D1337950

Carnivore Conservation

Carnivores are the focus of intense attention and resources in conservation biology. It is often argued that, because carnivores are at the top of the food chain, if they are protected, then other taxa will also be afforded adequate protection. Carnivores are also charismatic and compete with humans for dwindling space and environmental resources. In the past 10 years, theoretical and empirical studies on carnivores have developed very quickly. This volume reviews and summarises the current state of the field, describes limitations and opportunities for carnivore conservation, and offers a conceptual framework for future research and applied management. As such it will be of interest to students and researchers of conservation biology, mammalogy, animal behaviour, ecology, and evolution.

JOHN L. GITTLEMAN is Professor of Biology at the University of Virginia, Charlottesville. He is the Editor of *Carnivore Behavior, Ecology and Evolution* and co-Editor of *Animal Conservation*. His research focuses on macroevolutionary problems related to speciation, extinction and the evolution of biodiversity.

STEPHAN M. FUNK is at the Zoological Society of London.

DAVID W. MACDONALD is Director of the Wildlife Conservation Unit at the University of Oxford, and A.D. White Professor at Cornell University. He is also a Fellow of Lady Margaret Hall at Oxford. He has worked on carnivore biology for over 25 years, and has over 300 published papers. He has twice won the Natural History Writer of the Year award.

ROBERT K. WAYNE is at the Department of Organismic Biology, Ecology and Evolution, University of California, Los Angeles. He is co-Editor of *Animal Conservation*, and Editor of *Molecular Ecology*. His research uses molecular genetic approaches to address questions in evolution, ecology and conservation.

Conservation Biology

Conservation biology is a flourishing field, but there is still enormous potential for making further use of the science that underpins it. This new series aims to present internationally significant contributions from leading researchers in particularly active areas of conservation biology. It will focus on topics where basic theory is strong and where there are pressing problems for practical conservation. The series will include both single-authored and edited volumes and will adopt a direct and accessible style targeted at interested undergraduates, postgraduates, researchers and university teachers. Books and chapters will be rounded, authoritative accounts of particular areas with the emphasis on review rather than original data papers. The series is the result of a collaboration between the Zoological Society of London and Cambridge University Press. The series editor is Professor Morris Gosling, Professor of Animal Behaviour at the University of Newcastle upon Tyne. The series ethos is that there are unexploited areas of basic science that can help define conservation biology and bring a radical new agenda to the solution of pressing conservation problems.

Published Titles

1. *Conservation in a Changing World*, edited by Georgina Mace, Andrew Balmford and Joshua Ginsberg 0 521 63270 6 (hardcover), 0 521 63445 8 (paperback)
2. *Behaviour and Conservation*, edited by Morris Gosling and William Sutherland 0 521 66230 3 (hardcover), 0 521 66539 6 (paperback)
3. *Priorities for the Conservation of Mammalian Diversity*, edited by Abigail Entwistle and Nigel Dunstone 0 521 77279 6 (hardcover), 0 521 77536 1 (paperback)
4. *Genetics, Demography and Viability of Fragmented Populations*, edited by Andrew G. Young and Geoffrey M. Clarke 0 521 782074 (hardcover), 0 521 794218 (paperback).

Carnivore Conservation

Edited by

JOHN L. GITTLEMAN
University of Virginia, Charlottesville

STEPHAN M. FUNK
Institute of Zoology, London

DAVID W. MACDONALD
University of Oxford

and

ROBERT K. WAYNE
University of California, Los Angeles

CAMBRIDGE
UNIVERSITY PRESS

THE ZOOLOGICAL
SOCIETY OF LONDON

PUBLISHED BY THE PRESS SYNDICATE OF THE UNIVERSITY OF CAMBRIDGE
The Pitt Building, Trumpington Street, Cambridge, United Kingdom

CAMBRIDGE UNIVERSITY PRESS
The Edinburgh Building, Cambridge CB2 2RU, UK
40 West 20th Street, New York, NY 10011-4211, USA
10 Stamford Road, Oakleigh, VIC 3166, Australia
Ruiz de Alarcón 13, 28014 Madrid, Spain
Dock House, The Waterfront, Cape Town 8001, South Africa

http://www.cambridge.org

First published 2001

Typeface FF Scala 9.75/13 pt   *System* Poltype® [VN]

A catalogue record for this book is available from the British Library

Library of Congress Cataloguing-in-Publication Data

Carnivore conservation / edited by John L. Gittleman ... [et al.]
    p. cm. – (Conservation biology series ; 5)
  Includes bibliographical references (p.).
  ISBN 0-521-66232-X – ISBN 0-521-66537-X (pb)
  1. Carnivora. 2. Wildlife conservation. I. Gittleman, John L. II. Conservation biology series
(Cambridge, England) ; 5.
QL737.C2 C333 2001
333.95'9716 – dc21   00-045534

ISBN 0 521 66232 X hardback
ISBN 0 521 66537 X paperback

Transferred to digital printing 2003

Contents

Contributors

MARC BEKOFF
Department of Environmental,
Population and Organismic Biology,
University of Colorado, Boulder, CO
80309-0334, USA

LUIGI BOITANI
Department of Animal and Human
Biology, University of Rome 'La
Sapienza', Viale Università 32,
00185-Roma, Italy

URS BREITENMOSER
Institute of Veterinary Virology,
University of Bern, Länggass-Str. 122,
CH-3012 Bern, Switzerland

CHRISTINE BREITENMOSER-WÜRSTEN
KORA, Thunstrasse 31, CH-3074 Muri,
Switzerland

DAVID M. BROWN
Department of Organismic Biology,
Ecology and Evolution, University of
California, 621 Charles E. Young Dr,
Los Angeles, CA 90095-1606, USA

JANINE BROWN
Conservation & Research Center,
National Zoological Park, Smithsonian
Institution, 1500 Remount Road, Front
Royal, VA 22630, USA

LUDWIG N. CARBYN
Canadian Wildlife Service, 4999-98
Ave., Edmonton, Alta., Canada

TIM W. CLARK
School of Forestry and Environmental
Studies, Yale University, New Haven,
CT 96511, USA and Northern Rockies
Conservation Cooperative, Box 2705,
Jackson, WY 83001, USA

SARAH CLEAVELAND
Center for Tropical Veterinary
Medicine, University of Edinburgh,
Roslin, Midlothian, Scotland,
EH25 9RG, UK.

SCOTT CREEL
Department of Biology, Montana State
University, Bozeman, MT 59717, USA

NANCY CREEL
Department of Biology, Montana State
University, Bozeman, MT 59717, USA

EDUARDO EIZIRIK
Laboratory of Genomic Diversity,
National Cancer Institute, Frederick,
MD 21702, USA

CHRISTINE V. FIORELLO
Center for Environmental Research and
Conservation, MC 5556, Columbia
University, 1200 Amsterdam Ave., New
York, NY 10027, USA

LAURENCE G. FRANK
Field Station for Behavioral Research,
University of California, Berkeley, CA
94720, USA

STEFANIE FREITAG
Scientific Services, Kruger National
Park, Private Bag X402, Skukuza 1350,
South Africa

TODD K. FULLER
Department of Forestry and Wildlife
Management and Graduate Program in
Organismic and Evolutionary Biology,
University of Massachusetts, Amherst,
MA 01003-4210, USA

STEPHAN M. FUNK
Institute of Zoology, The Zoological
Society of London, Regent's Park,
London NW1 4RY, UK

ELI GEFFEN
Institute for Nature Conservation
Research, Faculty of Life Sciences, Tel
Aviv University, Ramat Aviv 69978,
Israel

ERIC M. GESE
National Wildlife Research Center,
Department of Fisheries and Wildlife,
Utah State University, Logan, UT
84322-5295, USA

JOSHUA R. GINSBERG
Wildlife Conservation Society, 2300
Southern Boulevard, Bronx, NY
10460-1099, USA

JOHN L. GITTLEMAN
Department of Biology, Gilmer Hall,
University of Virginia, Charlottesville,
VA 22903, USA

MATTHEW E. GOMPPER
Center for Environmental Research &
Conservation, Columbia University,
1200 Amsterdam Ave., New York, NY
10027, USA

JOGAYLE HOWARD
Conservation & Research Center,
National Zoological Park, Smithsonian
Institution, 1500 Remount Road, Front
Royal, VA 22630, USA

WARREN E. JOHNSON
Laboratory of Genomic Diversity,
National Cancer Institute, Frederick,
MD 21702, USA

M. KAREN LAURENSON
Centre for Tropical Veterinary
Medicine, Department of Tropical
Animal Health, University of
Edinburgh, Easter Bush, Roslin,
Midlothian, EH25 9RG, UK

GINA M. LENTO
Laboratory of Genomic Diversity,
National Cancer Institute, Frederick,
MD 21702, USA

GORDON LUIKART
Laboratoire de Biologie des Populations
d'Altitude, CNRS UMR 5553, Université
Joseph Fourier, BP53, F-38041
Grenoble, Cedex 9, France

DAVID W. MACDONALD
Wildlife Conservation Research Unit,
Department of Zoology, University of
Oxford, South Parks Road, Oxford
OX1 3PS, UK

GEORGINA M. MACE
Institute of Zoology, The Zoological
Society of London, Regent's Park,
London, NW1 4RY, UK

DAVID J. MATTSON
U.S.G.S. Biological Resources Division,
Forest & Rangeland Ecosystem Science
Center, Colorado Plateau Field Station,
PO Box 5614, BLD24, Northern Arizona
University Flagstaff, AZ86011, and
Department of Fish and Wildlife
Resources, University of Idaho,
Moscow, ID 83844, USA

BRIAN J. MILLER
Denver Zoological Foundation,
Conservation Biology Department, City
Park, Denver, CO 80205-4899, USA

M. GUS L. MILLS
Scientific Services, Kruger national
Park, Private Bag X402, Skukuza 1350,
South Africa

STEPHEN J. O'BRIEN
Laboratory of Genomic Diversity,
National Cancer Institute, Frederick,
MD 21702, USA

DAVID PAETKAU
Department of Zoology, University of
Queensland, St Lucia, Queensland
4072, Australia

ANDY PURVIS
Department of Biology, Imperial
College, Silwood Park, Ascot, SL5 7PY,
UK

RICHARD P. READING
Denver Zoological Foundation,
Conservation Biology Department, City
Park, Denver, CO 80205-4899, USA

MELODY ROELKE-PARKER
Laboratory of Genomic Diversity,
National Cancer Institute, Frederick,
MD 21702, USA

PAUL R. SIEVERT
Department of Forestry and Wildlife
Management, University of
Massachusetts, Amherst, MA
01003-4210, USA

CLAUDIO SILLERO-ZUBIRI
Wildlife Conservation Research Unit,
Department of Zoology, University of
Oxford, South Parks Road, Oxford,
OX1 3PS, UK

GÖRAN SPONG
Department of Animal Ecology, EBC,
Norbyvägen 18D, Uppsala University,
752 36 Uppsala, Sweden

CURTIS STROBECK
Department of Biological Sciences,
University of Alberta, Edmonton,
Alberta, 76G 2EG, Canada

FIONA SUNQUIST
Department of Wildlife Ecology and
Conservation, University of Florida,
Gainesville, FL 32611-0430, USA

MELVIN E. SUNQUIST
Department of Wildlife Ecology and
Conservation, University of Florida,
Gainesville, FL 32611-0430, USA

PIERRE TABERLET
Gordon Luikart Laboratoire de Biologie
des Populations d'Altitude, CNRS UMR
5553, Université Joseph Fourier, BP 53,
F-38041, Grenoble Cedex 9, France

MICHAEL D. THOM
Wildlife Conservation Research Unit,
Department of Zoology, University of
Oxford, South Parks Road, Oxford
OX1 3PS, UK

ALBERT S. VAN JAARSVELD
Conservation Planning Unit,
Department of Zoology & Entomology,
University of Pretoria, Pretoria 0002,
South Africa

PETER M. WASER
Department of Biological Sciences,
Purdue University, West Lafayette, IN
47907, USA

ROBERT K. WAYNE
Department of Organismic Biology,
Ecology and Evolution, University of
California, 621 Charles E. Young Dr,
Los Angeles, CA 90095-1606, USA

DAVID E. WILDT
Conservation & Research Center,
National Zoological Park, Smithsonian
Institution, 1500 Remount Road, Front
Royal, VA 22630, USA

ROSIE WOODROFFE
Ecology & Epidemiology Group,
Department of Biological Sciences,
University of Warwick, Coventry,
CV4 7AL, UK.

Acknowledgements

We are extremely grateful to the many individuals and institutions that contributed to the Meeting at which the chapters appearing in this book were first presented. Funding was received from The Zoological Society of London and Wildlife Conservation Society. Morris Gosling, Georgina Mace, and Linda DaVolls kindly helped with many logistical problems. The overseeing of the large editorial process of producing this volume was invaluably carried out by The Zoological Society of London. Tracey Sanderson at Cambridge University Press expertly gave us advice on pulling together all of the chapters into something coherent. Last, no edited book is produced without the criticisms, suggestions and effort of reviewers. Each chapter here was evaluated by at least two referees; we thank the following individuals for their time and assistance: Rob Atkinson, Todd Fuller, Jay Gedir, Joshua Ginsburg, Matthew Gompper, Lauren Harrington, Dom Johnson, Roland Kays, Devra Kleiman, David Maehr, Laurie Marker, Francoise Messier, Brian Miller, Richard Reading, Philip Riordan, Claudio Sillero, Daniel Simberloff and Mike Thom.

Why 'carnivore conservation'?

JOHN L. GITTLEMAN, STEPHAN M. FUNK,
DAVID W. MACDONALD AND ROBERT K. WAYNE

At present carnivore biologists are at an especially tough crossroads. Species are going extinct at a rate 100 times the natural background rates. With only around 5% of the planet's land surface protected in some form, continued habitat loss will produce much greater extinction rates, with potential disappearance of up to half of the world's species (May *et al.*, 1995; Pimm *et al.*, 1995). Critical conservation decisions are needed, at split-second timing, about which species to save, the best way to protect them, and how to divide resources for protecting and managing some taxa over others; indeed, the science of conservation biology is often characterized as the 'crisis discipline' (Soulé, 1985). Our intuition is to protect what we know and like. This is difficult with carnivores given that no one has a neutral position with them – they are loved or hated. On the one hand, carnivores are viewed as beautiful, powerful, and majestic; carnivores *are* 'megacharismatic'! It is unsurprising that visitors to the London Zoo recently indicated (Carvell *et al.*, 1998) that five out of their top 10 most popular animals are carnivores (Sumatran tiger, Persian leopard, Asiatic lion, meerkat, otter). On the other hand, carnivores are seen as the personification of evil, as exemplified in Theodore Roosevelt's description of the wolf as 'the beast of waste and desolation'. Such extremes in our perception of carnivores will continually work in favor and against conserving them.

The pressing issue is how to give carnivores priority, financially and intellectually, over other taxa when undertaking conservation measures. Carnivores *are* very expensive, not to mention labor intensive – radio collars, helicopters, laboratory costs all add up to millions of either dollars or pounds to carry out any successful conservation project on even a single species. Are carnivores this special? Relative to the disproportionate costs for doing conservation on them, can we rationalize that carnivores are worth it?

CARNIVORES ARE SPECIAL, OR ARE THEY?

In terms of species diversity, carnivores are not that unusual. From 4629 species (among 1135 genera) in the class Mammalia (Wilson & Reeder, 1993), the 271 species in the order Carnivora rank as the fourth largest group behind the Rodentia (2021), Chiroptera (925), and Insectivora (428). Size is not everything, though. The relevant question is: What do carnivores represent in terms of biodiversity, both historically and at present? Another picture emerges, one that gets at the question of whether carnivores are a special case for saving, even at high costs.

Evaluating what to preserve for the future necessarily involves understanding the past. The geological record shows that carnivores have often faired relatively well compared to other taxa. In contrast to many other mammals, carnivores have not gone through dramatic fluctuations in species numbers. General patterns of extinction versus origination rates in genera of Pleistocene carnivores suggest relatively high rates of new species appearances (Simpson, 1953; Gingerich, 1984). More detailed study of the fossil record over the past 44 million years in North America shows that carnivore species have remained relatively stable, despite a decline in herbivore diversity after the middle Miocene (Van Valkenburgh & Janis, 1993). That is, carnivores do not reveal directional evolutionary patterns, but rather are characterized by repeated evolution of certain ecomorphs such as cat-like and bone-cracking species (Martin, 1989). For example, sabertooth-like species have evolved independently at least three times (Felidae, Nimravidae, Creodonta), hyena-like species at least twice (Canidae, Oxyaenidae), and large dog-like predators at least four times (Canidae, Amphicyonidae, Ursidae, Hyaenidae). Such trends are consistent with predictions for extinction rates (McKinney, 1998) – of all mammalian orders, 24% of all Carnivora species are 'threatened' and, based on branching (birth–death) models measuring the number of species extinctions per unit time (McKinney, 1998), show the third lowest projected extinction rate (behind bats and rodents). Further, if we calculate the mean extinction rate of carnivores relative to the initial number of species in the group, the projected duration of the order is 2486 years, a fairly healthy duration relative to the median of 1179 years across mammals as a whole. So, despite the attention carnivores have received, extinction rates are not especially dire for the group as a whole. The causal reasons for this 'resilience' are obviously important to understand in conservation biology and emphasize the need to learn more about why some carnivores evade extinction (see Weaver *et al.*, 1996). However, an extremely important qualifier is needed here – even though carnivores may fair relatively *well as a whole*, historical and current patterns of extinction clearly indicate that large, carnivorous species with restricted ranges are highly threatened. Unless prompt measures are taken, carnivores may be represented largely by raccoons, coyotes, red foxes, and common weasels.

Extinction vulnerability among species is frequently expressed by particular biological traits (e.g., Terborgh, 1974; Wilson, 1987; Gittleman, 1994; McKinney, 1997; Purvis *et al.*, 2000b). These include: the number of species in a monophyletic group; species with narrow geographical ranges; species with only one or a few populations, small population sizes, or declining population sizes; species with low population densities; species requiring large home ranges; species with large body sizes; species with little genetic variability; species with specialized niche requirements; and species that are harvested or hunted by people. In many ways, these are exactly the characteristics that reflect the biology of carnivores!

Undoubtedly, the underlying reason for carnivores withstanding such multiple extinction risks is their tremendous variability, both within and across species. The observed range for an array of important biological traits is greater for carnivores than any other mammalian order (see Eisenberg, 1981; Gittleman, 1989, 1996; Macdonald, 1992), including:

- Body sizes range from the 100 g least weasel to the gigantic 800 kg or so polar bear.
- Reproductive rates are as low as one offspring every seven years, as in some black bears, to as high as three litters per year with eight young in a litter, as in some populations of mongooses.
- Carnivores are found in virtually every habitat or vegetational zone, from short grassland (meerkat) to sparse woodland (dwarf mongoose) to desert (fennec fox) to thick tropical forest (kinkajou) to oceanic waters (sea otter).
- Home ranges may be fairly small (0.55 km²: coatis; 0.20 km²: red foxes) to extremely large and non-defensible (1500–2000 km²: wild dogs), with worldwide geographical ranges lying between the restrictive island forms (e.g., island gray fox or Cozumel coatimundi) to the massive distribution of nearly 70 million km² of the red fox.
- Social structure ranges from spatially solitary individuals, with only brief encounters during breeding (ermine) to those species that form monogamous pair bonds (golden jackal) to those that live in extended social groups with as many as 80 individuals (spotted hyena).

Of course, overlaid onto this interspecific variation is considerable variation and flexibility within species. For example, in gray wolves adults weigh from 30 to 80 kg, litter size varies from one to 11, and populations are found in every vegetational habitat except tropical forests and arid deserts.

Within such variability lies our answer to the question of whether carnivores are special – a resounding 'yes', especially when we think about variation in carnivores explicitly in the context of conservation. This point is more compelling if we think in terms of species lists and classification schemes. Species are often classified into the following categories: indicator species, those that reflect critical environmental damage; keystone species, those that play a pivotal role in ecosystems; umbrella species, those that require large areas and thus, if protected, will protect other species; flagship species, those popular species that attract much attention; and, vulnerable species, those species most likely to become extinct. Each classification informs whether a species or taxonomic group are pivotal in terms of conservation status and the relative attention they receive for protection and management. It is quite remarkable that not only do many single carnivore species fit *all of these labels* but that there are entire carnivore clades that match these criteria. In the end, we suggest that an important reason why carnivore conservation should receive resources and attention, even perhaps disproportionately so, is that carnivores are renaissance taxa, involving a synthesis of conservation problems, causal factors and solutions.

SUCCESSES AND PROBLEMS – WORKING TOGETHER IN CARNIVORE CONSERVATION

Many classic examples of successful conservation biology involve carnivores. Problems of genetics, reintroduction, management, animal behavior and behavioral ecology, ecology, and policy all use carnivores as basic test cases in the literature. This is not surprising. Motivation for using carnivores obviously relates to their megacharismatic status, though equally important is their intrinsic variability. If we can sort out complex carnivores, we are bound to solve problems of other taxa. Importantly, we need to recognize a unique feature of our successes – collaboration. Take two examples. First, synthetic studies in ecology, population biology, behavioral ecology, and wildlife management have been critical for showing significant responses of carnivores to losses in prey (see Berger, 1998). Even experimental approaches to how prey respond to different kinds of olfactory and auditory stimuli in predators are informative for restoration of predator–prey communities. Secondly, the previously antagonistic relationship between field biologists and molecular geneticists is now coalescing into powerful conservation science. Studies on canids, felids, single species analyses of cheetah, black-footed ferrets, and tigers, all involve scientists in different disciplines showing that population fragmentation, genetic uniformity, and habitat loss are interrelated to such a degree that adopting only one approach is bound to miss important elements (Wayne, 1996).

These successes in more academically-related fields reveal an important contribution that we as carnivore conservationists need to acknowledge – conservation organizations studying other taxa are using carnivore-based studies. For instance, fundamental conservation goals are constructed from IUCN/SSC Action Plans on mustelids and viverrids (Schreiber et al., 1989), otters (Foster-Turley et al., 1990), procyonids and ailurids (Glatston, 1994), canids (Ginsberg & Macdonald, 1990), felids (Nowell & Jackson, 1996), African wild dog (Woodroffe et al., 1997a), and Ethiopian wolf (Sillero-Zubiri & Macdonald, 1997). Occasionally, single-species conservation goals of carnivores are achieved so well that they leave a unique influence – in Nepal's Chitwan National Park, the population density of tigers is the highest in the world because of an unusually forceful blend of protection in the park, anti-poaching policies, and monitoring by governmental and non-governmental organizations (Dinerstein et al., 1999). In turn, this has had a lasting effect on restoring ecological processes in the park, community-based ecotourism, and significant increases in the population density of other species (e.g., one-horned rhinoceroses). Newsletters formed from canid and small carnivore specialist groups are tremendous sources of information for conservation and management. These all are substantial contributions of which we as carnivore conservationists should be proud and continue to develop in conservation biology at large.

These successes should not disguise biases that have crept into conservation programs for carnivores. Two problems are particularly vexing. First, considerable attention in carnivore conservation has been focused almost exclusively on large carnivores. Indeed, large carnivores are often the textbook examples for prioritizing which species to save in conservation biology. As examples, large carnivores galvanize public opinion toward the greater goal of habitat conservation and the cessation of wildlife trade; the protection of large carnivores requires enormous reserves which protect other species; large carnivores occupy the top trophic levels of most food chains and offer stability to food webs; and large carnivores often are easy to breed in zoos and are notable successes in reintroduction programs. In sum, large carnivores reflect many critical problems of and solutions to carnivore conservation in general. The dilemma is to what extent is it effective to use large carnivores as a model and whether we should rethink this approach? For example, population decline and extinction are more likely with large species such as giant pandas and Siberian tigers. Conservation of large carnivores also affect negatively the conservation of other equally endangered species – cage space is expensive in zoos and conservation monies are limited. Despite large carnivores being used as symbols of conservation, they remain difficult to study and even detailed investigations of them often provide little return to basic science or consequential issues in conservation biology. For some large species there seem to be as many researchers working on them as there are individuals left in the wild. Should we redirect efforts toward other smaller and more abundant species, whose future is more certain and whose study provide potentially more significant lessons? In essence, we have made large carnivore species our symbols, bred them in zoos, displayed them to the public as synonymous with conservation. What do we do if our best efforts fail?

The other serious problem is what might be referred to as the 'human–carnivore interface'. Many examples could be given for this problem but generally the issue is that carnivore conservation becomes human-based, anthropocentric. For example, there is now whole-scale control of coyotes and foxes in the US and of cheetahs and jackals in Namibia. Should we be controlling carnivores in such situations, or should we adopt a strict ecological view, advocating complete uncritical protection? We need to work toward more balanced and flexible guidelines that allow control when it is effective and does not endanger populations.

These are only two problems that emphasize modern difficulties of carnivore conservation. There are many others that are just as pressing. The point is that we begin to assess what these problems are, what methods have worked and failed, and begin to focus on preserving carnivore species into the future.

PRIORITIZING PRESENT AND FUTURE PROBLEMS

Rare, elusive, dangerous – carnivores are difficult to study. This means collaboration is essential for successful conservation science. As problems become more severe, and they will, we must increase levels of collaboration to bring about quicker and more effective work. This was a primary motivation for organizing a Meeting, held 20 and 21 November, 1998 at The Zoological Society of London, in which we assembled for the first time a group of distinguished international researchers solely devoted to *carnivore conservation*, the result of which is the proceedings published herein. While acknowledging that not everything can be covered in this volume, the overarching goal was to assess where we are so to better carry out future studies of carnivore conservation. Within the framework of specific problems, we asked participants to organize chapters with reference to some broad issues:

- Are the methods and approaches effective in answering the question(s) at hand and to what extent are these carnivore-specific?
- How can we better assess which carnivore populations and species are more vulnerable?
- Might carnivore conservation be better served by prioritizing geographical areas or ecological communities rather than species-by-species (taxonomic) approaches?
- How can the science of carnivore conservation develop more effective means of communicating with the public and general environmental organizations, particularly when addressing human–carnivore conflicts?

In essence, the contributions in this volume lay out what we have accomplished thus far. More importantly, we hope that the papers here will help decide in which direction we head for carnivore conservation. As George Schaller (1996: p. 9) put it, 'We cannot ease the burdens of the past, but we can atone by assuring the carnivores of the future'.

PART I

Problems

2

Past and future carnivore extinctions: a phylogenetic perspective

ANDY PURVIS, GEORGINA M. MACE AND JOHN L. GITTLEMAN

INTRODUCTION

In many ways carnivores are an enigma for conservation biology. Consider the following. (1) The history of carnivores shows periods of devastating extinction rates, with 352 genera going extinct relative to the 129 currently living (McKenna & Bell, 1997; for comparison, primates and artiodactyls have lower rates but rodents much higher). (2) Episodes of extinction often have been followed by taxonomic replacement, as evidenced by saber tooths replacing themselves at least four times (Van Valkenburgh, 1999). (3) Of 4761 living mammal species spanning 11 orders, five have significantly more threatened species than expected (artiodactyls, insectivores, primates, perrissodactyls, sirenians) but carnivores are not one of them nor does any carnivore family have an unusually high level of threatened species (Mace & Balmford, 2000). (4) Carnivores carry the dubious distinction of facing more types of threat (e.g., habitat loss, effects of introduced species, rarity) than any other mammalian order yet no single carnivore taxon is unusually threatened. (5) Many carnivore species are the ultimate symbols of conservation biology (tigers, black-footed ferrets, red wolves, giant pandas) despite the high costs and extreme difficulties of conserving a single population or species of carnivore. Amazingly, the price tag for saving a single carnivore species (e.g., red wolf) may be over 4.5 million US dollars (Wayne & Gittleman, 1995). (6) Carnivores, and especially large felids, generally come out top in animal popularity in television polls, animal magazines, and among zoo visitors (Balmford et al., 1998; Carvell et al., 1998; Serpell, 1991).

Carnivores thus represent extremes of problems in conservation biology. Within one lineage, some species such as the red fox or the raccoon have virtually no risk of extinction – indeed, many are now considered pests in our largest cities – while others such as the giant panda and black-footed ferret are rapidly heading toward inevitable extinctions. Such diversity between closely related taxa suggests that carnivores are an ideal taxon to study the patterns and processes of extinction. In the words of Jared Diamond (1984a: p. 824), 'We need to understand why extinctions did befall some beasts in some places at some times, but not other beasts in the same places nor similar beasts at other places or times'. In this chapter, we show how taxonomic comparisons can identify what factors produce these disparate species conditions. We begin by stepping back from contemporary problems of carnivore conservation to consider historical patterns of extinction, using both available fossil data and new phylogenetic approaches. We then use this information to identify those carnivore taxa that were influenced by past extinction events. This guides our analysis of those factors that are important to consider as we develop plans to conserve present-day carnivore species.

PAST AND PRESENT TRENDS IN CARNIVORE EVOLUTION

The landscape of today's carnivores is very different from that which existed millions of years ago. Imagine small hyaenas with narrow, sharp teeth, and elongate, slender limbs; a gigantic mustelid (*Aelurocyon*) the size of a leopard; and bears of small size, living in social groups, having a meat-eating diet. These are only a fraction of the carnivore diversity that is now missing. Gone completely are the precursors to present-day carnivores, the 'miacoids' (a paraphyletic group including the stem canoids and stem feloids; Flynn & Galiano, 1982), and the creodonts that had three to four times more taxa than all of the 'modern' Carnivora. In relative terms, we now only have 129 extant genera compared to 352 fossil genera (McKenna & Bell, 1997). What characteristics of fossil carnivores contributed to their extinction? Recently, a number of excellent reviews have revealed important details about carnivore evolution (Flynn, 1996; Hunt, 1996; Janis *et al.*, 1998; Van Valkenburgh, 1999; Van Valkenburgh & Janis, 1993; Werdelin, 1996). The following is a brief summary of this literature that reveals general patterns in the fossil record, hints about causal factors influencing past extinction, and potential hypotheses for future risks of carnivore extinction.

Comparisons between studies of fossil carnivores and studies of extant species should take into account the different methodologies used in each case. Two types of palaeobiological information are used in our discussion. One is descriptive, essentially general information about the size, diet and overall ecomorphology of fossil forms emerging during the history of carnivores. The other uses these data to assess taxonomic turnover, duration of different ecomorphs, and what characteristics were likely to be more successful in the adaptive radiation of carnivores. Although the latter type study is more directly comparable to neontological study, it almost exclusively deals with medium (> 7 kg) to large body sizes as reflected in the availability and completeness of the fossil record. Therefore, statements regarding size are relative whereas our studies of living species are more complete. However, we believe that these comparisons reveal some important general trends that apply across the very different assemblages.

In general, three mammal groups have an exceptional fossil record including the Rodentia (about 560 genera), Artiodactyla (about 500 genera), and Carnivora (about 350 genera). All of these taxa represent a diversity of ecological types and have a worldwide distribution. Patterns of origination and extinction represent a general pattern that forecasts the more detailed examination of the carnivores presented below. Essentially, over a 65 million year period throughout the Cenozoic, there is a remarkable parallel between origination and extinction – one event follows another representing a dynamic equilibrium (Gingerich, 1984). However, this long-term, stable pattern seems to change in the Pleistocene. Towards the end of the Pleistocene extinction rates of large mammals greatly exceed their appearance, roughly 56% of the large mammal species disappeared in the Rancholabrean-3 (Gingerich, 1984; Webb, 1984). Relative to the overall pattern of origination–extinction balance in Cenozoic mammals, this disappearance is an unusual event. Possible explanations are that there has not been enough geological time for replacement or, more seriously this may be a direct consequence of human activity affecting many large mammal populations and so they will not be replaced.

The most informative studies of carnivore history derive from the fossil record in North America where detailed evolutionary changes can be assessed among the major carnivore clades. Herbivores in North America reveal consistent trends in morphological diversity, mainly showing increasing cursoriality and dental specializations for grazing. In contrast, carnivores do not appear to have directional evolutionary trends (Van Valkenburgh & Janis, 1993). Rather, carnivores are characterized by the repeated evolution of specific ecomorphs such as bone-cracking and cat-like (predatory) species. This pattern suggests one historical clue about carnivore evolution. As described below, certain ecomorphological characteristics are repeated, presumably because they are successful. These are also characteristics that give carnivores the unique flexibility to utilize different foods. Indeed, it is thought that the progenitors of the Carnivora, the 'miacoids', were more successful than their carnivorous (and more abundant) counterparts, the Creodonts, precisely because of a greater heterodont dentition (a single pair of cutting teeth; the carnassials, comprising the upper fourth premolar and the lower first molar) that permitted a more diversified diet.

Three major events have occurred in the history of modern carnivores. (1) During the last 10 million years or so of the Eocene (43–33 Ma), there were two rapid and dramatic cooling events. At this time many North American creodonts went extinct (39.5–45.9 Ma) and the 'miacoids' flourished (35.5–39.5 Ma) along with some new predators including the first canids, nimravids and mustelids (Janis *et al.*, 1998). The causes for this turnover remain unclear other than some unknown morphological feature that helped to utilize the new potential prey species that appeared with the major climatic change. (2) In the late Oligocene (34–24 Ma), the nimravid/hyaenodontid creodonts were replaced in North America by canids, amphicyonids, and ursids, and in the Old World by hyaenids, amphicyonids, and ursids. Both transitions are poorly represented in the fossil record though they are clear with regard to one feature (Van Valkenburgh, 1999): the absence of cat-like nimravids and true felids. This 'cat gap' shows that many caniform species evolve felid type morphologies indicative of hypercarnivory including reduced snouts, enlarged canines, and a reduction in crushing molars. This reflects an overall recurrent pattern in which there is always some hypercarnivorous taxa; whenever a taxon filling this niche goes extinct it is rapidly filled in kind. (3) The final turnover of carnivore taxa occurred during the late Miocene (5–8 Ma), involving a major extinction event that eliminated 60–70% of Eurasian genera and 70–80% of North American genera. Replacement taxa comprised large mammalian predator guilds including large felids, wolf-like and fox-like canids, bone-cracking hyaenids, and omnivorous ursids. In addition to carnassial teeth that allowed for dietary flexibility, these carnivores also revealed diversity in locomotor types with more efficient cursorial hunting. In sum, the rise-and-fall of carnivore taxa in the fossil record suggest that large carnivores have gone extinct many times. Each predator dynasty has been represented by only a few subfamilies lasting around 10 to 15 million years. Replacement has frequently involved hypercarnivorous morphotypes, and competition among carnivores has been excacerbated by habitat and food shortage that may have driven some extinction events. In terms of correlates of extinction risk in the fossil record, small size seems to be advantageous. The primary macroevolutionary change in the carnivores is the overall decline in the number of large, highly carnivorous predatory species. These

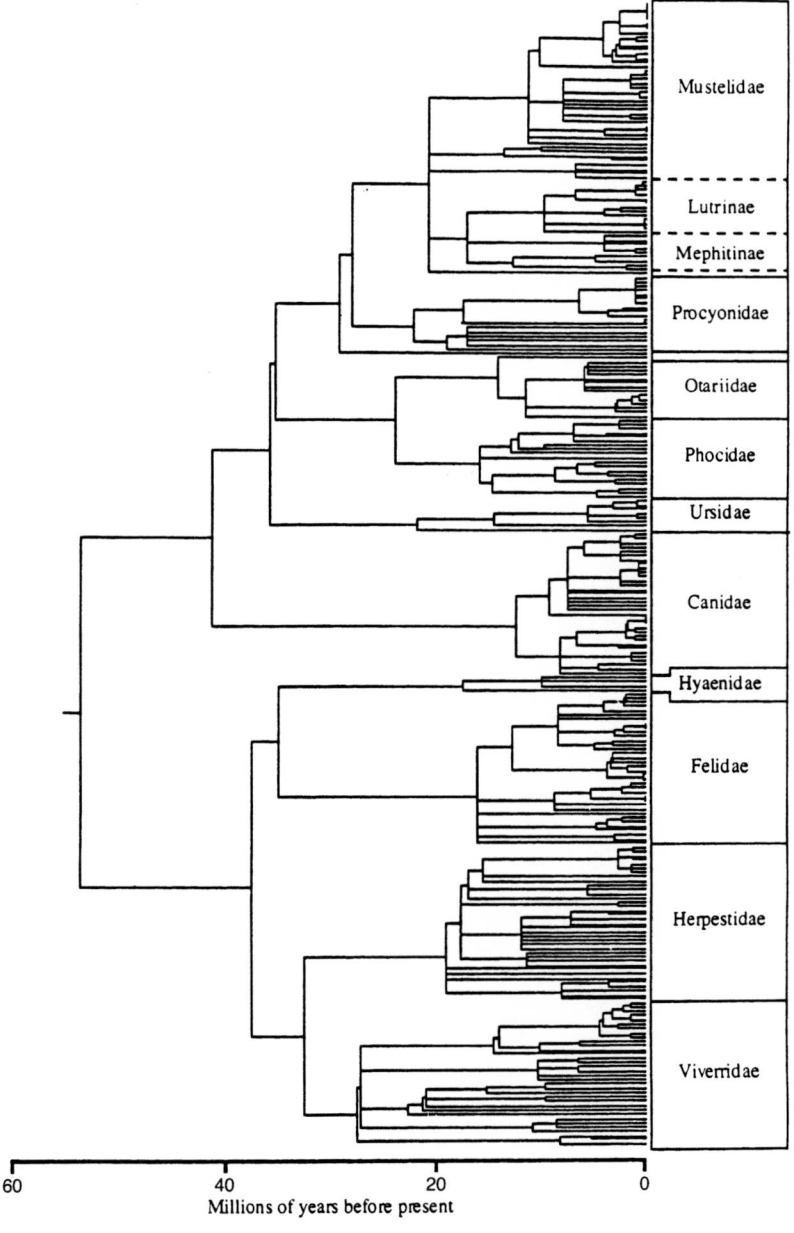

Mustelidae

Lutrinae

Mephitinae

Procyonidae

Otariidae

Phocidae

Ursidae

Canidae

Hyaenidae

Felidae

Herpestidae

Viverridae

60   40   20   0
Millions of years before present

Figure 2.1. A composite tree for all 271 extant species of carnivore, including estimated time of divergence based on molecular and fossil measures. The tree is comprised of all forms of phylogenetic hypothesis (morphology, molecular data, behavior, ecology), analyzed by matrix representation with parsimony. See Bininda-Emonds *et al.* (1999) for further details of methodology, data and analyses.

patterns are also consistent with studies of late Pleistocene extinctions where islands have been carved off of continents showing that large, carnivorous, habitat specialists are more susceptible to extinction than small, herbivorous, habitat generalists (Hope, 1973). One underlying explanation for these differences is that small mammals, particularly herbivorous habitat generalists, are more abundant and therefore survive better (Diamond, 1984a).

Today, the modern order Carnivora comprises 271 extant species (Wozencraft, 1993). Phylogenetic analyses (Flynn 1996; Werdelin 1996; Bininda-Emonds *et al.* 1999) indicate that: (1) the Carnivora are monophyletic; (2) two major clades exist within the Carnivora (see Figure 2.1), the Feliformia (cats, hyenas, civets, mongooses) and the Caniformia (dogs, bears, weasels, skunks, raccoons); and (3) well-substantiated clades exist within the Caniformia (Canoidea–Arctoidea), there is monophyly of the Feloidea (Herpestidae, Hyaenidae, Felidae, Viverridae), and monophyly of the Pinnipedia within the Arctoidea. Notably, carnivores range in body size over four orders of magnitude – greater than any other mammalian order – and are extremely diverse in physiology, behaviour, geographic distribution, life histories and, despite their ordinal name, dietary ecology. A relevant feature for our study is that species richness differs greatly among carnivore families ranging from 65 species in the Mustelidae to only four hyaenids (see Gittleman & Purvis, 1998). Such variation is essential to assessing phylogenetic patterns of extinction risk in carnivores.

USING PHYLOGENIES TO ESTIMATE EXTINCTION RATES

Ideally, the discussion above of the carnivore fossil record would lead us directly to the historical patterns of extinction and their causes. Unfortunately, however, palaeontological information does not provide sufficient detail for studying extinction in relation to the kinds of forces and biological characteristics that are important and measurable in extant taxa. Nevertheless, recent statistical approaches do permit reconstructions of birth and death processes when accurate phylogenies (either of fossils or extant taxa) are available. We briefly review these approaches to illustrate the kinds of analyses and inferences that can be performed and we then apply these techniques to the carnivores.

With fossils

Armed with a complete phylogeny of a clade, indicating the timing of every speciation and every extinction, it is a simple matter to measure the ob-

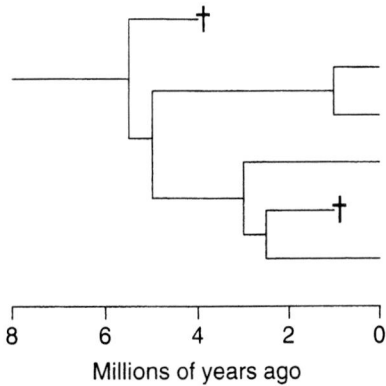

Figure 2.2. Phylogeny of clade showing all speciation and extinction events. Calculation of rates from such a phylogeny is trivial.

served rates of speciation and extinction. The top panel in Figure 2.2 shows the history of a small clade. The total length of phylogenetic branches in the clade is 20 Ma. These branches involve five speciation events and two extinction events, so the per-lineage rates of speciation have been $5/20$ Ma^{-1} and $2/20$ Ma^{-1} respectively.

Unfortunately, the fossil record is not complete, so cannot be taken at face value when analysing speciation and extinction rates. The earliest occurrence of a species in the fossil record is a biased estimator of its true time of origin (Norell, 1993) – in all likelihood, the species will have been around for some time before we first see it. Similarly, it will probably have been around for a while after our last record of it, so the last occurrence is a biased estimator of its date of extinction. Statistical techniques are now available that model and allow for these effects (Smith, 1994; Foote, 1997), which place the estimation of vital rates on a firmer footing. Additionally, species-level analysis of the fossil record is made harder by non-random incompleteness and problems of species definitions (see Smith, 1994; Sepkoski, 1998). Consequently, data are commonly analysed at some higher taxonomic level. Incompleteness then becomes less of a problem, although it is likely that an individual species has not been found as a fossil, the logic of sampling suggests that more and more inclusive taxa are more and more likely to be represented in the known record. Palaeontological genera and families are often more stable taxonomically and easier to identify than are species. There is also some evidence that higher taxa such as these can be useful proxies for the real underlying species-level patterns and processes (Gaston, 1996; Roy et al., 1996b), though the usefulness of non-monophyletic taxa is a matter of debate. By assuming particular distributions of species among higher taxa, it is then possible to extrapolate from origination and extinction rates of, say, families to obtain estimates of species origination and extinction rates (Raup, 1991). We do not wish to dwell on such analyses here, interested readers can refer to Stanley (1979), MacFadden (1992), or Caughley & Gunn (1996).

Extant taxa

Perhaps surprisingly, phylogenies of only extant species contain information that allows estimation of past rates of speciation and extinction (Harvey *et al.*, 1994; Kubo & Iwasa, 1995). As an illustration, consider the results of simulations in which clades are 'grown' under a simple null model where the per-lineage probabilities of speciation and extinction per unit time are constants. This model has a long history in macroevolutionary studies (see reviews by Purvis, 1996, and Mooers & Heard, 1997). When speciation is more likely than extinction (a necessary condition to avoid the 'gambler's ruin': Raup, 1991), the model produces exponential growth over time. A plot of the logarithm of the number of extant lineages against time is therefore expected to yield a straight line, the slope of which estimates the difference between speciation and extinction probabilities. Figure 2.3(a) shows the history of an artificial clade (simulated using the Bi-De program: Rambaut *et al.*, 1996) in which the per-lineage speciation probability per unit time, $\lambda = 0.1$ and the corresponding extinction probability, $\mu = 0$. Initially, the increase in lineage number is uneven – although the probabilities are fixed, the process is stochastic so the observed rates can vary. As the clade becomes large, the law of averages makes the observed per-lineage rate of increase increasingly constant, and the slope of the semi-logarithmic plot settles down to 0.1.

When $\mu > 0$, two lines can be drawn (Figures 2.3b–d). One (the thick line in the figures) is the number of lineages extant at each moment in time. As before, this becomes a straight line with slope equal to $\lambda-\mu$. However, in the absence of fossil information, we know only about lineages with extant descendants. The thin lines in Figures 2.3(b) to (d) show how the numbers of such lineages have changed over time (the vertical distance between the two represents then-extant lineages with no living descendants). These lines are also straight for much of the time axis, with a slope that estimates $\lambda-\mu$, but they curve up at the right-hand side (Harvey *et al.*, 1994). The degree of upturn reflects λ – the slope of a tangent to the line at the present estimates the speciation rate. In practice, maximum likelihood algorithms (Nee *et al.*, 1994b) give more precise estimates than ruler and pencil, and can even give statistical confidence intervals (Purvis *et al.*, 1995), but Figures 2.3(a–d) serve to illustrate the point. The pattern of lineages through time, i.e. the temporal spacing of lineage splits in a phylogeny of only extant species, allows estimation of the net rate of clade growth and the speciation rate, hence permitting estimation of the extinction rate.

What happens when this procedure is applied to carnivore phylogeny of Bininda-Emonds *et al.* (1999)? As Figure 2.4 shows, the lineages-through-time plot is a straight line with no upturn – the maximum likelihood estimate of the extinction rate is 0. Yet many genera are known to have gone extinct during carnivoran history, so why the mismatch? There are various possibilities. An initial methodological possibility is that the branch lengths are incorrect and therefore the observed pattern of extinction is wrong. Undoubtedly, some error exists in the carnivore branch lengths. However, out of a total of 211 nodes in the composite tree, 150 were estimated with divergence times and 73 of these involved at least one estimate from both fossils and molecules. Tests for errors in median dates were acceptable, with only 12 having unusually high 'coefficients of variation'. Given this, we would not anticipate that the errors that may exist in the branch lengths would consistently leave strong patterns.

Three macroevolutionary explanations are more informative. First, the rate of clade growth could have been declining over time, perhaps in a logistic fashion (Sepkoski, 1991). This hypothesis is testable because, under exponential growth, the product of N_t (the number of lineages at time t with extant descendants) and t (the time between successive splits in the phylogeny) should be constant. If, rather than being constant, $N_t \cdot t$ changes as the clade grows, that would suggest time- or density-dependence of the growth rate. However, $N_t \cdot t$ shows no trend over time for the carnivore phylogeny (Figure 2.5; note that the observations were grouped into successive tens because polytomies in the phylogeny artifactually cause many of the $N_t \cdot t$ to be estimated to be zero).

A second possibility is that different clades within the order have very different rates of clade growth, with the straight line in Figure 2.4 emerging as an artifact. This explanation is unlikely, species are distributed fairly evenly among carnivoran families and temporal-branching patterns follow an exponential model, thus there is little evidence for marked variation in rates among clades within the Order (Bininda-Emonds *et al.*, 1999).

Our preferred explanation for Figure 2.4 is that carnivores have suffered a mass extinction in the geologically recent past. Figure 2.6 illustrates the argument via simulated patterns. As before, the top line is what actually happened. The lines below it are the inferred history based on a phylogeny of (moving from top to bottom) 100%, 50%, 10% and 1% of the species,

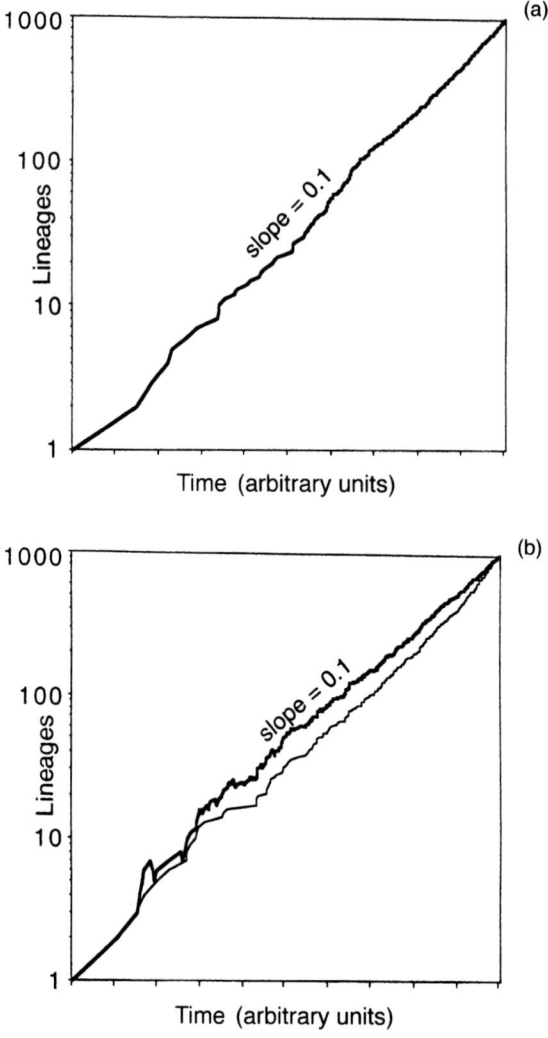

Figure 2.3(a–d). Lineages-through-time plots showing patterns from different speciation and extinction rates.

sampled at random. If the phylogeny is complete, the upturn is clearly visible. However, if only a small fraction of the species are represented, the line starts to plateau, not to climb. At some intermediate fraction, a straight line results. The simplest way to get a straight line when extinction is not zero is to sample only part of the clade's diversity. Given that all extant species *are* present in the carnivore phylogeny, the sampling process has

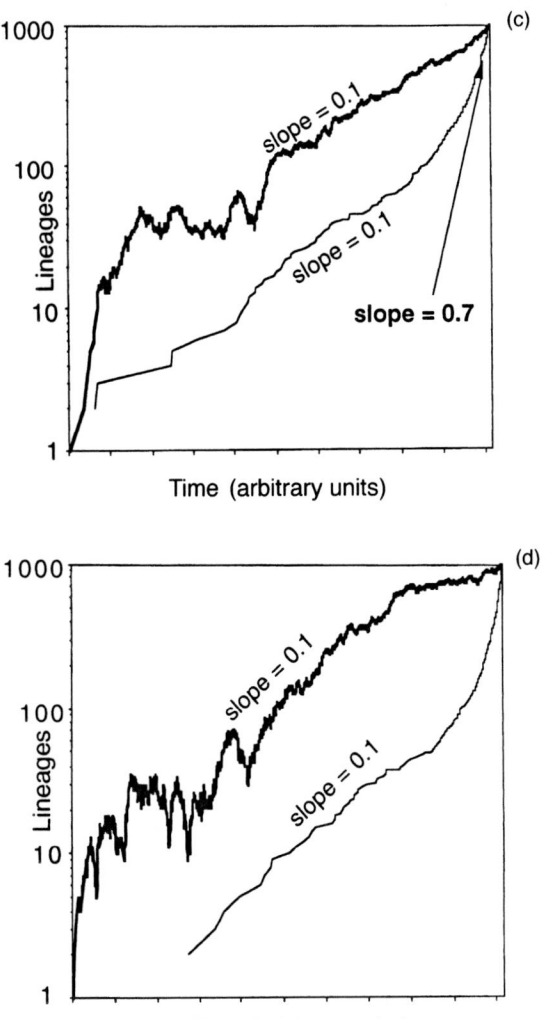

already happened – the species in the upturn have gone extinct in the geologically recent past.

The severity of the mass extinction can be estimated provided that extinction is assumed to be random and that mean species longevity (not at times of mass extinction) is known. Estimations of typical mammalian species longevity lie within the range 5 Ma down to 1 Ma (Stanley, 1979; May *et al.*, 1995), implying per-lineage extinction rates of 0.2 Ma^{-1} to 1 Ma^{-1}, respectively. Our simulations indicate that a pattern like that in Figure 2.4 is

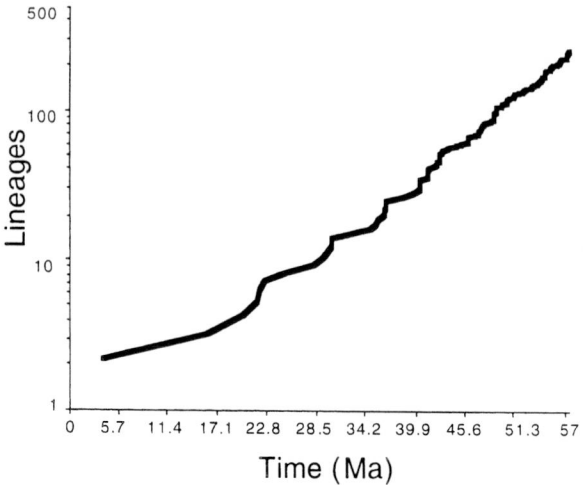

Figure 2.4. The lineages-through-time plot for the carnivore phylogeny.

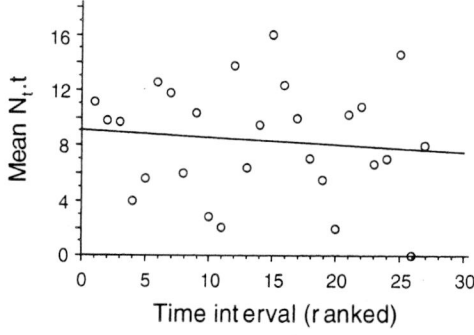

Figure 2.5. Test of density-independence in rate of clade growth in carnivores. Regression equation: $Y = 9.13 - 0.05X$, $r^2 = 0.01$, n.s. See text for explanation.

produced by a mass extinction that culls, respectively, two-thirds to five-sixths of species. These are approximate figures, doubtless with wide confidence intervals (the sampling theory for the estimate has not been worked out), but they are strikingly high. Such a high figure could reflect an incorrect assumption. It could result if, for instance, carnivore species have longer typical durations than other mammals, or if they are more taxonomically 'lumped'. However, carnivores have long been thought to have high turnover rates (Simpson, 1953; see also Van Valkenburgh & Janis 1993), and the low mean number of species per genus in carnivores relative

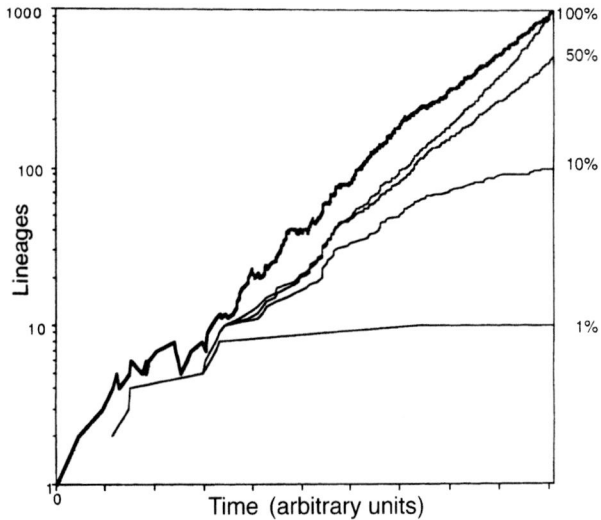

Figure 2.6. Lineages-through-time plots in which only a fraction of the clade's diversity (indicated at the right) is sampled at random. The thick line represents what actually happened and the thin lines represent the inferred history based on all, 50%, 10% and 1% of the species sampled at random. See text.

to other orders suggests that there is little lumping at that taxonomic level. Extinction would also be overestimated if the date estimates in the carnivore tree have a systematic bias, with young clades being thought too old. If the date estimates are reasonable, then the extinction has been severe.

Estimates of the number of species extinctions from the phylogeny are not easy to relate to known numbers of species extinctions in the fossil record. For example, a simple analysis of the North American data analysis of Kurtén & Anderson (1980) suggests that we may be able to account for only half to a third of all those predicted by the model. Nevertheless, as we have discussed above, the model assumptions are unlikely to be violated to such a degree that errors of this sort could be a result. It is certain that the fossil record underestimates species extinctions, especially those that have short persistence times and restricted distributions but it seems unlikely that this will be the only explanation. It may also be that different species definitions could lead to non-comparable results. Further investigations of both the regional fossil record and phylogenetic relationships should reconcile information from the two different sources. In sum, information from the fossil record does not match well with our estimates of carnivore extinction from the phylogeny. There are possible biases in both estimates which need further investigation before we can resolve the issue.

EXTINCTION AND FOSSILS

We can now begin to consider how patterns of extinction risk among contemporary species compare to those of the past. This is an important question that has interest whatever answer it gives. For example, if it transpires that the same taxa, or species with the same biological traits, had unusually high extinction rates in the past and in the present, then this would be rather good evidence for strong biological determinants of extinction risk that were significant even when the structure of communities, environmental conditions and threat processes had changed markedly. If, on the other hand, it emerged that the factors that were associated with high threat levels in extant mammals were different from those of extinct species, then it would indicate that the current extinction spasm is very different in kind from the past. We know that anthropogenic effects have altered the rate of species extinctions but this would support the idea that it had also changed their nature.

There are several ways to approach this question, but here we adopt a simple one. We first assume that extinct and extant members of the same families share characteristics that relate to their extinction risk, i.e. the biological determinants of vulnerability are shared within evolutionary clades. This seems to be a reasonable assumption since body size and other life history traits are related to extinction risk (Laurance, 1991; Gaston & Blackburn, 1995) and these same traits are significantly correlated with phylogeny (Gittleman *et al.*, 1998). If it turns out that taxa that had unusually high rates of extinction in the fossil record are those that are most threatened today, this would be strong circumstantial evidence that certain biological characteristics evolve repeatedly, but tend to be associated with high extinction rates. However, the lack of an association here can result simply from poor resolution of the method and would be inconclusive.

We collated information from McKenna & Bell (1997) on the carnivore genera that are recorded in each epoch and sub-epoch during the Cenozoic. These data simply indicate the presence of a genus in any location. We then use changes in generic counts between time periods as a measure of diversity or success of each. We chose to analyse this information for the period from the early Miocene onwards, as this is the point at which all extant families are known in the fossil record.

We use two indices for each family within the Carnivora to compare the past to the present trends in extinction. The first is a measure of generic persistence and is the median persistence time of extant genera as indicated by the number of sub-epochs present. This measure is weighted towards the more recent taxa but because it is based around extant genera is less sensitive to sampling biases of different families during different periods.

Secondly, we take the period from the early Miocene to the Recent and estimate mean turnover of genera within families across all sub-epochs, using the equations:

$$\text{mean Turnover } (T) = N_{t+1} / N_t$$

where N_t is the number of genera recorded in each sub-epoch from the early Miocene to the late Pleistocene and N_{t-1} is the number of genera in the sub-epoch straight afterwards. Many measures of taxonomic turnover have been developed (see Stanley 1979; Oliver & Pedder 1994; Russell 1998), all of which are weakened by averaging across species, genera and families. Despite this, such broad estimates remain useful indicators of extinction patterns and are appropriate for the purpose here (Gilinsky 1998). The problem with the measure used here is that it is an average over all time periods and does not differentiate success at different times, as clearly evident from our depiction of carnivore history; however, it does control for different generic numbers in families.

As Figure 2.7 shows there is no simple relationship across families with either of these fossil measures. With so crude an analysis this is perhaps unsurprising, and particularly as there is rather little variation among families in the extinction risk of extant taxa. Other than the hyaenids and ursids, the families all have threat levels of between 21% and 44%. However, the plots show some interesting patterns. The Ursidae with the highest turnover rates are currently the most threatened family and the hyaenids, which have the lowest threat rates, have the longest persistence times. The lack of any general relationship may result from inadequate and uneven sampling of the fossil forms. For example, the smaller bodied species (such as viverrids) and the marine species (seals) all have a relatively poor fossil record. A cross-family analysis may well be inadequate here and we suspect that detailed studies of particular groups within well-studied fossil faunal assemblages will be a more productive approach.

EXTINCTION AND PHYLOGENY

As species are lost from a clade, so is biodiversity – but there is not a one-to-one relationship between the two. An obvious measure of the 'biodiversity cost' of a species extinction is the amount of unique evolution that is lost,

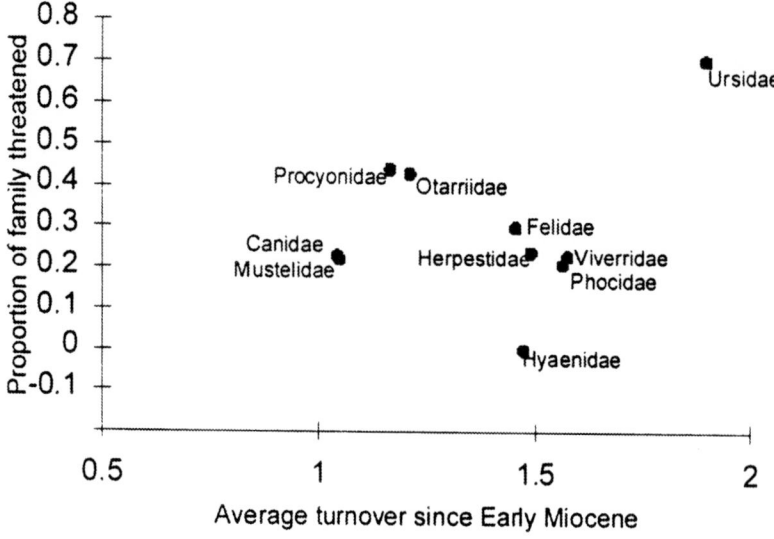

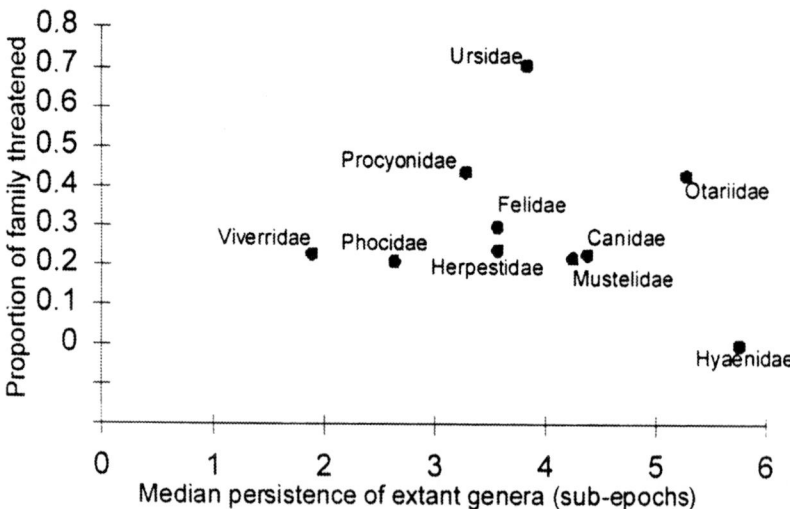

Figure 2.7. The relationship between measures of extinction vulnerability in extinct and extant members of carnivore families.

i.e., the length of time for which that species has been distinct from its closest living relative – the length of the branch on the phylogeny that leads to it and it alone. Most extant species have living close relatives, so the first species extinctions tend to incur only a small biodiversity cost. However, as more species are lost, survivors are less and less closely related to each other, so the biodiversity cost of successive extinctions tends to increase. Another way of putting it is that higher taxa like genera and families are unlikely to be lost at first, because their loss requires extirpation of *all* constituent species.

Nee & May (1997) modelled this process assuming that extinction is phylogenetically random. They showed that a high proportion of species can become extinct before many main branches are lost from the phylogenetic tree – in one case, loss of 95% of the species still left 81% of the tree's total branch length intact. An implication of their result is that evolutionary recovery from the current mass extinction could be rapid in geological time. Most niches would harbour at least some survivors, which might radiate rapidly to restore pre-extinction levels of species-richness and character diversity. The restoration would be particularly quick if speciation rates show density-dependence.

However, the shape of the curve relating biodiversity cost to species extinctions depends upon the phylogenetic pattern of extinction. Nee & May's (1997) simulations assumed that extinction is random with respect to phylogeny (the so-called 'field of bullets' scenario). If, instead, extinction risk is clumped – such that close relatives tend to have unusually similar levels of risk – biodiversity cost increases more rapidly as species are lost. Conversely, if extinctions are overdispersed, biodiversity cost remains low for longer.

The list of mammalian species known to have gone extinct since 1600 (Groombridge, 1992) yields mixed messages. On the one hand, the extinctions are distributed among orders roughly in proportion to the numbers of extant species those orders contain (among 20 orders, $r_s = 0.757$, $P < 0.01$; see Table 2.1), perhaps indicating randomness. Conversely, there is one marked clump. Among the 59 extinctions were all five species in the insectivoran family Nesophontidae. These cave-dwelling shrew-like species were endemic to islands in the Caribbean, and presumably all went extinct when humans and associated introduced species arrived there (Nowak, 1991). Quantitative analysis of the list indicates that historical extinction has been significantly non-random (Russell *et al.*, 1998).

Interpreting the list of known recent extinctions is complicated by two problems. First, the list is rather short (in particular, it contains only two

Table 2.1. *Numbers of species in mammalian orders according to a recent compilation (Wilson & Reeder, 1993), and numbers thought to have gone extinct since 1600 (Groombridge, 1992)*

Order	Number of species	Number extinct since 1600
Rodentia	2015	29
Chiroptera	925	6
Insectivora	428	5
Marsupialia	272	10
Carnivora	271	2
Artiodactyla	220	3
Lagomorpha	80	2
Perissodactyla	18	1
Sirenia	5	1
Others (average number)	1–233, median 7	0

carnivores). While good news in one sense, this limits statistical analysis. More importantly, there are biases. For a species to be listed, we have to be aware of its existence, probably biasing the list towards species extirpated by direct human exploitation and away from those that fell victim to introduced species or habitat loss. For instance, the list contains few species from tropical moist forests (Groombridge, 1992), yet probably around half this habitat has been lost since 1600.

A larger and more complete data set is provided by the living fauna (see Purvis *et al.*, 2000a for a fuller analysis). The pattern of perceived extinction risk among living species permits an assessment of whether risk is phylogenetically random. In particular, if currently threatened species were to go extinct, would we pay a higher biodiversity cost than expected under the 'field of bullets' scenario? We use a simple analysis, using numbers of extirpated genera as our index of biodiversity cost.

The IUCN (Baillie & Groombridge, 1996) assigned a category of threat to 252 carnivore species (the other 19 were classed as data deficient, and we have excluded them from what follows). Of these, 84 were listed as being at some risk of extinction (i.e., not in the lowest category of risk). If all 84 species were to go extinct, 34 genera would be lost entirely. How many genera would we lose if 84 randomly selected species went extinct? We simulated the process to answer this question, recording the numbers of genera lost in each of 100 trials (Purvis *et al.*, 2000a). The mode and median are 29 genera, and the observed value of 34 was equalled or exceeded in only seven trials (i.e., $P = 0.07$). These simulations suggest that we stand to pay a greater biodiversity cost than expected under random extinction.

This non-random pattern of extinction risk seems to be the rule rather than the exception (Russell *et al.*, 1998). If past extinctions have also been phylogenetically clumped, then more biodiversity has already been lost than random extinction would lead us to expect.

WHO GOES EXTINCT?

The apparent non-randomness of extinction risk poses an obvious question: why are the threatened species at risk of extinction? In this section, we take a comparative approach to this question, assessing the evidence for a range of hypotheses that are commonly found in the literature (see reviews in Soulé, 1983; Simberloff, 1986; Lawton, 1995; McKinney, 1997). Each hypothesis makes a prediction that certain attributes of species will correlate with extinction risk on the average.

(1)   Small populations are more likely than large ones to die out, other things being equal. Demographic stochasticity, local catastrophes, slow rates of adaptation, mutational meltdown and inbreeding are all more serious for small populations. Species' total population size can be figured extremely roughly as the product of their geographic range and their typical density. Small ranges and low population densities are therefore likely to confer enhanced extinction risk.

(2)   Island endemics are very likely to have small ranges. Additionally, they may have evolved in isolation from predators and competitiors, perhaps making them particularly vulnerable to effects of introduced species.

(3)   Species at higher trophic levels live at lower population densities and therefore are more prone to extinction risk.

(4)   Species with 'slow' life histories – small litters, slow growth rates, late sexual maturity and long interbirth intervals – cannot recover quickly from population bottlenecks and are more vulnerable to population extinction in the face of increased density-independent mortality, e.g., hunting.

(5)   Social structures involving complex mating displays (e.g., lekking), vocalisations or co-ordinated group defence render their bearers more vulnerable to extinction.

(6)   Species where individuals range widely are particularly vulnerable to edge effects, if density-independent mortality is high at the edges of nature reserves.

(7)   Last but not least, large body-size correlates with many of the extinction-promoting traits above. Larger species tend to have low

population densities, slower life histories, more complex social systems and larger home ranges. Additionally where carnivores are concerned, humans may be less tolerant of, and so more likely to persecute, larger species, providing a possible direct link between size and extinction risk.

This is not intended to be an exhaustive list. Perhaps surprisingly, rigorous comparative tests of these hypotheses are rare. Woodroffe & Ginsburg (1998), in a detailed study of 10 carnivore species, demonstrated that home range size was a more important predictor of population extinction risk in nature reserves than was population density. Large home ranges, together with low latitudinal limits, were also implicated as correlates of vulnerability to logging among 27 primate species. Large body size and, independently, slow life history were found to correlate with high extinction risk in a large-scale comparative analysis of birds (Bennett & Owens, 1997).

We tested the hypotheses above using data for fissiped carnivores that we collected from the literature and which will appear elsewhere (see Purvis *et al.*, 2000b for a more complete analysis). Estimates of extinction risk were taken from the IUCN Red List (Baillie & Groombridge, 1996), coded on a six-point scale. Among categories of the IUCN list, we lumped conservation-dependent with vulnerable, and extinct in the wild with extinct; results were not markedly affected by using instead a three-point scale, with lower risk, conservation dependent and vulnerable species being classified as intermediate. A wide range of potential correlates of extinction risk were tested, most with sample sizes of well over 100 species: geographic range; island endemicity; body mass; life history (gestation length, litter size, age at sexual maturity, interbirth interval); diet (trophic level, whether or not vertebrates comprised the main part of the diet); activity timing; space use (day range length, home range size, mean population density and ecological population density); and social structure (sociality and group size). Phylogenetically independent contrasts (Felsenstein, 1985; Pagel, 1992; Purvis & Rambaut, 1995) were used to account for similarities among related species inherited from common ancestors rather than acquired independently, and regression through the origin (Garland *et al.*, 1992) applied to the independent contrasts to test which variable in turn was a significant predictor of extinction risk.

As Table 2.2 shows, small geographic range was far and away the most important single predictor of extinction risk, with small litter size, low mean population density and diurnality also significant. When geographic range was controlled for by multiple regression, the pattern of correlation

Table 2.2. Results of 1-predictor and 2-predictor regressions to predict extinction risk

Predictor	N_c	Sole	With range
Geographic range	166	-10.98^{****}	N/A
Island endemic?	166	3.72^{****}	0.79
Body mass	155	1.55	2.74^{***}
Gestation length		1.59	1.81^{*}
Litter size	132	-2.58^{**}	-0.82
Age at sexual maturity	83	0.16	-0.18
Inter-birth interval		?	?
Diet		?	?
Diurnal?	142	2.19^{**}	1.45
Day range length	43	0.62	0.85
Home range size	81	-0.13	2.92^{***}
Mean population density	59	2.06^{**}	-2.01^{**}
Ecological population density	62	-1.06	-1.61
Group size	80	1.55	1.29

N_c: number of contrasts; $^*P < 0.1$; $^{**}P < 0.05$; $^{***}P < 0.01$; $^{****}P < 0.001$; N/A: not available.

in other variables changes markedly (Table 2.2). Large size and large home range size become significant, whereas island endemicity, litter size and activity timing drop out. The endemicity result is particularly noteworthy, it appears that the high level of threat typical of island endemics is simply in proportion with their typically small geographic ranges. There is no evidence of any additional effect that might be attributed to evolutionary isolation.

To test all the competing hypotheses concurrently required a multiple regression with model simplification. Initially all variables are included as predictors, and the predictor with the lowest marginal reduction in variance is dropped at each step until all remaining predictors are significant at the $P = 0.05$ level. Table 2.3(a) shows the model that results. Because of missing values in the data set, it is possible that predictors discarded early in the model simplification might invade the final model. Indeed, home range size displaces group size and population density to produce the second model in Table 2.3(b).

In both models, geographic range is the most important predictor of extinction risk. This, however, is not really strong evidence that small range predisposes species to extinction risk, because restriction to a small range is a criterion by which species can be assigned high extinction risk (Mace & Lande, 1991). Species feeding primarily on vertebrates are more at risk in both models, as are those with long gestation periods. Both of these are in

Table 2.3 (a). Regression equation resulting from model simplification

	Geographic range	Population density	Gestation length	Age at sexual maturity	Group size	Vertebrate-eater
Coefficient	− 0.430	− 0.218	0.245	− 0.879	0.355	1.810
t	− 7.24	− 3.35	4.34	− 2.65	2.29	7.23
P	< 0.001	0.002	< 0.001	0.01	0.03	< 0.001

N: 55 species, 48 contrasts; $r^2 = 78.9\%$.

Table 2.3 (b). Regression equation resulting from reintroduction of home range size and subsequent model simplification

	Geographic range	Home range	Gestation length	Age at sexual maturity	Vertebrate-eater
Coefficient	− 0.424	0.169	2.59	− 1.040	0.830
t	− 8.42	2.44	4.45	− 3.24	2.57
P	< 0.001	0.02	< 0.001	0.002	0.01

$N = 75$ species, 64 contrasts; $r^2 = 62.5\%$.

line with theory, but the association of risk with early sexual maturity is precisely the opposite of what theory would predict. Perhaps early maturity is associated with unstable population dynamics (Diamond, 1984b). Alternatively, the finding could reflect a difference between wild and captive populations. In the wild, it is common for the onset of reproduction to be inhibited by older members of the population. Such inhibition is presumably less common in captivity. If disproportionately many of the highly threatened species have been studied only in captivity, where social structure does not delay reproduction, then threatened species might seem to mature earlier.

The difference between the two regression models is that the first contains population density and group size, whereas home range size replaces these in the second. These three variables are intercorrelated in a way that simply explains the difference – knowing the population density and group size for carnivore species lets you make a good guess at its home range size. It is hard to say which aspect of space use or social structure is most important in predisposing lineages to the risk of extinction, but all of the associations are as predicted by theory.

Interestingly, neither model contains body size. While it is true that large species are more at risk (at least when geographic range is controlled for), it seems that the risk is because of their ecology and life history, not because of their size *per se*.

CONCLUDING REMARKS

We began this chapter with a cautionary note – carnivore conservation is difficult, expensive and unpredictable. After examining with different approaches both long-term (phylogenetic) and short-term (comparative analyses of threat levels) patterns of extinction in carnivores, we are even more convinced that carnivores are indeed an engima for conservation biology. Nevertheless, our phylogenetic approaches leave us optimistic for a number of reasons. First, adopting different analyses whereby both fossil and extant lineages are examined in parallel for what factors increase the likelihood of extinction are complementary. Generally, carnivores of large size living in isolated populations and eating strict carnivorous diets have been and are lost to extinction more than small omnivorous, widely-distributed species. This is alone is not surprising. What is unusual is that the independent (multivariate) comparative trends of extant taxa, classified as to various levels of extinction risk by the IUCN Red List, verify many of the characteristics that were thought to increase extinction proneness. It is the carnivore species for which there appears a discrepancy between classifications of threat and predicted threat we should now be focusing more attention toward. Species such as the black-footed cat (*Felis nigripes*), brown hyaena (*Parahyeana brunnea*), sea otter (*Enhydra lutris*) and culpeo (*Pseudolopex culpaeus*), all posted as 'least concern' on the IUCN Red Data List, each possess a combination of characteristics that make them of greater concern for potential extinction risk.

Our analyses of both fossil evidence and comparative trends across the complete carnivore phylogeny also suggest that extinction episodes are fairly common in carnivores. Unlike other mammal groups (e.g., insectivores), carnivores have historically shown the ability to adapt to novel environmental problems, often responding to extinction with a new radiating lineage. A clue for why this is the case may come from what sets carnivores apart from most other mammals – extreme variability. Some factors such as geographic range, diet and home range size are implicated as being important for extinction risk in both carnivores and primates (Purvis *et al.*, 2000b). However, life history traits such as gestation length and age at sexual maturity seem to be influential in carnivores but not primates. Overall, primates show less variability in home range size, some life histories, geographic range distribution, and body size. As a group, carnivores may therefore be safe from complete elimination, short of some extinction event that will make all discussion mute, partly because of such variation. The cause for concern is when population decline, habitat destruction and other factors of threat winnow away this variation. Our models reveal that the safest carnivore species are the European badger (*Meles meles*), mink (*Mustela vison*) and raccoon (*Procyon lotor*). A goal for carnivore conservation is to ensure that the entire landscape of carnivores is not as homogeneous as this group.

Acknowledgements

This work was supported by the Royal Society, the NERC (grant GR3/11526) and The Institute of Zoology (The Zoological Society of London). We thank the Editors for inviting us to the Meeting, Bob Wayne for his editorial expertise, and two anonymous referees for many helpful suggestions.

3

Interspecific competition and the population biology of extinction-prone carnivores

SCOTT CREEL, GÖRAN SPONG AND NANCY CREEL

It is sometimes argued that the major problems in conservation are not biological in nature, but have more to do with economics and other human affairs. It is true that humans have major impacts on the distribution and abundance of most large mammals, and it would be foolish to ignore human impacts on the conservation of carnivores (Childes, 1988; T. Clark *et al.*, 1996a). However, it is also true that some carnivores fare much better than others, despite facing similar human impacts. Thorny conservation problems remain for many carnivores, even after they are ensconced within a protected area.

A well-studied example illustrates this point. In Serengeti National Park, lions (*Panthera leo*) attained densities of 140 adults/1000 km² and spotted hyenas (*Crocuta crocuta*) attained densities higher than 1000 adults/1000 km², but African wild dogs (*Lycaon pictus*) never exceeded 15 adults/1000 km² (Frame *et al.*, 1979; Packer, 1990; Burrows, 1995; Hofer & East, 1995). It is unlikely that these differences in density are due to differences in the impact of humans. Hyenas, lions and wild dogs were all killed under predator control operations prior to the establishment of the National Park in 1951. No data tell how many of each species were shot in Serengeti, but data from Kruger illustrate the general point that all carnivores were commonly killed by game wardens in the first half of the twentieth century. Between 1903 and 1927 (when Kruger Park was established), wardens shot a minimum of 1272 lions, 1142 wild dogs, 660 leopards (*Panthera pardus*), 521 hyenas, and 269 cheetahs (*Acinonyx jubatus*) (Smuts, 1982). In the Serengeti region, lions were also hunted by the Maasai tribe, and were hunted by tourists over most of the current park's area until 1959, when the park was enlarged (Turner, 1987). Lions and hyenas are still hunted by tourists in game reserves bordering Serengeti, but wild dogs are not (Leader-Williams *et al.* 1996). Overall, the intentional impact of humans on wild dogs in Serengeti appear similar to the impacts on lions and hyenas.

Wild dogs have large home ranges and thus might be incidentally killed when they leave the park (Frame *et al.*, 1979; Woodroffe & Ginsberg, 1998). However, hyenas with home ranges near the center of the park also leave the park frequently (22% of 50 wet-season fixes: Hofer & East, 1995), where they are frequently killed by snares (Hofer & East, 1995). Lions at the center of the park are rarely snared, but snaring is thought to be common for lions with ranges near the park border (C. Packer, pers. comm., M. Turner in Schaller, 1972). In contrast, no wild dogs are reported to have died by snaring in Serengeti (Arcese *et al.*, 1995; Burrows, 1995), though they are snared elsewhere. Wild dogs rarely scavenge (Kruuk, 1972b; Creel & Creel, 1995), so they are weakly attracted to ungulate carcasses or the remains of slaughtering at snare lines. In contrast, spotted hyenas scavenge 20–50% of their meals and lions scavenge 10–30% of their meals, so their attraction to snare lines may be stronger (Gasaway *et al.*, 1991; Schaller, 1972; Packer, 1986). While snares are dangerous for all carnivores and can be a serious problem, their impact on wild dogs is not unusually strong.

Wild dogs are killed by viral diseases, including rabies and canine distemper, and domestic dogs are thought to be a significant reservoir from which viral diseases spill over to African wildlife (Roelke-Parker *et al.*, 1996), so viral epidemics could be considered an indirect human impact on protected carnivore populations. In some wild dog populations, mortality due to infectious diseases is severe (Gascoyne *et al.*, 1993b; Alexander & Appell, 1994). In other populations, infectious diseases have little impact on mortality (van Heerden *et al.*, 1995; Creel *et al.*, 1997b). As with other human impacts, spotted hyenas and lions are also vulnerable to viral diseases, and mortality can be severe. In a 15-year study of spotted hyenas in Kalahari, 43% of mortality was attributed to rabies (M. Mills, 1991). In Serengeti, canine distemper virus killed approximately 30% of the lion population in a period of nine months, with substantial mortality in the hyena population (Alexander *et al.*, 1995; Roelke-Parker *et al.*, 1996).

To return to our original point, measurable human impacts on wild dogs in Serengeti were similar to human impacts on the other carnivores, but wild dogs were always 10 to 100 times less common than lions and hyenas, and ultimately disappeared. Similar patterns of population density have been found everywhere that the three species have been studied – lions outnumber wild dogs threefold to 21-fold, and spotted hyenas outnumber wild dogs eightfold to 122-fold (Creel & Creel, 1996). For reasons that have to do with ecology, wild dogs are less common than sympatric large carnivores, and this makes wild dogs prone to local extinction. What ecological process underlies this effect? Is it important for carnivores in general? In this chapter, we propose that interspecific competition has particularly strong effects on carnivores, because it often leads to direct aggressive interactions (Palomares & Caro, 1999). While interspecific competition is taxonomically widespread, the stakes are unusually high for carnivores, due to their behavioral and morphological adaptations for killing. It would be surprising to see a goldfinch kill a chickadee over a seed, but it is commonplace to see a wolf kill a coyote, or a coyote kill a swift fox (Carbyn et al., 1994; Peterson, 1995).

MECHANISMS AND COMPLEXITIES

A relationship in which a pair of species competes for the same prey but one also preys on the other has been dubbed intraguild predation, and theory suggests that intraguild predation can have strong effects on the population dynamics of the subordinate competitor (Holt & Polis, 1997). Simple models of intraguild predation make some predictions that are not obvious, but seem valid for large carnivores, and are important for their conservation. For example, it is reasonable to assume that interspecific competition will be weak where prey density is high, and that high prey density will consequently reduce the likelihood of competitive exclusion (Karanth & Sunquist, 1995). However, an intraguild predation model suggests that competitive coexistence is more likely when prey density is low or intermediate (Holt & Polis, 1997). This prediction is closely linked to the model's assumption that the intraguild prey (i.e., the smaller predator) is more efficient than the intraguild predator at exploiting basal prey species for which both predators compete. This assumption is probably true in some cases. For example, African wild dogs are highly efficient hunters of prey that are also taken by lions (e.g., wildebeest: Fanshawe & Fitzgibbon, 1993; Creel & Creel, 1995), and lions often kill wild dogs (Scheepers & Venzke, 1995; Creel & Creel, 1996; Mills & Gorman, 1997). These patterns fit the assumptions of intraguild predation models, and as the models predict, the density of wild dogs is highest at intermediate prey densities (both within and across ecosystems: Mills & Gorman, 1997; Creel & Creel, 1998; and see below).

As intraguild predation models assume, interspecific competition in the real world is usually asymmetric, meaning it does not have an equal impact on each competitor (Lawton & Hassell, 1981). Asymmetric competition between carnivores can affect the subordinate competitor in three main ways. First, the dominant competitor may kill the subordinate competitor outright. This has been recorded for at least 97 pairs of carnivores, involving 54 victim species and 27 killer species (Palomares & Caro, 1999). Second, the dominant competitor may steal food from the subordinate, which can produce surprisingly large increases in the time and energy that must be expended to obtain sufficient food (Gorman *et al.*, 1998). Third, the subordinate competitor may simply avoid the dominant competitor to reduce the risk of fatal encounters or food loss (Mills & Gorman, 1997). Avoidance might involve occupying different habitats, or hunting at different times; both of these can be considered incipient character displacement. If the dominant competitor monopolizes areas of high prey density, then spatial avoidance might reduce prey availability for the subordinate species. This would increase the energetic costs of hunting (see below), and would reduce population density if the subordinate competitor was prey-limited.

Over evolutionary time spans, dominant competitors can affect subordinates by character displacement, imposing a selection pressure that alters the subordinate's morphology, behavior or ecology. In this chapter, our focus is on the immediate ecological and demographic impacts of competition, which operate on time scales relevant to conservation. That said, the boundary between evolutionary and ecological changes is not completely clear in some cases. For example, the tendency to cache food in trees (T. Bailey, 1993) or bury it (Malcolm, 1980), can respond to competition on both ecological and evolutionary time scales. For this chapter, it seems reasonable to assume that subordinate competitors generally act to reduce the impact of competition as much as possible, and the remaining impacts are unavoidable costs or trade-offs.

Conceptually, it is well-established that asymmetric competition can affect the distribution, demography and population dynamics of the weaker competitor (Lawton & Hassell, 1981; Tilman, 1986). In practice, it remains difficult to show that interspecific competition affects the dynamics of a given population, particularly when the species involved live at relatively low densities, are long-lived and are not particularly amenable to experiments. Demographic compensation can mask effects on survival and reproduction. If a competitive relationship is old (in the evolutionary sense), then mechanisms to reduce the impact are expected, and these mechanisms can make the competition subtle, despite its strength (Lack, 1971; Abramsky *et al.*, 1986). Active avoidance is a good example – a dominant competitor can have a strong effect on the density and distribution of a subordinate competitor even though the species now interact rarely. All that is required is an evolutionary history of consistent losses during aggressive interactions. To detect this sort of competitive relationship requires data on the spatial distributions of both competitors, and an independent measure of habitat quality for each of the competitors. Obtaining such data for large carnivores is a substantial undertaking.

In this chapter, we review studies of direct killing, food stealing and active avoidance, to assess how common these types of competition are, and how they affect demography and density. We then discuss their implications for conservation. Finally, we discuss how interspecific competition and habitat fragmentation might interact for species with large geographic ranges but low population densities, as is the case for many large carnivores.

ACTIVE AVOIDANCE

Direct killing and food stealing have obvious costs. When one sees a coyote kill a fox, the interpretation is relatively unambiguous. (That said, we still know little about the quantitative effects of intraguild predation on population dynamics. In the absence of experiments, it is difficult to establish whether mortality is additive or compensatory.) Spatial and temporal avoidance of competitors is more subtle, and its costs for the subordinate competitor are harder to measure. Despite these problems, active avoidance has been documented for a phylogenetically and geographically broad set of carnivores. There are also good counter-examples, where little or no evidence for active avoidance has been found for potentially strong competitors (e.g., kit foxes and coyotes, discussed below). Differences in methodology make it difficult to compare studies, but differences in habitat type may affect the need for active avoidance, and its utility. Most of the clear examples of active avoidance come from open habitats, and there is at least one example (the tiger and leopard) in which avoidance occurs in the more open of two parks but not in the more wooded (McDougal, 1988; Karanth & Sunquist, 1995).

In Asia, patterns of prey selection and space use have been studied for sympatric tigers (*Panthera tigris*), leopards and dholes (*Cuon alpinus*). McDougal (1988: p. 609) noted that tigers commonly kill leopards in Royal Chitwan National Park (Nepal), and stated that 'leopards are not common in habitat where tiger density is high. They are most prevalent on the peripheries of the park, sandwiched between prime tiger habitat on the one side, and cultivated village land on the other, dependent on both natural prey and domestic livestock'. While this pattern suggests that leopards are excluded from prime habitat by tigers, it is also possible that leopards are more able than tigers to exploit livestock as prey, and adjust their use of space accordingly. Karanth & Sunquist (1995) studied prey selection by tigers, leopards and dholes in Nagarahole National Park. While they did not present direct data on habitat selection or space use, they suggested that, unlike Chitwan, the distribution of leopards was not affected by tigers in Nagarhole. Prey densities are higher in Nagarahole than in Chitwan, and they attributed differences in competitive exclusion between the parks to Nagarahole's higher density of very large prey, particularly gaur (*Bos gaurus*), which are preferred by tigers. Because large prey were available in Nagarahole, tigers took prey averaging 66 kg, compared to 23 kg for leopards. Gaur comprised 45% of Nagarahole tiger kills, but less than 4% of leopard kills (Karanth & Sunquist, 1995). Dholes in Nagarahole preyed little on gaur, taking the smaller sambar (*Cervus unicolor*) in 90% of their kills (mean kill mass = 35 kg). Similar to their conclusions for leopards, Karanth & Sunquist suggested that the distribution of dholes was not constrained by larger carnivores. Johnsingh (1979, in Fox, 1984) concurred that niche separation in prey size, habitat type and diurnal patterns of activity reduces competition between dholes and big cats. This example illustrates that differences among ecosystems in the availability of prey can affect patterns of spatial avoidance, by altering overlap in prey selection. Also, note that this example does not support the prediction from intraguild predation models that competitive coexistence is more likely where prey densities are moderate or low (perhaps because some of the data are qualitative or indirect).

In Europe, Palomares *et al.* (1995; 1996) investigated spatial relationships among carnivores in Doñana National Park in southwestern Spain. Lynx (*Felis pardina*) preyed primarily on rabbits (*Oryctolagus cuniculus*), but opportunistically killed smaller carnivores, notably common genets (*Genetta genetta*) and Egyptian mongooses (*Herpestes ichneumon*). In Doñana, radiocollared lynx were found in a single habitat, Matasgordas, 85% of the time. In contrast, radiocollared mongooses and genets were found in areas other than Matasgordas 95% of the time. The densities of mongooses and genets were 2–10 times higher in the neighboring habitats not used by lynx. Because all of these carnivores prey heavily on rabbits, and rabbit densities are significantly higher in the habitat preferred by lynx, it is likely that mongooses and genets would occupy Matasgordas if possible, but their distribution is constrained by avoidance of lynx (Palomares *et al.*, 1995). An alternative explanation (which is viable for many cases of apparent avoidance) is that smaller carnivores and lynx differ in some unmeasured aspect of habitat preference. This alternative seems unlikely in the lynx–mongoose–genet example, because there is a sharp boundary to the area used by lynx, and this boundary aligns almost exactly with the boundary of the mongoose and genet distributions – the exclusion is almost complete.

In the Americas, interspecific avoidance has been studied with several species of canids, in several places. Mech (1977a) noted that the borders of adjacent wolf (*Canis lupus*) territories are often slightly separated, forming a buffer zone between territories that reduces the likelihood of interpack encounters. Fuller & Keith (1981) suggested that coyotes (*Canis latrans*) in Alberta use these buffer zones to reduce the likelihood of fatal encounters with wolves. In Manitoba, coyotes may have avoided wolves in late winter, and the densities of coyotes and wolves varied inversely, but coyotes sometimes trailed wolves actively in order to scavenge (Carbyn, 1982). Also in Manitoba, coyotes were more likely to follow wolves than vice versa (Paquet, 1991). Interactions between wolves and coyotes highlight the complexity of ecological relationships between carnivores. They potentially compete for prey such as ungulate fawns, and wolves often kill coyotes, but wolves can also provide coyotes with the remains of large carcasses, which are an important food source (particularly in winter). Paquet (1992) suggested that coyotes avoid wolves in areas where scavenging opportunities are rare, but overlap in areas where scavenging from wolves was more common. Coyotes must optimize the trade-off between a large but rare cost (being injured or killed) against a smaller but more frequent benefit (scavenging opportunities).

On a broader spatiotemporal scale, the removal of wolves from a region can allow range expansion by coyotes, as has been shown for the Kenai peninsula of Alaska and the prairies of the central US and Canada (Thurber *et al.*, 1992). The distribution of wolves may have a cascade effect on competition between coyotes and red foxes (*Vulpes vulpes*), because the ratio of coyotes to foxes is consistently lower where wolves are present than where wolves are absent, in the midcontinent of North America (Peterson, 1995).

Johnson *et al.* (1996) reviewed several other studies of sympatric populations of wolves, coyotes and foxes, some of which show avoidance and some of which do not. While it is likely that patterns of avoidance vary among ecosystems (dependent on variables such as habitat type and prey density), it is notable that many of the studies that did not detect avoidance did not use powerful methods. Some studies show clear-cut interspecific territoriality, as between grey zorros (*Dusicyon griseus*) and culpeo zorros (*Dusicyon culpaeus*: Johnson & Franklin, 1994b). In this case, interference competition by the larger culpeo seems to be the dominant factor determining the distribution of the smaller grey zorro. In contrast, North American red foxes and grey foxes (*Urocyon cinereoargenteus*) are often found with overlapping ranges, using overlapping sets of habitat types (see Johnson *et al.*, 1996). Such data will arise if the species genuinely do not avoid one another, but care is needed, because they can also be compatible with spatial or temporal avoidance on a scale finer than that used in the analysis. As Durant (1998: p. 371) notes, 'averaged over a long period ... a system might appear to have little spatial heterogeneity, yet at any particular moment might be extremely heterogeneous'.

Despite this caveat, there are cases, such as kit foxes (*Vulpes macrotis*) and coyotes in California, where competitors show strong overlap in habitat selection and diet, even at a fine spatiotemporal scale, and despite the fact that the dominant competitor often kills the subordinate (White *et al.*, 1994; White & Garrott, 1997). Perhaps the distribution of prey precludes active avoidance in these cases. In other words, one can avoid a dominant competitor only if there are locations that lack competitors but hold sufficient resources. Otherwise, one is constrained to face the competition.

Studies of cheetahs and wild dogs in competition with lions and spotted hyenas illustrate the importance and complexity of fine-scale active avoidance. Between 1987 and 1990, 95% of cheetah cubs in Serengeti died before the age of 14 months, and 73% of this mortality was due to predation – predators killed 69% of all cubs (Laurenson, 1995a). Over a period of 20 years, 91% of cubs that were killed fell prey to lions and spotted hyenas (which killed 79% and 12% of the cubs, respectively). It would be difficult for demographic shifts to compensate for this level of mortality, so cub predation is likely to affect population dynamics. Following this expectation, the density of cheetahs across ecosystems is inversely related to the density of lions, after controlling for variation in the density of prey (Laurenson, 1995a).

Based on anecdotal data, Eaton (1974: p. 84) wrote: 'Lions, when hunting the same prey as cheetahs, caused the cheetah to stop and move away. The mere presence or even sound of lions often resulted in cheetah changing their activities and moving away from the lions'. The diet of cheetahs overlaps substantially with the diets of lions and spotted hyenas (Kruuk, 1972b; Scheel & Packer, 1991; Caro, 1994), so the threat of direct competition ('encounter competition': Schoener, 1983) may affect the distribution of cheetahs relative to their prey. Point scans of cheetahs, lions, spotted hyenas and Thomson's gazelles showed that the gazelles were less common in the vicinity of cheetahs than in the vicinity of lions or hyenas (Durant, 1998). This pattern was not likely to arise from gazelles avoiding cheetahs, because hyenas and lions frequently kill gazelles (Kruuk, 1972b; Schaller, 1972), and hyenas and lions greatly outnumber cheetahs (Caro &

Durant, 1995). Consequently, the selection pressure on gazelles to avoid lions and hyenas should be stronger than the pressure to avoid cheetahs. Although cheetahs were not consistently found in areas with low densities of competitors, Durant (1998) suggests that the association of cheetahs with areas of low gazelle density is caused by avoidance of larger competitors. In support of Durant's explanation, cheetahs were likely to be moving, and unlikely to be hunting, when competitors were present.

In the Southern Kalahari, brown hyenas (*Hyaena brunnea*) are behaviorally subordinate to spotted hyenas, which easily displace brown hyenas from carcasses (M. Mills & M. Mills, 1982). In 19 direct interactions, brown hyenas were chased or harassed in 14 cases, killed in one case and avoided contact in four cases. To test whether active avoidance affected the distribution of brown hyenas, M. Mills & M. Mills (1982) used transect counts of both species in two areas of similar size and habitat type. They found a significant pattern of avoidance – spotted hyenas outnumbered browns by a factor of 8 in one area, while browns outnumbered spotteds by a factor of 2 in the second area. They also suggested that brown hyenas den in sand dunes (93% of 40 dens) to avoid the riverine habitats favored by spotted hyenas (69% of spotted hyena dens were in riverine areas).

For African wild dogs, avoidance of lions is clear-cut. Lions commonly kill wild dogs, accounting for 26% of 116 known-cause deaths in three ecosystems (Moremi: McNutt, 1995; Kruger: van Heerden *et al.*, 1995; Selous: Creel & Creel, 1998). An attempted reintroduction of captive wild dogs into Etosha National park failed when lions killed 6 of 13 dogs in a period of weeks (Scheepers & Venzke, 1995). Hyenas rarely kill wild dogs, but steal carcasses at substantial rates in some ecosystems (e.g., hyenas ate at 73% of 124 wild dogs' kills in Serengeti: Kruuk, 1972b; Fanshawe & Fitzgibbon, 1993). Given these pressures, wild dogs would do well to avoid larger competitors – the problem is to avoid competitors while maintaining adequate rates of encounter with prey.

Across ecosystems, there is a significant negative correlation between the densities of wild dogs and lions, and between the densities of wild dogs and hyenas (Creel & Creel, 1996). This pattern might be due to active avoidance, although it could also arise through purely demographic impacts (e.g., wild dog survival might be low where lion density is high). Two studies suggest that, within an ecosystem, wild dogs actively avoid lions, but do not avoid hyenas. In Kruger National Park, impala (*Aepyceros melampus*) comprised 73% of 79 wild dog kills (Mills & Gorman, 1997). In accord with this prey preference, the most preferred habitat (out of six types) was the same for wild dogs and impala. However, the second and third most-preferred habitats for wild dogs were the least preferred for impala, and the second most-preferred habitat for impala was least preferred for wild dogs. Overall, the habitat preference ranks of wild dogs did not correlate with those of impala. This pattern was explained by avoidance of lions, with a significant negative correlation between the ranked habitat preferences of wild dogs and lions (M. Mills & Gorman, 1997). The ranked habitat preferences of wild dogs and hyenas were not significantly correlated, but the three habitats least used by hyenas were also the three habitats most preferred by wild dogs, and the correlation between wild dog habitat preferences and an index of hyena density tended to be negative ($r = -0.61$).

Wild dogs also avoided lions in the Selous Game Reserve. For an area of 2600 km², there was a negative correlation between the spatial distributions of wild dogs and lions (Figure 3.1). The only significant habitat preference by wild dogs was for deciduous woodland, which held a low density of prey, but was little used by lions (Table 3.1). Data from 1352 encounters with prey during 3271 km of wild dog follows showed that only 3.75 prey were encountered per km, significantly lower than prey encounter rates in other habitat types, which ranged from 4.6 to 16.4 prey/km (Table 3.1). The energetic costs of preferentially using habitats with low prey density are discussed below.

Direct encounters between wild dogs and lions are rarely seen, perhaps because wild dogs can detect lions from a distance and avoid them. As mentioned previously, the strength of selection on avoidance behavior will depend on the cost of failing to avoid an encounter. For wild dogs, encounters with lions are clearly costly. Dogs were killed in two of 19 encounters, abandoned food in four and travelled several kilometers at a trot in some retreats (Figure 3.2; also see M. Mills & Gorman, 1997). Wild dogs exhibited clear avoidance in 17 of 19 natural encounters with lions in Selous. If dogs were resting when they detected lions, they moved off in single file, regardless of the time and temperature. If dogs encountered lions when travelling, their behavior changed immediately and markedly. From a loose group following several lines of travel, they congregated in single file, adopted a more linear travel route, moved at a steady trot and frequently glanced back. The dogs sometimes mobbed a lion group before retreating, by approaching very cautiously to within 20 m (standing on their hind legs if the view was obstructed) and alarm barking.

In response to audiotape playbacks of lion roars broadcast from 300 m or less, wild dogs tried to locate the lions within one minute, by approaching carefully or scanning in the direction of the roars. Failing to locate any lions, their next movement was directed at least 90° away from the roars in 11 of 12 trials with six different packs. Collectively, natural encounters and

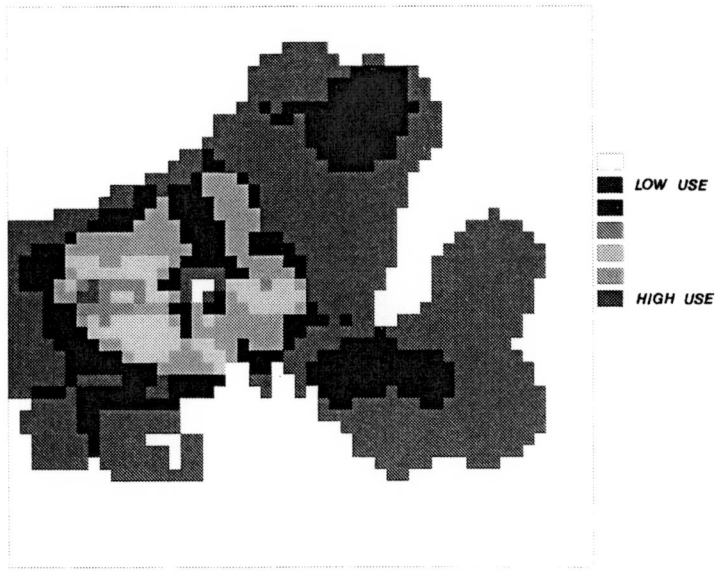

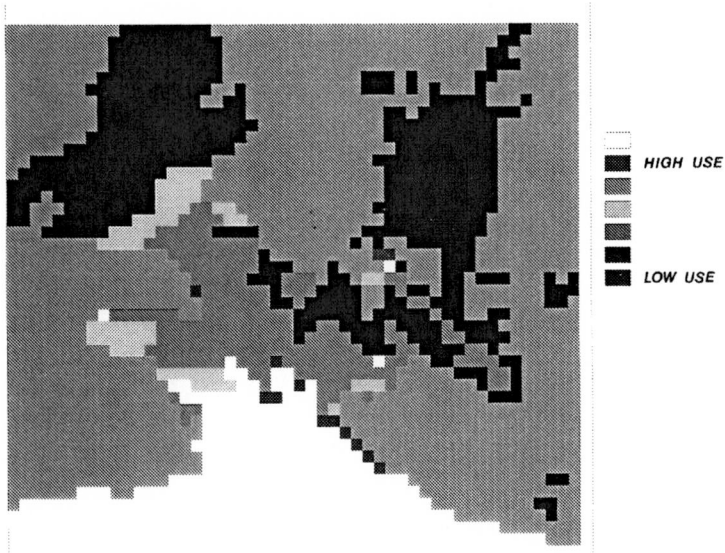

Figure 3.1. Space-use distributions of (a) wild dogs and (b) lions in the northern Selous Game Reserve. Plot frames show 37° 45' E to 38° 30' E and 7° 15' S to 7° 55' S. Light areas indicate little use, dark areas indicate heavy use, and white areas indicate no data.

Table 3.1. Encounters between wild dogs and prey as a function of habitat type[a]

| Habitat type | Wild dog | | | | | Lion preference rank |
	preference ratio[b,c]	prey herds encountered	prey individuals encountered	kilometers traveled in habitat	prey individuals per km travelled	
Deciduous woodland	1.30 ± 0.13*	422 (31%)	5818 (25%)	1552 (47%)	3.75	4
Thorn woodland	1.20 ± 0.27	370 (27%)	5908 (25%)	825 (25%)	7.16	2
Long grass	1.07 ± 0.20	235 (17%)	5183 (22%)	458 (14%)	11.31	5
Short grass	0.94 ± 0.28	297 (22%)	6462 (27%)	393 (12%)	16.4	3
Riverine thicket	not measured[d]	28 (2%)	199 (1%)	43 (1%)	4.63	1
Totals		1352	23 570	3271		

[a]Entries give raw numbers, then percentage in brackets.
[b]Preference ratio = % of radiolocations within a habitat type divided by % of homerange covered by that habitat type.
[c]Mean ± S.E.
[d]Riverine thicket covered < 1% of the wild dog study area, making its preference ratio highly sensitive to measurement error.
* $P < 0.05$ for single-point t-test comparing preference ratio with 1.

(a)

(b)

Figure 3.2(a,b). An encounter between African wild dogs and a male lion. Wild dogs are often killed in such encounters. (Photograph: G. Spong.)

playbacks showed clear-cut avoidance in 28 of 31 cases (χ^2 with Yates correction = 18.6, $P = 0.001$).

Wild dogs do not obviously avoid spotted hyenas. As noted above, there was not a significant negative correlation between the habitat preferences of the two species in Kruger. In Selous, the spatial distributions of wild dogs and hyenas actually show a significant *positive* association (Cramer's $V = 0.71$, $\chi^2 = 40.8$, df $= 13$, $P < 0.001$; this contradicts our prior report that the distributions were uncorrelated, which used a less powerful method: Creel & Creel, 1996). It is interesting to speculate why wild dogs avoid lions but not hyenas. Perhaps wild dogs would avoid hyenas if they could, but they simply cannot. Hyena densities are typically two to four times higher than lion densities (Creel & Creel, 1996), so hyenas cover the landscape more continuously than lions. Second, hyenas are well-adapted for scavenging. They are able to detect kills by smell from distances up to 4.2 km, and by hearing from distances up to 10.5 km (M. Mills, 1991). Given their extraordinary ability to locate kills, there may be no 'holes' in the hyena distribution, particularly for ecosystems with high hyena density. This explanation is supported by studies of food stealing (discussed below). Finally, if hyenas also avoid areas of high lion density, then the association between wild dogs and hyenas could be caused by a shared pattern of lion avoidance.

It is difficult to quantify the costs of active spatial avoidance for subordinate competitors. In some examples, subordinates are displaced into habitats where their rates of encounter with prey are low. If hunting is energetically costly, then a reduction in prey encounter rates is likely to carry a cost in terms of reproduction or survival. A rough sketch of the costs of avoidance is possible for African wild dogs. By avoiding lions, wild dogs in Selous encountered only 3.75 prey individuals for each kilometer they travelled in deciduous woodland, while the average prey encounter rate for other habitats was 9.88 prey/km (Table 3.1). All else equal, wild dogs would have to increase hunting effort by a factor of 2.6 to maintain average foraging success while restricting their movements to deciduous woodland. Gorman *et al.* (1998) estimated that hunting increases a wild dog's metabolic rate to 25 times the resting rate, with an average daily expenditure of 5.2 times basal metabolic rate (BMR). Independent data suggest that the maximum sustainable long-term expenditure of energy is six to seven times BMR (Hammond & Diamond, 1997). Consequently, it is not likely that wild dogs could maintain the 2.6-fold increases in foraging effort that would be required by a complete retreat to deciduous woodland.

FOOD STEALING

Although many studies report that one carnivore scavenges from another, quantitative estimates of food loss are not common. Good data come from cheetahs in Serengeti, which lost 12.7% of 110 carcasses to spotted hyenas and none to lions (Caro, 1994). Caro estimated that cheetahs lost an average of 65% of the edible meat on each carcass stolen, with a total loss of 9.2% of the edible meat they had captured. Serengeti cheetahs eat 1.4 meals/day (Caro, 1994), so a loss of 9% to kleptoparasites equates to one meal lost per eight days. Only 23–47% of hunts are successful (Caro, 1994), so a cheetah in Serengeti must make about two to five extra hunts per week, together with the travel and search required to produce a hunting opportunity, in order to replace meat lost to hyenas. Although the costs and risks of hunting have not been quantified directly, it is plausible that this additional foraging effort could affect reproduction or survival.

Hyenas also scavenge from African wild dogs, and the level of kleptoparasitism has been measured in five studies (Table 3.2). These data provide an unusual opportunity to compare populations. First, one can compare the rate of kleptoparasitism for a given pair of species across ecosystems, and try to identify variables that affect its intensity. Second, one can qualitatively assess kleptoparasitism's impact on population dynamics.

For hyenas stealing from wild dogs, the rate of kleptoparasitism varies among studies by a factor of 43 (Table 3.2) – in this case, the fact that both species are found in an area does not reveal much about the impact of interspecific competition. In the Selous Game Reserve and Kruger National Park, hyenas took little food from wild dogs (Table 3.2: M. Mills & Biggs, 1993; Creel & Creel, 1995). In contrast, hyenas took from 60–86% of wild dog kills in the Serengeti National Park and the Ngorongoro Conservation Area (Kruuk, 1972b; Fanshawe & Fitzgibbon, 1993). Surprisingly, hyena population density by itself is not a strong predictor of the intensity of kleptoparasitism across ecosystems (Table 3.2). Logically, high densities of kleptoparasites should increase the likelihood that carcasses will be found quickly enough to make kleptoparasitism worthwhile, and within a single ecosystem, it is likely that changes in the density of hyenas would alter the rate of kleptoparasitism. We know of no data that can test this hypothesis directly. In the absence of such data, it is intriguing that the proportion of carcasses stolen does not show a tight relationship to hyena population density across ecosystems. This observation suggests that some aspect of the environment can modify the intensity of competition between hyenas and wild dogs (Table 3.2). Two possibilities are plausible. First, scavengers can probably locate carcasses more easily in open habitats (by sight, sound or smell). Second, vultures also locate carcasses more quickly in open habitats (Figure 3.3), and carnivores often watch vultures, using rapid descents as a cue to locate carcasses. Vultures congregate more quickly and in larger

Table 3.2. The frequency and outcome of competition between wild dogs and hyenas at wild dog kills

Population	Number of wild dog kills	Number (%) with hyenas	Number (%) at which hyenas ate	Hyena population density (adult/km^2)	Thickness of habitat	Congregation of vultures at kills[b]	Status of dog population
Selous, Tanzania[g]	404	76 (18%)	14 (2%)	0.32	Wooded	Slow, few	Large, stable
Kruger, S. Africa[f]	52	Frequent	0 (0%)	0.14	Wooded	Slow, few	Large, stable
Serengeti, Tanzania[d]	62[a]	53 (86%)	53 (86%)	0.17	Open	Rapid, many	Gone
Ngorongoro/Serengeti, Tanzania[c]	62	46 (74%)	37 (60%)	0.82 – 1.43	Open	Rapid, many	Gone
Aitong, Kenya[e]	43	18 (41%)	Rarely	Moderate	Intermediate	Moderate, varies with migration	Gone

Sources: [c] Kruuk, 1972b; [d] Fanshawe & Fitzgibbon, 1993; [e] Fuller & Kat, 1993; [f] Mills & Biggs, 1993; [g] Creel & Creel, 1995.
[a]Original data excluded kills of gazelle fawns.
[b]Vulture congregation from pers. obs., Mills (pers. comm.), Holekamp (pers. comm.).

(a)

(b)

Figure 3.3. Vultures and other avian scavengers serve as a cue for kleptoparasites to locate kills made by other carnivores. In open ecosystems where vultures rapidly congregate in large numbers, direct competition between carnivores at carcasses may be more intense. Typical congregations in (a) Serengeti, where hyenas commonly steal carcasses from wild dogs. (Photograph: S. Creel.) (b) Selous, where competition at carcasses is less common. (Photograph: S. Creel.)

numbers in open habitats than they do in woodland, and this is likely to improve the ability of terrestrial kleptoparasites to locate kills. Both of these factors might affect the intensity of interference competition between hyenas and wild dogs (Table 3.2).

There is considerable variation in the time that dogs are able to feed before losing a kill, and this is related to the numbers of wild dogs and hyenas present. For wild dog kills on the shortgrass plains of Serengeti, the ratio of wild dogs to hyenas ranged between 1 dog : 10 hyenas and 9 dogs : 1 hyena, with a median ratio of 2 dogs : 1 hyena (Fanshawe & Fitzgibbon, 1993). Impressively, Serengeti wild dogs were outnumbered by hyenas at 30% of their own kills. The duration that wild dogs retained a kill increased as a function of the dog-to-hyena ratio, suggesting that wild dogs lost carcasses on which they would have continued to feed, if undisturbed. At wild dog kills in Selous (where the habitat is more closed and hyenas are less dense), the ratio of dogs to hyenas was much higher (median = 7.4 : 1, min = 0.63 : 1, max = 20 : 1), and hyenas only outnumbered dogs at one of 336 kills (Creel & Creel, 1996). As expected, wild dogs rarely lost food to hyenas in Selous (Table 3.2), and the duration of feeding was not affected by hyenas. These data illustrate how dramatically the intensity of kleptoparasitism between a pair of carnivores can vary, depending on population density and properties of the environment.

It is difficult to estimate how much food has been lost when a carcass changes hands. Although wild dogs lost many carcasses to hyenas, sometimes soon after the kill, Kruuk (1972b) suggested that the dogs generally lost portions that they would not have eaten anyway. Kruuk (1972b) and, later, Serengeti researchers (Fanshawe & Fitzgibbon, 1993) considered wild dogs less well equipped than hyenas or lions to eat skin and bones, but their observations were generally made at kills with competitors present, which complicates the problem of estimating what might have been eaten by individuals that were not disturbed by competitors. In Selous, where competitors were generally absent, wild dogs often ate long bones and ribs, and the skins of prey smaller than wildebeest (*Connochaetes taurinus*). For impala, wild dogs are capable of eating everything except gut contents and small portions of the skull, pelvis and vertebrae (even skulls and hooves are sometimes crushed and eaten completely). Van Valkenburgh (1996) reviewed videotapes of lions, hyenas, wild dogs and cheetahs consuming 33 carcasses (of five species) in the Masai Mara National Reserve, and found that wild dogs and hyenas ate similar proportions of skin and bone, while both ate a smaller proportion of muscle than lions. Thus, wild dogs are capable of eating most of the food they lose to hyenas (a conclusion also supported by the observation that wild dogs rarely give up carcasses without a struggle).

In contrast, cheetahs ate a much higher proportion of muscle than the other carnivores, and ate little skin and bone. A study that did not account for this difference would probably overestimate the impact of kleptoparasitism on cheetahs (unlike Caro's study above).

Two lines of evidence suggest that interference competition has an effect on wild dog population dynamics, though it is difficult to determine whether lions or hyenas drive the effect, because the densities of lions and hyenas correlate positively (Stander, 1991a). First, across ecosystems in which population densities have been measured, there is a significant negative correlation between the densities of wild dogs and lions, and between the densities of wild dogs and hyenas (Creel & Creel, 1996). Second, longterm data for Serengeti show that wild dogs were most abundant in the 1960s and early 1970s, when lions and hyena densities were relatively low (Kruuk, 1972b; Schaller, 1972; Frame et al., 1979). Wild dogs declined significantly over the next two decades, as lion and hyena numbers doubled, tracking increases in prey density (Burrows, 1995; Hanby et al., 1995; Hofer & East, 1995; Sinclair, 1995). The Ngorongoro Crater provides another natural experiment – wild dogs appeared just after a crash in the lion population in the mid-1960s (Estes & Goddard, 1967; Packer et al., 1991b). As the lion population rebounded (increasing fivefold between 1965 and 1980) wild dogs disappeared and remained absent.

In Northern Botswana's Chobe National Park, spotted hyenas commonly challenge lions for their kills (Cooper, 1991). Hyenas located 79% of 135 lion kills and actively trailed the lions in 10% of their hunts. Hyenas obtained meat at 21% of the lion kills they located, and obtained skin and bones at a further 59% of the kills. Combining the probability of locating a kill with the probability of feeding once present, hyenas obtained food from 63% of all lion kills. Cooper (1991) estimated that hyenas took 17% of lion kills before the lions were satiated, stealing about 20% of the food that the lions would have eaten in the absence of kleptoparasitism. Male lions strongly affected the outcome of contests over kills. When male lions were present, hyenas never obtained food. When male lions were absent, hyenas attempted to take food on 82% of occasions, with an outcome that depended on the ratio of lionesses to hyenas. Singletons and pairs of lionesses invariably lost their kills, but the probability of retaining a carcass increased up to a group size of 10 lionesses, at which point carcasses were never lost.

In Kruger National Park, leopard kills were detected by spotted hyenas in 54% of all cases, but were rarely stolen (Bailey, 1993). Leopards on the ground are easily driven from a kill by lions or wild dogs (Schaller, 1972; Creel & Creel, 1995), but they rapidly cache kills in trees or hide them in thickets, where they are presumably less vulnerable to scavenging. In Kruger, leopards tree-cached 84% of their kills, and were then able to feed without losses for an average of 2.4 days (Bailey, 1993). When leopards and hyenas encountered one another in Kruger, direct aggression was rare (T. Bailey, 1993). A different pattern of interactions was seen for leopards in northeastern Namibia (Stander *et al.*, 1997). There, only 12% of leopard kills were visited by other carnivores (hyenas, wild dogs, lions and other leopards), and tree-caching was rare, dangerous and ineffective. Of 90 kills dragged into thick cover, only 9% were located by other carnivores, and only 1% was lost. Namibian leopards rarely cached in trees, but of three cases, two kills were stolen and one leopard was killed by lions. Stander *et al.* (1997) reviewed several leopard studies and concluded that tree-caching is used primarily when the habitat is open and the rate of interaction with other carnivores is high, so that kills cannot be dragged to thick cover without a high probability of detection by would-be scavengers. Here again, we see that environmental differences can mediate the effects of interspecific competition among carnivores.

In North America, interactions between cougars (*Puma concolor*) and other carnivores at kills have been described for three sites. In Yellowstone National Park, bears visited 33% of 58 cougar kills, taking an average of 16 kg of the biomass edible for the cats (Murphy, 1998). This equates to a loss of 0.6 kg of food per day, or 17–26% of the cougars' daily energy requirements. In Glacier National Park, 47% of 15 cougar kills were detected by grizzly bears (*Ursus arctos*), which displaced the cats in 27% of cases, though quantitative estimates of food loss were not possible (Murphy, 1998). In the Frank Church Wilderness, 40% of 33 cougar kills were visited by coyotes, but no coyote kills (out of 24) were visited by cougars (Koehler & Hornocker, 1991). The number of carcasses that changed hands was not reported, but two coyotes were killed while trying to scavenge from cougars. These data illustrate a further complexity in the ecological interactions between carnivores. An intraguild predator may not only compete with and prey on a subordinate competitor, but may also be a source of scavenging opportunities. For some pairs of species, the subordinate competitor trades a risk of being killed against the energetic benefits of scavenging (e.g., coyotes scavenging from mountain lions, or jackals scavenging from hyenas: Koehler & Hornocker, 1991; Kruuk, 1972b). For other pairs of species, subordinate competitors do not attempt to scavenge from dominant competitors (e.g., cheetahs and wild dogs very rarely attempt to scavenge from lions or hyenas: Caro, 1994; Creel & Creel, 1996). Perhaps, if the dominant competitor is sufficiently well-armed relative to the subordinate, then the risk of injury may be too high to make scavenging worthwhile. Alternatively, the feeding behavior of the subordinate species may require uninterrupted feeding, rather than quick snaps at a carcass between aggressive encounters. Obviously, the impact of interspecific competition will be greater for subordinate competitors that do not gain scavenging opportunities from the dominant competitor, if all else is equal.

DIRECT KILLING

Two recent reviews discuss intraguild predation in carnivores (Palomares & Caro, 1999). Briefly, Palomares & Caro (1999) found 97 pairs of carnivores for which intraguild predation has been well documented, involving 54 species of victim and 27 species of killer, This tally excludes studies that simply found one carnivore in the feces of another, which might be due to scavenging. Within the order Carnivora, intraguild predators often do not eat their victims, so the killing is not predation in the normal sense. For 21 pairs of species, the intraguild predator commonly ate its victim in 10 cases, but 11 cases are better viewed as extreme interference competition, because the killer did not normally eat the victim (Palomares & Caro, 1999). The percentage of annual mortality due to intraguild predation ranged widely, from only 4% (martens preyed on by coyotes) to 87% (kit foxes preyed on by coyotes: Thompson, 1994; White & Garrott, 1997). Taking the average of 24 studies, the proportion of known deaths due to intraguild predation was 38% (Palomares & Caro, 1999). This value seems high at first glance, but we do not know the degree to which this mortality is additive in most cases, so the impact on population dynamics is hard to estimate. For many pairs of species, victims were generally juveniles. If juveniles have low reproductive value (as they usually do) then demographic compensation is likely to absorb the mortality with no change in population dynamics. If prime-aged females are commonly killed, a stronger effect on population dynamics is expected. With these caveats in mind, we present two case studies to show that direct interspecific killing can have strong effects on population dynamics.

Wolves commonly kill coyotes, and wolves may limit the distribution and abundance of coyotes (Peterson, 1995). However, the diets of wolves and coyotes show many dissimilarities, so an inverse relationship between their densities might also be due to differences in habitat selection unrelated to competition. The reintroduction of wolves into Yellowstone National Park provides an experiment that resolves this ambiguity. Yellowstone was declared a National Park in 1872, but wolves and coyotes were actively persecuted until the mid-1930s. From 1908 to 1935, at least 132 wolves and 4156 coyotes were killed by the government (Phillips & Smith, 1996). The government abandoned predator control in 1935, just as the wolf was eradicated. After a long absence, 31 wolves were released in Yellowstone in 1995 and 1996 (Phillips & Smith, 1997), and the population now numbers about 130. Of these, 40–50 wolves reside in the park's Northern Range, where coyotes were under study prior to the wolf reintroduction (Gese *et al.*, 1996a). Prior to wolf reintroduction, the coyote population on a study site in the Northern Range held 80 coyotes in 12 packs (Crabtree, 1998). Post-wolf, this population was reduced to 36 coyotes in 9 packs. The mean size of surviving packs dropped by 40%, from 6 to 3.6 adults and yearlings. Wolves were seen to kill 13 coyotes, with 19 additional deaths inferred from coyote carcasses with injuries and tracks attributable to wolves. For the post-wolf period, 25–33% of annual mortality has been attributed to wolves (Crabtree, 1998). We can infer that coyote mortality due to intraguild predation is largely additive, because population size has declined by 55% over a period of three years.

In a second example, coyotes are the intraguild predator instead of the prey. The San Joaquin kit fox is an endangered subspecies of kit fox, for which populations declines of 60% to 80% were observed in the 1980s and 1990s (White & Garrott, 1997). Studies at eight locations (36 years of observation) suggested that kit foxes were limited mainly by fluctuations in prey abundance and/or interspecific competition with coyotes. Using data from eight locations, White & Garrott (1997) estimated the proportion of each kit fox population that was killed by coyotes. Across populations, adult fox survival rates varied from 0.35 to 0.75 (mean = 0.55), and the proportion of deaths due to coyotes varied from 50% to 87% (mean = 70%). The proportion of kit fox populations killed by coyotes ranged from 24% per year to 63% per year (mean = 39% per year). White & Garrott (1997) showed that the mortality due to intraguild predation was largely additive. In low density kit fox populations, annual mortality due to factors other than predation is about 20%, so that mortality above 20% per year cannot be compensatory. Figure 3.4 shows the relationship between annual survival and annual offtake by coyotes, if mortality due to predation is completely compensatory up to 20%, and additive thereafter. The data for actual kit fox populations fall just below this line, demonstrating that intraguild killing is mainly an additive force of mortality (Figure 3.4). In three studies, mortality due to coyote predation was almost completely additive (points falling on the line in Figure 3.4). Across all eight studies, mortality due to coyote predation was 88% additive, suggesting that intraguild predation is a strong limiting factor for kit foxes.

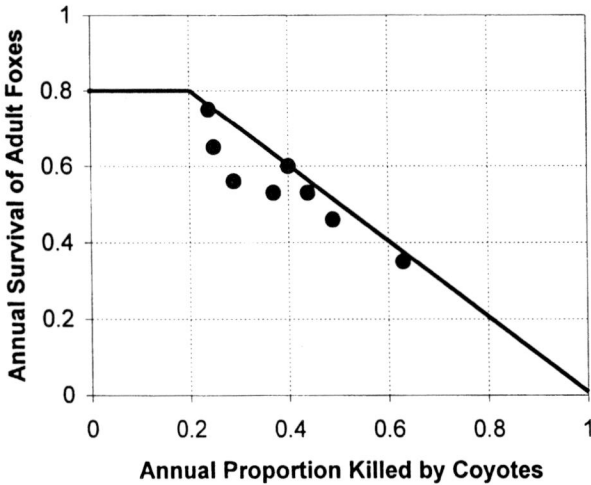

Figure 3.4. Annual survival rates for kit foxes as a function of coyote predation, in eight studies. The heavy line shows the relationship expected if mortality due to predation is compensatory up to 20%, and additive thereafter. Data from White & Garrott (1997).

IMPLICATIONS OF INTERSPECIFIC COMPETITION FOR CONSERVATION

Interspecific competition among carnivores can limit spatial distributions, constrain habitat selection, reduce prey encounter rates, reduce food intake (or require increased hunting effort) and increase mortality rates. For some species, a negative correlation between the densities of competitors suggests that these impacts limit the population density of the subordinate competitor (e.g., cheetahs limited by lions in Serengeti: Laurenson, 1995a). In a few cases, temporal changes in population densities support the correlational evidence (e.g., changes in the densities of wild dogs, lions and hyenas in Serengeti and Ngorongoro: Creel & Creel, 1996). In a handful of cases, humans have eliminated or reintroduced a dominant competitor, providing an experimental test of the effects of competition on population density (e.g., wolves and coyotes in Yellowstone: Phillips & Smith, 1996; Crabtree, 1998).

Taken together, the data suggest that large carnivores commonly affect the distribution, demography and population dynamics of medium-sized carnivores, which in turn limit smaller carnivores. What does this imply for conservation? The most important implication is that networks of protected areas with different ecological properties will be needed to preserve complete carnivore guilds. A protected area that provides ideal ecological conditions for wolves will probably be less than ideal for coyotes. Parks that support high densities of lions will probably support low densities of wild dogs, or none at all (Vucetich & Creel, 1999). In some parts of the world, policies are in place that produce this type of network. In Tanzania, National Parks allow photo-tourism but do not allow hunting. The National Parks are generally situated in areas of spectacular ungulate density, and therefore support high densities of lions and hyenas (which correlate with lean season ungulate density: van Orsdol *et al.*, 1985; Stander, 1991a). Because of the density of dominant competitors, wild dogs do not fare well in most of the National Parks. In contrast, Tanzanian Game Reserves (which are used for trophy hunting instead of photo-tourism) are generally located in areas that support lower densities of ungulates, and thus lower densities of lions and hyenas. As expected, wild dogs fare better in the Game Reserves. A similar pattern has been suggested for cheetahs in southern Africa.

The ubiquity and strength of competition between carnivores is cause for concern that subordinate competitors may drop out of ecosystems that favor dominant competitors. Particularly in nations that have a low proportion of public land under protection, carnivore conservation would benefit from using the principle of complementarity to assign conservation priorities, rather than relying on the principle that all species will flourish where apex carnivores persist. This highlights the importance of combining ecosystem-level and species-level approaches to carnivore conservation. Landscape and ecosystem level analyses are necessary to identify priorities for limited conservation dollars. But if interspecific competition among carnivores is common and strong, then conservation efforts will be more successful if they are guided by a knowledge of the ecological relationships between carnivores. Those concerned with the conservation and management of carnivores have had a long-standing focus on predator-prey relationships (Fuller, Chapter 8). Interactions with prey can affect the density and distribution of predators, so prey availability is important in assessments of the viability of carnivore populations. The same logic should be applied to ecological relationships between carnivores and potential competitors. Predation is only one of the ecological interactions in which carnivores participate – a priori, it is not obvious that a carnivore population will be limited by prey availability (bottom–up), or by competition and intraguild predation (top–down). The largest carnivores in an ecosystem are likely to be limited from the bottom up, but some very large carnivores are subject to surprisingly strong competitive interactions (e.g., spotted hyenas and lions: Cooper, 1991). For specific populations, evaluating the relative strengths of competition and prey limitation is likely to produce a good set of conservation actions.

Some factors mediating interspecific competition and its effects on population dynamics

The density of the dominant competitor is probably the strongest factor influencing the impact of competition on the subordinate competitor, but counter-examples exist, where some measure of competition (e.g., food loss) does not covary with the density of the dominant competitor in a one-to-one fashion. Several factors may modify the impact of competition on carnivore populations.

Open habitats appear to intensify competitive interactions. Many of the best examples of competition and intraguild predation come from savannas, prairies, open woodland and semi-deserts. This pattern may arise simply because field studies are biased toward open study sites and competitive interactions are easier to observe in open habitats. However, cases in which a pair of species has been studied in open and closed habitats support the hypothesis that visibility exacerbates competition between carnivores. High densities of flying scavengers may exacerbate interference competition at kills, especially if the habitat is also open. Flying scavengers locate carcasses more rapidly than those on the ground, and descending vultures can be detected over great distances, particularly if the habitat is open. Where birds serve as a cue for terrestrial scavengers, competitive interactions between carnivores will be more common.

The density of prey may also mediate the effects of competition among carnivores, but this effect is complex. Standard competition theory predicts that competition will weaken as prey density increases, because food is less likely to be limited. This line of logic assumes that live prey are the resource for which carnivores compete – but that assumption may be false. If competition is for carcasses (rather than live prey) then increasing prey density might not decrease the strength of competition. In this case, an increase in prey density may actually increase the strength of competition, by supporting a higher density of competitors. For example, the densities of lions and hyenas are positively correlated to the density of ungulates (van Orsdol *et al.*, 1985; Stander, 1991a). For wild dogs, this implies that food loss to hyenas and the risk of predation by lions will increase as the density of prey increases (Creel & Creel, 1998). These costs might be offset by increased hunting success, or they might not – logic alone cannot resolve the direction of the net effect. This requires empirical work, and we need more field studies before drawing general conclusions. In the case of wild dogs, the net effect of increasing prey density is probably negative, once above a minimum density.

The effect of competition also depends on whether or not the subordinate competitor can scavenge from the dominant competitor. In some cases, such as cheetahs in competition with lions and hyenas, the subordinate species rarely scavenges from the dominant, so the presence of the dominant species has purely negative impacts (Laurenson, 1995a). In other cases, such as coyotes in competition with wolves, the smaller species gains foraging opportunities that could partially or completely offset the risk of intraguild predation (Paquet, 1992).

INTERACTION BETWEEN INTERSPECIFIC COMPETITION AND HABITAT FRAGMENTATION

If competition is an important force limiting carnivore populations, a question arises. Why is competition now relevant to conservation, when these species have presumably been competing for eons? The answer involves changes in the landscape due to habitat fragmentation. Humans and large carnivores do not coexist well. Consequently, carnivores are confined to protected areas more completely than many other species, and this makes carnivore distributions very patchy. Prior to fragmentation, the intensity of competition might have ebbed and flowed at a particular location, but neighboring populations were available to recolonize areas that returned to favorable ecological conditions after a period of competitive exclusion. For example, the recent crash of the Serengeti lion population due to a canine distemper epizootic (Roelke-Parker *et al.*, 1996) might provide an opportunity for wild dogs to recolonize Serengeti, if there was a nearby source of immigrants. Because there is now no sizable wild dog population in that area, recolonization by more than a few individuals is unlikely.

This line of reasoning suggests that populations of subordinate competitors have always waxed and waned as the intensity of competition varied. With isolation, recolonization of populations that fail becomes less likely. Maintenance of sink populations by immigration is also less likely. As marginal populations blink out, the odds of immigration into the remaining populations become steadily lower, until the only populations remaining are those for which dominant competitors are ecologically limited at tolerable densities. If this scenario is correct, then a focus for conservation of many extinction-prone carnivores should be on areas in which interspecific competition is not intense, and which have good connectivity to other populations.

4

Strategies for carnivore conservation: lessons from contemporary extinctions

ROSIE WOODROFFE

INTRODUCTION

Many carnivore species have suffered dramatic declines over the past few hundred years. While some (such as the Falkland Islands wolf, *Dusicyon australis*) have become extinct and others (such as the black-footed ferret, *Mustela nigripes*) are extinct in the wild (Baillie & Groombridge, 1996), the range contractions suffered by some formerly-widespread species are perhaps more impressive (Figures 4.1 (a)–(d)). Species such as lions (*Panthera leo*), brown bears (*Ursus arctos*) and tigers (*Panthera tigris*) once occupied large tracts of land, yet today are restricted to isolated reserves, or to inhospitable areas where human densities are low.

There can be little doubt that the declines illustrated in Figures 4.1 (a)–(d) were caused, directly or indirectly, by people. Most carnivores come into conflict with people because of their predatory habits: red foxes (*Vulpes vulpes*) kill chickens, lions kill cattle, brown bears kill sheep, stoats and weasels (*Mustela* spp.) kill gamebirds. Some of the larger predators may even, occasionally, kill people. For these reasons the majority of carnivore species have been persecuted for hundreds, or even thousands, of years. Such persecution has not been a haphazard affair; in many cases it has been well-planned, nationally organised and, most critically, state-funded. The Emperor Charlemagne employed professional wolf-hunters as early as 800 AD (Boitani, 1995), while, in Norway, bounties were payable upon brown bears as recently as 1972 (Sørensen *et al.*, 1999). Similar measures were instigated by colonial governments to reduce predator numbers in Africa (e.g. Kenya Game Department, 1958), in Asia (e.g. Phythian-Adams, 1939) and in North America (e.g. Leopold, *et al.*, 1964). Persecution was not even confined to areas of human habitation: wolves (*Canis lupus*), mountain lions (*Puma concolor*) and African wild dogs (*Lycaon pictus*) were all persecuted inside designated National Parks, with the aim of protecting 'game'

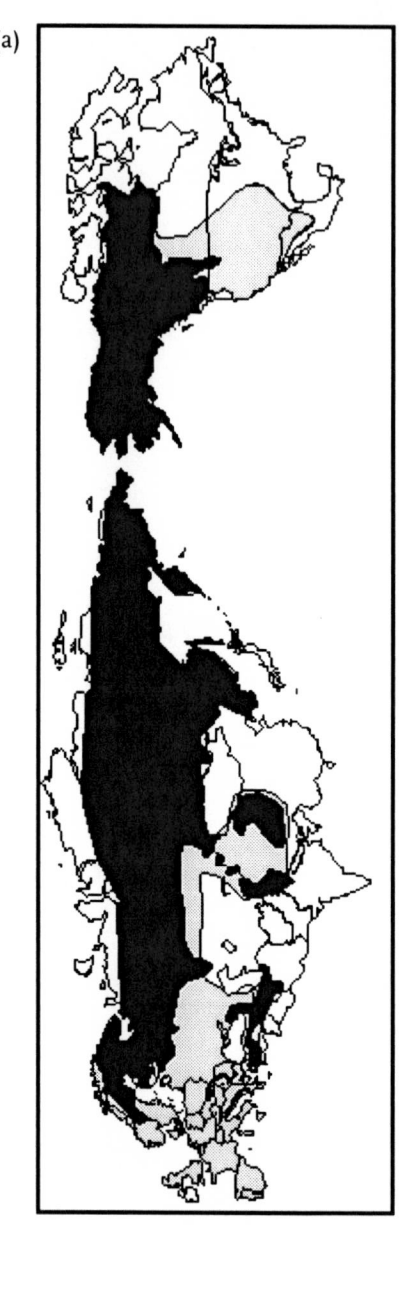

(b)

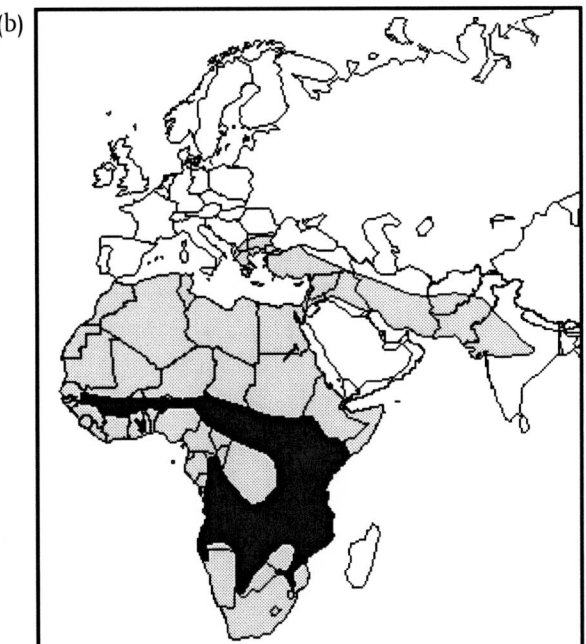

(c)

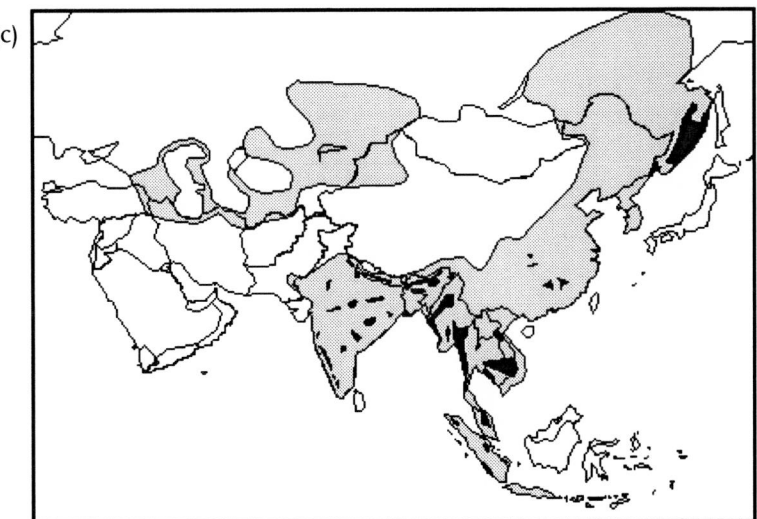

Figure 4.1. Past and present distributions of large carnivores; light shading indicates approximate historic distributions, heavy shading gives approximate current distributions. (a) (*opposite*) Brown bears; (b) lions; (c) tigers; (d) (*overleaf*) African wild dogs. Data sources: (a) Servheen *et al.* (1999); (b) Nowell & Jackson (1996); Pocock (1930); (c) Nowell & Jackson (1996); Seidensticker *et al.* (1999); Dinerstein *et al.* (1997); (d) Woodroffe *et al.* (1997).

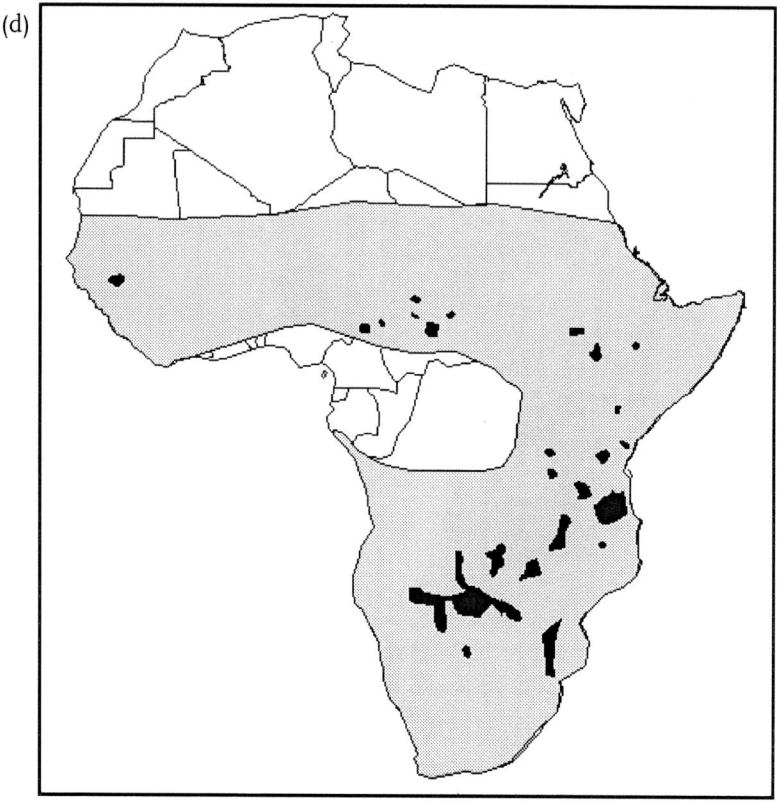

(d)

Figure 4.1 *(cont.)*

populations (Bere, 1955; Phillips & Smith, 1996). Habitat loss, exploitation for fur and prey depletion have all contributed to past carnivore declines, but this widespread, active persecution by people seems to have been the most important cause of most species' historic declines.

Today, most of the larger predators receive some degree of legal protection. Nevertheless, both legal and illegal persecution continue, and even nominally protected populations are affected by burgeoning human populations. Table 4.1 shows that contact with human activity – primarily on reserve borders – is a major cause of carnivore mortality, even in protected areas. For example, many Iberian lynx (*Felis pardina*) radio-collared in Doñana National Park, Spain, were killed on neighbouring private land (Ferreras *et al.*, 1992), while conflict with livestock farmers caused nearly half the lion deaths recorded in Etosha and Nairobi National Parks (Rudnai, 1979; Stander, 1991b).

Most large carnivore species are still declining and population extinctions continue. Perhaps most worrying is the loss of populations from protected areas (Woodroffe & Ginsberg, 1998). Clearly, active conservation measures are needed if further declines are to be avoided.

While the need for carnivore conservation cannot be disputed, it is vital that conservation action be based upon a realistic understanding of the causes of decline. For example, no amount of captive breeding can bring about the recovery of wild populations if they have declined as a result of habitat destruction. Likewise, the establishment of protected areas may be ineffective if those areas are too small to support viable populations (Woodroffe & Ginsberg, 1998), or too large to be guarded adequately (e.g. Leader-Williams & Albon, 1988). Management can only halt a decline if it is aimed at the causes of that decline – and this demands a knowledge of what those causes are (Caughley, 1994).

In this chapter, I investigate threats to carnivore species by analysing the causes of recent population extinctions. This approach is valuable because local extinctions are part of the process of species decline (Caughley, 1994); 'species extinction' simply describes the loss of a species' last population. There have been many, many population extinctions in the last few hundred years, but relatively few species extinctions; this means that statistical analyses can be performed at the population level which are impossible for entire species. While in some cases, frequent local extinctions are balanced by rapid recolonisation (Hanski, 1997), the vast majority of carnivore population extinctions have been irreversible components of ongoing decline processes.

Possible correlates of species' vulnerability

Populations of different species become extinct at different rates. Comparing these rates allows one to identify general characteristics of extinction-prone species. Not only can this approach provide a better understanding of the underlying causes of extinction, it may also help to direct conservation efforts towards little-known species likely to be at risk.

In the past, several factors have been postulated as predictors of species' and populations' vulnerability to extinction. Small populations are expected to be more extinction-prone than larger populations, partly because deterministic declines will drive them to extinction more rapidly, and partly because stochastic processes operating in populations of all sizes (such as small and large-scale variation in environmental conditions, variation in individuals' reproductive output and genetic drift) may hinder the recovery of very small populations (Soulé, 1987; Lande, 1988). Low reproductive rates will reduce population resilience and may also make extinction more likely. For these reasons, local extinctions are expected to occur most often among species that live at low densities (and hence are likely to live in small populations), and those which are large bodied (and hence likely to breed slowly and to live at low densities).

Recent models of habitat destruction predict that good competitors should also be extinction-prone (Nee & May, 1992; Tilman et al., 1994). This counter-intuitive prediction derives from the expectation that good competitors should be relatively poor dispersers. If this is the case, then, as habitat is destroyed, good competitors may find themselves unable to colonise new patches and therefore experience deterministic extinction (Nee & May, 1992; Tilman et al., 1994). In contrast, species that are poor competitors but good colonists – termed 'weedy' by Nee & May – should be less vulnerable to extinction, and may even rise in abundance in fragmented habitats (Nee & May, 1992; Tilman et al., 1994; Hannon, 1996). Since new data suggest that competition is an extremely important factor structuring carnivore communities (Seidensticker et al., 1990; Thurber et al., 1992; Johnson & Franklin, 1994b; Creel & Creel, 1996; Mills & Gorman, 1997; Durant, 1998), one might expect to find consistent differences in the vulnerability of competing carnivore species.

In this chapter, I compare carnivore species' vulnerability to extinction using three different 'currencies': (i) the size of reserve that species need for local persistence; (ii) the human population density that they can tolerate; and (iii) their extinction dates, relative to sympatric species. In each case, I use measures of species vulnerability to test the hypotheses outlined above, and to investigate the causes of local extinction. Each analysis was restricted to the species for which I could obtain sufficient data. For this reason, some species are represented in all three analyses, others appear in only one or two, and many poorly-known species are perforce omitted entirely. Nevertheless, the range of species considered should be wide enough to have some general implications for the conservation of large carnivores.

COMPARING THE MINIMUM AREAS SPECIES NEED TO PERSIST

Local extinctions are most easily documented in isolated areas. Human activity converts natural landscapes into mosaics of cultivation, grazing and urbanisation, isolating patches of natural habitat within a matrix of human-altered habitat. Species that can neither exploit nor cross this matrix become isolated in small populations, occupying the remaining fragments of

Table 4.1. The proportion of adult deaths caused by contact with human activity, recorded by studies of large carnivores inhabiting protected areas

	Proportion of mortality caused by people %(n)	Species total %
African wild dog		
Hwange National Park	81 (31)	
Kruger National Park	47 (19)	
Moremi Game Reserve	7 (15)	
Selous Game Reserve	25 (4)	
Various Zambian Reserves	75 (36)	61
Grey wolf		
Algonquin Provincial Park	54 (26)	
Glacier & Banff National Parks	95 (60)	83
Lion		
Etosha National Park	25 (4)	
Nairobi National Park	54 (31)	
Serengeti National Park	33 (27)	50
Tiger		
Royal Chitwan National Park	67 (3)	67
Iberian lynx		
Doñana National Park	75 (24)	75
Spotted hyena		
Masai Mara Game Reserve	61 (18)	
Serengeti National Park	42 (38)	49
Black bear		
Great Smoky Mountains National Park	50 (2)	
Pisgah Bear Sanctuary	90 (21)	87
Grizzly bear		
Selkirk Mountains	71 (7)	
Yellowstone National Park	89 (250)	89

All data come from intensive studies, of which all but two used radio-telemetry to local dead animals. The figures in brackets give the total number of mortalities recorded due to all causes. Causes of death include legal hunting outside reserves, poaching and persecution as well as accidental killing through road accidents and snaring. Data on Yellowstone grizzly bears also includes legal hunting and control of problem animals inside the park, which together account for 130 of the 250 deaths (Peek *et al.*, 1987). Primary data sources are given in Woodroffe & Ginsberg (1998).

Table 4.2. Results of logistic regressions (Cox, 1970) on the presence and absence of large carnivores in protected areas falling within their historic ranges

Species	Region	Number of reserves	Change in deviance	Critical reserve size	Density (adults/100 km^2)	Home range size (km^2)
African wild dog	East Africa	46	26.59 ***	3606 km^2	2.4 (4)	823.1 (12)
Grey wolf	Western Canada	44	19.82 ***	766 km^2	1.1 (9)	684.6 (11)
Dhole	India	71	30.59 ***	723 km^2	10.6 (1)	68.8 (2)
Lion	East Africa	32	17.61 ***	291 km^2	16.2 (12)	121.4 (59)
Tiger	India & Nepal	154	39.1 ***	135 km^2	3.6 (3)	16.9 (3)
Snow leopard	India, Nepal & Pakistan	30	21.09 ***	116 km^2	4.6 (6)	29.3 (2)
Jaguar	Central America	28	29.98 ***	69 km^2	6.8 (2)	18.8 (5)
Spotted hyena	East Africa	37	20.22 ***	179 km^2	74.5 (6)	34.9 (12)
Black bear	California	45	13.05 ***	36 km^2	58.0 (3)19.8 (32)	
Grizzly bear	Western Canada & North-West USA	54	18.48 ***	3981 km^2	2.0 (5)	773.8 (36)

Asterisks denote the significance values associated with each term. Critical reserve size is the area for which the regression model predicts a 50% probability of population persistence (see Figure 4.2). Wherever possible, data on population densities and home range sizes are taken from the regions for which critical reserve sizes were determined. Population density refers to the density of adults, averaged across studies; home range size refers to the mean area used by each adult female, or social group for social species. Figures in brackets give sample sizes. Primary data sources are given in Woodroffe & Ginsberg (1998).

natural habitat. Several studies have shown that isolated populations are more likely to go extinct if they occupy small habitat patches than if they inhabit larger patches (Brown, 1971; Bolger *et al.*, 1991; Newmark, 1995). The size of habitat patch that a population needs to persist will therefore provide one measure of that species' proneness to extinction in human-altered landscapes.

Woodroffe & Ginsberg (1998) investigated the impact of habitat fragmentation on large carnivores by comparing their persistence in reserves of varying size. For each species, they chose a geographic region in which the species was formerly widespread, but now persisted primarily in protected areas. As shown in Table 4.2, their analyses found that all of the species disappeared from small reserves, but persisted in larger ones.

While Table 4.2 shows a strong statistical effect of reserve size for all species, there is clear variation in the size of the reserves in which the species have persisted (Figure 4.2). Woodroffe & Ginsberg (1998) derived a measure of critical reserve size by using the fitted regression models to predict the area at which populations persisted with a probability of 50% (Figure 4.2). This measure is analogous to an LD_{50}, the dose of a drug which, when administered to experimental subjects, kills exactly half of them. As shown in Table 4.2, critical reserve sizes range from 36 km^2 for the American black bear (*Ursus americanus*), through to 3981 km^2 for its congener, the grizzly bear (*Ursus arctos*). These results suggest that species vary considerably in their ability to survive in isolated habitat patches.

This variation in species' proneness to local extinction could be due to several factors. If, as discussed above, small size predisposes populations to extinction, then species' critical reserve sizes should vary according to their typical population density. This is because the size of the populations at the time they were isolated will depend upon population density and the area of the reserves. However, carnivores inhabiting small reserves may be exposed to risks beyond those associated with small population size. As described above, most protected predator populations come into contact with human activity on reserve borders. Such contact is often fatal (Table 4.1). Deaths occur at rates high enough to cause local population decline (Woodroffe & Ginsberg, 1998), so that carnivore populations may be unable to persist on the edges of reserves unless they are continually replenished by immigrants from the safer reserve core. This 'edge effect' will be most marked for small reserves, which have relatively high perimeter:area ratios, and for wide-ranging species, which are most likely to travel beyond the borders of protected areas (Figure 4.3). If such edge effects are sufficiently powerful to accelerate (or even cause) local extinction, then, for a

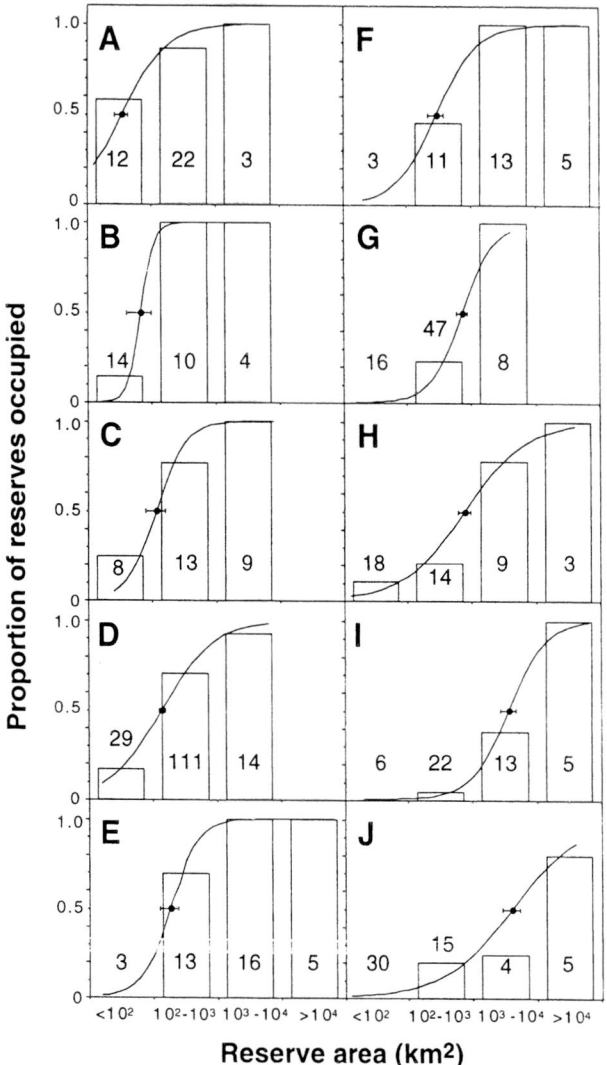

Figure 4.2. The proportion of reserves of varying size in which 10 species of large carnivores have persisted, redrawn from Woodroffe & Ginsberg (2000). Population persistence is related to reserve area for all species (Table 4.2). Curves show the probability of persistence predicted by logistic regressions fitted to the binary data; filled circles show the critical reserve sizes (± SE) for which the models predict a 50% probability of population persistence. Numbers in bars indicate sample size within each category (note that data were analysed using a continuous distribution of reserve areas). Species: A, black bear; B, jaguar; C, snow leopard; D, tiger; E, spotted hyena; F, lion; G, dhole; H, grey wolf; I, African wild dog; J, grizzly bear. (Reproduced by permission of Cambridge University Press.)

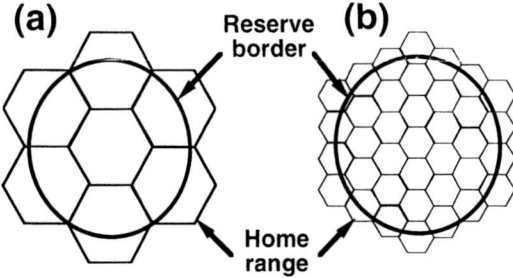

Figure 4.3. The effect of ranging behaviour upon the magnitude of edge effects. Shaded circles denote protected areas; hexagons denote home ranges. (a) When a species ranges widely, a large proportion of the population is exposed to reserve borders, and therefore to persecution; here six to seven home ranges incorporate border areas (86% of the population). (b) When home ranges are smaller, the proportion of the population exposed to the border is also smaller; here it is 49% (18/37 home ranges). Note that the size of the two populations could be identical if species (a) lived in groups, or occupied home ranges with a high degree of overlap. (Reproduced from Woodroffe & Ginsberg, 2000, by permission of Cambridge University Press.)

reserve of a given size, wide-ranging carnivores should disappear before those which restrict their movements to smaller areas. Species' critical reserve sizes should therefore reflect their ranging behaviour, rather than their population density.

Woodroffe & Ginsberg (1998) tested these competing explanations for the extinction of isolated populations, by comparing species' critical reserve sizes with average local population density and average female home range size within the regions for which they investigated population extinction (Table 4.1). After controlling for the potentially confounding effects of phylogeny (Harvey & Pagel, 1991), they found that average female home range size was a good predictor of critical reserve size, while population density had no significant effect (Figure 4.4; overall, $F_{2,7} = 20.6$, $P < 0.005$; effect of density, $t = 0.82$, $P > 0.4$; effect of home range size, $t = 4.00$, $P = 0.005$).

These results show that, in a reserve of given size, wide-ranging species are more likely to disappear than those with smaller home ranges, irrespective of population density. This suggests that mortality on reserve borders may indeed be an important contributor to local extinctions – even for populations which are nominally protected inside reserves. These results do not imply that stochastic effects cannot cause the extinction of small populations of carnivores, they simply indicate that edge effects have a much greater impact.

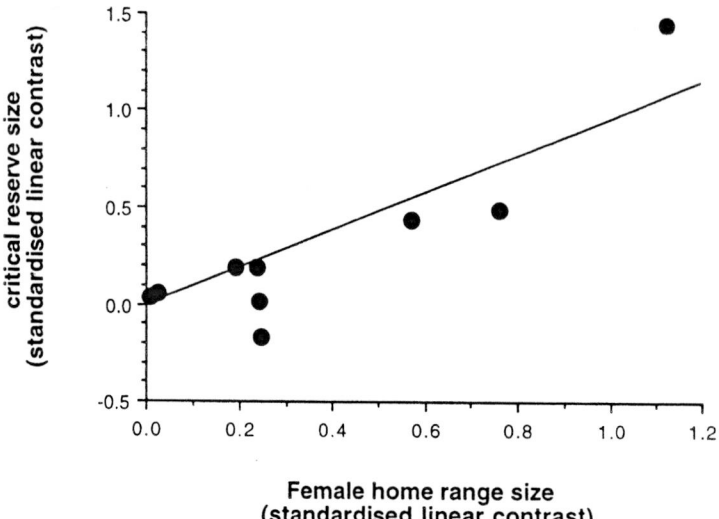

Figure 4.4. The relationship between phylogenetically independent contrasts in log (critical reserve size) and log (female home range size) calculated for 10 species of large carnivore. $r^2 = 0.84$, $F_{1,8} = 42.1$, $P < 0.005$. The effect remains strong after controlling for the (non-significant) effect of population density ($t = 4.00$, $P = 0.005$). (From Woodroffe & Ginsberg, 2000, reproduced by permission of Cambridge University Press.)

COMPARING SPECIES' RESPONSES TO HUMAN POPULATION DENSITY

Species' responses to rising human density may provide another common currency by which their proneness to extinction can be compared. As people become more numerous, they modify their surroundings in a variety of ways. This process creates new environments hostile to many wild species. Because large carnivores are tolerated by few human societies, they are likely to be especially sensitive to the growth of human populations.

I compared carnivore declines with human density using a combination of historical and contemporary data (Woodroffe, 2000). For each species, I selected a region and a point in time when carnivore populations had disappeared from some parts of that region, but had persisted in others (Table 4.3). I then compared the distribution of predators with human population density across regional sub-divisions (states, districts or provinces) at these times; an example is given in Figure 4.5.

I compared the persistence of carnivore populations with local human density using logistic regression, the same technique as that used to investigate critical reserve sizes. Most species had persisted only in areas of low human density, disappearing from more densely populated areas (Table 4.3, Figure 4.6). I characterised the human densities that each species could tolerate, by using the regression models to predict critical human densities in much the same way as critical reserve sizes were calculated (see above). For some species, I was also able to derive additional, independent estimates of critical human density by averaging local measures of human density for the years in which extinctions were recorded. These independent estimates are similar to those calculated from logistic regressions for the same species in the same regions (Table 4.3), suggesting that the regression method gives a realistic prediction of critical human density.

Critical human densities vary substantially between species (Table 4.3). Species such as grizzly bears and African wild dogs have very low critical densities (Table 4.3; Figure 4.6 A, B & C); such species are likely to have become locally extinct as a direct result of human persecution, in habitats which had, thus far, been modified relatively little in other ways. Other species, such as leopards (*Panthera pardus*) and spotted hyaenas (*Crocuta crocuta*), have much higher critical human densities (Table 4.3; Figure 4.6 H, I). These species appear better able to adapt to habitats modified by people – perhaps because they avoid people through nocturnal activity, and can survive on small prey and by scavenging when natural prey are depleted.

Limits to the explanatory power of human density

While human density is broadly associated with carnivore declines, human impact varies between regions as well as between species. For example, estimates of critical human density are substantially different for African wild dogs in eastern and southern Africa, and for cheetahs (*Acinonyx jubatus*) in Africa and India (Table 4.3, Figure 4.6 A, C). Furthermore, carnivores may persist in regions with substantially higher human densities than those found to be associated with extinction elsewhere. For example, the distribution of wolves in North America in 1900 suggested a critical human density of 13.0 people/km² (Table 4.3); yet, today, wolves persist in Cantabria, Spain (99 people/km²), Abruzzo, Italy (118 people/km²) and Rajasthan and Gujarat, India (129 and 211 people/km² respectively; Delibes, 1990; Jhala & Giles, 1991; India Network Foundation, 1999; Instituto Nacional de Estadística, 1999; Istat, 1999). Clearly, 13 people/km² is not an upper limit for wolf persistence in all regions.

This regional variation in particular species' ability to coexist with people indicates that, taken alone, human density gives rather inaccurate predictions of extinction probabilities. While this may limit the extent to

Table 4.3. Logistic regressions of carnivore persistence upon human population density

Species	Location	Date	Change in deviance	n	Critical human density[a]	Mean density at extinction[b]	Data sources
African wild dog *Lycaon pictus*	Kenya	1990	29.3***	41 districts	6.3 people/km² (4.1–9.6)	—	1,2
	Southern Africa	1996	22.1***	48 districts	0.7 people/km² (0.5–1.0)	—	1,3,4,5
Wolf *Canis lupus*	USA	1900	36.5***	49 states	13.0 people/km² (11.0–15.3)	13.5 people/km² (11.2–15.8)	6,7
Cheetah *Acinonyx jubatus*	Kenya	1986	11.7***	24 districts	16.5 people/km² (9.8–27.2)	—	2
	India	1901	0.42	10 states	not calculated (88.4–150.8)	120 people/km²	8,9,10
Lion *Panthera leo*	India	1901	1.27	8 states	not calculated	26.0 people/km² (13.4–38.6)	9,10,11,12
Mountain lion *Felis concolor*	USA	1900	36.1***	43 states	11.7 people/km² (9.7–14.1)	13.3 people/km² (10.7–15.9)	7,13
Jaguar *Panthera onca*	Brazil	1987	18.1***	21 states	17.3 people/km² (12.8–23.3)	—	14,15
Leopard *Panthera pardus*	Kenya	1986	8.4**	41 districts	958 people/km² (497–1857)	—	2,16
Spotted hyena *Crocuta crocuta*	Kenya	1990	22.6***	36 districts	79.5 people/km² (57.0–111)	—	2,17
Grizzly bear *Ursus arctos*	USA	1910	7.5*	16 states	4.2 people/km² (3.1–5.7)	4.2 people/km² (2.6–5.8)	7,18
Black bear *Ursus americanus*	USA	1970	2.6	48 states	not calculated	—	7,19

[a] Human population density at which logistic regression predicts a 50% probability of carnivore extinction (presented ± SE); calculated only where regression models showed a statistically significant effect of human density. [b] Mean human population density in the year in which extinction occurred (± SE). *** $P < 0.001$; ** $P < 0.005$; * $P < 0.01$.

Sources: [1]Woodroffe et al. (1997); [2]Hamilton (1986a); [3]Central Statistics Office (1992); [4]National Planning Commission (1992); [5]Central Statistics Office, Zimbabwe (1992); [6]Young & Goldman (1944); [7]U.S. Census Bureau (1999); [8]Divyabhanusinh (1995); [9]India Census Commissioner (1901); [10]India Census Commissioner (1941); [11]Pocock (1930); [12]Talbot (1959); [13]Nowak (1976); [14]Swank & Teer (1989); [15]Government of Brazil (1998); [16]Hamilton (1986b); [17]Mills & Hofer (1998); [18]McCracken (1957); [19]Herrero (1972).

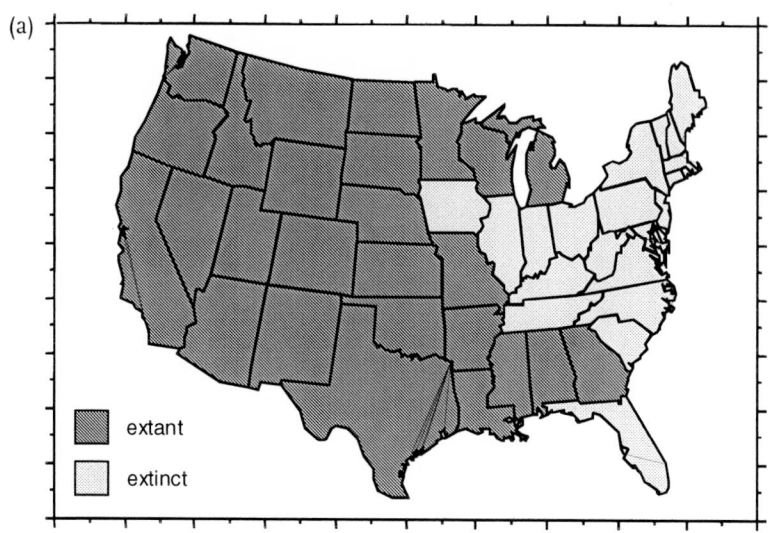

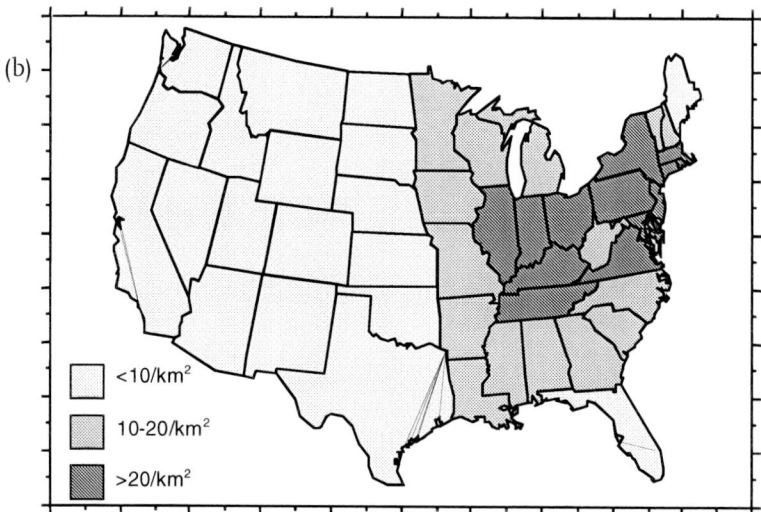

Figure 4.5. Distribution of wolves and people across the coterminous United States in 1900. (a) States in which wolves had been eradicated (light shading) and those in which one or more populations persisted (heavy shading). Heavy shading does not imply that wolves were continuously distributed across the entire state. Data from Young & Goldman (1944). (b) Human population density on a statewide basis; some states categorised as having densities > 20 people/km² contained as many as 200 people/km². Data from U.S. Census Bureau (1999).

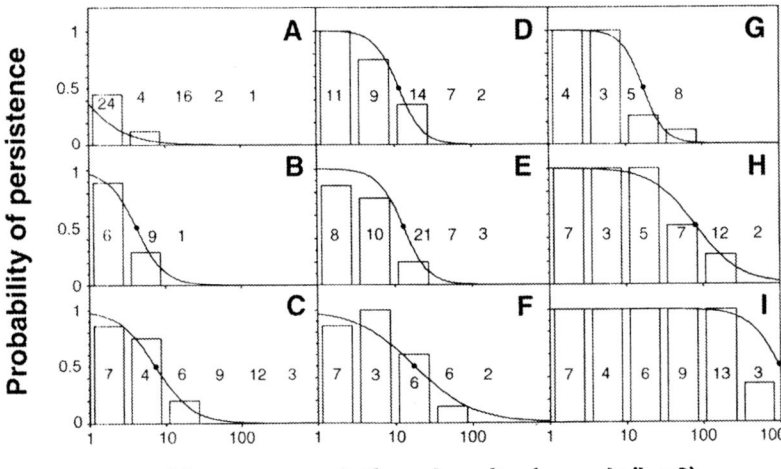

Human population density (people/km²)

Figure 4.6. Relationships between human density (measured on a logarithmic scale) and carnivore persistence. Each histogram gives the proportion of states, counties or districts still occupied by the carnivore species; curves show the logistic regression models fitted to the binary data. Solid circles indicate the human densities at which the regression models predict a 50% probability of local extinction. Numbers in bars indicate sample sizes within each category (note that data were analysed using a continuous distribution of human densities). Statistical analyses and data sources are presented in Table 4.3. Reproduced from Woodroffe (2000). Species and locations: A, African wild dog, southern Africa; B, grizzly bear, United States; C, African wild dog, Kenya; D, mountain lion, United States; E, wolf, United States; F, cheetah, Kenya; G, jaguar, Brazil; H, spotted hyena, Kenya; I, leopard, Kenya.

which critical human density can be used as a common currency to compare different species' proneness to extinction, it may provide a powerful tool for understanding the causes of local extinctions, by allowing comparison of human impact on the same species across different regions.

Regional variation in critical density estimates might be caused by phenotypic variation among carnivores; however, it is more likely to reflect both people's willingness to tolerate predators, and their ability to kill them. Local attitudes to predators will be shaped by a combination of cultural, economic and legal forces, all of which will strongly influence carnivores' chances of persistence. For example, cheetahs were tolerated or even prized in India until the arrival of British colonists, who not only hunted them, but introduced local noblemen to the sport (Divyabhanusinh, 1995). Similar differences were apparent in Africa. In nineteenth-century Cape Province, lions persisted some 30 years longer in black homelands than in

white-dominated areas, despite higher human densities (Skead, 1987).
This may reflect cultural differences, or simply variation in people's access
to firearms.

Regional and international trade in carnivore skins, bones and other
body parts may also encourage local people to kill predators. This probably
explains recent increases in killing of tigers by Indian villagers who have,
historically, shown an impressive ability to co-exist with predators despite
high human densities (Kumar & Wright, 1999). Likewise, government-
sponsored carnivore control programmes (e.g. the payment of bounties)
increase local 'demand' for dead predators and magnify human impacts
upon carnivore populations.

If local attitudes to predators play such an important rôle, then cultural,
political and economic change will be as important as human population
growth in determining carnivores' future. Human populations are projected to stabilise during the twenty-first century (US Census Bureau, 1998),
but this is expected to occur as a result of economic development and cultural change. For this reason, slowing of human population growth may
not result in a diminishing threat to predators. As economic demands rise,
and agriculture becomes intensified, carnivore habitats will be modified,
prey bases will be depleted, and people will, perhaps, become still less tolerant of predators. Despite these concerns, political and cultural change may
also permit the reversal of carnivore declines. The ongoing recovery of
wolves in North America is testament to the possibilities – while the process is difficult and highly controversial, the fact remains that wolf numbers are increasing, despite continuing growth in human populations
(Mech, 1995; Thiel & Ream, 1995; US Census Bureau, 1999).

COMPARING SPECIES' EXTINCTION DATES

Species' proneness to extinction can also be compared directly, using historical records of local extinctions. An example is given in Table 4.4. This
shows extinction dates for several large carnivore species, for various states
in India. These data suggest that, over the course of the nineteenth and
twentieth centuries, extinctions followed a consistent sequence – in most
states, lions disappeared first, followed by cheetahs, then dhole (*Cuon alpinus*). While tigers, wolves and leopards may be in urgent need of conservation in modern India, they remain widespread relative to other species
which formerly occurred there (Table 4.4). Further examples are provided
in Tables 4.5–4.7, which present similar data for North America, Africa and
Europe. Again, some species consistently disappear early (e.g. jaguars,

Table 4.4. Extinction dates for large carnivores in various states in India

Location	Lion	Cheetah	Dhole	Tiger	Leopard	Wolf
Bihar	1814[a]	1890[c]	—	s.p.[e]	s.p.[a]	s.p.[g]
Punjab	1834[b]	~1900[c]	—	~1930[f]	s.p.[a]	—
Utter Pradesh	1866[b]	1919[c]	1975[d]	s.p.[e]	s.p.[a]	s.p.[g]
Rajasthan	1872[b]	1860[c]	1966[d]	s.p.[e]	s.p.[a]	s.p.[g]
Madhya Pradesh	1872[b]	1967[c]	s.p.[d]	s.p.[e]	s.p.[a]	s.p.[g]
Gujarat	s.p.[a]	1940[c]	—	—	s.p.[a]	s.p.[g]
Tamil Nadu	—	1860[c]	s.p.[d]	s.p.[e]	s.p.[a]	—
Karnataka	—	1895[c]	s.p.[d]	s.p.[e]	s.p.[a]	s.p.[g,h]
Maharashtra	—	1896[c]	s.p.[d]	s.p.[e]	s.p.[a]	s.p.[g]
Orissa	—	1939[c]	s.p.[d]	s.p.[e]	s.p.[a]	s.p.[g]
Andhra Pradesh	—	1957[c]	s.p.[d]	s.p.[e]	s.p.[a]	s.p.[g]
Nagaland	—	—	1931[d]	s.p.[e]	s.p.[a]	—
Sikkim	—	—	1931[d]	—	s.p.[a]	—
Assam	—	—	1953[d]	s.p.[e]	s.p.[a]	—

's.p.' denotes that a species is still recorded as present in that state; dashes indicate either that no data are available, or that the species never occurred in that state. Note that, for most states, extinctions follow a consistent sequence – lions disappear first, followed by cheetahs, then dhole. While tigers, wolves and leopards may be in urgent need of conservation in modern India, they remain widespread relative to other species which formerly occurred there.

Sources: [a]Nowell & Jackson (1996); [b]Pocock (1930); [c]Divyabhanusinh (1995); [d]Jonsingh (1985); [e]Dinerstein et al. (1997); [f]Burton (1933); [g]Shahi (1983); [h]Karanth (1987).

Panthera onca, and grizzly bears in North America, lions in Africa), while others have experienced few extinctions detectable at this, rather large, spatial scale (e.g. black bears in North America, leopards in Africa).

A cardinal index of species' proneness to extinction

The data presented in Tables 4.4–4.7 can be used to compare multiple species' proneness to extinction. I derived indices of species' vulnerability by using a modification of Boyd & Silk's (1983) method for calculating dominance scores from social interactions. This technique, developed for behavioural research, calculates an individual animal's position on a continuum of social status, using the proportion of pairwise interactions that that individual 'wins'. I used the same technique, substituting species for individuals, counting 'interactions' as the number of times two species occurred at the same location, and 'wins' as the number of times one species became locally extinct before the other.

Using Boyd & Silk's (1983) technique, I calculated species' cardinal extinction scores by an iterative process. For t species, I aimed to calculate

Table 4.5. Extinction dates for large canivores in various areas of North America

Location	Jaguar	Grizzly bear	Wolf	Mountain lion	Black bear
California	1860[a]	1922[c]	1924[f]	s.p.[k]	s.p.[m]
New Mexico	1910[a]	1927[c]	1960[g]	s.p.[k]	—
Texas	1948[b]	1890[c]	after 1944[f]	s.p.[b,l]	s.p.[m]
Arizona	1949[a]	1935[c]	1960[g]	s.p.[k]	—
Northern Mexico	1955[a]	1957[d]	1980[g]	s.p.[l]	s.p.[m]
Nebraska	—	'long ago'[c]	1915–1941[f]	1903[k]	—
Kansas	—	'long ago'[c]	1915–1941[f]	1904[k]	—
Oklahoma	—	'long ago'[c]	after 1944[f]	1905[k]	—
North Dakota	—	1897[c]	1938[f]	1958[k]	—
South Dakota	—	1897[c]	1915–1941[f]	1958[k]	—
Utah	—	1923[c]	1915–1941[f]	s.p.[k]	—
Oregon	—	1931[c]	after 1944[f]	s.p.[k]	s.p.[m]
Montana	—	s.p.[e]	1941[f]	s.p.[k]	s.p.[m]
Wyoming	—	s.p.[e]	after 1944[f]	s.p.[k]	—
Maine	—	—	1740[f]	1906[k]	s.p.[m]
New Jersey	—	—	1820[f]	1830–1840[k]	—
Vermont	—	—	1821[f]	1881[k]	s.p.[m]
Connecticut	—	—	1842[f]	1800–1820[k]	—
Pennsylvania	—	—	1890[f]	1871[k]	s.p.[m]
New York	—	—	after 1890[f]	1890[k]	s.p.[m]
Kentucky	—	—	1894[f]	1863[k]	—
Tennesee	—	—	1895[f]	after 1900[k]	—
Florida	—	—	1895[f]	s.p.[l]	s.p.[m]
Newfoundland	—	—	1911[h]	mid-1800s[h]	s.p.[h]
West Virginia	—	—	1900[f]	1887[k]	s.p.[m]
Indiana	—	—	1908[f]	before 1850[k]	—
North Carolina	—	—	1911[f]	1886[k]	s.p.[m]
South Carolina	—	—	1911[f]	1916[k]	—
Virginia	—	—	after 1915[f]	1882[k]	s.p.[m]
Delaware	—	—	after 1915[f]	before 1800[k]	—
Alabama	—	—	~1921[f]	after 1921[k]	—
Georgia	—	—	1927[f]	1925[k]	s.p.[m]
Mississippi	—	—	932[f]	before 1900[k]	s.p.[m]
Arkansas	—	—	after 1944[f]	1920[k]	s.p.[m]
Louisiana	—	—	after 1944[f]	1931[k]	s.p.[m]
Wisconsin	—	—	1958[i]	1905[k]	—
Minnesota	—	—	s.p.[j]	1875[k]	s.p.[m]

's.p.' denotes that a species is still recorded as present in that area; dashes indicate either that no data are available, or that the species never occurred in that area.
Sources: [a]Guggisberg (1975); [b]Russell (1971); [c]McCracken (1957); [d]Trevino & Jonkel (1986); [e]Peek *et al.* (1987); [f]Young & Goldman (1944); [g]U.S. Fish & Wildlife Service (1996); [h]Banfield (1974); [i]Wydeven *et al.* (1995); [j]Mech (1995); [k]Nowak (1976); [l]Nowell & Jackson (1996); [m]Bauer & Bauer (1996).

Table 4.6. Extinction dates for large carnivores in various parts of Africa

Location	Lion	Wild dog	Cheetah	Striped hyena	Leopard
Natal (South Africa)	1870[a]	1930[a]	1930[a]	—	s.p.[a]
Eastern Cape (South Africa)	1879[b]	1925[e]	1889[b]	—	s.p.[b]
Tunisia	1891[c]	—	1960[c]	—	—
Algeria	1893[c]	1989[f]	s.p.[c]	s.p.[g]	s.p.[c]
Southern Namibia	before 1934[d]	after 1934[e]	s.p.[c]	—	s.p.[c]
Aïr et Ténéré (Niger)	1935[c]	1950[f]	s.p.[c]	1990[g]	s.p.[c]
Morocco	1940[c]	—	1950[c]	s.p.[g]	s.p.[c]
Laikipia (Kenya)	s.p.[d]	~1980[d]	s.p.[d]	s.p.[d]	s.p.[d]

's.p.' denotes that a species is still recorded as present in that area; indicate either that no data are available, or that the species never occurred in that area.
Sources: [a]Smithers (1983); [b]Skead (1987); [c]Nowell & Jackson (1996); [d]Frank (1998a); [e]Shortridge (1934); [f]Woodroffe *et al.* (1997); [g]Mills & Hofer (1998).

Table 4.7. Extinction dates for large carnivores in various areas in Europe

Location	Wolf	Lynx	Brown bear
Britain	1743[a]	before 0AD[c]	1000[e]
Denmark	1772[b]	—	3000 BC[e]
lowland Germany	before 1817[b]	—	after 1800[f]
German Alps	1847[b]	before 1900[c]	1912[f]
Austrian Alps	before 1900[b]	1800–1900[c]	1900[f]
Swiss Alps	before 1900[b]	1880–1900[c]	1916[f]
Italian Alps	before 1900[b]	1930[c]	s.p.[g]
French Alps	before 1908[b]	1900–1950[c]	1937[f]
French Pyrenees	before 1939[b]	1960[c]	s.p.[g]
Norway	1940[b]	s.p.[c]	s.p.[g]
lowland Spain	before 1950[b]	s.p.[d]	1600–1700[f]
Spanish Pyrenees	1950–1970[b]	1960[d]	s.p.[g]
peninsular Italy	s.p.[b]	'long ago'[c]	s.p.[g]
Greece	s.p.[b]	1970[c]	s.p.[g]

's.p.' denotes that a species is still recorded as present in that area; dashes indicate either that no data are available, or that the species never occurred in that area. Note that, while wolves and lynx tend, on average, to disappear before bears, this is not always the case. Since bears tend to disappear first from lowland areas (e.g. Britain, Denmark, lowland Spain), their apparent invulnerability may reflect the preponderance of mountainous areas in the dataset.
Sources: [a]Harting (1880); [b]Delibes (1990); [c]Breitenmoser & Breitenmoser-Würster (1990); [d]Rodriguez & Delibes (1992); [e]Curry-Lindahl (1972); [f]Couturier (1954); [g]Council of Europe (1989).

scores P_1, P_2, P_3, ... P_t from initial estimates $\hat{P}_1$, $\hat{P}_2$, $\hat{P}_3$, ... $\hat{P}_t$. If species i became extinct before species j M_{ij} times, at a total of N_{ij} locations where both species occurred historically (and where one or both species subsequently became extinct), then

$$P' = \frac{\sum\limits_{j=1}^{t} M_{ij}}{\sum\limits_{j=1}^{t} \frac{N_{ij}}{P_i + P_j}}$$

Each iteration generates new estimates of P_1, P_2, P_3, ... P_t which are substituted into the next iteration, until the process converges at stable estimates. Convergence does not occur when one species is always either the first or last to become locally extinct (analogous with individuals that either win or lose every encounter, Boyd & Silk, 1983), or when allopatric species are used (even if both allopatric species co-occur with another, more widely-distributed species). For this reason, I calculated extinction scores separately for the large carnivore guilds inhabiting Europe, Asia, North America and Africa.

Species that were always the last to become extinct (or never became extinct) were necessarily excluded from the iterative process, to permit convergence of estimates for other species. These species were assigned extinction ranks of zero; this would have been the rank calculated from the formula above, since the numerator $\sum_{j=1}^{t} M_{ij} = 0$ for such species.

I standardised extinction scores from different continents by re-scaling them such that the most extinction-prone species in each guild had a score of 100. These adjusted estimates were then log-transformed for analysis.

I compared species' standardised, log-transformed extinction scores with average body size, population density, female home range size and female dispersal distance (Table 4.8). As in the analysis of critical reserve size, I removed possible confounding effects of phylogeny by calculating independent contrasts (Harvey & Pagel, 1991), using a composite phylogeny for the Carnivora (Bininda-Emonds et al., 1999), and the comparative analysis program C.A.I.C. (without altyering branch lengths, Purvis & Rambaut, 1994). In the few cases in which independent data were available for the same species on different continents (e.g. brown bears in Europe and North America), I treated the continental data as though they came from sister-species. Repeating the analyses using only one continent for each species had no effect on the outcome of analyses.

These analyses show that, after controlling for phylogeny, species'

Table 4.8. Possible correlates of species' cardinal extinction scores

Species	Log (1 + extinction score)	Body size (kg)	Population density (adults 100 km^{-2})	Home range size (km^2)	Dispersal distance (km)
Panthera leo					
Africa	2.0	115	13.5[a]	121.4[a]	30[o,p]
Asia	2.0	110	14.3[b]	50[b]	—
Panthera onca					
North America	2.0	65.6	6.8[a]	18.8[a]	—
Panthera pardus					
Africa	0	36.1	4.9[b]	18.2[f,g]	—
Asia	0	30	6[b]	8.7[h]	—
Panthera tigris					
Asia	0.16	130	3.6[a]	16.9[a]	12[q]
Lynx lynx					
Europe	1.971	18.1	3.8[b]	216[i]	41[i,r]
Acinonyx jubatus					
Africa	0.496	38	2.1[c]	833[j]	—
Asia	1.441	—	—	—	—
Puma concolor					
North America	0.79	48	2.7[b]	274[k,l,m,n]	33[k,l,m,n,s]
Ursus arctos					
North America	1.977	298.5	2[a]	774[a]	27[t]
Europe	1.521	150	2.2[d]	—	—
Ursus americanus					
North America	0	110.5	58[a]	19.8[a]	14[n]
Canis lupus					
North America	0.85	30	1.1[a]	684.6[a]	93[u,v,w,x,y,z]
Europe	2.0	25	6.4[e]	—	—
Asia	0	25	—	—	—
Lycaon pictus					
Africa	1.224	22	2.4[a]	823.1[a]	40[aa,ab]
Cuon alpinus					
Asia	0.55	11.5	10.6[a]	68.8[a]	—

Standardized extinction scores are calculated from the relative extinction dates of sympatric carnivore species, by the procedure described in the text. Ecological data are taken from the regions for which extinction scores are calculated. Home range sizes and dispersal distances are given for females only; dispersal distances include only those animals which dispersed.

Sources: Body sizes; Nowell & Jackson (1996), Servheen *et al.* (1999), Ginsberg & Macdonald (1990). Ecological data: [a]Woodroffe & Ginsberg (1998); [b]Nowell & Jackson (1996); [c]Gros (1998); [d]Council of Europe (1989); [e]Delibes (1990); [f]Mizutani (1993); [g]Stander *et al.* (1997); [h]Seidensticker *et al.* (1990); [i]Breitenmoser *et al.* (1993b); [j]Caro (1994); [k]Hemker *et al.* (1984); [l]Ross & Jalkotzy (1992); [m]Logan *et al.* (1986); [n]Maehr (1997); [o]Hanby & Bygott (1987); [p]Pusey & Packer (1987); [q]Smith (1993); [r]Schmidt (1998); [s]Beier (1995); [t]Glenn & Miller (1980); [u]Mech *et al.* (1998); [v]Potvin (1987); [w]Gese & Mech (1991); [x]Ballard *et al.* (1987); [y]Fritts (1983); [z]Berg & Kuehn (1982); [aa]Fuller *et al.* (1992b); [ab]McNutt (1996).

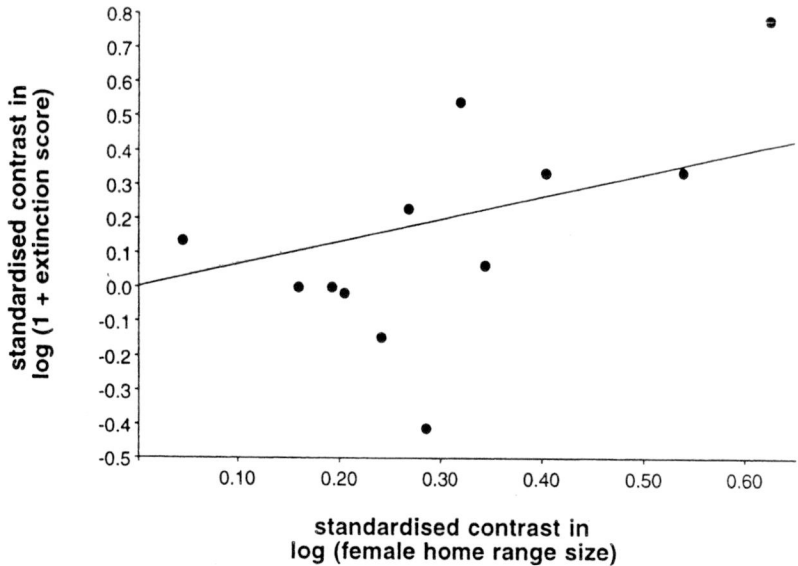

Figure 4.7. The relationship between female home range size and carnivore species' relative proneness to local extinction, calculated by comparing the extinction dates of sympatric species. $r^2 = 0.42$, $F_{I,II} = 7.97$, $P < 0.05$.

relative proneness to extinction is correlated with female home range size (Figure 4.7; $F_{I,II} = 7.97$, $r^2 = 0.42$, $P < 0.05$ using log-transformed data and forcing regression line through the origin). In contrast, neither body size nor population density or female dispersal distance are related to extinction score (all data log-transformed, all regressions forced through origin; body size: $F_{I,II} = 0.06$, $r^2 = 0.005$, $P > 0.8$; population density: $F_{I,II} = 1.4$, $r^2 = 0.11$, $P > 0.25$; dispersal distance: $F_{I,6} = 3.92$, $r^2 = 0.4$, $P < 0.1$).

Pairwise comparisons among species

The analysis presented above yields general conclusions about the characteristics of extinction-prone species. However, further examination of the data presented in Tables 4.4–4.7 may provide additional information concerning the process of local extinction.

For example, brown bears have tended to disappear before wolves in North America (bears first at 11/13 locations), suggesting that bears are more extinction-prone than wolves. However, while all native wolf populations have been driven to extinction in the coterminous United States, some bear populations have persisted. Weaver et al. (1996) attributed this difference to variation in the two species' behaviour in the mountainous areas where bears still persist – in winter, wolves follow their prey to lower altitudes, where they are more likely to encounter human activity, while bears remain at high altitude in their hibernacula. This may explain local variation in the relative extinction dates of wolves and bears – bears tend to disappear first from lowland areas (bears first at 8/8 locations in Europe and North America), while wolves disappear first from mountainous areas (bears first at 6/17 locations in the same regions; $\chi^2 = 9.24$, df $= 1$, $P < 0.005$). This effect of habitat upon the two species relative proneness to extinction probably underlies the lower extinction score calculated for brown bears in Europe, relative to North America (Table 4.8); most lowland bear populations had disappeared from Europe before historical records began, and relict mountain populations dominate the data set (Table 4.7).

A similar difference exists between the extinction sequences of wolves and mountain lions in North America. Again, wolves disappear first from mountainous areas (wolves first in 16/21 states) but persist longer than lions at lower altitudes (wolves first in 7/22 states, $\chi = 8.5$, df $= 1$, $P < 0.005$).

COMPARISONS BETWEEN DIFFERENT CURRENCIES

These three currencies generate broadly similar estimates of species vulnerability (Figure 4.8). Some species, such as grizzly bears and African wild dogs, appear consistently extinction-prone – they require large areas to persist, are intolerant of human activity, and often disappear from areas where other predator species persist (Tables 4.1, 4.3 and 4.8). Likewise, species such as American black bears appear consistently resilient, surviving in small areas, at high human densities, often for decades after other species have disappeared (Tables 4.1, 4.3 and 4.8).

Comparisons with red book status

Interestingly, such classifications of species' proneness to extinction do not necessarily reflect their red data book status (Figure 4.9). For example, the tiger appears to be a relatively resilient species, surviving in small reserves (Table 4.2) and in regions from which lions, cheetahs and dhole have long since disappeared (Table 4.4). This apparent resilience, which has been attributed to tigers' rapid reproduction and dietary flexibility (Sunquist *et al.*, 1999), may be surprising, given tigers' classification as an endangered species (Baillie & Groombridge, 1996). Likewise, brown bears' classification as lower risk (least concern) may be unexpected, given their apparently extreme vulnerability to extinction.

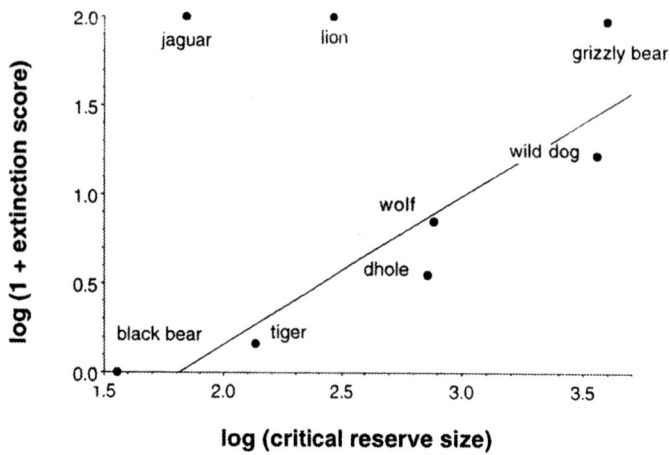

Figure 4.8. Correlation between species' critical reserve sizes and their extinction scores, calculated from relative local extinction dates outside protected areas. While the two measures of vulnerability are closely correlated for 6 of the 8 species ($r_6 = 0.91$, $P < 0.005$), lions and jaguars appear much more vulnerable outside than inside protected areas. If lions and jaguars are included, $r_8 = 0.36$, $P > 0.2$.

In fact, such discrepancies are far from surprising, and simply reflect the difference between a species' past record of extinction and the future it is likely to face. Tigers are a primarily tropical species. Most remaining populations occupy threatened forest habitat in nations with high-density, rapidly-growing human populations (Figure 4.1(c)). By contrast, brown bears have a circumboreal distribution, with most remaining populations occurring at high latitudes where human density is extremely low (Figure 4.1 (a)). Thus, by virtue of their distribution, brown bears are currently less likely to become globally extinct than are tigers, despite the their having been more extinction-prone over the last few hundred years.

Intraspecific differences in estimates of extinction-proneness

While most estimates of vulnerability are in broad agreement, some species which are found to be extinction-prone by one measure appear relatively invulnerable according to another (Figure 4.8). For example, lions have almost invariably disappeared before sympatric species, both in Asia and in Africa (Tables 4.4 and 4.6), yet protected populations persist well in relatively small reserves (Table 4.2). This difference may reflect the different environment experienced by lions inside and outside protected areas. Lions' ability to kill large prey means that they can cause substantial economic damage through depredation on livestock (Frank, 1998b). This, and

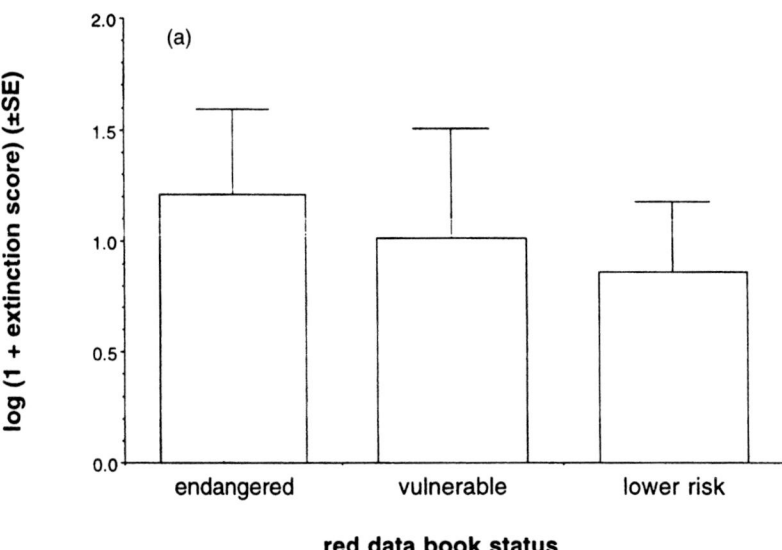

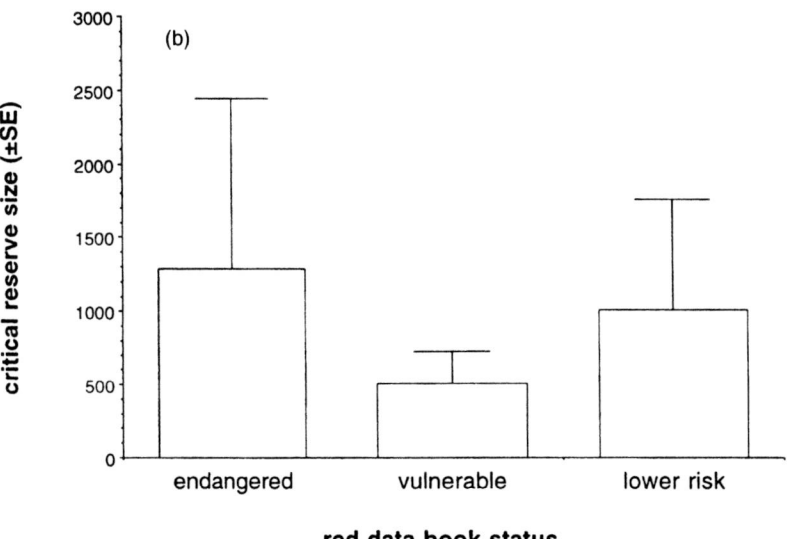

Figure 4.9. Relationships between species' red data book status and measures of their past proneness to extinction. Neither relationship is statistically significant. (a) Critical reserve size: $F_{2,7} = 0.13$, $P > 0.8$; (b) log(1 + extinction score): $F_{2,11} = 0.23$, $P > 0.8$. Data on red book status from Baillie & Groombridge (1996).

their occasional attacks on people, probably explains the intense persecution that they experienced in historical times. In Africa, other sympatric predators cause less severe damage and are not, therefore, so heavily persecuted (Frank, 1998b). When protected from such persecution, however, lions' relatively small home ranges, naturally high population densities and rapid reproduction allow them to persist well in comparatively small areas. Jaguars, likewise, appear much more extinction-prone outside than inside reserves (Figure 4.8). This, too, might reflect jaguars' dangerous penchant for killing livestock when living outside protected areas (Hoogesteijn et al., 1993; Nowell & Jackson, 1996).

Species' proneness to extinction may also vary regionally. For example, wolves appear consistently extinction-prone in North America, requiring large reserves (Table 4.2), and disappearing early (Table 4.5), when human density is still relatively low (Table 4.3). By contrast, in Europe and Asia wolves appear much less vulnerable, persisting longer (Tables 4.4 and 4.7), and tolerating higher human densities (see above). This difference might reflect a more intensive and widespread effort at wolf control in North America than in Eurasia. Alternatively, it might reflect behavioural differences – European wolves have adopted more secretive and nocturnal activity patterns where they coexist with people (Vilà et al., 1995; Ciucci et al., 1997). One might speculate that gradual development of human populations, and sustained local persecution over many centuries, allowed European wolves to adapt their behaviour to coexist with people. By contrast, human immigrants to North America confronted the local wolves with a rapidly-expanding, sophisticated and well-armed human population which may have driven them to extinction before such adaptation could occur.

BIOLOGICAL CHARACTERISTICS OF EXTINCTION-PRONE SPECIES

In general, species are expected to be extinction-prone if they are large-bodied, live at low densities, breed slowly and have limited dispersal abilities (see above). In fragmented habitats, wide-ranging behaviour may also be a disadvantage for predator species (Figure 4.3). Can the analyses presented above be used to identify biological characteristics that make some species particularly extinction-prone?

Ranging behaviour
The only general characteristic of extinction-prone species appears to be wide-ranging behaviour, which is associated with both large critical reserve

size (Figure 4.4) and with high extinction scores (Figure 4.7). A previous analysis also linked carnivores' relative home range size with their red data book status (Woodroffe & Ginsberg, 2000). As discussed above, wide-ranging behaviour is expected to predispose carnivores to local extinction where they rely upon protected areas to escape persecution (Figure 4.3); finding such a relationship outside reserves is, perhaps, more surprising. It is possible that extensive ranging brings animals into more frequent contact with human activity, even where no areas are formally protected. For example, it is wolves' winter home range expansions into human-dominated areas that is believed to explain their relative vulnerability at high altitudes (see above, Weaver et al., 1996). However, Harcourt (1998) has recently shown that species vulnerability is also correlated with home range size among forest-dwelling primates, which rarely leave forest and do not come into conflict with people. This suggests a more subtle explanation for the association with ranging behaviour, which might also apply to threatened carnivores.

Competitive ability

My analyses provide no support for Nee & May's (1992) suggestion that poor competitors, with good dispersal abilities, should be less extinction-prone. Indeed, the (non-significant) relationship between mean dispersal distance and extinction score suggests that good dispersers are, if anything, relatively more vulnerable to extinction ($F_{1,6} = 3.92$, $r^2 = 0.4$, $\beta = 2.06$, $P < 0.1$). Several species that are poor competitors and good colonists – such as African wild dogs, cheetahs and dhole – are disappearing from protected areas, while the competitors that dominate them, such as lions, spotted hyaenas and leopards, persist in the same areas (Nowell & Jackson, 1996; M. Mills & Gorman, 1997; Venkataraman, 1995; Woodroffe et al., 1997; Durant, 1998).

This discrepancy is probably due to Nee & May's (1992) definition of degraded habitat, and competitors' use of such habitat. Inferior competitors avoid areas frequented by dominant competitors, which are often highly aggressive to them (Thurber et al., 1992; Johnson & Franklin, 1994b; M. Mills & Gorman, 1997; Creel & Creel, 1996; Durant, 1998). In contiguous habitat, inferior competitors tend to occur at low densities, occupying areas of low prey abundance avoided by superior competitors (Johnson & Franklin, 1994b; Creel & Creel, 1996; M. Mills & Gorman, 1997; Durant, 1998); this behaviour allows competing species to co-exist. In fragmented landscapes, however, such avoidance behaviour often leads inferior competitors into human-altered 'edge' habitat (Seidensticker et al., 1990; Thurber et al., 1992; Woodroffe et al., 1997; Durant, 1998). This degraded matrix between patches is not simply uninhabitable (as in the Nee–May model), but actively hostile (animals crossing it, or living adjacent to it, suffer high mortality). Thus, while some 'weedy' carnivores (*sensu* Nee & May, 1992) may indeed thrive in fragmented habitats (e.g. Thurber & Peterson, 1991), other inferior competitors, less able to withstand persecution, may suffer extinction rates systematically higher than those of superior competitors.

Behavioural flexibility

Finally, species' vulnerability to extinction appears to be associated with their ability to adapt their behaviour to survive in human-altered habitats. While such flexibility is extremely difficult to measure (and therefore to analyse formally) it is a clear characteristic of resilient species. Predators such as leopards, wolves and spotted hyaenas capture wild ungulates when they are available but, at high human densities, may subsist on a diet of garbage, carrion and (in the case of leopards) pet dogs (Boitani, 1992; Nowell & Jackson, 1996; M. Mills & Hofer, 1998). Similarly, American black bears may switch from eating fruits and nuts to scavenging at rubbish dumps. In contrast, extinction-prone species such as lions, jaguars and African wild dogs respond to human presence by expanding their diets to include livestock – a behavioural response unlikely to promote coexistence with people. Other species, with middling extinction scores, seem to have a more neutral response to human presence. Since India contains 50% of the world's tigers and 15% of the world's cattle (Nowell & Jackson, 1996), it is remarkable that tigers appear to take relatively few livestock (by contrast, depredation on cattle is a serious problem in the one area of India still occupied by lions).

Behavioural flexibility involves more than a change of diet. As discussed above, Eurasian wolves become more secretive where they coexist with people. By contrast, ranch managers in Kenya describe the remarkable (and often fatal) curiosity and boldness that African wild dogs show when they come into contact with humans. Social flexibility (such as that described in Chapter 19), may also help some species to survive in the face of persecution.

CONSERVATION MESSAGES

These results suggest several important messages for those who plan and implement carnivore conservation.

People are still the biggest threat to carnivore populations

As discussed above, conflict with humans has been the primary cause of historic carnivore declines. The data presented here suggest that such conflict still underpins ongoing declines and local extinctions. Large predators disappear from areas where human densities are high (Table 4.3). Worse still, even nominally protected populations are affected – contact with human activity is the major cause of mortality among carnivores inhabiting protected areas (Table 4.1). This mortality, which is most severe where high densities of people live around reserves (Harcourt *et al.*, in press), appears high enough to cause local extinction. Discrepancies between species' relative vulnerability inside and outside protected areas (Figure 4.8) suggests that human impact may be especially severe for some species. While there are biological correlates of extinction proneness among carnivores, these seem to relate primarily to the extent and nature of species' contact with people.

These findings indicate that carnivore conservationists must seek to resolve the serious problem of conflict between predators and people. While measures such as disease control, captive breeding and reintroduction may be appropriate for some species in some areas, reducing the numbers of predators shot, speared, poisoned and trapped by people is the single most pressing need to halt global carnivore declines.

Protected areas must be managed to protect entire guilds

One way to minimise conflict between people and predators is to set aside areas for each, providing parks and reserves where carnivores are unmolested, and allowing people to kill any animals that stray beyond their borders. This approach might prove successful if the areas set aside for carnivores were planned and managed appropriately, to protect all of the species concerned.

A first, and major, concern is that reserves should be sufficiently large. As discussed above, some species – especially those that suffer aggression from superior competitors – live at low densities and range widely. Such species require very large areas to persist, especially when they are subjected to mortality on reserve borders (see above). As an example, Woodroffe *et al.* (1997) suggested that reserves in excess of 10 000 km^2 would be necessary to protect populations of African wild dogs.

It is important to bear in mind that the reserve area that a species needs to persist cannot be predicted from either its body size or its trophic position. Parks that are large enough to support the largest-bodied top predators – which might be expected to act as effective 'umbrella' species – may not support viable populations of their smaller competitors, which may live at lower densities and range more widely. Planning of protected area networks must therefore consider the densities and ranging behaviours of all species to be protected. This may be difficult for little-studied species, such as striped hyaenas (*Hyaena hyaena*) and honey badgers (*Mellivora capensis*), which seem to occur at low densities and may range widely.

Interspecific competition also means that parks must support a mosaic of high and low prey densities if complete carnivore guilds are to be protected. Species such as African wild dogs and cheetahs survive by using areas of low prey density avoided by superior competitors (Creel & Creel, 1996; M. Mills & Gorman, 1997; Durant, 1998); little-known species such as dhole, subject to similar competition (Venkataraman, 1995), may show similar ecological characteristics. If reserves – even large reserves – were actively managed to maximise prey densities, such poor competitors would almost certainly suffer.

Management cannot stop at reserve borders

Realistically, it is quite unlikely that many large tracts of land will be set aside to protect wide-ranging predators, especially in developing countries. It is more probable that carnivore conservation will have to depend upon networks of smaller reserves (where 'smaller' reserves still cover several thousand square kilometres), buffer zones and private and communal lands. In such landscapes, the effectiveness of conservation efforts will depend primarily upon managers' ability to resolve conflicts between people and predators, to minimise the numbers of carnivores killed – intentionally or accidentally, directly or indirectly – by people.

Such management is likely to draw heavily upon traditional methods of livestock husbandry. A few projects (e.g. Frank, 1998a) are piloting this approach, working with local people to refine husbandry practices, in an attempt to minimise livestock losses and, therefore, incentives to kill predators. The success of this approach depends absolutely upon people's willingness to participate – only if people see a benefit to tolerating predators are conservation efforts likely to succeed in the longer term (Dinerstein *et al.*, 1999). The declines mapped in Figure 4.1 (a)–(d) are testament to people's ability to eradicate predators. With ever-improving technologies, they will complete the job unless they are encouraged to do otherwise.

Acknowledgements

Much of the research for this chapter was carried out while I was visiting the Field Station for Behavioral Research, University of California at Berkeley. I would like to thank Laurence Frank, Josh Ginsberg, Sandy Harcourt, Sarah Durant, Pej Rohani, David Earn and Paul Paquet for extremely valuable discussions.

5

Alien carnivores: unwelcome experiments in ecological theory

DAVID W. MACDONALD AND MICHAEL D. THOM

INTRODUCTION

Humans are responsible for an extinction crisis which currently accounts for species losses between 100 and 1000 times faster than background rates (Balmford, 1996). Coblentz (1990) identified three categories of human-induced environmental perturbation: (1) over-exploitation of resources; (2) pollution; and (3) the introduction of exotic species. Habitat destruction is probably the most important cause of extinction (King, 1985; Diamond, 1984b, 1989), having in the past for example accounted for up to half of continental bird extinctions (Diamond, 1989). However the translocation or introduction of animals to new ecosystems, and the various impacts that result, is a contender for the second most important cause of extinction and endangerment of native animals (e.g. King, 1985; Soulé, 1990; Williamson, 1999). Of the 941 vertebrate taxa now in danger of extinction, 18.4% are threatened in some way by introduced species (Macdonald et al., 1989).

There is a litany of global catastrophic declines and extinctions which are wholly or partially attributable to the effects of introduced species. Among the most notable of these is the ecological collapse brought about in New Zealand since the arrival of humans 1000 years ago, which includes the extinction of 43% of the frog fauna and 40% of the avifauna (Towns et al., 1997). Most of these extinctions can be attributed to predation by introduced mammals (Towns & Daugherty, 1994). Similarly, Australia's small mammal population has been decimated since European settlement, largely through predation by introduced cats, foxes and dogs (Diamond, 1989; Smith & Quin, 1996). Another global impact of introduced species is the numerous island bird extinctions resulting from predation by introduced rats *Rattus* spp. and cats *Felis catus* – introduced predators are implicated in about half of all island bird extinctions (Diamond, 1984b, 1989; Milberg & Tyrberg, 1993; Balmford, 1996).

However, the evidence linking introductions to extinctions is often circumstantial (Frankel & Soulé, 1981; Ebenhard, 1988), and extinction is only the most extreme outcome of diverse consequences of introductions, which may cause changes in the distribution, abundance, behaviour, or evolution of the native fauna (Simberloff, 1981). In fact, any introduction alters the receiving environment, if only by altering the species composition. This leads to what is perhaps the most insidious threat, namely the homogenisation of global flora and fauna (Atkinson, 1996). These less dramatic impacts merit careful study as they may reveal the processes which threaten species and offer a unique, if lamentable, opportunity for testing several general ecological principles.

Biological invasions have always occurred (Lodge, 1993). Indeed, species have probably always been introduced to new ranges with assistance from others. The difference now is that the perturbation, which has in the past arisen through natural processes when an immigrant arrives has, recently, increasingly been the responsibility of humans.

Most of the effects of introduced carnivores can be classified into four categories: competition; interbreeding; predation; or disease. We will review instances of each category below, but first we should briefly consider the general importance of introductions globally. This topic has been reviewed by several authors, notably Lindemann (1956), Elton (1958), Diamond (1984b) and Kauhala (1996). Ebenhard's (1988) review is the most comprehensive – he tabulated 118 introduced mammal species worldwide, representing 30 families and eight orders. In this analysis, the order Carnivora comprises 23 (19%) of the 118 introduced mammalian species, more than three times its share of the total mammalian fauna (Ebenhard, 1988). Five carnivore families are represented amongst introduced faunas around the world, the exceptions being the Hyaenidae and Ursidae. Ebenhard (1988) judged the 10 most important introduced mammals, on the basis of number of individual successful introductions, to include the domestic cat *Felis silvestris catus* (at number 2), and the domestic dog *Canis familiaris* (at equal 9th). These species have been introduced 59 and 19 times respectively; in total there have been at least 151 introductions of carnivores.

REVIEW OF ALIEN CARNIVORE IMPACTS
Competition

Interspecific competition occurs when individuals of one species suffer a reduction in abundance, fecundity, survivorship or growth as a result of resource exploitation or interference with another species (Begon *et al.*, 1990). The effect on population dynamics can, in turn, lead to changes in species distribution and evolution. Although competition generally occurs only when two or more species obtain resources from a supply which is insufficient, it can occur even when resources are not in short supply, if the animals nevertheless harm each other (Connell, 1983). Competition may operate by exploitation or interference (Schoener, 1983). Exploitation competition occurs when a species affects the population dynamics and carrying capacity of another by depleting their mutual resources, but without the two species interacting directly (Ricklefs, 1990). Interference competition involves two or more species reacting directly to each other (Begon *et al.*, 1990) and, because aggression is conspicuous, this is the easier of the two types of competitive interaction to observe (Case & Gilpin, 1974; Schoener, 1977, 1983; Ebenhard, 1988).

Because one species is usually disadvantaged in competition (Schoener, 1983), the concern is that introduced species will prevail over native fauna. A classic example is the Argentine ant (*Linepithema humilis*) which invaded southern North America, driving out all native ant species as it progressed (Elton, 1958). However, such proven cases are relatively rare (Ebenhard, 1988), and documented severe impacts are few (Frankel & Soulé, 1981; Simberloff, 1981).

Interference competition

Interspecific aggression is common amongst carnivores. For example, lions (*Panthera leo*) kill leopards (*Panthera pardus*) and cheetah (*Acinonyx jubatus*) (Schaller, 1972; Laurenson, 1994). Tigers (*Panthera tigris*) also kill leopard (Seidensticker, 1976), and leopards in turn kill cheetah, serval (*Felis serval*) and African golden cats (*Felis aurata*) (Kingdon, 1997; Hart *et al.*, 1996). Wild dog (*Lycaon pictus*) numbers are limited by interference by lions and spotted hyaena (*Crocuta crocuta*) (Creel & Creel, 1996), and may, in turn, have limited the past distribution of the Ethiopian wolf (*Canis simensis*) (Gotelli *et al.*, 1994).

While intra-guild aggression emerges as a general rule for carnivores, firm evidence of competition remains elusive. However, Hersteinsson & Macdonald (1992) suggest that the southern limit of the distribution of the Arctic fox (*Alopex lagopus*) is determined by interspecific competition with the red fox (*Vulpes vulpes*). Indeed, Frafjord *et al.* (1989) observed that in six of seven interactions between these species, red foxes attacked Arctic foxes. An inadvertent experiment to test competition is provided by the introductions of Arctic foxes to over 450 islands during the early nineteenth century

(E. P. Bailey, 1992): on islands where red foxes were absent the Arctic foxes flourished (West & Rudd, 1983), but where red foxes occurred the Arctic species disappeared (Evermann 1914; Bower & Aller, 1917). On two of the islands, Adugak and Uliaga, control of Arctic foxes by introduction of sterile red foxes was attempted. The trial was an apparent success, with the Arctic species disappearing from both islands rapidly after release of red foxes (E. P. Bailey, 1992). The relative roles of exploitation and interference competition are not clear. A second natural experiment which revealed intra-guild competition between carnivores is the increase in pine marten (*Martes martes*) numbers in Scandinavia following a decline in red foxes due to sarcoptic mange (Lindstrom *et al.*, 1995). This circumstantial evidence of interference competition was substantiated by observations of foxes actively pursuing, killing, and occasionally eating, pine martens, leading to the conclusion that in Scandinavia, pine marten numbers are limited by intra-guild predation by foxes.

The relevance of competition exerted by introduced species is illustrated by the natural re-invasion of red foxes to the northern dune system of Holland, from which they had been extirpated in the Middle Ages (Mulder, 1990). When foxes returned to this region during the 1970s, stoats (*Mustela erminea*) declined. By 1985 the stoat had become extinct throughout the region, and foxes were once again widespread. Since foxes find mustelids unpalatable (Macdonald, 1977), it seems that they killed stoats in a form of aggressive competition. A parallel finding arose in California, where red foxes are not endemic, but to which State they were translocated, and where they now kill endangered San Joaquin kit foxes (*Vulpes macrotis mutica*) (Ralls & White, 1995). Indeed, the introduced red foxes have replaced the kit fox in some parts of its range. Native coyotes (*Canis latrans*) also attack and kill kit foxes however, and separating the impacts of the introduced red fox and the native coyote is difficult. Furthermore, coyotes are also hostile to red foxes and indeed, the interactions of these two species illustrate that the tenor of interspecific encounters is likely to be situation-specific. Coyotes are generally aggressive toward red foxes (Sargeant & Allen, 1989), and may limit their numbers in some areas. That foxes perceive coyotes as a threat was demonstrated by Voigt & Earle's (1983) observation that foxes avoid raising young in areas inhabited by coyotes, but do rear litters in the interstices between coyote ranges. Nonetheless, the level of hostility between these species is variable and coyotes are most likely to attack foxes in the presence of food (Gese *et al.*, 1996c), giving the impression that coyotes harass red foxes when contested resources are at stake, rather than because of any inherent aversion.

Perhaps the most devastating impact of an alien carnivore through competition is that of the American mink (*Mustela vison*) on the European mink (*M. lutreola*). The latter species was formerly widespread in Eastern Europe, where its American congener was released in large numbers during the early twentieth century to bolster the fur-trapping industry. Since the beginning of the twentieth century the European mink has slowly disappeared from most western countries. In the last few decades the decline has accelerated markedly, and at present it is definitely known to occur only (and declining sharply) in some isolated areas of Russia, Belarus, Ukraine, in Spain and in the extreme south-west of France. There seems little doubt that the European mink is Europe's most endangered mammal, and heading for extinction. Various hypotheses have been proposed to account for the European mink's decline, including pollution, human persecution, habitat change, decline of prey availability, competition or hybridisation with other predators, and disease (Maran *et al.*, 1998). To test these, our team undertook a radio-tracking field study in Belarus, where the two mink species live as part of a guild of semi-aquatic mustelids that also includes polecats (*M. putorius*) and otters (*Lutra lutra*). A pilot study in captivity revealed that male American mink dominated both male and female European mink, and female American mink dominated European females (Maran *et al.*, 1998). In the field, we classified an aggressive interaction between radio-tracked animals as an incident during which one individual moved to a location where another animal was present, from which one of them was then displaced to a habitat not usually frequented. During two years of radio-tracking in Belarus, involving 3500 hours spent radio-tracking 9 European mink, 40 American mink and 6 polecats, we documented 12 episodes during which an American mink displaced the European species. In five of these cases, the American mink chased the European mink for 200–500 m and in all 12 cases the European mink were driven away. These encounters were initiated by American males, and directed toward European mink of either sex. The sharp decline of the European mink in the study area coincided exactly with the arrival and explosive increase of the American mink. The evidence for interference competition seems compelling.

Not all instances of competition between introduced and native carnivores are decided in favour of the invading species. Despite their threatened status (only around 1000 individuals survived in 1995), Iberian lynx (*Felis pardina*) have been shown to limit numbers of the feral Egyptian mongoose (*Herpestes ichneumon*) (Palomares *et al.*, 1995). Egyptian mongooses, along with native genets (*Genetta genetta*) avoid areas used by the larger Iberian lynx (Palomares *et al.*, 1996), and therefore may minimise the frequency of competitive interactions by segregation of habitat.

Exploitation competition

Exploitation competition has been cited as a major impact of alien introductions (e.g. Clout & Lowe, 1997), but is notoriously difficult to prove (Ebenhard, 1988). Simberloff (1981) considered competition to be a negligible consequence of species introductions, with his meta-analysis revealing competition effects in only 6% of introductions. In contrast, de Vos (1977) described competition as an inevitable consequence of any introduction. Among 504 mammal introductions, Ebenhard (1988) found only 45 instances (9%) of potential competition. Of these, 51% probably resulted in altered abundance of a native species. In only three of these cases was the introduced species a carnivore (*Felis silvestris catus, Mustela vison, Mustela sibirica coreana*), and for only two of these three species was any evidence of actual competition with 'ecological effects' observed. Whether this reflects a genuine rarity of competitive interactions, or merely the difficulty of obtaining clear evidence of competition, is uncertain. For example, introduced American mink are expected to compete with sable (*Martes zibellina*) as are introduced Siberian weasel (*Mustela sibirica coreana*) with the native Japanese subspecies (*M. s. itatsi*) (Lever, 1985). Similarly, the feral cat in Australia may compete with native marsupial 'cats' *Dasyurus geoffroii* and *D. maculatus*, insofar as they share both prey and denning resources throughout their geographic ranges (Dickman, 1996b). However, in each case, critical evidence of competition is lacking. In one case exploitation competition has effectively been demonstrated. Petren & Case (1996) used experimental enclosures to show that the displacement on Pacific islands of a native species of gecko lizard by an invading species was due to differential food-harvesting ability, rather than direct aggression (although aggression is also a component of interactions between the species). However, examples such as this are unfortunately rare.

Overlap in resource use is not necessarily evidence of competition. For example, Sidorovich *et al.* (1998) sought evidence of exploitation competition between American and European mink, along with polecats and otters, in Belarus. The possibility of such competition emerged from a study in Estonia where European mink are now extinct and American mink widespread. To look for evidence of dietary overlap, Maran *et al.* (1998) analysed scats collected prior to the European mink's demise. During the first few years of the American mink's invasion of Estonia the two species had eaten somewhat different diets. However, this was probably because of habitat separation between them in those early years. Subsequently, after the European mink had gone, American mink in those places ate a diet very similar to that of the previously occupying European mink. This similarity in niche raises the possibility of exploitation competition between the two mink species.

Against this background, Sidorovich *et al.* (1998) suggested that the decline in European mink numbers might have been caused at least partly by changes in food availability, because of either (1) declining prey populations or (2) increased competition for food. This situation could arise if: (1) the European mink is more specialised than the American mink or other potential competitors (the polecat and otter); (2) the European mink is dependent on prey which has declined; or (3) there is a large overlap in diet with other predators which have increased in numbers. Having analysed the remains of over 4000 predator faeces, Sidorovich *et al.* (1998) calculated the niche breadth and overlap for the four species using, respectively, Levin's and Pianka's indices. In the Belarussian study sites, the niche breadth indices were: for American mink 4.31; for European mink 3.33; otter 2.57; and polecat 2.37. Generally, there was a large overlap of prey selection between European mink and both American mink and otter (and least overlap between polecat and otter). There was therefore no doubt of the potential for competition between the introduced American mink and native European mink. Despite this beguiling finding there was no evidence that this potential for competition was fulfilled. American mink had the most varied diet, and the polecat the most specialised. The European mink was clearly no more specialised than other members of this guild. Furthermore, the European mink's most important prey, the brown frog (*Rana temporaria*) was very abundant despite also being eaten in large numbers by all other members of the guild. Indeed, brown frogs were of major importance in the diet of all these carnivore species, at almost all times of year, and in virtually all habitats. The greatest similarity in diet (potential for competition) occurred between European mink and otter, closely followed by the European mink/American mink combination. In fact, the American mink showed a large diet overlap with all three other species. Frogs were least important to American mink and polecat during the winter, for European mink the reverse was true. The possibility of competition for food therefore exists, but on present evidence cannot be used to explain the decline in European mink.

The question of whether American mink is competing with native mustelids arises in the southern American cone, where introduced mink are now sympatric with the southern river otter or huillin (*Lutra provocax*)

(Medina, 1997; Previtali *et al.*, 1998). In Alaska native American mink coexist with river otters (*Lutra canadensis*) by exploiting different habitats and dietary niches (Ben-David *et al.*, 1995). This problem can potentially be resolved by comparing the diet of the two species in sympatric and allopatric populations, as Clode & Macdonald (1995) did with mink and otters in Scotland.

Outcomes of competition

If an alien species does compete with members of the native fauna, at least three outcomes are possible: (1) the loser may be forced out of part of its geographical range, or suffer a decline in abundance, or become extinct; (2) the loser may alter its spatial or temporal behaviour to avoid the other species; or (3) either or both competitors may coexist having undergone a niche-shift, and perhaps ultimately character displacement.

Extinction, extirpation and range contraction Competitive exclusion leading to extinction appears to be very rare. Simberloff's (1981) meta-analysis of introductions worldwide concluded that of seventy one extinctions across several taxa, only three could be attributed to competition. However, it might be premature to dismiss the impact of competition, as the consequences of some introductions may not yet have run their full course (Frankel & Soulé, 1981). Herbold & Moyle (1986) suggest that two species may more likely co-occur on a continent for centuries, rather than years, before competitive exclusion drives one to extinction. Although not an example from the Carnivora, there are lessons to be learnt from the case of the tuatara (*Sphenodon punctatus*) which appears to have been extirpated from some of New Zealand's offshore islands as a result of competition with introduced kiore (*Rattus exulans*) (Cree *et al.*, 1995). Juveniles, while making up 22–30% of the tuatara populations on two kiore-free islands, were absent from five of six islands where kiore were present. Even with such a dramatic impact on recruitment, the effect on tuatara numbers was not obvious, although historical records show that, over 150 years, the tuatara has become extinct on at least four islands that it shares with the introduced kiore. The impact of competition from an alien species may be inconspicuous but, over a long time, severe.

The dingo (*Canis familiaris dingo*), introduced to the Australian continent 3500–11 000 years ago, may have displaced by competition both the thylacine (*Thylacinus cynoecephalus*) and the Tasmanian devil (*Sarcophilus harrisi*) on the mainland (de Vos *et al.*, 1956; Burbidge & McKenzie, 1989; Lever, 1994; Smith & Quin, 1996). If so, this may be the only instance of extinction of a continental native species by an introduced carnivore via exploitative competition.

Interspecific competition leading to a change in abundance of the native competitor is cited by Simberloff (1981) in only 51 of 854 (6%) introductions. Boitani (1983) suggested that the decline in wolf numbers in Italy may be partially due to competition with stray dogs (see also Macdonald & Carr, 1995). Dogs, although frequently killed by wolves, are at an advantage under the protective shadow of people.

Coexistence, and avoidance of competition   Competing species may coexist by altering their resource-use to alleviate their conflict. An example is provided by the invading American mink and native otters in Europe. In Sweden, Erlinge (1969, 1972) demonstrated an overlap in the diet of these species, particularly in winter when he suspected they competed for food. Bueno (1996) concluded that the two species could coexist in Spain because of the otter's preference for aquatic prey, and the mink's ability to specialise on terrestrial species. Clode & Macdonald (1995) demonstrated that mink shift their diet in the face of competition with otters. They studied mink and otters on Hebridean islands off the West of Scotland and found that mink living sympatrically with otters tended to shift their diet to focus on more terrestrial prey species, while the diet of otters did not change (Figure 5.1). The fact that the alien species in this interaction appears to be the poorer competitor does not necessarily indicate that the native otter suffers no ill-effects of the new arrival. On smaller islands, where terrestrial prey were unavailable, there was tentative evidence that the mink was unable to change its diet and avoid competition with the otter (Clode & Macdonald, 1995). The two species do coexist on small islands, and are thus forced to share resources, however the effect of this competition on either species is unknown.

In some cases, competition is 'avoided' through coincidence rather than alteration of habitat or resource use. The raccoon dog (*Nyctereutes procyonoides*) introduced into Finland in the 1950s, provides an excellent example of coincidental absence of competition. In Finland, the raccoon dog shares many food resources with native European badgers (*Meles meles*) and red foxes, and may also share den sites with the badger (Kauhala et al., 1993, 1998; Kauhala, 1995). However, despite the similarity in the diet of raccoon dog and badger, numbers of both have increased substantially in Finland since the 1960s (Kauhala, 1995). The two species share resources which are abundant in summer; in winter, food is short and this might be expected to lead them into competition, however both hibernate and hence avoid the

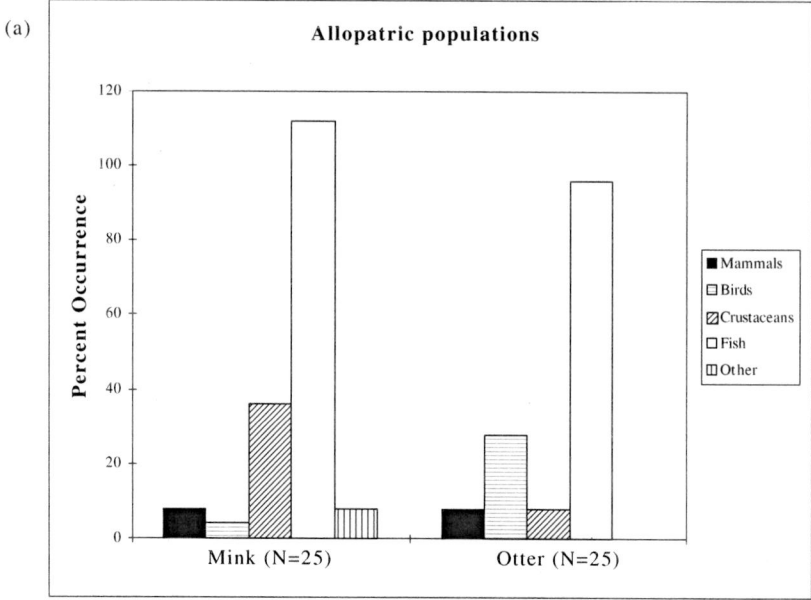

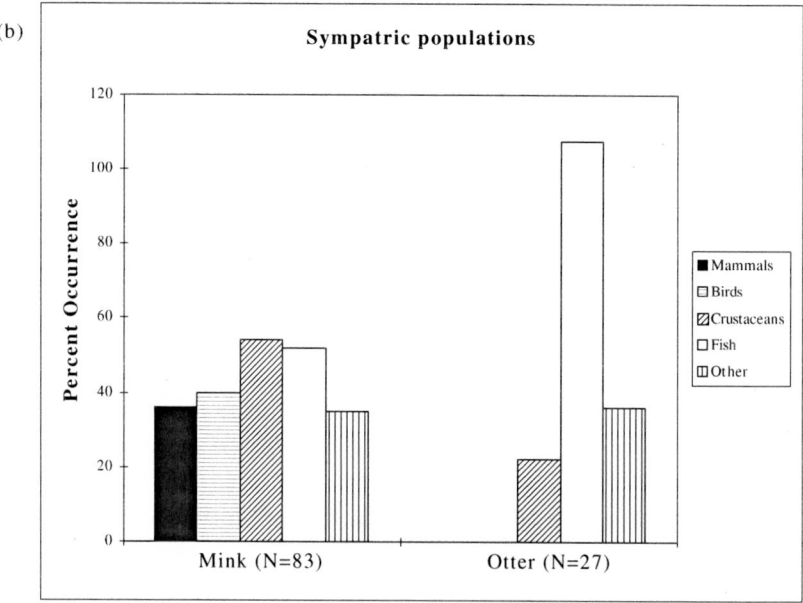

Figure 5.1. Diet composition of mink and otter in (a) allopatric populations and (b) a sympatric population in the Western Isles, Scotland. Because more than one prey item can occur in each scat, percentage occurrences can sum to greater than 100%. Redrawn from Clode & Macdonald (1995).

period of limited resources (Kauhala, 1995). There is also, therefore, no winter competition between either species and a third generalist sympatric carnivore, the red fox.

The distinction between an absence of competition and avoidance of competition is an important one. Firstly, avoidance of competition by alteration of behaviour affects the evolutionary trajectory of a species. Secondly, and linked to this, it may sometimes be difficult or impossible to determine whether two species not currently competing were formerly in competition (the 'ghost of competition past', Connell, 1980). This is relevant insofar as the vacant niche concept has been used as a justification for many introductions which later proved devastating (see Herbold & Moyle, 1986; Pimm, 1991).

Character displacement   The long term coexistence of competing species that undergo a behavioural or niche shift to minimise overlap in the use of contested resources can lead to morphological changes. Ecologically similar species that remain in competition can develop morphological differences as a consequence of resource separation (Abrams, 1996). Generally this phenomenon, character displacement, is demonstrated by a greater degree of interspecific morphological differentiation in sympatry than in allopatry. For example, sympatric mustelids have been shown to adjust, precisely, dental dimensions to accommodate coexistence (Dayan *et al.*, 1989). Among larger carnivores, divergence in body size is particularly important, as prey are difficult to partition except by size (Van Valkenburgh & Wayne, 1994). It has also been observed that where sympatric species specialise on the same resource, or where divergence is constrained by other factors, convergence of morphological characters may occur (Van Valkenburgh & Wayne, 1994; Abrams, 1996). The conditions which determine whether divergence or convergence occurs are not entirely understood. However, some form of character displacement is a potential outcome of animal introductions. Although species survival is not necessarily threatened by this situation, the resulting change in the evolutionary course of one or both species is clearly a perturbation arising from species introduction.

Our own work on European and American mink in Belarus has revealed evidence of character convergence taking place over a period of less than ten years. Body size of these species, as well as the European polecat (*Mustela putorius*), was studied over a 10-year period before and after invasion by American mink. On arrival in the study area, American mink were larger than same-sex individuals of each of the native species. After the American mink's arrival, their mean body size decreased, while the resident male and female European mink and female polecat increased in absolute mass, length, and relative mass (Sidorovich et al., 1999) (Figure 5.2). These results suggest a strong character convergence in response to the invading exotic by the residents, as well as in the invading species itself. The data are consistent with the hypothesis that the resident species are responding to direct aggression from American mink, insofar as larger individuals are better able to withstand attack from the invaders. Why the initial influx of American mink were particularly large is unknown, but robustness may be advantageous at the front of an invading wave. Subsequently, perhaps the same selective pressures that had formerly selected for relatively small size in European mink are also operating on subsequent generations of American mink.

Character convergence in sympatry has also been observed among native Canidae. Three species of jackal are sympatric in Eastern Africa, namely the side-striped jackal (Canis adustus), the golden jackal (C. aureus) and the blackbacked jackal (C. mesomelas). Instead of the expected body-size divergence to allow prey partitioning, convergence was observed for the three species (Van Valkenburgh & Wayne, 1994). In this case, body size divergence is probably restricted by the richness of other carnivore taxa in the region – there are at least 20 smaller and seven larger carnivores in the study area (Van Valkenburgh & Wayne, 1994). This result is relevant to a natural invasion of C. mesomelas, which in Zimbabwe has recently spread into the range formerly exclusive to C. adustus (A. Loveridge, pers. comm.). The result has been an increase in body size of female C. adustus in the area, and a consequent reduction in sexual dimorphism of the latter species. Whatever the long-term result of this change, it illustrates, as does the character convergence between American and European mink in Belarus, the rapidity with which competition may cause character displacement following the invasion of a new competitor.

Interbreeding

Hybridisation occurs in nature among sympatric carnivore species, and can lead to the production of fertile and viable hybrids (Lynch, 1995). For example, sable (Martes zibellina) and pine marten (M. martes) commonly hybridise, as do the European and steppe polecats, (Mustela putorius and M. eversmanni) (Grakov, 1994; Lynch, 1995). Similarly, the grey wolf (Canis lupus) hybridises with the coyote in some parts of the United States and Canada (O'Brien & Mayr, 1991). In all these cases the parent taxa remain separate species, despite interbreeding, because of their morphological and

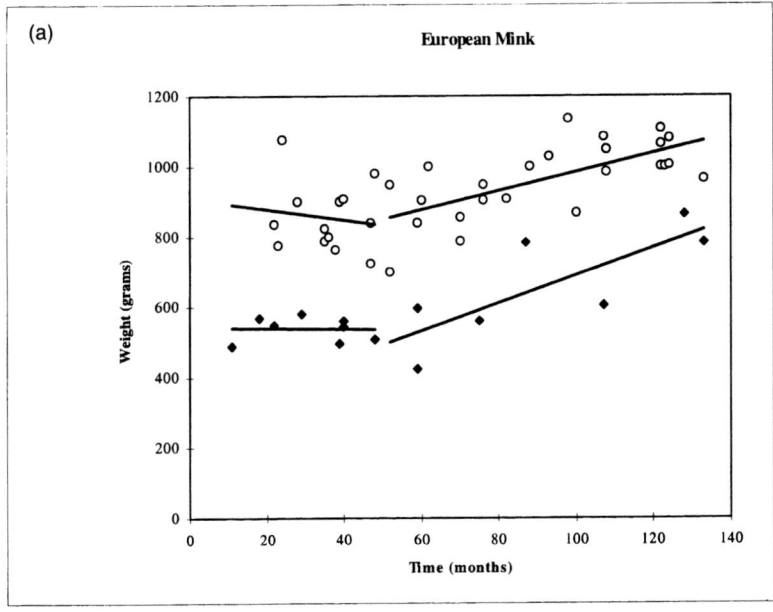

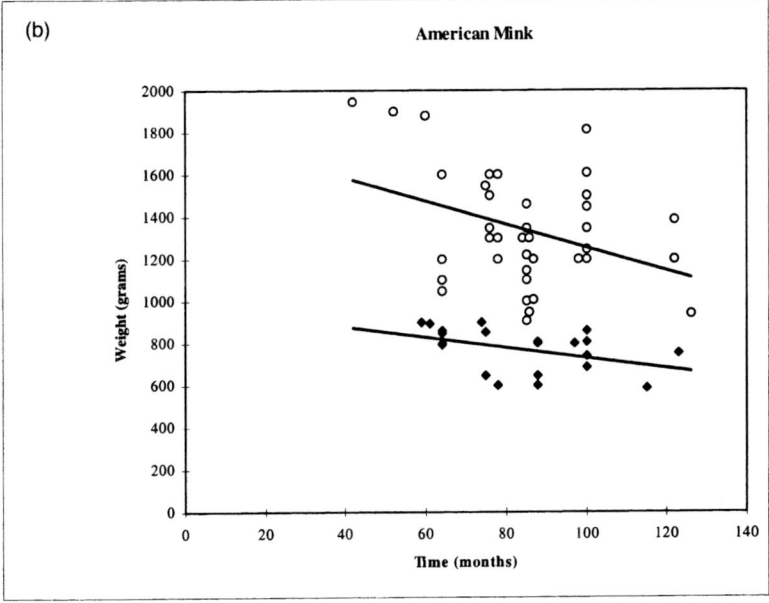

Figure 5.2. Weights of (a) European mink and (b) American mink between January 1986 and February 1997 in Belarus. Separate trendlines in (a) are fitted to data before and after the time of arrival of American mink at the study site (April 1990). Males: open circles, females: closed diamonds. Redrawn from Sidorovich *et al.* (1999).

phylogenetic distinctiveness (Lynch, 1995), and because introgression (gene flow from the hybrids back to the parental populations) is at current rates insufficient to cause homogenisation of the gene pool. These groups of species which can interbreed but are taxonomically distinct for other reasons, have been called syngameons (Lynch, 1995), a term originally used to describe interbreeding taxa of plants. Some members of syngameons are prevented from reproducing simply by geographical separation. Introductions or habitat changes sometimes allow these species to come into direct contact, and it is in these cases that hybridisation becomes a conservation problem (Lynch, 1995).

Hybridisation involving introgression may be a significant cause of lost genetic diversity, and yet the majority of cases may go undetected, largely because hybrids and parentals are often indistinguishable on phenotypic characters (Rhymer & Simberloff, 1996). Indeed, if interbreeding has occurred for a long period of time, as is the case with dingoes and wild domestic dogs in Australia, and feral cats and wildcats in Scotland, there may be no reliable basis for phenotypic or genetic comparisons (Daniels *et al.*, 1998). Even more difficult to detect are cases where interspecific matings do not lead to introgression, such as when embryos are resorbed, or offspring are sterile (Rhymer & Simberloff, 1996).

The problem for native species is particularly acute where their own population is small and, worse, when the population of the introduced species outnumbers their own (Simberloff, 1996). There is evidence, however, that *naturally* scarce or isolated populations are less susceptible to the demographic and genetic problems of small population size than other taxa (Balmford, 1996 and references therein), so this risk may apply only to populations which have been reduced by other agencies.

One of the most frequent sources of introgression of alien alleles into native carnivore populations is through contact with domesticated descendants. The domestic cat, dog, and ferret are all implicated in hybridisation with their native ancestors. This raises a morass of biological, ethical and legislative issues, illustrated by the worldwide distribution of the domestic cat (Fitzgerald, 1988), and its sympatry with the wildcat. The 'domestic cat lineage' is thought to have diverged from the other felid branches between 6–10 million years ago (Masuda *et al.*, 1996; Johnson & O'Brien, 1997). Hybridisation with feral house cats threatens the genetic integrity of European (Hubbard *et al.*, 1992) and African wildcats, with which introgression has been detected in even remote regions (Stuart & Stuart, 1991). Domestication of the cat probably occurred between 4000 and 5000 years BP (Todd, 1978; Price, 1984). Fossil evidence suggests that the wildcat was isolated in Britain after the last ice age, approximately 10 000 years ago (Yalden, 1982), whereas the domestic cat reached Britain between 2000 and 3500 years ago. As soon as the domestic cat arrived in Britain it is likely that interbreeding with the wildcat began, and in the intervening years the distinction between the two forms has become obscure (Daniels *et al.*, 1998; Reig *et al.*, 2001). As a result, there is a long history of contention over the difference between a wildcat and a wild-living domestic cat. Many authors have invoked a range of morphological and pelage characteristics to separate these animals. Pelage and body size are typically used in field identification guides (e.g. Easterbee, 1991; Macdonald & Barrett, 1993), while other authors have used gut length, body measurements, skull morphometrics, and genetic and immunological differences to discriminate between wildcats, wild-living domestic cats and hybrids (Daniels *et al.*, 1998 and references therein). The main difficulty in distinguishing between wildcats and feral domestic cats is that no wildcat population exists which has been isolated from contact with domestic cats, to provide a baseline against which to compare contemporary animals (French *et al.*, 1988a). The earliest museum specimens in Scotland, for example, date from the mid-nineteenth century, when both wild and domestic forms had potentially been interbreeding for more than 1800 years.

In an attempt to differentiate between wildcats and wild-living domestic cats in Scotland, Daniels *et al.* (1998) assessed 333 wild-living cats for pelage colouration and various morphological characters. On the basis of limb bone size and intestine length two relatively distinct groups could be identified. Group 1 cats typically had short intestines and long limb bones, while the Group 2 animals exhibited longer intestines and shorter limb bones, characteristics usually associated with domesticated animals. The Group 1 cats were similar to those traditionally associated with wildcats, although they displayed a greater range of pelage variables than are usually described. There was also a degree of geographical separation between the animals, with Group 1 cats generally occurring on colder, drier land with poor forestry or agriculture potential. This suggests that these Group 1 cats may be better adapted to colder, less fertile lands, characteristics which could be attributable to a longer period of evolution in the Scottish highlands. The existence of these two groups was subsequently confirmed by an independent methodology (three-dimensional skull morphometry) by Reig *et al.* (2001), but there is no evidence of differences in their behaviour, ecology and genetics (Daniels *et al.*, in press), parasites (Delahay *et al.*, 1998) or pathogens (Daniels *et al.*, 1999). In summary, there is no definitive method for distinguishing between wildcats and wild-living domestic cats in Scotland. This case illustrates a major conservation issue due to introgression.

A parallel example concerns the Ethiopian wolf (*Canis simensis*) the world's most endangered canid (Sillero-Zubiri & Macdonald, 1997). Analysis of mitochondrial DNA has shown the Ethiopian wolf to be a distinct species, but closely related to the grey wolf (Gotelli *et al.*, 1994). A genetic analysis of individuals from the principal surviving population, in the Bale Mountains, revealed that seven individuals, which had been identified as abnormal on phenotypic grounds, all showed evidence of hybridisation with dogs. At least one of these animals may have produced viable offspring. Approximately 17% of the sampled population were phenotypically abnormal, suggesting that hybridisation may be having a substantial effect on the genetic character of the species. Dog alleles were found in a second population which has no sympatric domestic dog population, indicating introgression into that population has taken place (Gotelli *et al.*, 1994). In the Bale Mountains, dogs outnumber wolves 10:1. Two criteria make hybridisation particularly threatening: firstly, the wolf population is small; and, secondly, it is vastly outnumbered by the non-native species with which it interbreeds.

Grey wolf populations have also suffered from hybridisation with domestic dogs in Italy (Boitani, 1992), the former USSR, Portugal, Spain (Blanco *et al.*, 1992 and references therein) and the United States (Simberloff, 1996). In Italy, interbreeding between dogs and wolves is considered to be the major threat to wolf survival (Boitani, 1992), and is greatly facilitated by the vast numbers of stray dogs in this country, perhaps up to 800 000 (Boitani *et al.*, 1995). In Australia, the dingo is known to interbreed with domestic dogs (Newsome & Corbett, 1982). Although some authors have argued that domestic dogs, dingoes and hybrids may be relatively accurately distinguished on the basis of phenotypic characters (Newsome & Corbett, 1982), others insist these populations can only be described as 'dingo-like wild canids' (Jones, 1990). Like the wildcat in Scotland, the absence of a pure population of dingoes means no baseline is available against which to compare potential hybrids.

The so-called Samson foxes of Sweden may be symptomatic of a further consequence of introductions. These animals, which lack guard hairs and reputedly show behavioural differences to red foxes, appeared in Finland in the early 1930s (Lampio, 1982). Since then their frequency in the Finnish population has fluctuated from below 1% of the annual fox kill to over 50% in some areas (Lampio, 1982). One hypothesis is that they arise through introgression with English red foxes imported into Scandinavia (de Vos *et* al., 1956, cf. Voipio, 1990). Further examples of interbreeding among native and introduced carnivores are common. The Florida panther (*Felis concolor coryi*) which has declined to about 40 individuals (O'Brien *et al.*, 1990), is already affected by introgression. One population consists of hybrids between the panther and an unknown Latin American subspecies released illicitly, and another population contains individuals which display these hybrid traits, with occasional movement occurring between the two groups (Simberloff, 1996).

Among the Mustelidae, Kamchatka sable (*Martes zibellina kamschadalica*) was introduced to Siberia to improve fur quality of Siberian sable and cross-breeding has occurred between them (Lindemann, 1956). Siberian weasels (*Mustela sibirica coreana*) were imported to Japanese fur farms from Korea some time after 1930. Fur farm escapees then began hybridising with the native Japanese subspecies *M. s. itatsi*, which they are apparently replacing (de Vos *et al.*, 1956). In Britain, the polecat (*Mustela putorius*) is threatened by hybridisation with feral populations of its domesticated descendent, the ferret *Mustela furo* (Lynch, 1995; Davison *et al.*, 1999).

Interbreeding between species may occur without introgression if, for example, offspring of one or both sexes are sterile, or when matings produce no offspring. Such interbreeding can nevertheless threaten the parental populations. Female European mink breed with males of the introduced American species. Matings between these two species do not produce viable offspring, and the embryo is resorbed before parturition (Rozhnov, 1993; Ternosvkij, 1977). However, as male American mink are thought to become reproductively active earlier in the year than males of the European species, early inter-specific pregnancies might pre-empt the European males' reproduction and render the female European mink reproductively abortive for that year. In fact, evidence from the field does not suggest this is a major threat, although cross-breeding between European mink and polecats may be more serious (Davison *et al.*, 1999).

Predation

The most obvious impact of introduced carnivores is reduction in range or population size of native animals by predation (Dickman, 1996). Ebenhard (1988) classified the effect whereby introduced predators cause extinction or population decrease in native species as 'abundance-shifting predation', and specified three categories of impact as: (1) none; (2) co-existence through dynamic equilibrium; and (3) extinction of the prey species or population. Evidence for native predators regulating, or at least limiting, their native prey is mounting (e.g. Lindstrom *et al.*, 1994, 1986) but still not abundant (see review in Macdonald *et al.*, 1999). Evidence for lack of impact of introduced carnivores on their prey is largely meaningless. We will discuss Ebenhard's second two categories below.

Stable coexistence

Regulation of prey by predators is known to occur in native predator–prey systems, for example among the cyclic (Framstad *et al.*, 1997) and non-cyclic (Reid *et al.*, 1997) rodent populations of northern Scandinavia. Although much speculated upon (e.g. Lever, 1994; Ebenhard, 1988), empirical evidence of native prey regulation by introduced carnivores is all but absent. Clearly, it is a feasible outcome, as illustrated by the invasion of Isle Royale by wolves which quickly reduced and then stabilised the moose population (Klein, 1968). In Australia, feral cats and foxes may regulate introduced rabbit (*Oryctolagus cuniculus*) populations during periods of drought (Newsome *et al.*, 1989) and dingoes probably regulate native rodent and macropod prey populations after drought or wildfire (Corbett & Newsome, 1987). However, examples such as these are scarce. This may be because a stable relationship between introduced predators and native prey is itself an uncommon outcome of species introductions, but more likely, the difficulty which has been encountered in attempting to demonstrate prey regulation in any system also pervades the study of predator introductions.

Major impacts

Frankel & Soulé (1981) conclude that there are no cases of extinction of vertebrates on continents in recent history attributable solely to predation by animals other than man. Nonetheless, examples of major declines, local extirpations and island extinctions of native prey due to introduced carnivores are common. Ebenhard (1988) lists 10 carnivore species (out of his 23 introduced carnivore species) that have had ecological effects on native prey. Of these, the cat ranks highest, with 38 introductions involving an ecological impact on native prey (out of a total of 59 recorded introductions). In equal second place are dogs and small Indian mongooses, each having impacted on native species in 10 of the locations where they have been introduced (out of 19 and 14 introductions respectively). The remaining carnivores lag behind, the next highest being *Mustela erminea* with only three introductions resulting in damage to native prey.

However, this analysis may be rather misleading, as each introduction may impact on one or dozens of native species, and is still only recorded as a single introduction with ecological impacts. Lever (1994) listed 153 native prey affected by introduced carnivores, of which 42% had been affected by cats, 25% by small Indian mongooses, and 11% by feral dogs, with the remaining 22% primarily affected by various foxes and mustelids. Jackson (1977) estimated that 61 bird taxa had been extirpated or become extinct due to introduced predators (including rats). Cats and mongooses were responsible for 33 and 9 of these cases, respectively. Concentrating solely on extinctions, Frankel & Soulé (1981) estimate that of approximately 58 island species which have become extinct in recent history, about 25 succumbed to introduced predators. In some cases, New Zealand and Australia being the major examples, separating the impacts of the various introduced carnivores may be impossible (Redhead et al., 1991). In New Zealand, 43% of anurans and 40% of birds have become extinct in the last 1000 years, and most of the impact has been attributed to predation by introduced mammals (although this includes non-carnivores, such as *Rattus* spp.) (Towns & Daugherty, 1994).

Several case studies illustrate these impacts.

Small Indian mongoose

The small Indian mongoose (*Herpestes auropunctatus*) has had its greatest impact in the West Indies. It was introduced to 29 West Indian islands in the late nineteenth and early twentieth centuries (Steadman et al., 1984; Henderson, 1992). Since then, it has been responsible for 7–12 extinctions and 12–13 local extirpations of amphibians and reptiles on the smaller islands. On large islands, although no extinctions have been attributed to the species, it is implicated in a number of range constrictions and dwindling populations of the local herpetofauna (Henderson, 1992).

In the Virgin Islands, mongooses have caused the 'near extinction' of ground nesting quail doves (*Geotrygon mystacea*), and the local extirpation of the lizard *Ameiva polops* (Philobosin & Ruibal, 1971). They are also responsible for the extinction of the Jamaican least parque, the diablotin, and possibly the endemic wren *Ferminia cerverai* in Cuba (King, 1985). There is strong evidence that mongooses cause severe range restrictions of ground-nesting birds to uninfested parts of islands (Baldwin et al., 1952; King, 1985).

On the Hawaiian islands, mongooses are reported to have caused extirpation of small seabirds from infested islands. The Newell shearwater (*Puffinus newelli*), for example, disappeared from Maui, where it had previously nested in ground burrows, but survives on mongoose-free Kauai (Baldwin et al., 1952).

Feral cat

Predation by cats is a major cause of the extinctions around the world, and they have been ranked as the second most important introduced predator of island birds, accounting for 26% of all predator-related island extinctions of birds (King, 1985). A famous instance is the extinction by a single cat of the final population of the Stephen Island Wren (*Xenicus lyalli*), the only flightless songbird in the world (Diamond, 1984b). Subsequently, cats caused a further 12 extirpations of native birds from this island (King, 1984).

Dramatic evidence of the impact of cats is provided for the West Indian island of Pine Cay in the Caicos Islands. Between 1974 and 1978 cats eliminated a population of some 15 000 native iguana (*Cyclura carinata*), and went on to affect the populations of two other native reptiles, *Anolis scriptus* and *Leiocephalus psammodramus* (Iverson, 1978). The impact of domestic cats has been similarly devastating on subantarctic islands (Johnstone, 1985). On Marion Island, for example, cats are probably responsible for the local extirpation of diving petrels, which are abundant on the neighbouring, cat free, Prince Edward Island. On Marion Island in 1975–76 the estimated 2137 cats were calculated to kill about 455 120 birds annually. On Macquarie Island, a smaller population of 375 cats was estimated to take 56 000 rabbits, 47 000 Antarctic prions and 11 000 white-headed petrels annually (Jones, 1977). Finally, Pascal (1980) calculated that the cats on the subantarctic Iles Kerguelen accounted for 1.2 million birds annually. Many species of birds have disappeared from New Zealand's offshore islands since cats were introduced between 1840 and 1931 (Veitch, 1985). Good experimental evidence of their impact is provided by the rapid recovery of many bird species following the elimination of cats from these islands (Veitch, 1985).

Red fox

Red foxes are widely claimed to have devastated native Australian prey, but much of the evidence is circumstantial (Redhead *et al.*, 1991), and confounded by the simultaneous introduction of feral cats (Dickman, 1996b; A. P. Smith & Quin, 1996). Nonetheless, the presence of foxes appears to be the primary factor determining the success or failure of macropod reintroductions – these were successful at 8% of sites without predator control, but 82% of predator-free sites (Short *et al.*, 1992; see also Kinnear *et al.*, 1988).

American mink

The value of the American mink (*Mustela vison*) as a farmed fur-bearer has resulted in it being transported around the world. Invariably, farmed mink have escaped, and in some cases they were deliberately released with the intention of establishing a locally harvestable crop. Almost everywhere that American mink have established beyond their native range they are now at least suspected of damaging local native wildlife (Macdonald *et al.*, 1999).

The spread of mink in British riparian systems has received considerable attention in the literature (e.g. Linn & Chanin, 1978a,b; Strachan & Jefferies, 1993). This concern has focused not only on game and fish industries (an impact which Birks & Dunstone, 1991, conclude is negligible), but also on British native wildlife (Linn & Chanin 1978a; Birks & Dunstone, 1991; H. Clark, 1991). Predation by mink has been associated with terns coalescing in larger colonies in the west of Scotland (Craik, 1995, 1997) although apparently not in the Sound of Harris (Clode *et al.*, 2000). Burrow nesting shelduck (*Tadorna tadorna*) are vulnerable to the arrival of mink on islands in Loch Lomond (Bignal, 1978). The ability of mink to dive for fish and crabs as well as scavenge on the shoreline enables them to avoid periods of food shortage on off-shore islands and so makes them superior island-hopping colonists (Birks, 1990). Other species may be at risk from mink predation. The native white clawed crayfish (*Astacus astacus*) is a preferred food for mink (Smal, 1991) in some areas. This species is declining with the spread of introduced signal crayfish (*Pacifasticus leniusculus*) and the introduced crayfish plague, *Aphanomyces astaci*.

In the West Highlands of Scotland, mink severely affect recruitment of island-breeding native seabirds. All eggs and chicks at many seabird colonies are killed, and the damage is particularly severe because mink are able to swim to islands three or more kilometres off shore, where dense seabird colonies congregate to avoid land predators. The mink's habit of surplus killing may also contribute to the devastation caused to seabird colonies. After one or more years of mink-caused whole-colony breeding failure, affected islands are abandoned by breeding seabirds. The long-term effects are that whole areas lose all breeding seabirds, and numbers of seabirds decline. Between 1987 and 1998 on the West of Scotland black-headed gulls decreased by 49%, common gulls by 39% and herring gulls by 37%, common terns by 48% and Arctic terns by about 58%. The belief that mink were responsible was supported by the observation that annual mink removal was followed by successful seabird breeding in numerous areas following several years of failure or absence (Craik, 1998).

Mink also threaten inland water birds. Ferreras & Macdonald (1999) found that for coots the number of chicks hatched per pair was significantly fewer in stretches of river occupied by resident mink, although there was no such effect on moorhens. This may be because coots nest on the water surface, usually among emergent vegetation (Chanin & Linn, 1980) whereas moorhens often build their nest on branches of trees and shrubs above water level (K. Taylor, 1984) and are perhaps therefore less accessible to the mink.

The American mink is heavily implicated in the drastic decline in Britain of the water vole (*Arvicola terrestris*) (e.g. Lawton & Woodroffe, 1991). Where they occur, water voles are initially an important component of the mink's diet, but they can be largely exterminated within one breeding season (see Strachan *et al.*, 1998). The mink's impact appears to be particularly due to the intensive foraging activities of breeding females within *c.* 1.5 km either side of her den (Macdonald & Strachan, 1999). As the mink radiate from their initial strongholds, water voles become more isolated and fragmented, the population decreases and the risk of local extinction due to either demographic or environmental stochasticity increases. The devastating impact of mink on water voles may be exacerbated by concomitant habitat loss which increases the water vole's susceptibility to predation (e.g. Barreto *et al.*, 1998).

Disease

Many factors usually combine to cause population decline, however infectious disease may be the agent which finally extirpates a threatened population (Pain, 1997). Infectious diseases introduced with exotic animals can spread rapidly through immunologically naïve native populations, an event which often arises when wildlife comes into contact with domestic animals (Schrag & Wiener, 1995; Simonetti, 1995).

Diseases have been inadvertently transported around the world in their hosts. For example, plains bison (*Bison bison*) infected with tuberculosis and brucellosis, were transported from Montana in 1907 as part of a conservation effort to bolster the Canadian population; the pathogens they took with them now threaten the remnant native Canadian wood bison (*B. b. athabascae*) population (Woodford & Rossiter, 1994). In the Caribbean island of Grenada, the rabies epizootic is maintained by the introduced feral small Indian mongoose (Everard *et al.*, 1974), deliberately introduced in a futile attempt to control introduced rats and cane toads (Lever, 1985).

Importantly for conservation, diseases which can occur naturally and sustainably in large populations may eradicate a small or genetically homogenous group of individuals. For example, the endangered Florida panther, which probably numbers fewer than 50 animals in the wild, suffers feline panleukopenia, caused by the feline parvovirus (Roelke *et al.*, 1993b), a disease that is commonplace in domestic cats (Macdonald *et al.*, 1998b).

As mentioned earlier, domestic or feral dogs and cats are among the primary vectors of wildlife disease, largely because the great mobility that people confer on domestic animals, and the relatively high density at which they often occur (Macdonald, 1982; Pain, 1997). Examples of conservation problems due to introduced domestic carnivores include the remnant populations of Ethiopian wolf threatened by rabies spread via domestic dogs (Sillero-Zubiri *et al.*, 1996). The domestic dog is probably also a vector in the rabies epizootics of African wild dogs (Macdonald, 1993; Perry, 1993; Woodroffe *et al.*, 1997), and may be the primary host of rabies in side-striped jackals (*Canis adustus*) (Cumming, 1982; Rhodes *et al.*, 1998). Indeed, there is evidence to suggest that feral domestic dogs may be important in 'driving' the cycle of wildlife rabies in several regions (Macdonald, 1982). Domestic dogs may also be a vector involved in sustaining visceral leishmaniasis among native canids, for example, in Brazilian populations of crab-eating zorro (*Cerdocyon thous*) (Macdonald & Courtenay, 1993). Mainka *et al.* (1994) found giant pandas (*Ailuropoda melanoleuca*) rescued from the Wolong Reserve in China were positive for canine parvovirus antibodies. Introduced canine distemper completely eliminated black-footed ferret (*Mustela nigripes*) in Wyoming (Thorne & Williams, 1988) – the likely source was ranch or feral domestic dogs (Williams *et al.*, 1988). Wild felids are also affected by diseases carried by domestic dogs. For example, as many as 1000 lions (*Panthera leo*) in the Serengeti National Park, Tanzania, amounting to one third of the population, succumbed to an epidemic of canine distemper virus (CDV) in the mid-1990s (Roelke-Parker *et al.*, 1996; Pain, 1997). CDV prevalence in the domestic dogs of local villages increased markedly the three years preceding the lion epidemic. Although direct dog–lion interactions are unlikely, the disease may be transmitted via an intermediate route, most likely spotted hyaenas. Approximately 30 000 domestic dogs surround the Serengeti (Roelke-Parker *et al.*, 1996). Other wild felids that may be affected by transmission of diseases by free-ranging domestic cats include the European wildcat (McOrist *et al.*, 1991; Artois & Remond, 1994; Delahay *et al.*, 1998; Macdonald *et al.*, 1998) and the Iriomote wildcat (*Felis iriomotensis*) (Mochizuki *et al.*, 1990).

Goltsman *et al.* (1996) describe the fate of the Mednyi arctic fox (*Alopex lagopus semenovi*), a subspecies endemic to the Commander Islands. The fox was historically present at extremely high densities, and even after extensive fur trapping in the nineteenth and twentieth centuries, Mednyi Island retained a stable population of about 600 individuals. However, from the mid-1970s, the island suffered a dramatic population crash, with cub mortality at over 95%, and from 1978 onwards the total population fluctuated between 60 and 100 animals. This decline was associated with an outbreak of otodectic mange, caused by the mite *Otodectes cynotis*, probably introduced to the islands by visiting dogs. The impact on Mednyi Arctic foxes suggests that they had no natural immunity to the parasite.

Long-term impacts

The long-term impact of exotic species is a global homogenisation of ecological communities (Soulé, 1990; Lodge, 1993). Lodge (1993) states that the number of documented introductions for most countries is in the range of 10^2–10^4 species – clearly mankind's contribution to the process of homogenisation is already huge, and its reverberations through communities may be far-reaching (Towns *et al.*, 1997). The shifting relations in the guild of semi-aquatic mustelids in Eastern Europe caused by the arrival of the American mink and the extirpation of the European mink is a case in point (Sidorovich *et al.*, 1998). In this case, the relationship between the introduced American mink and native prey is manifestly different to that of the usurped European mink – water voles have declined in the face of predation by the invading species (Macdonald & Strachan, 1999), doubtless also shifting the dynamics of the rodent herbivore community through changed relationships between water voles and root voles.

Interaction with other factors

A general rule in invasion success holds that for every 10 species imported to a country, one will appear in the wild, one in ten of those will become established, and for every 10 established introductions, one will become a 'pest' (Williamson & Brown, 1986; Williamson, 1993; Williamson & Fitter, 1996). The tens rule does not hold for every group of introductions in which it has been investigated, but as a generalisation, it seems to provide a fair indication of the impact of introduced species (Williamson & Fitter, 1996), although Carnivora are more successful as invaders than predicted by the rule (Boitani, Chapter 6). Below we consider some of the factors that can interact with introduced species to mediate their impact. To allow us to illustrate broadly the type of interactions we wish to highlight despite deficiencies in the literature, we have included a number of examples which do not come from the Carnivora.

The importance of introduced prey

The impact of introduced predators may be mediated by introduced prey. Whereas an introduced predator may be constrained by the availability of natural prey, its numbers may increase dramatically when sustained by an *r*-selected introduced prey, a phenomenon called 'hyperpredation' (Smith & Quin, 1996). The endemic parakeet *Cyanoramphus novaezelandiae*, co-existed with feral cats and wekas (*Gallirallus australis*) for over 60 years on subantarctic Macquarie Island until rabbits were introduced to the island in 1879. Rabbits sustained tremendous growth in the cat population, which duly extinguished the parakeet by 1890, along with an indigenous rail (*Rallus philippensis*) (Taylor, 1979). The introduction of the rabbit was the 'single factor that spelt doom for so many small and medium-sized birds on Macquarie Island' (Johnstone, 1985: p. 109). Similarly, in Australia, Redhead *et al.* (1991: p. 305) speculate that introduced rabbit and mouse 'essentially support a whole system of predators in Mediterranean Australia'.

A reverse example, on Pine Cay in the West Indies, involves feral cats that had minimal impact on the native reptile fauna while feeding on abundant introduced rats, but which largely eliminated the native iguana (and the rats) when their numbers were boosted during an influx of construction workers (Iverson, 1978). Amaresekare (1994) found that predation by cats and mongooses (*H. auropunctatus*) appeared to limit black rat (*Rattus rattus*) numbers in Hawaii, and the abundance of rodents may have ameliorated the impact of these introduced predators on native prey. This is close to the principle on which biological control relies.

Isolation

Frankel and Soulé (1981: p. 13) concluded that there is no clear-cut case of continental extinction of an indigenous species due to competition with an introduced form on a continent (see also MacArthur, 1972). In general, introduced species are more likely both to survive and to produce significant ecological impacts on islands than in mainland areas (Simberloff, 1995).

Since the arrival of man – another invading species – in the West Indies about 4500 years ago, there have been 37 extinctions of non-volant mammals. This approximates to one extinction every 122 years, and is therefore double the average for the last 20 000 years of one extinction per 267 years. Many of these extinctions are the result of direct exploitation, habitat destruction and the introduction of *Rattus* and *Herpestes* (Morgan & Woods, 1986).

Of the 93 species and 83 subspecies of birds that have become extinct since 1600, 93% were island forms (King, 1985). In 1978 the rate of bird extinction was estimated at one species or subspecies per 3.6 years (King, 1985). During the 1000 years of Polynesian occupation of New Zealand, at least 32 species of large birds became extinct, through direct predation by humans, the effects of fire and predation by rats and dogs (King, 1984). Of the approximately 77 species of bird and mammal that have become extinct in recent history, 53 were insular forms (Frankel & Soulé, 1981).

Historically, then, introductions to islands have at least the potential to cause more damage to native ecosystems than do introductions to continents, and why this is so is at least partly unresolved (Simberloff, 1995). Although priority might sensibly be attached to island invasions, it would be unwarranted to dismiss continental introductions as insignificant.

Habitat change

Introduced species do not act independently of other agents – their interaction with habitat loss is particularly important (Atkinson, 1996). Habitat disturbance may be an important factor determining whether invading species take hold. For example, Case (1996) found the most important correlate of successful bird invasions was the number of native extinctions in the past 3000 years, which he considered to be a surrogate for human activity and habitat destruction. Habitat change can even be the primary cause of an invasion. The spread of coyotes (*Canis latrans*) into the former range of the red wolf (*Canis rufus*) provides a convoluted conservation dilemma that may stem originally from humans creating the agricultural conditions that facilitated the coyote's invasion (Wayne & Jenks, 1991; Wayne, Chapter 7). Habitat change is also a confounding variable when identifying the impacts caused by invading species. Herbold & Moyle (1986) in their criticism of Simberloff (1981), point out that in many of the cases where 'no effect' of an introduced species was found, the habitat in question was already extensively modified. For example, considering the extinctions of birds on New Zealand islands, the effects of cats and habitat modification are hard to disentangle (Veitch, 1985). Macdonald & Strachan (1999) argue that the impact of American mink on British water voles has been exacerbated by the linearisation and fragmentation of riparian habitat through agricultural intensification. Many successful introduced species prosper in modified habitats while, coincidentally, indigenous species are dwindling in the face of habitat loss, as was demonstrated by Simberloff (1995) in the Mascarene Archipelago, where introduced bird species occupy lowland anthropogenic habitat whereas native species persist in upland native forest. Modified habitat may also favour introduced species, enabling them to outcompete their native counterparts. Petren & Case (1998; 1996) concluded from experimental work on gecko lizards that human alteration of the physical environment facilitated the invasion of the exotic species, and enhanced competitive displacement of the native gecko.

Interspecific differences

Ebenhard (1988) calculated that cats, dogs, and small Indian mongooses, probably the three most damaging introduced predators, had been introduced 59, 19 and 14 times respectively, far in excess of the number of introductions of most other carnivores. Nevertheless, numbers of introductions alone do not explain interspecific differences in predation effects – of the four mongoose species which have been introduced widely, one (*H. auropunctatus*) has preyed catastrophically on native species wherever it was introduced, while the others have had little obvious impact. Four of five introduced canids have had an effect on native prey populations, the exception being the racoon dog (Kauhala *et al.*, 1998). Ebenhard (1988: p. 35) classified the domestic cat as 'the most dangerous predator ever introduced', having been implicated in 38 cases of abundance-shifting predation. It is also evident that domestic cats are inimical to competitors, threaten wildcats through hybridisation (Daniels *et al.*, 1998), and pose ominous threats for disease transmission (e.g. Macdonald *et al.*, 1998).

CARNIVORE INTRODUCTIONS AS NATURAL EXPERIMENTS

Although there are undoubtedly undesirable introductions, there are potential benefits to be gleaned from those carnivore introductions that have already taken place. We suggest that the data on carnivore introductions should be viewed not just as a catalogue of historical mistakes and a warning for the future (a beneficial lesson in itself), but also as a source of information providing the opportunity to investigate ecological theory to an extent which would otherwise be neither practical nor ethical. Although it is not our intention here to analyse such data exhaustively, we have attempted to highlight some cases from which general lessons emerge, and provide brief examples below.

Lodge (1993: p. 133) noted that 'the characteristics and ecological impact of exotic species may provide clues to long-standing issues in the study of community assembly'. One such key issue in community assembly, which remains largely unresolved, concerns the relative importance of competition and predation in community composition (Wiklund *et al.*, 1998). The introduced Hawaiian avifauna has been modelled in terms of good competitors surviving and poor competitors failing to establish, but this is conditional on the successive filling up of niches – as more introduced species became established, the chances of further introductions surviving decreased (Moulton, 1993). Some introductions may also provide the opportunity to compare communities before and after invasions, and the answers clearly vary with circumstances. For example, whereas Maran *et al.* (1998) found that dietary competition with American mink was not a convincing explanation for the decline of European mink in Estonia, Clode & Macdonald (1995) found evidence of just such competition between American mink and otters in the Hebrides (although in this case competition appeared to favour the native otter).

Predation is thought to be a key factor in prey regulation (Sinclair, 1989), and understanding this relationship is of fundamental importance both for conserving rare species and controlling overabundant predators (Macdonald *et al.*, 1999). Perturbation experiments provide the best evidence for identifying limited or regulated prey populations (Sinclair, 1989), and carnivore introductions fall into this category. Craik (1998) removed feral mink from some seabird colonies in Scotland, causing a demonstrable increase in chick survival. Sinclair *et al.* (1998) used models based on threatened extant populations and attempted reintroductions in Australia, New Zealand and the Pacific islands to explore various aspects of the dynamics of predation. For example, they demonstrated that in Australia some native species such as the black-footed rock-wallaby (*Petrogale lateralis*) behave as classic alternate prey in the presence of feral predators, conforming to the expectations of a Type II functional response, while others, such as the brush-tailed bettong (*Bettongia penicillata*) were able to survive at low densities, and showed a Type III response to predation. Aside from their theoretical interest, such analyses provide practical guidance for management of rare or re-introduced populations.

Experimental investigation of invasions is rare (Williamson, 1999: p. 8), and as a result, much of the potential research value arising from species invasions is wasted. Petren & Case (1996) provide an example of how experimental methods were used to examine the mechanisms underlying competitive exclusion of a native species by an invader. They used large-scale enclosure experiments to eliminate direct aggression as the driving force behind the displacement of native gecko lizards by an invading species in the Pacific islands, and attribute this effect to exploitation competition. An experimental approach such as this is invaluable in ensuring capitalisation on the research benefits arising from species introductions.

FUTURE RESEARCH

There are many aspects of species introductions which merit further research, many of which have been outlined in the section about carnivore introductions as natural experiments and elsewhere in this chapter. As was

outlined earlier, the ecological changes which take place following an introduction present a unique opportunity to examine questions which might
otherwise be unanswerable. For example, to what extent can competition
and predation alter the composition of an ecosystem? What are the differences in individual species or ecosystems that allow some carnivore species, for example the raccoon dog, to be introduced with little apparent
impact (Kauhala, 1995) while others, such as the American mink, wreak
destruction wherever they appear (Macdonald *et al.*, 1999)? Perhaps the
most important practical problem to be resolved, however, is how the impact can be mitigated of those carnivores which have already been introduced. This is particularly important in key areas such as New Zealand and
Australia, where control of a small number of key carnivores would benefit
large numbers of native species. Traditional methods of control, such as
trapping and poisoning, are slow and labour intensive. However, important
research into novel control methods, such as immunocontraception (Tyndale-Biscoe, 1994; McCallum, 1996) is currently under way. Besides, the
practical problem of removing or controlling feral predators is the more
theoretically complex issue of managing the ecological repercussions of
removing what may be established exotic species, and the restoration of the
remaining habitat. New Zealand is among the leaders in the field of carnivore control and subsequent restoration (see Veitch, 1985; Towns *et al.*,
1997), but for the most part efforts there, as elsewhere, have been directed
at providing small insular refugia for threatened species. This is certainly
an essential first step, however in the future it may be possible to consider
larger-scale reclamation of ecosystems from the effects of introduced carnivores.

A further, more controversial practical issue is the matter of free
ranging and feral domestic species, particularly dogs and cats. These two
species are responsible for many conservation problems, through predation, hybridisation, and the spread of disease, and methods of managing
them to minimise their interactions with native species would be invaluable.

CONCLUSIONS

Introduced carnivores are a major threat to native species worldwide, primarily through predation, hybridisation and the spread of disease. Little is
known about the nature of the interactions between native and introduced
species in many cases. The problem of quantifying the impacts of introduced carnivores is confounded by the number of contributing factors, such as

habitat loss, human persecution and pollution, which have begun to take their toll within the same time frame as most carnivore introductions. Unravelling the conservation importance of introduced carnivores is further complicated because many of the threatened species involved may be extinct before the importance of introduced species in their decline can be assessed, let alone remedied. The likelihood that some populations of European mink and water vole are likely to be extinct before the detail of their interaction with American mink is revealed illustrates this point (Maran *et al.*, 1998; Strachan *et al.*, 1998), and highlights the importance of urgent research and action in dealing with alien carnivores.

6

Carnivore introductions and invasions: their success and management options

LUIGI BOITANI

Exotic species have always had a prominent position in conservation biology, especially since their introduction has been implicated as one of the four most important causes of species extermination – the 'Evil Quartet' (Diamond, 1984b). Indeed, introductions are increasingly more frequent in recent centuries with the number of documented introductions in most countries ranging from 100 to 10 000 species (Lodge, 1993). Today, invasion ecology and the impact of exotic species on natural ecosystems have become popular topics in conservation biology. In spite of a paucity of extensive data, a large scientific literature and at least 15 volumes have been published on this topic (see Lodge, 1993; Williamson, 1996). Most recently Shigesada & Kawasaki (1997) published a monograph on the mathematics of invasions and Bright (1999) published a general book on invasive species.

Exotic species have a decidedly negative reputation for themselves and are considered one of the most urgent matters to be dealt with on the conservation front (Drake et al., 1989; Bright, 1999). Despite this interest, there is still some confusion about the use of terms. The concepts of introduced, invasive, and pest species are associated with the problem of exotic (or alien) species, but are not always used consistently, leaving room for ambiguity. The terms introduction and invasion are often used as synonyms even though they express different concepts depending on the spatio-temporal scale, the causal contexts, and the consequences of the processes being described. A possible reason for the confusion of terms is that ecologists working on very different taxa (plants, invertebrates, vertebrates) may use the terms with slightly different perspectives and assumptions.

As I use the term, introduction applies only to the movement of species directly or indirectly brought out of their range by people into areas in which they did not occur in historic times. The term more strictly indicates only the anthropogenic movement of a species, even though it often implies also a minimum level of success in establishing a self-sufficient population. Lever (1994) used the term naturalization to indicate the processes of introduction and establishment in the wild. The rather vague meaning of 'historic times' allows for some overlap with the term 're-introduction' which applies to species brought back through human intervention to areas from which they were exterminated (Breitenmoser *et al.*, Chapter 12).

I refer to invasion as the process that follows an introduction and consists of an introduced species establishing itself as a reproducing population and spreading (invading) through the new environment. The word 'invasion' implies infringement, irruption, conquest of new territory, with some important effects on the existing system. Invasions may also occur naturally. Changes in the ranges of species are normal on both local and broader scales. Following the end of the last Ice Age, some 10 000 years ago, species were redistributed almost everywhere on this planet. On such broad spatial scales and when mass exchanges of biota occur across continents and islands, the most appropriate term is always invasions.

Invasive species may then become pests if they damage natural habitats (OTA, 1993) or if their presence conflicts with the interests of people to the point by causing damage to human health or economic, social or aesthetic interests (Putman, 1989; Williamson, 1996).

Traditionally, the study of introductions and invasions has been directly related to their bearing on human interests and only recently has there been more organized efforts to develop general theories on animal invasions (Crawley, 1986; Rejmanek & Richardson, 1996; Williamson, 1996). In particular, it would be useful to be able to predict the outcome of a species' introduction, and systematic searches have been done to discover the characteristics of successful naturalized species invaders and communities more easily invaded. However, despite the numerous empirical and theoretical studies, most proposed generalizations suffer from too many exceptions and lack of statistical testing. With the exception of some large taxa, there has been little success in achieving reliable predictiveness regarding invasions (Hengeveld & van den Bosch, 1996; Kareiva, 1996). Still, among mammals, the introduced species that successfully established new populations appear to have in common *r*-selected traits, high dispersal rates, high genetic variability, phenotype plasticity, large native ranges, eurytopy, polyphagy, and human commensalism. The most 'invasible' environments are described as early successional, disturbed, climatically matched with the native habitat of the invading species, and having a low diversity of native species and no predators (Diamond & Case, 1986; Lodge,

1993). Although they do not have predictive power, these generalizations are nevertheless useful as a framework for hypothesis testing.

Holdgate (1986) and Williamson (1993) proposed the ten-ten rule for successful invaders: it states that about 10% of introduced species are successful invaders and that about 10% of these become pests. The tens rule has several exceptions and Williamson (1996) points out that it could be conveniently expanded to 5 or 20%, and invites people to test it among different species.

Carnivore species have been introduced in many areas of the world and their effects have been often documented – especially when negative. This review analyses existing data on carnivore introductions against the theoretical generalizations given above and summarizes options for management.

WHICH SPECIES AND WHERE

The information published on carnivore introductions is scattered and largely incomplete. Ebenhard (1988) pointed out that there was no formal or even customary way of reporting introductions in the past and available information was more often available when linked to other aspects of the introductions that appeared of greater interest to ecology and conservation, such as the negative effects of introduced species on local fauna.

A few reviews on mammal introductions are available (deVos et al., 1956; Deems & Pursley, 1983; Ebenhard, 1988; Atkinson, 1989; Lever, 1994; Nummi, 1996) and they have been integrated in Table 6.1 with various other data (Lavrov, 1970; Bonner, 1984; Baker, 1986; Usher, 1988; Chapuis et al., 1994; Long, 1995). A minimum of 29 species of carnivores belonging to five families have been introduced into a great variety of environments. Information on the number of introduced species is likely to be reasonably accurate, because almost all introduced carnivore species are conspicuous and of some ecological and economic importance, therefore, they could hardly go unnoticed. On the other hand, the number of introductions reported in the literature (173) is definitely too low. There are two main reasons for this low number. The first and most important is that data on introductions that failed are lacking, and failed introductions were probably the majority (deVos & Petrides, 1967; Ebenhard, 1988). The second is that the number of reported introductions does not account for repeated introductions in the same or contiguous areas and in successive years. For example, there are only about 68 published records of cat introductions (see Table 6.1), but of 131 major islands and island groups in the world, cats

Table 6.1. List of introduced carnivores. Number of introductions are given for continents (C), continental shelf islands (S) and oceanic islands (O) and for introductions causing negative effects (abundance shifts) on native prey species through predation. Number of introductions are obtained only from published records for each introduction. See text for comments and references

Species	Introductions (with ecological effects)				Pests
	C	S	O	Total	
Canidae	16(2)	5(4)	18(11)	39(17)	
Dog (*Canis lupus familiaris*)	2	1	16(10)	19(10)	Yes
Dingo (*Canis lupus dingo*)	1(1)			1(1)	Yes
Coyote (*Canis latrans*)	7			7	Yes
Arctic fox (*Alopex lagopus*)		1(1)	1(1)	2 (2)	Yes
Red fox (*Vulpes vulpes*)	4(1)	2(2)	1	7(3)	Yes
Patagonian grey fox (*Dusicyon griseus*)		1(1)		1(1)	No
Raccoon-dog (*Nyctereutes procyonoides*)	2			2	No
Procyonidae	2	1(1)	3	6(1)	
Raccoon (*Procyon lotor*)	2	1(1)	1	4(1)	Yes
Kinkajou (*Potos flavus*)			1	1	No
Coati (*Nasua nasua*)			1	1	No
Mustelidae	8(1)	12(2)	9(4)	29(7)	
Stoat (*Mustela erminea*)		2(1)	2(2)	4(3)	Yes
Weasel (*Mustela nivalis*)			3(2)	3(2)	Yes
American mink (*Mustela vison*)	4(1)	5	1	10(1)	Yes
Ferret (*Mustela putorius furo*)	1	3(1)	3	7(1)	Yes
Siberian weasel (*Mustela sibirica*)		1		1	No
Polecat (*Mustela putorius*)	1			1	No
Sable (*Martes zibellina*)	1			1	No
Striped skunk (*Mephitis mephitis*)		1		1	No
Stone marten (*Martes foina*)	1				
Viverridae	3	6	21(10)	30(10)	
Lesser Oriental Civet (*Viverricula indica*)		1	3	4	No
Oriental civet (*Viverra tangalunga*)			1	1	No
African civet (*Civetticus (Viverra) civetta*)			1	1	No
Palm civet (*Paradoxurus hermaphroditus*)			2	2	No
Masked palm civet (*Paguma larvata*)		1		1	No
Small Indian mongoose (*Herpestes auropunctatus*) (includes *H. javanicus*)	1	3	11(10)	15(10)	Yes
Egyptian mongoose (*Herpestes ichneumon*)		1	1	2	No
Indian grey mongoose (*Herpestes edwardsi*)	2		2	4	No

Table 6.1. (*cont.*)

Species	Introductions (with ecological effects)				Pests
	C	S	O	Total	
Felidae	4(1)	7(2)	58(35)	69(38)	
Cat (*Felis catus*)	3(1)	7(2)	58(35)	68(38)	Yes
Yaguaroundy (*Felis yaguaroundii*)	1			1	No
Totals	33(4)	31(9)	109(60)	173(73)	

are absent from only 13 of them (Atkinson, 1989). In Alaska, arctic and red fox introductions started as early as 1750 for fox farming. Red foxes have been recorded on at least 455 islands, and arctic foxes on 77 islands, but there is no specific information on introductions (Bailey, 1993). In Table 6.1 all Alaskan islands are considered together as one item. The same is true for red fox introductions to the Kuril Islands (Litvinenko, 1993).

Cats and dogs are among the 10 mammalian species with the highest number of reported introductions (Ebenhard, 1988; Lever, 1994), followed by the American mink and the small Indian mongoose. Some species have been introduced only once or in only one area, and they succeeded in establishing viable populations; for example, coati on the island of Juan Fernandez (Torres & Aguayo, 1971), *Viverra civetta* on the island of Sao Tomé (Feiler, 1984), the polecat in Norway (Nummi, 1996), and the striped skunk on Prince Edward Island in Canada (Deems & Pursley, 1983). Other species have been introduced repeatedly in different, relatively close areas. For example, the coyote has been introduced into Florida and nine other states of the United States by hunters (Deems & Pursley, 1983).

Introduced species were classified by Ebenhard (1988) in groups depending on whether the introduction occurred on a continent, a continental shelf island, or an oceanic island (see Table 6.1). About 19% of all recorded introductions of carnivores occurred on continents, about 18% on continental shelf islands and about 63% on oceanic islands. Since most of the failed introductions are not known, it is impossible to state if this pattern is caused by different introduction efforts into the three groups of areas or it reflects different failure rates of introductions. However, the fact that this pattern grossly reflects the percentages for all recorded introductions of birds and mammals (Ebenhard, 1988) may indicate that introduction efforts have been different in the three groups of areas.

Introductions have occurred throughout human history and the earliest carnivore introduction is likely to have been dogs brought to the Americas by the Paleo Indians. Similarly, the dingo was brought to Australia by Aboriginal explorers as early as 3000–5000 years BC. The rate of introduction of carnivores accelerated in the nineteenth century. Most introductions for the fur industries took place between 1850 and the early decades of the twentieth century (Lever, 1985), whereas accidental introductions of cats and dogs to islands peaked during World War II as a result of military activities (Atkinson, 1989). A decrease in new introductions in the last 50–70 years is probably not the result of growing conservation awareness or ecological concerns, but merely due to the fact that most of the islands have already been colonized by the domestic species most frequently associated with people or by other alien species of potential economic interest.

A topic lying on the margins of the subject of introductions by humans is natural invasions – the more or less rapid expansion of the range of some species beyond their usual range (Diamond & Case, 1986). It is probably a good idea to distinguish the invasions of domestic and feral species from those of wild species. For some domestic and feral species, the process of invasion is almost predictable, such as for dogs and cats which have progressively invaded all available environments and for which the causes of their self-introduction are sufficiently known. These invasions are not limited to insular environments or those contiguous to human activity, but also extend to protected areas on continents where the natural conditions are relatively better preserved and should, therefore, be able to withstand these invasions (Usher, 1988). Smaller reserves and those that are located in the vicinity of urban areas appear to be more easily invaded (I. Macdonald et al., 1988; Loope et al., 1988). Good examples are the stray and feral dogs that compete with wolves even inside many Italian protected areas (Boitani, 1983). There have been some recent cases of natural invasions by wild carnivores, for which the ecological mechanisms are only inferred. A notable case is the coyote, which was present only west of the Mississippi River in the United States and west of Ontario's Lake Nipigon in Canada until 1900. Since then, coyote populations have expanded eastward to states and provinces on the east coast, helped by the eradication of wolves and extended habitat alteration (Deems & Pursley, 1983). An east–west expansion has been found for the red fox in North America (Soulé, 1990). A rather sudden invasion of the Canada lynx (*Lynx canadensis*) occurred in Minnesota in the early 1970s where it lasted for only a few years. It is presumed that the Canadian population of lynx had reached a high, forcing most of the dispersing individuals south (Mech, 1973d). Another puzzling case is that of the golden jackal (*Canis aureus*) which, in the last 20 years, has increased its range in Europe, conquering the Balkans up to the eastern fringes of the Alps.

THE REASONS FOR INTRODUCTIONS

Carnivores have been introduced for a great variety of reasons that fall into four main categories: (1) domestic animals have turned feral after arriving with humans (e.g., cats and dogs); (2) animals have escaped from captivity such as many furbearers imported for fur farms (e.g., arctic foxes and American mink); (3) animals have been accidentally transported as commensals or trapped in containers (e.g., small Indian mongoose in Florida, where it disappeared); and (4) most frequently, animals that were deliberately released for several reasons, primarily economic, recreational, or for biological control. The fur industry centered on red and Arctic foxes, sable, and American mink was mainly responsible for introducing these species into continental Eurasia and most of the northwestern and northeastern islands of the Pacific (Bailey & Kaiser, 1993). Animals were kept in cages and enclosures or released on the islands and then cropped regularly. However, when the fur industry collapsed at the end of the 1930s, many islands were abandoned by people, but several carnivore populations survived. American mink were brought to Europe for fur farms in the 1920s, but a few free-living animals were seen soon afterwards (see Kauhala, 1996, for a review). An eagerness to improve the economic and recreational return for hunters and trappers led to many carnivore introductions, sometimes involving an enormous number of released animals, such as the nearly 19 000 mink, 10 000 raccoon-dog and 1200 raccoons released into various hunting grounds in the former USSR from the 1920s to the 1940s (Lavrov, 1970). Red foxes were introduced to Australia mainly for hunting purposes and not for rabbit control (Lever, 1985).

Most species of *Mustela* and the Viverridae family were introduced as a form of biological control in an attempt to reduce irruptive populations of rabbits and rats. *Viverra* spp. is the source of civet, a valuable substance in perfume production, and for this reason *V. tantalunga* has been introduced onto several islands of the East Indies including Celebes (Groves, 1976). The Patagonian grey fox was introduced to Tierra del Fuego to control rabbits, as were the stoat, weasel, and ferret to New Zealand. However, there is not a single case in the literature of an introduced carnivore managing to control a pest species and not becoming a bigger pest itself. A paradigmatic case is described by Easteal (1981) and Nellis (1989) in Jamaica. Ants (*Formica omnivora*) were imported in an attempt to control introduced rats that were causing serious damage to sugar cane. As the ants turned out to be ineffective against the rats, toads (*Bufo marinus*) were introduced to control the ants and the rats. The effect was negative, as the toads ate many other things besides ants. At this point the small Indian mongoose was introduced. And once again the effect was negative, as the mongoose nearly eliminated the rat problem but ate several other species as well, and caused even more widespread damage to local ecological systems.

More recently, the introduction problem acquired a new dimension through the activity of 'animal liberation' groups who released animals kept in captivity by the fur industry. The first cases occurred some 15 years ago (Baker, 1986), but continue, and recently thousands of American mink were released in Italy, France, and Great Britain. In Italy these introductions have led to new established populations.

SUCCESS OF INTRODUCED CARNIVORES

Determining whether an introduction succeeds or not has been debated using scanty empirical evidence and much theoretical modeling, but we are far from agreeing on a general pattern. Support for the role of biotic and abiotic factors comes from correlational results which cannot prove a causal relationship. Moreover, most of the data feeding these analyses are incomplete for vertebrates (such as on failed introductions), thus weakening most results. To evaluate the success of introductions, we require data on the ratio of established populations to the total number of introductions. To understand the mechanisms of the success, we need to know at least a series of ecological and logistic factors that played a potential role in determining the outcome of the introduction attempt. Unfortunately, little data is available on carnivore species for both of these aspects, the only exception being the well recorded history of introductions into a few countries or large islands. But even these cases do not provide sufficient information for robust generalization. The factors suggested as potential predictors of successful invaders are not supported by empirical studies, forcing discussion to remain qualitative. I review some of the most common arguments on what makes a successful invader and discuss their applicability to the existing data on carnivore introductions.

In broad terms, all successfully introduced carnivores possess, to some degree, all of the characteristic life history traits and factors considered common among successful invaders (deVos *et al.*, 1956; Erlich, 1986; Lodge, 1993; Williamson, 1996). All are medium-small size, generalists or opportunistic predators, highly fecund, and very flexible in their habitat requirements. Williamson (1996), using data from the full taxonomic range of invaders (from plants to invertebrates and vertebrates), discussed some factors considered to be predictors of invasion success: the intrinsic rate of natural increase, modes of reproduction, genetic structure, abundance and range in native habitat, taxonomic isolation, and climatic matching (that is, the similarity between the climate of the original range and that of the introduction area). Although these characteristics seem to apply well to the successful introductions of the American mink, raccoon-dog, raccoon, lesser Indian mongoose, cats, and dogs, it is easy to find notable exceptions. As an example, the American mink succeeded even in areas where it was not ecologically distinct from other species in the invaded community (Macdonald and Thom, Chapter 5) and several species of the family Viverridae do not have particularly large ranges nor are they especially abundant in their native habitats. In summary, even though we do not have further support from failed introductions, it seems that current models are poor predictor of carnivore introduction success.

The concept of a vacant niche is often mentioned as a factor explaining invasion success (see Williamson, 1996, for a review). The argument, however, though intuitive and appealing, provides little help in understanding invasion success. Using the phrase 'vacant niche' to mean a new functional role played by the species in the new community, does not help explain the mechanisms of an invasion, as a vacant niche can only be defined *a posteriori* (i.e., after a species has colonized the new community). For example, the American mink and raccoon-dog were able to find 'vacant niches' in the rich carnivore guilds of Eurasia.

Elton's (1958) conclusion that invasibility decreases with increasing numbers of extant species would appear supported in broad terms by the high number of successful carnivore invasions on islands compared to continents (Table 6.1). Although those numbers cannot be compared against true introduction attempts and may only show that introductions onto islands occurred more frequently than on continental areas, it may be argued that the relatively higher vulnerability of island systems may have played a significant role in determining introduction success. Lack of effective antipredator behavior among island fauna is an important factor of island vulnerability (Diamond, 1984b) and thus, for carnivores, a potential factor influencing introduction success.

The biological and ecological characteristics of successful carnivores (high intrinsic growth rate, r, and hardly any intrinsic mechanisms of population control) and the ecological features of most introduction areas (high environmental variance and low complexity) have often provided the perfect setting for irruptive population increase. What are the critical factors that trigger an irruptive population increase? Why some introduced carnivores that had all the potential characteristics of a successful invader, did not succeed in establishing an irruptive population or disappeared in a short time? If we were able to compare detailed descriptions of the species' life history traits against the ecology of invaded areas, we could hope to get useful insights into the mechanisms of a successful invasion. For carnivores, we have scanty data to build upon that are not sufficient to draw any conclusion. As examples of successful irruptions, the American mink colonized all of Sweden (except the northern mountains) in just 35 years (Gerell, 1967); the raccoon colonized most of east and west Germany in less than 45 years (Lutz, 1984); and the raccoon-dog spread throughout central Europe at an average annual speed of 40 km (sometimes up to 120 km/y) (Kauhala, 1996).

There are also examples of introduced carnivores that all had potentially successful traits, yet did not build up populations rapidly. As examples of unsuccessful irruptions, the weasel, despite being freed in large numbers in New Zealand, occur only intermittently on the North and South Islands (R. Taylor, 1984); the stone marten escaped into the wild in Wisconsin in late 1960s and currently only a small breeding population exists in southeastern counties (Long, 1995; A. Wydeven, pers. com.), outside the range of fisher (*Martes pennanti*) and American marten (*Martes americana*); in a dry coastal habitat in central Italy, introduced small Indian mongooses disappeared after existing for 20–30 years at very low densities (Carpaneto, 1990). The reasons for these failures are not known, but King (1983a) pointed out the crucial role of habitat stability and prey population fluctuations in determining population densities of mustelids, particularly the smaller species. Competition with introduced stoats (*Mustela erminea*) and cats for introduced and fluctuating small rodent populations may be one of the factors influencing the relative low densities of the weasel in New Zealand. Competition with American mink and other mustelids may also be the crucial factor preventing the stone marten in having an irruptive population increase in Wisconsin. The highly seasonal fluctuations of rodent and insect populations in a Mediterranean maquis may have affected reproduction and survival of the mongooses in Italy. Studying what limits population increases and expansions of otherwise successful species would probably provide better insights on invasion processes than studying cases of irruptive populations.

The size of an island appears to have an important effect on irruptive population increases. Extremely successful population irruptions have occurred, such as feral cats on Marion Island (290 km^2), which grew from just five colonizers in 1949 to more than 2200 animals in 1975 (Rensburg *et al.*, 1987). All feral cats on the main island of the Kerguelen archipelago (Grand Terre, 6500 km²) descend from a pair introduced in 1956; and despite an intensive control program between 1973 and 1977, they numbered more than 10 000 in 1984 (Chapuis *et al.*, 1994). As an example of failures, a viable cat population could not be established on the 7 km² island of Saint Paul, that support a healthy rabbit population. Similar failures were found on some Aleutian Islands, where foxes disappeared naturally (helped by trapping in a few cases) from 387 of 455 islands that once harbored introduced populations (Bailey & Kaiser, 1993). The main reason for the lack of survival was apparently reproductive failure and starvation after the destruction of accessible bird colonies (E. Bailey, 1993). It is interesting to note that arctic foxes disappeared from most islands smaller than roughly 500 ha, although there are some exceptions, including an island as small as 32 ha (E. Bailey, 1993). Inbreeding and, most likely, the demographic stochasticity of very small populations may have played a critical role in the extinction of many fox populations, especially since environmental stochasticity is naturally high on many islands. The role of demographic stochasticity is rarely given much importance in the literature on invasive carnivore species, possibly because most invasive species quickly reach high numbers and densities. However, among carnivores, this factor could be important in small introduction areas.

Scientific data on carnivore introductions are not of good enough quality nor complete enough to support useful statistical analyses. The tens rule (Williamson, 1996) suggests 10% success of imported species escaping to the wild, and 10% success of introduced species becoming established with viable populations. These rates cannot be evaluated for carnivores as we do not know the real number of imported and introduced species. Anecdotes provide contrasting evidence: if the case of the red fox in the Aleutian Islands is considered, the rule appears confirmed as the fox population persists in about 15% of the islands (Bailey & Kaiser, 1993), but if the case of the American mink in European countries is considered, the rule is not validated (Kauhala, 1996). There is also the anything but trivial problem of classifying cats and dogs as true species or in a category by themselves because of their evolutionary advantage in terms of ecological and behavioral plasticity.

In discussing deviations from the tens rule, Williamson (1996) presents the case of mammals on continental islands (Ireland and Newfoundland), where rate of success in establishing viable populations almost reaches 100%. Strikingly impoverished fauna on the islands and good ecological match to the new communities are indicated as key factors for the high success. In broad terms, carnivore introductions may tend toward a similar pattern. The analysis of a large number of introduction cases may provide useful insights on the general trends and correlates of success. It could also provide interesting grounds for speculation, but would not significantly improve either our understanding of the intricate mechanisms of introduction success or our predictive power.

EFFECTS OF INTRODUCED CARNIVORES

The popular wisdom on carnivore introductions is that they are among the worst ecological disasters on earth, but the truth is much more complex. The impact of introduced carnivores has ranged from nil to minor economic, cultural and ecological loss through to true unmitigated disaster. The story of the lighthouse keeper's cat on St. Stephen's Island in the Cook Strait has been taken as the paradigm of the impact of introduced carnivores on endemic fauna. After its arrival in 1894, in less than a year this cat alone exterminated the tuatara (*Sphenodon punctatus*), a rare endemic lizard, and the island's endemic bush wren (*Xenicus lyalli*) (Greenway, 1967). There are several descriptions of the ways in which endemic species, populations and communities have been negatively affected by introduced carnivores (see Lever, 1994, for a review; Macdonald and Thom, Chapter 5). At least 25 reptile (Ebenhard, 1988), 96 bird and 30 mammal (Lever, 1994) species have been ecologically affected by introduced carnivores. Almost 45% of all successful carnivore introductions have had a negative effect on native species (i.e., lowered their population density) and have become a pest. This percentage is much higher than the 10% predicted by the tens rule. The most common kinds of impact are summarized.

Effects on native populations through predation

Introduced carnivores have been among the most common causes of population reductions and even extinction of native prey species on islands (Diamond, 1984b), but it is not always so. Ebenhard (1988) distinguished three main types of effect through predation. First, there may be no particularly negative effect, as shown by the cases of cats on Campbell Island (Dilks, 1979), and small Indian mongooses in Hawaii (Fischer, 1948); both species were feeding on rats without reducing their numbers. Second, predation may strongly reduce native species and bring them to extinction as shown by the 33 bird species extinguished by cats and nine by mongooses (Jackson, 1977). Third, a sort of balance may be established between introduced carnivore and native preys so that coexistence is possible. This appears to be the case for the American mink and the native prey populations in Britain and Sweden, despite the heated debate between high impact (Lever, 1978) and almost no effect at all (Chanin & Linn, 1980). Gerell (1985) has shown that minks kill only a minor portion of the breeding birds, even on small islands off the Swedish Baltic coast, and that coexistence has become possible in just two decades (but see Macdonald and Thom, Chapter 5, for a more detailed discussion).

At least a third of all cases of decline in Australia's endangered species are due to cats, the dingo, and the red fox (Burbidge & Jenkins, 1984; Dickman, 1996a), and Ebenhard (1988) concluded that the cat is by far the most dangerous introduced carnivore for native preys (see Table 6.1). Cats are known to prey on mammals as large as *Solenodon cubanus* and birds such as *Branta sandvicensis* (Ebenhard, 1988). Ascension Island provides an excellent example of how cat predation eliminated almost all the nesting grounds of seabird species from the main island (Ashmole *et al.*, 1994).

Dogs and canids are generally a good second in the list of introduced carnivores with negative impact, the only exception being the raccoon-dog. For example, on many small Aleutian Islands in Alaska, fox introductions caused drastic reductions in numbers of several seabird species (E. Bailey, 1993). The potential scale of the impact is shown by the example of Shaiak Island (Alaska) where two red foxes devastated a colony of 156 000 nesting seabirds – all eggs and nestlings were killed and cached all over the island (Bailey & Kaiser, 1993). The best evidence of the negative effect of introduction is the contrast in population densities and species diversity between islands with foxes and those without, and the striking increase in seabird numbers that occurs after foxes are eradicated (E. Bailey, 1993). A similar effect was found in Australia, confirming the negative impact of foxes on several Australian mammal species (Dickman, 1996b). The effect of the dingo has also been clearly demonstrated by comparing areas where the species is present with areas where the species has been eliminated. Densities of red kangaroos (*Macropus rufus*) and emu (*Dromaius novaehollandiae*) have increased and remained higher in dingo-free areas than in areas where dingoes still occur. In central Australia at least 10 medium-sized mammals have been eliminated by dingoes (Corbett, 1995). However, dingoes have also had a positive effect, as shown by the role they apparently play as controllers of cats and foxes, reducing their numbers and their impact on smaller preys (Dickman, 1996b).

Only four species of the introduced Mustelidae appear to cause negative effects, especially stoats in New Zealand (Atkinson & Cameron, 1993). The raccoon has not yet been found to have negative effects on its new European range, but on Queen Charlotte Islands in northern Pacific and other islands off the coast of British Columbia, the raccoon has been responsible for the destruction of more than 95% of many seabird colonies (ancient murrelet, *Synthliboramphus antiquus*; Cassin's auklet, *Ptychoramphus aleuticus*; rhinoceros auklet, *Cerorhinca monocerata*) (Harfenist & Kaiser, 1997). Of the nine species of the Viverridae family, only the small Indian mongoose has been found to have a consistently negative impact wherever it has been introduced (Ebenhard, 1988; Lever, 1994; Nellis, 1989).

Prediction of the possible effects of introduced predators is complicated by the lack of understanding of the subtleties of the defence behavior of potential preys. As an example, the opposite population trends of the highly endangered takahe (*Porphyrio mantelli*) and its closest and successful relative, the pukeko (*Porphyrio porphyrio*), show how small differences in behavioral responses to mammalian predators, such as the introduced stoat, can make to the difference between survival and extinction (Bunin & Jamieson, 1994).

Effects on ecologically similar species through competition/displacement

Several examples of competition between native and introduced species have been reported (deVos & Petrides, 1967), but quantitative studies of ecological competition through interference or exploitation between introduced carnivores and native species are limited. The potential competition of the American mink and the otter in Europe has been highly debated. Although dietary overlap has been estimated to be from 40% (Wise *et al.*, 1981) to 55–75% (Erlinge, 1969), Dunstone & Ireland (1989) concluded that there was no substantial evidence to support the view that mink were significant predators of native preys and competitors to native carnivores. However, Clode & Macdonald (1995) have shown that there is some competition for food between otter and mink and that the outcome might depend on local patterns of food types, quantity and availability. The American mink has also been indicated among the most important causes of extinction of the European mink (*Mustela lutreola*) (see below, and Macdonald and Thom, Chapter 5) and a competitor of the Siberian weasel (*Mustela sibirica*) and other native mustelids (Ebenhard, 1988). Cats have been shown to compete with dunnarts (*Sminthopsis macroura*), and dingoes are supposed to have been among the causes of extinction of the thylacine (*Thylacinus cynocephalus*) and Tasmanian devil (*Sarcophilus harrisii*) from mainland Australia (Dickman, 1996b). Despite these cases, conclusions on the effects of ecological competition between introduced carnivores and native species remain qualitative and weak.

Effects on community structure and ecosystem functions

Clear examples of invaders having an effect on processes at ecosystem level are very rare, and Vitousek (1990) believes that very few successful invasions alter large-scale ecosystem functions. Inductive reasoning may conclude that when an introduced carnivore has a dramatic impact on insular native species populations it could probably deeply affect the simple structure of those resident communities, but there is no extensive documentation to show the presence or absence of the effects it could have on all components of the systems.

Vitousek (1990) suggested that invaders can change ecosystems where they: (a) differ substantially from natives in resource acquisition or use; or (b) alter the trophic structure of the invaded area; or (c) alter disturbance frequency and/or intensity. It could be argued that many examples given in this chapter fall into one or more of these categories, but evidence to support the view of a major ecosystem impact is mostly anecdotal and incomplete.

Introduced predators can induce boom and bust cycles, and interact in complex and unpredictable ways with the native environment and other introduced species. For example, the Macquarie Island parakeet (*Cyanoramphus novaezelandiae erythrotis*) was not affected by cats and weka (*Gallirallus australis*) predation until the rabbit was introduced (Taylor, 1979). Rabbits provided the cats with an abundant year-round food supply, allowing the cats to increase and prey on the parakeet. On Dassen Island, off the coast of South Africa, cats predate on rabbits and scavenge on seabird carcasses in a sort of sustainable equilibrium; if the cats were removed, increasing rabbit numbers would damage the vegetation, and if rabbits were removed the cats would start preying on seabird colonies (Apps, 1984). A good indication that the effects of introduced predators are enhanced by fragmentation of native vegetation and prey species is provided by cats and dingoes in Australia (Corbett, 1995).

Attempts at biological control by introducing the small Indian mongoose have often led to extensive ecological disturbance to native ecosystems. The species is a generalist and attacks many non-target species causing disturbance throughout communities. In the Caribbean, small Indian mongooses prey on rice rats (*Oryzomys* spp.) and roof rats, but the latter have been more successful in escaping predation by becoming more arboreal, with the result that preying on nesting birds has increased (Nellis, 1982, 1989).

Effects on genetic integrity through hybridization

Hybridization and genetic introgression between subspecies are the most common genetic effects of introduced species (see Wayne & Brown, Chapter 7). Among carnivores, several good examples show these effects. In Japan, the native subspecies of the Siberian weasel (*Mustela sibirica itatsi*) was swamped by introduced Korean weasels (*M. s. coreana*), and in western Siberia *Martes zibellina* hybridized with *M. z. kamtshadalica* and lost its original genome (deVos & Petrides, 1967). In Britain, hybridization occurred between polecat and ferret, whereas it has not been found between polecat and American mink (Lever, 1994). The introduction of red foxes into Sweden from Britain has been linked with the appearance of the 'Samson foxes' which have an inherited deficiency of guard hairs (deVos & Petrides, 1967).

The highly endangered small population of Florida panthers (*Felis concolor coryi*) was recently found to consist of hybrids between the original population and animals from the Latin American subspecies (O'Brien *et al.*, 1990). Genetic introgression has also been one of the main issues in the reintroduction program of the red wolf (*Canis rufus*) in the southeastern United States. MtDNA haplotypes of both wolf and coyote have been found in the red wolf, showing that hybridization between these animals occurred in the past (Wayne & Jenks, 1991; Wayne & Brown, Chapter 7).

Hybridization may be more difficult to classify when feral and domestic animals are involved, but invading feral and house cats are known to threaten the genetic integrity of both European wild cats (*Felis silvestris*) (Hubbard *et al.*, 1992) and African wild cats (*F. s. libyca*) (Stuart & Stuart, 1991). Hybridization with feral dogs threatens the wild populations of the endangered Ethiopian wolf (*Canis simensis*) (Gottelli *et al.*, 1994), and also some wild gray wolf populations in North America and Europe. In Italy, however, even though wolf-dog hybridization has been observed (Boitani, 1983), there is no evidence yet of genetic introgression (Randi *et al.*, 1995; E. Randi, pers. com.). Lower offspring viability and perhaps also unknown behavioral barriers may limit the negative effects of these crosses.

An interesting aspect of 'genetic competition' between American and European mink in Europe has been described by Simberloff (1996). Since the American mink is larger than the European mink and becomes sexually active earlier, it is likely that female European mink mate with male American mink. The embryos are resorbed, however, and there are no viable offspring, leading to further decline of the European mink.

Effects on population viability through importation of parasites and diseases

Introduced carnivore species often carry parasites and diseases for which they can provide a reservoir and a means of transportation into invaded areas. However, there is not a single study proving that any introduced carnivore has been responsible for introducing any parasites or diseases to new areas. Rabies has been found in mongooses in several areas of introduction, including most Caribbean islands and Hawaii (Nellis, 1989). Small Indian mongooses also are known to carry leptospirosis, canine distemper, canine hepatitis, feline panleukopenia and other diseases potentially dangerous to other animal species (Nellis & Everard, 1983); and raccoon-dogs are known to carry *Trichinella*, mange and other potentially dangerous diseases, but all these diseases were already established in the invaded areas.

The effect that introduced carnivores, as potential reservoir for diseases, may have on native species is supported by indirect evidence. For example, the raccoon-dog is also a rabies vector and rabies outbreaks in Finland have been connected to the arrival and increase of this species (Kauhala, 1996). Feral dogs in Italy are affected by parvovirus and *Echinococus granularis* which are now also found in wild wolves (V. Guberti, pers. com.). Nevertheless, there is no evidence of any native species having been threatened by any parasites or diseases carried by introduced carnivores.

Effects on behavior of prey species

This is a rather neglected effect with few published examples. The elegant study on lava lizards (*Tropidurus* spp.) of the Galapagos Islands is one of these. Differences in the wariness of lava lizards on various islands were found to be positively correlated with the presence of cats and not correlated with other potential predators such as rats and snakes (Stone *et al.*, 1993). Predation pressure selected anti-predator response and eliminated the rarer tame behavior often found in island animals.

Nellis (1989) reported that roof rats became almost completely arboreal in St. Croix in response to predation by mongooses. On the same island, the bridled quail dove (*Geotrygon mystacea*) was known to nest on the ground before being thought extinct in 1921, and now nests in low trees and has increased to moderate densities (Nellis & Everard, 1983). The behavioral responses of native species to introduced carnivores are very difficult to predict because their full range of behavioral flexibility is unknown. However, this area of research appears to be of crucial importance to improve our ability to predict the outcome of animal introductions.

MANAGEMENT OPTIONS FOR INTRODUCED CARNIVORES

Management of introduced species has been debated from several perspectives. For example, the idealists call for the eradication of all introduced biota; the pragmatists try to give priority to the worst cases and those technically most feasible; and the animal rights organizations oppose control of exotic species because they are focused on the welfare of each individual animal and its right to live (Soulé, 1990). The removal or control of introduced carnivores has significant ethical and emotional implications, particularly when they are dogs or cats. In this complicated social context, finding a rational and consistent approach to management of introduced carnivores is not easy. The problem arises when the introduced species becomes a pest, that is, when the introduced population adversely affects (or it is perceived to affect) a resource of some value to people. Not all introduced carnivores are necessarily pests, as suggested by the American mink in Scandinavia which Gerell (1985) claimed no longer had to be controlled because an equilibrium had been established. But the many examples mentioned above have shown that a substantial share of all successful carnivore introductions have had a negative impact on ecosystems and endangered species, and a full range of control measures has been applied with diverse practical results.

The goal of many control programs is eradication of the introduced population, but this is often an impossible task. Bomford & O'Brien (1995) listed six criteria that are essential for eradication: (a) the rate of removal must exceed the rate of increase at all population densities; (b) immigration must be prevented; (c) all reproductive animals must be at risk; (d) animals must be detectable even at low densities; (e) discounted cost-benefit analysis must favor eradication over control, that is, the quantitative analysis should also estimate the present value of a future benefit at a chosen interest rate; and (f) there must be a suitable socio-political environment. It doesn't take more than a superficial analysis of these criteria to realize that, with a few exceptions for small islands, it is impossible to meet them for most pest carnivores. Not only do the latter all have the characteristics of successful invaders, but they also quickly become trap and bait shy, learn new tricks, spot new dangers and, at least with dogs and cats, can often count on huge reservoir domestic populations (Boitani et al., 1995).

These obvious remarks shed a pessimistic light on carnivore eradications. A notable case of an attempted and abandoned eradication program involved the American mink in Britain in 1970 (King, 1983a).

The most common solution adopted has been direct control of populations by shooting, trapping and poisoning. However, most successful invaders are successful because they have a very high rate of increase and any reduction in numbers is generally short lived. A short-term reduction may be enough to provide some relief from predation pressure, but such efforts must be repeated every year, implying high costs and a great deal of commitment. Direct killing succeeded in eliminating pest carnivore populations only on small islands, such as foxes from some of the Aleutian islands in Alaska (E. Bailey, 1993) and cats from Little Barrier Island (2817 ha) in New Zealand (Veitch, 1985), but failed on larger islands as with cat control programs on the main island of the Kerguelen (Chapuis *et al.*, 1994) and on Macquarie Island (Brothers *et al.*, 1985). In mainland areas, direct killing has never achieved any long-lasting result. Moreover, removal of target animals has to be carefully planned in relation to the biology of the species and the boom and bust characteristics of the populations involved. In New Zealand it has been found that intensive stoat trapping was more effective when concentrated in periods of population irruption following an increase of *Mus musculus* (R. Taylor, 1984).

More sophisticated approaches have been tested on carnivores with interesting, though local results. E. Bailey (1993) obtained good results in controlling introduced arctic foxes on some Aleutian islands using sterile male red foxes (but see Schmidt, 1985, for a few caveats of the technique). Red foxes apparently eradicate arctic foxes through competitive exclusion, at least on small islands, and there are indications that this is also the outcome when the two species compete in mainland areas (Hersteinsson *et al.*, 1989). Similar competition has been reported for coyote and wolves, and one of the reasons why the coyote succeeded in invading much of the eastern United States may have been the extermination of the wolf from those areas.

Cat control has been successful on Marion Island using the host-specific, contagious disease feline panleucopaenia (van Aarde, 1984). The virus did not eradicate cats, but was effective in reducing the population at an average annual rate of around 30%. However, a pathogen is unlikely to drive its host to extinction and continuation of population control by a combination of techniques appears to be necessary (Dobson, 1988).

Use of a species-specific contagious virus is also being planned to spread a contraceptive agent with the aim of sterilizing the population. Fertilization would be immunologically suppressed using recombinant virus vectors that can transmit immunogens inducing a specific immune response against reproductive proteins, but a vector virus has not yet been identified for carnivores (Creagh, 1992).

Poisons (sodium monofluoroacetate and diphacinone) have been widely used with carnivores, although there are several drawbacks. Animals may become resistant if exposed continuously to sub-lethal dosages following problems of bait acceptance, poison shyness or when it is difficult to ensure that a lethal dose is delivered. Populations may acquire a genetic resistance by selection of resistant strains. It is also difficult to poison the target species selectively. Recently, a potentially useful case of secondary poisoning was revealed in stoats and weasels that had preyed on mice killed by Talon 20, a powerful rodenticide (Alterio *et al.*, 1997). Although non-selective, it may prove a useful additional tool to support an integrated control program. A very promising control tool is a cat-specific poison that is being developed and tested in Australia (B. Richardson, pers. com.). If it will maintain its expectations of specificity and efficiency it may well be the optimal solution for cat control.

Fencing was widely applied to control dingoes in Australia and it soon became clear that it had a limited effect without other forms of control (Corbett, 1995). Fencing supported by intensive year-round trapping was the approach used to control stoats, cats and ferrets on the Otago Peninsula in South Island, New Zealand (Atkinson, 1996).

In short, there is no one means that is more effective than others in controlling introduced carnivores. It is interesting to note that New Zealand's vast experience on this topic has converged into a strategy to cope with pest animals based on a few solid actions: island eradication wherever possible; actions on mainland 'islands', highly managed sites for pest control and habitat restoration; and national pest species plans that are coordinated and integrated, as well as long-term programs that have clear objectives and tested techniques. Although more than 100 mammal populations have been eradicated from the 700 islands in New Zealand, success was possible only in smaller islands and those where re-invasion could be excluded (Veitch & Bell, 1990). Eradication is an urgent necessity for many natural areas but, in spite of the many calls for action (Coblentz, 1990; Temple, 1990), it is bound to remain a utopian goal for most of the introduced carnivores until new technologies are available.

The best control measure remains prevention and it can never be too obvious to repeat that all efforts should be made to prevent introductions and to develop contingency plans for prompt eradication of the first propagules.

CONCLUSION

In general, carnivore introductions have been an important threat to conservation worldwide and there should be no question that every effort should be made to control introduced populations and to avoid new releases. Control and prevention, however, appear to be most effective when they are understood and supported from the public. Most introductions have been carried out by people seeking personal benefits and the public became aware of the negative impact of introduced species only when the effects were obvious on some economic or health aspect. The public is rarely aware of the effects on the environment and a specific communication effort should be aimed at reducing this ignorance. A change in people's attitude is necessary and this could be achieved through an information campaign showing the results of previous researches.

Carnivore introductions are likely to continue, although at a lower rate than in past years. A decreasing trend seems possible for the release of wild species, but it does not seem likely for domestic animals. Free-ranging cat and dog populations continue to thrive and accompany the spread of human settlements into natural habitats. The study of their impact on ecosystem processes is urgent, and there is a need for more experimental work.

Experimental design has been rare in research on invasion ecology, and especially on carnivore introductions which are often described only through anecdotal reports. The possibility of manipulating introduced carnivore populations offers an especially good opportunity for study and experimentation on predation processes in simple and complex predator–prey systems, and for study and identification of possible keystone species. More generally, research on carnivore introductions is an excellent opportunity for insights on community structure and functioning. As pointed out by Vitousek (1990), there is a need for an approach that integrates population level processes into ecosystem level ecology to broaden our understanding of introduction patterns and make them more useful for study of ecosystem processes.

Study of failed introductions appears a neglected and yet promising research area. New carnivore introductions into an island may not be a frequent opportunity in the near future, but the spreading of an invasive carnivore throughout a continent (for example, the raccon-dog in Europe) and the fate of the animals freed by animal rights organizations could be interesting and useful case studies.

I doubt that carnivore introductions will ever be able to contribute significantly to the theoretical models of introduction and invasion and that their success could ever be predicted with reasonable accuracy. The number of ecological and behavioral variables that should be considered for a predictive model, their variation at population and, most important, at individual level, and the largely unpredictable role of people, would all require building a most complex model. Even if such a model could be built for a single case, it seems difficult that it could have a more general validity. Nevertheless, more and better field studies on carnivore introductions and invasions are necessary, and they should be a matter of priority for sound ecosystem conservation.

Acknowledgements

I thank John L. Gittleman, David W. Macdonald, and two anonymous referees for their critical comments on drafts of this paper. I am grateful for financial support from the Zoological Society of London enabling me to present this paper to the London symposium.

7

Hybridization and conservation of carnivores

ROBERT K. WAYNE AND DAVID M. BROWN

INTRODUCTION

Mating between individuals from different species (interspecific) or between individuals from differentiated populations of the same species (intraspecific) can be a source of evolutionary novelty and potential or an undesirable homogenizing influence on adaptively divergent populations. In plants, hybridization between individuals from distinct stocks or species has resulted in the creation of new reproductively isolated species that have unique adaptations (Arnold, 1997). A well documented phenomena in plants is transgressive segregation in which hybrids have trait values exceeding that of the parental forms (Rieseberg *et al.*, 1999). Such traits may better adapt individuals to new environments. Thus, hybridization can be a source of evolutionary innovation producing uniquely adapted forms. In animals, transgressive segregation also may occur (Rieseberg *et al.*, 1999) but here less substantive evidence exists for a significant role of hybridization. However, vertebrate hybrids may be better adapted to intermediate habitats or changing climate conditions (Grant & Grant, 1992; Quattro *et al.*, 1992; Dowling & Demarais, 1993), and, may present new favorable gene combinations (e.g. Hedges *et al.*, 1992; R. J. O'Neill *et al.*, 1998). Furthermore, hybridization may allow traits of higher adaptive values to move from one species to another (e.g. Parsons *et al.*, 1993). Consequently, hybridization, if occurring naturally, does not necessarily impact negatively on the persistence of species or populations. Hundreds of hybrid zones have been reported in vertebrates and most are small in width relative to the geographic range of hybridizing species (Hewitt, 1988). Therefore, the consequences often are minimal and new genetic combinations potentially can be sorted by natural selection to improve the level of adaptation in populations.

Hybridization is a concern if it is due to anthropogenic changes in habitat or species composition. This includes habitat destruction or succession, predator control or introduction of non-native forms. For example, the introduction of non-native species that hybridize with native relatives may result in their local genetic extinction (Simberloff, 1996), especially if the non-native forms are domesticated or otherwise adapted to the presence of humans. Examples include domestic cats, in areas where the European or African wild cat is present (Stuart & Stuart, 1991; Hubbard *et al.*, 1992; McOrist & Kitchener, 1994; Daniels *et al.*, 1998) or ferrets introduced to British hunting estates to control rabbits, which hybridize with native polecats (Davison *et al.*, 1999). Introductions may not be intentional, such as in France where American minks that have escaped from fur farms may hybridize with native European minks (Maran & Henttonen, 1995).

Equally insidious are examples in which habitat alterations or predator control efforts have increased rates of gene flow between otherwise reproductively distinct species. A well documented example is the extirpation of the gray wolf in the Great Lakes region of the United States and Canada which was followed by replacement with the smaller coyote (Moore & Parker, 1992). Conversion of forest to agriculture or urban environments also favored the coyote which then began to hybridize with dwindling populations of gray wolves (Lehman *et al.*, 1991) or an endemic North American wolf-sized canid (Wilson *et al.*, 2000). The asymmetry in population size may have led to hybridization of gray wolves and coyotes because when abundant, gray wolves will actively kill coyotes in their territories (Peterson, 1995; Phillips & Smith, 1997; Creel *et al.*, Chapter 3).

Hybridization, even when caused by humans, may not endanger the genetic integrity of native species. For example, hybrid offspring may not survive to reproduce or the geographical extent of hybridization may be limited. If hybrids are less fit, natural selection may help to preserve the genetic integrity of species or geographic factors may effectively limit the spread of hybrids. Therefore, a specific criteria for conservation is needed such that species or populations under threat can be identified. A taxon is endangered by hybridization if it is diagnosably distinct and is threatened with loss of unique characteristics through hybridization that is due to anthropogenic causes. The diagnosably distinct criteria refers to the possession by all members of a species or population of unique characteristics that indicate common ancestry or adaptive divergence (Cracraft, 1983; Avise & Ball, 1990; Crandall *et al.*, 2000). These characteristics should be heritable and preferably should be those that have a direct relationship to fitness and environmental differences between taxa (see Crandall *et al.*, 2000). Molecular characters commonly are used to distinguish populations (Avise, 1994), but these differences often refer to selectively neutral polymorphisms and are less consequential for conservation although they may indicate common ancestry and 'evolutionary potential' (Moritz, 1994, 1999).

The second part of the definition applies to the loss of unique characteristics. This is important because limited hybridization may not substantially affect the genetic characteristics of a widely distributed species. Moreover, hybrids often may be adaptively inferior (Hewitt, 1988). If hybrids are less fit then selection will constrain the width of hybrid zones and only features that improve fitness will be carried into the range of hybridizing species (Hewitt, 1988). Finally, to be of conservation concern, hybridization should owe its origin to anthropogenic causes, otherwise, it cannot be considered a departure from the natural process of evolution and is part of the historical legacy of a species (Dowling *et al.*, 1992; Dowling & Demarais, 1993). However, any human actions that alter the historic densities of species such that hybridization occurs would be regarded as anthropogenic in origin. These actions range from indirect effects on abundance caused by habitat loss and alteration to direct extirpation of carnivores.

An important category of hybridization that is not a conservation concern is hybridization that is a natural consequence of range expansion. For example, eastern Canadian black bears have increased their geographic range west with the expansion of forests after the last glaciation (Wooding & Ward, 1997). A hybrid zone has formed between genetically distinct eastern and western populations in the Western Rockies of Canada. Similarly, west and east European brown bear populations meet to form a hybrid swarm in Romania (Kohn *et al.*, 1995). Such hybrid zones likely are due to a natural cycle of forest expansion and contraction with glacial periods. Therefore, they may not merit conservation concern unless the process is substantially altered by habitat changes or other human activities. Another category of hybrid of less concern is that between populations that are negligibly differentiated and have few apparent diagnosable characteristics. For example, genetic characterization of the red wolf has not revealed any diagnosable characteristics (Roy *et al.*, 1996a). Similarly, although red wolves are morphologically intermediate between gray wolves and coyotes, no single feature uniquely characterizes them (Nowak, 1979). Their intermediate status may be a result of a hybrid origin, and as such the taxon does not merit concern for hybridization under the above definition (for an alternative view see Wilson *et al.*, 2000).

Occasionally, hybridization may be the only means to save taxa facing imminent extinction. For example, the dusky seaside sparrow, a unique dark form of the seaside sparrow found near Cape Canaveral (USA), was reduced to just five males before being removed from the wild for captive breeding (Avise & Nelson, 1989). Hybridization with another subspecies was the only means to save the taxon. Similarly, the Florida puma (*Puma concolor coryi*) is a morphologically unique population of about 50 individuals that was suffering the effects of inbreeding depression to such an extent that extinction seemed likely (Roelke *et al.*, 1993a). Consequently, a plan to introduce individuals from a genetically similar population in Texas was devised (Hedrick, 1995). In captivity as well, stocks often are so small that interbreeding with other populations is the only means to prevent severe fitness declines. Therefore, hybridization may be the lesser of two evils.

Finally, feral domestic species, even if they have had a long history of wild existence, such as the dingo, New Guinea singing dog or Native American 'Carolina dog' are of less conservation priority than their wild progenitors. Therefore, our criteria for concern focuses on species with a long natural legacy. Similarly, our criteria excludes hybridization between rare breeds or stocks as a conservation concern. Preservation of unique breeds and varieties created by artificial selection needs a special defense with regard to conservation of biodiversity because these organisms do not have a natural legacy independent of human manipulation.

In this chapter, we review studies of inter- and intraspecific hybridization between carnivores including wild and captive populations (Table 7.1). We begin with case studies of large terrestrial carnivores. The width of hybrid zones depends in part on dispersal distance (see Hewitt, 1988) and thus, in large carnivores, hybrid zones may be substantial relative to the geographic range of the hybridizing species. We follow with a discussion of case studies in hybridization of small carnivores and marine mammals. Finally, we review cases of substantial hybridization in captivity. The majority of cases of hybridization in the wild do not merit conservation concern because they are geographically limited or do not involve endangered species or distinct populations. Therefore, we conclude the review with discussion of specific endangered species threatened by hybridization.

HYBRIDIZATION IN THE WILD
Large carnivores
Felidae

Florida panther (*Panthera concolor coryi*)   The Florida panther is a highly endangered cat distributed in the Big Cypress Swamp and Everglades National Park in southern Florida (USA). This subspecies has been considered distinct because of diagnostic characteristics such as a kinked tail and cowlick (Wilkins *et al.*, 1997), and was placed on the endangered species list in 1967. Genetic analysis of multiple independent loci demonstrated convincingly that the population was greatly reduced in genetic variation relative to other panther populations (Roelke *et al.*, 1993a). Lower heterozygosity was associated with a suite of physiological problems; sperm viability was 18–38 times lower than other panther subspecies, 56% of males were cryptorchids, and cardiac defects and disease incidence were high (Roelke *et al.*, 1993a). Florida panthers also had a high incidence of stress lines in cortical bone indicating growth arrest (Duckler & Van Valkenburgh, 1998). The Florida panther is the clearest example yet of a natural carnivore population suffering from inbreeding depression.

Based primarily on mitochondrial DNA (mtDNA) evidence, the Everglades population is a genetic mix of two otherwise distinct subspecies (Roelke *et al.*, 1993a). The non-native subspecies is South or Central American in origin and derives from panthers introduced into the Everglades by the National Park Service between 1956 and 1966 (O'Brien *et al.*, 1990). The population in the Big Cypress Swamp likely represents a pure population of the authentic Florida subspecies and numbers fewer than 30 individuals (Hedrick, 1995). A captive breeding program probably would not be viable with such a limited and inbred founding stock. Moreover, as might be expected, inbreeding depression is less severe in the genetically mixed Everglades population, and consequently the influx of genes from outside Florida may have increased the probability of survival of the entire population. The US Fish and Wildlife Service recognized the importance of hybridization and has begun augmentation of the Florida population with individuals from Texas (Derr, 1999a). In pre-Columbian times, the Texas population likely was connected by gene flow with those throughout the Gulf and genetic tests have shown them to be closely related to Florida panthers (Roelke *et al.*, 1993a; Culver *et al.*, 2000). The introduction of panthers from Texas has already had measurable effects on viability of the Florida population (see Chapter 10). Therefore, genetic exchange is acceptable considering the substantial evidence for inbreeding depression, the genetically mixed status of the population and the availability of individuals from a closely related wild population (Hedrick, 1995). Such limited hybridization improves the likelihood of persistence and allows an opportunity for natural selection to retain unique adaptive features of the Florida population.

Indian tiger (*Panthera tigris*)   In 1978, in an attempt to increase the population of tigers in the Dhudhwa tiger preserve of Northern India, a Siberian–Indian tiger hybrid was introduced. Recently, individuals having the characteristic pelage pattern of a Siberian tiger (*P. altaica*) were reported in the preserve. Hairs from two individuals had diagnostic microsatellite and mtDNA polymorphisms (Shankaranayanan *et al.*, 1997;

Table 7.1. Location, cause and study citation of hybridizing carnivore species

Hybridizing species		Location	Cause	Reference
HYBRIDIZATION IN THE WILD				
Large Carnivores				
Felidae				
Florida panther (*Felis concolor coryi*)	X	S. or C. American panther (*F. concolor*)	Everglades National Park, USA	Captive panthers released into the Everglades	Roelke *et al.*, 1993
Indian tiger (*Panthera tigris tigris*)	X	Siberian tiger (*P. tigris*)	Uttar Pradesh, India	Siberian tiger markers found in two wild individuals	Shankaraarayahan *et al.*, 1997
Canidae				
Gray wolf (*Canis lupus*)	X	**Coyote (*Canis latrans*)**	**Great Lakes region of N. America**	**Low wolf density leads to mating with coyotes**	**Lehman *et al.*, 1991**
Red or Algonquin wolf (*Canis rufus* or *C. lycaon*)	X	**Coyote (*Canis latrans*)**	**Southeast United States & Great Lakes region**	Relatively low red/Algonquin wolf density leads to mating with coyotes. Naturally occurring? Diagnosable characters?	Wayne & Jenks, 1991, Wilson et al., 2000
Swift fox (*Vulpes velox*)	X	Kit fox (*Vulpes macrotis*)	Narrow hybrid zone in northwest Texas, USA	Naturally occurring at end of Rocky Mountains	Mercure *et al.*, 1993
East African wild dog (*Lycaon pictus*)		South African wild dog (*Lycaon pictus*)	Southeastern Tanzania to Botswana	Naturally occurring	Girman *et al.*, 1993; Girman & Wayne, 1997
Ursidae				
W. European brown bear (*Ursus arctos*)		E. European brown bear (*Ursus arctos*)	Eastern and Western European clades mixed in Romania	Naturally occurring due to expansion since last glaciation	Kohn *et al.*, 1995, Taberlet *et al.* 1995
W. North American black bear (*Ursus americanus*)		E. North American black bear (*Ursus americanus*)	Eastern and Western American clades mixed in Cascade Mountains	Naturally occurring due to expansion since last glaciation	Wooding & Ward, 1997

Small Carnivores

Mustelidae

Species		Species	Location	Notes	Reference
Sable (*Martes zibellina*)	X	Pine marten (*Martes martes*)	Pechero-Ilych National Park, Ural Mountains, Russia	Naturally occurring, not extensive	Yazan & Knorre, 1964
European mink (*Mustela lutreola*)	X	American mink (*Mustela vison*)	Europe	Speculative. *M. vison* males mate with *M. lutreola* females, embryos are reabsorbed	Maran & Henttonen, 1995
European mink (*Mustela lutreola*)	X	European polecat (*Mustela putorius*)	Europe	Speculative, range expansion of polecat during 1900s swamped mink populations	Maran & Henttonen, 1995

Marine mammals

Species		Species	Location	Notes	Reference
Antarctic fur seal (*Arctocephalus gazella*)	X	sub-Antarctic fur seal (*Arctocephalus tropicalus*)	Marion Island, Antarctica	Naturally occurring, not extensive	Hofmeyr *et al.*, 1997
New Zealand fur seal (*Arctocephalus forsteri*)	X	Australian fur seal (*A. pusillus doriferus*)	Sub-Antarctica	Naturally occurring, not extensive	Brunner, 1998
Harp seal (*Phoca groenlandica*)	X	Hooded seal (*Cystophora cristata*)	Gulf of St. Lawrence	Naturally occurring, not extensive	Kovacs *et al.* 1997

HYBRIDIZATION BETWEEN WILD AND DOMESTIC TAXA IN THE WILD

Felidae

Species		Species	Location	Notes	Reference
European wild cat (*Felis silvestris*)	X	Domestic cat (*Felis catus*)	Britain	Feral domestic cats mating with wild cats	Hubbard *et al.*, 1992
African wild cat (*Felis libyca*)	X	Domestic cat (*Felis catus*)	Southern Africa	Feral domestic cats mating with wild cats	Stuart & Stuart, 1991

Canidae

Species		Species	Location	Notes	Reference
Ethiopian wolf (*Canis simensis*)	X	Domestic dog (*Canis familiaris*)	Ethiopian highlands	Low wolf population and high dog density leads to extensive interspecific mating	Gottelli *et al.*, 1994
Gray wolf (*Canis lupus*)	X	Domestic dog (*Canis familiaris*)	Europe	Low wolf population and high dog density may lead to interspecific mating, not extensive	Vilà & Wayne, 1999

Table 7.1. (cont.)

Hybridizing species		Location	Cause	Reference
Small Carnivores				
European polecat (*Mustela putorius*)	X	Britain	Feral ferrets mating with wild polecats	Davison *et al.*, 1999
Domestic ferret (*M. furo*)				
HYBRIDIZATION IN CAPTIVITY				
Asiatic lion (***Panthera leo persica***)	X	Zoos	Founders of captive population bred with African subspecies	O'Brien *et al.*, 1987
African lion (***Panthera leo leo***)				
Indian tiger (*Panthera tigris tigris*)	X	Zoos	Interbreeding of subspecies in zoos	Shankaraarayahan *et al.*, 1997
Siberian tiger (*Panthera tigris altaica*)				
South African cheetah (*Acinonyx jubatus jubatus*)	X	Zoos	Interbreeding of subspecies in zoos	Marker-Kraus, 1997
East African cheetah (*Acinonyx jubatus raineyi*)				
Sand cat (*Felis margarita scheffeli*)	X	Zoos	Interbreeding of subspecies in zoos	Sausman, 1997
Sand cat (*Felis margarita harrisoni*)				
Mexican wolf (*Canis lupus baileyi*)	X	Zoos	Interbreeding of subspecies in zoos	Hedrick *et al.*, 1997
Gray wolf (*Canis lupus*)				

Taxa in bold are those threatened by hybridization as defined in this chapter.

Shankaranayanan & Singh, 1998) and had genetic markers that otherwise were present only in pure Siberian tigers. This suggests that progeny of the introduced Siberian–Indian tiger persist and may be contributing to the gene pool of the population. However, given the limited number of tigers in the reserve and tenuous status of the population, removing any suspected hybrids would threaten the persistence of the population. Moreover, the ecological role of the tiger is still being maintained by the hybrids and because of the extreme isolation of the population, successful migration to other areas is unlikely.

Canidae

Gray wolf (*Canis lupus*)   The gray wolf and coyote (*C. latrans*) are phenotypically, ecologically and genetically distinct species (Mech, 1970; Nowak, 1979; Lehman *et al.*, 1991; Roy *et al.*, 1994a, 1996a). However, mtDNA and nuclear microsatellite analysis of canids from eastern Canada and Minnesota (USA) found that some wolves have genetic markers otherwise unique to coyotes (Lehman & Wayne, 1991; Roy *et al.*, 1994a). The presence of coyote-like alleles in wolves likely reflects interspecific hybridization at a level of two to three hybridization events per generation (Roy *et al.*, 1994a). Hybridization between the two species has occurred because of the changes in the relative abundance of the two species. Wolf numbers decreased dramatically by 1900 due to predator control programs and habitat loss. With the disappearance of wolves, coyote numbers increased especially in disturbed areas (Moore & Parker, 1992). Prior to that time, wolves were more abundant and coyotes rare or absent, therefore wolves could readily find same-species mates (Lehman & Wayne, 1991). Consequently, the occurrence of hybridization likely is due to a human induced changes in the relative abundance of the two species.

The gray wolf in the Great Lakes (USA) region fulfills our criteria for a species of conservation concern due to hybridization. The gray wolf is diagnosably distinct and human-induced hybridization is affecting the genetic integrity of the species over a wide area. Increased protection might come in a very simple form; wolves when abundant kill coyotes within their home range as in Yellowstone National Park (Robbins, 1997). Thus, if gray wolves are numerically abundant, hybridization between the two species may not occur. In fact, genetic evidence suggests that hybridization may no longer be occurring in Minnesota where wolves now are abundant. However, hybridization may be ongoing in Ontario and Quebec where the species is less protected (Lehman *et al.*, 1991).

Red wolf (*Canis rufus*). An older hybrid zone between gray wolves, red wolves and coyotes may exist in the southcentral United States (Mech, 1970; Wayne & Jenks, 1991; Roy *et al.*, 1994a, 1996a). However, molecular genetic analysis suggests that the origin of the red wolf was due to repeated hybridization events between gray wolves and coyotes possibly beginning as long as several thousand years ago but increasing in historic times (Wayne & Jenks, 1991; Jenks & Wayne, 1992; Roy *et al.*, 1994a, 1996a; Reich *et al.*, 1999). The red wolf may be a hybrid form that is not diagnosable with molecular markers and morphologically cannot be distinguished from wolf–coyote hybrids (Wayne, 1992). Therefore, under the definition used here, the red wolf should not be regarded as threatened by hybridization (Jenks & Wayne, 1992; Wayne, 1992). However, new results suggest that the red wolf and Great Lakes wolf ('Algonquin wolf') share a common ancestry independent of the gray wolf (Wilson *et al.*, 2000). If the Algonquin wolf is diagnosably distinct and hybridization is not from natural causes, then it is threatened by hybridization.

African wild dog (*Lycaon pictus*) The African wild dog was once abundant throughout much of sub-Saharan Africa, but habitat disturbance and disease severely reduced populations to currently no more than several thousand individuals (Ginsberg & Macdonald, 1990; Woodroffe *et al.*, 1997). The most severe losses occurred in west and east African populations (Woodroffe *et al.*, 1997). East (Kenya and Tanzania) and southern (Zimbabwe, Botswana, Namibia and the Republic of South Africa) African populations have phylogenetically distinct mtDNA sequences and are significantly differentiated in microsatellite allele frequencies suggesting a long history of isolation (Girman *et al.*, 1993; Girman & Wayne, 1997). However, a large admixture zone exists from southeastern Tanzania through Botswana. Therefore, natural hybridization is occurring between east and southern African subspecies across a wide area. According to our criteria, the endangered east African subspecies cannot be considered threatened by hybridization. However, an east African captive breeding program should be developed so that their unique characteristics are represented in captivity (Girman *et al.*, 1993; Girman & Wayne, 1997).

Swift fox (*Vulpes velox*) and kit fox (*V. macrotis*) The kit and swift foxes are the smallest New World canids and frequent the arid lands of the American West (Mercure *et al.*, 1993). Swift foxes live on the eastern side of the Rocky Mountains and are morphologically and genetically distinct from kit foxes found on the western side (Dragoo *et al.*, 1990; Mercure *et al.*,

1993). Interbreeding between the two taxa occurs only in the south, in northwest Texas, where the mountains are lower (Packard & Bowers, 1970). Because the taxa are diagnosable, hybridization has the potential to cause the loss of distinct traits. However, the kit and swift foxes have much lower dispersal abilities than large canids, dispersing on average only 11 km (O'Farrell, 1987). Consequently, the hybrid zone is very narrow and may be stable overtime due to a balance between selection and dispersal (Hewitt, 1988). Because the hybrid zone is not geographically extensive, it appears not to be increasing and hybridization does not reflect anthropogenic habitat changes, we do not regard either taxon as threatened by hybridization.

Ursidae

Brown bear (*Ursus arctos*)   Control region sequences of European and North American brown bears show a striking pattern of phylogenetic subdivision (Taberlet & Bouvet, 1994; Waits *et al.*, 1998). Three sequence clades are found in Europe representing bears from: Scandinavia, France and Italy; from Greece and the Balkans; and from Russia and Romania. The existence of distinct western and eastern European clades probably is due to isolation and divergence in different Ice Age refugia. A hybrid zone, where both haplotypes from west and east European clades are found, exists in Romania (Kohn *et al.*, 1995). However, morphologic differentiation between the molecular groupings is not apparent. In North America, where four control regions sequence clades were discovered, nuclear microsatellite and ancient DNA analyses did not support the groupings based on mtDNA sequences (Paetkau *et al.*, 1998b; Leonard *et al.*, 2000). Thus, the mitochondrial clades of the Old and New World may not represent independent and distinct evolutionary units. Moreover, the development of a hybrid zone in Romania probably should be regarded a natural process of admixture as distinct populations expanded from Ice Age refugia. Therefore, although diagnosable, eastern and western populations will not lose traits consequential to fitness as a result of hybridization that in addition, should be considered a natural process. Under the definition used in this paper, they are not threatened by hybridization.

Black bear (*U. americanus*)   A similar situation exists with North American black bears where eastern and western mitochondrial control region clades are distinct but intergrade in the western cascades of Canada (Wooding & Ward, 1997). Here, as in European brown bears, black bear populations isolated during the Ice Ages may be merging naturally as forests expand across Canada. Morphologic differentiation is not apparent between eastern and western clades (Wooding & Ward, 1997). Therefore, as above, we do not regard either population as threatened by hybridization.

Small carnivores
Mustelidae

Reports on hybridization between small carnivores are poorly documented and often border on speculation. Very limited hybridization between sable (*Mustela zibellina*) and pine martin (*Martes martes*) may occur in the Ural Mountains (Yazan & Knorre, 1964). Potentially more significant is hybridization of the European mink (*Mustela lutreola*) with American mink (*M. vison*) or with the European polecat (*M. putorius*) in western Europe (Maran & Henttonen, 1995; Simberloff, 1996). The European mink is declining precipitously across many areas of Europe whereas the American mink, the descendents of minks imported for fur farms, have expanded their geographic range. A coincidence in the appearance of American minks and local declines of the European mink led to suggestions that hybridization may be the cause. However, hybrids have not been described. One theory is that the population decline may be mediated by infertile matings between larger American minks and female European minks which once mated by them, cannot successfully mate with conspecifics due to the presence of the copulatory plug (Maran & Henttonen, 1995). Hybrid embryos are thought to be reabsorbed. However, this scheme is hypothetical and competition between the two forms may explain the decline. Until further studies are done, the European mink, although in decline, cannot yet be considered endangered by hybridization.

Marine mammals

Reports of hybridization in natural populations of marine mammals are few and often anecdotal (Hofmeyr *et al.*, 1997; Brunner, 1998). The scarcity of reports might reflect the difficulty of comprehensively sampling areas where hybridization is occurring rather than indicating its overall rarity. Only three reports find hybridization supported by morphologic or molecular methods. On the sub-Antarctic Marion Island, hybrids between the Antarctic fur seal (*Arctocephalus gazella*) and sub-Antarctic fur seal (*A. tropicalis*) represent only about 0.02% of the population even though both species breed in the same areas (Hofmeyr *et al.*, 1997). Nine of 137 measured skulls of the New Zealand fur seal (*A. forsteri*) are morphologically divergent, leading the author to suggest that they are hybrids with the Australian fur seal (*A. pusillus*) (Brunner, 1998). Comprehensive surveys such as that done on Marion Island are needed to assess the significance of hybridization in the wild between *A. forster* and *A. pusillus*.

Harp seals (*Phoca groenlandica*) and hooded seals (*Cystophora cristata*) may hybridize in the Gulf of St. Lawrence (Kovacs *et al.*, 1997). A single pup had intermediate morphologic features and nuclear and mtDNA markers suggested that the pup was a hybrid of the two species. The pup was normally weaned by a hooded seal mother and appeared viable before it took to sea. Thus, this report suggests hybridization has the potential to genetically impact these species. In general, much more research needs to be done to assess the importance of hybridization in marine carnivores.

HYBRIDIZATION BETWEEN WILD AND DOMESTIC SPECIES
Felidae
European wild cat (*Felis silvestris*)   The most dramatic example of hybridization between wild and domestic felids involves the European wild cat. A morphologic and limited genetic study of wild cats in Scotland suggests that hybridization may be common and influencing the phenotypic characteristics of wild cats (French *et al.*, 1988; Hubbard *et al.*, 1992). However, hybrids are difficult to distinguish from pure wild cats leading to a concern that introgression has been overwhelming. Pure wild cats can be distinguished from domestic cats (French *et al.*, 1988; Daniels *et al.*, 1998), therefore as defined in this chapter, this species is clearly threatened by hybridization which has occurred over a wide area. The process will likely continue unless the number of domestic cats allowed to roam freely is controlled. Similarly, African wild cat/domestic cat hybrids have been documented in South Africa (Stuart & Stuart, 1991). Although not as extensive as hybridization in Scotland, hybridization has the potential to affect the genetic integrity of wild populations unless domestic cat numbers are limited and habitats suitable for the wild cat are preserved. In general, hybridization between wild and domestic species is not natural by definition and thus there is a strong mandate to take corrective action.

Canidae
Ethiopian wolf (*Canis simensis*)   The highly endangered Ethiopian wolf inhabits the Ethiopian highlands above 3000 m and numbers fewer than 500 individuals. Recent and historic declines in population size has led to loss of genetic variation; no mtDNA polymorphisms were found and heterozygosity in microsatellite loci was sharply reduced (Gottelli *et al.*, 1994). The domestic dog (*C. familiaris*) is closely related to the Ethiopian wolf and microsatellite and phenotypic analysis showed that in the Bale

Mountains, feral domestic dogs hybridize with them. Minimally, 15% of wolves are hybrids. In one case, a single litter had at least two fathers, one of which was a domestic dog (Gottelli *et al.*, 1994). In contrast to gray wolves, female Ethiopian wolves mate with males from outside their natal pack including stray domestic dogs. Thereafter, females return to their natal pack to potentially raise hybrid offspring. Consequently, the hybrid offspring can be socialized as wolves and may more successfully propagate in the wild (Vilà & Wayne, 1999). Although hybridization was observed in only one disturbed locality, given the limited area inhabited by Ethiopian wolves, continued habitat fragmentation and human population increase, the situation is dire and demands action (Wayne & Gottelli, 1997).

Gray wolf (*Canis lupus*) Since the origin of dogs from gray wolves, perhaps over 100 000 years ago, dogs may have interbred with wolves several times (Vilà *et al.*, 1997). Dog breeders and many North American cultures occasionally cross their dogs with wolves to improve vigor (Schwartz, 1997) and many thousand wolf–dog hybrids exist in the United States (García-Moreno *et al.*, 1996). In the wild, hybridization between gray wolves and dogs is likely to be most frequent near human settlements where the density of wolves is low and feral and domestic dogs are common (Boitani, 1982, 1983; Bibikov, 1988; Blanco *et al.*, 1992). Italian wolves could have been numerically augmented through hybridization with dogs following a severe bottleneck before 1980 (Boitani, 1983) and wolves on a high Arctic island show morphologic signs of interbreeding with Eskimo dogs (Clutton-Brock *et al.*, 1994). Recently, Butler (1994) suggests that European wolf populations are mainly hybrids between dogs and wolves. Although such inferences are based on anecdotal evidence, the genetic integrity of wild wolf populations is a real concern among conservationists (Boitani, 1983; Blanco *et al.*, 1992). However, a review of morphologic and genetic evidence suggests that hybridization had not materially affected the genetic composition of gray wolf populations (Vilà & Wayne, 1999). Offspring of male dogs and female wolves rarely survive because males dogs show limited parental care whereas the reverse cross might not often occur because sperm production and estrus cycle are not coincident in dogs and gray wolves (Vilà & Wayne, 1999). However, additional genetic analysis and observational studies of wolf–dog hybrids in the wild are needed to confirm these hypotheses.

Small carnivores

European polecat (*Mustela putorius*) The European polecat was largely eliminated from Britain by the end of World War I due to persecution by gamekeepers and survived only in a small corner of Wales (Davison *et al.*, 1999). Populations began to rebound there after the popularity of game reserves declined (Birks & Kitchener, 1999; Langley & Yalden, 1977). However, the domestic ferret (*Mustela furo*) which was introduced as late as the fourteenth century, achieved a wide distribution and survived throughout Britain (Thomson, 1951). Based on mtDNA analysis, pure ferrets likely persist only in Wales and elsewhere ferret–polecat hybrids or ferrets are common. Morphologic analysis suggests introgression is extensive. Because the European polecat is phenotypically distinct and has unique mitochondrial sequences it can be considered diagnosable. The species is clearly threatened by hybridization in Wales and can potentially receive protection under law in Britain (Davison *et al.*, 1999). However, the mtDNA study suggests that genetic difference among species in the polecat group (*M. putoris*, *M. furo*, *M. nigripes* and *M. eversmanii*) are small and could be considered a single holarctic polytypic species. Therefore, although significant, the magnitude of genetic and phenotypic divergence between the European polecat and the domestic form is small. With the exception of this qualification, the observed threat to the British European polecat is substantial and would support conservation action including translocation, removal of domestic ferrets, restoration of habitats for polecats and captive breeding.

HYBRIDIZATION IN CAPTIVITY

Hybridization between carnivore subspecies and between species, even from different genera, is reported in the zoo literature (e.g. Melish & Foster-Turley, 1996). The most significant reports from a conservation perspective are cited in Table 7.1. However, this is not an exhaustive account and hybridization only merits concern if it threatens the genetic integrity of diagnosably distinct endangered taxa (see Introduction).

Perhaps the most disturbing report of hybridization concerns Asiatic and African lions. The lion formerly had a geographic range including most of sub-Saharan Africa, parts of the Middle East and southwestern Asia (O'Brien *et al.*, 1987a). The lion now is found outside of Africa only in the Gir Forest of India, with a population of less than 250 individuals (O'Brien *et al.*, 1987a). No allozyme polymorphisms are evident in Gir Forest lions, contrasting sharply with levels of polymorphism of nearly 10% in most African lion populations. African lions contain three unique alleles not found in Asiatic lions and are thus diagnosably distinct. However, zoo lions identified as Asiatic are polymorphic for African lion alleles (O'Brien *et al.*, 1987a). Two of the five founders of the Asiatic lion breeding program likely are descendents of the African subspecies. Much of the breeding program may have been compromised by unintentional interbreeding between the two subspecies. This illustrates the importance of certifying the origin of individuals through genetic techniques if possible (e.g. O'Brien, 1994a,b; García-Moreno et al., 1996).

Of 79 sand cats (*Felis margarita*) in captivity in 1974, 21 are thought to be hybrids of the Arabian subspecies (*F. margarita harrisoni*) and the Pakistan subspecies (*F. margarita scheffeli*) (Sausman, 1997). However, the subspecies are connected through intermediate populations and the pelage and size characteristics show a clinal distribution (Hemmer et al., 1976). Thus, individuals from different subspecies may not always be diagnosably distinct especially if the populations are in close geographic proximity. Therefore, although interbreeding of cats from widely separated populations is not advised, some hybrids that derive from crosses between individuals from closely situated subspecies may be suitable for captive breeding. More attention should be given to geographic origin in future breeding plans.

South and east African cheetahs are genetically very similar although the east African population appears to have higher levels of genetic variation (O'Brien et al., 1987b). The two populations represent different subspecies (*Acinonyx jubatus jubatus* and *A. j. raineyi*) but in the absence of significant genetic divergence and definitive diagnosable characteristics, interbreeding between the two subspecies may not threaten the loss of consequential characteristics. In fact, the reproductive success of hybrids is significantly greater than matings within a subspecies (Marker-Kraus, 1997). Between 1990 and 1994, hybrids represented 13% of the breeding population but were responsible for 24% of births. Conceivably, given the low levels of genetic variation in each cheetah population (O'Brien et al., 1987b), hybrids may show hybrid vigor. Therefore, as in the puma example above, the benefits of crossing two populations, even those having some measure of distinction, may be outweighed by the increased fitness of hybrids.

The Mexican wolf is a diagnosably distinct endangered subspecies of gray wolf that was native to parts of Mexico and the southwestern U.S. (García-Moreno et al., 1996). Currently, only a few individuals, if any, still exist in the wild and thus planned reintroduction programs must use captive-raised wolves. However, only in one captive population, designated the certified lineage, are all the founders ($n = 4$) known to be derived from a wild population of Mexican wolves. Two captive populations were founded from individuals of uncertain ancestry and were not included in the Species Survival Plan breeding program of the American Association of Zoological Parks and Aquariums. To preserve genetic diversity and reduce inbreeding depression, these two captive populations should be included in the breeding program if it could be shown they are not hybrids of dogs, coyotes or northern gray wolves.

Analysis of hypervariable microsatellite loci found that uncertified Mexican wolves do not have genetic markers from dogs or coyotes. Moreover, the three captive populations are genetically more similar to each other than to any other populations of dog or wolf-like canid and shared two unique alleles. Therefore, the two uncertified lineages are not hybrid populations and should be included in the breeding program (García-Moreno *et al.*, 1996; Hedrick *et al.*, 1997).

CONCLUSIONS

Hybridization in the wild between endangered carnivore subspecies or between related carnivore species is limited. Subspecies commonly are allopatric so there is little opportunity for mating. However, related carnivore species often are sympatric and can potentially mate (Table 7.1; Van Valkenburgh, 1988). For example, three jackal species coexist in east Africa and do not hybridize although they may do so in captivity (Wayne *et al.*, 1989). Character divergence may be sufficient to maintain genetic isolation in the wild. In fact, hybridization is a threat to the genetic integrity of species generally where there are dramatic changes in habitats or relative abundance that reduce mating opportunities (e.g. gray wolves and Algonquin wolves in the Great Lakes region), or involve introductions (ferrets to Britain, South or Central American pumas to Florida, Siberian–Indian tiger hybrids to India). The high density of domestic species in many areas similarly has threatened the genetic integrity of wild species (e.g. Ethiopian wolves and domestic dogs, Scottish wild and domestic cats).

In several instances, there is need for action as the loss of unique characteristics is likely and the long-term persistence of distinct populations is threatened (Table 7.1). The most effective strategy for decreasing rates of hybridization in the wild involves habitat restoration that allows native carnivores to increase such that heterospecific matings are less likely. For example, protection of the gray wolf in Minnesota has allowed the population to increase from less than 50 individuals at the turn of the nineteenth century to over 2000 individuals at the end of the 1980s (Ginsberg & Macdonald, 1990). The genetic evidence suggests hybridization no longer occurs in Minnesota reflecting the increased abundance of wolves (Lehman *et al.*, 1991). In Yellowstone National Park introduced wolves actively exclude and kill coyotes (Phillips & Smith, 1997; Creel *et al.*, Chapter 3).

In a few carnivore species urgent action is needed. For example, native polecats in Wales are severely threatened by the rapid expansion of ferrets which thrive in agricultural settings. Here, removal of domestic ferrets from the wild in combination with habitat restoration might be necessary. Also, a captive breeding program to ensure a reservoir of genetically distinct individuals for reintroduction is needed. Similar strategies might be necessary for the Scottish wild cat, Ethiopian wolf and European mink.

Hybridization in captivity between endangered species or subspecies does not appear to be common. Often considerable expense and effort is paid toward protected breeding of individuals from seemingly distinct populations. However, species may be oversplit and subspecies based on limited surveys. In large carnivores, molecular genetic surveys rarely have supported extensive subspecies designation (e.g. Lehman & Wayne, 1991; Roy *et al.*, 1994a; Paetkau *et al.*, 1998b). For example, only four genetic units were supported in North American brown bears whereas seven subspecies had previously been designated (Waits *et al.*, 1998). Similarly the 24 subspecies of North American wolf probably reduce to only three or four genetic units (Roy *et al.*, 1994a). Consequently, plans for interbreeding subspecies in zoos need to weigh the costs of increasing inbreeding depression against the benefits of separate breeding. If the level of distinction between subspecies is modest, as in the east and southern African cheetah, interbreeding may be an acceptable course of action.

Acknowledgements

We would like to thank B. Van Valkenburgh and Klaus Koepfli for helpful comments.

8

Carnivore demography and the consequences of changes in prey availability

TODD K. FULLER AND PAUL R. SIEVERT

INTRODUCTION

The conservation of carnivores ultimately depends on accurately assessing their distribution and abundance, and then making informed management decisions. If a carnivore population seems to be stable and 'in balance' with its habitat, then monitoring the *status quo* may be all that is required. If a population co-occurring with humans is the cause of conflict, it may be 'controlled', possibly over a long period of time, such that conflicts are reduced to an acceptable level (Harris & Saunders, 1993). Finally, decisions often are made to increase a carnivore population in number and distribution because of its rarity (Reading & Clark, 1996). A necessary prerequisite for any of these actions is a sound understanding of the factors that affect carnivore density, as well as the means by which populations change. Only with this information can one responsibly manage change and thus fulfill conservation goals.

Carnivore densities apparently vary from one to over three orders of magnitude within species, and though inappropriate study design has been proposed to explain a portion of this reported variation, biological reasons for any such variation remain undecipherable in some cases (e.g., Smallwood & Schonewald, 1996, 1998). Clearly, some populations have been reduced to very low or even zero density through excessive natural (e.g., disease, starvation) or human-caused (e.g., bounties) mortality, and mortality of a wide variety is certainly an important component of density changes. However, the potential densities that populations reach are more generally understood to be a reflection of resource abundance, and in the case of carnivores, this usually means prey resources. For many carnivores, short- and long-term changes in prey abundance and availability, as well as geographic variation in food resources, are the major natural forces that influence population viability. For example, some populations of giant pandas (*Ailuropoda melanoleuca*) in China experienced a major decrease in density when their bamboo food resource all flowered and died off simultaneously (Schaller *et al.*, 1985). Similarly, it makes sense that brown bear (*Ursus arctos*) densities are highest in coastal Alaska where runs of salmon are seasonally superabundant, and lowest north of the Arctic Circle where the growing season is short and both plant and animal biomass are low (McLellan, 1994).

An understanding of the relationship between carnivore density and food resources is important because it affords us the opportunity to predict what a population size will be or could be. It may, in concert with other information, explain what is present versus what is possible. Although there are a number of carnivore species, especially those whose populations are cyclical, for which we have a good understanding of effects of food on demography, there are others of which we are surprisingly ignorant. For example, black bears (*Ursus americanus*) are one of the most studied carnivores in North America and their densities have been reported to vary 40-fold (summary in Garshelis, 1994). However, there has been no quantitative analysis undertaken that would give us a capability to predict potential black bear density based on food resources. Similarly, little quantitative synthesis of mountain lion (*Puma concolor*) demography has occurred since Seidensticker *et al.* (1973) proposed that the 'density of [the] breeding population, was set by a vegetation-topography/prey number-vulnerability complex'. These limits in our understanding of the demography of some species may be the result of logistical constraints or simply a missed opportunity.

Here we qualitatively review examples of the effects that changes in food abundance have on carnivore demography, including density, reproduction, survival, and behaviour. We consider food to be a density-dependent factor (*sensu* H. Smith, 1935, p. 889; Lack, 1954), and more specifically a limiting factor (*sensu* Krebs, 1985, p. 341) that can be investigated comparatively and experimentally. We also discuss a myriad of factors that confound our understanding of the relationships between food abundance and carnivore demography. By extolling the virtue of understanding of the factors that influence carnivore demography and applying them to a wider array of species, we outline ways of more effectively conserving carnivore populations.

EVIDENCE OF DEMOGRAPHIC RESPONSE

Carnivore food resources may vary in time (temporal variation) or space (geographic variation). The consequent effects on carnivore demography are best understood for cyclic species, particularly felids, canids, and mustelids and their prey in boreal forest ecosystems, as a result of using consistent methodologies in relatively simple ecosystems. In addition, synthetic analyses of carnivore demographic studies from the past 30 years are now being conducted and making their way into the literature. From these sources of information the consequences of changes in prey availability on carnivore demography can be more quantitatively evaluated.

Abundance and density

Within a given area, changes in relative abundance (usually numbers of individuals trapped each year) of cyclic species such as lynx (*Lynx lynx*; Elton & Nicholson, 1942), arctic foxes (*Alopex lagopus*; Hersteinsson et al., 1989), red foxes (*Vulpes vulpes*; Lindstrom, 1989), and stoats/short-tailed weasels (*Mustela erminea*; King, 1989) have long been known to vary with the relative abundance of their major prey. In fact, subsequent studies have demonstrated that densities of these and other carnivores, such as coyotes (*Canis latrans*), bobcats (*Lynx rufus*), and least weasels (*Mustela nivalis nivalis*), and pine marten (*Martes martes*) often are highly correlated with densities of prey from the same or previous years (e.g., Raymond & Bergeron, 1982; Erlinge, 1983; Knick, 1990; Korpimaki et al., 1991; Jedrzejewski et al., 1995; Kaikusalo & Angerbjorn, 1995; O'Donoghue et al., 1997; Jedrzejewski & Jedrzejewski, 1998). Within the Serengeti area of East Africa, spotted hyenas (*Crocuta crocuta*) more than doubled their population during a 20-year period when migratory wildebeest (*Connochaetes taurinus*) numbers increased more than 2-fold (Hofer & East, 1995). In Sweden, numbers of weasels (*Mustela nivalis*) apparently dropped by half when the vole population crashed (Erlinge, 1974).

Across a wide range of species, data indicate that carnivore density are positively correlated with prey density. Snow leopard (*Panthera uncia*) densities may be highest where blue sheep (*Pseudois nayaur*) densities are highest (Oli, 1994). A more quantitative comparison of lion (*Panthera leo*) densities in two areas indicated higher numbers where year-round prey were more numerous (Hanby et al., 1995). Similarly, coyote densities were 'many times' denser in a coastal area with abundant food than in adjacent areas without similar resources (Rose & Polis, 1998). Comprehensive analyses across geographic ranges indicate that wolf (*Canis lupus*; Keith, 1983; Fuller, 1989; Messier, 1995), lion (Van Orsdol et al., 1985), and European badger (*Meles meles*; Kruuk & Parish, 1982) density, and cheetah (*Acinonyx jubatus*; Laurenson, 1995; Gros et al., 1996) and leopard (*Panthera pardus*; Stander et al., 1997) biomass are strongly correlated ($r^2 \geq 0.72$, 0.76, 0.82, 0.61, 0.72 respectively) with prey biomass or lean season biomass (e.g.,

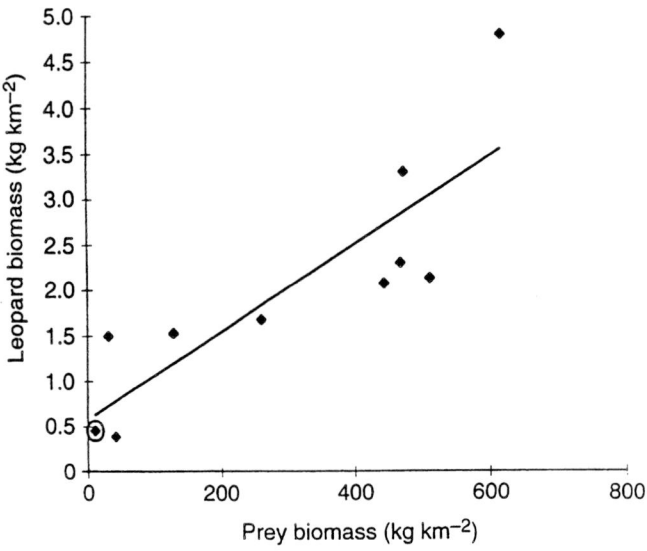

Figure 8.1. Leopard biomass correlated with medium-sized prey (15–60 kg) biomass in 11 arid/eutrophic savanna conservation areas in Africa (from Stander *et al.*, 1997).

Figure 8.1). Even polecat (*Mustela putorius*) density was strongly correlated ($r^2 = 0.91$) with numbers of common frogs (*Rana temporaria*), the major prey of polecats in Poland (Jedrzejewski & Jedrzejewski, 1998). These results strongly support the idea that food resources determine a large component of carnivore density.

Reproduction and neonate survival

There are ways populations may adjust to prevailing food supplies, and thus be observed to have densities directly related to food abundance. Changes in reproductive output are certainly a major potential response (Figure 8.2), and the relationship of nutrition and both reproduction and recruitment is well-documented in mammals (Sadleir, 1969). Since recruitment to a population is a function of the proportion of productive females, litter size, and offspring survival, these demographic parameters should be higher when prey is relatively abundant. Indeed, as snowshoe hare (*Lepus americanus*) numbers decline in Alberta (Canada), so did condition of lynx and coyotes, as indexed by fat reserves (Brand & Keith, 1979; Todd & Keith, 1983). As a consequence, for the lynx the average first age of reproduction has been shown to increase as food supplies decrease (Saunders, 1961).

Figure 8.2. Pregnant African wild dog feeding on a male Thomson's gazelle (*Gazella thomsoni*) (photo by T. K. Fuller). Changes in prey availability usually result in changes in carnivore reproduction (e.g., fecundity, litter size, pup survival), and thus potential for populations to increase.

Similarly, lynx and coyote pregnancy rates in Alberta were lower when hares were less abundant (Brand & Keith, 1979; Todd & Keith, 1983). In Sweden, red fox pregnancy rates declined with the abundance of voles (Englund, 1970; Lindström, 1989), in England female common weasels (*Mustela nivalis*) failed to breed during the year of lowest vole numbers (Tapper, 1979), and, in California, kit fox pregnancy rates were decreased when prey was scarcest (White & Ralls, 1993).

Also, lynx and coyote litter sizes in Alberta were smaller when hares were less abundant (Brand & Keith, 1979; Todd & Keith, 1983). In Sweden, red fox litter sizes also declined with the abundance of voles (Englund, 1970; Lindstrom, 1989). Litter sizes of wolves in Alaska were significantly smaller in areas with low prey densities (Boertje & Stephenson, 1992), as well, and analysis of a number of studies indicates that significantly fewer wolf pups occur in litters where ungulate biomass is lower (Fuller, 1997).

In addition to affecting the per capita reproductive output of female carnivores, changes in food abundance can also affect the survival of newborns. For example, post-partum mortality of lynx kittens during the population decline in hares was mainly responsible for a lack of recruitment to the lynx population in Alberta (Brand & Keith, 1979). For wolves, determination of pup survival during the first 5 months of life is difficult, but for 3

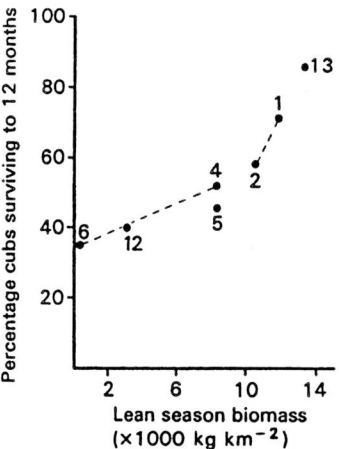

Figure 8.3. Percentage of African lion cubs surviving to 12 months and lean season prey biomass recorded in seven lion studies (from Van Orsdol *et al.*, 1985). Dotted lines connect points from different studies carried in the same study site.

studies survival was higher where biomass was higher (Fuller, 1989). Also, the number of pups per pack in fall for a number of studies was positively correlated with ungulate biomass (Fuller, 1989, 1997). The percentage of lion cubs surviving to 12 months was positively correlated with lean season biomass of prey (Figure 8.3). Finally, the number of surviving wolverine (*Gulo gulo*) cubs in Norway was positively related to small rodent abundance (Landa *et al.*, 1997). As a reflection of poorer nutrition and perhaps higher susceptibility to death of juveniles during years of food scarcity, Lindström (1983) found that growth of red fox juveniles was slower when voles were scarce. A similar phenomenon was observed by Messier (1987) in wolves.

Some changes in the overall number of offspring recruited result from changes in all reproductive parameters. For lions in east Africa, there was no difference in litter first age of reproduction, or interbirth interval, but survival of cubs during the first year was more than twice as high where prey was higher (Hanby *et al.*, 1995); overall, lion reproductive rates are consistently higher where food was higher (Van Orsdol *et al.*, 1985, Packer *et al.*, 1988). Also, decreased stoat populations during years of prey scarcity in New Zealand are for the most part due to decreased survival of neonates because nearly all females are pregnant regardless of prey densities (King, 1983b; 1989: p. 159).

There also have been several experimental attempts to document the change in neonate recruitment with changes in food availability. In a supplemental feeding experiment of arctic foxes in Swedish Lapland, Angerbjorn et al. (1991) found that there was a larger number of cubs at dens in summer when those dens were provisioned with extra food prior to parturition. Lindström (1989) also found an increased number of active red fox dens when foxes were provisioned.

Adult mortality

Another consequence of relative food shortage is increased mortality of adult carnivores. In the Yukon and Northwest Territories of Canada, the highest mortality of lynx, including starvation, occurred during lowest hare numbers (Poole, 1994; O'Donoghue et al., 1997). In addition, starving lynx are possibly more susceptible to human-related mortality (e.g., trapping) during hare scarcity (Brand & Keith, 1979). For territorial wolves, food shortages often result in increased trespassing into adjacent pack territories, the consequence of which is increasing mortality due to intra-specific strife (Mech, 1977b). Famine also has been suggested as the cause of a decline in coati (*Nasua narica*) numbers in Panama (Gompper et al., 1997).

Behaviour

In addition to direct demographic changes in carnivore populations when prey availability changes, indirect changes in behaviour also occur that contribute to observed changes in carnivore densities. One of the most common behavioural changes is in the size of carnivore home ranges or territories. For example, individual pack territory size was negatively correlated with white-tailed deer (*Odocoileus virginianus*) density in Wisconsin (Wydeven et al., 1995). A comparison of several studies (Fuller, 1989) also indicated that mean wolf territory size is negatively correlated with density of deer. In Europe, smallest wolf ranges were reported from regions where red deer (*Cervus elaphus*) were common (Okarma et al., 1998). Other species exhibit this same trend. As lagomorph populations declined in Idaho (USA), the average size of bobcat home ranges increased fivefold (Figure 8.4; Knick, 1990), and lynx home ranges increased threefold with a decline of snowshoe hares in the Yukon (Ward & Krebs, 1985). In east Africa, lions in areas of lower prey availability had larger home ranges (Hanby et al., 1995), and overall lion range sizes are inversely correlated with lean season biomass (Van Orsdol et al., 1985). Weasels in Poland had larger home ranges during the crash year of rodents (Jedrzejewski et al., 1995), and feral ferrets (*Mustela furo*) in New Zealand increased home ranges sizes nearly threefold when rabbit densities were reduced 99% (Norbury et al., 1998).

However, home ranges did not decrease in size with decreases in prey

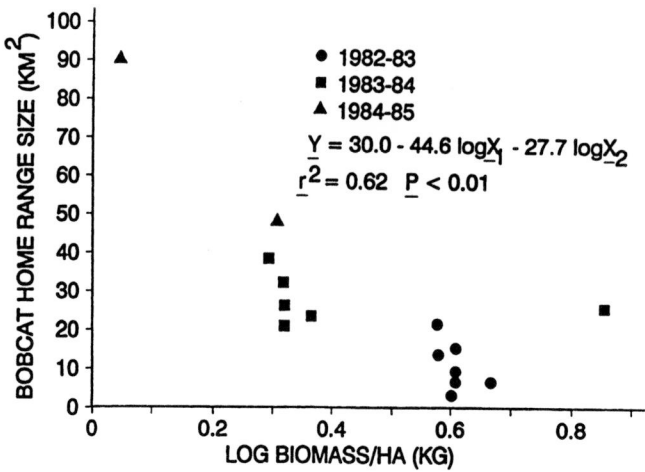

Figure 8.4. Relationship between the annual range sizes (Y) of female bobcats in Idaho, USA, and jackrabbit (X_1) and cottontail (*Sylvilagus nuttallii*; X_2) densities within their range (from Knick, 1990). Female range sizes were 95% minimum convex polygons. Lagomorph densities were determined from fecal pellet plots. (Reprinted by permission of The Wildlife Society.)

abundance for kit foxes (*Vulpes macrotis*) in California (White & Ralls, 1993). The maintenance of large and relatively non-overlapping home ranges in kit foxes may be an adaptation to drought-induced periods of prey scarcity that are episodic and temporary in that region of the country. Similarly, territory size did not decrease with lower earthworm (*Lumbricus* spp.) biomass, though the numbers inhabiting a territory were lower (Kruuk & Parish, 1982) and their spatial organization possibly reverted from strictly territorial clams to the more general mustelid pattern of solitary, single-sex ranges (Kruuk & Parish, 1987).

Another behavioural consequence of changes in prey availability is an increase in the number of transient and dispersing individuals in a population. Although coyotes in one study area in the northern Great Basin area of the western United States did have larger home ranges and territory sizes when abundance of cyclic black-tailed jackrabbits (*Lepus californicus*) was low, this change did not occur in a second study area – in that area, the proportion of transient coyotes increased when prey declined and the reason for this difference was not clear (L. Mills & Knowlton, 1991). In North America, the long-range dispersal of lynx during cyclic lows in snowshoe hare numbers is best known from documented 'invasions' of lynx into areas bordering the southernmost parts of their geographic range (Mech, 1973b). Relatively more lynx become nomadic, daily travel rates are higher, and more lynx disperse long distances when hare numbers are at their nadir (Ward & Krebs, 1985). In addition, during periods of prey scarcity, increased numbers of transient lions (Packer, 1986), wolves (Messier, 1985), and feral ferrets and cats (*Felis catus*; Norbury *et al.*, 1998) have been observed. These transients and dispersers have lower reproductive output and often also have higher mortality rates.

Finally, it has been proposed that at relatively high prey density, social regulation acts to regulate carnivore numbers (i.e., a Type II numerical response; e.g., Currier, 1983; Pech *et al.*, 1992; Messier, 1994, 1995). However, demographic evidence for this hypothesis is weak. First, for some species there have not been any quantitative assessments of carnivore densities in relation to prey density (e.g., mountain lions; Currier, 1983). Secondly, plots of prey density versus log-transformed predator density (e.g., moose *Alces alces* and wolves; Messier 1994, 1995) certainly show an asymptote (considered to reflect a type II response); however, a plot without such a transformation shows no evidence of predator densities levelling off at high prey densities (even though such a plot has a poorer fit of residual values; F. Messier, pers. comm.). Finally, single data points at high prey densities could indicate a levelling off caused by social factors (e.g., red foxes and feral cats versus European rabbits *Oryctolagus cuniculus*; Pech *et al.*, 1992), but just as easily it could reflect that, biologically, the numerical response of the predators cannot be fulfilled in six months, or before cyclic prey populations crash (Angerbjörn *et al.*, 1999). Clearly more research on carnivore populations at very high prey densities needs to be done to assess the veracity of the social limitation hypothesis.

CONFOUNDING FACTORS

Although variation in prey biomass explains a substantial amount of the variation in carnivore density, and the underlying factors affecting such a relationship seem sound, none of the correlations are so good that one's confidence in predicting carnivore density should be unrestrained. The statistical confidence of many food and density-related demographic parameters often is wide, and additional important factors can influence carnivore density on a site-specific basis, or even at a species level. Thus, it is also important to understand the factors that confound the relationship between prey biomass and carnivore numbers.

Methodology

Carnivores are notoriously difficult to census with great accuracy and precision. They are logistically difficult to count because of their low density (and thus large home range size), sometimes complicated social organization, and general elusiveness. Also, not all studies estimate a species' density in identical manners. Finally, a stated concern has been that biologists have inappropriately extrapolated carnivore density estimates from small, high density areas to larger study areas without considering potential density variation (Smallwood & Schonewald, 1996, 1998). This might be especially true of species for which little data has been gathered, but for well-studied wolves, study area size is a consequence of density, not the other way around (Fuller & Murray, 1998).

In addition, estimates of food availability might be poor. Although ungulate prey are more easy to survey than are carnivores, it is expensive to get precise estimates, and methodologies can vary greatly. Also, inappropriate extrapolations might be more common than we think. If more than one prey or food type is important for a carnivore species, then estimates of total food may be even more suspect when comparing across studies.

Biology

Even the definition of food density can be difficult to formulate. In ungulate-rich environments, calculating the numbers of appropriate-sized prey, given the carnivore assemblage present, might be a factor critical to accurately predicting potential carnivore density. For African lions, 'lean season' prey biomass, not maximum annual biomass, correlates most closely with lion density across habitats (Van Orsdol et al., 1985). Also, abundance is not always the same as availability. For example, in Ontario (Canada), wolves can prey on white-tailed deer or moose, but deer are preferred (Forbes & Theberge, 1996). Where moose are more common, but total prey biomass is equal to areas where deer are most common, wolves are generally less abundant; thus, wolf densities correlate with deer biomass but not moose biomass. On Isle Royale, wolf numbers correlate with numbers of old ($\geq$ 10 years) moose (Figure 8.5), and not with total numbers of moose (Peterson et al., 1998). Even though deer (*Odocoileus* spp.) and moose may occur throughout the distribution of mountain lions in the American West, only those adjacent to adequate stalking habitat (Seidensticker et al., 1973) may be available to them.

The use of alternative foods also can make understanding predator relationships to food difficult. Differences in the relative importance of a particular prey species in the diet may result in different demographic

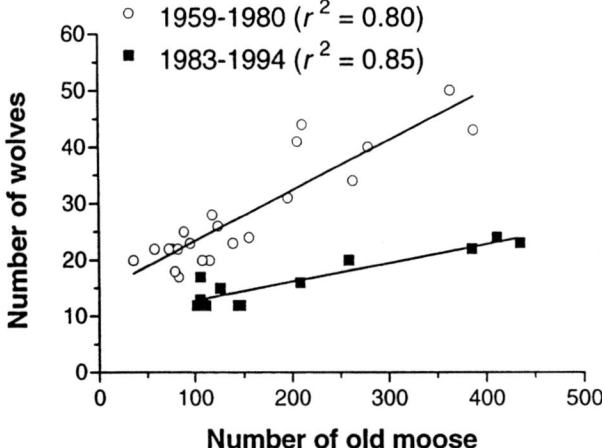

Figure 8.5. Relationship between abundance of moose $\geq$ 10 years old and size of the wolf population on Isle Royale, Michigan, USA from 1950 to 1980 and 1983 to 1994 (from Peterson et al., 1998).

responses to changes in prey abundance. For example, red foxes in Spain had smaller litter sizes after a decline in European rabbit numbers due to haemorrhagic disease only where rabbit consumption was relatively high (Villafuerte et al., 1996). Also, supplemental farm carrion as an alternate food source resulted in coyotes in farmland Alberta to maintain relatively high, and less variable, reproductive rates while in adjacent forested areas where hares declined, so did coyote reproductive rates (Todd, 1985). In general, responses to changes in primary prey seem moderated where alternate prey is more abundant (Angelstam et al., 1984). In Massachusetts, there were no identifiable changes in black bear cub production or survival in years when nut crops failed because nearby corn fields and other natural foods were sufficient alternatives (McDonald, 1998). Although several American marten (*Martes americana*) populations have declined following declines in prey abundance (Thompson & Colgan, 1987), in coastal Alaska, martens switch from rodents to alternative food sources in years when rodent numbers are low to maintain their body weight (Ben-David et al., 1997) and, presumably, their density. These and related demographic results also indicate that carnivore species with a relatively narrow diet breadth are affected more profoundly by changes in food resource availability.

Time lags in carnivore response to changes in prey often need to be taken into account for predictive purposes. Lynx and coyote numbers correlate most closely with numbers of hares the previous winter (O'Donoghue *et al.*, 1997), and on Isle Royale, Peterson & Page (1988) suggested that wolf population changes may lag three to five years behind changes in moose density. Thus, longer term views of predator and food resources often are needed. Relatively more ungulates (low wolf: ungulate ratio) occur than might be expected where wolf populations are newly established or newly protected (Fritts & Mech, 1981; Wydeven *et al.*, 1995) Conversely, relatively fewer ungulates (high wolf: prey ratio) occur where prey is declining and wolves are protected (Fuller, 1989, 1990).

Competition and predation are inherent complicating factors in multiple-predator systems. For example, Karanth & Sunquist (1995) hypothesized that occurrence and relative densities of large carnivore species in tropical forests may depend largely on relative densities of the different size classes of potential prey. In areas of differing prey species composition or densities, there should be different densities of predators. Although there may be spatial exclusion of leopards by tigers (*Panthera tigris*) in some areas (Seidensticker *et al.*, 1990), in other areas where large prey are available, leopards may be higher relative to tigers because competition is lower. Where both large and medium-sized prey are absent, leopards may be favoured over tigers (Rabinowitz, 1989). In the community of ungulates and carnivores in some portions of boreal North America, carnivore numbers are limited by ungulate abundance which, in turn, is limited not by low quality vegetation, but by the predators themselves. Unless brown bear and wolf populations decline substantially simultaneously, ungulate populations are viewed as residing in a low equilibrium (sometimes referred to as a predator 'pit') that they cannot escape (Messier, 1994; National Research Council, 1997).

Regarding interspecific predation, lions and spotted hyenas are a major cause of death for African wild dogs (*Lycaon pictus*; Creel & Creel, 1996). In Kruger National Park, it is at first counterintuitive that wild dogs live at their lowest density where the density of impala (*Aepyceros melampus*) and kudu (*Tragelaphus strepsiceros*), their major prey, are highest (Figure 8.6). This, however, is because lions heavily use such areas (Mills & Gorman, 1997), and wild dogs avoid them. Similarly, cheetah biomass, adjusted for prey biomass, is inversely related to lion and spotted hyena density, likely due to increase mortality of cheetah cubs due to predation (Laurenson, 1995a). Cheetahs exhibit local avoidance behaviour in both space and time with respect to lions and hyenas in the Serengeti (Durant, 1998).

Parallel circumstances occur in communities of smaller carnivores. Population trends of kit foxes are strongly influenced by food availability,

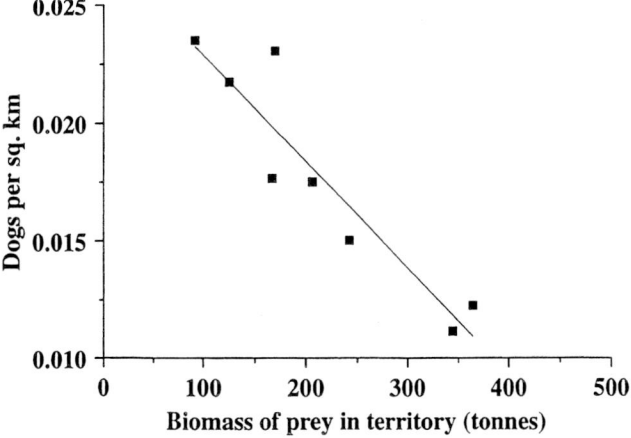

Figure 8.6. The relationship between African wild dog densities in different territories in Kruger National Park, South Africa, and the biomass of prey (Impala and kudu) in those territories (from Mills & Gorman, 1997, reprinted by permission of Blackwell Science, Inc.).

but coyote-related mortality on foxes act in concert with prey abundance to keep foxes at lower densities than they might otherwise attain (White & Garrott, 1997; Cypher & Spencer, 1998). Coyote predation is the major mortality factor for swift foxes (*Vulpes velox*), but expansion of competing red foxes into areas occupied by swift foxes may also limit swift fox populations (Sovada et al., 1998), much like as has been proposed for kit foxes (Ralls & White, 1995). For mongoose species in the Serengeti (i.e., *Helogale parvula*, *Mungo mungo*, and *Herpestes sanguineus*), annual variation in population density for a single population apparently was not driven by food, but rather predation (Waser et al., 1995). Conversely, smaller carnivores often undergo population increases subsequent to the removal of large, dominant carnivores (but see Wright et al., 1994); this phenomenon has been termed 'mesopredator release' (Soulé et al., 1988).

Finally, a number of other non-food-related factors can play significant roles in affecting carnivore densities. For example, jackrabbit density is a partial determinant of coyote density in the American West, via litter size and percentage of females breeding, but human-caused mortality confounds the relationship (F. Clark, 1972). The same is true where wolves are heavily exploited (Peterson et al., 1984; Ballard et al., 1987; Gasaway et al., 1992, p. 39). Genetic concerns, especially for small populations, also may be important. Inbreeding of the small wolf population on Isle Royale (Peterson et al., 1998) has contributed to keeping numbers low, even in the face of relatively abundant moose. In addition, diseases have played an important role in the demography of some carnivore populations, regardless of density of prey (Peterson *et al.*, 1998; Murray *et al.*, 1999).

RESEARCH AND CONSERVATION

The comparative approach to understanding carnivore ecology is not new (Gittleman & Harvey, 1982) and continues to make progress (Gompper & Gittleman, 1991). From a conservation perspective, intra- and interspecific comparisons are an efficient way to understand more about the population phenomena of species whose very existence may depend on quick action. The question arises, then, as to what more information is needed by carnivore conservationists with respect to the relationship between carnivore density and food supplies.

For some species (e.g. wolves, African lions, lynx) enough consistencies across studies and years make the general relationship between density and food quite reliable for conservation use. This is not to say that substantially more research is not needed. Additional investigations into the mechanisms of change that can actually be manipulated in a wildlife management sense seem very appropriate for these species. Certainly, human-caused mortality, and poaching or illegal killing in particular, is one of these research areas. Similarly, prey enhancement plans or even competitor reduction plans may be important for endangered populations. We may even have overlooked important aspects of carnivore adaptation to changes in prey availability that should be investigated. For example, lynx behavioural adaptations to prey population cycles make it essential to understand which animals have been or are being monitored. Breitensmoser *et al.* (1993) hypothesized that for cyclic lynx in the Yukon, a core number of females maintains stable territories through the hare cycle that they share with female offspring. Lynx increase their reproductive output during the increase phase, and maintain higher hunting success by hunting in family groups when hares are declining and low (O'Donoghue *et al.*, 1998). Monitoring lynx whose behaviour is different from these animals might result in a different view of how lynx respond to changes in prey abundance.

For species whose variation in density is not well explained, primary population density data need to be collated, evaluated and, where sufficient, synthesized. We suspect that such efforts will be difficult because of the lack of a large number of good, comparable density estimates, and simultaneous and adequate estimates of the availability of appropriate prey. Still, the effort needs to be made because the benefits are numerous.

Although the use of carnivore demographic data for conservation planning needs to be carried out carefully, such data can allow for assessment of population status and potential. Given food availability estimates, extant populations can be judged as to their nearness to the hypothetical ratio of predators to prey, or conceivably the biological carrying capacity, that the data suggest. For those populations that seemingly are reduced from their potential population levels, the ecological factor relegating them to that state can be identified and moderated if need be. For those populations 'exceeding' the potential, one can either anticipate a subsequent decline in carnivore numbers, or propose to enhance food resources such that a higher density of carnivores could be supported in the long run. Such approaches are especially important when restoring or anticipating restoration of populations to their former range (Mladenoff & Sickley, 1998), or when planning to conserve isolated populations over the long-term (Litvaitis et al., 1996). In relation to minimally viable populations, one can one can use demography/food information to assess the potential of an area to support a specified number of carnivores.

From a modeler's perspective, actually explaining 60–80% of the variation in real data sets is an excellent achievement. The simple relationship between carnivore density and prey biomass has in itself provided this degree of explanation for a number of species, and thus further elucidation of such relationships for other species seems a most fruitful area of investigation for carnivore conservation.

SUMMARY

For many carnivores, variation and long-term change in prey abundance and availability are major natural forces that shape the potential for a population to increase, but the understanding of the nature of this relationship varies by species. Relevant demographic data for carnivores are available for some species in areas where food resources vary in time (i.e., cyclic species; temporal variation) or space (geographic variation). Here we qualitatively review examples for numerous species for which changes in food abundance have been related to their demography, including density, reproduction, survival, and behaviour. We also discuss several factors that confound the understanding of these relationships.

For at least six carnivore species, densities are positively correlated with prey densities or biomass. This is because relatively higher food resources result in carnivores in better physical condition, and thus affect carnivore reproduction by lowering the average age of first reproduction, increasing litter size, and lowering mortality rates of juveniles. In combination, these changes increase net reproduction and thus carnivore density. Relatively poor food resources often increase mortality of adults due to starvation or intra-specific strife, cause carnivores to use larger home ranges, and result in increased rates of movements, including dispersal and transient behaviour; these result in lower densities. Poor density estimates of carnivores or their food supply, time lags in demographic responses, competition with other carnivores, and mortality due to disease and humans can confound this relationship. Because prediction is important, assessment of carnivore food resources and their demographic effects clearly are important because they allow one to more confidently predict the potential of an area to support a carnivore population. These are essential components of carnivore conservation planning.

Acknowledgements

We are grateful to the conference organizers for their support and enthusiasm, and to F. Messier and M. Gompper for helpful comments on the manuscript.

9

Human–carnivore interactions: adopting proactive strategies for complex problems

MARC BEKOFF

> Those who complain of the 'inconsistencies' of animal lovers understand
> neither the complexity of attitudes nor how rapidly they have developed.
>
> *(Mighetto, 1991: p. 121)*

HUMAN–CARNIVORE INTERACTIONS

There is no doubt that humans influence the behavior of carnivores and many other animals.[1] For carnivores, even within protected areas, conflicts with humans are usually the most important cause of mortality in adults (Woodroofe & Ginsberg, 1997b, 1998). The problems we face in the area of human–animal interactions raise numerous difficult and troubling questions (Bekoff, 1998a; Mack, 1999). For example, do animals have rights and if so, what responsibilities does this entail? How *should* humans treat other animals? What *ought* we do? Can we do whatever we please to other animals? Should we interfere in animals' lives when we have spoiled their habitats or when they are sick, provide food when there is not enough food to go around, or translocate them? Should our interests trump theirs? Should we be concerned with individuals, populations, species, or ecosystems? Should we let animals be and not intentionally interfere in their lives? Negotiating the path between what we do and what we ought to do can be a difficult journey with many twists and turns, and there are many junctions where it would be a good idea to pause for deep reflection.

The ethics of science and society's involvement with carnivores are important, complex, and ultimately subjective (for balanced discussion of all sides of these issues see Bekoff, 1998a; see also Aitken, 1997; Dawkins, 1998; Webb, 1999). Emotions and passions run high. The editors of this volume recognized this and wanted a contribution that would highlight just how complex and multidimensional these issues are. However, they faced the dilemma that, because judgments in this sphere are so highly personal, whoever they selected would likely show a bias for his or her view (limited space did not allow for more than one paper in this area). My intentions here are to draw attention to some of the issues that are central to carnivore conservation, and inevitably, my view is more visible than others. However, this does *not* mean that other views are less important nor that any of the editors of this volume share my opinions. What is important is that we all agree that ethics is an essential element in any discussion of conservation biology, as it is in any other sphere of science.

Animals depend on our goodwill and mercy. Each person chooses to be intrusive, abusive, or compassionate and each is responsible for her or his choices. Science, including conservation biology, is not value-free (many chapters in Clark *et al.*, 1995b and Caro, 1998; see also Dietz & Stern, 1998; Estes, 1998; Seidensticker *et al.*, 1999). Indeed, dealing with personal sentiments and emotional conflicts makes questions about what we ought to do extremely difficult (Bekoff, 1995; see Estes', 1998 quotation below). Complicating the situation is the fact that values and sentiments change with time and are sensitive to demographic, political, and social-economic variation, and personal whims.

THE REAL WORLD

My perspective on human–animal relations warrants against claiming that any species of animals is special, especially without informed argument on a case-by-case basis (Bekoff, 1998b). All individuals count, no animals are resources or property with whom we can do what we please, their lives matter very much, and they should be firmly entrenched in our moral community. Nonetheless, there are numerous and incredibly difficult practical issues at hand in the real world (e.g. Clark *et al.*, 1995b; Noss *et al.*, 1997; Estes, 1998; Minteer, 1998; for discussion of carnivores see Reading & Clark, 1996; Wikramanayake *et al.*, 1998; Berger, 1999; Terborgh *et al.*, 1999) and species-egalitarianism is problematic when we try to make decisions about how to interact with other species (French, 1995). For example, there is no doubt that carnivores, as predators, play a vital role in the maintenance of the biodiversity, stability, and integrity of various communities (Berger, 1999; Crooks & Soulé, 1999; Terborgh *et al.*, 1999). As such, it has been argued that we have special responsibilities to retain or restore top predators so that biodiversity is not jeopardized (Terborgh *et al.*, 1999). By focusing on carnivores we are taking into consideration numerous other species. However, because different species may play varying roles in maintaining the integrity and biodiversity of different ecosystems, developing a consistent ethic is difficult when adjudicating conflicts of interests concerning what humans should do (for discussion see French, 1995). While species-egalitarianism is a reasonable starting point, we ultimately may come to rank them when we can only give special protection to some but not all species (French, 1995). But species ranking should be done carefully and in an informed way. *We simply cannot please every individual even if we wish to do so and how we decide who to help requires serious reflection on our part.*

SOME AREAS OF CONCERN

In this chapter I will discuss some basic issues that center on the use and exploitation of carnivores and discuss some representative studies (see also Schaller, 1993, 1998; Clark *et al.*, 1995b; Bekoff & Jamieson, 1996; Cooper & Carling, 1996; Newmark *et al.*, 1996; Reading & Clark, 1996; Taylor & Dunstone, 1996; Tobias, 1998b; Varner, 1998). I will try to provide some guidelines for proactive decision-making especially concerning how much human intervention is permissible. I will argue that when humans do intervene we must do so by treading lightly with respect and compassion. For many questions about how animals should be treated by humans there are no 'right' or 'wrong' answers. However, there are better and worse answers. Open discussion on all sides will help us make progress. Perhaps it will turn out that in some cases what we think is the 'right' action to take is not when the big picture is carefully analyzed. A major goal is to stimulate discussion among interested parties so that all pertinent issues are discussed and competing agendas are given due consideration.

INDIVIDUALS VERSUS SPECIES: TROUBLED WATERS FOR CONSERVATION BIOLOGISTS?

While many people believe that it is permissible to trade off individual lives for the good of their species, others maintain that it is individuals who count in our moral decisions. For example, in reintroduction or captive breeding programs, difficult dilemmas arise for those who believe that individuals' lives can be trumped for the good of their species or for some supra-individual entity. Difficult decisions need to be made that may result in the death of numerous individuals by a variety of means (trapping, transporting, starving, being hit by cars, being shot) because other conspecifics may survive and eventually establish a sustainable population. Those who argue that it is individuals who count, that our moral decisions should be based on an individual's own characteristics and not on the species to which it belongs (Bekoff, 1998b,c), likewise have to make difficult decisions that might ultimately result in the loss of a species, or perhaps a species that remains in a perilous state.

One important question concerns if and how human interests in individuals and species can be brought together in a practical way that is not narrowly prescriptive as speciesism. Speciesists make decisions about how humans are permitted to treat other animals based on an individual's species membership rather than on that animal's individual characteristics. For example, all and only humans, or all and only mammals, might constitute protected groups.

Clearly, how individuals and species are traded off raises numerous difficult issues (Sober, 1986; Aitken, 1998; Albrecht, 1998; Callicott, 1998; Crisp, 1998; Estes, 1998; Jamieson, 1998; Varner, 1998; also see essays in Zimmerman et al., 1993). The difficulty of assigning relative value to different animals is vexing, and it also is impossible to dispense with emotional conflicts and personal biases. For example, if given the choice, I would choose to kill the last wolf (or mated pair of wolves) rather than my dog companion, Jethro (Bekoff, 1995). I realize my decision counters the one that many people interested in species preservation and biodiversity would make, but it is impossible for me to discount Jethro's friendship and his trust in me. Would others who choose their own companion animal over a wild animal make the same choice if someone else's companion animal was involved? I raise these issues because I think they need serious discussion in debates about the value of individual lives versus the value of populations and ecosystems. (I thank an anonymous reviewer for suggesting this example.)

Although conservation biologists need to deal with individuals, populations, species, and ecosystems, they usually are more concerned with populations, species, and ecosystems. Also, they often are troubled when making decisions about the relative value of individuals, versus species, populations, and ecosystems. It is worth quoting Estes (1998: p. 1157) on this issue for he poignantly and succinctly gets to the heart of the matter in his discussion of whether or not to rehabilitate oiled wildlife, specifically California sea otters (*Enhydra lutris*).

> The differing views between those who value the welfare of individuals and those who value the welfare of populations should be a real concern to conservation biology because they are taking people with an ostensibly common goal in different directions. Can these views be reconciled for the common good of nature? I'm not sure, although I believe the populationists have it wrong in trying to convince the individualists to see the errors of

their ways. The challenge is not so much for individualists to build a program that is compatible with conservation – to date they haven't had to – but for conservationists to somehow build a program that embraces the goals and values of individualists because the majority of our society has such a deep emotional attachment to the welfare of individual animals ... As much as many populationists may be offended by this argument, it is surely an issue that must be dealt with if we are to build an effective conservation program.

Sober (1986) presents a detailed discussion of the difficulties of assigning value to supraindividual entities, and argues that it is the maintenance of ecological balance and diversity that is holistically valued, not an ecosystem itself. The difference between those who value individuals and those who value balance and diversity centers on the unit of value; both are monolithic.

Some of the main issues concerning trade-offs among individuals, populations, species, and ecosystems are highlighted when considering reintroduction programs.

REINTRODUCING SPECIES

One area of great interest that also generates a lot of controversy centers on whether or not animals should be reintroduced into areas where they lived in the past (for detailed discussion of carnivores see Reading & Clark, 1996 and Linnell et al., 1997; see also Hartwig, 1998 and many chapters in Clark et al., 1995b and Caro, 1998). A major question centers on when and whether it is permissible to override an individual's life for the good of its species – when can individuals be traded off for conservation gains? Berger (1998) notes that the addition, range expansion, and restoration of carnivores to different ecosystems alters selection pressures on animals, especially prey that have evolved in their absence. Researchers have an obligation to know about the effects on life history strategies, demography, behavior, and animals' lives (Clark et al., 1995b; Reading & Clark, 1996; Sutherland, 1998; Caro, 1999a).

What are some of the events surrounding reintroduction? Consider the reintroduction of gray wolves (Canis lupus) into Yellowstone National Park. Frequently, wolves who are to be reintroduced into an area are taken from another area where they still live and moved to the new area. Is it permissible to move individual wolves from areas where they and other wolves have thrived and place them in areas where they might not have the same quality of life, for the perceived good of their species? Is it permissible to interfere in large ecosystems that have existed in the absence of the species to be reintroduced, and is it permissible to remove animals from an ecosystem in which they play an integral role? What are the effects of removing individuals and placing them elsewhere in *each* location? The vital importance of predators to the integrity of ecosystems (Berger, 1999; Terborgh *et al.*, 1999) demands that these questions be given serious consideration.

What about other predators who might now experience increased competition for food? For example, reintroduced wolves and perhaps their offspring are killing coyotes (*C. latrans*) very frequently in Yellowstone (Crabtree & Sheldon, 1999). Before the wolves were reintroduced, coyotes did not compete with wolves. What about prey who now will be eaten when in the past, in the absence of wolves, they would not have been preyed upon?

There are other questions that need to be considered. Sometimes, individual wolves are kept in captivity for purposes of breeding. They live in cages and they never get to roam. Is this permissible for the good of the species?

JINXED LYNX? A CASE STUDY

In Colorado, Canadian lynx (*Lynx canadensis*) were reintroduced into areas where they once lived. Individuals were released beginning in February 1999. This project, like others of its type, is highly controversial (e.g. Kloor, 1999; Lynx Summary, 1999; RMAD, 1999), and brings to light some concerns about reintroduction efforts. Critics believe that it is hasty and ill-planned. Supporters have also agreed that the program was rushed (Kloor, 1999), but that they had to begin reintroducing lynx when they did because of the presence of adequate snowshoe hares, the major food supply for lynx.

Colorado represents the southernmost portion of the lynx's historical range. Individuals were trapped in Alaska and Canada and released into a rather different ecosystem in Colorado with an expectation of at least 50% mortality. The reintroduction of lynx is justified by some people because the animals 'will be killed anyway by trappers.' This line of reasoning simply buys into a system that supports the exploitation of animals. Just because an animal might die in one way does not justify killing them in another way. There was likely not enough food for translocated animals (Human, 1999) and John Seidel, the former head of the Colorado Division of Wildlife (CDOW)-based project viewed the reintroduction as 'an experiment of sorts' (Human, 1999: p. 10A). As of August 1999, 41 lynx were released, five had starved to death, one was shot, and two were hit and killed by cars (one near the Vail ski area after traveling about 150 miles north). The status of 12 others was unknown.

The importance of rigorous science and public support (especially of special interest groups) in reintroduction programs cannot be emphasized too strongly (Reading & Miller, 1995), and this is what the Colorado lynx program lacked. No public opinion survey was performed and there were problems with the validity of the food surveys that were conducted (Kloor, 1999; see also Scott *et al.*, 1999 for a discussion of habitat quality). Experts were concerned from the start about the lack of food, but their concerns were ignored. On Seidel's own admission the program was rushed. Having qualified staff and making uncompromising efforts to know the fate and behavior of reintroduced animals is also essential for judging the success of a program (see for example, Gustafson & Brocke, 1998; for discussion of the difficulties of lynx reintroduction in the Swiss Alps and the necessity of a long-term vision, see Breitenmoser, 1998a). For example, it is necessary to know if lynx show enough flexibility in behavior to allow them to adapt to ecosystems differing in climate, vegetation, and food resources (for consideration of possible genetic differences between predators of cyclic prey, see Breitenmoser *et al.*, 1993a). It also is essential that suitable habitat be protected indefinitely, and this is questionable for the Colorado project (Whipple, 1999).

The Colorado program highlights some major issues centering on reintroduction programs. Of course, not all of them are managed in this manner, and some seem to be on the path to sustainable populations (see below). Nonetheless, it needs to be asked if it is unethical to: (i) perform reintroduction experiments when it is believed at the start that half the animals will die from starvation; and (ii) undertake reintroduction programs simply to prevent species from being listed as endangered or threatened under the Endangered Species Act. In the United States, when the Endangered Species Act is invoked and a species is listed, local control over land use is trumped by federal control, and critical habitat must be protected. Some people understandably want to keep the federal government out of local concerns. One way to keep the federal government out is to attempt to reintroduce animals to keep their numbers up. In Colorado, it was noted 'If we don't begin work on this reintroduction, the federal government will take the lead within the next several years' (Seidel, 1998: p. 1). Action to list the lynx as endangered or threatened was to have taken place in June 1999, but was postponed until January 2000. Along the same lines, 'Idaho officials acknowledge granting permission to relocate lynx is partly an effort to block possible Endangered Species Act restrictions in the state.' (*Bozeman (Montana) Daily Chronicle* (September 12, 1998: p. 5).

In addition to poor biology and the lack of sociological studies, there also are political questions that need to be answered. The Vail ski area, which gave $200 000 to the lynx project, wants to expand its skiing terrain into probable lynx habitat (Hansen, 1999). Information gathered from inquiries made under the Freedom of Information Act showed that there were some questionable relationships between Vail Associates and the Colorado Division of Wildlife. Vail Associates (VA) wanted to know 'If VA takes the lead in securing funding for lynx (and) wolverine recovery, would the state be willing to release VA from all further obligations for preserving and protecting lynx habitat in the Vail ski area expansion zone?' (Hansen, 1999).

Rushing into translocation and reintroduction programs for carnivores (and other animals) because of political and other pressures is ill-advised. It is difficult to learn anything useful from poorly planned projects (Seddon, 1999). Individuals needlessly suffer, and conflicts arise that could have been avoided with better pre-release planning. While I and other critics of the lynx program support well-planned reintroduction programs, these projects are highly visible and failure can spell doom for future efforts.

IS MORE BETTER, IS BETTER ALWAYS GOOD?

Reintroduction programs also raise other questions. For example, it is not clear that species preservation and conservation always have to be valued, why 'more is better', why each species matters so much, why biodiversity should be conserved, or if we can improve nature (Sober, 1986; Hettinger, 1996; Elliot, 1997: especially pp. 97ff; Aitken, 1998; Caro, 1998). Sober (1986) correctly notes that in the absence of knowledge about the consequences of the extinction of different species, the argument from ignorance raises the possibility that the extinction could be beneficial as well as deleterious.

Personal attitudes, including human short-sightedness and greed, also often inform views on these controversial subjects (Berger, 1994; Tobias, 1998a,b). Ludwig *et al.* (1993) call into question the idea that we can manage animal populations in a sustainable way. They argue that science is probably incapable of predicting sustainable levels of exploitation of an animal population, and even if it were possible to make such predictions human shortsightedness and greed would prevent us from acting on them.

Given that even many experts are skeptical of attaining the goals of many captive breeding, translocation, or reintroduction efforts, specifically establishing healthy and self-sustaining animal populations (for discussion see Rabinowitz, 1991; Schaller, 1993; Bekoff, 1995; Clark *et al.*, 1995b; Beck, 1996; Reading & Clark, 1996), it is important to reassess what we are doing and why we are doing it. Indeed, these are very difficult projects and there are many factors that are beyond the control of the scientists who so dearly want to them to succeed. Rabinowitz (1991: p. 165) notes that many captive breeding programs 'provide no comprehensive management of captive populations and no follow-up programs to reintroduce the young to the wild.' Furthermore, he points out that 'the proper techniques of reintroduction are rarely used.' Reading & Clark (1996: p. 296) stress 'It is clearly desirable to improve approaches to reintroduction.' Waples & Stagoll (1997: p. 120) argue that 'Human responsibility does not end until the animal has properly integrated into the wild population and no longer requires or seeks human care and attention.' In many ways human responsibility never ends because we control land use and are obligated to protect habitat indefinitely.

I raise the questions I have, not because I am against all reintroduction and translocation programs. Indeed, some well-planned efforts look to be on the road to yielding sustaining populations (gray wolf recovery seems to be going faster than predicted, Bangs *et al.*, 1998; red wolves, *C. rufus*, on the Alligator River National Wildlife Refuge in northeastern North Carolina, Brian Kelley, pers. comm.) and they can serve as models for future efforts. In all fairness, the lynx project is going smoothly as of January 2000. However, a challenge for which there are no easy solutions is to determine if and how the lives of the 15 individuals who are known be dead can be traded off against the 14 or so individuals known to be alive (12 are missing). Perhaps if the project does ultimately result in a sustaining population of lynx in Colorado some who are reluctant to make this trade-off would be more willing to do. Clearly, one would hope for a mechanism providing an adequate ethical review balancing individual animal harms against the benefits on a population level. Animal experiments are in some countries assessed by a classification system for ranking the degree of animal pain and distress (e.g. the Animals (Scientific Procedures) Act in the UK; Orlans, 1997). However, in reintroductions, where even the estimation of individual survival probabilities is inherently complex and difficult, any such classification system is likely to allow for subjectively interpreted margins, driven by personal agendas or variations in ethical approaches, and thus providing only pseudo-objectivity.

I ask these questions because the issues are not as clear as some people want them to be. When trying to conserve species or restore ecosystems we need to be concerned with the animals who are involved, not only our own human-centered goals. How are costs and benefits balanced for the humans and other animals involved? Populations and ecosystems that have continued to develop in the absence of predators will also be changed. It may turn out in some cases that it would be wrong to try to regain what was lost. It might be impossible to recreate what once existed simply because times have changed and we cannot recreate what once was. Certainly, patience is required (Hettinger, 1996).

AFRICAN WILD DOGS

A highly visible and disputed example of the possible effects of human interference into wild populations concerns the plight of African wild dogs (*Lycaon pictus*; for summaries see Woodroffe *et al.*, 1997; Woodroffe, 1999a,b; see also East & Hofer, 1996; Stearns & Stearns, 1999: pp. 147–87). Interference into the lives of wild dogs involved vaccinating them against rabies and canine distemper. While some scientists maintain that handling the dogs and inoculating them was indirectly responsible for their decline because the handling weakened the dogs' immune system making them less resistant to stress, others, using the same data, conclude just the opposite. Here we have an example of scientists, all of whom care deeply about African wild dogs, not being able to discern what caused their decline. This is because the problems are so incredibly difficult. Should the researchers interfere and possibly cause animals to die or let nature take its course? If the rabies and distemper were introduced by domestic dogs (*C. familiaris*) that would not have been there in the absence of people, are we more obligated to try to help the wild dogs than if the rabies and distemper were natural?

Clearly, human presence poses a serious threat to wild dogs. Woodroffe & Ginsberg (1997b: p. 73) conclude that '61% of recorded adult mortality is caused directly by human activity' (see also Creel & Creel, 1998; Woodroffe & Ginsberg, 1998). Furthermore, correlation and causation can be conflated in studies of the effects of human intrusion. Ginsberg *et al.* (1995) concluded that disease was responsible for the collapse in the population of wild dogs living in the Serengeti-Mara region and that handling was correlated but not causally related to mortality.

Kirkwood (1992) considers such questions as whether we should intervene on behalf of free-living wild animals, and if so, to what extent and how it should be done (see also Tuyttens & Macdonald, 1998). While Kirkwood recognizes that there are many different views, he claims that 'Most would probably agree that when wild animals are harmed by man's very recent (in evolutionary terms) changes to the environment (such as oils-spills, power lines, roads, and environmental contamination) there is a reasonable case, on welfare grounds, to intervene' (p. 143). Kirkwood calls for 'an international code on intervention for wildlife welfare to provide guidance on ethics, methods and standards' (p. 151).

DO WE HAVE DIFFERENT OBLIGATIONS TO DOMESTIC AND WILD ANIMALS AND NATIVE AND NON-NATIVE SPECIES?

Do we have different obligations to domesticated animals and individuals belonging to wild and/or endangered species? Certainly domestic animals do not suffer less than wild relatives when their lives are compromised. Indeed, because of their unique relationship with humans, they might even suffer more (Bekoff, 1995). Does it matter if the domesticated animals are feral? What about wolf–dog hybrids? Does their hybrid status place them in a different category than pure wolves with respect to our attitudes and actions toward them?

Consider this example. Historically, burrowing bettongs (*Bettongia lesueur*) and western barred bandicoots (*Perameles bougainville*) were eradicated in Western Australia. At Shark Bay in Western Australia, feral cats (*Felis catus*) were poisoned to protect these two endangered native mammals that had been reintroduced to the site (Short et al., 1997). Should this have been done and if so, why? What if the feral cats were wild animals? Would this make a difference in our decisions and why so? These difficult questions demand principled consideration, even if final decisions are more pragmatic than principled (Sober, 1986), or based on emotional responses (Bekoff, 1995).

Another current debate centers on predation by non-native red foxes (*Vulpes vulpes*) on native colonies of endangered California least terns (*Sterna albifrons*) and clapper rails (*Rallus longirostris*; Gosselin, 1999). Foxes were introduced about 100 years ago to areas where they are killing these birds. Various conservation organizations are trapping and killing the foxes to keep them from destroying the native birds. Should the foxes, which are now resident, be killed? Is 100 years residency enough to grant them a strong claim to occupancy? Are there any statutes of limitations that apply so that the doing–undoing cycle does not repeat itself elsewhere? Whereas some claim that it is permissible to kill the foxes for the good of the native birds because the foxes are making the landscape 'less wild and less diverse' (Gosselin, 1999: p. 6), it seems reasonable to favor non-lethal solutions in this and other situations (see below). The foxes do not *have* to be killed. The extant foxes are caught in a bind and should not have to pay with their lives because humans previously translocated their ancestors.

Inevitably, these are difficult questions, and answers will definitely reflect individual moral stances and biases. Some might argue that domesticated animals are merely human inventions, they are artificial, and that their lives are not worth as much as wild animals, whereas others might argue that a life is a life and one cannot make assessments of value using the variables of 'wild' and 'domesticated.' One might also argue that human-engineered animals are not that different from human-managed ecosystems. Neither is 'natural' or 'wild' (recognizing that there are very few places on earth that have not been influenced by humans). Katz (1996) has argued that restored and redesigned ecosystems are anthropocentrically designed human artifacts. The natural integrity of ecosystems is compromised by human intervention.

There also is the issue of the value of native versus introduced species, some favoring native species, whereas others arguing that it is wrong to kill non-native individuals so that natives can live. They believe that other more humane solutions will emerge if non-natives are not killed. For example, non-natives may be moved to another area so that native species can survive, or non-natives can be controlled using non-lethal methods (see below).

WHAT IS NATURAL?

Much of the present discussion centers on the difficult question 'what is natural?' Sober (1986: pp. 179ff) claimed that 'The environmentalist's lack of concern for humanly created organisms and environments may be practical rather than principled' (p. 180). He also noted that 'If we are part of nature, then everything we do is part of nature, and is natural in that primary sense.' Debates about the question 'what is natural?' are essential if supposed differences between what is natural and what is not are used to implement differential policies toward animals who are placed in one or the other category. If the distinction is more practical than principled, so be it, but it is important for people to parse out the reasons underlying their choices.

ADOPTING PROACTIVE STRATEGIES – TOO LITTLE TOO LATE?

Clearly, humans and many other animals, including carnivores, are going to cross paths with cross purposes. Everyone realizes that the issues are extremely complex, and allowing human interests always to trump the interests of other animals is not the solution if we are to resolve the numerous and complex problems at hand. Excellent examples of what needs to be done on a wider scale are provided in the interdisciplinary essays on tiger conservation in Seidensticker *et al.* (1999).

Norton (1987) correctly worries about the length of time that it takes to develop theories that can be applied in conservation decision-making. During these deliberations, numerous species will be lost and our decisions simply may come too late. This is especially so for *k*-selected carnivores. Nonetheless, some grounding is needed for why we do what we do. Nature is complex and there are few quick fixes or easy answers to most of the problems with which we are faced. Facile solutions such as killing 'problem' animals, as supported by the National Audubon Society (Gosselin, 1999) and such organizations as Wildlife Services in the United States, generally have not worked in the past, and there is little reason to think they will work in the future. Indeed, the proof that they have not worked is the fact that there still are numerous problems with coyotes and other predators. Formally called Animal Damage Control, Wildlife Services slaughters hundreds of thousands of animals annually using leghold traps, snares, poisons, explosives, and aerial shooting. In 1996 about 82 000 coyotes and numerous other carnivores were destroyed. During the last 50 years, about 3.5 million coyotes have been killed.

Killing coyotes and other predators has not been very effective in eliminating problems. Sometimes simple non-lethal alternatives are possible. For example, in an on-going study of an endangered subspecies of the loggerhead shrike *(Lanius ludovicianus mearnsi)* on San Clemente Island, California, nest predation by island foxes (*Urocyon littoralis*) is controlled by using the 'invisible fence' system (a buried wire that produces a current that delivers a shock to an animal who crosses over it wearing a collar) and not killing them (David Garcelon, pers. comm.).

Concerning carnivores, we still need more detailed basic biological information on all species (e.g. behavioral ecological, physiological, genetic), especially for those we want to relocate and reintroduce or manage in other ways (Clark *et al.*, 1995b; Reading & Clark, 1996; Berger, 1998; Caro, 1998). Intraspecific variability also needs to be given close attention. Not only are most carnivores very difficult to study, but pertinent issues concerning their conservation and management are extremely difficult with which to deal. Thus, even with focused discussion among experts, solutions to existing problems are not readily forthcoming (Seidensticker *et al.*, 1999; Terborgh *et al.*, 1999). Habitat loss is considered by most conservation biologists to be the biggest threat to animal and plant biodiversity (Noss *et al.*, 1997; Wilcove *et al.*, 1998). Even if humans want to protect resident animals, or reintroduce animals to the wild or relocate them to suitable habitats, there might not be any place for them to go. One of our major obligations to other animals should be the indefinite preservation of protected habitats.

It is a privilege to study other animals and to share their lives with them. As we learn more about how we influence other animals we will be able to adopt proactive strategies. Part of learning entails changing our practices and asking 'would we do what we did again, or have we learned something that can make other animals' lives better?' Being proactive means that we will increase the chances of extricating ourselves from the repetitive cycle of doing something to a species that leaves it in a perilous or endangered state and then having to undo what our ancestors did to remedy the situation. Often, in remedying the situation, the animals become imperiled once again (for example, the wolves reintroduced into Yellowstone National Park are predictably traveling outside of the park and their status as 'endangered' is being called into question). There may be a limited number of cycles through which many species will be able to go. There also may be a limited number of cycles that can be funded or tolerated by humans.

Hettinger (1998) stresses that we need to find a positive role for humans in nature and that too much emphasis on the notion of wildness may prevent this. I agree. While we often value wild nature because of its autonomy from humans, realistically there are few, if any places on Earth, that are totally free of human intervention. Being an intimate part of the nature does not mean that we cannot help nature along in some instances.

ETHICAL ENRICHMENT

It is in the best traditions of science to ask questions about ethics; it is not anti-science to question what we do when we study other animals (Bekoff & Jamieson, 1996; Bekoff & Elzanowski, 1997). Ethics can enrich our views of other animals in their own worlds and in our different worlds, and help us to see that variations among animals are worthy of respect, admiration, and appreciation. The study of ethics can also broaden the range of possible ways in which we interact with other animals without compromising their lives. Ethical discussion can help us to see alternatives to past actions that have not served us or other animals well. In this way, the study of ethics is enriching to other animals and to ourselves. If we think that ethical deliberations are stifling and unnecessary hurdles over which we just jump in order to achieve our goals, then we will lose rich opportunities to learn more about other animals and also ourselves. The application of ethical enrichment, similar to the utilization of environmental enrichment, is a two-way street. Our greatest discoveries come when our ethical relationship with other animals is respectful and not exploitive. While animals are unable to consent to or to refuse our intrusions into their lives, it is useful to ask what they might say if they could do so.

There is a continuing need to develop and improve general guidelines for research on free-living and captive animals. These guidelines must take into account all available information. Professional societies can play a large role in the generation and enforcement of guidelines, and many journals now require that contributors provide a statement acknowledging that the research conducted was performed in agreement with approved animal care regulations. Guidelines should be aspirational as well as regulatory (Jamieson & Bekoff, 1996). In the future the challenge is to make guidelines more binding, effective, and specific. Many projects concerning the management of wild carnivores are not subjected to the same degree of scrutiny as are more 'academic' efforts (there is no Institutional Animal Care and Use Committee, for example, and a greater reliance on in-house appraisals and approval). More than one conservation biologist has told me that they would like to see more rigorous internal assessments of the methodological and ethical aspects of non-academic research protocols.

SCIENCE, VALUES, CONSERVATION, AND CARNIVORES

Dunlap (1988: pp. 175–6) noted that 'The role and impact of "science" in the transformation of American ideas is ambiguous at best.' While many people respect scientists and endow them with special powers to fix things when they break, scientists and science will not alone be able to deal effectively with the many difficult and puzzling problems that arise when humans and other animals interact (Nader, 1996; Wolch & Emil, 1998). Personal and cultural values influence how individuals view and study other animals (e.g. Asquith, 1996; Bekoff & Jamieson, 1996; Kellert, 1996; McCrindle, 1998; Rasmussen, 1999) and the choices they make. It is essential to involve local people when researchers 'go into the field' in countries other than their own and to respect local customs and beliefs about matters of conservation. Common sense also plays a large role in our decision-making and political and sociological components cannot be ignored. Most decisions about conserving species are multidimensional matters. In the future, scientists and other people involved in conservation efforts will have to reconcile how values and facts are to be factored into the deliberations in which they partake and the choices they make.

Animal advocacy often places a priority on individuals, especially among animal rightists, whereas animal welfarists take a utilitarian position and favor decisions where the presumed costs to the animals are less than the presumed benefits to humans (Regan, 1983; Singer, 1990; Bekoff, 1998a). In conservation biology, often the interests (and rights) of individuals are traded off against perceived benefits that accrue to higher levels of organization, including populations, species, and ecosystems (see below). Biocentrists and anthropocentrists often clash because the issues are so difficult and are so highly driven by social and personal views. In the end. all levels or organization need to be considered in any deliberation about humans' interference in nature, and because of this, very difficult issues need to be dealt with (Webb, 1999). Achieving win–win situations for the humans and animals who are involved will be very difficult but we should never stop trying.

Carnivores and all other animals will undoubtedly benefit from open discussions about human–animal relationships. Trade-offs always need to be made, and Hettinger (1996: p. 8) has poignantly argued that 'it is far from clear that we can improve nature in non-utilitarian ways.' How costs and benefits to humans, other animals, species, populations, and ecosystems are balanced, and how value is assigned to various costs and benefits, represent some of the most difficult challenges facing those who want to improve human and animal lives and the health of ecosystems. Because humans will forever try to help other species in one way or another, it is important to make each and every effort count because we may have fewer and fewer opportunities to make positive contributions in the future.

Carnivores are closely linked to the wholeness of many ecosystems. By paying close attention to what we do to them, and why we do what we do where and when we do it, we can help maintain the integrity and vitality of individuals, species, populations, and ecosystems. *When animals lose, we all lose.*

ENDNOTE

1. For general references see Goude (1994), Noss *et al.* (1997), Neumann (1998), Tobias (1998a), Wilcove *et al.* (1998), and Wolch & Emel (1998); see also Caro (1999b). For recent studies of the influence of human activities on carnivore behavior see Harris (1979, 1981), Harrison (1993; 1997), Travaini *et al.* (1993), Laurenson & Caro (1994), Bekoff & Jamieson (1996),

Clutton-Brock (1996), Newmark *et al.* (1996), Putnam (1996), Cypher (1997), Olson *et al.* (1997), Jackson & Ahmad (1997), Breitenmoser (1998a), Dexter & Meek (1998), Estes (1998), Fleming *et al.* (1998), Liss (1998), Schaller (1998), Wikramanayake *et al.* (1998), and references therein.

Acknowledgements

I especially thank two anonymous referees and Schuyler Greenleaf along with Dan Blumstein, John Reed, Colin Allen, Joel Berger, Roger Fouts, Dale Jamieson, Jeff Masson, Carron Meaney, Bill Merkle, Rich Reading, Elliott Sober, Dan Whipple, and Susan Rodriguez-Pastor for comments on an ancestral version of the entire manuscript or on specific sections of an earlier draft. I also thank Jim Estes for allowing me to quote from Estes (1998) and David Garcelon for informing me of his research on loggerhead shrikes.

10

The control, exploitation, and conservation of carnivores

WARREN E. JOHNSON, EDUARDO EIZIRIK AND GINA M. LENTO

INTRODUCTION

Human history and evolution have been shaped and influenced by interactions with a diverse array of carnivores. Our cultural and historical heritage reflects a long-term relationship with carnivores, and our attitude towards them often mirrors our interaction with the natural world. Until recently, we viewed carnivores primarily as adversaries to be avoided or killed, reflecting a history of predation and competition. With time, as we improved our ability to use tools and to 'tame' the environment, perspectives changed and carnivores became a resource or a commodity. Carnivores were exploited for subsistence, medicine, commercial profit, and recreation. Individuals and populations were also persecuted or 'controlled' as vermin or pests because of concerns over personal safety, disease transmission, or depredation on livestock and pets. However, during the last few decades, often under the guise of conservation, an increasing number of more noninvasive or less prejudicial uses of carnivores have become important and more emphasis has been placed on protecting and restoring previously decimated carnivore populations.

The goal of this chapter is to explore the interactions among the control, exploitation, and conservation of carnivores, to review instances in which they can be compatible, and to relate similarities among each type of activity. We will summarize several case studies and discuss some of the insights for conservation that have been garnered from the exploitation of carnivores. Finally, we will make recommendations for conservation and management actions, and address research areas which can improve our understanding of how best to coexist with carnivores without sacrificing their existence while meeting our own needs.

GENERAL PROBLEMS

Our coexistence with carnivores depends on the availability of appropriate physical space, our tolerance for sharing this space, and our desire and propensity to extract advantage or enjoyment, whenever possible, directly from these animals. The analysis of these extractive activities can be performed from many perspectives, encompassing philosophical, moral, historical, economic, ecological, and social issues. In this chapter we will not debate many of these social and ethical questions, but will start from the premise that the maintenance of viable populations of all extant carnivore species is desirable and that the exploitation of many carnivores will likely exist for the foreseeable future. In addition, we will focus on direct, as opposed to indirect, impacts that exploitative or control activities can have on long-term viability and health of carnivore populations and on carnivore conservation. In this context, much of the discussion is encompassed within the rubric of population dynamics, but to a lesser extent will also include behavioral and genetic considerations. A detailed discussion of these parameters is beyond the scope of this chapter, but they are essentially interrelated measures of recruitment (reproduction) and loss (mortality) to a population and estimates of total and effective population size (density, dispersal, and other behavioral considerations). The interactions among these parameters are complicated and have rarely been considered collectively in carnivore populations in an empirical fashion. These interactions include density-dependent recruitment, compensatory and additive mortality, and density dependent immigration and emigration.

A significant difficulty in objectively measuring the effects of exploitation is the almost universal problem of obtaining appropriate biological and demographic data from these elusive species. During the last decade there have been substantial improvements in population sampling and estimation techniques and there has been increased understanding of the parameters that are important in maintaining viable populations. For many species, increased monitoring and regulation of control and exploitation activities have provided additional data on these populations. However, for most species and populations, neither the needed baseline data have been collected nor have populations been monitored for sufficient periods of time to make broad conclusions about how to distinguish between detrimental and non-detrimental effects. These problems are exacerbated when dealing with the more secretive, rare, and widely dispersed species or those that inhabit aquatic environments.

REVIEW OF CASE STUDIES

Carnivore exploitation: the hazard of being economically valuable

Many carnivores have been exploited commercially in some fashion, with the degree of effort depending predictably on supply and demand. There have been major differences in this process among species, populations, and administrative divisions (regions, countries, etc.) that are illustrative of how carnivores respond to commercial harvest and how exploitation and conservation are generally linked. Several commonly hunted and trapped carnivores of the Americas are good examples.

There are three common fox or zorro species found in the southern cone of South America that are persecuted to varying degrees in large portions of their ranges. The culpeo zorro (*Pseudalopex culpaeus*) and the South American gray zorro (*P. griseus*) occur east and west of the Andes from northern Chile and Argentina south to the island of Tierra del Fuego. Azara's zorro (*P. gymnocercus*) is distributed in southern Brazil, Uruguay, Paraguay, and northern Argentina (Ginsburg & Macdonald, 1990). Although some of these zorros are exploited for subsistence purposes or recreation, the major factors motivating their harvest are economical and commercial (Ojasti, 1993).

During the 1970s, foxes, or zorros, were among the South American species most commonly traded for their pelts. From 1976 to 1979, around 3 600 000 gray zorro (including Azara's zorro) and 32 000 culpeo zorro pelts were exported from Buenos Aires, Argentina (Figure 10.1) with a total value of over 80 million US dollars (Mares & Ojeda, 1984). From 1982 to 1984 these figures had dropped to around 70 000 gray zorro and 4500 culpeo zorro (Ojasti, 1993). This decline was probably the result of a combination of reduced demand in the foreign markets of the United States, Germany, and the rest of Europe, stricter enforcement of regional wildlife legislation, and reductions in population densities in the most readily accessible areas.

In South America the exploitation of zorros is also related to the pelt trade of other species, such as the wild cats (McMahan, 1986; Bowles, 1996), and to the dynamics of national and international markets, laws, and regulations for these species. Trade in carnivore skins in South America focused initially on the jaguar. Later it shifted to zorros and smaller cats such as the ocelot (*Leopardus pardalis*), margay (*L. wiedii*), tigrina (*Oncifelis tigrina*), and eventually to the Geoffroy's cat (*O. geoffroyi*), as laws regulating exploitation and trade changed both nationally and internationally in the 1960s and 1970s. Traders and professional trappers were motivated to develop new wildlife products for trade and to find trade routes through coun-

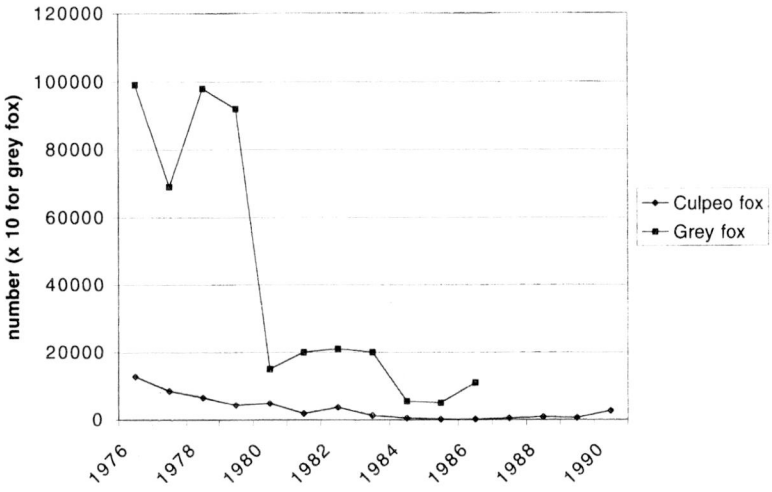

Figure 10.1. The number of culpeo zorro and grey zorro (including Azara's zorro) pelts exported from Argentina between 1976 and 1990. (Data compiled from Mares & Ojeda, 1984; IUCN, 1988; Novaro, 1995).

tries with the least enforcement and the most relaxed laws (N. Smith, 1976). Since the inception of the CITES (Convention on the International Trade in Endangered Species of Wild Flora and Fauna) agreement, the Appendix status of several species of cats and zorros had to be changed because of either unsustainable trade or the difficulty in distinguishing products from protected and unprotected species (McMahan, 1986; Bowles, 1996).

Although the international pelt trade has apparently continued to decline since the 1980s, national and international trade still occurs, especially in Argentina (e.g. Iriarte & Jaksic, 1986; Johnson & Franklin, 1994a; Novaro, 1995). However, most cats and foxes are not hunted or trapped by professionals, but instead by small farmers, ranchers, or herdsman or while spotlighting for other economically important species such as the European hare (*Lepus capensis*). There are still significant economic incentives for continued exploitation of these carnivores. The money derived from the sale of a few pelts a year can represent an important contribution to the annual income of many rural residents (Novaro, 1995). These factors, along with cultural traditions, maintain trapping throughout most of southern South America, despite restrictions. Culpeo zorro hunting, in particular, is also continued in an attempt to reduce sheep depredation (Bruggers & Zaccagnini, 1994).

Although the legal trade of pelts in South America has been credited with leading to overexploitation of target and non-target populations and species, the impact of harvest on the South American zorro populations is unknown, mostly because there have never been any broad estimates of population sizes. Many of these zorro species appear to be able to tolerate high levels of exploitation (Broad *et al.*, 1988). Although estimates of maximum sustained yields of South American zorros that are based on empirical data and theory do not exist, Pils *et al.* (1981) estimated that 50% of pre-harvest red fox (*Vulpes vulpes*) populations could be harvested. The ability of South American zorros and many other canid species to tolerate such high mortality is likely due in part to their generally high intrinsic growth rate, but may also be a function of the heterogeneous spatial distribution of hunting pressure. This in effect creates 'source' populations of non-exploited zorros and 'sink' populations of exploited animals (Pulliam, 1988). Novaro (1995) concluded that, in spite of unsustainable trapping levels on some ranches (46 to 73%), zorro densities remained relatively stable because of recruitment from other areas. Although zorro populations have also benefited from the establishment of National Parks and other protected areas, this does not always guarantee their protection. In Torres del Paine National Park, 45% of the documented mortality in grey and culpeo zorro resulted from poaching (Johnson & Franklin, 1994a).

In large part because of difficulties in regulating and limiting the legal hunting, trapping, and trade of zorros in South America, in many areas the most effective conservation tool for maintaining carnivore populations has been to completely ban these activities. In contrast, the characteristics and history of carnivore exploitation in North America have been somewhat different. In Canada and the United States a combination of greater monetary resources, stronger governmental agencies, broader and stricter enforcement of hunting and trapping limits, and better population estimates have led to the management of carnivore populations through legal harvest. Hunters and trappers, as well as a large segment of the general public, are often active participants in the process of managing populations to ensure long-term viability. This has led to the maintenance of legal hunting and trapping of numerous carnivore species including foxes, cats, bears, and mustelids.

The North American raccoon (*Procyon lotor*) is one of these species that have been heavily harvested. Raccoons originally occurred throughout most of North America in a wide variety of habitats, and were introduced into the Soviet Union and Germany in the middle 1930s (Sanderson, 1987). The raccoon is the most economically important furbearer in the United States based on total revenue (Shieff & Baker, 1987). Records of total harvests of raccoon in the United States were begun in 1934–35, when about 400 000 animals were traded. Since then numbers have fluctuated greatly, reaching over five million per year in the early 1980s. Harvests in Canada have been relatively lower, reaching a peak of almost 200 000 in early 1980s (Obbard et al., 1987). Raccoon populations were relatively low in North America prior to the early 1940s, but have since greatly increased in numbers, possibly due to increased habitat availability. Their distribution has expanded into new habitats and geographic ranges, and raccoons have become one of the principal species of nuisance wildlife in urban settings (Williams & McKegg, 1987). Populations have withstood high hunting pressure (as a result of increases in pelt values), although locally numbers have decreased periodically due to disease (i.e. distemper or rabies). There is the potential of density-dependent compensation among mortality sources during the harvest season (W. Clark et al., 1989), limiting demographic effects of hunting and trapping in a healthy population. In heavily hunted areas, however, legal harvest can account for over 75% of mortality (W. Clark et al., 1989). In Iowa, light harvest levels affect juveniles more than adults, but these differences were not apparent under heavy harvest rates. Although raccoons have been stocked at times in certain states, there has been little active management of the species except in regulating the timing and length of the hunting and trapping season (Sanderson, 1987). Due to the relatively small impact of raccoons on humans and their current market value, raccoon populations are unlikely to become severely threatened in the near future.

As with their terrestrial counterparts, almost all species of marine carnivores or pinnipeds (i.e. Otariidae, Odobenidae, and Phocidae) have been exploited historically. During the late eighteenth and early nineteenth centuries sealing, like whaling, was a lucrative business, and many species were left balancing on the edge of extinction by the early 1900s. Today, many species have recovered to levels of abundance comparable to presealing levels, but several remain endangered or vulnerable, and few populations are currently exploited commercially.

Classic examples of historic over exploitation include the commercial killing of harp seal (*Phoca groenlandica*) and hooded seal (*Cystophora cristata*) pups for their pelts, two species with high tolerance and natural tameness towards humans (Riedman, 1990). An annual quota for 30 000 hooded seal pups remained in effect for several years in Canada and Norway, being reduced to 15 000 in 1974. Commercial hunts were prohibited in Canada by 1989, but continued longer in Norway (Reeves & Ling, 1981;

Riedman, 1990). In the southern hemisphere, adult fur seal (*Arctocephalus* sp.) were among the most-exploited highly prized species for their pelts. Several hundred thousand skins per year from multiple species of fur seals and sea lions (*Otariidae*) were imported into Europe at the peak of the sealing period (Mattlin, 1978). South American fur seals were hunted for oil and pelts as late as 1990 (Riedman, 1990).

The New Zealand (NZ) fur seal (*Arctocephalus forsteri*) exemplifies many of the trends common to commercial exploitation of marine carnivores. It had one of the widest ranges of any fur seal, ranging from southern Australia and islands south of Tasmania to the South Island of New Zealand, the lower North Island and all of the NZ subantarctic islands. Occupation of New Zealand by the Maori in the 1500s and their subsistence use is suggested to have caused a historical extirpation of this species in the North Island, which is currently being recolonized. Many populations of NZ fur seals from the subantarctic islands were exterminated and most others were markedly reduced by sealing in the 1800s. For example, in the New Zealand region, one sealing gang alone reported a take of over 60 000 NZ fur seal skins from Antipodes Island, and during one season prior to 1815, over 100 000 NZ fur seal skins were reported taken from Macquarie Island (Mattlin, 1987). Populations of NZ fur seals at both islands were exterminated by 1820 (Shaughnessy & Fletcher, 1987). Today, the estimated population of NZ fur seals at Antipodes Island is 1100 individuals (Mattlin, 1987), and 1200 individuals at Macquarie Island (Shaughnessy & Fletcher, 1987). The current estimated number of NZ fur seals across their entire range is 66 000 individuals (Mattlin, 1987; Shaughnessy & Fletcher, 1987).

The NZ fur seal is now recovering from historic exploitation as evidenced by the reestablishment of rookeries in the upper South and lower North Islands of New Zealand in the last 15–20 years and an increase in numbers at original rookeries. However, incidental by-catch has become a significant problem in the last decade with several hundred fur seals being killed each year during normal fisheries operations when they get caught inside trawl nets or become entangled in set nets and long lines (Shaugnessy & Payne, 1979; Donoghue, 1997). There have been recent voluntary and legal attempts to reduce seal mortality from entanglement by including 'escape hatches' in trawl nets (Gibson & Isakssen, 1998), restricting the timing and length of tows, and using lights and acoustic harassment devices (S. Anderson & Hawkins, 1978; Geiger, 1985; A. Branson, pers. comm.). Maximum allowable annual by-catch quotas have also been set in some commercial fisheries (Woodley & Lavigne, 1993). For example, it was estimated that up to 300 NZ sea lions of a total population of less than

16 000 were killed per year in troll nets. These findings prompted the establishment of a maximum by-catch quota of 63 individuals (Gales & Fletcher, 1996).

By-catch mortality in pinnipeds can also be biased toward certain sex and age classes. For example, from 1987 to 1994, the females of reproductive age made up from 54 to 65% of the by-catch mortality in NZ sea lions (Slooten & Dawson, 1995). Although the specific demographic effects of this bias were unknown, the potential threat to viable populations prompted the New Zealand government to prematurely close the relevant fisheries in both 1997 and 1998, resulting in several million dollars of losses to the New Zealand fishing industry.

Carnivore control: the hazard of being considered a pest

In addition to commercial exploitation, some carnivores are also persecuted to reduce populations or to remove specific individuals. This phenomenon is demonstrated by the history of three of America's largest carnivores, the puma (*Puma concolor*), jaguar (*Panthera onca*), and brown bear (*Ursus arctus*). The puma is the most widely distributed terrestrial mammal in the Americas and one of the continent's most recognized carnivores. Throughout its range pumas are commonly associated with forested areas or with drier, more-open regions. Although deforestation, agriculture, and human settlement have altered and fragmented suitable puma habitat, large areas remain, many of which retain high puma densities. Pumas have also adapted to some human-altered habitats and can be found near many urban areas. Conflicts between humans and pumas are numerous in North, Central and South America, where puma have traditionally been hunted or trapped for sport, to reduce depredation on livestock, to increase ungulate populations, or to decrease negative interactions with humans (see Andersen, 1983; Bruggers & Zaccagnini, 1994; Reynolds & Tapper, 1996).

The history of puma persecution in southern South America demonstrates a pattern that has been common throughout much of the continent. In Patagonia, puma were common and often approachable when Europeans began settling the area in the 1800s (Hudson, 1895). By 1916, however, there were over two million domestic sheep just in southern Chile, and human and puma confrontations increased dramatically. For almost 100 years southern Chile and Argentina were dominated by large corporate ranches, some with units with as many as several thousand sheep. Professional puma hunters killed as many pumas as possible and did not focus on problem cats alone. Although some pelts were sold in international markets (Mares & Ojeda, 1984), most were killed to reduce depredation on livestock. As agrarian reform and economic conditions caused a shift in land use patterns toward smaller ranches, puma hunting became less organized, and occurred on a smaller scale. In spite of this persecution, puma still cause considerable economic losses in some areas, depredating on livestock such as cattle, sheep, horses, and goats and often killing several animals in one night (Bruggers & Zaccagnini, 1994; Cunningham *et al.*, 1995). In one of the few efforts to quantify this effect in South America, livestock ranchers in La Rioja Province, Argentina estimated annual losses of 20–27% of goats and 4–12% of cattle. In the sheep grazing region of Santa Cruz Province, Argentina, puma were held responsible for an estimated 23% of perinatal lamb mortality (Bruggers & Zaccagnini, 1994). Depredation on livestock can be especially acute in areas where geography and habitat provide refuge for sizeable puma populations.

The establishment of protected areas throughout the Americas and the enactment of stricter national and international laws regulating wildlife provided additional refuges for pumas and other large carnivores (e.g. Belanger, 1988). These reserves also often led to increased tensions between resource managers and ranchers, however. For example, after the creation of Torres del Paine National Park in southern Chile in 1959, guanacos (*Lama guanicoe*), one of the main prey items of the Patagonia puma, increased in number in the Park from around 100 animals to almost 1300 by 1988. During this period, puma densities also grew to 30/100 km^2, levels rarely seen in other populations (Franklin *et al.*, 1999). However, neither the National Park, nor general laws against killing pumas passed in 1972, were sufficient to protect this population from the persecution of ranchers and hunters. Seven of the eight adult pumas that were followed during a radio-telemetry study had home ranges that included sheep ranches. At least three of these preyed on sheep, sometimes killing as many as 2–14 sheep (0.5–2% of the herd) every other night before returning to the relative protection of the Park. Following a regional pattern, pumas were pursued after making kills by hunters and dogs, often becoming the victims of poaching within the National Park (Bruggers & Zaccagnini, 1994; Franklin *et al.*, 1999). Some attempts have been made in southern Chile to adapt to increased puma populations. These include shifts to other economic activities, such as the raising of cattle, that are less susceptible to puma (and culpeo zorro) depredation and the establishment of more service-oriented businesses.

Improved protection of pumas has also altered other aspects of their behavior. In addition to increased interactions with neighboring ranches, the growing puma population also became more visible and less secretive.

Sightings by Park visitors, Park staff, and field researchers increased dramatically from the early 1970s when they were rare into the 1980s. By the mid-1990s, weekly reports of day and night-time puma observations were the norm. The increase in puma numbers and sightings, plus a large rise in tourist visits culminated in 1998 when a puma killed a tourist. A Park guard subsequently killed the offending puma.

Human–puma interactions are also on a rise in North America, as well as documented attacks on people and pets (Anderson, 1983; Aune, 1991; Beier, 1991; Halfpenny et al., 1991). In Boulder County, Colorado, reports of puma interactions with humans increased from an average of one to two per year in the 1960s and 1970s to close to 100 in 1990 and became more common during daylight hours and around human population centers (Halfpenny et al., 1991). Puma populations are stable or increasing in several parts of the United States and in many areas hunting and/or animal damage control (of animals suspected to be killing livestock) are used to try to prevent conflicts. These efforts can be significant. Depredation control was the leading cause of death in an exploited population of puma in southeastern Arizona, contributing to a high estimated mortality rate of 45% (Cunningham, 1995). Management practices in other states vary, ranging from efforts to reestablish a healthy, viable population in Florida and hunting bans in California to hunting seasons of over 200 days a year in many western states of the United States and unlimited hunting in Texas. In eleven western states, the legal harvest of pumas from 1987 to 1990 ranged from an average of 82 animals per year in Washington to 330 animals per year in Idaho (Green, 1991). In most western states property owners may kill pumas that inflict damage as long as the kill is reported (Tully, 1991). During the same period an average of 290 pumas per year, or around 1160 pumas were culled by agencies in attempts to limit livestock depredation. Several states expended well over $US 100 000 in puma control efforts in addition to the resources provided by the Animal Damage Control section of the Animal and Public Health Information Service (APHIS) of the US Department of Agriculture (USDA) (Tully, 1991).

The puma is a success story compared with other large carnivores of North and South America such as the gray wolf, brown bear, and jaguar. In spite of persistent persecution and hunting as a game species, it has prospered in many areas with appropriate habitat, low human densities, and sufficient availability of appropriate prey items (Beier, 1995). Moreover, the puma has also been well studied compared with most other persecuted carnivores. There have been several studies documenting the demographics of harvested and non-harvested puma populations (Logan et al.,

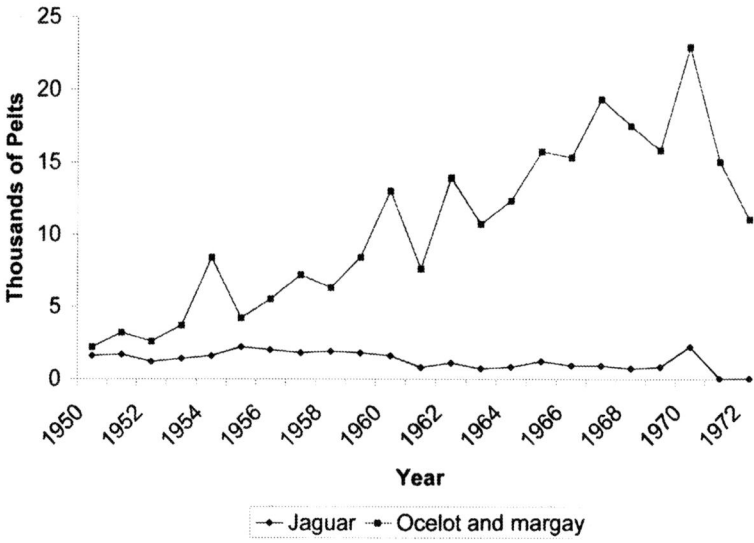

Figure 10.2. The number of jaguar, ocelot, and margay pelts exported from the Peruvian Amazon region between 1950 and 1972. (From Ojasti, 1993, reprinted by permission of the Food and Agriculture Organization of the United Nations.)

1986, 1996; Ross & Jalkotzy, 1992; Lindzey *et al.*, 1994). These studies have highlighted the importance of human-induced mortality on puma populations and have led to the suggestion that puma populations might be able to sustain an overall mortality rate of about 15% (Jalkotzy *et al.*, 1992) and that long-term viability may depend upon dispersal among populations (Beier, 1995).

The history of jaguar interactions with humans is both similar and different from that of the puma. Compared with the puma, the jaguar has a more limited distribution, both geographically and in terms of habitat requirements. Jaguar populations are therefore more likely to be isolated in smaller areas of suitable habitat and thus to be more vulnerable. Jaguars were originally found from southern Arizona, New Mexico, and Texas south throughout Central and South America to Patagonia in Argentina. Perhaps more so than the puma, the jaguar is a highly prized trophy species, but it has also been persecuted for its pelts and to reduce depredation on livestock (Quigley & Crawshaw, 1992; Rabinowitz, 1995). In the 1960s over 15 000 jaguar pelts were removed from the Amazon region yearly (N. Smith, 1976) (Figure 10.2). Although the pelt trade diminished significantly after hunting became illegal in 1967 and the CITES treaty was enacted in 1973, persecution continues throughout most of the jaguar's range.

Two of the best-documented examples of jaguar persecution resulted from pioneering work in the Pantanal (Crawshaw & Quigley, 1991) and Foz do Iguaçu regions of Brazil (Crawshaw, 1995). The Pantanal, a vast flood plain formed by the Paraguay River and its tributaries, still has a large jaguar population. With the creation of a large number of extensive cattle ranches, on which cattle generally roam freely, this population began to be routinely and systematically persecuted. On a typical ranch, a professional hunter killed 68 jaguars and 275 pumas from 1959 to 1966 (Crawshaw & Quigley, 1991). More recently the outlook for jaguars has improved. A recent decrease in ranching due to economic factors has apparently helped the jaguar population in this area rebound in numbers, and many ranches are looking to tourism as a viable alternative to ranching (P. Crawshaw, pers. comm.).

At Iguaçu National Park, in southern Brazil, at the boundary with Argentina and Paraguay, a different situation occurs. The Park protects the natural beauty of the Iguaçu waterfalls and one of the largest pieces of subtropical rain forest remaining in the continent. Compared with the Pantanal region, however, this is a smaller area of contiguous jaguar habitat, and is thus less likely to maintain viable population sizes. Park management is also complicated because there is no buffer zone with contiguous agricultural and livestock land. Illegal hunting for sport, subsistence, and commercial purposes occurs in and around the Park. For example, during a field study of seven radio-collared jaguars, three were killed by poachers (two opportunistically and one as a trophy), two by vehicles on the highway bordering the Park, and two others may also have been killed (Crawshaw, 1995). As with many other carnivores, the full dimensions of jaguar depredation on livestock in the area are unknown and are often overstated by ranchers. Most research suggests that a large portion of livestock losses attributed to jaguars through their range are actually due to other causes, and that if abundant levels of natural prey are maintained and husbandry practices are improved, mortality rates can be reduced significantly (Quigley & Crawshaw, 1992; Rabinowitz, 1995). The long-term viability of this population will probably depend upon its close monitoring and control, increased local outreach, and the maintenance of jaguar populations in adjoining protected areas in Argentina (Eizirik et al., 2000).

In North America, the brown or grizzly bear (*Ursus arctos*) is one of the carnivores most sensitive to control efforts. The dynamics of human–brown bear interactions have been well-studied in the Greater Yellowstone Ecosystem, where from 1973 to 1985, illegal kills accounted for 41% of the 101 known deaths, management action 35%, and road kills 6% (Knight *et al.*, 1988). Since 1995, humans have been responsible for over 90% of adult female mortality (Weaver *et al.*, 1996). These and other data have led to the hypothesis that brown bear populations cannot sustain human-caused mortality rates of over 5% annually (Bunnel & Tait, 1981). This is similar to estimates of the maximum sustainable levels of around 7% for bear populations in Sweden (Swenson *et al.*, 1994).

Bear studies have also demonstrated that patterns of exploitation, such as the preferential harvest of adult males, can affect the social structure and population growth, even in a fairly asocial species. For example, a harvested bear population in southwest Alberta, Canada had skewed age and sex ratios, with fewer females and older adults than expected (Wielgus & Bunnell, 1994; Wielgus *et al.*, 1994). These authors observed that the population had a growth rate close to zero and seemed to be decreasing in size. They hypothesized that this was partially due to increased mortality of older males and the disruption of the social (territory) system. This caused an increased influx of young males, a potentially higher frequency of infanticide, and decreased litter size as females avoided the prime feeding grounds that were occupied by the immigrant males.

The control of carnivore populations is not restricted to terrestrial species. Several species of pinnipeds habitually take fish from fishing gear being towed behind commercial vessels. Fishermen argue that seals cause significant damage to fishing gear while removing fish and reduce catch rates. Studies thus far have failed to demonstrate that pinniped foraging reduces the available biomass of commercially valuable fish species either during or outside the fishing season (e.g. Fay, 1982; Kajimura, 1984; Carey, 1992; Dix, 1993). However, some local fisherman perceive these 'pests' as threats to their livelihood and shoot seals (J. David, pers. comm.; S. Childerhouse, pers. comm.). As with terrestrial carnivores, seal populations can be hard to estimate. A common misconception is that seal populations are expanding as numbers of seals increase in local fisheries. However, these perceptions are often incorrect since seal distribution patterns shift periodically in response to trophic changes and prey availability, especially during the pupping season (Fay, 1982; Gales & Mattlin, 1997).

Nonextractive exploitation: the hazard of becoming an attraction

Although nonextractive exploitation of carnivores, by definition, implies no direct harvest of individuals, many of these activities have the possibility of disrupting populations indirectly by either decreasing recruitment or increasing mortality. It has been argued that uncontrolled ecotourism can potentially harm many carnivore species. For example, organizing marine mammal watching tours has rapidly developed into a multi-million dollar business in several countries (Constantine, 1999). Pinnipeds, due to their amphibious lifestyle, are vulnerable to disturbance by ecotourism activities both at sea and on land. Land-based operations are a particular concern as the time pinnipeds spend on land is primarily for resting, mating, and pup-rearing activities. These negative impacts have not been fully characterized, but may be especially important for already vulnerable populations (Boo, 1990; Wright, 1996; Lidgard, 1997). Direct provocation by tourists has been shown to delay birthing and to shorten lactation times (Lidgard, 1997; Constantine, 1999). Boat and aircraft noise may also affect some marine carnivores as disturbance often leads to increased attention to young and avoidance maneuvers.

Ecotourism also has the potential of disturbing terrestrial carnivores, especially those restricted to small areas. For example, many brown bear populations inhabit areas habitually visited by tourist. Extensive tourist activity increases the probability of conflict between humans and bear, especially during breeding. Habituation towards humans or human food sources will also often lead to confrontations. The political and social challenges of maintaining bear populations are considerable (see Mattson *et al.*, 1996a). However, in spite of the fact that encounters with brown bears can be fatal (Tough & Butt, 1993), current management policy in many protected areas is being geared towards the coexistence of brown bears and humans. This policy often results in the exclusion of humans from areas of high bear densities or from areas inhabited by females with cubs. Tolerance towards offending animals in more settled areas is usually more limited, however.

SYNTHESIS AND DISCUSSION
From exploitation to conservation: the route most traveled
Although the earliest bands of humans undoubtedly produced many localized effects on carnivore populations, it was probably not until groups cooperated more effectively and became more proficient at tool use that they had a large impact on carnivore populations. Although it has been hypothesized that humans played a role in the Pleistocene extinction of large mammals (Martin & Wright, 1967; Martin & Klein, 1998), it is unlikely that they were solely responsible for the removal of a carnivore species from large portions of its range (see Woodroffe, Chapter 4). More recently, however, several carnivore species became extinct by human actions, including the Falkland Island wolf (*Dusicyon australis*), the sea mink

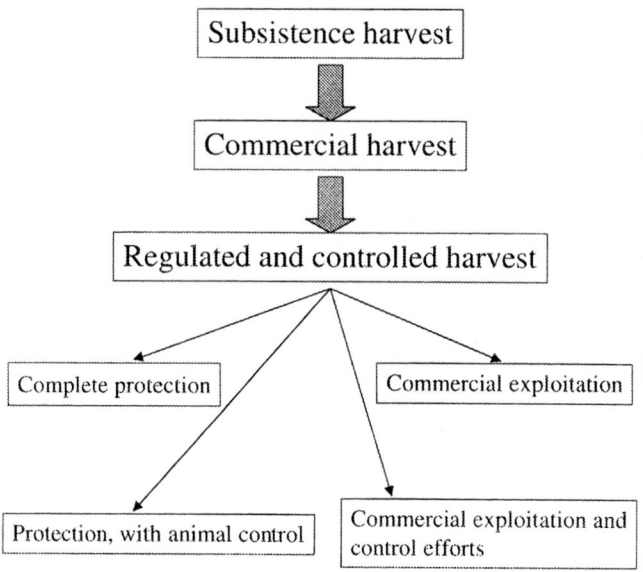

Figure 10.3. Commonly observed temporal pattern of transition from carnivore exploitation to conservation.

(*Mustela marcodon*), the Caribbean Monk Seal (*Monachus tropicalis*), and the Japanese sea lion (*Zalophus californianus japonicus*) (Nakamura, 1989; Groombridge, 1992; Malik *et al.*, 1997; Woodroffe, Chapter 4). Many other carnivores have experienced large reductions in numbers and distribution due to human exploitation and predator control. For example, the lion (*Panthera leo*) was gradually removed from Europe and the Middle East 2000–3000 years ago, in part due to sport hunting. Only a small, genetically very homogeneous population of Asian lions remains in the Gir forest of India (O'Brien *et al.*, 1987c). Brown bear (*Ursus arctos*) and grey wolf (*Canis lupus*) were totally eradicated from large parts of North America and most of Europe by 1900 through hunting and poisoning due to their threat on humans and livestock (Elgmork, 1994; Craighead & Vyse, 1996). Wolves were deliberately removed from all of the Japanese islands except Sahkalin in the nineteenth century (Dobson, 1994). Northern elephant seals (*Mirounga angustirostris*) were reduced to 10–20 individuals during the nineteenth century from overexploitation for their blubber (Hoelzel *et al.*, 1993), resulting in a dramatic reduction in genetic variation (Bonnell & Selander, 1974). Other examples include the grey seal (Mowat, 1984), sea otter (*Enhydra lutris*) (Garshelis 1987), river otter (*Lutra canadensis*)

(Melquist & Dronkert, 1987), lynx (*Lynx canadensis*) (Nowell & Jackson, 1996), wolverine (*Gulo gulo*) (Hash, 1987), and American marten (*Martes americana*) (Strickland & Douglas, 1987).

There are certain patterns of events and conditions that are common to the exploitation and persecution of most carnivores (Figure 10.3). First, humans establish efficient ways to access areas where carnivores are found, and this leads to habitat changes as people arrive and economic activities such as logging, mining, hunting, fishing, agriculture and manufacturing are introduced. Carnivores then come into growing conflict with humans leading to the commercial, subsistence, incidental, recreational, or indirect exploitation or persecution of the animals. Carnivores may also be killed in attempts to control the spread of diseases (e.g. Voigt & Johnston, 1992). With time, available habitat becomes more fragmented and isolated (e.g. Seidensticker, 1986). For most carnivores, this leads to an eventual reduction in population size, which in turn will either result in the local eradication of the population or in a decrease in the levels of exploitation. At any time during this pattern of events, management authorities may intervene and attempt to control exploitation and maintain viable populations. Historically, the first step of game management has been to restrict extractive activities by altering the time, place, or method of hunting or by controlling the number of hunters and their bag limit. However, eventually they will also include the establishment of refuges and protected areas, reintroduction efforts, and habitat restoration and improvement.

From exploitation to conservation: lessons to be learned

There is much that can be applied to carnivore conservation that comes from studies of their exploitation and population biology. The impacts of exploitation on the demographics of carnivore populations can be complex, and are affected by numerous ecological, economic, social, and legal factors (McMahan, 1986; Houji & Heli, 1986). Harvest levels tend to fluctuate greatly over time and vary considerably from one area to another. Historically, the largest impact has been on species exploited for commercial gain for their pelts and blubber. The impact has been especially acute on species with low densities, low birth rates, and high economic value. Species with larger population sizes, higher birth rates, lower economic value, and/or which are more elusive have fared better. Species in the latter category that have been heavily exploited or persecuted include the bobcat (*Lynx rufus*), raccoon (*Procyon lotor*), red fox (*Vulpes vulpes*), culpeo zorro, and coyote (*Canis latrans*).

The impact of carnivore control efforts is more difficult to evaluate. For some species, such as the coyote, dingo, and red fox, it is clear that extraordinary efforts are required to completely eliminate the species from an area. In these species, moderate harvest rates can easily be compensated for by productivity or immigration and broad, sustained control efforts are needed to maintain reduced populations (Harris & Saunders, 1993; Reynolds & Tapper, 1996). For others, such as the wolf, brown bear, puma, and jaguar, a combination of habitat destruction, reduction in prey availability, and direct persecution can eradicate populations.

Populations respond to human induced mortality from harvest or control in many different ways. One of the most common responses is to increase recruitment. In larger carnivores, increasing total reproductive output more often depends on changes in pregnancy rates, especially among juveniles, and not on changes in litter size (Brand & Keith, 1979; Pils *et al.*, 1981; Allen, 1984; Fritzell *et al.*, 1985; Johnson & Halloran, 1985; S. Harris & Smith, 1987). For example, in northern fur seals, harp seals, and southern elephant seals, sexual maturity in both sexes tends to occur earlier in new, less crowded rookeries or in rookeries reduced by hunting, leading to more rapid population growth rates (Riedman, 1990). However, in other species, there can be a slower recovery from intensive harvest, perhaps due to a high maternal investment in pups, a relatively rigid breeding cycle, and an extremely low frequency of viable multiple births (Estes, 1979). Immigration and emigration are also important density-dependent mechanisms that can compensate for heavy exploitation, especially in mobile species. In heavily harvested populations, emigration may become insignificant or be delayed due to the increased availability of resources (S. Harris & Smith, 1987). Conservation efforts to increase carnivore populations might thus benefit from actions that shorten generation times or increase reproduction rates.

Carnivore populations can also respond to exploitation through compensatory mortality, where harvest simply replaces other forms of death, thereby limiting its effect on populations (Errington, 1956). However, several carnivore studies have found that harvest can be more additive than compensatory. In a heavily harvested population, mortality in young raccoons (less than one-year-old) increased in an additive manner, compared with adults, that showed no increase in mortality (W. Clark *et al.*, 1989). Harvest mortality of lynx may also occur in addition to natural causes (Brand & Keith, 1979; Bailey *et al.*, 1986). For population managers managing species for which additive mortality is operative, even low levels of mortality from harvest or poaching can be detrimental.

In addition to affecting population size directly, demographic changes can also result from the preferential harvest of certain sex and age classes of carnivores. For example, adult male pinnipeds have been harvested for their penises and harp and hooded seal pups for pelts. As adolescent and sub-adult males remain vagrant and adult males tend to haul out singly or in isolated areas, hunting of males for penises predominantly impacts breeding males and can directly lower productivity. Other forms of harvest, such as for pups usually involves killing the mothers, as well (Reeves & Ling, 1981), and the loss of some proportion of pups due to orphaning. These age and sex shifts may cause further impediment to recovery in species with distant recruitment pools or small isolated breeding populations (Avise, 1996).

Similar patterns of demographic impact occur in terrestrial carnivores. A study on the effect of exploitation of martens along trap lines found biased age and sex ratios, with larger numbers of young animals recolonizing the intensively harvested locales from areas that had not been trapped (Quick, 1956). Similar age and sex bias was documented in a Japanese brown bear population (Mano, 1995). In many carnivore species, males disperse farther and in greater numbers than females, perhaps resulting in additional differences in genetic and demographic patterns (see review by Waser, 1996). For example in a study of urban red fox, 73% of male juveniles dispersed compared with 32% of females (Harris & Trewhella, 1988) and dispersed longer distances (Trewhella et al., 1988).

The effects of exploitation and control vary among species. For example, at least one species of fur seal and one species of sea lion are sympatric in several locations around the world. For every life history parameter measured for such pairs, the fur seal species is significantly more successful than the sea lion (R. Merrick, pers. comm.). Reasons for this phenomenon are unclear though this situation suggests that fur seals may be more adaptable to adverse conditions thus predisposing them to better recover from exploitation. Though fur seals and sea lions alike were heavily hunted in the Southern Hemisphere, more populations of fur seals may have survived in fragmented and isolated ranges as sealing became less productive per unit of effort. This may have been due to the type of habitat preferred by fur seals. Southern hemisphere fur seals, in general, prefer rocky outcroppings or near off-shore rock piles for haul outs and rookeries, whereas sea lions prefer flat, sandy beaches making them more exposed to capture (Lento, 1995).

Responses to exploitation and control may be hard to predict in many situations, however, because of complex interactions among the various population parameters. For example, most forms of carnivore exploitation are spatially nonrandom. The heavy harvest of certain populations, areas, or habitats can affect the demographic connectivity of populations and the ability of individuals to disperse, creating or exacerbating the complexity of a metapopulation structure and increasing the probability that harvest mortality may be additive instead of compensatory. Aspects of this scenario have been studied by some authors with source and sink population models incorporating differing rates of productivity, survival, and dispersal (Pulliam, 1988; Danielson, 1991). This approach has not often been tested empirically, but the role of metapopulations in the dynamics of puma (*Puma concolor*) and brown bear (*Ursus arctos*) populations has been discussed and modeled (e.g. Doak, 1995; Beier, 1995).

From exploitation to conservation: is it inevitable?

One of the controversies concerning exploitation is whether it can be sustainable (Malik *et al.*, 1997; Willers, 1994; Geist, 1994). The numerous ways in which North American wildlife has been used over the last 70 years has been offered as an example of exploitation that is both biologically and economically sustainable (Geist, 1994). The success of this exploitation has been attributed to several factors, including the public ownership of wildlife, the elimination of a commercial market for the parts of vulnerable wildlife species, the centralization of wildlife laws, and the prohibition of frivolous killing (Geist, 1988). The sustainability of exploitation is also facilitated when the commercial value of the wildlife products is low relative to alternative sources of income and when the investment by the hunter or trapper in terms of time and money is high (Lavigne *et al.*, 1996). Utilization is also more likely to be sustainable when supported by increased economic benefits derived from related activities. However, in much of the world, a legal market for wildlife products, either nationally or internationally, especially when valuable, promotes overexploitation and illegal poaching (Ehrenfeld, 1970). In pinnipeds, even the sustainability of activities such as ecotourism has been questioned. It has been suggested that because ecotourism obligates tour operators to be stakeholders in the welfare of the target species, this may act as a conservation measure in the long run, but only if it proves not to be a detriment to the animals (Constantine, 1999).

Genetic consequences of exploitation and control efforts also need to be considered. An important way in which exploitation affects populations is by reducing the effective population size (N_e), which can occur in several ways. First, overexploitation can lead to dramatic reductions in total numbers. Fluctuations in population size induced by cycles of overharvesting and recovery will decrease N_e to the harmonic mean of population size over time, which will be well below the maximum observed population size

(Frankham, 1995). In addition, nonrandom harvesting can skew sex and age ratios and can increase variation in family size (reproductive success). These factors, by reducing N_e relative to N, make it necessary to have a larger number of individuals to effectively conserve genetic diversity over time. For example, in an exploited lynx population with apparent male-biased harvesting, it was suggested that although populations appeared to have increased with harvesting, effective population size may actually have been reduced (Quinn & Thompson, 1987).

If effects of harvest are compensatory and random, instead of additive and nonrandom, harvesting could have little to no effect on genetic structure. However, exploitation can be a disruptive factor in populations if selection is biased towards certain age, size, experience, and behavioral classes. Theoretically, exploitation can thus also serve as a new agent of natural selection, resulting in an evolutionary pressure that might be in a different or opposite direction than previously encountered and which could affect (reduce) the fitness of the population.

From exploitation to conservation: towards the future

The lack of sufficient data with which to anticipate the impact of exploitation, population control, or conservation efforts is one of largest problems facing wildlife managers (Harris & Saunders, 1993; Singer, 1994; Weber & Rabinowitz, 1996). Field research is crucial to understanding the basic ecological requirements of exploited species. However, it is as important to monitor the results of management plans, not only to be able to modify these actions, but also to learn from them. To this end, long-term data sets are particularly valuable. Good data have an additional value when enacting management plans in that some of the paralyzing debates over policy decisions can sometimes be avoided when discussing facts instead of dealing with opinions. Since in the future we will probably be making active management decisions about all carnivore populations, we must accept the responsibility to do this as well as possible. This implies being able to collect the data with which to make these decisions.

To improve our ability to manage carnivore populations, both for harvest and conservation, research is needed in several areas. Some of the needed research efforts are species-specific while others are more broadly applicable. For many carnivores, basic information on life history parameters and population ecology is still unknown, much less how they will respond to exploitation. The management of many carnivores would benefit from research similar to that which has been conducted on wolves, which has allowed simple models to be written estimating sustainable mortality rates, and estimating the sizes of wolf populations given available ungulate biomass (Fuller, 1989). These data would then allow managers to predict the resiliency of carnivores to exploitation, factoring in their aptitude to alter their behavioral patterns, their capacity to compensate demographically to increased exploitation levels, and their ability to disperse across different habitats, distances, and barriers (Weaver *et al.*, 1996).

More research is needed to compare the effects of different exploitation techniques on different species. For example, whether hunters are using dog, baiting, stalking, or ambushing techniques will influence the age and sex of black bears that are killed (Litvaitis & Kane, 1994). Often only the number of animals harvested is taken into consideration when assessing whether populations are threatened or endangered. However, harvest based on reproductive potential but which does not consider behavioral aspects might severely disrupt relationships in regard to territories or social groups, as with wolves (Haber, 1996). More research is also needed to compare exploited and non-exploited populations of the same species, in similar environments.

One of the most fundamental tools missing for managers is a method of obtaining reliable population size estimates. Increased emphasis needs to be placed on the development of both direct and indirect methods of monitoring populations using new technologies. This should not only include demographic characteristics, but also genetic aspects such as reliable estimates of effective population size and the amount of gene flow among areas, as well as approaches that assess the prevalence of pathogens and their impact on the population. It is important to maintain the integrity of a functional ecosystem, since exploitation can change the ecological relationships in a community. For example, small carnivores have been shown to benefit from the loss of larger ones (see Johnson *et al.*, 1996).

More research is needed on the importance of heterogeneous harvest levels, which leads to source–sink dynamics. This is also related to the study of the utility of refugia. For example, the availability of refugia of adequate size has been shown to be important for the maintenance of black bear populations (e.g. Elowe & Dodge, 1989; Hellgren & Vaughan, 1989; Powell *et al.*, 1995). Models have predicted that the appropriate size of a refugium for bobcats depends on the intensity of the exploitation in the areas separating them (Knick, 1990). A similar model predicted that multiple refugia dispersed so that they permitted complete mixing among protected, and exploited populations would lead to sustainable harvests while minimizing the probability of extirpation (Joshi & Gadgil, 1991).

CONCLUSIONS

Our future interactions with carnivores will evolve along with shifts in our attitudes and values. Rarely will we have the luxury of deciding between interfering or not in natural systems. By altering one part of an ecosystem, invariably we are forced to make decisions of how to control other components. This often includes the control of carnivores. The need to control or the desire to exploit populations increases as wildlife populations come into greater contact with humans through habitat alteration and human encroachment into carnivore habitat. Experience with several carnivore populations demonstrates that they could potentially increase in size, as long as illegal harvest is prevented and adequate management practices, including harvest, are enforced.

From our knowledge of the effects of exploitation on carnivores, we can begin to classify the degree and type of management that each species will probably have considering the current exploitation and the goals of maintaining viable populations. Carnivores can be divided into several categories in regard to their most common pattern of control and exploitation (Table 10.1). To ensure their survival, some need to be completely protected and sometimes actively managed. These will generally be found in small, isolated populations and have specialized ecological requirements and/or low intrinsic growth rates. Species included in this category are small, rare terrestrial carnivores such as some small cats and the black-footed ferret (*Mustela nigripes*). These may or may not require large amounts of contiguous habitat, depending on their density, ecological requirements and dispersal abilities. Some species will also require a certain amount of control in addition to thorough protection. They will include some of the larger species such as tigers, brown bears, and jaguars, which in most portions of their range will unlikely ever to be common enough to permit controlled harvest, but which may require selective culling in certain populations, especially of 'problem animals' (see Mech, 1995). As predator control is also expensive, these activities will generally be restricted to agricultural areas. A second group of carnivores will be composed of species that may allow sustainable harvests in some areas. These populations will generally be large and distributed over broad geographic areas, and present rapid growth rates and some economic value. Not all of these species will necessarily be harvested, however, for ecological, political or social reasons, and any exploitation activity will surely require concurrent research, monitoring, and regulatory efforts. A subset of this group will have no intrinsic economic value and may be controlled as pests. These species

Table 10.1. Selective list of carnivore species and their most common category of exploitation throughout their distribution

---

Complete protection

Simien jackal	*Canis simensis*
Iberian lynx	*Lynx pardinus*
Giant panda	*Ailuropoda melanoleuca*
Giant otter	*Pteronura brasiliensis*
Australian sea lion	*Neophoca cinerea*
Auckland sea lion	*Phocarctos hookeri*
Mediterranean monk seal	*Monachus monachus*
Hawaiian monk seal	*Monachus schauinslandi*

Protection, but with selective animal control

Jaguar	*Panthera onca*
Tiger	*Panthera tigris*
Asiastic black bear	*Ursus thibetanus*

Commercial exploitation

Golden jackal	*Canis aureus*
Red fox	*Vulpes vulpes*
Racoon dog	*Nyctereutes procyonoides*
Canadian lynx	*Lynx canadensis*
Bobcat	*Lynx rufus*
Puma	*Puma concolor*
Raccoon	*Procyon lotor*
Siberian weasel	*Mustela sibirica*
American mink	*Mustela vison*
European pine marten	*Martes martes*
Sable	*Martes zibellina*
American pine marten	*Martes americana*
Fisher	*Martes pennanti*
Striped skunk	*Mephitis mephitis*
River otter	*Lutra canadensis*
Sea otter	*Enhydra lutris*
Baikal seal	*Phoca sibirica*
Caspian seal	*Phoca caspica*

Commercial exploitation and control efforts

Gray wolf	*Canis lupus*
Black-backed jackal	*Canis mesomelas*
Coyote	*Canis latrans*
Dingo	*Canis familiaris dingo*
Culpeo zorro	*Pseudolopex culpeus*
American black bear	*Ursus americanus*
Brown bear	*Ursus arctos*

(From Ginsberg & Macdonald, 1990; Nowak, 1991.)

include the dingo, coyote, and caracal. A final group of species is composed of those that generally are not exploited and which are not in need of conservation action.

The three goals of minimizing the potential adverse effects of carnivores, maximizing the benefits that they provide, but also ensuring their long-term conservation are often seen as conflicting with each other. However, each of these disciplines or approaches has much in common as they are all fundamentally involved in population management (Shea *et al.*, 1998). For example, the successful reestablishment and conservation of many carnivores will eventually require some control of these same species. Alleviating the anxieties of the local human community can be one of the most crucial steps in promoting the successful reintroduction of carnivores. The debate over sustainable use of carnivores can also add a scientific framework to discussions of wildlife management. The process of regulated exploitation and harvest may help enlist public support for conservation efforts, foster local participation in finding solutions to conservation and management problems, and provide monetary compensation to the community for the presence of a carnivore population.

Exploitation of wild animals is an emotional issue, which has been the subject of intense debate on moral, pragmatic, and economic grounds (see Robinson & Redford, 1991; Swanson & Barbier, 1992; Taylor & Dunston, 1996). Humans affect all wildlife species and all wildlife management has animal welfare implications (Taylor & Dunston, 1996). For wildlife utilization to be compatible with conservation, it will probably be through broadly based ecosystem and cultural approaches. This will lead to the creation of the necessary infrastructure and trained personnel for the management and preservation of all species in a multi-use, multi-species approach. Carnivore conservation will ultimately depend on the collective education of consumers of carnivore products, resource managers, and those communities living in closest proximity to and interacting most directly with these animals.

Acknowledgements

We would like to thank J. Slattery and several anonymous reviewers for providing helpful comments on this manuscript. We also appreciate the support of M. Carrère and L. Utz.

PART 2
Some approaches and solutions

11

Interdisciplinary problem solving in carnivore conservation: an introduction

TIM W. CLARK, DAVID J. MATTSON, RICHARD P. READING
AND BRIAN J. MILLER

INTRODUCTION

The goal of carnivore conservation is to reverse declines in populations and to secure remaining populations in ways that gain enduring public support. As noted by Minta *et al.* (1999: p. 374), 'Clearly, carnivore conservation rests both on reliable scientific information and informed public consent.' Inasmuch as these species play keystone roles in ecosystems, indicate the health of ecosystems, function as umbrella species, focus public attention on conservation, and garner support for broader conservation efforts, conserving carnivores pertains to more than just saving the animals (Terborgh *et al.*, 1999). But despite some progress, such as the wolf reintroduction into Yellowstone National Park, there are very few places where long-term conservation is assured. Although it is generally recognized that there are significant 'human dimensions' involved, little attention has been devoted to these factors (T. Clark *et al.*, 1996a; Weber & Rabinowitz, 1996; M. Miller, 1999). Ignoring or overlooking key variables can lead to inaccurate definitions of the problem, inadequate solutions, and continuing losses.

A more comprehensive, contextual, and rational approach to carnivore conservation is urgently needed. Concerned people must conceive of conservation as a process of human decision-making, upgrade this process to achieve better outcomes, and by this means change the human practices that threaten carnivores. The decision process is the whole sequence of actions and events by which human communities identify and solve problems that hinder achievement of their goals. Functionally speaking, they first recognize a problem, study it, debate and promote various alternatives, and prescribe new rules. Then there is a process of invoking the new prescriptions in specific contexts, enforcing sanctions, applying new standards, administering them, and resolving disputes. The decision process also encompasses monitoring or appraising the new rules, and finally terminating these rules and moving on to new ones. We review and analyze problems of carnivore conservation by looking at case studies, and we synthesize a view of some basic problems that must be overcome to achieve better outcomes. We introduce an interdisciplinary approach that, by focusing on the decision-process problem at hand, encourages users explicitly, empirically, and systematically to address the full set of variables at play in any context within the limits of time and resources. The approach is practical, effective, and widely applicable (see T. Clark *et al.*, 2000a).

CASE STUDIES IN CARNIVORE CONSERVATION: SURVEYING SOME CHALLENGES

Three cases, briefly discussed here, suggest the range of factors at play in actual conservation, most of which are human factors that are not formally addressed by biology-based, scientific approaches, but which nonetheless have real and profound consequences.

Grizzly bears

Ursus arctos horribilis exist in the contiguous United States as two large and three small populations totaling fewer than 1000 individuals. Grizzlies were listed as threatened under the US Endangered Species Act in 1975 because of threats posed by human-caused mortality and loss of secure habitat. Key management documents produced by the decision process include the Grizzly Bear Recovery Plan (US Fish and Wildlife Service, 1993), standards and guidelines specified in national forest management plans (most current grizzly bear range occurs within US national forests), guidelines for the management of 'problem bears,' and a recent conservation strategy. The ability of these prescriptions to achieve bear conservation is debatable, as is the adequacy of the decision process (Wilkinson, 1998).

Stark differences in perspective exist about the quality of the decision process. This takes the form of intense and public debate about the adequacy of the science, the goals of recovery, the openness of the debate, the adequacy of implementation, and the ability of appraisals to provide realistic feedback. There is excessive, unproductive conflict and growing litigation as evinced in news articles, scientific journals, and court proceedings. The grizzly bear recovery process falls short of meeting recognized decision standards, such as being factual, comprehensive, rational, integrative, inclusive, timely, contextual, and ameliorative (see Lasswell, 1971). Critics charge that government researchers and managers chronically overlook and underestimate human dimension issues and that no mechanism exists to integrate such information (Primm, 1993, 1996; Mattson & Craighead, 1994; Mattson 1995, 1996, 1997; Mattson *et al.*, 1996a,b).

Many value-based decisions are miscast as science-based issues, such as the number of bears and the quality of habitat conditions that are sufficient for recovery and related specification of time frames and appropriate levels of risk (see Shaffer, 1992). Even where scientific information has had a legitimate role to play, its influence has sometimes been overridden by other values, such as power interests, agency loyalty, or responsiveness to elected officials (see Mattson & Craighead, 1994; Mattson, 1996, 1997; Mattson *et al.*, 1996a). Recovery is defined in terms of certain demographic goals and habitat conditions, and thus couched in the language and outlook of science. The debates reveal little appreciation that all scientific models and projections are based on human value judgments about acceptable risks, probabilities, and thresholds, and that management activities are based on multiple and often competing values at different scales of society. Specific demographic and habitat-related objectives vary, sometimes substantially, from one bear population to another. Even when the goals are agreed upon, goals are only descriptions of what needs to be done, and the means of achieving them may remain in dispute. Habitat goals are expressed in terms of security from humans and specified in terms of maximum acceptable road densities and minimum percentages of bear range that are free of roads. Research is underway to justify demographic goals in terms of population viability analysis.

Grizzly conservation draws diverse participants, holding differing perspectives and values, interacting in complex and shifting situations, using diverse strategies, and seeking different outcomes. The Grizzly Bear Recovery Plan (US Fish and Wildlife Service, 1993: p. 29) acknowledges the influence of human contextual factors in a diagram taken from Kellert & Clark (1991), but study of these factors has been consistently *ad hoc* and incomplete.

Without exception, state and federal agency officials whose primary job is grizzly bear management were trained in traditional wildlife biology. Many were also involved in wildlife research at some point in their careers. Although top-level managers with more diverse backgrounds substantially set and influence the implementation of grizzly bear policies, biologists-turned-managers hold primary responsibility for developing programs, writing documents, and guiding their implementation. According to Mattson *et al.*, (1996a), most agency officials seem to be highly responsive to power incentives, particularly individuals both inside and outside the agencies who control resources and opportunities for them. As a consequence, a complex human social dynamic holds sway over the decision process. Grizzly bears are at additional risk because of antagonism among participants and because all parties are held hostage to larger political agendas such as states' rights and national political issues (Primm, 1993, 1996; Mattson *et al.*, 1996a; Primm & Clark, 1996). No study to date has provided an adequate basis for anticipating and minimizing conflict or systematically improving the decision process in this case.

Jaguars

Panthera onca once ranged from the southwestern United States throughout the Americas to northern Argentina (Seymour, 1989). The species is threatened by the fur trade (now largely curbed via CITES), habitat destruction, direct persecution, and declining prey (Weber & Rabinowitz, 1996).

The decision process surrounding jaguar conservation is complex and less than effective. It is highly fragmented and, in many ways, under-organized. There is a general lack of biological knowledge to use in decision-making, and the level of uncertainty sometimes heightens the conflict that arises when people try to clarify and resolve issues. It is very costly to undertake the long-term studies needed because of the animals' low population densities and extensive ranges, and there are relatively few trained biologists to undertake the work (Mares, 1991). Most planning is therefore local and may be driven more by opinion than systematically gathered, reliable knowledge.

There is presently no unified strategy across the species' range, although efforts are underway to address this problem. Some nations, such as the United States and Mexico, list the jaguar as endangered, while others, such as Bolivia, permit hunting. Different groups and nations operate from different goals and problem definitions; indeed, one may see a problem where another sees a desired situation. In addition, administrative boundaries may not reflect informal levels of authority and control within a region. For example, large areas may be controlled by groups that have little to do with official government wildlife policy (e.g., drug cartels or political rebellions). The major threats to jaguars are rooted in human activities, which derive from the values that people place on resources. First, deforestation rates in Latin America vary by country, but are generally regarded as among the highest in the world. Second, human population growth, poverty, devaluing economies, and social conflicts have pushed people further into the tropical forests, and their main activity there is subsistence agriculture. Third, many tropical countries have huge international debts, and 45% of the economies in all developing countries rely on resource extraction, a situation that leaves little alternative but to destroy the environment for short-term survival. Fourth, economic globalization (e.g., treaties like NAFTA) encourages conversion of habitat to fulfill growing export markets for fruit and produce (Soulé & Noss, 1998). Fifth, persecution of jaguars is usually not penalized, even in countries that protect them. There are typically few individuals assigned to enforcement, salaries for such jobs are barely livable, and the legal system in rural areas does not yet play a strong role in society (Shaw, 1993). Sixth, jaguars are largely viewed as a threat to livestock. When humans expand their activities deeper into the forest, they kill the natural prey of jaguars and introduce domestic animals so that jaguars must either leave or switch to eating livestock. Relocating problem animals does little to address the conditions that created 'stock-killers.'

Finally, the perspectives of participants are problematic. Latin American wildlife agencies are often understaffed, causing delays in information gathering and planning. Because political appointees staff these agencies, important wildlife positions may go to people who are not necessarily well trained in biology, much less management of decision processes (Mares, 1991). In general, Latin American agencies are organized as top–down bureaucracies with complicated rules and slow processes that may impede conservation. In addition, if there is a national political change, entire institutions may be replaced. Economic devaluation is a constant threat, and unstable economies make it difficult to commit long-term funding to conservation.

Mongolian gray wolves
Canis lupus management follows a classic pattern of controlling large carnivores for the sake of game management (National Research Council, 1997). The postulate is that killing predators will result in larger numbers of prey species, a theory that prevails among game biologists everywhere. The literature refers to wolves as 'enemies' of wild ungulates (e.g., Dulamtseren, 1970) and the 1997 Red Book lists wolves as a threat to several species. Although wolves remain distributed throughout most of Mongolia, they have declined in recent decades and are actively hunted by herders and biologists alike (Reading *et al.*, 1998).

Wolf management is ostensibly based on biology, but since little is known about Mongolian wolf biology or the impacts of wolves on livestock or wild ungulates, non-biological considerations drive management. Biologists and wildlife managers do know that wolves occasionally kill both livestock and wild ungulates. Nomadic herders, who kill wolves to reduce livestock depredation, hold significant political power based on tradition.

The vast majority of decision makers and biologists in Mongolia also advocate killing wolves (Reading *et al.*, 1998). As result, wolves are persecuted through bounties in the name of livestock and game protection. For example, the recently formed Association for Protecting Livestock from Wolves holds periodic wolf hunts that are encouraged by the local press. Moreover, a traditional Mongolian myth holds that the person who kills a wolf gains that animal's strength and prowess.

The single most important factor influencing wolf management is the values and perspectives of Mongolians, from nomadic herders to managers to biologists. Wolf control is not a contentious issue – current practices are not questioned and little, if any, study of wolf management or alternatives occurs. The widely accepted goal of reducing wolf numbers appears to be succeeding (Reading *et al.*, 1998). Changing wolf management from active, widespread control activities to conservation will therefore require changing perspectives and underlying values – an extremely important, non-biological challenge. Furthermore, wolves are little studied, and non-biological considerations receive almost no attention.

Finally, wildlife biologists and managers in Mongolia receive traditional wildlife management training, several traveling to Russia (or the former Soviet Union) for higher university degrees. Until the last few years, coursework was strongly disciplinary, usually restricted to zoology, with little to no broader training in other fields, not even ecology. As a result, wolf management remains strongly traditional, with little to no questioning of management goals, underlying assumptions and theories, or methods. This traditional biological approach dominates despite a paradigm shift underway among biologists elsewhere in the world toward a more ecosystem-scale view of wolves and large carnivores in general (Clark *et al.*, 1999).

Other cases

Overall, carnivore conservation worldwide has been characterized, at least until recently, by insufficient attention to human dimensions, with disastrous consequences in some carnivore conservation programs (e.g., Lopez, 1978; Reading & Clark, 1996). For example, wolves reintroduced into the Upper Peninsula of Michigan (USA) during the mid-1970s were all killed within eight months of release. Hook & Robinson (1982: p. 382) examined local attitudes following the release and concluded that 'the wolf's future in Michigan depends upon the attitudes of Michigan residents.' Similar problems were encountered in the Florida panther (*Puma concolor*) recovery program following an experimental release of five panthers into Osceola National Forest in 1988 (Beldon *et al.*, 1990) and in the recent reintroduction of Mexican wolves (*C. l. baileyi*) in Arizona, where some released animals were shot. Other cases did attend to contextual factors. For example, a proposed red wolf (*Canis rufus*) reintroduction was moved from a site with an antagonistic public to a site with a more supportive public (Moore & Smith, 1991). The reintroduction of gray wolves into Yellowstone National Park was proceeded by massive attention to local values and attitudes. Both of those programs appear to be succeeding (D. Smith & Phillips, 2000; Kelly & Phillips, 2000).

SYNTHESIS AND ANALYSIS: THE NATURE OF CARNIVORE CONSERVATION

The cases show many of the 'real-life' factors at play in carnivore conservation. In addition to the obvious biological problems of carnivores, including small populations at low densities in fragmented, declining habitats, there are weaknesses in the decision process – the process by which people conceive of problems and carry out solutions – that greatly influence whether conservation programs will be effective. We briefly examine three of these.

Bounded professional perspectives

Most threatened carnivores live in human-influenced environments, and no amount of fixing of the biological problems will help unless the human-caused conditions of carnivore endangerment are also remedied. Professional attitudes too often dismiss the decision process and its effect on carnivores simply as 'politics' – an impediment to scientific or rational progress – and then discount it as a subject for scientific study or management by someone else. But, as Weiss (1989: p. 117) points out, 'Difficult to measure and understand is not the same as unimportant.'

Narrow, discipline-based, technical perspectives of those who figure most prominently in carnivore management and conservation – managers, researchers, government agency leaders, non-governmental advocates – may lead to defining problems and crafting solutions largely in technical terms of animal behavior, population dynamics, or habitat relations or in terms of data collection and modeling. To compound matters, most carnivore work is done through bureaucracies, which, according to Finlayson & McMahon (1994), exacerbate conservation crises with their cumbersome structures and operating procedures. When professionals act on narrow disciplinary and bureaucratic definitions and constructions, there is a mismatch or incongruence between their actions and the actual conservation challenge.

The predominance and persistence of less-than-comprehensive

methods for solving complex societal problems, we suggest, is the result of the 'bounded rationality' of all participants. Herbert Simon (1983), who coined this phrase, noted that because the world is more complex than we can comprehend, we filter out important stimuli and operate on simple representations of reality. These filters, largely subconscious, consist of each individual's values, disciplinary training, epistemology, parochial interests, organizational allegiances, and other factors. People bounded by biology tend to define conservation problems as biological constructs. Bounded rationality can impede creative thinking and effective problem solving (A. Miller, 1999). This is most clear in the Mongolian wolf case, where the extremely narrow training of wildlife researchers, the fact that many managers have no biological background at all, and the strength of the bureaucracy mean that there is no consideration of alternatives to the widespread, very traditional (even folklore-based) belief that wolves should be killed. Jaguar conservation, on the other hand, will likely be impeded by the large number of varied perspectives, many of which are likely to be bounded by geopolitical focus, biological training, organizational allegiances, and other parochializing factors.

Neglect of decision process

Decision-making is always much more about values than it is about 'facts' as scientifically understood. It is in this arena that science, analysis, and politics merge. Through this process people must mesh their different perspectives, values, and interests to find solutions to shared problems (T. Clark & Brunner, 1996). Whether acknowledged or not, all knowledge and all choices have social, political, and value content and consequences. Decisions cannot be made by neutral decision makers because neutral human beings do not exist (Primm & Clark, 1996).

Every carnivore conservation effort entails decision-making processes through which people attempt to solve a problem and choose a course of action. By knowing how the decision process is set up and how to judge whether it is working well, participants can insist on good practices and fix poorly performing processes. Decision processes can be empirically studied and managed for the benefit of carnivores as well as humans. Good problem solving entails asking, for instance, why certain prescriptions for carnivore conservation are developed and maintained, what information exists, what alternatives should be considered, and what their consequences might be, not only for the animals but also for people and institutions, how the process will be evaluated, corrected, and ended. Answering such questions requires an explicit conception of the decision-making process, adherence to recognized standards of decision making, and skill in managing decision processes.

Although the decision process for jaguar conservation on an international scale is just beginning to coalesce, grizzly bears are the subjects of a highly organized decision process, in which certain weaknesses have been institutionalized. Few participants 'see' the full decision process they are caught up in, acknowledge standards against which their activities could be measured, or envision any way to intervene to reduce the conflict, build trust, and clarify and secure common interests. Instead, they see only that one agency or group is deliberately thwarting the aims of another, or that that there are dueling data sets with participants siding with one or the other, or in more personal terms, that one side is 'right' and the other 'wrong.' These myopic views lead to interactions and debates focused on technical matters and protection of personal, professional, or organizational 'territory' (Mattson & Craighead, 1994).

Improving the decision process is not merely a matter of getting the biological science right and having decision makers act on it. Many scientifically-trained professionals dismiss value-related considerations as beyond their scientific and professional responsibility and fail to recognize the values that underlie their ostensibly 'neutral' science or policy perspective. Assuming their work to be value free, many simply expect that their recommendations will be heeded by the public and decision makers (see Brewer & Clark, 1994). But this presumes a relationship between science and society that is little supported by abundant case material and hard-won experience. This presumed relationship – an uncritical and trusting client (society) depending on the authoritative input of expert, value-free, objective science – is deeply institutionalized in the professional norms, social institutions, and organizations that select professionals to do conservation work. Copeland & Lewis (1997: p. 304) call these assumptions into question by asking whether scientists are adequately trained with the necessary skills to ensure the effective use of their information. Failure to recognize, disclose, and compensate the value positions of all participants can heighten and personalize conflict by shifting the focus of debate to the competence or integrity of those people or organizations at odds. It is rare to find programs that explicitly and systematically address the perspectives and values of participants, rarer still for professionals in carnivore conservation to examine their own standpoints relative to the decision processes in which they participate.

Underappreciation of the human context

Human perspectives and practices are the ultimate cause of the problems that grizzly bears, jaguars, wolves, and many other carnivores face – perspectives that value other things more highly than carnivores and practices that are directly and indirectly destructive. Human-caused mortality is often driven by markets for pelts and other body parts, fear, livestock conflicts, usurpation and destruction of habitats, and human depredation on their prey. Even the severity of diseases, such as canine distemper or rabies, has been worsened by reductions in carnivore populations and interactions with domesticated animals. At the same time, conservation-oriented perspectives and practices are the source of solutions to these problems. Saving these species comes down to changing what, how, when, and where humans do the things they do. The numbers and distributions of people and livestock, the presence, size, and extent of human disturbances, and the outlooks and activities of people living near carnivores become very important. The task of an interdisciplinary professional is to be realistic about the context.

Local contexts are always nested within larger contexts. Programs to reduce human disturbances must be based on information and a humane concern for all those affected by the program. Such programs are contingent on the actions of many people and organizations, such as game or conservation departments or non-governmental environmental groups, which in turn are affected by factors such as the culture, structure, and resources of organizations, relations among subcultures, and external relations with powerful elites. The policies that guide organizations that manage wildlife (and human action with regard to wildlife) are the outcomes of larger political and economic processes devoted to determining who gets what resources, how, when, and where (Lasswell, 1971; Kellert *et al.*, 1996). At the largest scale, carnivore conservation problems are enmeshed within global social phenomena, such as the growing interdependence of human communities and economies and the global movement of people, goods, and services.

Based on the consensus that prevails with regard to exterminating Mongolia's wolves, it is legitimate to ask why conservation efforts should be initiated there or how they could possibly succeed. Yet, the overriding contextual factor is the global losses of this species and large carnivores in general that compel conservationists both within and outside the country to persuade Mongolia to change. The jaguar case is a prime illustration of the large number of contextual factors, from regional rebellions to international debt structures, that influence people's perspectives and practices as they pertain to large carnivore populations.

CONCLUSIONS AND RECOMMENDATIONS FOR CONSERVATION: A PRACTICAL INTERDISCIPLINARY APPROACH

Carnivore conservation poses a complex, interdisciplinary challenge because diverse human factors threaten many species and their habitats worldwide. We maintain that the overall goal is to conserve carnivore communities, while gaining lasting public support. Many problems prevent achievement of this goal, among them traditional professional perspectives, underappreciation of the human context, and disregard for the central role of the decision process. While the biological sciences are necessary for conservation, they are not sufficient. The best way to address these three problems is with an interdisciplinary approach (although we agree that 'decades of lip service to the idea of interdisciplinary research have so far not succeeded in generating much activity,' Finlayson & McMahon, 1994: p. 48).

The interdisciplinary approach we recommend offers a way to move forward practically. It is a means of organizing knowledge for thought and research and of integrating it to solve problems. Its categories serve as a 'checklist' of variables to address in any conservation project, thus enabling users to construct a realistic map of the social context and decision process and to use it to define and solve problems. It is rational, integrated, and comprehensive. As Brewer & deLeon (1983: p. 22) noted, 'other approaches may appear to offer simpler or easier solutions, but each usually turns up lacking in important ways – not the least of these being their relative inability to help one think and understand, and hence to become a more humane, creative, and effective problem solver.'

Recommendations

We recommend viewing conservation as a 'systems' challenge – a system of decision making. The four tasks demanded of professionals by this system are to: (1) establish a standpoint for yourself; (2) carry out problem solving; (3) ensure an adequate decision process; and (4) understand (map) the context of the problem at hand. These systems components are briefly described and interrelated below and in Figure 11.1. A fuller description and examples are in Lasswell (1971) and T. Clark (1997).

1.   **Establish a standpoint for yourself.** All problem solvers have a perspective (or standpoint) in relation to the problem – a product of their training, personal values, and other factors. The purpose of clarifying one's own and others' standpoints is to recognize and avoid

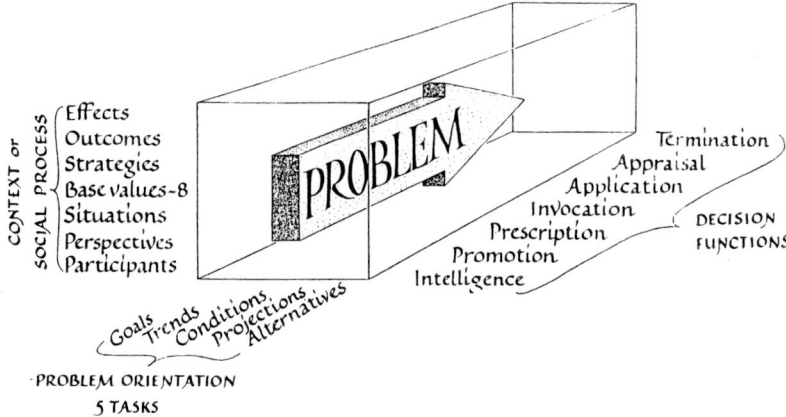

CONTEXT or SOCIAL PROCESS
- Effects
- Outcomes
- Strategies
- Base values-8
- Situations
- Perspectives
- Participants

PROBLEM

Termination
Appraisal
Application
Invocation
Prescription
Promotion
Intelligence
} DECISION FUNCTIONS

Goals Trends Conditions Projections Alternatives

·PROBLEM ORIENTATION
5 TASKS

Figure 11.1. A systems view of carnivore conservation. The three axes show the key dimensions of interdisciplinary problem solving: (1) carrying out the five tasks of problem solving; (2) appraising the adequacy of the decision process; and (3) understanding (mapping) the context of your work and the conservation problem at hand. Also, but not illustrated here, is the task of clarifying your own standpoint (see text for explanation).

unconscious bias and to position oneself to solve problems in the common interest (see T. Clark & Wallace, 1999). Narrow, partisan standpoints will be less effective in resolving complex problems than those that are as free as possible from parochial interests, cultural bias, ideology, disciplinary rigidity, and fixed bureaucratic loyalties. In the broadest terms, we recommend that carnivore conservationists identify themselves as citizens of the global community who strive to maintain environments for a healthy, sustainable future for humanity and all life forms, and that they commit themselves to improving decision processes toward this end.

2.   **Carry out problem solving.** How we characterize problems largely determines how we respond to them. Too frequently, conservationists commence 'biological' solutions before they define conservation problems and their context fully. If we miscast or under-represent what is involved in a problem, we virtually guarantee the misallocation of resources and increase chances of failure.

Problem solving requires five interrelated tasks. Wallace & Clark (1999) offer explanations of the tasks and a worksheet. First, conservation goals, both biological and social, must be specified. Without biological sideboards, socio-economic forces may compromise the ecological health of the ecosystem under study. Although goals may be somewhat general at first, over the course of a conservation program clarification of the social and biophysical context will demand greater specificity in the goals. Second, the history of the situation must be examined to determine if trends in events and processes are moving toward or away from the specified goals. Third, factors, relationships, and conditions behind the trends must be understood, including the complex interplay of factors that affected prior decisions. Models, both qualitative and quantitative, are often useful at this stage. Fourth, trends must be projected into the future, based on past trends and conditions. Problem solvers should attempt to project a wide variety of possible future trajectories and outcomes, rather than accurately predict the 'correct' future. Finally, developing management and policy alternatives is the last task. Interdisciplinary approaches cultivate creativity in inventing new alternatives in policies, rules, norms, institutional structures, and procedures. Alternatives must be evaluated relative to the specified goals.

3. **Ensure an adequate decision process.** Carnivore conservation is concerned with establishing *who* will make decisions about *how* we use resources. Conservationists must successfully influence this process if they expect to save species and their habitats.

Seven activities characterize the decision-making process: data gathering; recommending; setting rules; enforcing rules; administration; monitoring; and ending the rules (Table 11.1). Clark & Brunner (1996) describe these steps, give examples, offer questions to ask about each activity, and list standards for each of the seven functions. There are standards to judge the adequacy of each decision-making function, and applying these can raise the quality of conservation by identifying poorly performing processes and providing a basis for appraisal and upgrading (see Lasswell, 1971). The decision-making process should be, first of all, rational, comprehensive, and integrated. The biophysical and social information included in decision-making should be dependable; if not, some measure or description of uncertainty (or risk) is needed. The decision process should be open and accessible to those with something to contribute or something at stake in the program. Openness also refers to the extent to which the process is available to scrutiny. It should also be inclusive – 'selective omission' may serve personal or special interests and cause unproductive conflict. Timeliness is essential – the lag between perceiving a problem and taking effective action should be as short as possible, and obsolete or ineffectual management practices and policies should be

Table 11.1. *Decision process in carnivore conservation is made up of seven interrelated activities or functions*

- *Planning (Intelligence Function)* is the gathering, processing, and disseminating of information key to decision making. Is intelligence being collected on all relevant components of the problem and its context from all affected people? To whom is intelligence communicated? How would you like to see the planning function carried out?
- *Recommending (Promotion Function)* is the advocacy of specific policies and proposals. Which groups (official or unofficial) urge which courses of action? What values are promoted or dismissed by each alternative and what groups are served by each? How would you like to see the recommendation function carried out?
- *Setting Rules (Prescription Function)* is the authoritative statement of policies. Will the new prescriptions harmonize with rules by which the key organizations already operate, or will they conflict? What rules does the conservation partnership set for itself? What prescriptions are binding (these are easier to determine if they are written down)? How would you like to see the prescription function carried out?
- *Enforcing Rules (Invocation Function)* is the process by which participants determine what needs to be done according to their interpretation of the rules. Is initial implementation consistent with prescription? Who should be held accountable for following the rules? Who will enforce the rules? How would you like to see the enforcement function carried out?
- *Administration (Application Function)* is implementing the final prescription. Will disputes be resolved by people with authority and control? How do participants interact and affect one another as they resolve disputes? How will final implementation be carried out? How would you like to see the application function carried out?
- *Monitoring (Appraisal Function)* is evaluating performance in terms of goals. Who is served by the program and who is not? Is the program evaluated fully and regularly? Who is responsible and accountable for success or failure? By whom are one's own activities appraised? How would you like to see the monitoring function carried out?
- *Ending Management (Termination Function)* is ending or changing a policy, usually according to conditions set forth earlier. Who should stop or change the rules? Who is served and who is harmed by ending a program? How would you like to see the termination function carried out?

dealt with promptly. Decision processes should also be honest, flexible, and efficient.

4. **Understand the context of the conservation problem.** The human factor is too easily overlooked, ignored, or viewed as a constraint to the central biological task of carnivore conservation. The context, or the social process surrounding the conservation task, is central to understanding the problem and finding a permanent solution. This social process must be 'mapped' realistically, and the rich and pervasive patterns of people influencing or affecting each other must be detailed.

Table 11.2. *Some questions to ask in mapping the social process or context of carnivore conservation problems*

- *Participants.* Who is participating? Identify both individuals, groups, and organizations. Who would you like to see participate? Who is demanding to participate but not currently a participant?
- *Perspectives.* What are the perspectives of those who are participating, of those you would like to see participate, and of those making demands to participate? What would you like their perspectives to be? Perspectives include:
  - *Demands,* or what participants or potential participants want, in terms of values and organization.
  - *Expectations,* or the assumptions of participants about past and future.
  - *Identifications,* or on whose behalf demands are made.
- *Situations.* In what situation do participants interact? In what situations would like to see them participate?
- *Base Values.* What assets or resources do participants use in their efforts to achieve their goals? All values can be used as bases of power. What assets or resources would you like to see participants use to achieve their goals?
  - *Power* is having the ability to make, influence, and carry out decisions.
  - *Enlightenment* is the gathering, processing, and disseminating of information and knowledge.
  - *Wealth* is to have money or its equivalent (e.g., the production, distribution, and consumption of goods and services).
  - *Well-being* is to have safety, health (physical and psychic), and comfort.
  - *Skill* is the acquisition and exercise of capabilities in vocations, professions, and the arts.
  - *Affection* is to have intimacy, friendship, loyalty, other warm relationships.
  - *Respect* is to show and receive deference.
  - *Rectitude* is participation in forming and applying norms of responsible conduct (i.e., to have ethical standards).
- *Strategies.* What strategies do participants employ in their efforts to achieve their goals? Strategies can be considered in terms of diplomatic, ideological, economic, and military instruments. What strategies would you like to see used by participants in pursuit of their goals?
- *Outcomes.* What outcomes are achieved in the continuous flow of interaction among participants? Outcomes can be considered in terms of changes in the distribution of values. Who is indulged in which values? Who is deprived of which values? Outcomes also refer to the ways in which values are shaped and shared. The ways in which values are shaped and shared are called practices or institutions. How are practices changing? How would you like to see practices change? What is your preferred distribution of values?

Clark & Wallace (1998) describe the human social process and how to map it (Table 11.2).

One practical concern for problem solvers is to decide just how focused or broad their inquiry should be. The interdisciplinary framework described here encourages problem solvers to be both comprehensive and selective simultaneously by focusing on the decision-making process in real-life contexts – the appropriate focus of inquiry should be the decision process, large or small, and how well it is working. Jaguar conservation, for instance, will necessarily involve international components, while the grizzly bear case is clearly more regional in scope.

Applying the interdisciplinary approach

There are several ways to apply this interdisciplinary, analytical approach to actual conservation situations (Clark, 1997, 1999). First, 'cooperative problem solving' through an adequate decision process is a way of getting participants or communities to seek their common interests. Many conservation programs attempt this approach with varying success, but few explicitly use an interdisciplinary approach. Problem-solving efforts can be set up by participants, coordinators, or decision makers to help groups integrate their knowledge to solve complex problems of all kinds (Willard & Norchi, 1993). This design seeks to explore the problem at hand, its context, and find enduring solutions in an integrative manner.

Second is 'prototyping,' which is a small-scale, trial change in a social or policy system. It has been used successfully in Australia in an endangered species case (T. Clark *et al.*, 1995a,b) and elsewhere. The primary goal of 'experimenting' with problems and solutions is to get information on relevant factors and to learn how to solve problems. Thus, the effort should include an explicit protocol for learning and integrating lessons across experiences and later scaling up through pilot studies and eventually full-scale applications.

Third, 'workshops for capacity building' improve basic problem-solving knowledge and skills. This approach is being actively used in the Greater Yellowstone Ecosystem; five two-day workshops focused on grizzly bear conservation, wolf recovery, and general problem-solving skills were carried out in 1999. Diverse people can be involved in workshops, even those at odds with one another. Team efforts are particularly rewarding. Workshops can help participants avoid approaches that are overly technical, parochial, or ones that tend to promote special interests. The challenge is to learn how to orient to complex problems using knowledge and methods from many disciplines and to integrate that knowledge for practical purposes. Of utmost importance are workshops to build a shared problem definition of conservation problems, improve cooperation among participants, enhance the capacity of participants to be effective through group action and discussion, establish priority areas for conservation, and open up opportunities to experiment and learn.

FUTURE RESEARCH AND NEEDS

First, not only must biological and ecological research on carnivores continue, but it should be directed toward conservation needs.

Second, social science research is also needed to conserve carnivores. In general terms, such work would examine the social and decision processes at play in any given case. Specifically, this would entail detailing via sociological, anthropological, psychological, or political science methods the practices, culturally and technologically complex though they may be, that hinder (or support) carnivore conservation. The perspectives behind the practices must also be determined, including the identities, the expectations, and demands of people involved. The play of values and institutions in society must be examined. T. Clark *et al.* (2000) reviews the range of biological and social science methods frequently used or available to conservation efforts.

Third, an interdisciplinary approach that synthesizes reliable information is needed to systematically integrate biological and social knowledge into a unified conservation program.

Fourth, applications of the interdisciplinary method described above (cooperative problem solving, prototyping, and workshops for capacity building), as well as more traditional conservation programs, should be systematically documented and studied for the purpose of learning what has worked and what has not. This kind of comparative learning approach is the only basis for genuine adaptive management. Comparisons should be carried out at regular intervals at professional meetings or by an oversight body. The lessons should be published and distributed widely. Participants in other programs can evaluate their utility, incorporate them as appropriate, and report results back to professional colleagues and societies. An endless repetition of this approach – field work using interdisciplinary and disciplinary approaches, comparison of results, lesson finding, distribution of lessons, refining methods, and new field work – offers a way to learn continuously and to improve carnivore conservation across the worldwide community.

Acknowledgements

We would like to thank Denise Casey for critically reviewing the manuscript and five anonymous reviewers for their comments on an earlier draft. This work was supported, in part, by grants to the Northern Rockies Conservation Cooperative (Cathy Patrick, Gil Ordway, Ted Smith and the Kendall Foundation, Fanwood Foundation, New-Land Foundation), Denver Zoological Foundation, and Yale University.

12

Assessment of carnivore reintroductions

URS BREITENMOSER, CHRISTINE BREITENMOSER-WÜRSTEN,
LUDWIG N. CARBYN AND STEPHAN M. FUNK

INTRODUCTION

Human beings have long translocated wild animals, often to gain extra game for hunting or trapping, sometimes just by mistake. Soon after human activities had led to the local extinction of species, the first attempts were made to bring these same species back to where they once belonged. In the nineteenth century, for instance, roe deer and red deer were reintroduced to different sites in western Europe. To restore carnivore populations, however, became conceivable only when the general attitude towards predators changed in the middle of the twentieth century. The first known reintroduction programmes for carnivores date from the 1940s. Carnivores, however – in particular the large ones – cause conflicts with human interests, and a reintroduction project often provokes a long-lasting controversy among interest groups. So why do we reintroduce carnivores? The most obvious objective may be to save a threatened species. But, with a few exceptions, the carnivores reintroduced so far do not belong to the globally threatened species (Tables 12.1 and 12.2). On the contrary, many of them are still widespread and rather common. Carnivores are reintroduced because we now judge their eradication as a wrongdoing that must be corrected, and because we consider predators to be important components of local ecosystems. Consequently, the majority of projects were carried out in North America (Table 12.3), where predators became locally extinct in the past, and where contemporary society has recently developed a strong interest in conservation. But this is a typical assessment of our 'enlightened' urban society estranged from nature, not always shared by people who live in and from the countryside and who have to share resources with carnivores once they are back.

Reintroductions in general and carnivore reintroductions in particular are extremely lengthy, costly and complex processes (IUCN/SCC, 1998).

Table 12.1. Overview of carnivore reintroduction projects

Family	Species		Continent	Number of projects	Outcome of projects				
					s	f	u	?	r
Canidae	*Canis lupus*	Wolf	North America	6	3	2			1
			Asia	1	1				
	Canis rufus	Red wolf	North America	2	1			1	
	Lycaon pictus	Wild dog	Africa	10	2	7		1	
	Vulpes velox	Swift fox	North America	2	1	1			
Felidae	*Acinonyx jubatus*	Cheetah	Africa	10	1	5	1	1	2
	Felis concolor	Puma	North America	1		1			
	Felis silvestris	Wildcat	Europe	4	1	2		1	
	Lynx canadensis	Canada lynx	North America	2			1		1
	Lynx lynx	Eurasian lynx	Europe	11	4	4		3	
	Lynx rufus	Bobcat	North America	1	1				
	Panthera leo	Lion	Africa	3		1	1	1	
			Asia	1		1			
	Panthera pardus	Leopard	Africa	3	3				
	Felis serval	Serval	Africa	1	1				
Hyaenidae	*Hyaena brunnea*	Brown hyaena	Africa	4	2	2			
Mustelidae	*Enhydra lutris*	Sea otter	North America	6	3	2		1	
	Lutra canadensis	River otter	North America	20	13		2	1	4
	Lutra lutra	River otter	Europe	4	2	1			1
	Martes americana	Pine marten	North America	39	18	7	14		
	Martes pennanti	Fisher	North America	22	10	6	5	1	
	Mustela nigripes	Black-footed ferret	North America	5	1	1			3
Ursidae	*Ursus americanus*	Black bear	North America	2	1				1
	Ursus arctos	Brown bear	Europe	4		1		1	2
			North America	1				1	
Total				165	69	44	24	13	15

For details see Appendix 12.1. Outcome of projects according to expert opinions – s: success; f: failure; u: unknown (the development of the population has not been followed, or the outcome has not been documented); ?: uncertain (cannot be judged as too little time has passed since the releases, or the development of the population is still unclear); r: running (phase of releasing animals still going on).

Table 12.2. Number of species reintroduced per Carnivora family in relation to the number of species listed on the Red List of the IUCN maintained by the World Conservation Monitoring Centre (www.wcmc.org.UK/species/animals/animal_redlist.html). Some species are on that list because important populations or subspecies are endangered

	Number of extant species	Species on the Red List	Number of reintroduced species		Number of reintroduction projects of species	
			listed	not listed	listed	not listed
Canidae	32	19	3	1	14	7
Felidae	35	27	3	6	12	25
Viverridae	70	27	0	0	0	0
Hyaenidae	4	2	0	1	0	4
Mustelidae	63	25	1	5	2	91
Procyonidae	18	12	0	0	0	0
Ursidae	9	7	0	2	0	7
Total	231	119	7	15	28	134

Table 12.3 Basic features of carnivore reintroduction projects in relation to the Carnivora families

	Number of projects					
	Canidae	Felidae	Hyaenidae	Mustelidae	Ursidae	Total
Total number of projects	21	37	4	96	7	165
Continent of release: North America	10	4		92	3	109
Europe		15		4	4	23
Africa	10	17	4			31
Asia	1	1				2
Origin of animals: wild caught	6	16		82	6	110
captive-bred	10	2		6		18
mixed	4	4		3		11
unknown	1	15	4	5	1	26
Year of project start: up to 1950		1		2	1	4
1950–1959		1		18	1	20
1960–1969		5		14		19
1970–1979	6	13	1	12		32
1980–1989	6	10	3	36	1	56
1990–1999	9	7		14	4	34
Animals released: unknown	1	6	1	1	1	10
2–10	9	14	3	10	4	40
11–20	3	11		15	1	30
21–50	5	3		27		35
51–100	1	2		19		22
101–200	1	1		12		14
201–500				11	1	12
500–1000	1			1		2
Median	11.5	13	5	45.5	6.5	29
Range	3–942	2–129	4–7	2–845	2–254	2–942
Success of projects: successful	8	11	2	47	1	69
failure	10	14	2	17	1	44
unknown		3		21		24
uncertain	2	6		3	2	13
running	1	3		8	3	15

The Mexican wolf reintroduction programme is expected to cost over US$ 7 000 000 over nine years (USFWS, 1996). The number of carnivore reintroductions has significantly increased over the past decades, but most efforts fail (Schaller, 1996). The problems met in reintroduction programmes of carnivores differ from those of herbivore or bird reintroductions in three aspects: (1) the eradication of carnivores was often the result of conflict with humans (menace to humans, competition for game, and predation on livestock), and their reintroduction causes massive controversy; (2) although many carnivore species are rather tolerant towards habitat quality – as long as there is an adequate prey base – a viable population needs extended living space as the abundance of carnivores is generally low; and (3) carnivores are elusive animals, and to monitor the progress and the success of a reintroduction programme is a difficult, expensive and long-lasting task. There is a clear need to improve the efficiency of carnivore reintroductions. However, any evaluation must go beyond the identification and ranking of factors affecting success of specific projects. Alternative conservation strategies, such as the protection of remaining wild populations elsewhere, must be compared to reintroductions in the light of their cost-benefit ratio.

Ideally, a quantitative analysis of carnivore reintroductions similar to the analyses of Griffith et al. (1989), B. Beck et al. (1994) and Wolf et al. (1996) of animal translocations would provide a critical assessment of the factors determining the success of carnivore reintroductions. Such an analysis is however difficult, because reintroduction programmes are often poorly documented and lack external evaluation and peer-review. The definition of success and even of short- and long-term aims are highly variable, monitoring is often inadequate, and for many carnivore reintroductions, the release phase ended too recently to allow for a conclusive judgement. Furthermore, the scope of carnivores reintroduced – ranging from black-footed ferret (*Mustela nigripes*) to brown bear (*Ursus arctos*) – makes every inter-specific quantitative comparison a questionable venture. Therefore, we have restricted our review to a qualitative identification of patterns emerging from previous attempts to reintroduce carnivore populations. We review seven early and recent projects, which represent a broad spectrum of species and geographical regions, to highlight some specific problems of carnivore reintroductions and their assessment. We will compare the idealistic propositions of the *IUCN Guidelines for Reintroductions* (IUCN/SCC, 1998) with the practical reality of the reintroduction projects and discuss some improvements in the procedures of carnivore reintroductions.

RECENT REVIEWS OF CARNIVORE REINTRODUCTIONS

Since the 1980s, several publications have reviewed the man-made restoration of carnivore populations, either in the form of general reviews of reintroductions or in specific analyses of carnivore projects. Griffith *et al.* (1989) and C. Wolf *et al.* (1996, 1998) have quantitatively analysed the outcome of reintroduction projects in birds and mammals and found that the success of reintroduction projects mainly depends: (1) on the habitat quality of the release area; (2) the position of the release site relative to the historic distribution; and (3) the number of individuals released. They concluded that reintroductions of omnivorous species are generally more successful than those of herbivores or carnivores. Reading *et al.* (1997) made a quantitative study of the influence of valuational and organisational aspects on the success of rare species translocation. Only a few of the respondents to their questionnaire identified valuational or organisational problems in their programmes. The analyses of the 131 translocations stress the importance of public support and effective public relation/education programmes to the success. Further, reintroduction projects had greater public support than restocking projects. Stanley Price (1991) recommended that reintroduction programmes should be incorporated into national and international conservation strategies, have a multidisciplinary, scientific approach, and that all results should be published. Yalden (1993) stressed the importance of the need for the fullest understanding and support from the people in the area where carnivores are to be reintroduced and suggested a concentration on projects promising a substantial conservation benefit. Kleiman *et al.* (1994) reviewed the criteria that should be met prior to a reintroduction and provided a framework for the initial decision-making, including a list of topics to be addressed as the project proceeds. They stressed the importance of defining goals and objectives to be able to measure the success of a project. Reading & Clark (1996) underlined the complexity of carnivore reintroduction projects, which should address not only biological–technical aspects, but also valuational (public relation, public support) and organisational (co-operation, qualification of personnel and leadership, etc.) aspects. Hein (1997) presented a list of points to be considered at an initial stage of a translocation, and suggested improvement of the data collection, the estimation of costs and benefits, research on listed species (experimental design), and exploration of alternative methods. He called for a more scientific, experimental approach, as each reintroduction project provides an opportunity to learn. This latter statement was supported by Soorae & Stanley Price (1997), who furthermore pointed out that carnivore reintroductions are always long-term, expensive projects, which typically are initiated by NGOs, but will ultimately require the commitment of GOs. Miller *et al.* (1999) discussed some general guidelines for reintroducing carnivores with the focus on biological–technical aspects. They concentrated on carnivores because of their high value to ecosystem health. Carnivores have been extirpated from many ecosystems, fragmentation rendered natural re-colonisation difficult, and carnivores are disproportionately harder to re-establish by means of reintroductions than other taxa. All the reviews published since 1989, came to the conclusion that the outcome of carnivore reintroductions is influenced by a complex and interwoven network of variables, which can be categorised in: (1) biological–technical; (2) organisational; and (3) valuational aspects (Reading & Clark, 1996; IUCN/ SSC, 1998). In 1989, the IUCN Reintroduction Specialist Group (RSG) was founded in order to promote a more rigorous approach in reintroductions. Recently, the RSG published *Guidelines for Reintroductions* (IUCN/ SSC, 1998), listing the requirements (see Table 12.4) that should be met for reintroduction projects. Besides the release procedures, there are important biological and socio-economical pre-release and post-release activities, which will influence the success of a project.

SYNTHESIS AND ANALYSES
Characteristics of carnivore reintroductions

We considered 165 carnivore reintroduction projects (Appendix 12.1) applying the IUCN/SSC (1998) definition of reintroductions – an attempt to re-establish a species within its historical range, while a translocation is the act of re-inforcement or supplementation or restocking, i.e. the addition of individuals taken from the wild to an existing population. We have only considered projects where animals were released to an area where the autochthonous population had become (virtually) extinct. The 165 carnivore reintroduction projects (Appendix 12.1) concerned 22 species from five families (Tables 12.1 and 12.2; basic statistics summarised in Table 12.3). Although this is a considerable number of projects, the amount and quality of information available is often insufficient for detailed analyses or for a comparison with other projects.

Of the projects, 58% concerned mustelids (Table 12.3), followed by felids (22%), canids (13%), ursids (4%) and hyanids (3%). The majority of the projects was realised in North America (Tables 12.1 and 12.3; Appendix 12.1: 78 projects in the USA, 31 in Canada). For Africa, we found 31 projects (25 in South Africa, 3 in Namibia and Zimbabwe and 1 in Kenya), for

Table 12.4. Checklist of requirements and activities for carnivore reintroduction programmes

Project number in Appendix 12.1	Mn 132–136	Vv 21	Luc 75	Ll 42	Lxc 38	Ua 162	Pl 54
Aims and objectives:							
1. Definition of general aims of re-introduction: why? (e.g. global conservation status vs. restoration of local/regional community structure, socio-economic reasons) and how? (e.g. self-sustainable population vs. metapopulation)	y	(y)	y	(y)	y	(y)	(y)
2. Definition of concrete short- and long-term objectives	y	(y)	y	n	n	(y)	[y]
3. Definition of success and of criteria, methods and timescale to measure it	y	(y)	y	n	(y)	n	(y)
Feasibility study (habitat and potential threats):							
4. Assessment of taxon. status of original population and of animals to release	/	[y]	y	n	(y)	(y)	n
5. Study of status, ecology, life history, physiology and disease susceptibility of wild populations to evaluate the species' needs and the release areas' suitability	y	n	y	n	(y)	(y)	y
6. Identification of historical reasons for the decline and assesment of their current status	y	(y)	y	(y)	(y)	y	y
7. Assessment of new threats (new guild structure, disease, human activity)	y	(y)	y	n	y	n	y
8. Habitat viability analyses of the release area and projected population range (environment, resources, carrying capacity, spatial characteristics, etc.)	y	n	y	n	n	(y)	[y]
9. Population modelling (PVA and metapopulation model) and sensitivity analysis of short- and long-term demography and distribution	(y)	n	y	n	n	n	[y]
Strategic pre-release activities:							
10. Research into previous comparable reintroductions and intensive contacts to experts	(y)	y	y	n	y	n	(y)
11. Construction of a multidisciplinary advisory team with expert knowledge for all aspects (biological, economical, sociological, political, legal)	(y)	n	[y]	n	n	n	n
12. Evaluation and development of efficient intra- and interorganizational structure (team structure, continuity over complete project, decision-making authority, etc.)	n	n	y	n	n	n	n
13. Guidelines for decision-making process in conflict situation (economic vs. biological decisions, transparancy and accountability)	n	n	(y)	n	n	n	(y)

#		C1	C2	C3	C4	C5	C6	C7
14.	Elimination or substantial reduction of previous causes of decline and of new threats	n	(y)	y	(y)	(y)	y	y
15.	Evaluation of source animals from wild populations: genetically related, viable population (must not significantly suffer from the loss)	/	n	y	y	y	(y)	y
16.	Evaluation of source animals from captive-bred stock: demographical and genetical healthy stock, trained individuals not habituated to humans (large carnivores!)	y	(y)	/	/	/	/	/
17.	Policies for intervention in relase phase (e.g. feeding, removal of released animals)	[y]	/	y	n	[y]	n	[y]
18.	Policies for mitigation of damages (compensation, husbandry practicies...)	y	/	y	[(y)]	y	y	y
19.	Long-term political support (involvement of all relevant GOs; cross-border co-operation; approval of all landowners)	y	(y)	(y)	n	[y]	(y)	y
20.	Adjustment of legislation to support the re-introduction	(y)	(y)	/	(y)	/	/	/
21.	Coordination with, and support from, national and international NGOs	[y]	(y)	[(y)]	n	(y)	n	n
22.	Consideration of health and quarantine regulations	y	(y)	y	y	y	y	y
23.	Socio-economical studies to assess impacts, costs and benefits to local human population	(y)	n	n	n	(y)	n	y
24.	Assessment of human attitudes, acceptance and support by local people	[y]	(y)	[y]	n	[y]	n	y
25.	Starting of public relations programmes on national/local level	y	(y)	(y)	n	y	[y]	y
26.	Involving local people and starting local education programme	y	[y]	y	n	(y)	[y]	y
27.	Securing adequate funding for all programme phases	y	n	y	n	(y)	n	y
28.	Policies for accurate/detailed record keeping on all aspects of project and for accessability of database for peer review (e.g. on World Wide Web)	n	n	y	n	[y]	(y)	y
29.	Design of monitoring programmes (demography, behavior, disease, genetics, ecosystem)	(y)	(y)	y	n	(y)	(y)	y
30.	Evaluation and design of long-term management programmes (incl. supplementation, culling, harvest etc)	(y)	n	y	n	n	n	[y]
31.	Re-evaluation and adjustments of 1 to 30	y	n	y	n	(y)	n	(y)

Release procedures:

#		C1	C2	C3	C4	C5	C6	C7
32.	Reduction of human activities where they are a threat to released animals	(y)	(y)	y	/	y	/	/
33.	Selection and capture of animals to release (representative demographic and genetic structure; no adverse effect on source population)	y	(y)	y	n	y	y	y
34.	Veterinary screening of wild or captive release stock and veterinary intervention if necessary	y	(y)	y	y	y	/	(y)

Table 12.4. (cont).

Project number in Appendix 12.1	Mn 132-136	Vv 21	Luc 75	Ll 42	Lxc 38	Ua 162	Pl 54
35. Animal welfare and ethical acceptability for capture, transport and release	y	(y)	y	n	(y)	y	y
36. Release strategy (transport, number, group composition, training and acclimatisation, timing, release patterns, supplementary feeding, vaccination, use of 'surrogate' species)	(y)	[y]	y	(y)	y	y	y
Post-release activities:							
37. Monitoring of all, or a sample of, released individuals (behaviour, space use, survival)	(y)	(y)	y	[y]	y	y	y
38. Monitoring of demography (mortality, reproduction) and population range	[y]	(y)	y	[y]	y	y	y
39. Analysis of causes of mortality (collection of carcasses)	y	(y)	y	[y]	y	y	y
40. Genetic surveillance	n	{y}	n	n	/	n	y
41. Health surveillance	y	n	y	[y]	/	n	y
42. Study of long-term adaptation of individuals and population to ecosystem	{y}	y	y	n	/	y	y
43. Monitoring of effects on ecosystem (predation, displacement)	(y)	(y)	y	(y)	n	(y)	y
44. Monitoring of carnivore/human relation (depredation, values, etc) and continuing of public relation activities (education, etc)	(y)	n	y	[(y)]	/	[y]	y
45. Intervention if necessary (feeding, disease control, re-capture, supplementation etc.)	y	n	/	n	y	[y]	y
46. Continuing of protective measures if necessary (habitat restoration, legislation etc)	(y)	(y)	y	[y]	/	[y]	y
47. Monitoring of demography in source population (if wild population)	/	n	y	n	n	n	n
48. Continuous updating of database	y	y	y	[y]	/	y	y
49. Making all data available for research and peer review (eg. World Wide Web)	n	n	y	n	/	(y)	y
50. Evaluation of population viability and decision for revision, rescheduling or discontinuation of the programme (according to criteria and time frame as defined in 3)	(y)	(y)	y	n	/	y	(y)

General assessment:

51.	Evaluation of effectiveness of strategic pre-release activities (organizational structure etc.)	(y)	n	y	n	n	n	{y}
52.	Evaluation of success of re-introduction techniques	[y]	y	y	n	n	y	y
53.	Evaluation of effectiveness with regard to global status of species (e.g. would protection of remaining populations had been more effective?)	y	n	y	n	(y)	(y)	(y)
54.	Evaluation of costs/benefits with regard to non-biological dimension (documentation, efficiency in time and resource use, inter- and intraorganizational relations, education, paradigmen shift in public attitudes)	n	[y]	[y]	n	n	[y]	(y)
55.	Publication in scientific, peer-reviewed literature	(y)	y	y	[y]	n	n	n
56.	Publication in popular literature	y	y	y	y	n	y	y

Adopted from the IUCN Guidelines for Re-introductions (IUCN/SSC, 1998)

y: requirements met at indicated stage of project; n: not met; /: not applicable; (y): partially established activities; [y]: a posteriori established activities and requirements; {y}: planned activities; Mn: *Mustela nigripes* (black-footed ferret); Vv: *Vulpes velox* (swift fox); Luc: *Lutra canadensis* (river otter); Ll: *Lynx lynx* (Eurasian lynx); Lxc: *Lynx canadensis* (Canada lynx); Ua: *Ursus arctos* (brown bear); Pl: *Panthera leo* (lion).

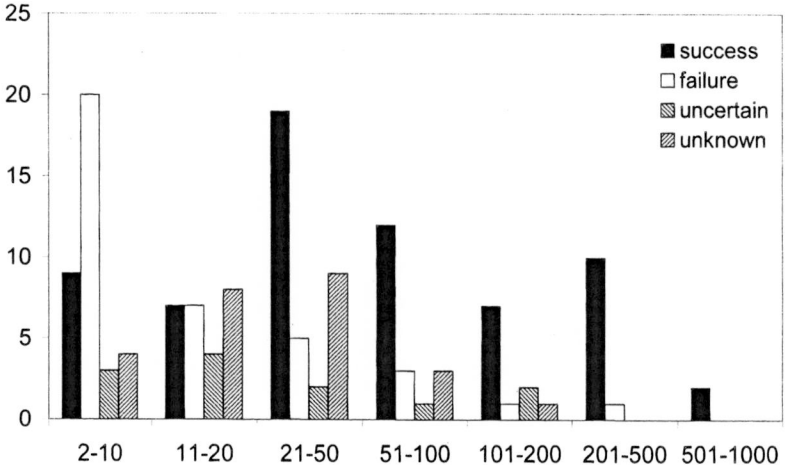

Figure 12.1. The number of carnivore reintroduction projects (Appendix 12.1) plotted against the numbers of animals released (*n* = 140). Projects with fewer animals released have a higher risk to fail than projects with > 20 animals released.

Europe, 23 (7 in Switzerland, 2 each in Germany, Poland, Austria, Italy and France, 1 each in Russia, Slovenia, Czech Republic, Spain, Sweden and United Kingdom), and for Asia 2 (India, Georgia). In North America, mainly mustelids (84%) have been reintroduced, whereas in Europe and Africa, felid projects dominated (65% and 55%, respectively). Canids have so far been reintroduced in the United States and in Africa, but not in Europe.

The origin of the animals has been documented for 139 (84%) projects. From those, the vast majority released animals caught in the wild, and only a few projects used captive-bred animals, or a mixture of wild-caught and captive-bred animals (Table 12.3). However, in canids, 67% of the projects used partially or exclusively captive-bred animals. Not surprisingly, reintroductions of endangered species or subspecies, such as the red wolf (*Canis rufus*), the Mexican wolf (*Canis lupus baileyi*), the African wild dog (*Lyacon pictus*) and the black-footed ferret (*Mustela nigripes*) were carried out with captive-bred animals (Appendix 12.1).

The number of animals released varied from two for the Eurasian lynx or the pine marten to more than 800 for the swift fox and the river otter (Appendix 12.1). Half of the projects were carried out with 25 or fewer animals (Figure 12.1). The median of the numbers of animals set free varied between 5 for hyanids to 45.5 for mustelids. Remarkably high numbers of animals released are documented for the swift fox (942 in Alberta, Canada; L. Carbyn, pers. comm.), the sea otter (412 in southeast Alaska, USA; Love,

1992), the river otter (843 in Missouri, USA; Hamilton, 1998), and the black bear (254 in Arkansas, USA; J. Clark & Smith, 1994).

The earliest documented carnivores to have been reintroduced were pine martens in the United States in 1934 and 1944, brown bears in Poland from 1938 to 1944, and lynx in East Prussia in 1941 (today Russia; Appendix 12.1). The early reintroduction projects predominantly concerned mustelids (Table 12.3), which are probably politically less problematic for translocation than larger carnivores. The highest number of mustelid projects however has only been realised in the 1980s. Cat reintroductions became prominent in the 1970s, essentially because of the projects to re-establish lynx populations in western Europe. Canid projects peaked in the 1990s. The total number of projects peaked in the 1980s (Table 12.3).

Successes and failures in carnivore reintroductions

Biological–technical aspects

According to the published opinions of the experts, 70 of the 165 (42%) projects were successful (Tables 12.1 and 12.3), 44 (27%) have failed, for 13 (8%), the outcome is still uncertain and for 24 (15%) the outcome was not documented. Fourteen projects (8%) are still running. Felids, canids and mustelids had a success rate of 30%, 38% and 50%, respectively (Table 12.3). The failure rate varied from 18% to 48% for these three families, and the proportion of projects with a yet uncertain outcome ranged from 3% to 16%. For several reasons, however, these results may be misleading and it is impossible to compare the results between studies, both for reintroduction projects and reviews, and from the information currently available to quantitatively analyse factors affecting success and failure.

First, many of the reintroduction projects are poorly documented. Important information, such as the origin of the animals or the fate of the released individuals, is often not reported or not known. For many projects, data have never been published in peer-reviewed journals or books. Internal reports, difficult to obtain from outside, are often the only documentation. Griffith *et al.* (1989) and Kleiman *et al.* (1994) emphasise the lack of documentation and argue strongly for improvement. B. Beck *et al.* (1994) report that they were able to collect reasonably complete information on fewer than 50% of projects that have reintroduced captive-born animals. It is most likely that published material conveys an over-optimistic impression in regard to the success rate of carnivore reintroduction projects, because small scale or *ad hoc* projects – which may have a higher failure potential – are hardly documented. The choice of units to include in comparative studies has been controversially discussed. Conservation related parameters appear to be best analysed in a multivariate approach or within families (Foose & Seal, 1992; Ginsberg, 1994). The available data, however, are not sufficient.

Secondly, key parameters influencing the outcome of reintroductions are difficult to quantify. How, for example, can the habitat quality be quantitatively compared between reintroductions of sea otters in Alaska and lions in South Africa? Or how can we compare the significance of numbers of animals released for brown bear and swift fox? Common sense tells us that suitable habitat must be available before reintroduction is considered, and habitat restoration programmes should be carried out before and during the release phase (IUCN/SSC, 1998). However, habitat requirements may not be known for extinct populations and reintroductions may also aim to determine habitat requirements and verify habitat suitability, as was claimed for the current reintroduction of Canada lynx (*Lynx candensis*) in Colorado (Seidel *et al.*, 1998). Moreover, reintroductions are influenced to a high degree by stochastic events. Even a well-prepared project may fail because of unforeseeable circumstances. The restoration of sea otters (*Enhydra lutris*) to the Pribilof Islands in 1968 failed because the 'abnormal extension of the icepack' killed all 55 released animals (Love, 1992). This statistical 'white noise', the small amount of quantitative data available and the obvious bias towards 'successful' projects, increases the likelihood of statistical errors type I and II, i.e. the likelihood that real correlations are not statistically significant or that correlations appear statistically significant without any true correlation.

Thirdly, the aims of the projects and criteria for how to measure their outcome have often not been defined. The success of a reintroduction project depends on the objectives and the point in time at which to measure it. The success rate of 42% given in Table 12.3 differs greatly from previous analyses. Using the project managers' assessment of success, Wolf *et al.* (1996) reported a success rate of 94% in 17 reintroductions of carnivorous mammals, which was significantly larger than in omnivorous mammals (two out of three), and herbivorous mammals (67%, $n = 57$). Across species, Wolf *et al.* (1998) found a success rate of 67%. In contrast, Schaller (1996) estimated a success rate of only 10% in Carnivora. B. Beck *et al.* (1994) defined success as the wild population reaching at least 500 individuals, free of human support, or where a formal genetic/demographic analysis predicted that the population would be self-sustaining. The authors found a success rate of 11% across animal species, but could only assemble sufficient information for 46 projects involving 39 mammal species. Most projects have not defined success at all, or used implicit definitions (survival and/or spread of released animals, successful reproduction) which did not assess the development and viability of the population. The lynx reintroduction in Austria was for long believed to be successful, although there was no evidence of a reproducing population (Huber & Kaczensky, 1998).

Woodroffe & Ginsberg (1997c) implicitly used survival, successful reproduction and persistence as criteria for success in wild dogs. On this basis, two out of nine reintroduction and population supplementation projects were judged as successful. The most successful reintroduction was in Hluhluwe-Umfolozi (South Africa) where 22 animals were released in 1980–81 and where the population has persisted with estimated numbers from 3 to 30 animals. However, the 960 km² park is too small to sustain a viable population in the long term (Woodroffe & Ginsberg, 1997c). Clearly, when applying the 500 animal rule of B. Beck et al. (1994), or any criteria for the probability of extinction or critical effective population size (Franklin & Frankham, 1998; Lynch & Lande, 1998), no wild dog reintroduction could be considered successful. These limitations apply to most of the 'successful' reintroductions of canids (e.g. red wolf, see Ginsberg, 1994), felids (e.g. lynx in the Swiss Alps, see below), mustelids (e.g. Eurasian otter, T. Sjöåsen & A. Angerbjörn, unpub. data), hyenas and ursids.

Fourthly, long-term monitoring of re-established carnivore populations is often inadequate. Monitoring is often limited to the release phase and some immediate years. A large proportion of programmes did not integrate long-term monitoring into the project design. Authors came to a conclusion about the success of a project often after only a few years. We lack the information to conclude how many projects once considered to be successful failed at a later stage or still will. Projects up to about 1980 have typically ended within one year of the releases (Slough, 1994). Since then long-term monitoring has been initiated for many of them, but few results have been comprehensively published and are often only available as internal status reports (e.g. Phillips, 1992; Henry & Lucash, 1998; Hamilton, 1998). Reintroduction projects in North America tend to be organised by government institutions and some form of monitoring is normally carried out. Approaches, intensity, quality and documentation varies greatly ranging from radiotracking and regular litter counts in the red wolf (Phillips, 1992; Henry & Lucash, 1998), to modelling the population in the Missouri river otter project (Hamilton, 1998), to the failure to monitor the status of released animals (Reading & Clark, 1996). In Africa, Asia and Europe, most projects tend to have been initiated by enthusiastic persons or organisations not qualified to make long-term commitments. Monitoring during the initial phase was often attached to research programmes and depended

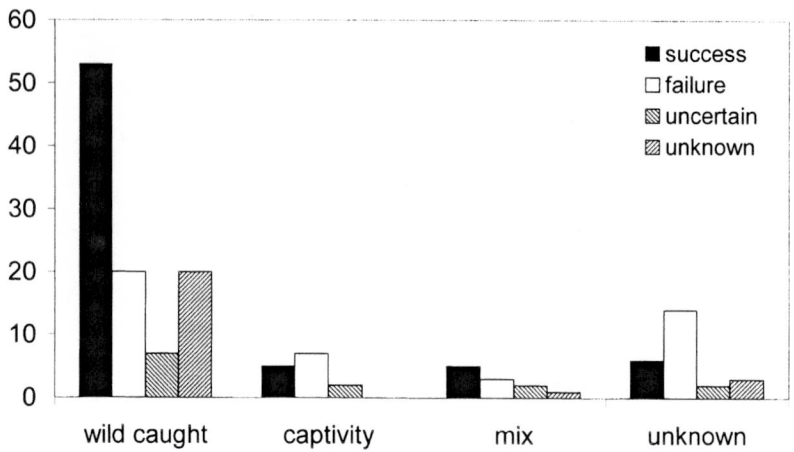

Figure 12.2. The number of carnivore reintroduction projects (Appendix 12.1) plotted against the origin of animals released (*n* = 150). Projects using wild caught animals have a higher ratio of success. This finding is however biased, as projects with smaller or more threatened species are more likely to use captive bred stock.

on outside funding. Such a project structure only exceptionally leads to a long-term surveillance, as was the case for the reintroduced lynx in Switzerland (Breitenmoser *et al.*, 1998).

Nevertheless, some general patterns emerge from previous failed (and from so far successful) reintroductions with regard to the biological/technical dimensions. Reintroductions have failed because of species, site or project specific reasons (e.g. insufficient release training of captive-bred wild dogs and disease; Woodroffe & Ginsberg, 1997c), or – more important – the use of inadequate founder animals or insufficient numbers. Figure 12.1 suggests that with an increasing number of animals released, the chance of success increases and the risk of failure decreases. Projects that used animals caught in the wild, or a mixture of wild-caught and captive-bred animals, tend to be more often successful than those carried out with captive-bred animals only (Figure 12.2). Successful reintroductions compensated for losses due to mortality and dispersal. Reported mortality rates for wild-caught animals were high, and even larger when using captive-born stock. For example, only 4 out of 49 captive-bred black-footed ferrets survived the first winter after release (T. Clark, 1994; Miller *et al.*, 1996). In the swift fox project, the proportion was 4–14% of captive-bred as against 47% of wild-caught animals (Carbyn *et al.*, 1994), and in Eurasian otters, 8–35% captive-bred and 60–67% of wild-caught animals survived (Sjöåsen & Sandegren, 1992). Population modelling of swift fox releases have indicated that an initial population of 100 individuals supplemented with 20 individuals/ year for 10 years had the lowest probability of extinction of all models tested. Single releases of 50 or 100 animals had a high probability of extinction due to demographic and environmental stochasticity (Ginsberg, 1994). Captive-bred large carnivores are particularly difficult to release, since they lack the experience of hunting, and associate food with humans. In Georgia, captive-born wolves were intensively trained to kill wild prey and to avoid humans prior to their reintroduction. This release procedure was considered successful after the third generation of wild-born wolves still showed the same behaviour as their hand-raised parents (Badrize, 1999). Woodroffe & Ginsberg (1997c), on the other hand, summarised the failure of a reintroduction of captive-bred wild dogs in Matetsi Safari Park, Zimbabwe, because the animals were unable to make a living in the wild. Herrenschmidt (cited in Breitenmoser & Breitenmoser-Würsten, 1990) reported the killing of two captive-bred Eurasian lynx released to the Vosges Mountains in France, when they tried to take chickens on a farm. Threatened taxa, where autochthonous populations cannot provide animals for translocations, may consequently generally be more difficult (but not less urgent) to reintroduce.

Valuational and organisational aspects

Evaluation of reintroduction success has traditionally been focused on the biological dimension. However, reintroductions may be well-designed with regard to their biological objectives, but may be ineffective or cause unexpected negative results (Reading & Miller, 1995; B. Miller et al., 1996; Reading & Clark, 1996). Only a very small number of projects have been scrutinised with regard to their valuational and organisational failures and successes (Reading & Clark, 1996). Reading & Miller (1995) and Reading & Clark (1996) have pointed out that the black-footed ferret programme may be regarded as biologically successful, but that this success was achieved in an ineffective and uneconomical way. The project has strained relations among many individuals and organisations involved, and may be doomed because the continuous fragmentation of suitable habitat (sponsored by federal agencies) is likely to lead ultimately to the species' extinction (Reading & Clark, 1996). Carnivores are typical 'flagship' species. Their high profile may increase conflict between persons or interest groups. Large carnivores potentially touch the interests of people living in the recovery areas and most often have an historical burden (Breitenmoser, 1998a). In successful reintroductions, human–carnivore conflict did not exist or had been reasonably solved. Such conflicts have been the major reason for carnivore persecution and, therefore, the most important factor for the survival of re-established populations (Moore, 1992). Human–carnivore conflicts are often the main argument against reintroductions, as in the case of Siberian tigers (Jackson, 1996). B. Beck *et al.* (1994) have identified public support and community education as a crucial factor correlated with success in reintroductions of captive-bred animals.

Carnivore reintroductions can have great cultural significance, as in the case of the Nez Perce Indians who entered into a co-operative agreement with US Fish and Wildlife Service in 1995 that gave the tribe state-wide recovery and management responsibilities for wolves throughout Idaho. Or, on the other hand, carnivore recovery projects may reinforce underlying socio-cultural conflicts, as with the lynx reintroduction in Switzerland. The return of the lynx is strongly supported by the majority of the urban population, while in the rural community of the mountains, strong opposition against the large predator persists even 30 years after its reintroduction (Breitenmoser *et al.*, 1999a). The recovery of the large predators is often regarded by the rural community as another point where they loose control over their traditional way of living. Human–carnivore conflict must be avoided not only through public information and involvement, but also through careful management of the released animals and the emerging populations. Such management has not only helped to maintain public support for the red wolf reintroduction project, but also local landowners now allow red wolves onto their land (Phillips *et al.*, 1995). However, the paradigm shift in human attitude towards carnivores should be interpreted very carefully and not be over-emphasised. Whilst over 90% of persons surveyed in Florida indicated support of the reintroduction of Texas cougars, inhabitants of areas near to the proposed release site showed strong opposition (Belden & McCown, 1996). Another very recent development is the increasing awareness of ethical considerations in release programmes and animal rights (see Bekoff, Chapter 9), and the increasing number of groups of animal lovers supporting certain species, such as wolves. Managing reintroduction projects may become a very complex process requiring sophisticated public relations strategies to balance pressure from different and often opposing pressure groups. The fair integration of divergent interest groups complicates the decision-making process anywhere, but it also impedes it, particularly when cultural differences are also involved. This is an aspect especially important in Europe and Africa. The Alps are a well suited range for the restoration of large carnivore populations. Besides the reintroduction programmes for lynx (Breitenmoser-Würsten *et al.*, 1998) and for brown bear (Gutleb, 1994; Rauer, 1997; Brugnoli, pers. comm.), wolves and bears recolonise the area spontaneously (Breitenmoser-Würsten *et al.*, 2001). The Alps stretch over 1000 km and touch seven countries, including seven different legislative and management systems and five different languages. In each country the distribution of large carnivores is relatively limited and cross-border co-operation is indispensable, but as a result of the cultural differences it is also very complicated and slow.

Project planning and success compared to the reintroduction checklist

Careful project planning, in the light of relevant guidelines (e.g. IUCN/SSC, 1998), is one piece of advice common to all reviews on reintroduction. Carnivore reintroductions are complex, long-lasting and, to a high degree, stochastic processes. Stochasticity in the form of unforeseeable events (e.g. extreme weather conditions, diseases, etc.) or unexpected responses from the animals or the human population to anticipated events can threaten even a well designed project. Obviously, small-scale projects (e.g. in regard to the number of animals released; Figure 12.1) are more likely to fail as a consequence of unfortunate coincidences. Still, there is no guarantee of success, and, even more perplexing, there is no guarantee of failure. Some reintroductions (e.g. the lynx reintroductions in the Alps, see below) have ignored almost all the rules and have nevertheless been 'successful'.

We have summarised the recent IUCN/SSC (1998) Guidelines for Reintroductions in Table 12.4 and have adapted and supplemented them to meet specifically the requirements of carnivore reintroductions. We have then chosen seven earlier or ongoing projects (Table 12.4) representing a wide taxonomic and geographic contrast, and have evaluated them by means of the checklist, or asked expert colleagues to do so. This brief inquiry by no means allows any quantitative analysis; it provides, however, an interesting qualitative comparison between some well-known projects. The reintroduction projects evaluated were the (for selected references see Appendix 12.1):

(1)   black-footed ferret (*Mustela nigripes*; Shirley Basin, Wyoming; Charles Russel NWR, Montana; Badlands NP and Buffalo Gap NG, South Dakota; Fort Belknap IR, Montana; Aubrey Valley, Arizona, all USA; project nos. 132–36 in Appendix 12.1; correspondents R. Reading and B. Miller), where some 870 captive-bred animals have been released since 1991;

(2)   swift fox (*Vulpes velox*; Alberta and Saskatchewan, Canada; no. 21; L. N. Carbyn), 942 wild-caught and captive-bred foxes released from 1983 to 1997;

(3)   river otter (*Lutra canadensis*; Missouri, USA; no. 75; D. Hamilton), 845 wild animals released from 1982 to 1992;

(4)   Eurasian lynx (*Lynx lynx*; western Swiss Alps; no. 42; U. Breitenmoser), 14–18 wild-caught lynx (of which at least 10 in the western Alps) released from 1971 to 1976;

(5)   Canada lynx (*Lynx canadensis*; South Fork, Colorado, USA; no. 38; R. Reading and B. Miller), 41 wild lynx released since February 1999;

(6)   brown bear (*Ursus arctos*; eastern Austrian Alps; no. 159; N. Gerstel and G. Rauer), 3 wild-caught animals released from 1989 to 1993;

(7)   lion (*Panthera leo*; Phinda, Natal, South Africa; no. 54; L. Hunter), 13 wild lions released from 1992 to 1993.

The reintroduction of Canada lynx into Colorado started only in 1999 and cannot yet be evaluated. Of two more projects, only preliminary results can be presented: in the area of brown bear releases in Austria, the number of individuals is estimated to be 13–16 (Rauer, 1997; Gerstel, pers. comm.); in the lion project in South Africa, the released animals have increased to 30. The reintroduction of black-footed ferrets to the Shirley Basin failed; no animals are presently known to be surviving (B. Miller, pers. comm.). The swift fox, river otter and Eurasian lynx projects were better designed and more successful and have produced self-sustaining populations up to the present.

For all projects, some general aims were defined at the beginning, and in most projects, concrete objectives, criteria for success and parameters to measure it were defined (Table 12.4). Most projects performed different kinds of feasibility studies, but some neglected to investigate the habitat quality and the carrying capacity of the projected population area. Only in one project – the river otter – was population modelling performed prior to the releases. With regard to the strategic pre-release activities, it is conspicuous that the biological/technical aspects (such as elimination of previous threats, mitigation of damage, health and quarantine considerations) were generally fulfilled, whereas the organisational aspects (interdisciplinary advisory team, intra and inter-organisational structure, guidelines for decision-making processes and conflict situations) were – again with the exception of the river otter project – neglected. Partly dismissed also was co-operation with national or international non-government organisations and the evaluation of long-term management programmes. Across the projects, the recommendations for the release and the post-release activities were generally respected. Only one project – the river otter – monitored the source population after the animals were translocated. A partial deficiency can also be identified in the later evaluation of effectiveness and costs, and in the publication of the results. All projects have produced popular articles (with the exception of the recently started Canada lynx reintroduction), but not all have made the data available for peer review or have published in scientific journals.

According to our checklist, the river otter reintroduction in Missouri (Melquist & Dronkert, 1987; Reading & Clark, 1996; Hamilton, 1998), together with the lion project in Natal (where an identically organised cheetah, *Acinonyx jubatus*, reintroduction was carried out; Hunter, 1998) were outstandingly well designed, as were the black-footed ferret projects, with exception of the first one in Wyoming (Reading & Clark, 1996; Vargas *et al.*, 1999; B. Miller, pers. comm.). In contrast, the re-establishment of the Eurasian lynx in Switzerland was exceptionally poorly done. Only a few individuals were released in the Swiss Alps, without co-ordination and at different sites (Breitenmoser *et al.*, 1998). Most of the recommendations in Table 12.4 were neglected. As a matter of fact, there was a second reintroduction at the same time, in the Jura Mountains (Breitenmoser & Baettig, 1992; Breitenmoser *et al.*, 1998), with even fewer animals. The animals, wild-caught in the Slovakian Carpathian Mountains, were set free without any pre or post-release activities nor any public information. Nevertheless, in both the Swiss Alps and the Jura Mountains, lynx populations have emerged from these releases and survived to the present. The lynx reintroduction in Switzerland was started in the early 1970s, when little experience (and no guidelines) was available for carnivore recovery programmes. However, these early projects offer an opportunity to assess the long-term effects of omissions in project design.

Lynx are not especially easy to reintroduce, as can be seen from the number of failed projects (Appendix 12.1). The population in the Swiss Alps has undergone several fluctuations, and its long-term survival is still not assured. At present, the population numbers about 100 individuals, with a relatively high abundance in the north-western Swiss Alps, but there is no clear evidence of emigration into unoccupied areas (Breitenmoser *et al.*, 1999). The limited expansion of the population seems to be a consequence of the natural and artificial fragmentation of the Alps together with the specific life history of the lynx. To overcome this deadlock and to link the small, isolated lynx occurrences of the Alps (Breitenmoser *et al.*, 2000), further reintroductions or translocations are necessary (Breitenmoser, 1998b). Such a strategy is however compromised through the ongoing controversy about the lynx, which is largely a result of the lack of any public information or public involvement in the early stages of the reintroduction and the lack of clear management guidelines up to the present. Meanwhile, illegal shooting of lynx is a considerable mortality factor and a threat to the small population (Breitenmoser *et al.*, 1998).

A poorly designed project will not necessarily fail. Omissions in regard to valuational/organisational aspects, communication and public involvement will, however, ultimately lead to conflict and controversy if, at a later stage of the project, further management decisions must be taken. Long-term support of government and non-government organisations, interest groups and the general public is very difficult to obtain if they have not been properly informed and involved from the beginning.

The importance of, and the struggle for, an adequate organisational structure can be illustrated by the swift fox recovery programme in southern Canada. The release programme in Canada began in 1983 and involved a large number of individuals and agencies (Scott-Brown *et al.*, 1987; Brechtel *et al.*, 1993; Carbyn *et al.*, 1994). The programme evolved through different stages. From its inception, the releases were not based on a single approved recovery plan. It had developed from a private initiative (1973) to a university project (1976) with some governmental support to an interagency co-operative programme (1984). Letters of agreement between governments, which had expired by 1989, were renewed to continue through to 31 March 1994, and extended to 1997. Documents that placed the programme in perspective initially were student thesis projects, done through the University of Calgary under the direction of Stephen Herrero, and subsequently agency reports (Carlington, 1978, 1980; Reynolds, 1983; Russell & Zendran, 1983). By 1989, a considerable amount of field work had been done. However, a general framework for the operation was still lacking. A newly-appointed Recovery Team developed options and a management strategy. Without the lengthy trial and error period, none of the information we now have on the responses of the foxes to different release techniques and to different environmental conditions, would have been available. When the Recovery Team was established in April 1989, a wealth of experience upon which to build a programme was available. The committee set out a schedule and maintained it throughout. From 1989 to 1992 the committee held 17 meetings, discussing organisational structures and policy as much as research. In the field, a greater emphasis was put on practical aspects relating to captive breeding, acquisition of wild foxes and broad-scale monitoring.

Was the Canadian Swift Fox Program a success? After a total of 942 foxes had been released over a period of 11 years, it was found that foxes could survive and reproduce in an area where they once flourished but had become extinct. Survival and reproduction to F_1, F_2 and subsequent generations was achieved. Continued monitoring revealed an increasing fox density in the core area. Despite these results, the prognosis of failure or success was still left open to question (Carbyn *et al.* 1994). One cannot be assured of long-term success, but, based on the thorough surveys and results from the 1996–97 winter surveys (Cotterill, 1997), the official delisting by COSEWIC (1978) from extirpated to endangered (Carbyn, 1998) was a step in officially acknowledging that the 15-year programme had achieved its objectives.

The shift from a private initiative to a university project and finally into the responsibility of governmental agencies of the swift fox programme is characteristic of many reintroductions and one explanation for organisational problems. Private institutions and universities are not adapted for long-term tasks, such as monitoring and continuous management. Even if a GO is in charge from the beginning, organisational and political problems may not be avoided. The Colorado reintroduction of lynx was, according to our checklist (Table 12.4), relatively well designed with regard to the biological/technical aspects, but it neglected important organisational recommendations, such as the creation of a comprehensive multidisciplinary advisory team. The early starvation of some of the released animals set off protests from animal welfare organisations, and the project was suddenly the centre of a major political controversy that went far behind the reintroduction of the lynx (Anon, 1999b). Negligence with regard to a peer review of the reintroduction project may have caused some errors in the release procedures. The lack of a multidisciplinary advisory board, including critical voices, however, has prevented the people responsible for the project realising the immense political dimension of this reintroduction. Wildlife biologists may be good scientists; however, focusing on their subject of interest, they tend to ignore the socio-cultural and political aspects of a carnivore reintroduction.

CONCLUSIONS AND RECOMMENDATIONS FOR CONSERVATION

As long as only relatively few reintroduction projects have been carried out for the same species under comparable environmental and socio-economical conditions, we have to rely upon general recommendations (e.g. the IUCN guidelines) and case-studies to prepare a carnivore reintroduction. From our review, we would like to emphasise the following recommendations to increase the chance of success for future reintroduction projects.

- *Learning from experiences.* Reintroduction projects must be better documented. The IUCN/SSC (1998) guidelines as adapted in Table 12.4 provide not only a checklist for the planning, but also for the documentation of a project. In order to learn not only individually by trial and error, but also as decision-making organisations, experience should be published to make it available to other people. It is important not only to advertise the success stories, because often the lessons from failures are more valuable. In order to publish conclusions from reintroduction programmes in recognised journals, the projects must follow a scientific approach and carefully designed monitoring to gain the relevant data. Aspects not qualified for scientific papers could still be distributed by means of the fast developing and powerful internet (World Wide Web; see for example USFWS's comprehensive and up-to-date information on the Mexican wolf reintroduction at http://ifw2es.fws.gov/MexicanWolf) or through periodicals such as the *IUCN Reintroduction Specialist Group* newsletter.
- *Long-term approach.* Most projects suffer from a too narrow time-frame at the beginning of the reintroductions. Often, the problems of a re-established population show up only after several generations. This is very obvious for possible inbreeding problems, and often the case for management problems. Carnivores have a high potential to cause conflict with human activities. The socio-economic problems may however only start when the population has reached a certain extension and number of individuals. Monitoring, often the first task cut when the budgets shrink, or even intensive population studies by means of telemetry could be as important 10 or 25 years after the reintroduction as they are for the released animals. We do not know a single case where financial support was ready for such a long time-span from the beginning of the project. Long-term surveillance of a population requires the commitment of government institutions, such as wildlife services. Such organisations must be incorporated from the very beginning of a project.
- *Human dimension aspects.* An obvious prerequisite for a successful reintroduction is that the reasons for the former decline are understood and that limiting factors are removed. This claim not only concerns the ecological resources, such as habitat or prey base, but also negative attitudes of people towards the carnivore species to be reintroduced. Especially, reintroductions of large carnivores like wolves, brown bears or large felids provoke conflict with interest groups, such as hunters and livestock breeders. Even in areas where the ecological conditions for the

return of large carnivores are favourable, the basic attitude of local people may still be negative (Breitenmoser, 1998a). Insignificant decreases in game densities or minor losses of domestic stock due to lynx can make illegal killing the most important mortality factor, as for the lynx in the Alps (Breitenmoser *et al.*, 1998). In the reintroduction of the Mexican wolf in Arizona and New Mexico, 4 out of 11 released individuals were shot (Parsons, 1998). Even smaller carnivores are often considered to be excessive killers of small game, fish or livestock and clandestinely persecuted. Viable carnivore populations need vast spaces, which they can rarely find within protected areas. Carnivore reintroductions will only be successful in the long-term if humans are willing to co-exist with the predators. It is not enough to gain the (urban) majority of public opinion in favour of a carnivore reintroduction; eventually the hunters and livestock owners living in and from the areas where the predators ought to live will decide the success or failure of the reintroduction. To gain the support, or at least the tolerance of local people, they need not only to be informed and educated, but should also be involved in decisions regarding the management of reintroduced populations. However, to involve local people, their values and basic believes must be understood. To gain this knowledge, human dimension research should accompany a reintroduction project from its initial phase.

- *Peer-reviewed examinations.* Reintroduction programmes have rarely received peer-reviewed evaluation despite its great potential to improve conservation efforts (reviews in Meffe *et al.*, 1998; Kleiman *et al.*, in press). An outstanding exception is the assessment of the conservation status of the southern sea otter population where the mechanism of the review process is outlined in detail (Ralls *et al.*, 1996). External evaluations should not only address whether or not specific aims and objectives have been achieved and analyse the reasons for the success or failure, but also help to refine project aims, definitions of success and methods used. Financial and structural limitations all too often have the result that reintroduction projects have to operate on the basis of compromises between ideal and achievable objectives, thus reflecting political, economic or organisational rather than biological policy. Here, peer review can potentially benefit projects greatly because reviewers can operate outside the thinking framework and socio-political context constricting institutions managing, or opposing, reintroduction projects. A first layer of review is now more commonly being implemented by creating expert panels to advise project managers.

Interdisciplinary workshops may not only be an ideal platform to gather and analyse all relevant data, but also to allow participation of and communication between all parties involved. Very often, outside observers have not only a more objective view and can help to define the underlying problems better, but also their judgement is more readily accepted by the people involved.

- *Definition of success.* We propose the application of the IUCN Red List Categories (IUCN, 1994) to assess success and failure at about 10 years (or an appropriate number of generations according to the turn-over of the species) after completion of the release phase. Any project where the population is assessed as 'critically endangered' should be considered as a failure because there is a very high risk of extinction in the wild in the near future, regardless of whether the population just 'hangs on'. 'Endangered' populations should be classified as 'uncertain'. The minimum standard for 'success' should be the 'vulnerable' category or better. Vulnerable populations still face a relatively high risk of extinction in the medium-term future and require ongoing management. The classification is based on an assessment where five criteria (population reduction, area of occurrence and occupancy, two criteria for population density and a quantitative analysis of the extinction probability) are carefully balanced with each other (e.g. Ralls *et al.*, 1996). A 'vulnerable' taxon / 'successful' reintroduction requires, for example, more than 250 mature animals. We also propose a first evaluation at five years after the completion of the release phase in order to estimate how likely it is that the programme will prove successful or a failure. This evaluation should be carried out using the same criteria, except the analysis of past population demography since this parameter is obscured by the population built-up following the releases (see Table 12.5).

- *Conservation value of carnivore reintroductions.* Despite the failure of many projects, carnivore re-establishment has been shown to be technically possible in almost all species where reintroductions have been attempted. The usefulness of reintroductions for carnivore conservation in a larger perspective should, however, be critically assessed *a priori* and *a posteriori* for all attempted and planned projects. The reintroduction of wild dogs into Hluhluwe-Umfolozi park would never have been realised under the consideration of the IUCN/SSC (1998) guidelines asking for the establishment of 'viable, free-ranging populations' or any probability for extinction criteria. In a comprehensive review of pros and cons of wild dog reintroductions,

Table 12.5. Biological criteria for project success or failure of reintroductions at about five/ten years after the completion of the reintroduction phase. Any projects that fall in between failure and success should be classified as uncertain. The guidelines follow the IUCN Red List Categories (IUCN, 1994; www.iucn.org/themes/ssc/redlists/ssc-rl-c.htm)

Criteria	Failure	Success
• Population reduction of x %, projected or suspected within the next 10 years or 3 generations, whichever is longer.	> 80%	< 20%
• Extent of occurrence estimated x km^2 or area of occupancy estimated to be y km^2, and estimates indicating two of the following: (1) severely fragmented or known to exist at only one location; (2) projected decline or extreme fluctuations in extent of occurrence, area of occupancy, habitat area or quality, number of locations or subpopulations, or number of mature individuals; (3) continuous, observed, inferred or projected decline in area, extend or quality of habitat.	x < 100, y < 10	$x \geq$ 5000, $y \geq$ 500
• Population estimated to number x mature individuals, and projected continuous decline in number of mature individuals, and population severely fragmented or all individuals in a single subpopulation.	x < 250	$x \geq$ 2500
• Population estimated to number x mature individuals.	x = 50	$x \geq$ 250
• Probability of extinction is x within y years or z generations, whichever is the longer.	$x \geq$ 50%, y = 10, z = 3	x < 20%, y = 20, z = 5

Woodroffe & Ginsberg (1997a) conclude that reintroductions are of limited value for wild dog conservation. Similarly, reintroductions of cat species may be of great local or regional significance, but they do not have a significant role on a global scale (Nowell & Jackson, 1996). In wild dogs, as well as in many other species, the species' range has contracted as a consequence mainly of habitat deterioration and/or fragmentation and persecution. Often, remaining habitat patches are too small or isolated to sustain populations over larger time-spans. Only continuous, intensive and expensive management of a network of small re-established populations as metapopulations increases the probability for long-term population persistence. Investing in the conservation of still existing larger populations may often be more effective. Whilst the swift fox was re-established in the Canadian prairie, the swift fox range has further decreased in the United States.

In a more regional perspective, carnivore reintroductions were carried out to restore local bio-diversity and ecosystems. This involved species not globally threatened, but locally extinct. Such reintroduction programmes are typical in industrialised countries. Here, changes in the attitude and in the distribution of people may have considerably increased the ecological conditions for the existence of the extinct taxon in certain, mostly mountainous areas (Breitenmoser, 2001). Some spectacular reintroductions (e.g. the recovery of the ibex, *Capra ibex*, in the Alps) have considerably increased the awareness of nature conservation of a large public. Carnivore reintroductions, however, which are often controversial, have a higher risk of backfire when they fail. Not only for the sake of the animals, but also for the maintenance of public support for nature conservation projects in general, carnivore reintroductions should only be undertaken when a high probability of success is likely.

FURTHER RESEARCH AND NEEDS

This and earlier reviews have shown that we lack the information to evaluate the decisive parameters defining the success of a reintroduction project. In most reintroductions, a more scientific and standardised approach would help to gain this information. However, reintroduction programmes are typically promoted by conservation or management organisations, and the scientific community is rarely involved at an initial stage; reintroduction projects would offer unique opportunities for scientific work on carnivores. The need for more objective documentation on the one hand, and the chance to do interesting basic research on the other, should lead to a fruitful co-operation between conservationists and scientists. Some of the questions that should be prioritised are:

- *Long-term development of the genetic structure of a reintroduced population.* Only recently, have projects started to include genetic aspects more rigorously into reintroduction planning and investigation of the genetic composition of reintroduced populations (Hedrick, 1995; Forbes & Boyd, 1997; Hedrick *et al.*, 1997; Serfass *et al.*, 1998). Most carnivore reintroductions were started from a relatively small number of individuals, which mainly came from one resource. Up to now, we do not have any studies showing the long-term effect of such inbreeding situations, but some of the early reintroductions of carnivores would today offer the time contrast needed to investigate such questions.
- *The ecosystematic aspects of carnivore reintroductions* should be given more emphasis. Ecosystems may have altered considerably when predators were removed. The conditions these species meet as they return may therefore be quite different from the time when they went extinct. This includes the effect of (large) carnivores on their (ungulate) prey populations. On the one hand, hunters often oppose carnivore reintroduction projects because they fear reduction of game populations, but on the other hand, the potential to control ungulate or other prey populations is used as an argument in favour of reintroducing predators. The impact of carnivores on their prey populations is generally not very well understood. Reintroduction projects offer opportunities to do experimental field work on predator–prey relationships. Such research would help to solve management problems related to carnivore reintroductions and at the same time would provide basic scientific information.
- *Definition of aims and success.* Many projects have no, or poorly defined, aims and, consequently, insufficient definitions of 'success'. As outlined above, the calculation of extinction probabilities using predictive demographic population modelling may contribute to the assessment of reintroduction success. However, the use of population models is rare and does not include predictive aspects (e.g. the Missouri river otter model; Hamilton, 1998). The adequacy of the extinction probability criterion is being controversally discussed. It is often not possible to apply this criterion, either because the frequency of disasters is unknown, or the available computer programs do not

incorporate some critical population parameters of carnivores, and habitat loss and persecution can not be considered (Ginsberg & Woodroffe, 1997a; Ralls *et al.*, 1996). Also the sensitivity of computer programs, such as VORTEX, to density-dependent parameters (Brook *et al.*, 1999; Lacy *et al.*, 1995; Lindenmayer *et al.*, 1995; Mills *et al.*, 1996) and critical threshold of effective population size for genetically secure populations (Franklin & Frankham, 1998; Lynch & Lande, 1998) is currently being critically discussed. Nevertheless, the development and use of predictive models will allow a more formal and critical measurement of the success of reintroduction programmes. The currently available predictive models are not suitable for modelling social and spatial organisation in some carnivores (Ginsberg & Woodroffe, 1997a). Moreover, specific parameters, such as density-dependence of reproductive success, are often not sufficiently known to allow predictive rather than retrospective, informative modelling. Therefore, both the development of demographically and spatially-structured models and intensified monitoring of population processes are required.

- *Human dimension research* in relation to carnivore reintroductions must be enhanced, using an interdisciplinary approach involving ecologists, anthropologists, psychologists and others. Conflict with human activities is often a major threat to carnivore recovery programmes. Habitat fragmentation and persecution of predators and their prey are key factors to carnivore conservation, and can not be separated from the human context. These conflicts tend to have far-reaching socio-cultural backgrounds that are often not really understood. Conservation projects are doomed to fail without a better understanding of the human dimension and without improvements in the attitude of human society towards conservation (Gipps, 1991). This applies especially to carnivores.

Acknowledgements

We are grateful to N. Gerstel, D. Hamilton, L. Hunter, B. Miller, G. Rauer, and R. Reading for the evaluation of the projects listed in Table 12.4. Special thanks to B. Miller and P. Jackson for their critical comments on the manuscript and improvement of the language.

Appendix 12.1. Reintroduction projects of carnivore species

ID	Species	Location	Country	Years	Numbers	Origin	Outcome	Selected references (WWW = WorldWideWeb)
	Canidae							
1	*Canis lupus*	Alaska	USA	1972	5 (2/3)	cap	failure	in Reading & Clark, 1996; in Woodroffe *et al.*, 1997
2		Upper Peninsula, Michigan	USA	1974	4 (2/2)	w.c.	failure	in Reading & Clark, 1996; in Woodroffe *et al.*, 1997
3		Georgia	Georgia	1974–79	22	cap	success	Badridze, 1999
4		Minnesota	USA	1975–78	107 (53/54)	w.c.	success	in Reading & Clark, 1996; in Woodroffe *et al.*, 1997
5		Central Idaho	USA	1995–96	35	w.c.	success	Fritts *et al.*, 1995; Forbes & Boyd, 1997
6		Yellowstone, C. Idaho	USA	1995–96	49	w.c.	success	Fritts *et al.*, 1995; Forbes & Boyd, 1997
7	*Canis l. baileyi*	Arizona	USA	1998	11	cap	running	Parsons, 1998; Parsons, 1999; WWW
8	*Canis rufus*	Alligator River NWR, N. Carolina	USA	1987–95	64 (32/32)	cap	success	Phillips, 1992; Reading & Clark, 1996; in Woodroffe *et al.*, 1997; Kelly, 1999; WWW
9		Southern Appalachians	USA	1991–92	37	cap	uncertain	Henry & Lucash, 1998; WWW
10	*Lycaon pictus*	Kalahari Gemsbock NP	South Africa	1975	5 (2/3)	w.c.	failure	in Woodroffe *et al.* 1997
11		Etosha NP	Namibia	1978	6	cap	failure	in Woodroffe *et al.* 1997

Appendix 12.1. (cont.)

ID	Species	Location	Country	Years	Numbers	Origin	Outcome	Selected references (WWW = WorldWideWeb)
12		Hluhluwe-Umfolozi P.	South Africa	1980–81	22 (12/10)	mix	success	in Reading & Clark, 1996; in Woodroffe et al. 1997
13		Matetsi Safari Area	Zimbabwe	1986	9 (5/4)	cap	failure	in Woodroffe et al., 1997
14		Etosha NP	Namibia	1989	5 (2/2/1)	cap	failure	in Woodroffe et al., 1997
15		Etosha NP	Namibia	1990	11 (4/7)	cap	failure	in Woodroffe et al., 1997
16		Klaserie Game Reserve	South Africa	1991	8 (6/2)	cap	failure	in Woodroffe et al., 1997
17		Venetia	South Africa	1992	14 (7/7)	w.c.	failure	in Reading & Clark, 1996; in Woodroffe et al., 1997
18		Madikwe Game Res.	South Africa	1995	6 (3/3)	mix	success	in Woodroffe et al., 1997
19	Vulpes velox	Tsavo NP	Kenya	1997	6 (2/4)	mix	uncertain	Soorae, 1997
20		South Dakota	USA	1980		uk	failure	in Reading & Clark, 1996
21		Saskatchewan, Alberta	Canada	1983–pres.	942	mix	success	see this chapter
	Felidae							
22	*Acinonyx jubatus*	Mkuzi	South Africa	1966	33	uk	failure	in Reading & Clark, 1996
23		Hluhluwe/Umfolozi	South Africa	1966–69	64	uk	failure	in Reading & Clark, 1996
24		Ndumu Game Reserve	South Africa	1971	14	uk	unknown	Hunter, 1998
25		Kruger	South Africa	1973		uk	failure	in Reading & Clark, 1996
26		St Lucia Reserve	South Africa	1978–81	18	uk	failure	Hunter, 1998

		Site	Country	Year	Count	Source	Outcome	Reference
27		Itala	South Africa	1979	13	uk	failure	in Reading & Clark, 1996
28		Bophuthatswana	South Africa	1981–84		uk	success	in Reading & Clark, 1996
29	*Felis concolor*	Phinda, Kwazulu-Natal	South Africa	1992–94	15	w.c.	uncertain	Hunter, 1998
30		Matusadona NP	Zimbabwe	1993–pres.	20	w.c.	running	Atkinson & Wood, 1995
31		Hluhluwe/Umfolozi	South Africa	1995–pres.	22	uk	running	Marker-Kraus, 1996
32		Northern Florida	USA	1988–89	7 (3/4)	w.c.	failure	in Reading & Clark, 1996
33	*Felis silvestris*	Augstmatthorn	Switzerland	1962–69	19	mix	failure	Stahl & Artois, 1991
34		Jura Mts of Vaud	Switzerland	1970–72	19	uk	unknown	Stahl & Artois, 1991
35		Franges-Montagnes	Switzerland	1975	5	uk	failure	Stahl & Artois, 1991
36		Bavaria	Germany	1984–89	129	cap	success	Stahl & Artois, 1991
37	*Lynx canadensis*	Adirondacks, NY	USA	1988–90	83 (48/35)	w.c.	uncertain	Brocke & Gustafson, 1992; in Reading & Clark, 1996
38		Colorado	USA	1999–pres.	96	w.c.	running	Seidel et al., 1998; Byrne & Shenk, 1999; WWW
39	*Lynx lynx*	Rominter Heide	Russia	1941	5 (2/2/?)	mix	failure	Festetics, 1980
40		Bavarian Forest	Germany	1970–75	5–7	mix	failure	in Reading & Clark, 1996; Thor & Pegel, 1992
41		Jura	Switzerland	1971–80	10 (5/5)	w.c.	success	Breitenmoser et al., 1998
42		Alps	Switzerland	1971–82	14–18 (8/6/4)	w.c.	success	in Reading & Clark, 1996; Breitenmoser et al., 1998
43		Gran Paradiso NP	Italy	1975	2 (2/0)	w.c.	failure	Boitani & Francisci, 1978
44		Kocevje	Slowenia	1976	6 (3/3)	w.c.	success	in Reading & Clark, 1996; Cop & Frkovic, 1998

Appendix 12.1. (cont.)

ID	Species	Location	Country	Years	Numbers	Origin	Outcome	Selected references (WWW = WorldWideWeb)
45		Alps	Austria	1977–79	9 (6/3)	w.c.	failure	in Reading & Clark, 1996; Huber & Kaczensky, 1998
46		Sumava	Czech Rep.	1982–89	17 (11/6)	w.c.	success	in Reading & Clark, 1996; Cerveny & Bufka, 1996
47		Vosges Mts	France	1983–92	16–18 (5/11/2)	mix	uncertain	in Reading & Clark, 1996
48		Jorat, Plateau	Switzerland	1989	3	uk	uncertain	Breitenmoser et al., 1998
49		Kampinoski	Poland	1993–95	5 (2/3)	cap	uncertain	Böer et al., 1995
50	Lynx rufus	Cumberland Is, Georgia	USA	1988–89	32	w.c.	success	in Reading & Clark, 1996; Nowell & Jackson, 1996
51	Panthera leo	Uttar Pradesh	India	1958		uk	failure	in Reading & Clark, 1996
52		Hluhluwe/Umfolozi	South Africa	1963		uk	failure	in Reading & Clark, 1996
53		Hluhluwe/Umfolozi	South Africa	1965	5	uk	unknown	in Reading & Clark, 1996
54		Phinda, northern Natal	South Africa	1992–93	13	w.c.	uncertain	Hunter, 1998
55	Panthera pardus	Mkuzi	South Africa	1975–76	6	w.c.	success	in Reading & Clark, 1996
56		Itala	South Afirca	1980–91	3	w.c.	success	in Reading & Clark, 1996

No.	Species	Location	Country	Year	Number	Source	Outcome	Reference
57		Songimvelo	South Africa	1990	7	w.c.	success	in Reading & Clark, 1996
58	*Felis serval*	Transvaal	South Africa	1981–83		uk	success	in Reading & Clark, 1996
	Hyaenidae							
	Hyaena brunnea							
59		Hluhluwe/Umfolozi	South Africa	1977	4	uk	failure	in Reading & Clark, 1996
60		Itala	South Africa	1980	5	uk	failure	in Reading & Clark, 1996
61		Transvaal	South Africa	1981–82		uk	success	in Reading & Clark, 1996
62		E shores of St.Lucia	South Africa	1983	7	uk	success	in Reading & Clark, 1996
	Mustelidae							
	Enhydra lutris							
63		Pribilofs	USA	1951–68	128	w.c.	failure	Love, 1992
64		South-east Alaska	USA	1965–69	412	w.c.	success	in Reading & Clark, 1996; Love, 1992
65		Washington	USA	1969–70	59	w.c.	success	in Reading & Clark, 1996; Love, 1992
66		Vancouver Is, B.C.	Canada	1969–72	179	w.c.	success	in Reading & Clark, 1996; Love, 1992
67		Oregon	USA	1970–71	93	w.c.	failure	in Reading & Clark, 1996; Love, 1992
68		San Nicolas Is, Calif	USA	1987–90	166	w.c.	uncertain	in Reading & Clark, 1996; Love, 1992
69	*Lutra canadensis*	Colorado	USA	1976–91	107 (42/42/23)	w.c.	uncertain	in Melquist & Dronkert, 1987; T. Beck, 1990; Hamilton pers. comm.
70		Minnesota	USA	1979	23	w.c.	success	in Melquist & Dronkert, 1987; Hamilton pers. comm.

Appendix 12.1. (*cont.*)

ID	Species	Location	Country	Years	Numbers	Origin	Outcome	Selected references (WWW = WorldWideWeb)
71		Arizona	USA	1981–83	46	w.c.	success	in Melquist & Dronkert, 1987; Hamilton pers. comm.
72		Alberta	Canada	1981	21	w.c.	unknown	in Melquist & Dronkert, 1987
73		Kentucky	USA	1982	11	w.c.	unknown	in Melquist & Dronkert, 1987
74		Kentucky	USA	1991–94	355	w.c.	success	Hamilton pers. comm.
75		Missouri	USA	1982–92	845	w.c.	success	in Reading & Clark, 1996; in Melquist & Dronkert, 1987a; Hamilton, 1998
76		Tennessee	USA	1983–94	487	w.c.	success	in Melquist & Dronkert, 1987; Hamilton pers. comm.
77		Pennsylvania	USA	1983–pres	112	w.c.	running	in Reading & Clark, 1996; Serfass, 1998; Hamilton pers. comm.
78		Kansas	USA	1983–84	19	w.c.	success	in Melquist & Dronkert, 1987; Roy, 1997; Hamilton pers. comm.
79		Indiana	USA	1995–pres	303	w.c	running	Hamilton pers. comm.; WWW
80		Iowa	USA	1984–91	261	w.c.	success	in Melquist & Dronkert, 1987; Hamilton pers. comm.

No.	Species / Location	Country	Period	Number	Source	Outcome	Reference
81	Oklahoma	USA	1984–85	20	w.c.	success	in Melquist & Dronkert, 1987; Hamilton pers. comm.
82	West Virginia	USA	1984–97	249	w.c.	success	in Melquist & Dronkert, 1987; Hamilton pers. comm.
83	Ohio	USA	1986–92	123	w.c.	success	Hamilton pers. comm.
84	Nebraska	USA	1986–91	159	w.c.	success	Andelt pers. comm.
85	North Carolina	USA	1989–96	267 (160/107)	w.c.	success	Spelman, 1998; Hamilton pers. comm.
86	Maryland	USA	1990–pres	80	w.c.	running	Hamilton pers. comm.
87	Illinois	USA	1994–97	346	w.c.	success	Hamilton pers. comm.
88	New York	USA	1995–pres.	153	w.c.	running	Serfass, 1998; WWW
89	*Lutra lutra* Schwarzwasser-Sense	Switzerland	1975	8(4/4)	w.c.	failure	Weber et al., 1991
90	Eastern England	UK	1983–89	18	uk	success	in Reading & Clark, 1996
91	south-central Sweden	Sweden	1987–92	36	mix	success	Sjöåsen, 1995
92	Cataluña	Spain	1995–pres.	~ 20	w.c.	running	Morán et al., 1998
93	*Martes americana* Alaska, southeastern islands	USA	1934–52	58 (10/16/32)	uk	success	in Reading & Clark, 1996; in Berg 1982
94	Silver Bow Co., Montana	USA	1944	12	w.c.	unknown	in Slough, 1994
95	Sibley Prov. Park, Ontario	Canada	1950–51	47 (31/16)	w.c.	unknown	in Slough, 1994
96	2nd College Grant, New Hampsh.	USA	1953	2 (1/1)	w.c.	unknown	in Slough, 1994
97	Stockton I., Wisconsin	USA	1953	5	w.c.	unknown	in Slough, 1994
98	Prince Albert NP, Saskatchewan	Canada	1954	24 (12/12)	w.c.	success	in Slough, 1994; in Berg, 1982
99	Lincoln Co., Montana	USA	1955	21	w.c.	unknown	in Slough, 1994
100	Porcupine Mt State Park, Michigan	USA	1955–57	29 (18/11)	w.c.	failure	in Slough, 1994; in Berg, 1982

Appendix 12.1. (cont.)

ID	Species	Location	Country	Years	Numbers	Origin	Outcome	Selected references (WWW = WorldWideWeb)
101		Liscomb Game Sanct., Nova Scotia	Canada	1956	12 (5/7)	w.c.	failure	in Slough, 1994; in Berg, 1982
102		Meager Co., Montana	USA	1956–57	9 (4/5)	w.c.	failure	in Slough, 1994
103		Parry Sound Dist., Ontario	Canada	1956–63	249 (155/94)	w.c.	success	in Slough, 1994; in Berg, 1982
104		Barb Lake, Manitoba	Canada	1960–61	11 (?/($\geq$ 3)	w.c.	failure	in Slough, 1994; in Berg, 1982
105		Tramping Lake, Manitoba	Canada	1961	2 (1/1)	w.c.	failure	in Slough, 1994; in Berg, 1982
106		Minago R., William L., Manitoba	Canada	1967–68	99	w.c.	success	in Slough, 1994; in Berg, 1982
107		Acadia For. Res. Stn., N.B.	Canada	1967–68	< 10	w.c.	failure	in Slough, 1994
108		Hiawatha Natl. Forest, Michigan	USA	1968–70	99 (62/37)	w.c.	failure	in Slough, 1994; in Berg, 1982
109		Duck Mt, Manitoba	Canada	1969	42	w.c.	success	in Slough, 1994; in Berg, 1982
110		La Poile River, Newfoundland	Canada	1975	3 (1/2)	w.c.	success	in Slough, 1994
111		White Mt. NF, New Hampshire	USA	1975	29 (20/9)	w.c.	unknown	in Slough, 1994
112		Nicolet Natl. Forest, Wisconsin	USA	1975–83	172 (121/51)	w.c.	success	in Slough, 1994; in Berg, 1982
113		Maine River, Newfoundland	Canada	1976–78	11 (7/4)	w.c.	unknown	in Slough, 1994
114		McCormick, Exp.For., Michigan	USA	1980	22 (9/13)	w.c.	success	in Slough, 1994; in Berg, 1982
115		Iron Co., Michigan	USA	1980–81	48 (21/27)	w.c.	success	in Slough, 1994; in Berg, 1982

	Location	Country	Years	N (M/F)	Source	Outcome	Reference
116	N.Black Hills, South Dakota	USA	1980–81	43 (25/18)	w.c.	success	in Slough, 1994; in Berg, 1982
117	Terra Nova NP, Newfoundland	Canada	1982–83	10	w.c.	success	in Slough, 1994
118	Northern Maine	USA	1983	76	w.c.	success	in Slough, 1994
119	Takhini R/Wheaton Lake, Yukon	Canada	1984–86	31 (16/17)	w.c.	unknown	in Slough, 1994
120	Haines Junction, Yukon	Canada	1984–87	51 (34/17)	w.c.	success	in Slough, 1994
121	Fundy NP, N.B.	Canada	1984–91	44 (27/17)	w.c.	success	in Slough, 1994
122	N Otsega, S Cheboygans C., Mich.	USA	1985	49 (25/24)	w.c.	success	in Slough, 1994
123	Braeburn, Yukon	Canada	1985–86	26 (14/12)	w.c.	unknown	in Slough, 1994
124	Takhini River, Yukon	Canada	1985–86	63 (42/21)	w.c.	success	in Slough, 1994
125	Northeast Lake Co., Michigan	USA	1986	36 (19/17)	w.c.	success	in Slough, 1994
126	Chequamegon NF, Wisconsin	USA	1987–90	139 (94/45)	mix	unknown	in Slough, 1994; Berg & Kuehn, 1994
127	Kejimkujik Natl. Park, Nova Scotia	Canada	1987–90	80 (45/35)	w.c.	unknown	in Slough, 1994
128	Chippewa Co., Michigan	USA	1989	20 (11/9)	w.c.	unknown	in Slough, 1994
129	Central Black Hills, South Dakota	USA	1989–90	39 (21/18)	w.c.	unknown	in Slough, 1994
130	Green Mt. Natl. For., Vermont	USA	1989–90	69 (52/17)	w.c.	unknown	in Slough, 1994
131	Riding Mt NP, Manitoba	Canada	1992–93	68 (37/31)	w.c.	success	Schmidt, 1993
132	Shirley Basin, Wyoming	USA	1991–94	228 (140/88)	cap	failure	in Reading & Clark, 1996; Vargas et al., 1999; Miller pers. comm.

Mustela nigripes

	Location	Country	Years	N (M/F)	Source	Outcome	Reference
133	Charles Russell NWR, Montana	USA	1994–pres.	163 (93/70)	cap	running	Buskirk et al., 1994; Stoneberg, 1995; Vargas et al., 1999
134	Badlands NP and Buffalo Gap NG, South Dakota	USA	1994–98	325 (189/136)	cap	success	Vargas et al., 1999
135	Fort Belknap IR, Montana	USA	1997–pres.	78 (43/31)	cap	running	Vargas et al., 1999
136	Aubrey Valley, Arizona	USA	1996–pres	79 (53/26)	cap	running	Vargas et al., 1999

Martes pennanti

	Location	Country	Years	N (M/F)	Source	Outcome	Reference
137	Nova Scotia	Canada	1955	12 (6/6)	cap	success	in Reading & Clark, 1996; in Berg, 1982

Appendix 12.1. (*cont.*)

ID	Species	Location	Country	Years	Numbers	Origin	Outcome	Selected references (WWW = WorldWideWeb)
138		Patricia portion of N Ontario	Canada	1956	25	w.c.	unknown	in Reading & Clark, 1996; in Berg, 1982
139		Wisconsin, Nicolet Natl Forest	USA	1956–63	60 (36/24)	w.c.	success	in Berg, 1982
140		Ontario, Parry Sound	Canada	1956–63	97 (37/60)	w.c.	success	in Berg, 1982
141		Michigan, Ottawa National Forest	USA	1958–63	61 (42/19)	w.c.	success	in Berg, 1982
142		Montana	USA	1959–60	36 (16/20)	w.c.	success	in Reading & Clark, 1996; in Berg, 1982
143		Oregon	USA	1959–63	24 (10/14)	w.c.	failure	in Reading & Clark, 1996; in Berg, 1982
144		Vermont	USA	1959–67	124 (19/16/89)	w.c.	success	in Berg, 1982
145		Idaho	USA	1962–63	39 (20/19)	w.c.	success	Powell & Zielinski, 1994; in Berg, 1982
146		Wisconsin, Chequamegon Natl Fst	USA	1966–67	60 (30/30)	w.c.	success	in Berg, 1982
147		New Brunswick, southern N.B.	Canada	1966–68	25 (10/15)	w.c.	failure	in Berg, 1982
148		Minnesota, Itasca State Park	USA	1968	15	w.c.	unknown	in Berg, 1982
149		Virginia, west	USA	1969	23 (6/10/7)	w.c.	success	in Berg, 1982
150		Maine, eastern Maine	USA	1972	7	w.c.	failure	in Reading & Clark, 1996; in Berg, 1982
151		Manitoba, Riding Mountain NP	Canada	1972–73	4	uk	failure	Powell & Zielinski, 1994
152		New York, Catskill	USA	1976–79	43 (19/24)	w.c.	unknown	in Reading & Clark, 1996; in Berg, 1982
153		Ontario, Manitoulin,Bruce Pennin.	Canada	1979–82	57 (27/30)	w.c.	unknown	in Berg, 1982
154		Oregon	USA	1981	13 (8/5)	w.c.	unknown	in Berg, 1982
155		Alberta	Canada	1984		uk	failure	Carbyn pers. comm.

			Years	Numbers	Origin	Outcome	References
156	Montana	USA	1988–91	45	w.c.	failure	Powell & Zielinski, 1994; Berg & Kuehn, 1994
157	Alberta	Canada	1990–91	20	uk	uncertain	Carbyn pers. comm.
158	Elk Is NP, Blackfoot RA, Ministik L. Bird Sanctuary, Alberta	Canada	1990–91	20	mix	success	Proulx et al., 1994; Powell & Zielinski, 1994
	Ursidae						
	Ursus americanus						
159	Interior Highlands, Arkansas	USA	1958–68	254	w.c.	success	Smith et al., 1990; J. Clark & Smith, 1994; Smith & Clark, 1996
160	Big S Fork, Tennessee, Kentucky	USA	1995–?	? (?/12)	w.c.	running	J. Clark, 1998
161	Bialowieza	Poland	1938–44	10	uk	failure	Swenson et al., 1999
162	Lower Austria, Styria	Austria	1989–93	3 (1/2)	w.c.	uncertain	Gutleb, 1994; Rauer, 1997; in Reading & Clark, 1996
163	Pyrenee Mts	France	1996	2 (0/2)	w.c.	running	Quenette et al., 1997; Moutou, 1995; Swenson et al., 1999
164	Brenta, Trentino	Italy	1999–pres.	5(2/3)	w.c.	running	Genovesi, 2000
165	Montana	USA	1990–pres.		w.c.	uncertain	in Reading & Clark, 1996

Note for row 162 the label *Ursus arctos* applies.

Years: period when animals were released. Numbers: Numbers of animals released (total (male/female/unknown sex)); Origin: cap = animals from captivity; w.c. = wild caught animals; mix = animals from captivity and from the wild; uk = origin of animals unknown (not documented). Outcome of the project according to experts' opinion.

13

Interactions between carnivores and local communities: conflict or co-existence?

CLAUDIO SILLERO-ZUBIRI AND M. KAREN LAURENSON

INTRODUCTION

Large carnivore populations are declining globally under the pressure of habitat degradation, hunting, disease and the commercial trade of body parts (Weber & Rabinowitz, 1996). Indeed, 22 of 30 large carnivore species reviewed by Fuller (1995) are already listed as 'endangered' either by the IUCN or the US Government. Even against this background, the Carnivora attract disproportionate interest from conservation biologists and more generally, intrigue a wide variety of people. This interest has arisen for a number of reasons, not least because of the elusiveness and powerful image of carnivores, but also as a result of their relative rarity and their role as predators. In this latter role, they frequently interfere with other animals, be they humans, other endangered wild species, game species or livestock. Carnivores are thus frequently perceived as competitors to humans and, historically, human and wild carnivore interactions have involved conflict and misunderstanding.

The rarity of carnivores arises from their life history and ecological characteristics, which makes them particularly susceptible to the damaging effects of incidental mortality, such as that caused by humans (e.g. Weaver *et al.*, 1996). Fundamentally, the biological niche of large carnivores at the top of the food chain means that they will always be less abundant than their herbivore prey. Furthermore, many species (the Canidae excepted) have low reproductive rates through small litter size and relatively delayed sexual maturity, and thus their productivity is inherently low and often cannot compensate for increased mortality rates.

In addition, because of their body size and high trophic position, large carnivores require extensive home ranges and large prey populations. Thus only vast, relatively intact ecosystems can support viable populations and it is fundamentally extremely difficult to maintain large areas for these large carnivores. As a consequence, these species are the first to suffer when human populations expand and cultivate previously untouched habitats. An example is the relative rarity of large carnivores in the densely populated areas of Europe and North America. Bears, wolves and lynx (*Felis lynx*) were rapidly extirpated from the British Isles as the human population spread, although they persisted longest in the less densely populated mountainous and northern areas.

With the strengthening of the conservation ethic during the twentieth century, carnivores are no longer perceived simply as '*vermin*', but as a key and valued part of the ecological community. This change, however, is not necessarily taking place among people living near wild carnivores. Carnivores that 'spill-over' from the edges of protected areas often come into conflict with human neighbours and sour the relationship between local communities and conservation agencies. However, policies and practices have recently moved away from 'fortress conservation' to embrace the idea of incorporating local communities in the process. This involves a broad spectrum of objectives and activities, from those designed to support a protected area or species, for example through community education, to those where rural development can be achieved through the use of resources, with the conservation of these resources a secondary benefit (Barrow *et al.*, 2000).

This move towards community-based conservation is clearly essential for carnivore conservation. If a problem exists between human and carnivore communities then seeking 'solutions' that do not involve local communities is futile, as only the human communities have the ability to ameliorate the situation by changing their own behaviour or that of the carnivores. Thus, as conservationists seek a tolerant co-existence between carnivores and humans there can be no movement towards this state without human involvement.

In this chapter we first look at the reasons why large carnivores frequently find themselves in conflict with humans and then examine the ways in which conservationists have tried to solve these problems. Here we refer particularly to the importance of community involvement in the various programmes. We refer to *communities* as the people of a local administrative unit (such as a municipality), of a cultural or ethnic group (such as a band or tribe) or of a local urban or rural area (such as the people of a particular neighbourhood or valley) (IUCN, 1991). Data for this chapter are derived from original research, literature review, and 25 replies to a questionnaire sent to scientists known to have been involved in carnivore research or conservation.

Although we have selected some good examples of conservation programmes for smaller carnivores, we generally focus on large carnivores
(maximum adult weight $\geqslant$ 15 kg) for a number of reasons. First, as top
predators, they are important indicators of ecosystem function and productivity and as such are vulnerable to habitat alteration and loss. Secondly,
they are thought to provide a protective umbrella for other wildlife species if
their conservation needs are ensured, and thirdly, they have a complex historical relationship with humans that continues to be the most influential
fact determining their fate. Lastly, they undoubtedly pose an enormous
conservation challenge (Fuller, 1995; T. Clark et al., 1996a). We deal only
with natural carnivore populations, excluding problems arising from introduced wild and domestic carnivores.

THE ISSUE: WHY PROBLEMS ARISE BETWEEN HUMANS AND LARGE CARNIVORES

Carnivores come into conflict with humans for a wide variety of reasons.
First and foremost, people see large carnivores as a direct threat to human
life and the real or perceived risk of predation on humans was cited as a
problem in 88% (n = 19) of questionnaire replies. Predation on humans is,
however, actually quite rare. Carnivores can also act as reservoirs of human
diseases, again threatening human life. On the other hand, carnivores can
be a resource for humans, with trade in body parts and skins for medicinal
use or clothing and fashion. It is when this trade becomes unsustainable
that a conservation problem develops. The greatest source of human–carnivore conflict is, however, competition for resources, whether this is for
land, man's domestic animals, his crops (e.g. European badgers – *Meles
meles*) or for prey species. In addition to these rational reasons for carnivore
persecution by humans, Hans Kruuk has suggested that the urge to kill
carnivores is in our instinctive behavioural repertoire, like the hatred for
snakes (Kruuk, 1976a).

The nature and degree of conflict between carnivores and communities
has been quantified to some degree by 74% (n = 19) of the projects that
replied to our survey. This quantification was assessed through informal
interviews in almost all cases, but structured questionnaires and government data were also used extensively. In some other cases, published literature was reviewed, data were collected directly by the project, or local
meetings were used to assess conflict between humans and carnivores.

Attacks on humans
In general, large carnivores are fearful of humans and prefer to avoid them.
However, in isolated cases an animal may be a poor hunter of natural prey

through inexperience, injury or old age and may then resort to attacking humans. If successful, this behaviour may become habitual. In other circumstances frequent human activities lead to habituation and a loss of this fear of humans. For example in tiger habitat, use of the area for tourism, illegal grazing, logging and firewood collection may have led to an increase in tiger attacks on humans (Singh, 1991).

Large cats, particularly tigers (*Panthera tigris*), lions (*Panthera leo*) and mountain lions (*Felis concolor*), account for most human deaths by predators. Lions, whether Asiatic or African, have been feared throughout history. Notorious man-eaters include those of Tsavo, Kenya which at the turn of the twentieth century killed over 100 workers constructing the East African railway (Patterson, 1907). Tigers have also become renowned man-eaters on occasion. Eight man-eating tigers and one leopard (*Panthera pardus*) shot by Jim Corbett in the early 1900s had killed and eaten nearly 1100 people (Corbett, 1957). Today, tigers are probably the most persistent problem species in this regard and are a continuing problem in parts of India where people and tigers use the same habitat. Somewhere between 36 and 100 people are killed each year in the Sundarbans mangrove forests in eastern India, although attacks may be considered to be relatively infrequent when considering some 35 000 people move through this area each year (Chakrabarty, 1992).

Increasing human use of areas previously used predominantly only by wildlife is a common underlying problem explaining increases in big cat attacks. In North America, for example, mountain lion attacks have apparently been increasing as suburban settlements encroach on their habitat (Foreman, 1992; Seidensticker & Lumpkin, 1992). Of 53 unprovoked attacks documented between 1980 and 1990, nine were fatal (Beier, 1991).

Human mortality from other species is rarer, but attacks are not uncommon. For example, there are relatively few reports of bears killing people, but in just one location in Norway, there has been a total of 50 serious confrontations between humans and polar bears since 1973 (Gjertz & Persen, 1987). Grizzly bears (*Ursus arctos*) are notoriously unpredictable and attacks on humans are reported yearly in North America (Herrero, 1985). Sloth bears (*Melursus ursinus*) have occasionally caused human injuries and casualties in Nepal's Royal Chitwan National Park (Nepal & Weber, 1995).

Interestingly, grey wolves (*Canis lupus*) seem to have a particular public relations problem in this regard. Although they have caused no human deaths in North America during the twentieth century (unlike mountain lions) and livestock losses are very low, wolves are often blamed for livestock attacks and are still widely and irrationally feared (Kellert et al., 1996).

Attacks on humans can have a significant negative impact on conservation efforts. For example, one of the major causes of park–people conflict in
Nepal's Royal Chitwan National Park comes from threats to human tres-
passers by tigers and sloth bears (Nepal & Weber, 1995). In the Gir Forest
in India, persistent attacks by Asiatic lions on humans (averaging 14.8 attacks and 2.2 deaths annually), although mostly outside the reserve, have
hindered support and fostered hostile attitudes among local peoples for
lion conservation (Saberwal et al., 1994). Following a drought in 1987–88,
the problem escalated through a combination of increased lion aggressive-
ness, villagers bringing their surviving livestock into their dwellings and a
higher incidence of attacks on villages closer to sites where lions were formerly baited for tourist shows.

Predation on livestock

Predation by carnivores on livestock is the root of a deeply ingrained hatred
for carnivores throughout the world, with every domestic species from
chickens to cattle being affected. Indeed, livestock predation was the most
frequently cited reason (40%) for problems between humans and carnivores in our questionnaire. A number of factors may contribute to this
situation. First, thanks to human protection, domestic animals exhibit little
effective anti-predator behaviour, making them particularly vulnerable to
predators (Kruuk, 1972a). In addition, livestock compete with wild herbivores for resources and thus can reduce the abundance or alter the distribution or behaviour of wild prey, thus changing the pattern of wild carnivore
predation.

Although farmers and ranchers consistently express the most negative
attitudes toward large carnivores (e.g. Kellert, 1985; Reading & Kellert,
1993), large carnivores often constitute a minor problem compared with
smaller carnivores such as jackals, coyotes or feral dogs. However, depredation by any predator of an individual's livestock can have severe emotional,
financial and political consequences (Mech, 1981). In some circumstances
livestock losses can have a significant impact on farmers' livelihoods. For
example, in the Gir Forest Sanctuary, India, 1900–2000 domestic animals
have been killed annually by lions in recent years (Singh & Kamboj, 1996).
Reported losses to snow leopards (*Panthera uncia*) in Annapurna, Nepal,
although only 2.6% of the total stock holding, represented almost a quarter
of the average annual Nepali per capita income (Oli et al., 1994). In another
study in Nepal, losses (18% of the livestock holding) amounting to half the
average annual per capita income were attributed to wolves and snow leopards (Mishra, 1997). Perhaps unsurprisingly, snow leopards are killed by

herdsmen in defence of their livestock and in local opinion total extermination of leopards was the only acceptable solution to the predation problem (Oli *et al.*, 1994).

Livestock predation by wild carnivores is by no means restricted to the developing world and is a common problem in Europe. Wolf and bear predation is a common complaint in parts of the Italian Abruzzo, where extensive grazing is practised (Cozza *et al.*, 1996). Brown bear (*Ursus arctos*) and wolverine (*Gulo gulo*) predation on free ranging sheep in Norway is also common (Landa & Tommeras, 1997; Sagor *et al.*, 1997).

The carnivore–livestock conflict, particularly in the developed world, has been exacerbated by a change in husbandry during the nineteenth century, especially in areas where these carnivores are recolonising or have been reintroduced. Domestic animals are now rarely herded or guarded by dogs whilst grazing and thus are more vulnerable to predation. Furthermore, stockmen have lost the tradition of coexistence with large predators and modern protective legislation for carnivores is not matched by a positive co-operative attitude (Breitenmoser, 1998a). Feral dogs also frequently contribute to problems of livestock predation, but the blame is often apportioned to their wild relatives (Cozza *et al.*, 1996).

Predation on game species

Throughout history, carnivores have been seen as competitors with humans for prey and as a result have been directly persecuted. For example, in the royal hunting preserves of the New Forest (UK) and Bialowiesa (Poland), wolves, bears and lynx were killed to protect populations of fallow deer (*Dama dama*) and red deer (*Cervus elaphus*). In our questionnaire survey, it is the second most common (15%) reason for conflict cited, and deer hunters are among the most vocal opponents to the reintroduction of lynx and wolves in Europe and North America. In Alaska, increasing grey wolf numbers are blamed by hunters for declining moose populations and the resulting reduction in hunting quotas (Gasaway *et al.*, 1992). In this case, diversionary feeding of wolves with carcasses from road kills has been attempted.

In Switzerland, particularly the northwestern Alps, the expansion of lynx after reintroduction has ceased, perhaps due to illegal killing. Hunters claim that the roe deer population has been substantially reduced by lynx and tension is growing as hunters find lynx-killed roe carcasses at deer yards in harsh winters (Breitenmoser *et al.*, 1999a).

In the UK, one of the main activities of professional gamekeepers who act as custodians of wild and reared game birds, is to reduce red fox, stoat

(*Mustela nivalis*), weasel (*M. putorius*) and mink (*M. vison*) numbers and thus increase the harvestable surplus of birds (Reynolds & Tapper, 1996). Conflicts between conservation and hunting interests often arise when protected species are killed illegally (e.g. wild cat *Felis silvestris* and pine marten *Martes martes* on grouse moors in Scotland).

Predation on other endangered wildlife

In somewhat exceptional circumstances, there may be a novel conservation dilemma when a rare carnivore has a significant impact on another endangered species. Proven examples of this are few, but some recent incidents have caused concern. Spotted hyaenas (*Crocuta crocuta*) have been blamed for predation of calves of endangered black rhino (*Diceros bicornis*) in Aberdare, Kenya (Sillero-Zubiri & Gottelli, 1991). Similarly, a study in Namibia suggested that calves born to dehorned rhino mothers that were sympatric with hyaenas were less likely to survive than calves of horned mothers in similar areas or calves of hornless mothers in predator-free areas (Berger & Cunningham, 1994). Hornless mothers were presumably less able to drive away predators.

Another example also comes from the Etosha National Park in Namibia, where three separate wild dog reintroductions failed to a greater or lesser extent when captive-bred wild dogs, which apparently had poorly developed anti-predator behaviour, were killed by lions (Scheepers & Venzke, 1995). Indeed, lion predation is thought to have an effect on free-living wild dog (and cheetah – *Acinonyx jubatus*) populations; there is an inverse correlation between the populations of wild dogs, or cheetahs, and lions (Laurenson, 1995b; Creel & Creel, 1996). When lion populations increased during the 1970s and 1980s in the Ngorongoro Crater in Tanzania, these mesopredators disappeared. In Namibia, the highest density of cheetahs is now found on ranch land, where hyaenas and lions have been controlled, rather than in protected areas (McVittie, 1979).

In Asia, it has been suggested that dholes (*Cuon alpinus*) in Alas Purwo National Park, Java, Indonesia were responsible for a decline in endangered banteng (*Bos javanicus*) (Hedges & Tyson, 1996) and that Asian wolves (*Canis lupus pallipes*) in Velavadar National Park, India, might limit endangered blackbuck antelope (*Antelope cervicapra*) (Yadvendradev, 1994). However, in both cases, although these predators did undoubtedly kill some of these species as prey, there were insufficient data to confirm that the prey population was limited by these predators and for further action to be taken (Yadvendradev, 1994; Sillero-Zubiri, 1996).

Consumptive use of carnivores

Humans can have a direct impact on carnivores by using them as a wild-life resource. This may take the form of harvesting fur-bearing animals, hunting for sport or the use of body parts. Until recently the fur trade concentrated on wild populations, particularly those of spotted cats such as lynx, leopard, jaguar (*Panthera onca*) and ocelot (*Felis pardalis*), but also arctic fox (*Alopex lagopus*) and South American zorros (*Pseudalopex spp.*). This harvest can have a serious impact on the density and demographic structure of these populations, with the impact of trophy hunting on lions particularly well documented (Smuts, 1976; Starfield *et al.*, 1981; Pusey & Packer, 1987; Creel & Creel, 1997).

More recently, the use of carnivore body parts for medicinal uses in Asia is seriously threatening tiger and bear populations. Indeed this illegal trade is regarded as having the greatest potential to reduce tiger populations in the shortest time (Kenney *et al.*, 1995) and with the increase in wealth in Asia this trade is considered to be driving tigers to extinction (Nowell & Jackson, 1996). The devastating effects of commercial poaching can also be exemplified by the dramatic decline of black rhinoceros and African elephants (Leader-Williams *et al.*, 1990).

Conflict over land

The ever-increasing demand for land for agriculture, timber, minerals, new roads, buildings and recreational development leads to wildlife habitat loss and fragmentation and results in increased contact and problems between large carnivores and people. Even when these fragmented habitats are protected, carnivores are not always safe, as their home ranges straddle boundaries. Human induced mortality in nominally protected populations of lions and grizzly bears reaches 50% to 89% (Woodroffe & Ginsberg, 1998). Furthermore, the greater edge effects in smaller protected areas can compound this human-caused carnivore mortality.

Carnivores and disease: reservoirs and victims

Wild carnivores can act as reservoirs of zoonotic diseases of humans, with rabies being the classic example. For examples, skunks (*Mephitis mephitis*) and racoons (*Procyon lotor*) act as rabies reservoirs in North America (Charlton *et al.*, 1988; Jenkins *et al.*, 1998), whereas in southern Africa, yellow mongooses (*Cynictis pencillata*), bat-eared foxes (*Otocyon megalotis*) and possibly jackals (*Canis mesomelas, C. adustus*) fill the same role (Bingham *et al.*, 1999; Chaparro & Esterhuysen, 1993; Thomson & Meredith, 1993). In Europe, where red foxes are rabies reservoirs, systematic attempts were made to control rabies by reducing the density of foxes, through providing bounties and employing fox control teams (Aubert, 1993). This approach appeared to have only limited success and it is only since oral vaccines were developed and extensively distributed that near eradication has been achieved, at least in France and Switzerland (Pastoret & Brochier, 1998). It should be noted, however, that even where wild carnivores are rabies reservoirs, domestic dogs still account for 98% of transmission events to humans.

With the removal of rabies as a limiting factor on the fox population and a consequent increase in fox numbers, the wider potential for foxes as a reservoir for another fatal disease of humans has emerged. *Echinococcus multilocularis*, which causes disseminated cysts in body organs in humans when the oocyst from carnivore faeces is ingested, is becoming a widespread problem in Europe. Preventing its introduction to the UK has been of particular concern when considering changes to the quarantine laws (MAFF, 1998).

On the other side of the disease–carnivore conservation dilemma, disease is increasingly seen as a problem for endangered carnivores (Macdonald, 1993). Indeed, the majority of examples of disease causing a conservation problem are in carnivores (e.g. Thorne and Williams, 1988; Gascoyne *et al.*, 1993b; Alexander *et al.*, 1994; Sillero-Zubiri *et al.*, 1996). In our survey, the threat of disease was cited as a real or potential conservation problem in over half (58%) of the replies with the spread of pathogens from domestic species being the most important potential source for problems. Rabies was the pathogen of greatest concern, although sylvatic plague and canine distemper in North American canids, the spread of feline immunosuppressive viruses to bobcats and pumas in North America, and tuberculosis in lions in South Africa were other highlighted examples.

THE IMPORTANCE OF COMMUNITY INVOLVEMENT IN APPROACHES TO CARNIVORE CONSERVATION

Although in the developed world sophisticated initiatives such as captive breeding or reintroductions are frequently involved in carnivore conservation, *in situ* conservation projects both in developed or developing countries involve a comprehensive approach including biological field research, identification of local human interests and cross-boundary co-operation (Weber & Rabinowitz, 1996). Clearly if carnivore conservation is impeded by a problem between local communities and carnivore behaviour, problem resolution must involve not only the carnivore that is the problem, but also the humans who are having the problem. Thus communication between wildlife managers and local communities is essential.

Community-based conservation initiatives have spanned a wide range of activities. At one end, some initiatives embrace the concept that the community is not the proprietor of the natural resources but is its neighbour, and thus that 'parks and people' must be good neighbours (Brandon & Wells, 1992). This leads to a focus of activities on park outreach strategies such as improving public relations and understanding through conservation education, and then progressing to revenue sharing, conflict resolution and community development strategies (Infield & Adams, 1999). At the other end of the spectrum, community-based conservation involves giving ownership to people living next to or with those resources, with a shift of responsibility and authority for land-use policy from central authorities to local communities (Hackel, 1999). This progression from outreach through revenue-sharing to ownership since the 1980s perhaps reflect both a maturation of the philosophy of community involvement in conservation, as well as political pressure locally, nationally and internationally (Barrow *et al.*, 2000).

The approaches we reviewed spanned this spectrum of philosophies, but could more clearly be separated into those that attempted to reduce or remove the problem and those that did not reduce the conflict but instead attempted to resolve the problem through improving tolerance. Here we discuss a number of these approaches.

INCREASE COMMUNITY TOLERANCE (ALTHOUGH SCALE OF PROBLEM MAY REMAIN)

Recognise the problem

As in many walks of life, simply listening to a grievance and recognising a community's problem can alleviate the problem through reducing tension. For example, when the socialist government was deposed in Ethiopia in 1991, the recognition that local residents' concerns were legitimate was enough to reduce the resentment and decrease the persecution of individual Ethiopian wolves *Canis simensis* (and mountain nyala *Tragelaphus buxtoni*). These species had been acting as 'surrogates' for the distant government officials (Gottelli & Sillero-Zubiri, 1992). Similarly, North American communities were directing aggression against grizzly bears rather than more distant elitist environmentalists, but the situation improved when a forum for complaint was estab-lished (Primm, 1996). Thus an improvement in local acceptance, mutual understanding and public opinion would greatly improve the outlook for many endangered large carnivore populations.

Recognition of a problem may require research into the ultimate and proximate causes of livestock or game mortality, in particular to determine whether losses to predation are additive or are compensated by reduced losses to other mortality agents. It may also be necessary to determine which species are causing problems and whether particular individuals are involved. In some cases (e.g. jaguar, zorros, red fox) actual losses are much lower than suggested by local communities. For example, investigation of lamb losses to red foxes revealed that foxes were the proximate and ultimate cause of only a very small percentage of lamb losses in Sutherland and Argyll in Scotland (Hewson, 1984). In other cases, investigations have supported the community's concerns and have led to action on the part of wildlife managers, such as the scale of livestock losses to lions in Bushmanland, Namibia (Stander, 1990).

Participation of local communities in management

Novel ideas for the co-management of habitat and wildlife with local communities are increasingly seen as the way forward for conservation, particularly *outside* protected areas. These frequently involve improving the economic benefits to communities from wildlife. Community participation in wildlife management might involve the design as well as management of protected areas, such as the creation of the Cockscomb Basin Jaguar Preserve in Belize (Rabinowitz, 1995) or the actual transfer of land and resource rights to local communities. In Canada's Western Arctic, polar bears have benefited from a co-operative wildlife management process established by the Inuvialuit Land Claim (Bailey et al., 1995). The Claim not only secured the title of the land and fishing and wildlife harvesting rights for the Inuvialuit, but also involved them in co-managing the resources with the government. As a result, the mutual distrust between government biologists and the Inuvialuit has been overcome. Now, in collaboration with the Inupiat of Alaska, a Polar Bear Management Agreement has been signed which aims to maintain a viable bear population by maximising the protection of females and cubs but exploiting male bears.

Improve economic benefits to the community

The philosophy that local communities should directly financially benefit from conservation underpins many of the recent strategies for community-based conservation (Hackel, 1999). Clearly where economic benefits are substantial, and given human nature, this is one of the most powerful ways of reducing negative perceptions of wild carnivores, or of wildlife more generally. There are a number of ways in which programmes have endeavoured to transfer economic benefits to local communities, particularly through ecotourism, hunting, employment and compensation for losses. Often, such strategies are used in combination. For example, the Communal Areas Management Programme for Indigenous Resources (CAMPFIRE) in Zimbabwe promotes conservation of wildlife by allowing communal landholders to receive direct income generated by hunting fees, game-viewing and curio sales (Child, 1996). In particular, large predators, previously persecuted for livestock losses, now have enhanced value for the local people as they command substantial hunting fees.

Another project that has attracted international attention and support is that of the Ranthambhore Foundation, which seeks to improve living standards of local people around the Ranthambhore Tiger Reserve. Since 1988 the Foundation has been working in seven villages surrounding the reserve, growing trees for firewood, timber and fodder for livestock, rehabilitating grazing lands and providing new cattle breeds, medical and family planning facilities. They also develop income-producing activities for women, and provide environmental education for all.

A recent initiative to create market outlets for hand-crafted products with wildlife motifs has kept herding families in Mongolia busy during winter months turning their camel, sheep and cashmere wool into gloves, hats and scarves, in return for protecting snow leopards (McCarthy & Allen, 1999). Careful market research led to the setting up of co-operatives, a local management system for collection and quality control and specific quality and style requirements. The most immediate conservation link enhances individual incentive by peer pressure through withholding from the whole community a bonus for meeting all contract terms if any individual herder illegally kills a snow leopard. Although recently set up with outside seed money, this enterprise aims to be self-sustaining in the near future and independent of international involvement.

Non-consumptive recreational use: ecotourism

Ecotourism has been a major growth industry over the last 20 years and there is no doubt that large carnivores are a major attraction for tourists planning a traditional wildlife safari. Estimates of the 'worth' of a major predator range from US $50 000 a year for a leopard in Londoloze Game Reserve in South Africa to US $128 750 for a male lion in Amboseli National Park in Kenya (Martin & de Meulenaer, 1988).

Traditionally, tourism is a good strategy for established conservation areas, with east and southern Africa abounding with examples where safari tourism has become a major source of income. In southern Africa, commercial farmers, whose precarious income from cattle farming has always been susceptible to drought, are increasingly turning to tourism as well as consumptive trophy hunting as an alternative source of income (Lambretchs, 1995). These private sector initiatives have led to the establishment of large, collaborative, private nature reserves and conservancies. These areas are added to the formally protected reserves at no extra cost to the taxpayer and provide relatively large areas for the protection and relocation of all major predators.

In addition, ecotourism is becoming an increasingly important land use for conservancies on communal lands (e.g. CAMPFIRE), provided that a share of the revenues and jobs go to the local community. A good example is that of tourists in Bushmanland, Namibia, where the use of baits and traps has been turned inside out; tourists pay to sit in traps, originally used to catch problem leopards, to get a close experience of a leopard approaching a bait (P. Stander, pers. comm.).

High-profile and visible carnivore species (e.g. lions, tigers, bears, wolves) may be capable of supporting a sustainable tourist trade, but this approach may be unsuitable for other more secretive species or those extremely sensitive to human pressure (e.g. pumas, leopards, snow leopards). Many visitors to areas renowned for their large predators (e.g. northern Rockies, east African parks, Indian parks) are attracted by the knowledge of the presence of these large predators, even if the chance of sightings may be minimal. Moreover, expectations may surpass reality and mass tourists (not unlike experienced naturalists!) may become disappointed if they do not see the elusive predators during their visit.

However, a note of caution regarding this approach is needed; the economic rewards of ecotourism may be low or not reach the expectations of the local community (Hackel, 1999). For example in the Bale Mountains in Ethiopia, home to the largest population of Ethiopian wolves and where income from tourism is often used to justify the presence of a park, the local community benefits in a limited way for a number of reasons. First, annual tourists number only in the hundreds, many use their vehicles rather than local horses to travel around the area, and those in vehicles usually use Addis Ababa-based guides rather than local guides. Although six to eight individuals offer themselves as private guides, they rarely work more than a few days each month, at a maximum monthly income of US $50. Often, these guides receive no income at all in a month. Furthermore, it was estimated that the revenue lost when tourists drove directly to one area of the park, without passing an entrance gate and paying fees, would have been insufficient to pay for the wages of scouts manning a new gate.

Furthermore, tourism is susceptible to changes in the global economy and to the political stability of a given country, and its growth is also unpredictable (Infield & Adams, 1999). Recurrent wars in the Democratic Republic of Congo exemplify this problem as they have stopped ecotourism in that country, which has also affected Uganda and Rwanda. The tourist market is also a limited one and thus if the trend to move from livestock raising to tourism continues to expand for private or communal lands, revenues may drop (Nowell & Jackson, 1996). Finally, there is a fundamental problem in distributing income from tourism (and other activities) to local communities – i.e. most foreign earnings go to tour operators or central government and only a tiny proportion may reach the community (Hackel, 1999). Overall, it would be unwise to justify carnivore conservation purely through the economic benefits accrued from tourism.

Consumptive recreational use (hunting)

Hunting of large carnivores in certain areas can increase the value of predators and this can be offset against livestock losses or other perceived problems (Stander, 1993). Hunting revenues can be substantial as hunters frequently spend more than tourists per capita, stay longer and also may require numerous local staff. Once more, however, most of the revenues raised from fees and licences go to government not to the local communities.

Edwards & Allen (1992) report that communities participating in Zimbabwe's CAMPFIRE programme earned about US $4 million, or US $400 per household, from hunting revenues. In some cases this was a doubling of the annual household income. In Namibia, most trophy hunting takes place on private game farms or hunting concessions rather than commercial cattle land or communal lands; for this to be a viable source of income for cheetah conservation, trophy hunting must be done on cattle ranches and the fees must increase.

A novel approach to hunting is the non-consumptive or 'green' hunts of large carnivores, similar to that attempted in Zimbabwe. There, hunters pay to dart and anaesthetise an elephant, take a cast of its tusks and a picture before reviving the animal. Arguably it is as challenging and even more dangerous than using firearms. However, it is uncertain whether hunters would generally consider this bloodless alternative to traditional hunting favourably and whether animals might be unduly stressed. Activities could be co-ordinated with the removal of problem animals or research projects, and perhaps also subsidise these.

Overall, trophy hunting may help carnivore conservation, but its economic benefit is limited and does not always reach the communities affected by problem animals. On the other hand, hunters and trappers are often extremely knowledgeable about wildlife, and a proportion consistently support carnivore conservation, notwithstanding problems associated with predators killing mutual prey (see above). Indeed, management programmes for areas with consumptive practices often target hunters and trappers. For example, education has long played a role in reducing conflicts between bear conservation and humans. Recently, male-selective harvest of grizzlies in the Yukon has been promoted with hunters (Smith, 1995) through educational videos and meetings with hunting outfitters and hunting guides. Hunters were taught how to identify adult males and avoid over-harvest of females. In many cases individuals with negative views can be influenced through peer pressure and self-policing by the other hunters.

Compensation for livestock losses

Improving tolerance of carnivores through compensation for losses may be achieved by direct compensation for livestock losses or through a 'conservation payment/ subsidy' for having carnivores on private land. In northwest USA, a NGO pays subsidies to farmers for having a breeding pair of wolves on their ranch (Hudson, 1993). Other alternatives include providing tax incentives to landowners and transferring user fees from recreation to landowners.

Direct compensation for livestock losses has proven to be a relatively widespread and sometimes inexpensive, but not always effective, means for relaxing opposition to carnivore conservation. In Italy for example, the local government compensates 100% of the value of livestock killed by wolves, bears and even feral dogs (Cozza *et al.*, 1996). This amounted to a modest 0.4–2.8% of total livestock subsidies in the region. In other situations, the financial compensation received by the villagers is often inadequate; for example it amounted only to 3% of the perceived annual loss in the Kibber Wildlife Sanctuary of Indian trans-Himalaya (Mishra, 1997).

For a compensation strategy to be successful, it is vital, through research, to establish a number of parameters. First, the carnivore species involved must be ascertained, as stock selection varies hugely with carnivore species. In Africa, lions take mostly cattle, hyaenas mostly sheep and goats, and jackals take mostly young lambs and kids. Jackals kill mostly in the day whereas lions and hyaenas are more likely to kill at night (Kruuk, 1980). Secondly, it is vital that the criteria for compensation are clearly laid out, to avoid abuses in the claim system. For example, a few farmers may take advantage of the situation to gain other subsidies. In Italy, farmers sometimes keep old or infirm sheep for headage payments, but these are more likely to be killed by predators (Cozza *et al.*, 1996).

Unfortunately, although in many circumstances a compensation system alleviates the direct losses to farmers, it does have disadvantages. First, it does nothing to reduce the problem and encourages and acknowledges a state of constant conflict. This is particularly true where only a few farmers suffer the vast majority of losses. Secondly, it does not encourage the improvement of management systems although this may be alleviated if compensation criteria are modified. Lastly, in some circumstances it can be extremely expensive. In Sweden, compensation payments to maintain the theoretical minimum viable population of bears are estimated at around US $7 per head per year (Sagor *et al.*, 1997).

People's participation in research/conservation activities

Our questionnaire survey revealed that local communities can be involved in the research and conservation of large carnivores in a number of ways. These can range from direct employment, through voluntarily providing information on carnivore sightings to using local knowledge to find or trap carnivores. Employment with conservation, research or hunting activities can provide an opportunity for 'poachers to turn gamekeepers'. Former poachers in Namibia and Zambia have been successfully employed on a research and conservation project as trackers and assistants (Owens & Owens, 1993; Cunningham & Berger, 1998). Reducing illegal killing of carnivores in this way could help regulation of wildlife consumption and the implementation of management strategies.

Many projects may employ one or two assistants, which may seem trivial but can actually be significant in the local economy. Our Ethiopian Wolf Conservation Programme is now the biggest employer in a local town (8–10 people and approximately US $10 000 a year in local wages). Staff have gained status and respect in the community and we continually receive requests for employment. With the responsibility of having individuals and their families dependant on this work, we sometimes feel like a welfare or job creation agency, not a conservation programme.

In Namibia, Phil Stander employed local Ju/'Hoan Bushmen trackers on a behavioural ecology study of lions and leopards. Stander adapted their traditional bow and arrows, used for many thousands of years for hunting, to dart and immobilise these large carnivores, finding this method more accurate, quicker and more flexible than traditional dart guns (Stander *et al.*, 1996). In addition, the Ju/'Hoan tracking skills were used to follow and reconstruct hunts for these predators, gaining valuable behavioural data on cryptic and nocturnal predators in a bush habitat where direct observation was impossible (Stander *et al.*, 1996).

Villagers and forestry guards in Thung Yai Naresuan Wildlife Sanctuary, Thailand were trained in line transect techniques for wildlife censuses. They then collected many kilometres of valuable transect data in this area of great conservation significance for carnivores, particularly cats and bears (Steinmetz & Mather, 1996). The project has revealed a tremendous potential for participatory wildlife surveys and the possibility of establishing a community-based wildlife monitoring process.

Improve the community's aesthetic and moral benefits
Conservation education

In many situations it is impossible to provide sufficient economic benefit to local communities to compensate for the resources that are lost by a protected area (Barrett & Arcese, 1995; Hackel, 1999). In these circumstances the most important way that public support can be gained for large carnivores and their conservation is through educational programmes, so that local people can relate positively with the species or habitats in question. Through community education local pride and awareness can be enhanced and effective conservation promoted. Thus by broadening people's *understanding* of the environment, their *concern* for the plight of endangered species and ecosystems can be heightened and this can lead them into *action* for long-term conservation (Dietz & Nagagata, 1995). Recommendations to involve the local community include targeting key groups with education programmes, building support through the use of spokes-people within the target groups, integrating human and ecological concerns and, if possible, designing species-specific education initiatives using the species as a flagship for other conservation concerns.

Policies for conserving large carnivores must emphasise all values represented by these species, not only their presumed ecological significance or their economic importance, but also the many emotional, intellectual and even spiritual benefits provided by these charismatic species (Kellert *et al.*, 1996).

Since 1996, the Ethiopian Wolf Conservation Programme has developed a community education approach with the chief aims of reducing wolf persecution by shepherds, diminishing road kills and, crucially, encouraging responsible dog ownership in order to reduce disease transmission and hybridisation with dogs (Sillero-Zubiri & Macdonald, 1997). A local primary school teacher is employed as Education Officer and works with the community and local schools. He visits human settlements inside the National Park to talk to people about dogs, disease and wildlife, and uses education materials in the Oromo language, designed locally to address specific problems. These activities prepare the ground for dog vaccination and sterilisation campaigns. Posters, stickers and road signs are widely used to reduce road kills, raise the profile of the wolves and gain their acceptance in local towns. At primary schools he lectures about wolf conservation and other environmental issues, organises art, literary and sport competitions, and supports various schools' activities, such as Nature Clubs, tree nurseries and football teams. Our approach appears to have been successful in raising the profile and acceptance of the Ethiopian wolf in the region and also in bringing attention to other wildlife conservation issues. Whether this awareness is translated into reduced conflict remains to be seen.

The success of efforts to involve local communities can be illustrated by a project in the prairies of North America on swift foxes, *Vulpes velox* (A. Moehrenschlager, pers. comm.). There, local farmers were unaware both of the presence of swift foxes on their land and of their conservation importance. The researchers built up trust and respect from the local community by participating in the lifestyle of the farmers, by helping them with cattle management, looking out for prairie fires and concurrently conducing education programmes in schools and giving talks to local landowners and natural history groups. They also involved farmers in the trapping and release of introduced foxes, named some individual foxes after locals, wrote biannual progress reports for the farmers and generally involved the community as active and crucial project participants. In time the farmers began to report swift fox sightings, brought canid carcasses for analysis and became so involved in the project that they were discussing sightings amongst themselves. When the Moehrenschlagers left the project, the landowners hosted a farewell party for the researchers.

The importance of the community education approach is highlighted by the results of our questionnaire. Despite 74% of our respondents being involved with carnivores primarily as researchers, 79% of projects had carried out some sort of education programme, even if only on a small scale. On those projects that were primarily aimed at species conservation, education was the cornerstone of their work. Just under half (47%) of these projects conducted education programmes in schools, whereas 52% carried out work directly with the local communities. The use of publicity material (posters, stickers, articles in the press, radio, etc.) was widespread. The success of this approach in these projects is, however, less clear. Just under half of respondents said that their educational efforts appeared to have limited success, as problems between carnivores and humans had not decreased, or that the projects were too new to assess. However, others could discern a change in attitude amongst the local community, or a direct decrease in the conflict. For example in northern Kenya, the killing of wild dog pups at dens ceased (K. Doherty, pers. comm.). It was also clear that education efforts would, in some cases, take a generation to come into effect as there were noticeable generational differences in attitudes to carnivores.

REDUCE PROBLEMS BETWEEN CARNIVORES AND PEOPLE
Fencing problem areas

Fencing reserves and their wildlife has been used traditionally as a way of reducing conflict with the surrounding communities. Although this is an old fashioned conservation approach and may be frowned upon by some, in many places it has proved very effective. Several public and private areas in South Africa are prime examples, such as Kruger or Pilansberg National Parks. In Nairobi National Park, Kenya, fencing has enabled wildlife, including large predators, to exist immediately adjacent to the city.

Fencing also serves to clearly delineate wildlife and human areas, potentially reducing arguments over primary land use. Unfortunately, permanent fence construction and upkeep are costly, thus precluding their use in poorer countries. More importantly fencing effectively cuts wildlife movement and may result in catastrophes during droughts or bushfires. Furthermore, the small size of many fenced reserves means that populations of carnivores with a small genetic pool will require active management.

However, when used to directly protect people or their livestock, electric fences can be relatively cheap and effective. Fences installed around camps in the Arctic effectively deterred bears in two traditionally troublesome areas (Follmann & Hachtel, 1990). Similarly electric fencing was perceived as the best short-term solution to protect people and livestock from large carnivores in India (Veeramani et al., 1996).

Change behaviour of groups in conflict
Reduce predation on humans

There are a number of strategies that can be used to reduce carnivore predation on humans. On one hand, humans can change their behaviour to avoid contact, and if contact occurs, to avoid attacks. In the Sariska Tiger Reserve in India, villagers have increased their security through adopting a variety of behaviours when encountering a tiger. They never get between a tiger and its prey, they distance themselves slowly when a tiger is seen and respect the tiger's behavioural repertoire by keeping quiet and looking it directly in the eyes when retreating (Galhano Alves, 1996). As a result, there has been only one human death in the last 10 years, an outsider who ignored these practices. A well-known and imaginative measure used to deter man-eating tigers is the use of masks by the Sundarban people when gathering wood, grass and honey. These masks, used back-to-front, seemingly reduce the chances of being attacked from behind as the tigers think that the human is facing them (Nowell & Jackson, 1996).

When a problem animal emerges, other preventive measures that can be adopted include fencing-off villages, sleeping indoors and, most importantly, staying inside at night. People can also avoid being alone in areas with a problem animal. However, in some circumstances, the only solution to this conflict is to kill a man-eater, as many re-offend once they have attacked humans. This must be done by an experienced hunter and can be a very difficult task.

Reduce demand for carnivores and their products

Whilst the trade in carnivore pelts and live animals for pets is no longer the major carnivore conservation problem, the use of cat and bear body parts by the Oriental medicine trade is one of the biggest challenges posed to the conservation of these species today. The illegal trade in tiger bone is extensive, with a high demand amongst ethnic Chinese and Koreans. Medicines are used mainly to help joint and muscle problems, but may also cure ailments ranging from alcoholism (bone ash) to toothache (whiskers) (Nowell & Jackson, 1996).

Legislation is obviously part of the solution and must occur at all levels of the administration. For example, in China, legislation on the use of tiger products has been recognised at higher levels of government through bans on importation and the national medical governing body has taken these products off their list of eligible medicines that can be prescribed and distributed.

Enforcement, however, is difficult when demand remains strong. In addition, an illegal medicinal trade is the most difficult to stop because the product is less visible, the consumer base is larger and more widespread, and consumer motivation is not at all obvious. In these circumstances, it is clear that the only fundamental solution is to decrease demand. Unfortunately, the illegality of the trade can make it difficult to learn enough about the demand for these products to take appropriate measures.

The success of reducing demand through publicity and education campaigns is illustrated by the long running campaign against the spotted cat fur trade by pressure groups in the West. For example, the world trade in cat skins has shown a steadily declining trend, falling from 450 000 pelts in 1980 to 100 000 by 1990 (WCMC, unpub. data). Moreover, the number of species being traded has declined by more than half. It has also been suggested that the small-scale illegal trade of fur coats in Nepal could be stopped with a small-scale publicity campaign aimed at tourists (van Gruisen & Sinclair, 1992). However, the effectiveness of this approach in the Asian medicine trade, a very different market, is not yet clear (Nowell & Jackson, 1996).

Nonetheless, publicity measures have been instigated in China and Taiwan, mainly by western organisations, and also by government, to try and alert the public to the tiger's plight and the illegality of the trade. For example, the Taiwan government has issued public awareness stickers to pharmacists who signed a written declaration that such products were not for sale in their shops. This approach must, however, go hand in hand with work to understand the dynamics of the market. Its effectiveness will also take time as cultural attitudes can take a generation to change.

Change livestock husbandry patterns

Livestock husbandry can be changed in a range of ways to reduce predation. In some situations, extensification of husbandry systems and the lack of supervision of livestock has led to an increase in this problem. For example this occurred in Europe during the twentieth century leaving sheep, cattle and equids vulnerable to lynx and wolf predation. In southern Africa, livestock losses to cheetah are concentrated on large-scale ranches, not in traditional pastoralist systems.

In some cases, measures are relatively simple and cheap. In northern Kenya, improved vigilance by herdsmen, the prevention of livestock from straying and returning herds to small thorn enclosures at night, drastically reduced livestock depredation (Kruuk, 1980). Avoiding areas where predators are known to be concentrated is also effective. Similarly in Namibia, enclosing calves at night and keeping calving cows and newborns under closer supervision was sufficient and also reduced calf mortality from other causes (Marker-Kraus et al., 1996). Allowing wild prey recovery in the range to reduce pressure on livestock has also proven effective.

In other cases more sophisticated or expensive solutions have been used. Guard dogs, donkeys or even llamas (Lama glama), have been trained to live with herds and identify with calves or sheep and deter most predators, even when humans are absent. Special livestock guarding breeds, originally used in Europe, have now been introduced to South America and Namibia (Marker-Kraus *et al.*, 1996). In Asia electric fences may temporarily stop cattle losses to leopard, tiger and dhole although resettling villagers from inside the forest areas may be the only permanent solution (Veeramani *et al.*, 1996).

In some cases it may be difficult to change livestock husbandry, either because farmers are resistant to change or because management options are too difficult or expensive to put into practice. There may simply be little incentive for farmers to change, if losses are relatively low, or else farmers may not be keen to take on perceived extra work.

Modify the behaviour of predators

Several non-lethal taste and olfactory agents that have an aversive effect on carnivores have been used to deter predation, but results so far are inconclusive. Lithium chloride, which induces vomiting and diarrhoea, has been widely tested as a conditioned taste aversion method to prevent canids from attacking livestock. Although strong aversion against sheep meat develops, individuals are likely to continue attacking sheep. Use of oestradiols produced a more generalised avoidance of eggs among mammalian egg predators, including red foxes, American badgers (*Taxidea taxus*) and coyotes (Nicolaus *et al.*, 1989). Similarly, volatile repellents attached to lambs were a potential method for decreasing wolverine (*Gulo gulo*) predation on sheep in Norway (Landa & Tommeras, 1997).

Dealing with problem animals

Livestock losses and attacks on humans are often caused by just a few individual 'problem' animals. The elimination of these individuals under controlled circumstances is usually acceptable to local communities as it apparently directly tackles the problem. Indeed, if problems involving just a few individuals are not resolved, animosity from the local community may increase and be extended to a whole species (Tilson & Nyhus, 1998). Actual disposal of problem animals may take the form of blanket or selective destruction or physical translocation.

Kill all predators or individual problem animals

Blanket reduction in predator populations such as the eradication of wolves across North America, or predators of game birds in the UK, has been used for centuries to reduce problems between carnivores and humans. However, the selective removal of problem animals can lead to increased success and better carnivore conservation. This approach of removing individual problem animals has been a long standing practice of conservation authorities in Canada. The Indian philosophy has been similar; they do not lay blame on the species as a whole, but seek to destroy the single individual responsible for the wrongdoing. With education, such an attitude may spread elsewhere.

For example, Stander (1990) investigated the options for alleviating this conflict during a study in northern Namibia, where long-term records of individual lions permitted the categorisation of stock-raiding lions as habitual 'problem animals' or 'occasional stock raiders'. He suggested management strategies for each group under varying conditions, with optimal solutions emerging as translocation for occasional stock raiders and elimination for problem animals. For this, co-operation between farmers and conservation authorities is required and success of the strategies depends on long-term monitoring of individual lions.

Similarly in the UK, some gamekeepers and shepherds now appreciate that only some individual red foxes are responsible for most predation on lambs or game birds. Thus, if they do not discern an appreciable problem, they will refrain from fox control, as indiscriminate killing may remove a 'good' fox and a fox that selects birds or lambs may move into the artificially created vacuum.

Due to demands for large carnivore trophy hunting a sustainable utilisation scheme may serve as compensation for livestock losses. Sport hunting cannot, however, replace the control of problem animals. It is only in the interest of local people not to kill carnivores indiscriminately if substantial returns from trophy are accrued.

Translocation of problem animals

Translocation of individual carnivores has been a standard management tool for decades in North America and Africa in response to livestock depredation. It is clear, however, that such operations must be carefully planned and carried out to avoid failure. Nevertheless there are virtually no data available on the success of this approach in the long term. Problem animals may offend elsewhere; they appear to consistently return to the site of capture over distances of up to 400 km. However, when stock raiding has not become habitual, Stander (1990) found that the act of translocation, even if the animals returned to its home range, was enough to prevent this habit developing and the animal re-offending. Survival of translocated animals may also be poor and in a saturated ecosystem, there are few areas where a translocated animal could fit into the social system. Researchers in

Namibia are currently investigating the fate and ranging patterns of translocated leopards and cheetahs (P. Stander & L. Hansen, pers. comm.). However, until further data are available, it would appear that unless there are large areas available where conflict potential is low, this strategy is unlikely to work and management efforts should concentrate on reducing the potential for problems or, where this is not practical, lethal control (Linnell et al., 1997).

Disease management in domestic or wild carnivores

The management of disease as a conservation problem is in its relative infancy, with almost no examples of a disease problem solved. Our questionnaire survey illustrates this observation. Despite the perception that disease was a problem, very few projects had instigated disease management strategies, let alone conducted investigations to assess potential problems. The only examples in this survey, apart from our project in Ethiopia, were that trapped canids on the North American prairies and translocated cheetahs in Namibia were quarantined before release and that domestic dogs were vaccinated in northern Kenya against rabies (A. Moehrenshlager, L. Marker and K. Doherty, pers. comm.). These actions, however, provide examples of the general disease management techniques that can be adopted, i.e. the direct vaccination of susceptible individuals, the vaccination of a putative domestic reservoir and the prevention of introducing disease with translocated animals.

Our questionnaire review also highlighted that in virtually all these situations, domestic dogs appear to be the reservoir for pathogens that are of concern in carnivore conservation. We could find no examples of felid or viverrid pathogens causing significant conservation problems in the wild. Indeed, only generalist canid pathogens (canine distemper and rabies virus) have hitherto caused severe conservation problems, with rabies being of greater importance than canine distemper virus (Macdonald, 1993; Roelke-Parker et al., 1996).

However, the fact that rabies virus causes significant economic losses as well as public health problems (e.g. the cost of rabies in Ethiopia may approach US $7.50 per household each year) actually helps the cause of carnivore conservation, as there is a degree of self-interest for local communities in controlling this problem and in co-operating with conservation managers. In Tanzania, with a little outside knowledge and advice, villagers have been so fearful of rabies and its concurrent losses that they have been motivated to sell cattle to generate cash to buy human vaccines for rabies post exposure treatment (S. Cleaveland, pers. comm.). Thus control of disease in carnivore conservation may give direct benefits to local communities.

The management of disease as a conservation problem does, however, offer an alternative path for community involvement and support. Disease control can be used more indirectly to benefit local communities and increase tolerance of carnivores. In the course of our work vaccinating dogs in Ethiopia, people have asked for additional veterinary assistance for livestock, which in many cases has been provided on the day by veterinarians and assistants participating in the dog vaccination campaign. As a result of such requests, dog vaccinators take some basic veterinary supplies with them and we have started a small scale 'revolving fund' to provide anthelmintics for livestock. Drugs are provided at cost price to remote communities and the income is used to buy more medicines. We try to ensure that recipients are aware that this help is only available because of conservation concerns for the Ethiopian wolf.

Disease control programmes also present extensive opportunities for two-way communication, monitoring and conservation education. By simply visiting houses and hamlets, explaining why we want to vaccinate dogs and that we work for the park authorities, we have increased the contact with the local community and pointed out that removing livestock and human deaths from rabies is a direct benefit of having Ethiopian wolves in their neighbourhood. Our village vaccination campaigns are inevitably attended by a horde of children and bystanders, gaining huge amusement from the extraordinary sight of people touching and paying attention to dogs. We take advantage of this otherwise aggravating situation during lulls in vaccination activities by turning our dogcatchers and vaccinators into educators. A lecture and story on rabies, dog husbandry and Ethiopian wolf conservation is followed, after another suitable spell of dog vaccinations, by fierce competition for prizes of stickers, posters and our dog management book.

Thus we have found in Ethiopia, that the threat of disease can be used to portray carnivore conservation in a positive light, give direct and indirect benefits to local communities and be used as a vehicle to improve wildlife manager–community relations.

CONCLUSIONS AND RECOMMENDATIONS FOR CONSERVATION

Community involvement in carnivore conservation is clearly fundamental in the long term. This involvement may be limited to improving knowledge in local communities through outreach and education programmes or to direct employment, but can also extend to communities obtaining economic benefits from carnivores through tourism or hunting revenue, and to sharing resource ownership and management decisions. The use of this approach has grown since the 1980s as conservation philosophy has shifted away from purely protectionist attitudes. However, there is a need to assess just how extensive such initiatives are and how successful and sustainable they have been in carnivore conservation.

Education

Education programmes associated with carnivore conservation programmes are the most frequent way in which conservation and communities interact. In the last 10 years such activities have been seen as the first step in outreach programmes around protected areas and indeed are increasingly becoming an integral part of the activities of conservation agencies. Even in carnivore research programmes, our review also illustrated that small-scale initiatives are frequently carried out by researchers who had no real conservation mandate or objectives.

The extensive involvement of education programmes is undoubtedly a welcome advance that must be continued, extended and encouraged if carnivores and local human communities are to be 'good neighbours'. However, their success is still open to question. In a few cases we reviewed, there was an obvious improvement in the situation. For example, in northern Kenya the killing of wild dog pups at dens stopped (K. Doherty, pers. comm.) and in Canada an education programme made farmers aware of the presence and conservation importance of swift foxes and the farmers became involved in monitoring them (A. Moehrenschlager, pers. comm.). However, in other situations, success is not always apparent, although this is not always necessarily because the programmes themselves may have failed, but because no attempt has been made to assess attitudes or behaviour. This is an important omission that must be rectified if such an approach is to be sustained in the future. Baseline information on attitudes to carnivores, habitats and conservation must be assessed so that changes can be quantified. Furthermore, the type and degree of human–conservation problems should also be measured to ascertain whether changes in attitude are translated into changes of behaviour.

Some programmes might also benefit from a more professional approach. Although many activities are undertaken as an extra activity, for example by people running research rather than conservation programmes, many programmes that are primarily conservation orientated do not always involve people and organisations who have extensive experience of implementing such programmes. Whilst recognising that budget and time limitations may be at the root of such omissions (in our own programme for example), we would urge such initiatives to seek adequate funding so that the wheel is not reinvented and personnel implementing the programme are professionally trained and have access to well designed methods and materials.

Economic benefit sharing

Carnivores clearly are attractive to both tourists and trophy hunters. Revenue sharing from these two activities has been carried out in a few conservation projects, such as that for jaguars in Belize (Rabinowitz, 1995) and those for lions, cheetahs and leopards in southern and east Africa (Stander, 1990; Marker-Kraus and Kraus, 1997). However, our review highlighted that the direct economic benefits of carnivores may be limited only to these two activities and that the situations in which these activities can have a real impact may also be limited. Although apparently successful in parts of southern and east Africa, factors affecting the success of these activities (such as the global economy, political instability and fashions in holiday destination choice) are often outside the control of the community. Furthermore, there is increasing concern and a realisation that the real economic benefits of such activities may never really compensate for the loss of the resource use by local communities (Martin & de Meulenaer, 1988; Hackel, 1999; Murombedzi, 1999). Nevertheless, even if benefits are limited, they can be used to partially offset the costs of conservation activities for local communities and still prove to be a useful incentive for conservation. With an unequal distribution of revenues between government, operators and communities and also within the community, finding ways to redistribute benefits is a priority. It may, however, be acceptable for government revenues to be directed to resolve the community's health, educational or other needs, or to be returned to conservation.

Local employment is also a form of economic benefit for communities. Unfortunately, although extensive, this activity can only ever be limited to a small proportion of the community. This can lead to the development of jealousies although competition for jobs could also raise the standard of employees.

There is also a danger that a price attached to wildlife can be compared to revenue that might be received from other activities (Geist, 1988). Thus, if a better economic alternative appears, communities might abandon conservation activities (Hackel, 1999) and even if not, they are still subsidising wildlife and are losing economic opportunities. Thus, although economic benefits for local communities should be sought wherever possible, we believe that using them as the only justification for conservation is inadvisable.

Management sharing

A shift of responsibility for carnivores and other conservation resources from central government to local communities is one of the more recently implemented activities for community-based conservation programmes and is one that appears to have at present, few examples in carnivore conservation. However, the signing of management agreements with the Canadian Inuit for polar bears is a striking example of how this activity can be innovatively put into practice. Other examples, such as in the CAMPFIRE project in Zimbabwe, have not been aimed specifically at carnivore conservation, but nonetheless have embraced it.

The success of such activities and management structures still remain to be tested as insufficient time has passed for them to have become well seated, robust and functioning without outside input (Infield and Adams, 1999). Nevertheless, communities that have a voice often feel more empowered and are sometimes more tolerant of a problem as a result of feeling they are at least being heard. The channels of communication do go some of the way towards changing situations.

The small number of examples in carnivore conservation of the transfer of resource responsibility/ ownership and economic benefits to communities might be explained in a number of ways. First, as community involvement in conservation is a relatively recent initiative, those involved with carnivores may have been late in initiating such programmes. There is little evidence to support or refute this idea, although there seems little reason to accept this explanation *a priori*. Furthermore, those involved with tiger and jaguars were certainly at the forefront of incorporating this approach two decades ago (Rabinowitz, 1986; Nowell & Jackson, 1996).

Secondly, the sheer difficulty of incorporating community interests in conservation may mean that few programmes have been attempted successfully, thus few are held up as examples to be cited. The difficulty of this ambitious approach is unquestioned and it is also apparent that a blueprint for a successful programme cannot easily be transferred to another situation. Programmes are also complicated to implement and administer. We suspect that this difficulty hampers any deeper involvement of communities in carnivore conservation.

Finally, although the rhetoric and philosophy of community-based conservation is compelling, it may be that such programmes are actually not suitable for many carnivore conservation programmes. Certainly there is an increasing awareness that community-based conservation that involves revenue and responsibility sharing is not a universal panacea for all conservation problems (Barrett & Arcese, 1995; Little, 1994; Safalsky, 1994; Hackel, 1999; Murombedzi, 1999). The economic needs of poverty-stricken and rapidly increasing populations in most developing countries are often foremost and it is almost impossible for their needs to be reconciled with conservation objectives. Even in developed countries, those that suffer from carnivores are often the poorer rural population, not the urban majority with a strong voice and conservation ethic. Thus in some situations such programmes simply cannot meet both community and conservation objectives. If this is the case, and we suspect that there are many scenarios where the economic benefits that can be obtained from carnivores can never outweigh the costs to the community from not using the habitat's resources, conservation must still have protectionism at its core.

Thus, instead of conservationists trying to implement unrealistically complex programmes, they should perhaps aim to use the most appropriate tools to build better relations with rural people (Hackel, 1999), as people must value habitats and species for reasons other than economic ones. The role of conservation education is thus vital in order to increase the moral benefits and tolerance of carnivores, and also wider conservation objectives. Such outreach programmes can be effective in improving attitudes to protected areas or species among local communities without materially altering the balance of costs and benefits associated with them (Infield & Adams, 1999; Hulme & Infield, 2000; Kangwana & Ole Mako, 2000). In addition, if some benefit for local people can be achieved through new financial opportunities and if other direct solutions can be found for problems between carnivores and people, then success is more likely.

Perhaps then, these examples of community involvement in carnivore conservation, which seem to incorporate primarily community outreach/ education programmes and the direct solving of some specific problems, reflect the reality of the way forward. When conservation issues are foremost but are confronted by extreme rural poverty, it may be that an improvement of mutual tolerance is the best that can be achieved.

Recommendations and the future

In our review we found several omissions and factors that might help progress in this area. For example, we found relatively little material on how to implement management activities such as problem animal control schemes or compensation schemes. Nowell & Jackson (1996) also recognised that few people work on ways of reducing human–predator conflicts. Thus there needs to be more recognition of the significance of this subject and development of the professional capacity to address it. Furthermore, it would appear that much of the information is found in the 'grey literature' which is not widely known or found through routine searching and is often not even available. We would urge conservation managers to disseminate their experiences to the widest audience possible through both the written and electronic media.

It is also clear that novel approaches to solving carnivore–community conflicts are still required, as many of the problems facing carnivore conservation now are the same as those that led to the extirpation of many species in developed countries over the last millennium. 'Knitting' for snow leopards, the use of face masks to deter tigers and the use of disease management activities as a vehicle for communication and co-operation are some examples from this review where innovative or previously unobvious solutions have emerged.

There are also some common factors that might give a better understanding of whether programmes have been successful. Our questionnaire highlighted the need for better quantification of human–predator problems and the benefits of carnivores, with improved methodology underpinning this need. Disease threats also need to be assessed carefully including improved methodology for risk assessments, as does the rigorous testing of management solutions to these problems. Without such baseline information, it is impossible to measure the success or otherwise of projects. This is relevant also to conservation education projects, where baseline assessment of peoples' attitudes and behaviours must be carried out.

FINAL SUMMARY

The downward population trends of many carnivore species are likely to continue with the expansion of human populations. Thus future conservation efforts will probably focus on increasingly fragmented populations within protected areas. However, since a large number of individuals are found outside those protected areas their fate depends very much on the aspirations of the local community. Many reports on wildlife conservation describe the local people as the '*problem*', seeing the local people as obstacles in the way of management and recovery of wildlife populations. However, local people are increasingly recognised not only as a problem, but also as a vital part of the solution (Kessler & Eastland, 1995). Thus it is of paramount importance that local people are involved in solving problems of carnivore conservation through the various approaches reviewed in the hope of reverting the traditionally, deeply based, negative views about large carnivores. Community involvement, however, requires strong partnerships, shared goals for both wildlife and human communities and shared responsibility. Each solution must be worked on a case-by-case basis, to fit a unique set of ecological, cultural and economic circumstances.

Realistically, in human-dominated landscapes where carnivores and people coexist there will only be, at best, an uneasy tolerance. Thus to conserve large carnivores, conservation policy must encompass and also wholeheartedly embrace a mixture of strategies, including protectionism, conservation education, public relations, community involvement and revenue sharing. Whilst some of the examples in this review have illustrated how steps have been taken along this path, conservationist must expand the number of innovative and imaginative solutions to carnivore–human conflicts in the future.

Acknowledgements

We would like to thank all the colleagues that replied to our questionnaire. The Ethiopian Wolf Conservation Programme is funded by the Born Free Foundation, with additional support from the Bernd Thies foundation, Dresden Zoo, Landesbund Fuer Vogelschutz, National Geographic Society, White Oak Conservation Center and Zoologische Gesellschaft für Arten- und Populationsschutz. Karen Laurenson is funded by the Wellcome Trust. Scott Newey, Jorgelina Marino and Simon Thirgood assisted with the literary review and they and an anonymous referee commented helpfully on the manuscript.

14

New methods for obtaining and analyzing genetic data from free-ranging carnivores

PIERRE TABERLET, GORDON LUIKART AND ELI GEFFEN

EMERGENCE OF NEW GENETIC TECHNOLOGIES USEFUL FOR CARNIVORE CONSERVATION

Many carnivore species are threatened or endangered and are difficult to study because of their low densities and secretiveness. In addition, many nature reserves and natural areas are too small to ensure the long-term persistence of carnivore populations. Thus, populations will often require monitoring (estimation of population size, sex ratios, dispersal rates, etc.) and management actions (translocations, habitat restoration, etc.) to ensure their persistence (Ballou *et al.*, 1994).

Recent advances in population genetics can help provide elusive information about population parameters, population ecology, and behavior that are critical for designing conservation programs. In the past few years, laboratory genetic techniques have greatly improved and now allow for DNA typing using a non-invasive sampling approach (Morin & Woodruff, 1996; Kohn & Wayne, 1997; Taberlet *et al.*, 1999) and microsatellite DNA markers (Bruford & Wayne, 1993). Major advances have also occurred in statistical methods for analyzing microsatellite data (Luikart & England, 1999) allowing the extraction of new information and providing more precise estimates of important demographic and genetic parameters.

Microsatellites represent a powerful tool for investigating population genetic issues, parentage, extra-pair copulations, structure and relationships between and within social groups, and for identifying individuals (DNA 'fingerprinting'), which is a prerequisite for many applications (forensics, dispersal rates, etc.). In this chapter, we will consider only these DNA markers. They present many advantages over the other available techniques. First, microsatellite alleles can be unambiguously assigned to loci, thus eliminating the errors associated with scoring non-allelic but similar-length bands as alleles from the same locus, which occasionally occur when using traditional microsatelite DNA fingerprinting methods. Second, the bands generated by microsatellite analysis are identified as alleles by their absolute size (number of base pairs) for each locus that allows combining data from separate gels and from separate data sets. Third, microsatellites evolve rapidly, and exhibit many alleles per locus, allowing accurate individual identification as well as the precise estimation of population genetics parameters. Finally, microsatellites are short in length and thus can often be recovered from ancient samples (museum skins) and from partially degraded DNA samples from hairs or feces collected non-invasively in the field. However, some subjectivity is still present in scoring microsatellite data because of occasionally ambiguous bands on gels and of the potential occurrence of null alleles (heterozygote genotypes which appear as homozygotes; Brookfield, 1996).

In this chapter, we present an overview of the recent improvements in genetic methods that can help to design and to implement a conservation program. After a brief discussion of five case studies using these new methods, we review recent technological improvements for producing reliable genotypes using non-invasive genetic sampling. Second, we discuss statistical issues and approaches for individual identification (DNA 'fingerprinting') and for estimating parameters such as population size, effective population size, and dispersal rates. Finally, we present new insights in parentage analysis and relatedness.

CASE STUDIES
Number, sex, and home range of Pyrenean brown bears (Taberlet *et al.*, 1997)

Pyrenean brown bears (*Ursus arctos*) are threatened with extinction. Management efforts to preserve them require a comprehensive knowledge of the number and sex of the remaining individuals. By combining field data on track sizes with microsatellite and Y-chromosome data, researchers showed that this dwindling population consists of at least one yearling, three adult males, and one adult female. Only 36 of 247 hair samples and 21 of 105 feces samples provided enough DNA for complete genetic typing at all polymorphic loci using the multiple-tube approach (seven independent polymerase chain reactions, PCRs, per locus per sample in this case). Because of the very low polymorphism (only six out of 24 loci were polymorphic with only two alleles), two individuals could not be resolved using the genetic data alone, but fortunately exhibited different track sizes. This study demonstrates that a non-invasive genetic approach alone will some-

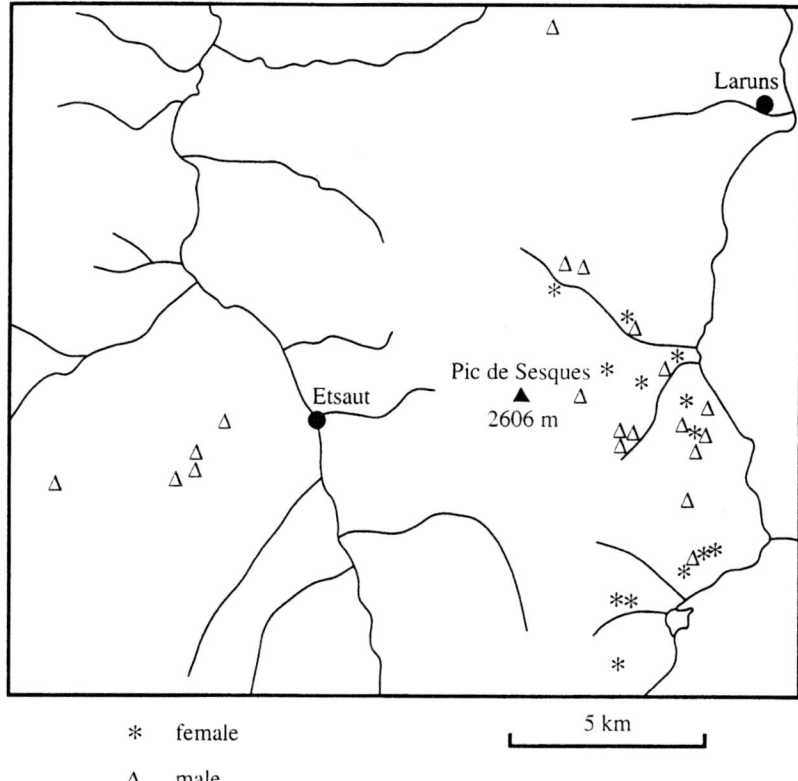

Figure 14.1. Approximate home ranges of two Pyrenean brown bears obtained by non-invasive genetic sampling and genotyping (from Taberlet *et al.*, 1997, reproduced by permission of Blackwell Science Ltd).

times not be sufficient to identify individuals in populations with reduced genetic variation. Additionally, home-range data were obtained via this non-invasive approach and showed that the female has as a smaller home range than males (Figure 14.1).

Mark-recapture experiments using genetic tags (Woods *et al.*, 1999)

A study of Canadian brown bears (*Ursus arctos*) demonstrated that it is potentially feasible to estimate population census sizes from genetic tags, without capturing or even seeing the animals. Clumps of hairs were repeatedly collected on barbed wire around baits, and analyzed. First, brown bear (versus black bear) samples were identified using a species specific mitochondrial DNA (mtDNA) test. Second, the sample sex was identified using a Y-chromosome test. Third, individual identification was carried out using six highly polymorphic microsatellites. Out of the 405 samples collected and identified as brown bear, 303 produced multi-locus genotypes leading to the identification of 54 genotypes (29 males, 25 females). The multiple-tube approach was not necessary, because several hairs from the same clump were used in the DNA extractions, increasing the amount of DNA and thus reducing the probability of genotyping errors. Nevertheless, two genotyping errors were found by reanalyzing all genotypes that had been detected only once and that differed by only a single allele from other genotypes that occurred many times. The six highly variable markers made it extremely unlikely that any two multi-locus genotypes would differ by only a single allele.

Estimating population size in coyotes (Kohn *et al.*, 1999)

A non-invasive genetic study using feces from coyotes (*Canis latrans*) provided precise estimates of population size. The study approach consisted of four steps: (1) feces were collected from the presumed target species (e.g., along trails of territorial boundaries); (2) DNA was extracted from feces, and species identity and sex was determined by mitochondrial DNA (mtDNA) and Y-chromosome typing, respectively; (3) hypervariable microsatellite loci were typed from the feces; and (4) the rarefaction analysis and a standard mark-recapture method were used to estimate population size.

In the first step of the coyote study, 651 recognizable carnivore-like feces were removed from 381 sites along six transects in a 15 km^2 area in the Santa Monica Mountains, California. From the 651 feces, 238 were randomly drawn for DNA analysis. Species-diagnostic mtDNA restriction sites identified 188 (79%) coyote feces (from the 238 total) and 115 of these were typed for the Y-chromosome marker and three canid-specific microsatellite loci. The Y-chromosome (SRY) analysis revealed a sex ratio of 1:1.14 (not significantly different from 1:1). The microsatellite analysis revealed 30 unique genotypes (an estimate of the minimum population size). The rarefaction estimate of population census size consisted of plotting the average number of unique genotypes, y, discovered as a function of the number of feces genotyped, x. The idea here is that the asymptote of this curve gives an estimate of the local population size. Thus, a curve defined by the equation $y = (ax)/(b + x)$ was fitted to the data. This process was repeated 1000 times to simulate a distribution of curve asymptotes. The mean and maximum/minimum of the distribution of possible curve asymptotes was taken as the population size estimate and the minimum/maximum estimates, respectively (Figure 14.2). For the coyotes, the (mean) population size estimate was 38 (CI$_{95\%}$ 36–41) and simulations suggested a minimum and maximum

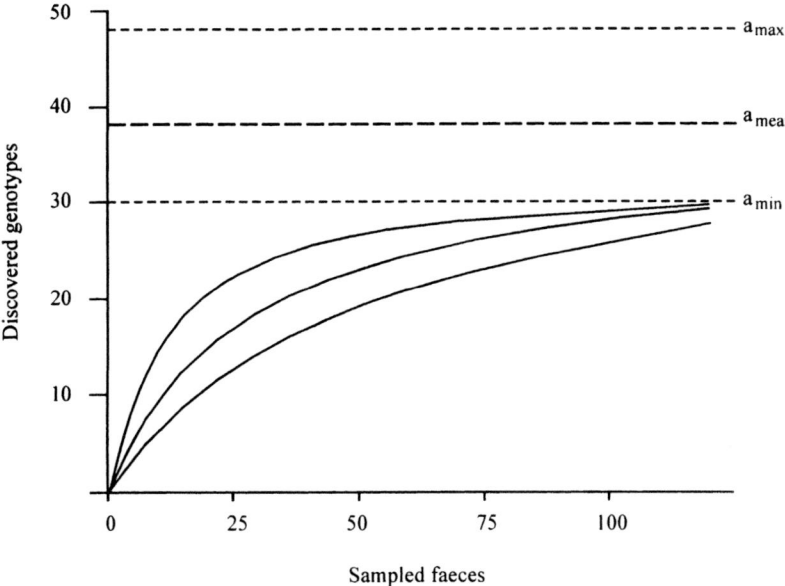

Figure 14.2. Genotype rarefaction curve for population size estimation based on non-invasive sampling using faeces. Plot of the average number of unique genotypes, *y*, identified as a function of the number of faeces analyzed, *x*; *a*, asymptotes and estimated population sizes (from Kohn *et al.*, 1999, reproduced by permission of The Royal Society; see text for further details).

possible size of 30 and 47, respectively. Similar estimates were obtained from the mark–recapture method.

Inbreeding in wolf packs (Smith *et al.*, 1997)

The purpose of this study was to determine whether incest is common in wolf packs. Individuals from the above populations were typed using 20 polymorphic microsatellite loci. Three social groupings (i.e., mother–offspring, sibling, and mated pairs) were defined based on behavioral observations in the field and from captive populations. Exclusion analysis was used to confirm suspected mother–offspring and siblings relationships. Pairwise relatedness values within each of the three social categories were calculated separately for each population. Relatedness values from each category were compared with the other two categories, and with known unrelated individuals in the captive populations, using randomization tests. The results confirmed what has been previously hypothesized based on behavioral observations – mated pairs in the wild are as related as any two unrelated individuals, and incestuous mating is avoided. Although wolves live in packs, which compose of family units, the results of this study suggest that adult offspring rarely replace a parent when the opposite sex parent is present.

Social structure, dispersal, and pack interactions in the African wild dog (Girman *et al.*, 1997)

Ninety-two African wild dogs (*Lycaon pictus*) from nine packs in the Kruger National Park, Republic of South Africa, were screened using 14 microsatellite loci and 402 bp of the mitochondrial control region. Each individual was subjected to a global exclusion search over all parental combinations. This approach enabled the authors to assign 25 (86%) of the pups to alpha males, and 47 (92%) to alpha females. Subordinates produced only about 9% of all pups. Pairwise relatedness values were calculated for several social categories. Alpha-ranked and within group adult female–male pairs were related at a level similar to unrelated individuals. This result implies that incestuous mating is avoided in the wild, and that opposite-sex siblings avoid dispersing to the same social groups. On the other hand, similar-sex siblings often disperse together (mean within group similar-sex relatedness was at the level expected for cousins). Using randomization tests, Girman *et al.* (1997) have demonstrated that social groups are composed of related individuals by comparing to a random collection of individuals. They also showed that both sexes disperse, mostly within or near territories of their close relatives.

NON-INVASIVE GENETIC SAMPLING

The first step in studying the genetics of free-ranging carnivores is to obtain tissue that will be subsequently used as a source of DNA. This sampling often represents the most challenging difficulty. Until recently, the only way to sample tissues was to immobilize the carnivore, either by trapping or by using a dart gun with a tranquilizing drug. This approach allows gathering of much additional information (home ranges via radio-tracking, age, morphometric measurements, physiological condition, etc.). But it also has some serious drawbacks in terms of cost and feasibility. Capturing large mammals in remote mountainous areas, or in dense forest may require extensive logistic assistance that can be too demanding financially for a conservation program. The sampling strategy that requires immobilizing the animal and taking blood or biopsy samples is called non-destructive sampling.

With the development of the PCR, it has become possible to use hair or feces as a source of DNA. Such a sampling strategy that involves tissues left behind by the animal is called non-invasive sampling. Scientists working in conservation biology and in ethology are particularly interested in non-invasive sampling techniques, because genetic studies can be initiated without having to capture, disturb, or even observe the animal. This interest was corroborated by the publication of many preliminary reports, the dominant idea being that non-invasive sampling has the same potential as blood or biopsy samples (e.g., Höss *et al.*, 1992; Taberlet & Bouvet, 1992; Morin *et al.*, 1993; Kohn *et al.*, 1995; Tikel *et al.*, 1996; Reed *et al.*, 1997). However, after 1995, some studies on primates and carnivores revealed risks of genotyping errors associated with the use of hair or feces. These technical difficulties might explain why, during the six or seven years after the initial development of these non-invasive methods, only a few comprehensive studies were published. Nonetheless, hairs or feces may represent the only sampling strategy to gather genetic information when capturing animals entails too much risk for an endangered carnivore population, or when it is logistically too expensive within a conservation program.

Genotyping errors associated with non-invasive genetic sampling

When using hairs, feathers, or feces from free ranging animals, the total amount of DNA available for the genetic typing can be very low. Under these circumstances, only one allele of a heterozygous individual might be detected (Gerloff *et al.*, 1995; Foucault *et al.*, 1996; Taberlet *et al.*, 1996, 1997; Gagneux *et al.*, 1997; Goossens *et al.*, 1998). This type of error, called 'allelic dropout', produces false homozygotes, and may be explained by sampling stochasticity – when pipetting template DNA in a very dilute DNA extract, sometimes only one of the two alleles is pipetted, amplified, and detected (Taberlet *et al.*, 1996). Allelic dropout might also be related to extreme DNA degradation, or differential amplification of one of the two alleles. A mathematical model has been developed to account for these stochastic events, and computer simulations have been performed to quantify error rates according to the amount of amplifiable template DNA (Taberlet, *et al.*, 1996). When using very low amounts of template DNA, there is a high risk of allelic dropout, even when positive PCRs are obtained in almost all experiments.

In our experience, when working with shed hairs or feces collected in the field, the amount of template DNA used per PCR is highly variable, and often in the range where allelic dropout occurs. In order to avoid this genotyping error, an ideal solution would be to quantify the DNA concentration of the extract, and to adjust the amount of template DNA per PCR.

But such an estimation is either technically difficult due to the very low DNA content of hairs, or even impossible when using feces because the DNA of the species under study is copurified with large amounts of prey or bacterial DNA. Furthermore, even if large quantities of DNA are obtained, it may be highly degraded and the number of intact target molecules may be in the range leading to allelic dropout. This could be especially problematic for obtaining long PCR products.

Another kind of genotyping error has been detected when using very little template DNA and when amplifying dinucleotide microsatellites – artefacts generated during the amplification process and then misinterpreted as true alleles because they have the same characteristic shadow band profile (Taberlet *et al.*, 1996). If such a 'false allele' occurs in a homozygous individual, then this individual might be recorded as a heterozygote, and if it occurs in a heterozygous individual, then the presence of three 'alleles' will allow the detection of the error. These artefacts generating false alleles are easily confused with sporadic contaminations. They generally occur in less than 5% of the PCRs (Taberlet *et al.*, 1996), but should not be disregarded as they can lead to erroneous genotypes.

To overcome all potential sources of genotyping errors (allelic dropout, false alleles, sporadic contaminations), researchers have proposed a multiple-tubes approach (Navidi *et al.*, 1992; Taberlet *et al.*, 1996) – amplification of each microsatellite is repeated independently several times, and the true genotype is deduced by analysing the set of all experiments (see Taberlet *et al.*, (1996) for precise guidelines; and the case study on the Pyrenean bear discussed above).

If the error rate is relatively low and if each multilocus genotype has a high probability of being sampled many times, then the single tube approach can be used. Errors can be detected by reanalyzing any multi-locus genotype that is found only once and/or that differs by only a single allele from another genotype that has been found many times (see the case study on the use of genetic tags discussed above).

A pilot study to assess feasibility

Based on the limitations outlined above, it is clear that some scientific questions cannot be assessed using non-invasive methodology exclusively, when the amount of DNA available is too low to obtain reliable results. Consequently, researchers should not switch to non-invasive sampling without evaluating whether the scientific question of interest is solvable using this approach. Not assessing the potential of non-invasive sampling prior to performing extensive genetic analyses can have dramatic conse-

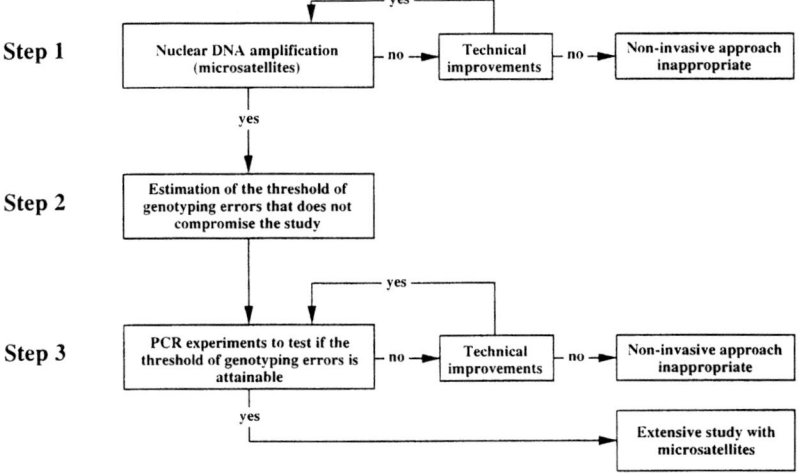

Figure 14.3. Flow chart diagram illustrating the three steps of the pilot study that should be carried out before any extensive study of microsatellite polymorphism based on a non-invasive genetic sampling (from Taberlet *et al.*, 1999, reproduced by permission from Elsevier Science).

quences such as the loss of time, money and energy invested in the collection of useless samples, and a delay of many years in the completion of an important study. Unfortunately, the disciplines in which non-invasive methods are most needed, ethology and conservation biology, are also the disciplines in which the consequences of genotyping errors may be most serious or in which the application may be most difficult (e.g., in paternity analysis or in endangered species with low levels of heterozygosity).

Conducting an appropriate pilot study is the best way to assess whether the non-invasive approach is feasible (Taberlet *et al.*, 1999; Taberlet & Luikart, 1999). As the amount of DNA that can be extracted from samples collected non-invasively greatly varies among species, it is not advisable to transfer an experimental protocol from one species to another. For example, in our laboratory, wolf feces provide much more amplifiable DNA than bear feces, and accordingly, the allelic dropout problem is much easier to avoid in wolves than in bears. The general goal of a pilot study should be to develop an experimental protocol that will produce results with a confidence level appropriate for the research question.

Such a pilot study should be carried out in three steps (Figure 14.3). The first step consists of amplifying at least one microsatellite locus using DNA extracted via the non-invasive approach. It is essential that the primers used have been shown to work on DNA extracted from tissue or blood samples from the study organism. This experiment will determine if the DNA extraction protocol provides good quality template DNA with few or no inhibitors of PCR. If this step fails, then the non-invasive approach is not appropriate, unless technical improvements are achieved (e.g., in sample preservation, DNA extraction, or PCR optimization).

The purpose of the second step is to estimate the maximum genotyping error rate that is compatible with the level of confidence required by the scientific question. This threshold of genotyping errors can be estimated either by using analytical equations, or by computer simulations. For example, if the goal of the study is to identify individuals using 10 loci with a confidence level above 95%, such a result will be obtained if the rate of allelic dropout is less than 0.005 per locus assuming 100% of the loci are heterozygous. Because allelic dropout does not affect homozygous genotypes, if 50% of loci are heterozygous, then the rate of allelic dropout will be less than 0.01 to achieve the goal of the above example. This means that even a very low level of allelic dropout can lead to erroneous genotypes and can compromise the entire study. It is important to realize that by increasing the number of loci, the error rate of a multilocus genotype will increase accordingly. Consequently, to increase the power to resolve individuals without increasing the error rate for multilocus genotyping, it would often be better to use loci with higher heterozygosity than to use more loci.

The third step is required to determine if it is technically possible, using non-invasive sampling, to achieve the threshold of genotyping errors estimated in the second step. Clearly, this corresponds to the most time-consuming, expensive, and technically challenging part of the pilot study. Because the error rate that must be quantified can be very low, the number of PCR amplification experiments to be conducted can be accordingly very large. For example, 200 amplifications of heterozygous individuals without allelic dropout are necessary to be sure ($P = 0.01$) that this kind of error does not occur in more than 1% of the PCRs. Such amplifications should be conducted on DNA extracted from each of several different samples in order to estimate the inter-sample variance in error rates.

Even after the three steps of the pilot study have been successfully passed, it is still not certain that the non-invasive approach represents the best choice. Indeed, if the multiple-tubes strategy has to be used, the total cost of the laboratory work can be 3 to 10 times higher than when using blood or tissue samples. Thus, the factors that must be weighed up are the field constraints involved in trapping the animals and the additional laboratory costs of using non-invasive sampling. Clearly, there are some situations where trapping of the species of interest is not conceivable,

particularly in the case of a small endangered population. In these conditions, the non-invasive sampling approach represents the only solution.

INDIVIDUAL IDENTIFICATION AND ESTIMATING
IMPORTANT POPULATION PARAMETERS

Individual identification

Accurate identification of individuals from their unique multi-locus genotype (DNA fingerprint) is a prerequisite for many uses of molecular
markers in carnivore conservation and population ecology. It is important
to analyze enough highly variable markers to resolve all the individuals in
the population of interest. A commonly used statistic for estimating the
power of a set of markers for resolving individuals is the probability of
identity (*PI*), which gives the theoretical probability of sampling two identical genotypes when randomly sampling from a population. *PI* is computed
per locus using allele frequencies from each locus and the following equation:

$$PI = \frac{n^3(2\ a_2^2 - a_4) - 2n^3(a_3 + 2a_3) + n(9a_2 + 2) - 6}{(n-1)(n-2)(n-3)} \tag{1}$$

where *n* is the number of individuals sampled, a_i equals $\sum_i p_i^i$, and *pj* is the
frequency of the *j*th allele (Kendall & Stuart, 1977; see also Paetkau *et al.*,
1998a). *PI* values for a set of independent loci can be combined to obtain
the overall (multi-locus) *PI* by simply multiplying together the *PI* from each
locus ($PI = PI\ 1 \times PI\ 2 \times PI\ 3$).

While *PI* formulae provide useful information, it is critical to recognize
that they may occasionally underestimate the true *PI* by 1–3 orders of magnitude. Such a bias may cause researchers to believe their set of markers
has more power for resolving individuals than it actually does. For example,
from a sample of 84 Scandinavian brown bears the theoretical *PI* (from
equation 1) for four microsatellite loci was one order of magnitude smaller
than the actual observed *PI* (Waits *et al.*, 2001). In a sample of 64 Montana
wolves, the theoretical *PI* for seven loci (data from Forbes & Boyd, 1997)
was three orders of magnitude lower than the actual observed *PI* (i.e.,
~0.001 instead of ~0.000001; Waits *et al.*, 2001). Why does this bias
occur? This (and related formulae) assume Hardy–Weinberg proportions
at all loci, linkage equilibrium and no shared ancestry (i.e., no close relatives in the population). Because these assumptions are violated in most
natural populations, the *PI* statistic will tend to under estimate the true

probability of finding identical genotypes. Thus, to ensure sufficiently high power to resolve individuals, researchers should either compute the actual observed *PI* (when approximately 50 individuals have been genotyped) or simply use one or more additional markers than are suggested as necessary by the theoretical formula. Of course, 'sufficiently high power' will depend on the specific research question and the severity of the consequences of failing to resolve all individuals. For example, if a researcher must reliably identify siblings or close relatives, they may compute the *PI* for siblings (Taberlet & Luikart, 1999) or compute *PI* using a formula that accounts for the presence of relatives in a population (see Balding & Nichols, 1994).

Example applications of *PI* in wildlife forensics are in identifying problem bears, or wolves that have killed live stock. In 1998, at Glacier National Park, Montana, a female bear was captured near the site where a hiker had been attacked. Before translocating or destroying the 'accused' bear, biologists conducted microsatellite DNA typing to verify that the bear's genotype matched that from the fresh bear scat which contained fragments of the hiker's clothing. An extremely important question here is how many microsatellite loci are necessary to be highly certain ($P < 0.0001$) that the genotype from the scat matches that of the killer bear. In this case, six loci were used and they provided a *PI* of < 0.0001 (L. Waits, pers. comm.) when using the following equation and the allele frequencies from the bear population:

$$PI = \sum p_i^2 + \sum (2p_ip_j) \tag{2}$$

where p_i and p_j and are frequencies of the *i*th and *j*th allele respectively. Equation (2) is different from equation (1) in that this one addresses the following forensic question – what is the probability of sampling a second genotype identical to the genotype 'already in hand' (i.e., from the scat in the above example)? Equation (1) gives the probability of sampling two individuals with identical genotypes and thus is more useful for evaluating the general power of a set of markers when planning a study requiring individual identification.

Minimum population size, census size, home range and sex ratios

Information on the minimum or census population size, home ranges and sex ratios is critical for population management and for predicting the long-term persistence of populations. Such information can be especially difficult to obtain for carnivores that are often elusive and at low population densities. Non-invasive genetic sampling of feces or hairs can help provide estimates of all of these parameters (e.g., Mills *et al.*, 1999). Furthermore, sample sizes can be larger and potentially less biased than those from traditional demographic capture–recapture methods because, for example, there is no problem with trap-shyness when collecting scats along a transect (Schwartz et al., 1998; Kohn et al., 1999). The case studies on the Pyrenean bear and the coyotes discussed above illustrate the potential advantages of combining genetic and demographic information to obtain more comprehensive estimates of these population demographic parameters.

Effective population size

Estimating and monitoring the effective population size (Ne) is essential in conservation programs because Ne influences the rate of inbreeding, loss of beneficial alleles and fixation of harmful alleles, and thus influences the risk of population extinction (Frankel & Soulé, 1981; Lande, 1994). It is important to monitor Ne even in populations for which the census size (Nc) is known because Ne can be much smaller than Nc (Frankham, 1995), and cryptic genetic bottlenecks (reduced Ne) can occur that are only detectable by directly monitoring Ne using molecular markers. Ne can be defined as the size of an ideal population (i.e., random mating, with even sex-ratio, non-overlapping generations, etc.) that experiences genetic change (e.g., loss of heterozygosity) at the same rate as the population of interest (Lande & Barrowclaugh, 1987).

It is possible to estimate the long-term historical Ne of a population using DNA sequence data or microsatellite allele frequency data (e.g., see Waples, 1991; Paetkau et al., 1998a). Estimates of long-term Ne may be informative when compared to estimates of the current Ne in order to determine if the Ne has recently declined, which may signal an increased risk of population extinction. However, estimating long-term Ne requires knowledge about mutation rates and dynamics, and the assumption of a constant-size population (mutation-drift equilibrium) for many generations in the past. Repeated estimates of current Ne can be used to monitor populations to detect population declines or cryptic genetic bottlenecks that could be caused by reproductive failures, sudden population fragmentation/isolation or the monopolization of reproduction by a few dominant males (Luikart et al., 1998b; Schwartz et al., 1998, 1999).

For the grizzly bears (Ursus arctos) on Kodiak Island, Paetkau et al. (1998) estimated the long-term historical Ne by using the contemporary microsatellite heterozygosities and assuming mutation rates of 0.0001–0.0002, a stepwise model of mutation (SMM), and a constant-size population. They obtained a Ne estimate of 106–532 using the following equation

(SMM; Ohta & Kimura, 1973):

$$He = 1 - (1 / (1 + (8 \, Ne \, \mu)^{0.5}))$$ (3)

where μ is the mutation rate. Such a small Ne is perhaps surprising when considering that the population census size (Nc) is estimated to equal 2842. Nonetheless, this gives a Ne/Nc ratio of 0.037–0.187 that is in reasonable agreement with the small Ne/Nc ratios estimated for many wildlife species (average $Ne/Nc = 0.11$; Frankham, 1995). A small Ne estimate of 239–1195 was also obtained for an island population of black bears (*Ursus americanus*) with a census size of 3000–10 000 bears (thus, $Ne/Nc = 0.024$–0.40). An important management implication of these low Ne/Nc ratios is that very large population census sizes will be required for long-term maintenance of genetic variation within bear populations. Relatively small and isolated populations, such as Yellowstone, will require translocations to maintain genetic variation.

Paetkau *et al.* (1998a) also estimated the 'current' or recent Ne for the grizzly bear population in the Yellowstone ecosystem by using the rate of loss of heterozygosity at eight microsatellite loci over the past 100 years (or ~ 10 generations). The rate of loss of heterozygosity was estimated at 0.8–4.0% per generation by subtracting the estimate of the contemporary He (0.554) from the estimate of historical He (0.67–0.70). The historical He in Yellowstone was assumed to equal that in contemporary populations in southern Canada and Montana, USA. Using the rate of He-loss per generation, an estimate of Ne can be obtained from the following equation (see also the equations of Harris & Allendorf, 1989):

$$Ne = 1 / (2 \times H_{LOSS})$$ (4)

The Ne estimate for Yellowstone was approximately 13–65. Given that the population census size is approximately 350, the estimated Ne/Nc ratio is 0.037–0.186 which is in reasonable agreement with the long-term Ne estimates for bear populations.

Another approach for estimating the current Ne is measuring temporal changes in allele frequencies (Waples, 1991). The amount of change (or drift) in allele frequencies between generations is inversely proportional to Ne, and can be converted into an estimate of Ne using a simple mathematical equation (Waples, 1989) or a more complicated maximum likelihood approach (M. Beaumont, unpub. data). This 'temporal method' promises to give more precise estimates of Ne than the heterozygosity-loss approach (Luikart *et al.*, 1998b), but might be less accurate because it is probably more sensitive to violations of several necessary assumptions, e.g., random sampling of individuals, no population substructure and no immigration (Waples, 1989, 1991). The consequences of violating these assumptions must be quantified by computer simulation modeling before the method can be used with confidence in many carnivore populations. Additional approaches for estimating and monitoring Ne (i.e., linkage disequilibrium and heterozygote excess) have been reviewed in Schwartz *et al.* (1998) and Schwartz *et al.* (1999).

Change in effective population size

In addition to estimating the current and long-term historical Ne, it is possible to detect recent changes in Ne (i.e., changes during the past few dozen generations (Cornuet & Luikart, 1996)). Detecting a severe reduction in Ne is critical in conservation because fragmented and declining populations may have an increased risk of extinction. Population declines generate a genetic signature (e.g., deficiency of rare alleles) that can be detected by analyzing 5–20 microsatellite loci from a single sample of approximately 30 individuals. For example, the microsatellite data from the endangered Mexican wolf reveals a signature of a recent bottleneck, whereas large wolf populations show no such signature (Luikart *et al.*, 1998a). Interestingly, such tests can identify declining or recently bottlenecked populations when nothing is known about the current or historical population size or effective population size.

New statistical tests for detecting recent increases in Ne have been applied to carnivore populations (Goldstein *et al.*, 1999). In populations of island foxes *Urocyon littoralis* with limited opportunity for long-term growth, the tests revealed no evidence of population growth. However, in the larger mainland population of foxes *U. cinereoargenteus* the test provided evidence for population growth. The test compares the observed distribution of branch lengths among microsatellite alleles (at a locus) with the distribution expected in a constant-size population. Populations that have been growing for hundreds of generations will tend to have genealogies of alleles with similar branch lengths at all loci. On the contrary, constant-size populations have genealogies of alleles with relatively variable branch lengths.

Inbreeding level of individuals

A recent statistical method could improve our understanding of the fitness consequences of inbreeding and outbreeding in carnivore populations (Coulson *et al.*, 1998). The method infers the relative inbreeding level of individuals using the information from the relative allele lengths of microsatellite alleles at heterozygous loci. The method infers that heterozygotes with alleles of different length are less inbred than heterozygotes with alleles of similar length (assuming a stepwise model of mutation (SMM); see for details Coltman *et al.*, 1998; Coulson *et al.*, 1998). The method has not yet been applied to carnivores, but in neonatal red deer (*Cervus elaphus*; Coulsen *et al.*, 1998) and harbour seals (*Phoca vitulina*; Coltman *et al.*, 1998), individuals that were outbred according to this method were significantly heavier at birth and survived longer than individuals that were relatively inbred. A classical statistic for quantifying individual inbreeding (heterozygosity, *Ho*) was not significantly related to birth weight or survival, in either the red deer or harbour seals, using the same genetic data. This new statistical method can potentially extract more information than *Ho* about an individual's inbreeding level. Thus, it may help researchers to quantify the fitness costs (and benefits) of mating between close relatives (and non-relatives) in natural carnivore populations.

Forensics applications

A serious threat to many carnivore populations is illegal killing by poachers. DNA markers can be used on tissue samples or body parts to determine the species of origin and even individual identity (see *PI* above). In addition, 'assignment tests' can be used to assign or exclude a population as the origin of an individual (Paetkau *et al.*, 1995; Rannala & Mountain, 1997). Such methods could be used to test if an animal came from a legal hunting area or a protected national park. Thus assignment tests could help reduce poaching and convict poachers. These methods require samples of approximately 20–40 individuals (from each candidate population) and 10–20 highly polymorphic loci ($He = 0.60$; Cornuet *et al.*, 1999) in order to exclude all non-source populations with high statistical certainty ($P < 0.01$ or $P < 0.0001$). The tests also require moderately high genetic differentiation among populations (i.e., $F_{st} > 0.08$). It is difficult to provide general guidelines as to the sample sizes of loci and individuals needed to perform such tests because their accuracy depends on several variables such as the mutation model of the loci and the number of candidate populations being considered.

Dispersal rates

Recent statistical advances allow improved estimation of past and current dispersal rates from genetic data. Traditionally, dispersal rates (averaged over many past generations) have been estimated using F_{st} (a statistic quantifying genetic differentiation among populations) and a model of dispersal such as the island model (e.g., Slatkin, 1987). Recent estimators of dispersal rates using maximum likelihood and coalescent theory generally outperform the estimators based on F_{st} (Beerli & Felsenstein, 1999). In addition, it now may be possible to estimate current dispersal rates directly from a single large sample of individuals (i.e., a sample of multi-locus genotypes) from each population of interest (see Chapter 22 in this volume).

Taxonomy and hybridization

Taxonomy and hybridization are important issues in conservation. A recent statistical method using microsatellites promises to help resolve between sets of populations that have become fragmented recently versus sets of populations that have been isolated for hundreds of generations (and thus may be locally adapted) and should be managed as separate units (O'Ryan *et al.*, 1998). Thus, microsatellites might be useful for identifying population units for conservation (e.g., management units and subspecies or evolutionary significant units). Another recently published coalescent method (Bertorelle & Excoffier, 1998) should improve our ability to detect and access hybridization events, e.g., in some canid species. The role of genetics in identifying taxa and detecting hybridization among carnivores has been thoroughly discussed by Wayne (1995)

PARENTAGE, RELATEDNESS, AND BEHAVIOURAL ECOLOGY
Parentage

Parentage analysis is used for deducing parent–offspring pairs in the population under study. The two main approaches currently in use are exclusion analysis and analysis of relative likelihood. Exclusion analysis is based on the Mendelian theory; alleles found in the offspring should be accounted for in the parents. A perfect match, across several loci, between an offspring and a pair of possible parents may indicate that these are the true parents. The probability of exclusion per locus (PE_i) can be calculated following Chakraborty *et al.*, (1988):

$$PE_i = (1 - a - b)^2 \tag{5}$$

where a and b are the frequencies of the alleles found in the offspring. The probabilities for all loci (n) combined can be calculated as (Chakraborty *et al.*, 1988):

$$PE(C) = 1 - \prod_{i=1}^{n}(1 - PE_i) \tag{6}$$

PE(C) is the probability of genetically excluding a randomly chosen male or female as being the father or mother of the given offspring. *PE(C)* is only a rough measure of the efficiency of a given exclusion analysis. It is not suitable for testing specific parental pairs.

The exclusion approach is powerful in searching for the true parents of an offspring. A global search over all parental combinations for each given offspring ideally should find only a single suitable parental pair. However, occasionally more than a single suitable pair of parents are located. To resolve this discrepancy, additional loci should be scored. Further exclusion of possible parents can be achieved by incorporating behavioral and morphological information (e.g., estimated age, presence in the study site, etc.) into the analysis (e.g., Girman *et al.*, 1997; Moritz *et al.*, 1997). The drawback of the exclusion approach is its great sensitivity to scoring mistakes and null alleles. Any mismatch, even on a single allele, excludes a pair of parents as being the putative parents. Such a mismatch, created by human error, may cause the misidentification of the true parents.

Another approach, which can be complementary to the exclusion analysis, is the relative likelihood of parentage (RLP; e.g., Aldrich & Hamrick, 1998; Prodöhl *et al.*, 1998). In this analysis, the likelihood that alleles in the offspring could have originated from a pair of possible parents is calculated. The calculations span all possible parent combinations for each offspring, and combine the information over all loci. To determine the most probable parents for each offspring a ratio between the likelihood of each specific parental pair and the total sum of likelihoods is calculated as shown in the equation below:

$$RLP = \frac{\prod_{L=1}^{n} P(Of|Fa_i, \ Mo_i) \cdot P(Fa_i) \cdot P(Mo_i)}{\sum_{j=1}^{m} \prod_{L=1}^{n} P(Of|Fa_i, \ Mo_i) \cdot P(Fa_i) \cdot P(Mo_i)} \tag{7}$$

where $P(Of|Fa_i,Mo_i)$ is the probability that alleles found in the offspring (*Of*) originated from the *i* parents (*Fa_i* and *Mo_i*), and $P(Fa_i)$ or $P(Mo_i)$ is the probability of selecting the alleles in the *i* father or *i* mother by random. The number of loci is donated by *n* and the number of possible or suitable parental combinations is donated by *m*.

The pair of individuals that are most likely the true parents of the focal offspring receive the greatest relative ratio. Marshall *et al.* (1998) developed Meagher & Thompson's (1986) and Thompson & Meagher's (1987) earlier work for the use with microsatellites. Their program CERVUS accommodates scoring and other errors associated with microsatellite data and population sampling (Slate *et al.*, 2000). Marshall *et al.* (1998) include slight modifications in the ratio calculations, yet the basic technique is similar. Since all possible combinations in a given population are considered, the technique is most powerful when all possible parents for each offspring are included. The method used in CERVUS is superior over exclusion analysis because it provides confidence for the paternity and single exclusions may not preclude the detection of the true parents (Marshall *et al.*, 1998).

Relatedness

The degree of relatedness between a pair of individuals in a population is determined by an index of relatedness. Indices of relatedness are useful in estimating the relative mean relatedness for social categories such as fa-ther–offspring, mother–offspring, siblings, within matrilines, etc. A simple index for calculating the relative degree of relatedness between any pair of individuals is the similarity index (SI). SI is calculated as $2S_{ab}/(n_a + n_b)$, where S_{ab} is the number of shared alleles between a pair of indi-viduals, and n_a or n_b are the number of alleles present in the a or b individ-ual. The values obtained with this measure are between 0 (non-related) and 1 (identical individuals). The advantage of this index is that it is simple and fast to compute for all possible comparisons within a given population. The drawback of this index is its sensitivity to the number of loci used, and to missing data.

Other more reliable indices of relatedness are available (Pamilo, 1990; Ritland, 1996). We favor the Queller & Goodnight (1989) index of related-ness. This index weighs each allele by its frequency in the population, so that rare alleles are given a relatively higher value. Overall, index values run between -1 and 1. In a sample that adequately represents a population in a Hardy-Weinberg proportions, the index values obtained for parent-off-spring or full sibs relationships approach 0.5. The pairwise relatedness (R) for any two individuals is calculated as follows:

$$R = \frac{\sum \sum (p_y - p')}{\sum \sum (p_x - p')} \tag{8}$$

The equation is summed over loci and allelic positions. P' is the population frequency of the allele present at the focal locus and allelic posi-tion, excluding the compared individuals. P_x is the frequency of the focal allele in the specific individual (i.e., 0.5 or 1 depending on whether the individual is a heterozygote or a homozygote). P_y is the frequency of the focal allele in the compared partner (i.e., 0 or 0.5 or 1). This index is not symmetric, thus comparing one individual over another would not yield the same index value for the reciprocal comparison (P_y over P_x). To accommodate for this discrepancy, the denominator values and numerator values are calculated for each of the combinations (P_y over P_x, and P_x over P_y), and are summed up prior to the division. This procedure yields an average estimate of relatedness between the two individuals concerned. Standard errors for the relatedness values can be estimated by jack-knifing over loci (Queller & Goodnight, 1989). This index can be used to calculate the relatedness within social groups or categories in the same manner.

In general, pairwise relatedness values vary greatly. This precludes the possibility of determining the level of relatedness between any two individuals without the aid of statistical confidence. Field *et al.* (1998) and Goodnight & Queller (1999) provide a log likelihood procedure for calculating the confidence for R values. Their aim is to calculate the probability that the genotypes of two randomly selected individuals in the population could occur if the two were related at a level x (e.g., full siblings), relative to the probability that they were related at a level y (e.g., cousins). The distribution of 10 000 such likelihood ratios, which were generated based on the observed allele frequencies, can be used for evaluating the statistical confidence for the observed pairwise R values. This approach allows categorizing R values into relatedness categories (e.g., full sibs or parent–offspring, non-related).

The Queller & Goodnight (1989) index of relatedness can also be used to estimate the number of loci needed for the parental analysis. By randomly selecting a cumulative number of loci, calculating the index of relatedness after each locus is added, and repeating the process a 1000 times, we can estimate the amount of variance contributed by each additional loci. The point of asymptote on a graph of the mean index of relatedness versus the number of loci used indicates the minimum adequate number of loci for the analysis. Adding additional loci beyond that point is not expected to significantly change the resolution of the analysis (e.g., Girman *et al.*, 1997).

In a given population of n individuals there are $n(n-1)/2$ unique pairwise relatedness (R) values. These R values are useful for testing patterns in the distribution of relatedness within and between social categories or groupings. For example, we can test whether individuals within social groups are more closely related (to each other) than to members of other groups, or whether individuals in neighboring groups are more closely related to each other than expected between randomly distributed individuals. Such simulations allow testing for social structures and patterns of dispersal within a population. This approach takes into account both the spatial arrangement of individuals or groups, and the composition of groups; thus testing whether the observed distribution of relatedness is different from the one expected from a random arrangement. The direction of significant deviations from random distribution may suggest an explanation for the observed distribution pattern (e.g., groups are family units, females migrate shorter distances than males, etc.). Examples for the use of the above method are in Girman *et al.* (1997), Smith *et al.* (1997), and De Ruiter & Geffen (1998).

Behavioral studies of mating systems and relatedness are important in conservation because they influence the effective population size and thus the rate of loss of genetic variation. They also influence the amount of genetic load (severity of inbreeding depression) and dispersal patterns (e.g. sex-biased dispersal). Furthermore, data on these parameters are necessary for modeling demography and assessing the probability of population persistence.

CONCLUSIONS AND RECOMMENDATIONS FOR CONSERVATION

The new developments in genetic methods outlined above clearly demonstrate that it is becoming easier to design and implement a comprehensive conservation program with a genetic dimension. Indeed, genetic data can provide important information (e.g., population history, mating system, and even population size and dispersal rates) that is nearly impractical to collect using only field approaches when studying elusive and secretive carnivores.

Nevertheless, genetic approaches are not without drawbacks. The technical difficulties of implementing non-invasive genetic sampling can be substantial. Furthermore, the possibility of extremely low polymorphism in small and endangered populations can limit the power of genetic inferences. Thus, it may not be advisable to reduce the field study component in favor of a genetic approach, especially if a thorough genetic pilot study has not been done. Indeed, field studies can also provide important information, such as age and body condition, that genetics cannot. Age, for instance, can be helpful when interpreting genetic relationships for reconstructing pedigrees in free-ranging populations (e.g., for resolving between father–offspring versus sibling–sibling pairs). Similarly, when mother–offspring pairs are known from field data, far fewer genetic markers will be necessary to assess paternity and subsequently to reconstruct pedigrees.

From these examples, it is clear that genetics and field investigations should be viewed as complementary rather than competitive. The best overall approach for carnivore conservation and management programs is to use both.

15

Applications of genetic concepts and molecular methods to carnivore conservation

WARREN E. JOHNSON, EDUARDO EIZIRIK,
MELODY ROELKE-PARKER AND STEPHEN J. O'BRIEN

INTRODUCTION

Carnivores, by nature, can be difficult to study. Some are elusive, rare, and often occupy inaccessible areas. The inability to collect sufficient amounts of reliable data on carnivore distribution, numbers, population structure, and habitat requirements is a severe impediment to conservation efforts and the lack of information is often the basis of many controversies. However, the tools available to study carnivore populations have increased dramatically in the last decade. Theoretical and technological advances in remote sensing, radio telemetry, population-size estimation, statistics, and modeling promise to play an integral and inescapable role in future conservation efforts. But some of the most exciting and influential changes are occurring in the realm of molecular genetics, taking advantage of improvements in biomedical technology and the ready access to large public databases. In addition, the development of more sophisticated analytical methods and a comprehensive body of theory in phylogenetics and population genetics is providing the means not only to describe current phenomena, but also to probe into the past evolutionary history of a species and to make predictions of its future fate. Molecular tools are not only providing data useful for management decisions, but are also serving as a catalyst for the integration of many of the disparate disciplines that contribute to carnivore conservation, including areas such as reproductive biology, physiology, immunology, demography, virology, behavior, ecology, and veterinary medicine. These molecular tools include allozyme electrophoresis, mitochondrial DNA (mtDNA), restriction fragment length polymorphisms (RFLP), polymerase chain reaction (PCR), direct sequencing, major histocompatibility complex (MHC) variation, minisatellites (DNA fingerprinting), and microsatellites. Many of these techniques and classes of molecular genetic markers are depicted in the examples that follow and are described at length elsewhere (e.g. Hoelzel, 1998).

The goals of this chapter are to review how molecular techniques and/or genetic approaches have influenced carnivore conservation. After reviewing some of the major issues in conservation genetics, we will focus on several case studies that demonstrate a direct link between molecular genetics and conservation actions in four broad areas: (1) the identification of conservation units; (2) the quantification and interpretation of genetic variation; (3) the determination of ecology, natural history, and identity; and (4) the characterization of pathogens in natural populations. Finally, we will discuss how future efforts will likely be enhanced and broadened by an increased focus on genetic issues and by an increased use of molecular methods.

MAJOR ISSUES IN CONSERVATION GENETICS

A major goal of conservation biology is the preservation of biological diversity and the maintenance of the evolutionary patterns and processes that generate it. To achieve these goals it is necessary to decide what to protect and then to determine how. The first question, of 'what to preserve' has centered on the determination of 'conservation units', or the species, subspecies, or populations that have historically shared a geographic area, show differentiation from other such units, and should therefore be managed independently (e.g. Ryder, 1986; Moritz, 1994). Discussions of 'how' have often centered on maintaining the general 'health' of these conservation units. In recent years molecular approaches and an emphasis on patterns of genetic variation have provided important insights into both of these questions, helping managers determine not only what to protect, but also what factors contribute to 'long-term health' and how to ensure the viability of natural populations.

Conservation genetics focuses in fundamental ways on biological and genetic diversity. Genetic variability is influenced by factors such as mutations, selection, gene flow, mating systems, and genetic drift. Gene flow, or the movement of genes within and among populations, occurs through dispersal and breeding and potentially also through extinction and recolonization of local populations; it potentially can homogenize the distribution of variability throughout a geographic area. Reductions in gene flow can lead to differentiation among populations as novel mutations accumulate, as local selection favors certain alleles, and as genetic drift leads to the retention of random subsets of the original variation. As long as a small fraction of the population is replaced by immigration, there will be no significant differentiation from neighboring areas. The amount of migration needed to avoid genetic differentiation has been debated and reviewed extensively (e.g. Slatkin, 1985; Hartl & Clark, 1997). However, the seminal rule by Wright (1931) often has been used as a benchmark, that one effective migrant every other generation is potentially sufficient to prevent extensive differentiation due to genetic drift.

Besides promoting differentiation, genetic drift is also responsible for loss of genetic diversity, especially in small populations. Lacy (1987), on the basis of simulation models, suggested that genetic drift was the most important factor leading to the loss of genetic variation, and proposed that this could be overcome by one or a few migrants per generation. An independent but often correlated phenomenon in small, isolated populations is inbreeding, the mating between relatives which tends to increase genetic uniformity by increasing the frequency of homozygotes (for a review see Templeton & Read, 1994).

The interactions among gene flow, genetic drift, inbreeding, selection, and mutation will vary due to many factors, and are strongly influenced by population size (N), or more exactly by the effective population size (Ne). The effective population size can be estimated in several ways. It is often defined as the size of an idealized population experiencing the same amount of genetic drift as the actual population under study (Lande & Barrowclough, 1987; Templeton & Read, 1994; Hartl & Clark, 1997). It can also be seen as the average number of individuals contributing equally to the next generation. Ne is often much smaller than the actual census size (Frankham, 1995), and is influenced mainly by demographic and behavioral factors such as population fluctuations over time, unequal sex ratios, and variable rates of reproductive success (unequal family sizes). More recently it has been recognized that a metapopulation structure (Lande & Barrowclough, 1987), from natural or induced fragmentation, can in some cases reduce the effective population size (Hedrick, 1996).

Because genetic drift and inbreeding get stronger as the effective population size gets smaller, considerable attention has been given to identifying Ne and the ratio of Ne/N in wild species (e.g. Frankham, 1995), as well as maximizing the Ne of managed populations (e.g. Lande & Barrowclough, 1987; Templeton & Read, 1994). In addition, debate has centered on the minimum (effective) population size needed to ensure the evolutionary potential of a population, i.e. the ability to evolve in response to environmental change. Based on the maintenance of genetic variability assuming a simple scenario, it has been suggested that an effective population size of

50 individuals would be acceptable for short-term conservation goals. In addition, it has been postulated that an *Ne* of 500 would be adequate to promote long-term survival and the reasonable evolutionary potential of a population (Franklin, 1980; Soulé, 1987). Others (e.g. Lynch, 1996) have argued that much higher numbers would be required, and it is important to realize that this parameter will vary in the context of different evolutionary histories and conservation scenarios.

Given a small effective population size, several issues relating to genetic variation of both populations and individuals become important. For example, although a matter of considerable debate, the reduction in genetic variability and the inbreeding common in small populations are generally assumed to increase the risk of extinction, especially in the harsher conditions faced by wild populations (see reviews by Lacy *et al.*, 1993; Frankham, 1995). Inbreeding has been hypothesized to decrease individual fitness through lower birth rates, higher death rates, and expression of various deleterious alleles, a phenomenon known as inbreeding depression (e.g. Laikre *et al.*, 1993, 1996). An increase in observed homozygosity can also impair the ability of individuals to respond to new disease challenges, climatic variation, and environmental change in general (Wright, 1977; O'Brien & Evermann, 1988; Jimenez *et al.*, 1994; Keller *et al.*, 1994).

Risk factors for small populations do not operate independently from one another. Ecological and anthropogenic effects on populations may be greater in genetically depauperate populations (Lacy, 1997). The synergistic relationship between a decrease in population size, an increase in the effects of genetic drift and inbreeding, and the subsequent loss of individual fitness and future demographic collapse has been called a mutational meltdown (Lynch & Gabriel, 1990; Lynch, 1996) or extinction vortex (Gilpin & Soulé, 1986). This is ultimately one of the events conservationists strive hard to avoid.

CASE STUDIES
Genetics and what to conserve

The first step in defining and characterizing units for conservation often is to determine the adequacy of commonly recognized specific and subspecific classifications, ideally with multiple morphological and molecular markers. This approach can provide insights into the historical patterns and processes leading to the formation and maintenance of these groups and helps provide an evolutionary framework within which more recent patterns can be interpreted. Within the carnivores, intraspecific and interspecific relationships among extant lineages have been well-studied in several families, including the Felidae, Canidae, Ursidae, and Phocidae (Table 15.1). Several examples depicting the use of phylogenetic approaches to carnivore conservation are summarized below.

Felidae

Although there are still some unresolved questions, most of the major evolutionary lineages among the 36–38 extant cat species have been identified (O'Brien et al., 1996; W. Johnson & O'Brien, 1997; Pecon-Slattery & O'Brien, 1998) (Figure 15.1). Based in large part on analyses of mtDNA sequence variation and variation in noncoding sequences on the X and Y chromosomes, eight major clusters or clades that likely reflect separate monophyletic evolutionary radiations have been identified. Four additional species did not consistently align with these groups and likely do not have closely related sister taxa. Within this framework, the uniqueness and taxonomic status of several endangered and/or virtually unknown species have been investigated. In some cases this has involved the reliance, in part, on museum material, as with the very rare and poorly known Bornean bay cat (*Pardofelis badia*). This cat, which was originally described from only eight incomplete specimens, had not been sighted in the wild since 1928. A specimen collected in 1992 provided the opportunity to confirm the uniqueness of this species, demonstrating that it was only distantly related to the Asian golden cat (*Profelis temmincki*), with which it formed a separate lineage. The study also affirmed that this specimen was derived from the same phylogenetic lineage as the type specimen of bay cat, collected in 1855 and currently held in the Museum of Natural History in London (W. Johnson et al., 1999b).

Similarly, the Andean mountain cat (*Oreailurus jacobita*) has never been held in captivity and has been seen only rarely in the wild. This species was originally described based on three skulls and 14 museum skins and, as with many small cats, can be easily confused with other species by casual observers. Phylogenetic analyses of DNA extracted from several historic and contemporary skins were used to resolve the evolutionary status of the Andean mountain cat, aligning it with the ocelot and margay and the other small spotted cats of the South American ocelot lineage (W. Johnson et al., 1998) (Figure 15.2). This research has also provided the basis for the definitive recognition of these cats using molecular methods. This is a tool that is currently being used in helping to establish conservation management plans and for recognizing occasions of natural hybridization, as between the pampas cat (*Lynchailurus colocolo*) and tigrina (*Leopardus tigrina*)

Table 15.1. Examples depicting the use of phylogenetic approaches to identify conservation units in carnivore species

Taxon	Method	Reference
Felidae (O'Brien *et al.*, 1996; Johnson & O'Brien, 1997; Pecon-Slattery & O'Brien, 1998)		
Leopard (*Panthera pardus*)	allozymes, mtDNA RFLP, minisatellites	(Miththapala *et al.*, 1996)
Tiger (*P. tigris*)	mtDNA sequences, microsatellites, MHC variation	(Wentzel *et al.*, 1999)
Tiger (*P. tigris*)	mtDNA sequences	(Cracraft *et al.*, 1998)
Puma (*Puma concolor*)	mtDNA sequences, microsatellites	(Culver, 1999)
Ocelot and margay	mtDNA sequences	(Eizirik *et al.*, 1998)
Iriomote cat	mtDNA sequences, microsatellites	(W. Johnson *et al.*, 1999b)
Geoffroy's cat	mtDNA sequences, microsatellites	(W. Johnson *et al.*, 1999a)
Asian leopard cat (*Prionailurus bengalensis*)	mtDNA sequences, microsatellites	(W. Johnson *et al.*, 1999b)
Canidae (Wayne 1995; Wayne *et al.*, 1997)		
Red wolf (*Canis rufus*)	microsatellites	(Roy *et al.*, 1994a)
Red fox (*Vulpes vulpes*)	allozymes, mt DNA	(Frati *et al.*, 1998)
San Joaquin Valley kit fox (*V. macrotis mutica*)	mtDNA sequences	(Mercure *et al.*, 1993)
African wild dog (*Lycaon pictus*)	mtDNA RFLP and sequences, microsatellites	(Girman *et al.*, 1993; Roy *et al.*, 1994b)
Gray wolf (*Canis lupus*)	mtDNA	(Wayne *et al.*, 1992)
Mexican wolf (*Canis lupus baileyi*)	microsatellites	(Garcia-Moreno *et al.*, 1996)
Channel Island fox (*Urocyon littoralis*)	allozymes, mtDNA, minisatellites	(Wayne *et al.*, 1991)
Ursidae (Goldman *et al.*, 1989; Shields & Kocher, 1991; Hanni *et al.*, 1994; Talbot & Shields, 1996; Zhang & Ryder, 1994)		
American black bear (*U. americanus*)	mtDNA sequences	(Byun *et al.*, 1997; Wooding & Ward, 1997)
Brown bear (*Ursus arctos*) in Europe	mtDNA sequences, microsatellites	(Taberlet & Bouvet, 1994; Taberlet *et al.*, 1995)
Brown bear in North America	mtDNA sequences, microsatellites	(Talbot & Shields, 1996; Paetkau *et al.*, 1998a,b; Waits *et al.*, 1998)
Polar bear (*U. maritimus*)	microsatellites	(Paetkau *et al.*, 1995)

Procyonidae		
N. A. raccoon (*Procyon lotor*)	allozymes	(Beck & Kennedy, 1980; Dew & Kennedy, 1980; Hamilton & Kennedy, 1987)
Mustelidae (Dragoo et al., 1993; Dragoo & Honeycutt, 1997; Flynn & Nedbal, 1998; Koepfli & Wayne, 1998)		
North American River otter (*Lutra canadensis*)	allozymes	(Serfass et al., 1998)
European Otter (*Lutra lutra*)	mtDNA RFLP	(Effenberger & Suchentrun, 1999)
Black-footed ferret (*Mustela nigripes*)	allozymes	(O'Brien et al., 1989)
Viverridae		
Yellow mongoose (*Cynictis penicillata*)	mtDNA RFLP, mtDNA sequences	(Van Vuuren & Robinson, 1997)
Phocidae (Arnason et al., 1995; Lento et al., 1995)		
Harbor seal (*Phoca vitulina*)	mtDNA, minisatellites, microsatellites	(Stanley et al., 1996; Kappe et al., 1997; Goodman, 1998)
Grey seal (*Halichoerus grypus*)	microsatellites	(Allen et al., 1995)
Hawaiian monk seal (*Monachus schauinslandi*)	mtDNA sequences, minisatellites	(Kretzmann et al., 1997)
Mediterranean monk seal (*M. monachus*)	mtDNA sequences	(Stanley & Harwood, 1997)
Otariidae		
S. hemisphere fur seals (*Arctocephalus* spp.)	mtDNA sequences	(Lento et al., 1994, 1997)
Steller sea lions (*Eumetopias jubatus*)	allozymes, mtDNA	(Lidicker et al., 1981; Bickham et al., 1996)
California sea lions (*Zalophus californianus*)	mtDNA sequences	(Maldonado et al., 1995)
Odobenidae		
Walrus (*Odobenus rosmarus*)	mtDNA RFLP, microsatellites	(Cronin et al., 1994; Andersen et al., 1998)

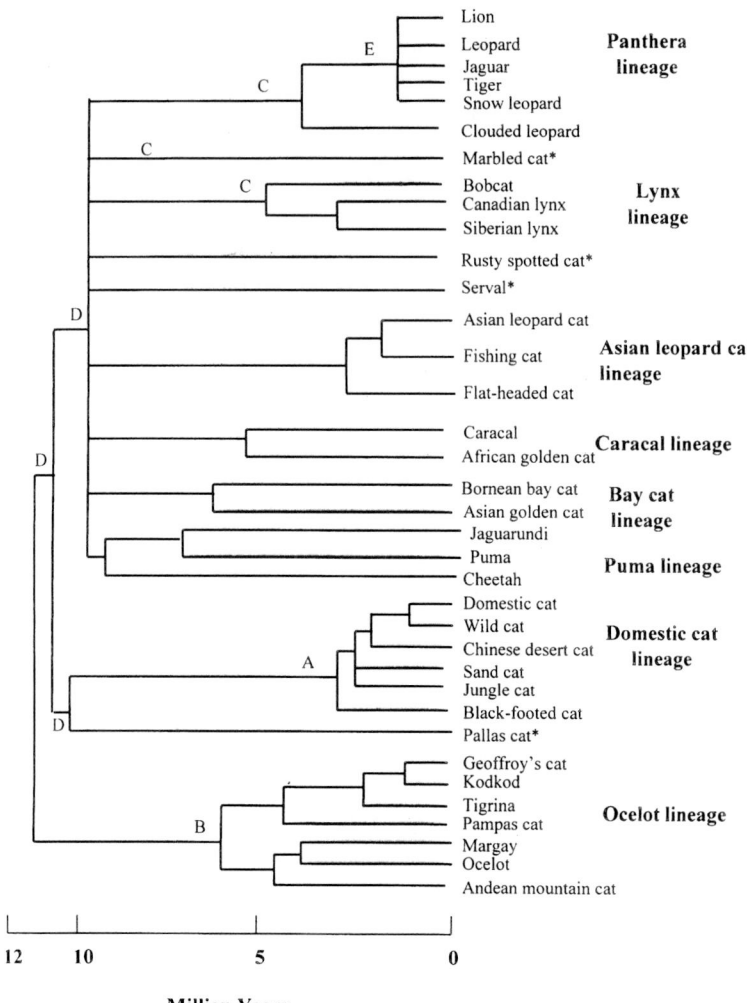

Figure 15.1. Phylogenetic relationships of Felidae based on sequence variation of mitochondrial DNA genes, variation among intronic sequences of *Zfy* and *Zfx* genes, mitochondrial DNA restriction length polymorphism variation, immunological distances, isozyme electrophosesis, karyology, and endogenous retroviruses. The nodes for all of the named lineages were supported by multiple analyses. Species labeled with asterisks did not consistently cluster in any lineage. A, indicates entry of endogenous retroviral families, feline leukemia virus and RD-114; B, indicates ocelot lineage species which have 36 chromosomes and share a metacentric chromosome C3 (Andean mountain cat not examined); C, indicates *Panthera* genus species and *Lynx* lineage species with the same karyotypes; D, indicates nodes supported by immunological distances, but not other methods; E, indicates *Panthera* genus species with the incorporation of a segment of mtDNA on nuclear chromosome F3.

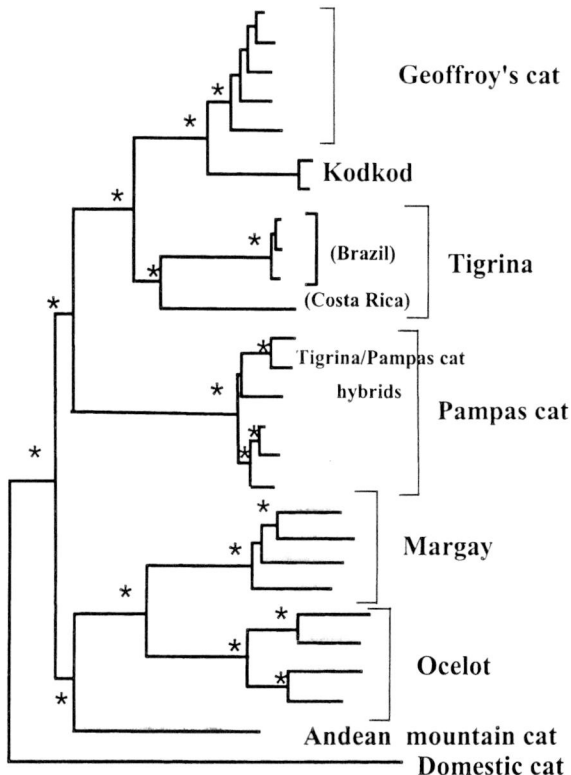

Figure 15.2. Depiction of the phylogenetic relationships among several individuals of the seven South American ocelot lineage species based on minimum evolution analyses of mitochondrial DNA (mtDNA) sequences (Neighbor joining tree constructed with Kimura distances). Nodes labeled by an asterisk were supported by at least 60% bootstrap support with minimum evolution (neighbor joining) and maximum parsimony analyses. Tigrina/pampas cat hybrids were individuals that had mtDNA haplotypes that were identical or similar to pampas cat haplotypes, but had Zfy sequences that were identical to sequences from southern Brazil tigrinas. (Modified from W. Johnson *et al.*, 1999a.)

(Figure 15.2) (W. Johnson *et al.*, 1999a). These results were used to estimate times of divergence among lineages and to demonstrate that sister taxa such as the kodkod (*Oncifelis guigna*) and Geoffroy's cat (*O. geoffroyi*) diverged relatively recently compared with ocelots and margays or with diverse lineages of Central American and Brazilian tigrinas (W. Johnson *et al.*, 1999a). The uniqueness of Central American tigrina, along with their endangered status, should provide an incentive for more aggressive conservation actions in the future.

After establishing the validity of commonly recognized species-level classifications, the next level of resolution in defining units for conservation emphasizes patterns of intraspecific variation and geographic population structure. Within the Felidae several studies have focused on the identification of phylogeographic patterns and broad-scale population structure (Table 15.1). These have generally not agreed with traditionally recognized trinomial designations. This is perhaps not surprising since the latter are often based on a very limited number of specimens, restricted geographic sampling, and morphological characters that undergo extensive individual variation, and are thus often not reliable indicators of historical population subdivisions. In most cases the number of groups that could be distinguished with molecular methods was lower than the number of previously recognized subspecies. For example, for both ocelots and margays, although current conservation efforts have generally been based on the eleven commonly recognized subspecies, phylogenetic analyses of mtDNA sequence variation identified fewer geographic regions that likely reflected historic faunal barriers to gene flow (Eizirik *et al.*, 1998). In addition, the geographic borders among these units, representing ancestral barriers to gene flow, are often different from what had previously been hypothesized. For example, current subspecies classifications of ocelot and margays consider populations on both sides of the Amazon river to be of the same subspecies, although the mtDNA genetic analyses suggest that they have been isolated for an appreciable length of time (Eizirik *et al.*, 1998).

Canidae

Conservation efforts of the 34 canid species have also benefited greatly from the use of molecular methods to help identify and characterize conservation units. For example, Darwin's fox (*Pseudolopex fulvipes*), found in Chile, was often considered a subspecies of the more common grey fox (*Pseudolopex griseus*). However, mtDNA markers demonstrated not only that Darwin's fox was unique, but that it was an older, ancestral species (Yahnke *et al.*, 1996). This classification immediately placed this species among the list of most endangered canids in the world, attracting renewed interest among conservationists. Darwin's foxes were only known to exist in two widely separated populations, one on the island of Chiloe in southern Chile, and the other in a patch of 'relict' old-growth forest in central Chile. Therefore, one of the first initiatives was an effort to identify other populations of Darwin's foxes in areas between the two known populations. This was done, in part, through trapping and by using mtDNA markers to determine if fecal samples collected in different areas were from Darwin's foxes. Although no other large populations of Darwin's foxes have yet been found, on the basis of these results the Chilean government is contemplating creating new protected areas designed to protect some of the few remaining patches of coastal old-growth forest.

Another advantage of using molecular markers and phylogenetic analysis in conservation efforts is that these approaches facilitate the recognition of hybrids. Hybridization has become a major conservation issue, especially in some families such as among canid species (see Wayne, Chapter 7). Mating among canid species has been reported both in captivity and in the wild, and is one of the major threats to the genetic integrity of several species, including the endangered Ethiopian wolf (Gottelli *et al.*, 1994) and the African wild dog (Girman *et al.*, 1993). The prevalence of hybridization in natural populations and its importance for conservation efforts is still being debated. The way that conservationists ultimately define and manage hybrids will have a large impact on the conservation of many populations, such as with the management of the red wolf, a naturally occurring mixture of gray wolf and coyote (Roy *et al.*, 1994a; Wayne & Gittleman, 1995; Wayne, 1995).

Ursidae

Molecular approaches have also added to our understanding of the phylogenetic and taxonomic relationships among the ursid species, including the giant panda (e.g. Goldman *et al.*, 1989, Shields & Kocher, 1991; Hanni *et al.*, 1994; Talbot & Shields, 1996). Some species in this group have also received considerable attention in terms of characterizing geographic partitions based on genetic variation at the intraspecific level (Table 15.1). The importance of these molecular genetic analyses for management and conservation efforts is demonstrated in the brown bear (*Ursus arctos*). Although over 90 subspecies of brown bears were once proposed in North America alone, currently only two to seven subspecies are commonly recognized (Hall, 1984). mtDNA sequence analyses have revealed, however, that there is considerable genetic differentiation among geographic regions in both Europe (Randi *et al.*, 1994; Taberlet & Bouvet, 1994; Kohn *et al.*, 1995) and North America (Talbot & Shields, 1996; Craighead & Vyse, 1996; Waits *et al.*, 1998), and that these observed subdivisions often do not correspond with these currently recognized subspecies classifications.

Pinnipeds

Among the three families of pinnipeds or marine carnivores, several studies have used molecular tools to address phylogenetic issues ranging from higher order relationships to intraspecific patterns of subdivision (Table 15.1). Because of their long and continued history of commercial exploitation, the recognition of distinct groups of these marine mammals has been of particular importance to these species. The habit of many of these species to return to the same locations for breeding not only leads to a greater probability of geographic differences among populations, but also increases the probability and ultimate evolutionary impact of local overexploitation. Among the 19 species of phocids, the harbor seal (*Phoca vitulina*) has been particularly well studied (Stanley *et al.*, 1996; Kappe *et al.*, 1997; Goodman, 1998). The high levels of geographic differentiation that were observed in this broadly distributed species, both on continental and regional scales, are probably going to be typical of many of the aquatic carnivores. For example, molecular genetic studies on several of the 14 species of the Otariidae have also revealed significant patterns of regional population differentiation, providing support for the separate protection and management of several subpopulations. These include southern-hemisphere fur seals (*Arctocephalus* spp.) (Lento *et al.*, 1994, 1997), Steller sea lions (*Eumetopias jubatus*) (Lidicker *et al.*, 1981; Bickham *et al.*, 1996), and California sea lions (*Zalophus californianus*) (Maldonado *et al.*, 1995).

Genetics and population health

Along with the delineation of different groups, molecular techniques have commonly been used to study the history, dynamics, and maintenance of genetic variation in natural populations (see review by Wayne & Koepfli, 1996). Measures of molecular diversity and use of the related theoretical concepts have been successfully applied to several carnivore species to assess threats to wild and captive populations, to predict the viability of endangered populations, and to plan management actions. Reductions in genetic variation are of particular concern to conservationists because these patterns sometimes have been associated with the manifestation of deleterious traits and loss of reproductive potential.

Northern elephant seals

Conservationists became concerned about the loss of genetic variation in endangered species with the first reports of a lack of variation in northern elephant seals (*Mirounga angustirostris*) at 24 allozyme loci (Bonnell & Selander, 1974). These results were later confirmed with additional allozyme loci and with mitochondrial DNA sequence data (Hoelzel *et al.*, 1993). The northern elephant seal was one of the first examples to demonstrate that the residual effects of past population history can be still carried in a population, with potential implications for its future viability. For the northern elephant seal, although present day population levels were high (over 100 000 in the 1970s) and the population was apparently healthy, the residual effects of eighteenth century overhunting, when populations were reduced to 10–20 individuals, were still visible in their genomes. Subsequently, other studies have also investigated levels of genetic diversity in several marine carnivores (e.g. Bickham *et al.*, 1996; Swart *et al.*, 1996; Kretzmann *et al.*, 1997; Andersen *et al.*, 1998).

Cheetah and lion

The issue of genetic diversity and conservation was brought to public attention mainly by the results of a series of studies on the cheetah (*Acinonyx jubatus*). The cheetah was shown to have extremely reduced levels of diversity at allozyme and MHC loci compared to other mammalian species, but moderate levels of variation in fast-mutating loci such as minisatellites and microsatellites (O'Brien *et al.*, 1985; Yuhki & O'Brien, 1990; Menotti-Raymond & O'Brien, 1993; Driscoll, 1998). These results led to the suggestion that cheetahs have undergone a severe population bottleneck around 10 000 years ago, to the extent that most of their genetic variation was lost, and that only fast evolving loci would have recovered some diversity during this period (Menotti-Raymond & O'Brien, 1993). This reduction in diversity was accompanied by physiological impairments, including a high incidence of sperm abnormalities (O'Brien *et al.*, 1985) and juvenile mortality. These animals also appeared to be susceptible to pathogens such as infectious peritonitis (Pfeifer *et al.*, 1983; Evermann *et al.*, 1984; O'Brien *et al.*, 1985; Heeney *et al.*, 1990), herpes virus (Junge *et al.*, 1991), *Helicobacter* sp. (Munson, 1993, 1996), and *Campylobacter* sp. (Skirrow, 1994). These combined results led to the hypothesis that poor male reproductive traits (especially the high incidence of abnormal sperm forms) and increased disease susceptibility were related to the extreme genetic uniformity of the cheetah. Interest in these results has prompted a lively debate over the relative importance of genetic variation in determining the probability of populations persisting over long periods of time and in establishing priorities for the conservation of endangered species (e.g. see review in Wayne & Koepfli, 1996; O'Brien, 1998).

The hypothesis that genetic variation was directly linked to reproductive traits and to disease resistance could not be directly tested in cheetahs because there are no living, outbred populations that can be used to make direct comparisons. However, other felid species have been used as a model to test the same principles. For example, a series of studies on three populations of lions (Gir forest in India, Ngorongoro Crater in Tanzania, and Serengeti in Tanzania) found correlations among estimates of molecular genetic variation (allozymes, MHC, mtDNA RFLP, and DNA fingerprints) and male reproductive traits. Estimates of ejaculate volume, spermatozoal motility, the number of spermatozoa per ejaculate (both absolute count and total motile), percentage of normal sperm, and serum testosterone levels decreased as estimates of inbreeding increased (Wildt *et al.*, 1987a; O'Brien, *et al.*, 1987a; Gilbert *et al.*, 1991). These results, along with comparisons among other *Panthera* genus species and among cheetah and puma populations (Table 15.2), provided further support for the hypothesis that reductions in molecular genetic variation from demographic bottlenecks could impair the reproductive potential of a population, and thus its long-term health, or viability.

Florida panther

Another well-documented example of the study of genetic variation comes from extensive studies of a small relict population of approximately 30 animals of the Florida panther (*Puma concolor coryi*). Pumas in this population had extremely reduced genetic variation, as measured by mtDNA RFLP variation, allozymes, mtDNA sequences, MHC, and microsatellite variation, compared with a group of 'nonauthentic panther hybrids' of partial Central American origin and with other puma subspecies (O'Brien *et al.*, 1990; Roelke *et al.*, 1993a; Culver, 1999). There is evidence that suggests that this dramatically reduced genetic variation, likely the result of drift and close inbreeding during the population reduction, was tied to several physiological impairments (Roelke *et al.*, 1993b). Comparisons among other puma subspecies and other cat species revealed that Florida panther males had relatively poor seminal quality traits (Table 15.2), including sperm concentration, number of sperm per ejaculate, percentage of normal sperm and percentage of acrosomal defects (Roelke *et al.*, 1993a). Acrosomal defects can be especially debilitating since these significantly decrease a sperm's fertilization potential (Howard *et al.*, 1991, 1993). In other carnivores, these traits are often linked with infertility (Wildt *et al.*, 1988; Howard *et al.*, 1990; Wildt *et al.*, 1987b).

The Florida panther population also displayed an unusually high incidence of cryptorchidism or the retention of one or both testicles within the body (56% of males examined between 1981 and 1992; Roelke *et al.*, 1993b). Cryptorchid males tended to have lower circulating testosterone concentrations and fewer motile sperm per ejaculate than normal males (Barone *et al.*, 1994). The incidence of cryptorchidism, a heritable trait,

Table 15.2. Comparisons of allozyme and microsatellite size variation with percentage normal sperm in felid populations

Species	Population	Allozyme		Microsatellite		Normal Sperm (%)	References
		% P	% H	No. alleles	% H		
African Lion	Serengeti	11.0	3.8	3.4	47.2	75.0	1, 2, 3
Tiger		10.0	3.5	5.6	38.1	65.3	4, 5, 6
Leopard		10.0	3.1	ND	ND	49.1	4, 6, 7
Clouded leopard		6.0	2.3	ND	ND	28.9	4, 5
African Lion	Serengeti	11.0	3.8	3.4	47.2	75.0	1, 2, 3
African Lion	Ngorongoro	4.0	1.5	2.9	42.5	50.0	1, 2, 3
Asian Lion	Gir Forest, India	0.0	0.0	1.3	7.6	34.0	1, 2, 3
Cheetah	East Africa	4.0	1.4	3.6	43.3	30.0	3, 8, 9
Cheetah	South Africa	2.0	0.04	3.4	43.8	30.0	3, 8, 9
Puma	South America	ND	ND	6.6	56.2		3, 8, 10
Puma	Texas and Idaho	9.8	4.1	2.5	32.9	14.0	3, 8, 10
Puma	Big Cypress, Florida	4.9	1.8	1.5	16.1	5.7	3, 8, 10

[1] O'Brien et al., 1987c; [2] Wildt et al., 1987a; [3] Driscoll, 1998; [4] Howard et al., 1993; [5] Newman et al., 1985; [6] unpublished; [7] Miththapala et al., 1996; [8] Roelke et al., 1993a.b; [9] Wildt et al., 1987b; [10] Barone et al., 1994.
Percentage of polymorphic allozyme loci (% P), average percent allozyme and microsatellite heterozygosity (% H), and average number of microsatellite alleles (No. alleles).

increased dramatically in the Florida population from 0% of pre-1975 births to 90% of the males born after 1990, in association with documented consanguineous mating (two males born in 1991 were bilaterally cryptorchid and thus sterile) (Roelke *et al.*, 1993b). During this same time period a second class of deleterious traits started to appear in the population. Between 1985 and 1998, atrial septal defects (ASDs), which are an abnormal fetal opening between the two atria in the heart, was found in 6 of 33 necropsied Florida panthers, including a juvenile female, the probable offspring of a backcross of the sire with his daughter. Providing support for the genetic basis of these traits, all three male panthers with ASDs were also cryptorchid (compared with 9 of 17 cryptorchid males without ASDs) (Cunningham *et al.*, 1999). These observations were consistent with the rapid rise of maladaptive traits due to fixation of deleterious alleles from incestuous mating.

Following the suggestions of a Florida panther working group (Seal 1994), eight female pumas from Texas (*P. concolor stanleyana*) were released in the southern Florida swamps in the spring of 1995 in an effort to reverse the patterns of both demographic and genetic reductions. Texas pumas were chosen because prior to the twentieth century, Texas populations were contiguous with that of the Florida panther. The occurrence of recent gene flow between these two populations was subsequently confirmed by extensive mtDNA and microsatellite analyses of pumas from throughout North and South America (Culver, 1999). Of the original eight Texas females, one was hit by a car, a second was shot, and a third has never bred. From 1995 to 1998, 12 kittens were born to the other five of the introduced Texas pumas (Land & Taylor, 1998). Field observations suggest that two of these kittens may have produced second generation offspring in 1999 (D. Land, pers. commun.). During this same time period, 26 kittens were born to seven resident panthers. Of these kittens, 7 (58%) of the 12 Texas kittens were subsequently captured and collared compared with 5 (19%) of 26 of the Florida panther kittens, suggesting that survival of hybrid kittens may be significantly greater than that of native panther kittens. This is consistent with the hypothesis that increased genetic variation is correlated with enhanced juvenile survival rates.

By June 1998, at least 12 (25.5%) of the 47 pumas being monitored in southern Florida (including released Texas females) were descended from Texas lineages. This suggests that the initial goal of having 20% of the breeding population being derived from Texas pumas is close to being reached. The initial success of this experiment bodes well for the long-term genetic health of this population and may help increase the overall population size. The Florida panther story remains to date one of the best examples of how a large amount of detailed knowledge about the genetic variation of a population can help focus conservation efforts, crystallizing support around the search for practical solutions to a conservation problem (Seal, 1994; Hedrick, 1995).

Captive populations

Patterns of genetic variation have been discussed and documented in more detail in captive populations than in wild ones (Lacy, 1993a). Carnivores are held in captivity for many reasons, but most are held for farming (for pelts), scientific study, conservation efforts, and pleasure. For some carnivores, especially those that are severely threatened in the wild, more individuals actually exist in captivity or semi-captivity than in the wild. Examples include the Siberian tiger (*Panthera tigris altaica*), black-footed ferrets (*Mustela nigripes*), Mexican wolves (*Canis lupus baileyi*), Asian lions (*Panthera leo persica*), and Amur leopards (*Panthera pardus orientalis*). Many captive populations show evidence of reduced genetic variability and of inbreeding depression. For example, inbred populations of captive wolves demonstrated reduced juvenile weight, longevity, and reproductive success, as well as a high incidence of a hereditary form of blindness (Laikre & Ryman, 1991). Similarly, inbreeding depression, evidenced by decreased juvenile survival, has been seen in most inbred captive populations studied (Ralls *et al.*, 1988, reviewed extensively by Lacy *et al.*, 1993). The importance of differences in selective pressures between wild settings and captivity has also been a matter of discussion, especially when considering the release of captive-born and captive-reared animals into the wild (Frankham *et al.*, 1986; Templeton, 1990).

Genetics and molecular ecology

Molecular techniques are being used increasingly to obtain insights into aspects of carnivore biology that traditionally have been difficult or impossible to determine, even through careful visual observations or with more recent advances such as radio-telemetry and remote-sensing technologies. The work performed so far has addressed issues of interest to conservation at the level of individual animals, groups, populations, or entire species. Several of the pioneering studies in the area described the relatedness among individuals in a population and the patterns of reproductive behavior and social structure (reviewed by Gompper & Wayne, 1996). Although attempts were made to verify patterns of social behavior using analyses of allozyme variation (e.g. Evans *et al.*, 1989), it was not until the advent of more rapidly evolving genetic markers that this became more feasible.

Social structure

An understanding of natural history traits such as social structure is often important in the development of successful conservation plans. Understanding of social interactions provides insights into what activities might potentially disrupt group cohesiveness and in turn influence effective population sizes. In addition, social interactions are one of the key variables in determining the impact of pathogens on populations.

In one of the first studies using multilocus DNA fingerprints on a carnivore population, it was demonstrated that female lions within a pride were always closely related to each other but not to the adult males. Males were either closely related to each other or unrelated (Gilbert *et al.*, 1991; Packer *et al.*, 1991). The study also showed that as the number of male coalition partners increased, it became more likely that they were all close relatives and that mating would be dominated by only a few individuals. These studies helped establish a direct link between the occurrence of infanticide in lions with the tenure of males and demonstrated how instability in lion prides can affect population growth rates. Studies on the structure of another large, social carnivore, the wolf, using similar methods, also revealed novel results. Multilocus DNA fingerprints demonstrated that the degree of relatedness between individuals from nearby wolf packs varied among three different regions (Lehman *et al.*, 1992). These results suggested that differences in management strategies in the areas, from total protection to managed hunting, were affecting the local demography, social structure and patterns of genetic variation.

In a similar study of social structure, Waser *et al.*, (1994a) found that male slender mongooses (*Herpestes sanguineus*) form loose coalitions, some which last as long as seven years. Using DNA fingerprinting data they confirmed that the males within these groups are often closely related and no one individual monopolizes matings with the females within the coalition's territory. More recently, Gompper *et al.* (1998) used multilocus DNA fingerprints to study the relationships among white-nosed coatis (*Nasua narica*). They determined that female coatis live in bands with other closely related females, that female dispersal is low, and that females are generally not closely related to those in nearby bands, unless a nearby band was formed by dividing into two. Males within the range of these bands were also generally more closely related to the members of the resident female band and to other males using the same area, but may roam widely during the breeding season (Gompper et al., 1997, 1998).

Natural history

Molecular techniques have been used to improve upon other centuries-old methods of studying wildlife, such as the use of animal signs to make inferences about their presence and behaviors. These techniques have traditionally been limited, however, by the inherent uncertainties of accurately identifying and interpreting these signs. The use of molecular techniques to interpret signs such as feces and hair left by wild animals has become possible with better methods of DNA extraction from a variety of tissues in various states of degradation and with improvements in the amplification of DNA products from minute amounts of material. These technologies, along with more accessible and complete databases, have great implications for the future of carnivore conservation.

For example, molecular methods are being adapted that will allow researchers to obtain a variety of data about individuals and populations from fecal samples including information on behavior, census and effective population size, movement and distribution patterns, genetic variation, and disease (see review by Kohn & Wayne, 1997). The application of these techniques to carnivore conservation is only beginning, but promises to be useful in multiple ways. In European brown bears, PCR-based analyses of fecal DNA has provided insights into the number, sex, and diet of bears in the Brenta mountains of northern Italy (Höss et al., 1992; Taberlet & Bouvet, 1992; Kohn et al., 1995). Fecal DNA has also been used to assign an individual identity to brown bear feces (Taberlet et al., 1996), to determine the species, sex, and individual identity of seal feces (Reed et al., 1997), and to estimate the populations size of coyotes (Kohn et al., 1999). Molecular markers have been designed to rapidly assign what species produced a scat (Foran et al., 1997a; Paxinos et al., 1997). Similar applications are being worked on to use hair samples, which can be collected from a variety of kinds of hair traps (Foran et al., 1997b; Gossens et al., 1998).

These techniques will also be useful in determining the characteristics of extinct populations (e.g. Roy et al., 1994b, 1996a) through the extraction of DNA from historic samples of hides and bones. This sort of information can influence current management decisions. For example, it is suspected that clouded leopards (Neofelis nebulosa) have either been extirpated from the island of Taiwan, or are very rare. A comparison of historic samples (from hunters, museums, etc.) with several mainland populations from China, Vietnam, and Malaysia would establish what mainland population most resembles the original (or remaining) clouded leopard population on the island and would help determine the best source of animals for reintroduction efforts.

Forensics

Another, perhaps more immediate benefit to many carnivores may be the application of forensic genetics to the management of domestic exploitation and the control of illegal international trade in pelts, bones, organs, and medicinal extracts. Although these products may be impossible to classify on the basis of appearance, they often contain DNA that can be amplified and compared to known 'reference' samples (e.g., Baker & Palumbi, 1994; Malik *et al.*, 1997). With adequate reference information from wild populations and a thorough understanding of their systematics and population structure, available genetic methods enable the identification of species, geographic origin and individual identity. In a similar sense, reference samples can help in the resolution of forensic cases. For example, in 1996, results of an analysis of 10 microsatellites amplified from the DNA of a single cat hair deposited at a human murder scene in Prince Edward Island, Canada assisted in the successful prosecution of the primary suspect (Menotti-Raymond *et al.*, 1997a,b).

These techniques can also be used with live animals to determine their sex, species, subspecies, or geographic origin. This information can not only be used in regulating trade of live animals, but also can be useful in the management of species. Throughout the world, there are captive individuals of many species that, because they are of unknown origin, are not included in the various breeding plans designed to preserve natural levels of variation. For example, determination of provenance can be of great utility in including as many individuals as possible in breeding programs. Statistical tools using the frequency of private alleles in microsatellite loci promise to be particularly useful in determining the origin of unknown animals, even when only 5–10 'voucher specimens' are available (Wentzel *et al.*, 1999).

GENETICS AND PATHOGENS

Molecular techniques are being increasingly used to study pathogens in natural populations of carnivores. These studies have generally used molecular methods either to address the evolutionary relationships among different viral strains, or to more accurately assess the extent of viral exposure in populations. For example, the molecular epidemiology of rabies has been studied in several carnivores in South Africa using sequence variation in two genomic regions of the virus (Nel *et al.*, 1993; von Teichman *et al.*, 1995). Evidence was found for two distinct viral groups. One group, made up of closely related virus isolates from domestic dogs, jackals, and bat-eared foxes, was related to European strains of rabies. The second group, which was more distantly related and phylogenetically distinct, contained isolates from mongooses (herpestids), and perhaps is unique to South Africa. These molecular data also showed that the rabies isolates from mongoose occasionally infected canid hosts as well. More recently, mtDNA analyses of the genetic structure of yellow mongoose populations (*Cynictis penicillata*) was compared with regional hot spots of the rabies biotype found in mongooses to support the hypothesis that rabies outbreaks reflected differences in population densities more than genetic or geographic partitioning (Van Vuuren & Robinson, 1997).

Molecular techniques were also used to determine that the canine distemper virus that affected at least 85% of the Serengeti lion population in 1994 was closely related to viral isolates from domestic dogs (Roelke-Parker *et al.*, 1996). This virus, which also affected uncounted numbers of leopards, hyaenas, and bat-eared foxes, clearly had an important impact on the carnivore community. These results not only provided added incentives to vaccinate the domestic dogs, which were the probable reservoir for the disease, but also highlighted the value of molecular tools in monitoring and determining the source of infectious diseases in wildlife species.

Molecular phylogenetic analyses of conserved viral genes were also used to study the different strains of the feline immunodeficiency virus (FIV), which have been found in domestic cats and several free-ranging endangered species such as the puma, leopard, cheetah, and lion (Brown *et al.*, 1994; Carpenter *et al.*, 1996). These analyses demonstrated that the different FIV strains tend to be found in a single host, and that these strains have deep evolutionary roots, implying that they have been infecting cat species for a long period of time. Although FIV causes immune cell collapse and eventual death in domestic cats, it does not appear to cause a similar disease in exotic cat species, which indicates that this system may be an example of a host–parasite interaction in which the lentivirus and host have accommodated to each other.

SYNTHESIS, CONCLUSIONS AND RESEARCH NEEDS

The examples in this chapter have documented various ways that molecular genetics can help carnivore conservation efforts. The application of molecular genetic methods to carnivore conservation can be distilled into a fairly clear recipe. First, it must be determined what to conserve. This can be integrated with a determination of the genetic health status of the population (e.g. measures of genetic variation), the establishment of basic natural history data (molecular ecology), and the collection of baseline data. Molecular methods can also help identify and monitor current and future threats such as genetic introgression and pathogens. Finally, based on the results of the previous steps, management plans that integrate molecular genetic data with insights from other disciplines should be established.

Molecular genetics has contributed the most towards the conservation of carnivores that were the focus of large, multidisciplinary efforts involving different research groups, government agencies, and non-government organizations, and which have had both *in situ* and *ex situ* research and management components. Invariably, these conservation efforts used several of the approaches outlined in this chapter to help establish and meet specific management goals. The most influential studies, both in terms of scientific and popular impact and notoriety, have involved felid and canid species, followed by studies on ursid and phocid species. Much less emphasis has been placed on hyaenids, herpestids, viverrids, otarriids, odobenids, procyonids, and mustelids, with the black-footed ferret being the one exception. The result is that our knowledge of the different carnivore families is very unbalanced. Even within each family, research efforts have concentrated heavily on a small number of species and on restricted geographic regions. This disproportionate treatment is due in large part to the focus of laboratories interested in carnivore conservation genetics, which tends to propagate further research efforts on the same and closely related species. Secondly, the popularity of domestic cats and dogs as pets and their use in medical research and as disease models increases their interest to both scientists and funding agencies, facilitating and channeling research efforts. Finally, the carnivores that have received the most attention are generally large species, which are often the most affected by human disturbances because of biological characteristics such as low fecundity, long generations, and low population densities.

As a result, research needs vary greatly among the various carnivore families. In the Felidae, Canidae, Ursidae, and pinniped families, continued focus is needed on phylogeographic issues and on the characterization of life history traits and measures of genetic diversity in the poorly known taxa and geographic areas. For pinniped species that are still being harvested, either directly or indirectly, this information is especially critical. Studies at rookeries and mating grounds also promise to provide interesting insights into mating and social behavior that will aid in future conservation efforts (e.g. Amos *et al.*, 1993). There is still much to be gained from continued in-depth studies of these marine carnivores and their use as models for integrating molecular genetics, comparative genomics, ecology, evolutionary biology, and conservation issues. Among the other terrestrial carnivores, some of the most basic studies of molecular genetic variation are needed. Our knowledge of the interspecific and intraspecific classifications of several of these groups, of their evolutionary history, and their distribution, ecology and role as disease vectors remains sketchy.

THE FUTURE ROLE OF GENETICS IN CARNIVORE CONSERVATION

Since their divergence from other mammals around 100 million years ago (Li *et al.*, 1990), carnivores have evolved many unique morphological, physiological, and behavioral characteristics; traits that are encoded in their genomes. The concept of preserving this genetic variability has become an increasingly key component in our considerations over conservation issues (e.g. Soulé, 1987). Many of our conflicts with carnivores are now resolved with at least a tacit understanding that it is important to conserve biodiversity or genetic diversity at several levels, including intraspecific patterns of variation.

With rapid technological advances in molecular techniques, it is becoming easier to measure and interpret variability (Chakravarti, 1999). Therefore, our ability to conserve carnivore populations in many cases is no longer limited by technological considerations, but is constrained by the lack of biological samples from which we can obtain the data needed to design more effective management plans. As demonstrated by the examples presented in this chapter, there are now good precedents for the utility of these biological resources. Increased effort and resources should be spent on the establishment of viable cell lines from representative animals of known origin which would be available for future, as yet unforeseen, research needs. These samples would then be the equivalent of the specimens collected by the Natural History Museums during the last century.

Similarly, our understanding of pathogens, and their effect on carnivore populations, requires much more attention and can benefit from better collections of biological materials. Although demographic issues are often the most pressing concerns with the conservation of carnivore populations and receive most attention, ultimately the long-term viability of these populations will also depend on controlling sources of mortality. The importance of disease in population dynamics and carnivore conservation, and the difficulty in tracking disease in natural populations, argue for the need for all scientists and managers to collaborate with epidemiologists and wildlife veterinarians and to collect, whenever possible, data for disease studies.

In the future, improved techniques will be developed for not only detecting but also counteracting the loss of genetic variability in populations. There will also be advances in methods to identify deleterious genes and to determine genetic uniqueness (Chakravarti, 1999). Maps of the entire genomes of carnivores, such as the cat and dog, will be completed and the determination of the function of most genes will eventually follow. This progress is being accompanied by advances in our ability to rapidly screen genetic diversity, incorporating new techniques which combine the fields of chip making, molecular genetics, and robotics. Our ability to directly intervene in populations will also improve with advances in genetic engineering, gene synthesis, and the possibility of duplicating and cloning whole genomes. These technologies will dovetail with improving methods for the viable preservation of genetic variation and potentially adaptive genotypes in collections of sperm, eggs and embryos, and with advances in assisted reproductive techniques (see Wildt, Chapter 16).

In this chapter we have shown ways that molecular techniques have aided conservation efforts and have suggested how molecular-based techniques will become even more important. As humans increasingly alter their environment, impacts on patterns of genetic variability will continue. The importance of species conservation has been generally accepted, as has the role of molecular genetics in the conservation of endangered species (Avise 1989; O'Brien 1994a,b, 1995). In an era of increased technological options and a greater degree of intervention in natural processes, it is crucial that we reflect on our alternatives. Ideally we will make decisions that will not only meet human needs and preferences in regard to carnivores, but also maintain the potential for persistence, adaptation, and continuing evolution of these populations in the millennia to come.

16

Role of reproductive sciences in carnivore conservation

DAVID E. WILDT, JOGAYLE HOWARD AND JANINE BROWN

INTRODUCTION TO PROBLEMS AND ISSUES
What is the role of reproductive sciences in conservation?

Reproductive biology as a discipline is misunderstood in the conservation world. It generally is perceived as a set of high-tech solutions for producing offspring, usually in *ex situ* (zoo) conditions. The field is largely regarded as a highly applied science. The reality is that it is the goal of the reproductive biologist to understand the diverse and wondrous ways that living organisms (including carnivores) reproduce. Reproductive biology is pri-marily a basic science geared to generating new knowledge about reproductive mechanisms. If done systematically and appropriately, then resulting data can have applied benefits.

The origins of this misunderstanding probably hail to the late 1970s and two significant events. The first was the birth of Louise Brown, the first human infant to be conceived in a Petri dish (by *in vitro* fertilization). The second was the emergence and routine use of embryo transfer in the cattle industry, making it possible to produce 'litters' of calves from genetically superior cows that normally produce only a single calf per year. The idea that the technology could easily be extrapolated and extended to endangered species was proposed by a few intrepid zoo directors who contemplated assembly-line propagation of endangered species. A small cadre of reproductive physiologists, hired by a few zoos, began applying cattle technologies to zoo-held species.

The first significant birth, a gaur calf born to a Holstein cow after interspecies embryo transfer, occurred at the Bronx Zoo in the early 1980s (Stover & Evans, 1984). This was followed by a few other significant, usually one-time births at other zoos, primarily in Artiodactyla. There were just enough of these 'gee-whiz' events to capture the public's imagination and to stigmatize the discipline as one with the sole purpose of enhancing offspring production.

It was normal that animal managers and the public might become excited about producing wildlife using hi-tech reproduction. The potential benefits were significant (Holt *et al.*, 1996; Wildt *et al.*, 1997). Among other advantages, gamete and embryo technologies (artificial insemination, embryo transfer, *in vitro* fertilization) could offer a means of moving genetic material while eliminating stresses associated with transporting animals long distances. Assisted breeding would overcome problems associated with mate incompatibility, mate preference and pair bonding. In combination with the ability to cryopreserve sperm and embryos, these techniques would help avoid genetic drift, facilitate germplasm transport and provide frozen 'insurance' in the event of catastrophes capable of wiping out small populations.

Why assisted breeding is not routine in wildlife

If there are so many advantages to assisted breeding in wildlife management, why does the potential of these techniques continue to be discussed rather than seeing products from their use? That is, why are these techniques not being used routinely to help manage and conserve endangered species, including carnivores? The simple answer is that species are different physiologically as well as phenotypically. Assisted breeding techniques that are effective in humans and cows are not necessarily applicable to bears and hyaenas. On the contrary, reproductive mechanisms among animal groups are highly varied (Wildt *et al.*, 1992). Without understanding the fundamentals of reproductive processes, assisted breeding can never become consistently successful.

The following example may be useful for illustrating the dilemma. In the late 1970s, our laboratory became interested in the high proportion of zoo-maintained cheetahs that failed to naturally mate. We speculated that this management problem could simply be solved by artificial insemination. Sperm were collected, processed and cryopreserved from cheetahs in Africa using cattle technology. In the United States, sperm were deposited into hormonally treated cheetahs, again using cow technology. This highly applied study failed because, although most cheetahs ovulated, none became pregnant (Wildt *et al.*, 1986). Nonetheless, failure forced us to realize that a cheetah was not a cow! In retrospect, it was logical that cow technology would likely be inefficient in the highly specialized cheetah that had evolved for reasons considerably different than its bovine counterpart. It was clear that we needed more basic research on the fundamental biology of the cheetah. This realization resulted in more than 15 years of research on many physiological aspects of cheetah biology, including multidisciplinary collaborations with field biologists, behaviourists, geneticists, nutritionists and veterinarians. More than a dozen papers resulted on the reproductive biology of the cheetah (see reviews Wildt *et al.*, 1993; Brown *et al.*, 1996). A great deal was learned not only about cheetah reproduction, but also about the integrative biology of the species, findings that are contributing to enhanced management of cheetahs in zoos. One by-product was more effective production of cheetah cubs by assisted breeding (Wildt *et al.*, 1997). Furthermore, the technology has been transferred to the field in a partnership with the Cheetah Conservation Fund, a non-governmental organization in Namibia. Guidance and assistance are being provided to create a genome resource bank (a frozen repository of sperm, blood products, tissue and DNA) to help protect the largest disease-free population of cheetahs in Africa. Sperm are being collected and cryopreserved from wild-caught cheetahs, and its biological competence has been proven by the birth of cubs in the United States using sperm transported intercontinentally (Wildt *et al.*, 1997). This is one example of what the reproductive sciences are all about – curiosity driven research and increasing intellectual capital, with the results beginning to pay dividends in management and conservation.

REVIEW OF OTHER CASE STUDIES
The value of fundamental reproductive studies in carnivores

A primary advantage to studying carnivores is the opportunity to discover new mechanisms in previously unstudied species. Almost all reproductive biologists today study reproductive systems of humans, livestock and laboratory animals. Therefore, conventional wisdom on how reproduction occurs mechanistically is precariously derived from only a few species. This provincial approach ignores thousands of wildlife (and hundreds of carnivore) species that hold a wealth of information about novel reproductive phenomena. This is a resource not to be taken casually because of its fundamental importance, as well as its potential economic, social and even human health value.

One illustrative example in carnivores is the phenomenon of teratospermia, a condition in which high proportions of pleiomorphic (or malformed) sperm are consistently ejaculated. Certain species, subspecies or populations within the Felidae family have a predilection for teratospermia. Cheetahs (in zoos and in the wild) routinely ejaculate *c.* 75% structurally-abnormal sperm (Wildt *et al.*, 1983; Wildt, 1994). Studies also have revealed that lions isolated in the Serengeti's Ngorongoro Crater as well as the remnant population of Asian lions in the Gir Forest Sanctuary of western India also are teratospermic (Wildt *et al.*, 1987). The most provocative example has involved the Florida panther, an isolated subspecies in southern Florida that consistently ejaculates more than 90% pleiomorphic spermatozoa (Barone *et al.*, 1994).

But what is the significance of the high number of malformed sperm observed in certain felids? Reproductive techniques, especially *in vitro* fertilization, have been used to examine teratospermia in the context of sperm fitness and function. Early studies determined that there was a direct relationship between the proportion of sperm in the ejaculate that were normally-shaped and *in vitro* fertilization success (see review Wildt, 1994). Thus, fertilization and embryo cleavage occurred readily in species such as the domestic cat and tiger (normospermic species), compared to teratospermic counterparts such as the cheetah, Florida panther and clouded leopard. Similarly, following *in vitro* culture was the discovery that the deformed sperm never advanced into the egg proper, either failing to bind or penetrate the glycoprotein coat of the egg (the zona pellucida) (Howard *et al.*, 1993). Of more basic interest was the function of the normal-shaped sperm from teratospermic ejaculates. Are these cells really normal? Again detailed technical studies revealed that structurally-normal sperm from teratospermic males were also compromised at the subcellular level, including in capacitation (Long *et al.*, 1996) and tyrosine kinase signalling (Pukazhenthi *et al.*, 1996) (biochemical processes that assist in achieving fertilization). More recent data reveal that normal-appearing sperm from teratospermic donors also have unstable chromatin (Anzalone *et al.*, 1998).

In addition to addressing a fascinating phenomenon (sperm morphology and function), these studies can have applied relevance. We suspect that the cause of teratospermia is related to a lack of genetic diversity within populations (in the case of the lion and Florida panther; Roelke *et al.*, 1993a) or entire species (in the case of the cheetah; O'Brien *et al.*, 1985), although a clear cause and effect relationship has yet to be established. Nonetheless, it is firmly documented that these sperm do not participate efficiently in the fertilization process. Furthermore, it is apparent that these studies of wild carnivores have provided a new perspective on teratospermia, a condition that commonly occurs in humans (Kholkute *et al.*, 1992). Therefore, these findings have human health implications and is one reason why this work in our laboratory has been supported by the National Institutes of Health.

Non-invasive hormone monitoring for tracking reproductive status

One of the most powerful, contemporary technologies for studying reproduction in carnivores is the measurement of gonadal hormonal metabolites in excreted faeces or urine. This technique has no direct role in propagating offspring, yet provides information on important reproductive characteristics such as onset of puberty, seasonality, oestrous cycle duration, time and type (induced versus spontaneous) of ovulation, pregnancy and even parturition prediction. Hormones drive reproductive success and give an indication of the reproductive status of an individual, a population or even a species. Until recently, the sole means of assessing endocrinological status in wildlife was through the measurement of hormones in blood samples collected invasively, usually at anaesthesia. In addition to disrupting an animal's normal routine, there was always the potential for the anaesthetic drugs to perturb normal endocrine patterns that, in turn, could influence subsequent processes, such as ovulation. However, assessing hormonal metabolite content in faeces or urine avoids these pitfalls and permits long-term sampling with no animal disturbance. Validation studies have proven that metabolite profiles generated from serial collections of carnivore faeces reflect physiological hormone fluctuations in the peripheral circulation (see review, Brown & Wildt, 1997).

Whether the predominant hormones are excreted in faeces or urine is species-specific and is determined during the validation process. Once this information is known, it is possible to design studies that assess hormonal-reproductive status acutely or long-term. Particularly valuable studies have been possible under controlled situations in zoos where frequent excreta collection is a simple by-product of routine enclosure cleaning. Projects have demonstrated that certain species (e.g., ocelot) produce consistent reproductive cycles throughout the year (Figure 16.1), whereas others (e.g., Pallas' cat) exhibit marked seasonality (Figure 16.2). The gonads of both male and female Pallas' cats generally are quiescent from late spring to early fall. Beginning in October, the male begins to excrete androgens in the faeces, no doubt reflecting the need for testosterone to stimulate spermatogenesis for the impending, but brief, breeding season. In late December, the female begins to excrete gradually increasing, baseline concentrations of oestrogen metabolites followed by cyclic and peak oestrogen patterns that coincide with maximal androgen excretion by the male (Figure 16.2). Particularly interesting is the brevity of peak gonadal activity in both sexes, with an abrupt and sharp decline to nadir hormone production by late February or early March. To-date, no other such felid species shows this remarkably brusque seasonality.

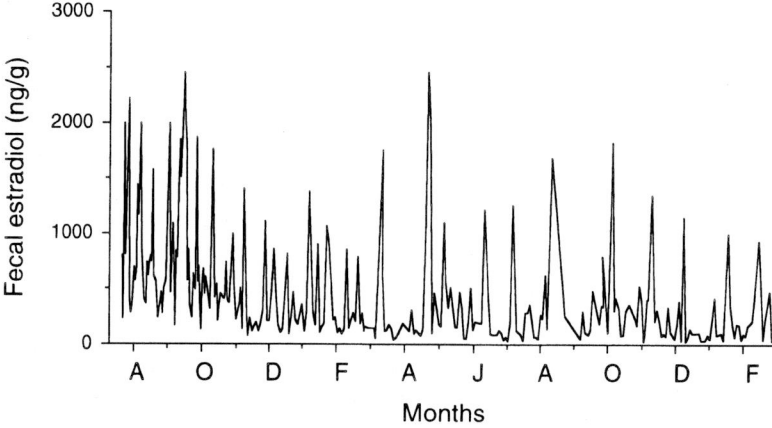

Figure 16.1. Faecal oestradiol profile in an adult female ocelot over time illustrating continuous patterns of ovarian activity.

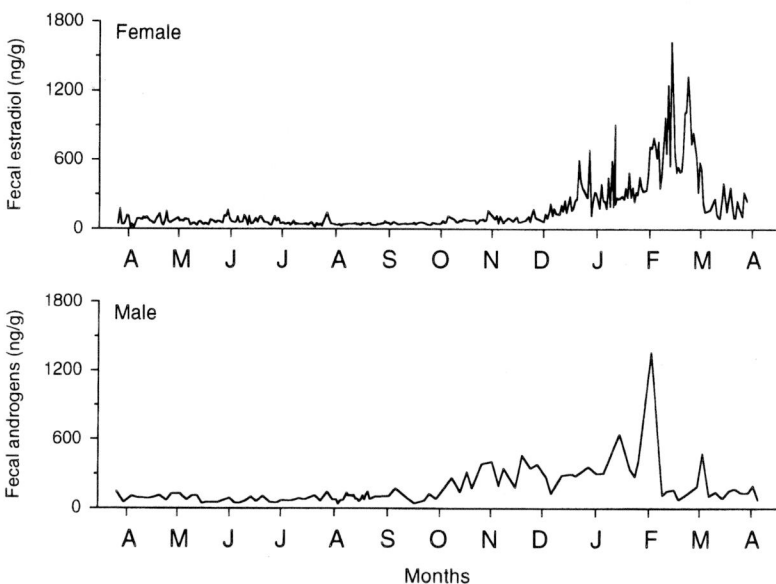

Figure 16.2. Profiles of faecal oestradiol excretion in an adult female (top panel) and faecal androgen excretion in an adult male (lower panel) Pallas' cat, illustrating marked gonadal seasonality.

Such fundamental data can have applied value, including discovering that excessive artificial light in one zoo (as a result of an early winter evening activity, 'Festival of Lights') disrupted the onset of normal seasonality in the Pallas' cat, thereby preventing breeding and kitten production (Swanson & Brown, unpub. data). Isolating this pair from the extended artificial lighting in the subsequent year resulted in normal hormone patterns in both the female and male, breeding and the production of kittens. Another example involves the impact of pairing behaviourally incompatible female cheetahs, a practice that can disrupt normal ovarian cyclicity (Wielebnowski, Wildt & Brown, unpub. data). Separating zoo-held cheetahs and maintaining them singly (as cheetahs normally live in nature) has been shown to provoke the reinitiating of normal reproductive cycles. Other management applications include: identifying the causes of reproductive inactivity (senescence, cystic ovarian follicles, retained corpora lutea); improving hormone treatments useful to increasing the efficiency of ovulation induction for artificial insemination; and assessing the level of physiological stress for the purpose of evaluating and improving habitats to enhance reproductive success and animal well-being. The latter possibility holds great promise because the technology has advanced to allow accurate measures of adrenal corticoid hormones, often used as indices of 'stress' (Graham & Brown, 1996).

There also is a growing database suggesting that non-invasive hormone metabolite monitoring has applications in field studies of carnivores. Creel et al. (1992) first described the potential of conditioning free-living dwarf mongooses in the Serengeti National Park to scent mark rubber pads, thus providing a source of urine for hormonal assessments. Especially important was the finding that alpha females in mongoose packs excreted more oestrogen that was a likely means of suppressing reproduction in subordinate counterparts. In contrast, the alpha male had indistinguishable androgen profiles from subordinates. A similar approach was extrapolated to African wild dogs of the Selous Game Reserve, but in this case using faeces (Creel et al., 1997a). It was possible to positively correlate wild dog reproductive behaviours and dominance hierarchies to oestrogen excretion (in the case of females) but not androgen excretion (in the case of males). Also of significance was that adrenal status could be assessed in wild dogs by measuring faecal corticosterone concentrations (Creel et al., 1997a). It was discovered that there was an 'hyperadrenal cost' to dominance status within the pack (alpha males and females excreted more corticosterone). This was intriguing because it was widely believed that reproductive suppression in subordinates was caused by social stress, a hypothesis that now could be rejected on the basis of non-invasive assessment of adrenal status. In sum, hormonal metabolite monitoring combined with conventional behavioural ecology is providing new, more holistic information of relevance to conservation.

Using reproductive technology to assist in the recovery of an endangered carnivore, the black-footed ferret

When sufficient fundamental knowledge was secured (after years of basic research), reproductive technologies became routine in propagating live-stock and humans. In theory, the same scenario should be possible for carnivores given a similar strategic approach. Although there is a growing number of examples using artificial insemination and even *in vitro* fertiliz-ation and embryo transfer to produce carnivore offspring (see reviews, Wildt *et al.*, 1998; Howard, 1999), there is one species in which reproduc-tive technologies are routinely contributing to recovery and reintroduction – the black-footed ferret, a small, charismatic mustelid whose original natu-ral range was the Great Plains (USA).

The black-footed ferret's survival was linked to the common prairie dog, its primary prey. As the Great Plains were settled and converted to farm-lands, the government actively supported massive poisonings of prairie dogs. The decline of black-footed ferrets followed and, by the 1960s and 1970s, the species was largely eliminated from its native range. No captive breeding programmes existed in zoos, and by the late 1970s the species was believed to be extinct. A small population was discovered in 1981, but even-tually suffered a disease epidemic that caused the death of all but 18 individ-uals. These last remaining ferrets were brought into captivity and were the subject of an enormous collaborative effort by the State of Wyoming, the U.S. Fish & Wildlife Service, zoos throughout North America, the Conser-vation Breeding Specialist Group and the American Zoo and Aquarium Association. With the production of ferret kits in Wyoming, animals were eventually distributed to six other facilities to assist in the breeding pro-gramme.

There have been two areas of focus at the Conservation and Research Center in northern Virginia. Originally, kits were produced solely by natu-ral breeding. Since 1986, our laboratory also has been involved in learning the basics about ferret reproductive biology while exploring the potential use of assisted breeding to accelerate offspring production. Original studies centered on the common European ferret and Siberian polecat as 'research models', developing an understanding about sperm physiology and opti-mal methods for processing, culturing and cryopreserving semen. These species also were valuable for developing a laparoscopic intrauterine artifi-cial insemination technique, because an early study demonstrated that vag-inal insemination was ineffective (Wildt *et al.*, 1989). Studies over time demonstrated that European ferrets and Siberian polecats could be pro-duced routinely by artificial insemination (Table 16.1; Howard, 1998). The

Table 16.1. Ferrets produced by laparoscopic intrauterine artificial insemination using fresh or frozen-thawed spermatozoa

Species	Sperm treatment	No. of pregnancies (%)	No. of kits	Mean (± SEM) no. kits/litter
European ferret	fresh	17/24 (70.8)	85	5.2 ± 0.5
	thawed	7/10 (70.0)	31	4.4 ± 1.0
Siberian polecat	fresh	1/1 (100.0)	6	6.0
	thawed	5/6 (83.3)	26	5.2 ± 1.2
Black-footed ferret	fresh	11/14 (78.6)	32	2.9 ± 0.4
	thawed	2/3 (66.7)	1	1.5 ± 0.5

Adapted from Howard, 1998.

significance of having assisted breeding technology available for black-footed ferrets eventually became highly apparent.

A 1995 survey of breeding records of the black-footed ferret species survival plan revealed that more than 50% of males failed to breed naturally, a serious problem in capturing all genetic diversity from the small number of original founders. Since then our research has focused on potential causes for this lack of reproductive fitness, which has revealed at least two problems. The first is a relatively high rate of sexual incompatibility that was expressed as excessive male aggression, apathy or poor copulatory positioning (Wolf et al., 1998b). Males with this problem that are from important, underrepresented genetic lineages have been relegated to being sperm donors for artificial insemination, and kits are now being produced routinely using fresh or frozen-thawed sperm (Table 16.1; Figure 16.3; Howard et al., 1997).

The second complication originated with the assumption that yearling males approaching their first breeding season were 'prime breeders'. Collaborators at the National Black-Footed Ferret Conservation Center (Wyoming) had already confirmed that yearling black-footed ferrets had comparable testes sizes and libido to two- and three-year-old counterparts. However, systematic semen collection and evaluation revealed that the yearlings produced sperm for a shorter time period during the breeding season than their older counterparts (Wolf et al., 1998b; Howard et al., 1999). Thus, yearlings were undergoing abbreviated spermatogenesis in the face of normal sexual drive, so mating was occurring often in the absence of sperm release, which resulted in a high rate of pseudopregnancy (sterile matings). When males were selected for breeding in 1998 on the basis of sperm production (rather than testes volume), whelping success increased 20%, and an additional 59 kits were produced by natural matings

Figure 16.3. Black-footed ferret kits produced by artificial insemination from genetically valuable parents.

(Howard *et al.*, 1999). Thus, a modest use of reproductive technologies boosted the efficiency of the black-footed ferret breeding programme. The result has been more animals available for the reintroduction programme in progress in five western USA states. This is the first example of consistent accelerated propagation of an endangered species, which for reintroduction programmes includes the use of assisted breeding technology.

CONCLUSIONS AND RECOMMENDATIONS FOR CONSERVATION

Reproductive biology is an important component of wildlife research. If 'to conserve' means 'to preserve and protect', then reproduction (the essence of continued species existence) is a high priority study area. However, reproductive biology now almost always has a high-tech nuance. There is a perception that the discipline largely involves assisted breeding techniques, such as artificial insemination, *in vitro* fertilization and/or embryo transfer. But reproductive biology does not have to be defined by, or closely linked to, technology. Furthermore, reproductive techniques often are most valuable for producing new knowledge, and not necessarily for generating offspring. This has been the case for several carnivores, especially for a diverse array of felids, a few canids and a mustelid, the black-footed ferret. Fundamental studies have revealed a wealth of information on topics ranging from gonadal–hormonal–behavioural relationships, to the early events of fertilization/embryogenesis, to factors influencing reproductive fitness. Especially interesting have been studies of novel phenomena (e.g., teratospermia) and the recognition of species-specific differences in reproductive mechanisms, even among closely related taxa. Powerful tools also have been developed: (1) to measure hormonal patterns from steroidal metabolites excreted in faeces or urine; and (2) to systematically collect and cryostore sperm, both of which are applicable to certain wild carnivore populations. Once this fundamental reproductive information is known, then wildlife can be propagated using improved natural or assisted breeding. The black-footed ferret, discussed earlier, is a prime example of how reproductive technologies are being used to enhance management and increase offspring numbers for reintroduction into nature.

FUTURE RESEARCH AND NEEDS

It is apparent that the reproductive sciences are more than high-tech assisted breeding procedures to be used for a 'quick fix'. Rather, this is a broad-based discipline that can range from recording reproductive behaviours to the use of the most sophisticated laboratory techniques for understanding the subcellular mechanisms that allow the creation of embryonic life. There is no doubt that the reproductive sciences can contribute to carnivore conservation, but the first priority always must be the production of new fundamental knowledge that eventually may have management application.

Wildt & Wemmer (1999) have summarized other high priority areas that will allow reproductive research to contribute more to wildlife management:

Studbooks as sources of reproductive data

Many zoo populations are tracked by means of a studbook, a catalogue of living (and deceased) animals. Such documents (in computerized and hard copy form) hold a wealth of information on specific life history variables. These data not only have a fundamental value (e.g., on pubertal onset, gestation/generation interval, seasonality and reproductive senescence), but also have relevance in simulation modelling for *in situ* situations, especially in the context of assessing stochastic events on population dynamics and extinction risks (Lacy, 1993b).

Partnerships among disciplines

Tools in reproductive technology become even more powerful when merged with other conservation biology disciplines. This chapter already has offered some illustrative examples in the context of animal behaviour and the simultaneous assessment of gonadal or adrenal hormones (see discussion of free-living dwarf mongooses and behaviourally incompatible cheetahs). Non-invasive hormone monitoring has enormous potential for other interdisciplinary studies of carnivores (e.g., stress versus disease, mortality and environmental conditions). Additionally, gamete technologies have had an application in understanding reproductive fitness, including in wild populations associated with suspected losses in genetic variation (e.g., Florida panther; Roelke *et al.*, 1993a). Certainly, wildlife managers should be confident about collaborating with reproductive physiologists to assess animal fitness (in nature or in zoos) while looking for links to other variables (behaviour, genetics, disease and nutrition, and environmental factors, including stress). Some fascinating connections can be expected. Also, when aggressive genetic intervention is necessary, managers should consider exploiting techniques such as artificial insemination or genome resource banking to accelerate population growth, infuse genetic vigor or simply to bank germplasm as 'insurance' for the future.

Consistently controlling reproductive capacity

As emphasized throughout this chapter, reproductive technology has an application in carnivore conservation, but there is a continuing problem with *consistency*. Newspapers occasionally report an anecdotal story of a tiger produced by *in vitro* fertilization or a bear by embryo transfer. But there are too few studies in which the technology has become so reliable that wildlife managers actually want to incorporate more reproductive technologies into recovery programmes. That is why the consistent success of artificial insemination in the black-footed ferret is so important as a model to demonstrate what is possible. More such examples are needed among carnivores, and are likely to be achieved only after fundamental reproductive mechanisms are established. Likewise, the ability to effectively *reduce reproductive capacity* is as serious as the more common interest of enhancing fecundity. Contraception in carnivores is a high priority for research. An ability to routinely and safely prevent reproduction is key to the successful genetic management of zoo populations, so that precious space is not wasted on generic animals.

Promoting the reproductive sciences

Finally, there is a need for educating the public, scientific colleagues in the conservation community and our fellow reproductive biologists in academia about the true role of this discipline in conservation biology. Hopefully this chapter will be a first step in this direction. Ironically, the greatest challenge is not necessarily with scientists in other conservation disciplines, but with conventional reproductive biologists in academia. What is interesting is the evolution of the reproductive sciences in universities, where it is much more common to find researchers exploring the molecular biology of a cell rather than enjoying the satisfaction of working with whole, living animals. The graduate student who has either experienced the thrill of handling a wild carnivore, or ever seen one in nature is a rarity. This problem possibly is prevalent in the other life sciences, but this can be rectified by those of us who are motivated to entice (and provoke) young scientists to become involved in integrative conservation biology.

17

Monitoring of terrestrial carnivore populations

ERIC M. GESE

INTRODUCTION

There is increasing concern about the status and distribution of terrestrial carnivore populations throughout the world (Schaller, 1996). Changes in land-use practices, habitat loss and fragmentation, sanctioned human persecution, declines in natural prey, disease, illegal poaching, and increased competition within carnivore guilds have brought about a general decline in several carnivore populations with some species now occupying a fragment of their former range. The continued loss of suitable habitat due to an ever expanding human population has placed the issue of conservation and protection of some carnivores as a top environmental priority and/or controversy for many agencies and organizations. Paramount to carnivore recovery, reintroduction, or development of management plans and policies, is having reliable and accurate information regarding the status, health, and well-being of the carnivore population of concern. One of the most commonly asked questions when dealing with carnivore conservation is: where are the animals, how many are there, and what is the population trend? These questions often place biologists and managers in the difficult position of determining the status of a carnivore population. Biologists need reliable methods that provide accurate data on the distribution, abundance, and population trend of a species in order to make informed decisions and recommendations to policy makers. Many carnivores are secretive, nocturnal, far-ranging, live in densely vegetated habitats or remote areas, or exist at extremely low densities, making censusing and monitoring a carnivore population very difficult, if not sometimes seemingly impossible.

Monitoring of a carnivore population may be performed at various levels of resolution. First, biologists may only need to know where a particular carnivore occurs (i.e., species distribution). Second, the biologist may need to know how many animals are in an area (i.e., species abundance).

Animal abundance may be assessed in two ways: relative and absolute abundance. Relative abundance uses indices of animal abundance (e.g., track counts, scent-post visitation rates) that can be compared over time or between areas, but of itself does not estimate animal numbers. In contrast, absolute abundance involves using methods to actually count animals and then estimate the number or density of animals in the population. With repeated sampling over time, both relative indices and absolute estimates of animal abundance can be used to monitor population trends. For many carnivore species this amount of information may be adequate. However, if the population trend indicates an increasing or declining population, then it may be important for the biologist to ask: why is the population changing? This final question involves examining the demographic processes of birth, death, emigration, and immigration that determines the persistence of a population.

The objective of this chapter is to describe the techniques that have been developed to census and monitor terrestrial carnivores and a discussion of the advantages and disadvantages of each technique. Many of the techniques described herein still need an in-depth evaluation as they pertain to accuracy and reliability in monitoring population trends of carnivores. This chapter will focus on terrestrial carnivores only (suborder Fissipedia); inclusion of aquatic carnivores would require an entire chapter to itself. Capture, handling, or immobilization procedures will not be discussed; the volume of literature is enormous and species specific (readers should consult Pond & O'Gara, 1994; Schemnitz, 1994, and references therein). Current methods for censusing or surveying wild carnivores range across the gradient of accuracy, reliability, and cost. I have included references of several studies that used, or attempted to use, a technique to determine species distribution or abundance. These references are only provided as examples of studies, and are not inclusive of all studies using that specific technique.

SOME CONSIDERATIONS BEFORE IMPLEMENTING A MONITORING PROGRAM

Before embarking on a large-scale effort to monitor a carnivore population, the biologist or manager should carefully consider what question(s) they are asking, and if an estimate of population size is needed, then one must decide on the precision and accuracy of the estimate required to answer that question (Lancia et al., 1994; Zielinski & Stauffer, 1996). For example, if a biologist is assessing the abundance of a very rare carnivore that numbers 50 animals in the wild, then even a slight decline in population size would be critical and surveys would need to be sensitive to even the smallest change in numbers. In contrast, a carnivore population numbering 5000 animals could use a survey with less sensitivity because a slight decline would not be catastrophic to that population. The precision (the measure of how close an estimate is to the expected value), accuracy (the measure of how close an estimate is to the true population size), power (the probability of rejecting the null hypothesis when it is in fact false and should be rejected), sample size, survey design, and the statistical assumptions of each method should be considered before implementing a monitoring program (Peterman, 1990; Reed & Blaustein, 1997; Van Strien et al., 1997). Macdonald et al. (1998a) provides a thorough review of statistical considerations when designing a monitoring program. Two major problems that a biologist must typically address when developing a monitoring program is observability or catchability of the animal (the probability is generally < 1) and the size of area to be sampled because time and money constrain sampling the entire area (Lancia et al., 1994, Macdonald et al., 1998a). In addition, the costs, logistics, manpower, and time constraints must all be considered before deciding on the usefulness of a particular method to monitor a carnivore population. These considerations sound quite intuitive and fundamental, but success of the project may hinge on careful examination, prior planning, and development of an appropriate study design (Skalski & Robson, 1992; Macdonald et al., 1998a).

METHODS EMPLOYED TO DETERMINE SPECIES DISTRIBUTION

Often biologists may only need to know if a species is present in an area. This fundamental question is needed to determine the presence and distribution of rare, threatened, or endangered species. Methods typically employed to determine species distribution include habitat mapping, questionnaires, interviews, sighting reports, or confirmation of a sign left by the species in question. Any survey method (direct or indirect measures) that provides an estimate of animal abundance provides distribution information as well. However, for discussion of those survey methods see under 'Methods of estimating animal abundance'.

Habitat mapping
Biologists should not necessarily race out into the bush and start looking for animals or signs of them. Careful consideration regarding the kind of suitable habitat required for a species followed by examination of habitat

maps or aerial photos (if available) can save time (e.g., Macdonald *et al.*, 1998a). Habitat suitability models have been developed for many carnivore species. With the continued development of satellite imagery, remote sensing, and Geographic Information Systems (GIS), areas containing suitable habitat for a particular species can be identified allowing for maximization of survey effort. Surveys can then be stratified by habitat types or land classes (Macdonald *et al.*, 1998a). In the UK, use of landscape data from the Countryside Information System (CIS), plus existing mammal records and knowledge of habitat requirements, were used to predict mammal distribution on a national scale (Macdonald *et al.*, 1998a). Use of GIS has also been instrumental in identifying potential habitat for restoration of carnivores (e.g., Mladenoff *et al.*, 1995; Mladenoff & Sickley, 1998).

Questionnaires, interviews, and sighting reports

One of the simplest methods of determining species distribution, and possibly gaining a subjective estimate of animal abundance, is collecting sightings and general impressions from various people in the field. Questionnaires, interviews, and sighting reports from hunters, trappers, rangers, mail carriers, tourists, guides, and field personnel have been used with some success to measure animal distribution, and sometimes animal abundance, of different species of Canidae (Lemke & Thompson, 1960; Allen & Sargeant, 1975; Harris, 1981; W. Clark & Andrews, 1982; Fuller *et al.*, 1992a; Fanshawe *et al.*, 1997), Felidae (Tewes & Everett, 1982; Erickson, 1982), Mustelidae (Fortenbery, 1970; Hillman & Linder, 1973; Powell, 1982; Strickland & Douglas, 1984; Slough & Smits, 1985; Melquist & Dronkert, 1987), Procyonidae (Kaufman *et al.*, 1976; W. Clark & Andrews, 1982), and Ursidae (Kolenosky & Strathearn, 1987). Questionnaires were successfully used in the UK to detect the presence of elusive carnivores, such as pine marten, *Martes martes* (Strachan *et al.*, 1996), western polecats, *Mustela putorius* (Birks & Kitchener, 1999), and wildcats, *Felis silvestris* (Balharry & Daniels, 1997).

More in-depth questionnaires or interviews with persons with intimate knowledge of the area and who spend considerable time in the field (e.g., trappers, game wardens, rangers, guides) not only may provide a range and status report (Kaufman *et al.*, 1976; Fuller *et al.*, 1992a), but may also be used to obtain a general, subjective estimate of abundance (e.g., Allen & Sargeant, 1975; Harris, 1981). Many agencies compile status reports using this method to access the relative abundance and distribution of carnivores, particularly in countries that are unable to invest the considerable resources more accurate population assessment requires (e.g., Fanshawe *et al.*, 1997).

Questionnaires have been used when agencies require a large-scale assessment of carnivore distribution (e.g., Fuller *et al.*, 1992a), or in circumstances when little is actually known about the biology of the species in question. This is especially useful for rare species that have a wide distribution. For example, surveys by park staff, field workers, and rangers provided a subjective estimate (absent, rare, common, uncommon) of the abundance of African wild dogs (*Lycaon pictus*) throughout the African countries (Fanshawe *et al.*, 1997). In North America, questionnaires are often sent to trappers and field personnel to monitor trends in furbearer populations (W. Clark & Andrews, 1982; Strickland & Douglas, 1984). Drawbacks of this technique include misidentification of carnivores, low response levels to the questionnaire, and concentration of animal sightings along roads or near human habitation (i.e., rare carnivores inhabiting areas of low human density may go undetected or unreported).

Presence of sign

Sightings of the carnivore species in question allow for direct confirmation of species presence. Spotlight surveys have been commonly used to detect rare or endangered nocturnal species, such as black-footed ferrets (*Mustela nigripes*). However, in the absence of visual confirmation of the animal itself, biologists may resort to surveys of animal sign to determine whether a species is present in a given area. Sign surveys have been used to determine species distribution of most carnivore groups, including several felids (Schaller & Crawshaw, 1980; Newman *et al.*, 1985), mustelids (S. Macdonald & Mason, 1982; Melquist & Hornocker, 1983; Richardson *et al.*, 1985; Melquist & Dronkert, 1987; Macdonald *et al.*, 1998a), ursids (Pelton & Marcum, 1977; Kohn, 1982), and canids (Sargeant *et al.*, 1993). Several different methods of sign surveys have been used, including counting tracks, scats, scratches, burrows or dens, and hair samples. For example, diurnal surveys for signs (scat, tracks, fresh dirt diggings) of black-footed ferrets have been conducted throughout the prairie ecosystem to locate remnant populations (Fortenbery, 1970; Hillman & Linder, 1973; Richardson *et al.*, 1985). Trained dogs have even been used to search for ferrets and their burrows (Dean, 1979). Tewes & Schmidly (1987) describe the use of predator calls to attract ocelots (*Leopardus pardalis*) in south Texas. Conspicuous burrows of American badgers (*Taxidea taxus*) and European badgers (*Meles meles*) have been used as an indicator of species presence (Macdonald *et al.*, 1998a). Surveys at bridges crossing over rivers have been used to determine presence or absence of river otters, *Lutra canadensis* (S. Macdonald & Mason, 1982; Melquist & Dronkert, 1987). Sprainting (defecation) surveys for otters (*L. lutra*) provide good distribution information in the UK, but appear to be unrelated to otter abundance (Conroy & French, 1987; Kruuk *et al.*, 1986). Schaller & Crawshaw (1980) identified tracks and scats to determine the presence and movement patterns of jaguars (*Panthera onca*). Hairsnares or hair tubes can be used to assess distribution through species identification by characteristics of the hair (e.g., Adorjan & Kolenosky, 1969; Moore *et al.*, 1974) or DNA techniques (Foran *et al.*, 1997a,b; Paxinos *et al.*, 1997; Kohn *et al.*, 1999).

Track plates
The use of track plates to determine carnivore presence is gaining in popularity, particularly for the detection of forest carnivores (e.g., Zielinski, 1995). This technique provides a reliable measure of species distribution or presence, but may be unreliable for determining relative animal abundance. Track counts in prepared beds have been used to estimate the extent of mink (*Mustela vison*) distribution, but not numbers of mink (Burgess & Bider, 1980; Humphrey & Zinn, 1982). Similarly, smoked track plates have been used to record tracks of weasels (Barrett, 1983; T. Clark & Campbell, 1983), marten, *Martes americana* (Barrett, 1983; Zielinski & Truex, 1995), and fisher, *M. pennanti* (Zielinski, 1995). A detailed description of tracking plates and the implementation of both enclosed track-plate boxes and unenclosed track plates is provided by Zielinski (1995). In general, track surfaces may be produced from smoked or carbon-sooted aluminum plates, contact paper (tacky, white paper), chalk, or ink. A visual and/or olfactory lure is used as an attractant to bring the animal to the tracking station and while investigating the attractant the carnivore leaves tracks on the tracking surface. Identification of tracks, getting the animal to step on the plate, transportation of the tracking plates, and protecting the track plates from the weather are all problems that require some prior planning when using this technique (but see Zielinski, 1995 and Zielinski & Truex, 1995 for suggestions).

Remote cameras
A relatively new method that is gaining popularity is the use of remote cameras set along trails, near bait stations, or nests. Remote cameras have been used successfully to detect several forest carnivores (Kucera *et al.*, 1995; Foresman & Pearson, 1998) and elusive or nocturnal felids (Joslin, 1982; Rappole *et al.*, 1985). The cameras are commercially available from several manufacturers (see a list in Kucera *et al.*, 1995). They can be set up to be triggered by an animal tripping a line, or activated remotely by

pressure-sensitive plates, motion or heat detectors, or breaking of an infrared beam. While these camera systems are mostly used to detect the presence of carnivores (Kucera *et al.*, 1995; Naves *et al.*, 1996; Foresman & Pearson, 1998), or identify predators at bait stations or nests (Savidge & Seibert, 1988), they could potentially be used to determine animal abundance if individuals can be identified by artificial tags (e.g., ear tags, radio collars) or natural features (e.g., pelage characteristics) and then applying mark–recapture estimators. Remote cameras have the added benefit that a permanent photographic record is available for examination by other researchers. Disadvantages of remote cameras are expense (although some systems are not too costly), getting animals to trigger the camera (similar to problems associated with track plates), and the time delay between photo acquisition and development of the film (i.e., results are not instantaneous). However, development of digital cameras that download images into a computer may negate this concern.

Some considerations for sign surveys

A problem with using sign to determine carnivore distribution is the proper and consistent identification of tracks, scats, burrows, and hair samples. Species identification from scats can be assisted by the use of fecal bile acid patterns detected by thin-layer chromatography (Major *et al.*, 1980; Johnson *et al.*, 1981). Examination of hair samples with a light microscope and comparison to a hair key (e.g., Adorjan & Kolenosky, 1969; Moore *et al.*, 1974) or reference collection can provide species identification. Recent advances in DNA techniques have opened the door for more accurate assessment of species identification and carnivore distribution based upon scat or hair samples (Foran *et al.*, 1997a,b; Paxinos *et al.*, 1997; Kohn *et al.*, 1999). It should be emphasized that most sign surveys only provide distribution information. However, these DNA techniques can also be used to identify individual animals allowing for estimation of population size (Kohn *et al.*, 1999). The amount of sign left behind by an animal does not appear to correlate with animal density for most carnivores (Messick & Hornocker, 1981; Melquist & Hornocker, 1983; Messick, 1987). Also, simply because observers fail to find sign does not necessarily indicate species absence.

Surveys for species presence used as measures of animal abundance

The previously discussed sign surveys can serve a dual purpose. In their most rudimentary form they provide distributional information, but with standardization of the methodologies and the amount of effort conducting the survey, sign surveys may also be used as an index of animal abundance.

For example, if certain areas or habitats are repeatedly surveyed over time and the number of hours of searching are recorded, then biologists may standardize their surveys to tracks/hour, scats/hour, etc., allowing for trend information over time or comparisons between areas.

METHODS FOR ESTIMATING ANIMAL ABUNDANCE

Once a biologist has determined that a carnivore is present in a particular area, the next question that may need to be answered is: how many animals are there and what is the trend in abundance? Biologists may monitor animal abundance by direct methods of counting the animals themselves, or indirectly by counting animal sign (Macdonald *et al.*, 1998a). Estimating animal abundance requires consistent and standardized application of a technique to be able to detect changes or differences with some degree of accuracy, precision, and power (Macdonald *et al.*, 1998a). Thus, for any of the following techniques, biologists must maintain a standardized study protocol for the survey or count that is used and consistently apply that protocol to all future surveys to allow for direct comparisons over time. Whether biologists use sign surveys, indices of relative abundance, or measures of absolute animal abundance, caution should be exercised when examining population trends for carnivores. Assessing rates of increase or decrease from trend data should be done carefully, taking into account the precision and accuracy of the methods used to determine population size estimates or indices of relative abundance. Biologists should be aware of the influence of other variables on survey results. Biologists should consider the characteristics of the animals themselves (e.g., behavior, size, color); the topography and vegetation where the survey will be executed; temporal factors; observer experience, ability, and fatigue; and the spatial distribution of the species concerned (i.e., widely distributed versus high density). Before embarking on population trend analyses, biologists and researchers should examine the assumptions and estimate the power of the survey technique in its ability to detect population changes; see Gerrodette (1987), Eberhardt & Simmons (1992), and Kendall *et al.* (1992) for more details.

Indirect methods
Scent-station surveys
One of the most common sign surveys utilized for indexing carnivore abundance in North America is scent-post or scent-station surveys. Scent-post surveys have been widely used to estimate the relative abundance of several canids (Linhart & Knowlton, 1975; Roughton, 1979; Sumner & Hill,

1980; K. Johnson & Pelton, 1981; Morrison *et al.*, 1981; Roughton & Sweeny, 1982; Conner *et al.*, 1983; Travaini *et al.*, 1996; Sargeant *et al.*, 1998), cats (Conner *et al.*, 1983), mustelids (Brown, 1969; Lord *et al.*, 1970; Humphrey & Zinn, 1982; Melquist & Dronkert, 1987; Hein & Andelt, 1995), raccoons (Sumner & Hill, 1980; W. Clark & Andrews, 1982; Conner *et al.*, 1983; Smith *et al.*, 1994), and bears (Lindzey *et al.*, 1977; Kohn, 1982). Scent-post or scent-station surveys involve placing a scented tablet (e.g., fermented egg extract, mackerel oil) or other attractant within a 1-m circular area of sifted dirt. Tracks left by an animal are identified to species, and presence or absence of the species is recorded. Typically, stations are spaced at a predetermined interval along roads or trails and then visited for three to four consecutive nights to record tracks; the sifted area is swept smooth after each night. Biologists should consider the movement patterns and home-range size of the species of interest when determining the spacing of the stations (i.e., close spacing for close-ranging species, increased spacing for larger species). The frequency of animal visitation to operable stations (i.e., not disturbed by wind, rain, vehicles) is used as an index of abundance. For details on this method and its application, see Linhart & Knowlton (1975), Roughton (1979), and Roughton & Sweeny (1982). Biologists interested in using scent-post surveys should consult Smith *et al.* (1994) and Sargeant *et al.* (1998) prior to implementation. While some biologists reported that scent-station surveys reflect changes in raccoon abundance, Smith *et al.* (1994) found no association between visitation rates and density of raccoons. Knowlton (1984) found a positive correlation ($r^2 = 0.79$) between coyote (*Canis latrans*) scent-station indices and estimated coyote density. Seasonal changes in habitat use and visits to multiple stations by a single animal can contribute to invalid correlations of animal density and visitation rates. Sargeant *et al.* (1998) makes several recommendations regarding sample unit specification and interpretation of scent-station surveys. Misidentification of tracks, problems with the weather (mostly wind and precipitation), wariness of animals in relation to the sifted substrate, and a fairly labor intensive technique are items to be addressed when considering scent-station surveys.

A variation of the scent-post survey that has been used to index dingo (*C. familiaris dingo*) populations is the activity index (Allen & Engeman, 1995, Allen *et al.*, 1996). This index of animal visitation simply uses a sifted dirt area on a road without any scent or lure to attract animals. The number of track sets crossing the sifted area is used to assess relative abundance and calculate a variance estimate (Engeman *et al.*, 1998).

Scat deposition transects

The rate at which scats are deposited along established roadways has been used as an estimate of relative abundance for some canid species, mainly coyotes (Clark, 1972; Davison, 1980; Andelt & Andelt, 1984) and wolves, *Canis lupus* (Crête & Messier, 1987). The general methodology involves designating transects or routes along a roadway, clearing all scats from the road, then returning and collecting all scats encountered two weeks later. The scat index is computed as the number of scats collected per transect per 14-day period (Davison, 1980). If transects vary in length, or the time periods vary in the number of days between collections, then the index can be standardized to scats/km/day. Scat deposition rates for coyotes were found to be correlated ($r^2 = 0.97$) with estimates of animal density derived from mark–recapture techniques using radioisotope tagging of feces (Knowlton, 1984). For long-term monitoring, scat transects should be conducted along the same routes at the same time of year to avoid introducing biases associated with differential prey digestibility (hence differential scat deposition rates) and seasonal changes in food items consumed (Andelt & Andelt, 1984). Misidentification of scats and heavy vehicle traffic on roadways can also be problematic when using scat deposition counts. Use of DNA techniques for identifying species from scats may alleviate the problems of misidentification (Foran *et al.*, 1997a,b) and identification of individual animals collected during scat deposition transects could potentially be used to estimate population size (Paxinos *et al.*, 1997, Kohn *et al.*, 1999).

Track counts along a transect

Tracks left by carnivores along river beds, dry washes, sandy fire breaks or roads, or on snow-covered roads and trails have been used as a relatively simple and inexpensive measure of relative animal abundance for several species of canids (Beasom, 1974a; Crête & Messier, 1987; Palomares *et al.*, 1996), felids (Anderson, 1981; Van Dyke *et al.*, 1986; Van Sickle & Lindzey, 1991, 1992; Smallwood & Fitzhugh, 1995; Beier & Cunningham, 1996; Stander, 1998), mustelids (Ruff, 1939; Quick, 1944; de Vos, 1952; Coulter, 1966; Priklonski, 1970; Fitzgerald, 1977; Powell, 1982; S. Johnson, 1984; Slough & Smits, 1985; Golden, 1986; Melquist & Dronkert, 1987), ursids (Pelton & Marcum, 1977; Stirling *et al.*, 1980; Kohn, 1982; Kendall *et al.*, 1992), and Egyptian mongooses, *Herpestes ichneumon* (Palomares *et al.*, 1996). Carnivores that occupy regions that receive snow have been monitored through the use of counting tracks along established transects within one to two days following fresh snowfall. Winter track counts along standard transects have been routinely used to index the relative abundance and population trends of marten (Slough & Smits, 1985), weasels (Ruff, 1939; Quick, 1944; Priklonski, 1970; Fitzgerald, 1977), and fisher (de Vos, 1952; Coulter, 1966; Powell, 1982; S. Johnson, 1984). Similarly, counts of tracks left by cougars (*Puma concolor*) along dry washes has been used to index animal abundance (Beier & Cunningham, 1996). Golden (1986) was able to conduct aerial track counts for wolverines (*Gulo gulo*) in unforested areas of Alaska. Ballard *et al.* (1995) reported good precision between line-intercept sampling of tracks and estimates of wolf density based upon radiotelemetry. This technique was repeatable, efficient, reasonably accurate, and relatively inexpensive. Biologists attempting transect counts of tracks should be aware of some pitfalls. Misidentification of tracks and low power to detect population changes can occur when using track counts (Van Sickle & Lindzey, 1991; Kendall *et al.*, 1992; Ballard *et al.*, 1995; Beier & Cunningham, 1996). Precision can be increased by increasing sampling effort (more transects), or increasing the length of transects if dealing with a far-ranging species (e.g., cougars: Van Sickle & Lindzey, 1991), although see Kendall *et al.* (1992). Much of the power of this estimator is dependent upon a high rate of encountering sign along the transects (Kendall *et al.*, 1992). When working in areas with snowfall, variables one must consider include the condition and consistency of the snow, variable depth of snow (i.e., no snow negates data collection), temperature, and time of year. Observer experience at interpreting tracks is also crucial for consistent and reliable monitoring.

Den and burrow surveys

Ground and aerial surveys for active dens have been conducted along transects as a method of indexing relative abundance of some carnivore species. Annual den surveys have been used to monitor populations of arctic fox (*Alopex lagopus*) in northern dry tundra (Macpherson, 1969; Garrott *et al.*, 1983), but appear to have little application in areas of coastal wet tundra (Anthony, 1996). Ground and aerial surveys for dens has been used to monitor kit fox (*Vulpes macrotis*) populations in desert environments (O'Farrell, 1987) and red fox (*V. vulpes*) populations on the prairie (Trautman *et al.*, 1974). The key to this survey technique is relatively open habitat with little vegetative cover and a carnivore species that makes conspicuous dens or burrows. These surveys can be relatively expensive (aerial searches) and/or labor intensive (ground searches). In general, this survey entails personnel walking or flying along a route or transect searching for active dens. The presence of feces or tracks at the burrow or den can assist in species identification. Ground surveys conducted along transects can also be used to calculate the density of dens if biologists record the perpendicular distance from the transect to the den (Burnham *et al.*, 1980). Conspicuous burrows dug by badgers have been used to indicate species presence, but there appears to be no correlation between density of burrows and animal abundance (Messick & Hornocker, 1981; Messick, 1987). This technique would probably not work well for indexing carnivores with large social units. No matter how large the pack, coyotes and wolves typically have one natal den to rear offspring (i.e., a pair of coyotes uses the same number of dens as a pack of seven coyotes). For animals that exist in packs or clans, the number of dens would more likely indicate the number of social units present across an area, but not the number of animals in each social unit.

Vocalization response surveys

For social carnivores that utilize long-range vocalizations (roars, howls, or whoops) to communicate, biologists have been able to use the response rate to simulated vocalizations as an estimate or index of relative animal abundance. Howling surveys for coyotes (Wenger & Cringan, 1977, 1978; Okoniewski & Chambers, 1984) and wolves (Harrington & Mech, 1982; Carbyn, 1982; Fuller & Sampson, 1988), roaring for lions, *Panthera leo* (Rodgers, 1974; Maddock *et al.*, 1996; Ogutu & Dublin, 1998), and long-distance whoops for hyenas, *Crocuta crocuta* (Ogutu & Dublin, 1998) have all been used as a technique for estimating animal abundance. Vocalization response surveys typically employ recorded vocalizations, although human imitation of sounds is sometimes effective. Traveling along roads or trails and stopping at predetermined intervals, vocalizations are produced and then observers listen for a specified amount of time for a response from the target species. The biologist may conduct the survey over several nights and use the vocalization response as a means of estimating the relative abundance of the carnivore species. Standardization and consistency of this method is needed for reliable and comparable results for trend analyses. Biologists should also be aware of the seasonal, social, temporal, and spatial factors that may influence carnivore vocalization rates (Laundré, 1981; Harrington & Mech, 1982; Walsh & Inglis, 1989; Gese & Ruff, 1998). For an accurate population census, biologists need to intensively survey the area of interest to obtain adequate coverage (Fuller & Sampson, 1988). In the Masai Mara National Reserve of Kenya, Ogutu & Dublin (1998) estimated that 20% of the study area had to be sampled to acquire reliable estimates of hyena and lion abundance.

Frequency of depredation complaints

The frequency of livestock depredation complaints may be useful as an indicator of relative abundance and population trend under the general belief that animal abundance is correlated to rates of livestock predation. Because this relationship has not been explicitly tested, biologists should be cautious of this technique as depredation rates are subject to changes in livestock stocking rates, habitat type, size of area used, husbandry practices, and environmental variables (Fritts, 1982; Lindzey, 1987; Mech et al., 1988a).

Some considerations when using indirect methods

Indirect methods provide only relative abundance, not absolute abundance, and must be applied consistently for any reliable comparisons between areas, habitats, or over time. Whenever indices of relative abundance are used, biologists should attempt to learn if the relationship between relative indices and absolute abundance is positively and monotonically related, or if the relationship is nonmonotonic. Is the relationship linear with a constant slope, or linear with a variable slope? Indices that are nonmonotonic to animal abundance are of little use in monitoring trends of a carnivore population. Comparison of an inexpensive indirect method to a more expensive direct method could prove worthwhile for calibration of the less expensive technique. During such a calibration, the techniques should be performed concurrently and may need to be conducted on a species-specific, habitat-specific, and seasonal basis. Unfortunately, few indices of relative abundance have been properly tested with a known carnivore population estimate. Of those that have been examined, results are mixed. Knowlton (1984) found positive correlations between scat deposition rates along transects and estimated coyote population density. Scent-post survey indices were also positively related to coyote density. In contrast, Smith et al. (1994) found no association between scent-station visitation rates and density of raccoons.

Direct counts

Direct counts involve the actual counting of animals themselves, in contrast to counting sign. These counts may use either dead animals (e.g., mortality samples, road kills, harvest reports) or live animals (e.g., trapping or sightings). The assumptions of direct counts and the estimators used to determine population size should be carefully reviewed (Caughley, 1977; Burnham et al., 1980; Skalski & Robson, 1992). Counts may involve total counts of the area, or a subsample of the area and extrapolation to the rest of the area of concern. Stratification of subsamples to different habitat types or land classes may increase the validity, usefulness, and precision of the surveys (Macdonald *et al.*, 1998a).

Harvest reports and pelt registration

A method of gaining insight into abundance (and certainly distribution) of a species is examination of harvest and trapping records. Current and historical harvest records can be a valuable resource in obtaining a general, if subjective, idea of animal distribution and abundance (Seton, 1909; Hewitt, 1921). In the Canadian provinces, mandatory pelt sealing reports has also been used to estimate furbearer population trends (Novak, 1987). In the UK, a decline in otter numbers was observed through a decrease in hunting success (Strachan & Jefferies, 1996). While detailed information from harvested animals can be used to construct models for population estimation (W. Clark & Andrews, 1982), harvest data alone is generally not a reliable estimate of population trends. Pelt prices, differential harvest methods, and environmental and social factors all influence harvest rates. W. Clark & Andrews (1982) speculated that harvest surveys may indicate population trends of furbearers with low commercial value because harvest trends would be less affected by management actions and fur prices. Other problems associated with the use of harvest records include hunters and trappers not keeping records, trappers having faulty memories, only some hunters submitting reports (usually successful hunters), and sometimes trappers will give inaccurate reports to avoid tax auditors (Sanderson, 1951a; W. Clark & Andrews, 1982). For rare species (e.g., coati, *Nasua narica*), fur harvest reports are generally unreliable for population trends (Kaufman, 1987), while harvest reports for abundant furbearer populations (e.g., long-tailed weasel, *Mustela frenata*) may be reliable measures of population trend (Hamilton, 1933; Barbour & Davis, 1974).

One method for estimating harvest rate and population size of bobcats (*Lynx rufus*) uses the total number of harvested animals, the sex-specific age distribution of the harvest, and the estimates of harvest effort over the span of years represented in the age distribution (Paloheimo & Fraser, 1981; Rolley, 1987). Interpretation of the sex and age structure of harvested samples is commonly used to assess changes in black bear (*Ursus americanus*) populations (Whelan *et al.*, 1978; Lindzey & Meslow, 1980; Kolenosky & Strathearn, 1987). However, when using harvest data, the validity of the underlying assumptions should be carefully evaluated (Gilbert *et al.*, 1978). Population trends of carnivores have been examined in relationship to past and current harvest records for many species of Canidae (Elton, 1942;

Chitty, 1950; W. Clark & Andrews, 1982; Erickson, 1982), Felidae (Elton & Nicholson, 1942; Erickson, 1982; Lindzey, 1987; Rolley, 1987; Quinn & Parker, 1987), Mustelidae (Hamilton, 1933; Barbour & Davis, 1974; W. Clark & Andrews, 1982; Powell, 1982; Linscombe *et al.*, 1982; Strickland & Douglas, 1984; Melquist & Dronkert, 1987), Procyonidae (Seton, 1909; Hewitt, 1921; Sanderson, 1951a; W. Clark & Andrews, 1982; Kaufman, 1987; Novak, 1987), and Ursidae (Whelan *et al.*, 1978; Lindzey & Meslow, 1980; DeMaster *et al.*, 1980; Kolenosky, 1987). However, no in-depth testing has been conducted to confirm the relationship between animal population density and reports of fur or animal harvest statistics.

Road mortality samples

The frequency of animal carcasses found on roadways has been proposed as a measure of population trend for some carnivore species, usually as an index of relative abundance. For example, the number of raccoons (*Procyon lotor*) and skunks (*Mephitis mephitis*) killed along roads have been used as measures of relative abundance (W. Clark & Andrews, 1982; Bartlett & Martin, 1982). While this technique is intuitively simple and appealing, differences in animal behavior and movements, habitat, traffic density, road surface, and road density likely influence kill rates of some carnivores; nor has the relationship between population density and road kill rate been adequately examined. However, Birks & Kitchener (1999) calibrated road kills of polecats with numbers estimated from intensive live trapping. Road mortality samples can be used to confirm species presence.

Spotlight surveys

Spotlight surveys are a cost effective method typically used for assessing the relative abundance of nocturnal animals. Estimates of relative abundance for nocturnally active carnivores, such as raccoons (Andrews, 1979; Frederickson, 1979; Rybarczyk *et al.*, 1981; W. Clark & Andrews, 1982), badgers (Hein & Andelt, 1995), kit foxes (Ralls & Eberhardt, 1997), red foxes (Weber *et al.*, 1991), black-footed ferrets (Campbell *et al.*, 1985), and skunks (Schowalter & Gunson, 1982; Rosatte, 1987), have been determined with spotlight surveys. These surveys usually involve two observers standing in the back of a truck being driven slowly (16–24 km/hr) along roadways, scanning the road and sides for animals using spotlights of > 500 000 candlepower. When an animal is detected, usually by eye shine, the driver stops the vehicle and the observers identify the animal (using binoculars or a spotting scope). The mileage and time of detection is recorded for each sighting. An index of animals/km is then calculated.

Spotlight counts can be used to estimate population size with line-transect methodology if the perpendicular distance to the sighted animal is recorded (Thompson et al., 1998). Transects need to be fairly lengthy (> 10 km), and because vegetative cover and topography can influence visibility (Whipple et al., 1994; Ralls & Eberhardt, 1997) which influences survey results, these variables should be considered in survey design (Ralls & Eberhardt, 1997). For a description of this technique in assessing fox abundance, see O'Farrell (1987) and Ralls & Eberhardt (1997). Surveys can be conducted over several nights (repeated counts) to obtain a measure of sampling error (Norton-Griffiths, 1975). Large samples with replication are needed to detect changes in population size with any statistical power (Ralls & Eberhardt, 1997). Surveys can be conducted seasonally and annually for population trend analysis. Spotlight counts do not work well in areas containing low densities of carnivores. Spotlight counts may also be used to acquire a relative estimate of the abundance of certain prey species at the same time (Barnes & Tapper, 1985; White et al., 1996), but Ralls & Eberhardt (1997) believed that spotlighting was not a sensitive method for assessing prey abundance.

Catch-per-unit-effort

Live trapping certainly gives a positive confirmation of species presence and hence distribution. The number of animals captured per trap-night can also be used as an index of relative abundance of carnivores. Live trapping is expensive and labor intensive, and can be ineffective in areas with low carnivore density. In addition, standardization of capture procedures and variation among individual trappers can cause problems with this methodology. This technique has been used to assess the relative abundance of coyotes (F. Clark, 1972; Davison, 1980; Knowlton, 1984), island gray foxes, *Urocyon littoralis* (Crooks, 1994), kit foxes (Cypher & Spencer, 1998), felids (Rolley, 1987), and some species of mustelids (Lindzey, 1971; Simms, 1979; Bjorge et al., 1981; King, 1981; Hein & Andelt, 1995). For weasels, the number of animals caught per trap-night appears to be linearly related to animal density (Caughley, 1977), but few experimental tests have been conducted for other carnivore species.

Capture–mark–recapture

A technique originally developed with small mammals and proving useful for estimating carnivore populations is capture–mark–recapture. While mark–recapture is fairly time consuming, labor intensive, and costly, it does provide a reliable estimate of population size (i.e., absolute abundance) for many carnivore species, including badger (Messick & Hornocker, 1981), ringtail (*Bassariscus astutus*) and coati (Kaufman, 1987), mustelids (Bailey, 1971; King & Edgar, 1977; Messick & Hornocker, 1981; Douglas & Strickland, 1987; Rosatte, 1987; Strickland & Douglas, 1987), bears (Pelton *et al.*, 1978; DeMaster *et al.*, 1980; Miller & Ballard, 1982; Kruuk, 1995; Miller *et al.*, 1997), canids (F. Clark, 1972; Todd *et al.*, 1981; Roemer *et al.*, 1994), felids (Schaller, 1972; Currier *et al.*, 1977; Mills *et al.*, 1978; Miller, 1980; Quinn & Parker, 1987), hyenas (Kruuk, 1972b; Sillero-Zubiri & Gottelli, 1993), and raccoons (Sanderson, 1951b). Mark–recapture can provide relatively accurate estimates of population size if sample sizes are adequate, data collection techniques are unbiased, and the basic assumptions for the population estimator are not violated (see Caughley, 1977; Wilson *et al.*, 1996; or Thompson *et al.*, 1998; and references therein for assumptions of various estimators). This method involves capturing and marking individuals, then recapturing a number of the marked individuals again and estimating population size based upon the ratio of marked to unmarked animals recaptured using one of several models (Pollock, 1981; Seber, 1982; Montgomery, 1987).

Marks employed to tag the animal include ear tags, radio collars, dyes, and physiological markers such as radioactive isotopes. 'Recapture' may involve actual physical recapture of the animal, resighting of the animal (Smuts, 1976; Todd *et al.*, 1981; Miller *et al.*, 1997), returns from trappers or hunters (Sanderson, 1951b), recapture via fecal analysis for a physiological marker, or a combination of these (e.g., Currier *et al.*, 1977). Kohn *et al.* (1999) estimated coyote population size by identifying individual animals through fecal DNA analysis combined with mark–recapture methodology. Several different models for population estimation (e.g., Petersen, Jolly–Seber, Schnabel) can then be used to calculate population size (Caughley, 1977; Jolly, 1982; Seber, 1982; Thompson *et al.*, 1998). Many of these models are now available on software for use on a computer (e.g., programs CAPTURE by White *et al.*, 1982; NOREMARK by White, 1996; EAGLES by Arnason *et al.*, 1991). If the area of interest or trapping effort is known, then density estimates can be derived. Researchers should review capture–recapture methodologies outlined by Caughley (1977) or Thompson *et al.* (1998) to assist in the study design prior to implementation. Various trapping designs have been used with mark–recapture estimators. Roemer *et al.* (1994) used a trapping grid to estimate population size of island gray foxes. A trapping web design was used to estimate numbers of Indian mongooses, *Herpestes javanicus* (Corn & Conroy, 1998). F. Clark (1972) captured and marked coyote pups at dens in the spring then recaptured them during late-summer trapping sessions.

The use of physiological markers has received increased interest as a means of marking animals and then using 'recaptures' of those marks to estimate animal abundance with mark–recapture estimators. The method involves capture of the animal, injection or oral dosing of the animal, then resampling the animal at a later date either by direct recapture and blood sampling, collection of labeled scats, or examination of hunter killed animals. Radioactive isotopes have been used to determine densities of black bears and other carnivores (Pelton & Marcum, 1977; Kruuk et al., 1980). Radioactive zinc has been used to estimate the density of European badgers by injecting the captured individuals, then detecting the isotope in feces and estimating the population size from the ratio of radiolabeled to normal feces (Kruuk et al., 1980; Kruuk & Parrish, 1982). Kruuk et al. (1993) used radioactive isotopes to mark otter spraints and then identify which otter deposited that spraint. With the added responsibility and permitting needed to handle and store radioisotopes, researchers have examined other compounds to serve as individual markers for carnivores. Knowlton et al. (1988) reported that oral doses of iophenoxic acid were detectable or traceable in coyotes up to 16 weeks post-ingestion. Johnston et al. (1998) tested the use of chlorinated benzenes as physiological markers for coyotes and found that injection or ingestion (oral dose) of some compounds were detectable up to 100 days later in feces and blood serum. Biomarkers have been used to estimate animal abundance in canids (Davison, 1980; Knowlton, 1984), mustelids (Kruuk et al., 1980; Kruuk & Parrish, 1982; Knaus et al., 1983; Melquist & Dronkert, 1987), raccoons (Conner et al., 1983; Conner & Labisky, 1985), and bears (Pelton & Marcum, 1977).

Direct counts by removal

For some species of furbearers, most often species that are considered pests, the removal method has been used to estimate animal abundance. The method has been used to estimate population size mainly on skunks (Skalski et al., 1984; Rosatte, 1987) and raccoons (Twichell & Dill, 1949; Fountain, 1975). Disadvantages of this technique is the lack of knowledge of what proportion of the population was missed or not captured, and how large an area was affected by the removal. Due to the economic importance of the furbearer species, intrinsic values, and/or the social and political ramifications, the removal method is rarely employed.

Transect, strip, or area sampling

In certain circumstances it may be possible for the biologist to directly count the number of animals along transects, strips, in quadrants, or within a defined area and estimate animal population size or density (Gates,

1979; Burnham *et al.*, 1980; Rao *et al.*, 1981; Bibby *et al.*, 1992a). While transect and quadrant surveys are commonly used for estimating populations of ungulates, some of the larger carnivores may be surveyed with this technique. Trends in relative abundance can be compared from direct counts; absolute abundance may be estimated if correction factors are available to account for problems with sightability (Samuel *et al.*, 1987). Population estimates can also be calculated by distance methods along line-transects (Burnham *et al.*, 1980). Software programs that will estimate population size using distance data along transects include DISTANCE (Buckland *et al.*, 1993; Laake *et al.*, 1993) and TRANSECT (Burnham *et al.*, 1980). Aerial surveys typically require a large carnivore occupying a relatively sparsely vegetated habitat that allows for maximum sightability. Aerial surveys have been used to estimate animal abundance of coyotes (Nellis & Keith, 1976; Todd *et al.*, 1981), brown bears, *Ursus arctos* (Erickson & Siniff, 1963), and polar bears, *Ursus maritimus* (Scott *et al.*, 1959; Prevett & Kolenosky, 1982). Air and ship censuses of polar bears have been conducted during the summer when bears are concentrated along the polar ice pack (Larsen, 1972).

The number of animals sighted can be affected by weather, vegetation, visibility, and observer experience and fatigue. Miller & Russell (1977) compared aerial transect-strip counts and ground counts of wolves and reported that the behavior of the animals, width of the survey strip, and visibility all contributed to unreliable estimates of wolves using aerial surveys over open tundra habitat. The use of ultraviolet, infrared, or thermal imagery photography has been proposed for enhancing sightability of polar bears (Lavigne & Øritsland, 1974) and cougars (Havens & Sharps, 1998) during aerial surveys. Ground surveys are practical for smaller carnivores or animals that can be readily viewed in open habitats. Population trends for coati were measured by making visual counts along walked transects (Kaufman, 1987). Hyenas were sampled by ground transects in Africa (Hanby & Bygott, 1979). In certain situations, the entire area of interest may be surveyed, and through repeated sampling and reobservation, the entire population may be counted. For example, the wolves on Isle Royale have been observed and counted for decades, with each wolf pack counted on the island each winter (Jordan *et al.*, 1967; Wolfe & Allen, 1973; Peterson *et al.*, 1998). However, the ability to count all individuals in a defined area is a rare circumstance, but correction factors from a radio-marked sample can be used for determining a more accurate estimation of population size.

Identification of individual animals

While the opportunity to directly observe carnivores may be considered rare, there are certain species living in national parks or reserves with open habitats that allow for direct observation and identification of all individuals in the study area. This technique has been used successfully in studies of large carnivores in Africa. Biologists studying African lions have been able to identify all individuals by using sketches and photographs so that all lions found could be positively reidentified by a combination of ear notches, vibrissae spot patterns, and other natural features (Pennycuick & Rudnai, 1970; Bretram, 1975; Hanby & Bygott, 1979). Similarly, identification of individual hyenas by distinct spot patterns, scars, and ear notches (East & Hofer, 1991) has been used to determine population size (Hofer & East, 1995). Throat patches have been used to identify individual European otters (Watt, 1993). Individual coyotes in Yellowstone National Park were identified through radio collars, ear tags, and unique phenotypic characteristics. Observation of these animals permitted determination of pack size, and hence population size (Gese et al., 1996a). Maddock & Mills (1993) censused African wild dogs by collecting photographs from tourists and other field personnel. They were able to identify 357 wild dogs from 26 packs by examining over 5000 photographs.

Common to studies using identification of individuals is relatively open habitat and a carnivore species that is readily observable and generally tolerant of human presence. In fact, the animals do not necessarily need to be marked for individual identification, as individuals may be resighted and identified indirectly. Track characteristics of cougars has been used in which tracks of individual animals were separated on the basis of characteristics and location. These individual tracks were then combined to provide a density estimate (Koford, 1976; Ackerman et al., 1981; Van Dyke et al., 1986; Smallwood & Fitzhugh, 1993). The main advantage of using characteristics of individual tracks for identification was that it entailed less effort than a large-scale trapping program, but the accuracy of this method in relation to changes in population size remains untested (Lindzey, 1987). While individual identification allows for a relatively complete count of animals, the time and effort for this type of monitoring avails itself only to particular situations and is often conducted in conjunction with behavior studies (e.g., East & Hofer, 1991; Gese et al., 1996a). Another method that is receiving increasing attention is the use of hairsnares to acquire hair samples from carnivores, then using DNA sequencing to identify individuals in the population (Foran et al., 1997b; Paxinos et al., 1997; Kohn et al., 1999).

Radiotelemetry

With the introduction of radiotelemetry back in the 1960s (Cochran & Lord, 1963), the ability to monitor secretive carnivores increased tremendously. This method allows researchers to estimate the home-range size or territory size of an animal. Combining territory size (and overlap) with the number of members of the social unit or pack, plus the percentage of radio-collared transients sampled from the population, density estimates can be derived for the population in question. Because canids tend to be highly social with well-defined territories, radiotelemetry is now widely accepted as a method to measure population size and density (e.g., Mech, 1973a; Fritts & Mech, 1981; Fuller, 1989; Gese et al., 1989). For more solitary carnivores, estimates of home-range size, the extent of inter- and intrasexual home-range overlap, and the proportion of transients in the population are used to estimate population density. This method has been used for felids (Hornocker, 1970; Seidensticker et al., 1973; Hemker, 1982; Rolley, 1987; Quinn & Parker, 1987), mustelids (Melquist & Hornocker, 1979; Hornocker & Hash, 1981; Magoun, 1985; Douglas & Strickland, 1987; Strickland & Douglas, 1987), ringtails and coatis (Lanning, 1976; Trapp, 1978; Russell, 1979; Lacy, 1983), and bears (Kolenosky, 1987). While radiotelemetry is very labor intensive and costly, this technique provides one of the best and reliable estimates of population density for many carnivores. Long-term studies using radiotelemetry provide the most reliable annual estimates of population density for several secretive, far-ranging, low-density carnivores, such as cougars (Hornocker, 1970; Seidensticker et al., 1973; Hemker, 1982), wolverine (Magoun, 1985), and lynx, Lynx lynx (Quinn & Parker, 1987). With the advent of satellite and GPS technology, more intensive monitoring of large and medium-sized carnivores will be possible (Ballard et al., 1998; Merrill et al., 1998), but systems for smaller carnivore species will require further technological development.

MONITORING ANIMAL POPULATION DEMOGRAPHICS

The previously described methodologies provide information on how a carnivore population may be doing numerically, but do not necessarily answer questions of why the population trajectory is up, down, or stationary. In order to do this, one must know the rates of survival, fecundity, immigration, and emigration that influences the persistence of a carnivore population. Thus, in this section I will attempt to summarize the important features that one may need to measure in order to understand these important demographic processes. There are entire books devoted to the analysis of animal population dynamics (e.g., Caughley, 1977; Royama, 1992), therefore I will not go into detail of the mathematics involved. Because most of the actual techniques used to measure survival, fecundity, immigration, and emigration are species specific, for the scope of this chapter I will only provide a listing of the various measures one may want to monitor. I strongly recommend that readers embarking on a study of carnivore population dynamics consult Caughley (1977), Royama (1992), Thompson *et al.* (1998), and White & Garrott (1990) during the design and planning stages of studies so as to maximize their effort in collecting the proper data needed for demographic analyses.

Fecundity

The fecundity rate of a female is the number of offspring she produces over an interval of time (Caughley, 1977). Measuring fecundity or reproduction is fairly involved and time consuming. However, there are several basic questions dealing with fecundity that biologists may wish to ask: (1) when does the breeding season start and how long does it last, both in terms of estrous and gestation?; (2) when are the young born?; (3) what proportion of the females in the population breed?; (4) how many young are produced?; (5) is there one (monestrous) or multiple (polyestrous) breeding seasons in a year?; (6) what is the sex ratio at birth?; and (7) what is the age of first reproduction? There are various techniques to answer these questions. For carnivores, collection of carcasses, recovery of tagged animals, and observations in the field or captivity may address some of the questions. More specifically, examination of ovaries (corpora lutea counts) and placental scar counts from recovered animals or hunter killed animals, the ratio of juveniles to females in harvest counts, and/or observation of litter size in the field will give some measure of reproductive output (e.g., age-specific fecundity). Behavioral observations of animals in the field or captivity, physical examination, or tissue histology may provide information on initiation and cessation of the breeding season, and age of first breeding or sexual maturity.

Survival

Measuring the survival rates of carnivores usually involves construction of a life table or estimation of survival from radiotelemetry data. Pertinent questions a biologist may consider when designing a study to address survival rates are: (1) what is the number of deaths in each age interval?; (2) what is the probability of dying in each age interval?; (3) does mortality vary between seasons?; and (4) what are the causes of mortality? Ages from animals collected from hunters and trappers can be used to construct life tables. Caughley (1977) presents detailed information on various models for life-table construction and survival analysis. Measuring radio-days and numbers of deaths during defined time intervals derived from radio-collared animals can be used to calculate daily and interval survival rates (Trent & Rongstad, 1974; Heisey & Fuller, 1985). Application and assumptions of various survival estimators using radio-collared animals is covered thoroughly in White & Garrott (1990). Popular software programs that will estimate survival rates include SURVIV (White, 1983) and MICROMORT (Heisey & Fuller, 1985). The statistical package SAS (SAS Institute Inc.) will also calculate survival rates using the Kaplan–Meier product limit estimator (White & Garrott, 1990).

Immigration and emigration

Measuring emigration and immigration from a carnivore population usually involves the capture and tagging of several individuals and the subsequent recapture or radio-tracking of those individuals. Monitoring of the movements of animals out of a marked population (e.g., dispersal) is a simpler task that monitoring movements into the population, because biologists can not predict where immigration will occur from outside the known study population. Thus, biologists typically assume that the rate of movement out (egress) of their study area is equal to the rate of ingress. This assumption is usually violated, particularly if one of the populations is receiving control or some form of management. Whether the population being studied is maintained as a source or sink is pertinent to understanding the system and carnivore population in question.

DISEASE MONITORING

A subject often overlooked when monitoring carnivores is the question pertaining to the role of diseases in population dynamics. With an increasing interface between carnivores and humans and their pets, livestock, and expanding development, the possibility of disease transmission continues to escalate. Rare or endangered carnivores exposed to disease agents can have dire consequences. Canine distemper caused a rapid decline in black-footed ferret numbers and almost caused the species to become extinct (Williams et al., 1988). Similarly, rabies has been implicated in the decline of African wild dogs (Woodroffe & Ginsberg, 1997b). Biologists beginning a study should investigate the possible need for a disease monitoring program and handling protocol (for animals and samples collected), especially if dealing with a plan to reintroduce a species, or a rapidly declining carnivore population. Physical examination of living animals, blood collection for serological analysis, and post-mortem examinations of animals collected from trappers or hunters and recovery of telemetered animals can be used in a disease monitoring program. Consultations with wildlife veterinarians affiliated to a diagnostic lab or university are recommended to recognize which diseases should be screened for and then design an appropriate monitoring program.

MODELS

Computer simulations have been used to model carnivore population dynamics. Models which take into account different levels or rates of demographic variables, such as survival, fecundity, age structure, etc. have been used to develop computer simulations of population trends of various carnivore species, including coyotes (Connolly, 1978; Sterling et al., 1983), river otters (Tabor & Wight, 1977; Mowbray et al., 1979), polar bears (Stirling et al., 1976), and black bears (Lindzey & Meslow, 1980). These models can then be used to simulate the population response when one or more demographic variables is manipulated. The use of simple population models has now expanded into more sophisticated models and software programs (e.g., VORTEX by Lacy, 1993b). Population viability analysis (PVA) and population and habitat viability assessment (PHVA) has been used to evaluate the outcomes of various management actions, environmental perturbations, and stochastic events on the population viability of a species over a predetermined period of time (Shaffer, 1981; Boyce, 1992; Reed et al., 1998) using life-history data in relation to environmental factors (e.g., Shaffer, 1983). Biologists using these models should consider the 'realism' of these models. A PVA or PHVA is only a model and may not actually reflect or predict population persistence, and thus should not be the primary tool for developing a conservation plan. Macdonald et al. (1998a) recommended that PVAs appear to be most useful to biologists by guiding management actions and identifying practical monitoring methods. The accuracy of the data inputted into the model, levels of uncertainty, as well as the sensitivity of the model should be evaluated (Reed et al., 1998). Some PVAs and PHVAs may actually be best used to raise questions and formulate hypotheses for future testing (Macdonald et al., 1998a; Reed et al., 1998).

SUMMARY

In closing, as with all of the techniques mentioned above, personnel utilizing these methodologies should seriously consider what questions need to be answered before starting a monitoring program. Careful thought and planning will save headaches down the road. Logistical, political, ethical, social, and economic considerations should be included in the planning process. Noninvasive techniques are becoming more prevalent for monitoring carnivore populations and will continue to be important for monitoring rare, threatened, and endangered species, particularly when capture and handling could jeopardize the health and welfare of a species. I encourage anyone planning to initiate a monitoring program to talk to other researchers and gain their insight into what works and what does not work. Regrettably, techniques that fail are usually not published. Often times, field personnel have valuable knowledge about particular aspects of carnivore monitoring that are not readily available in the published literature.

Acknowledgements

I greatly appreciate the helpful reviews of the manuscript by Ann Kitchen, Fred Knowlton, Russ Mason, and four anonymous reviewers, plus input and discussions with my graduate students and colleagues.

PART 3

Prospects for research and conservation

18

Changing landscapes: consequences for carnivores

MELVIN E. SUNQUIST AND FIONA SUNQUIST

INTRODUCTION

As we enter the twenty-first century, human activities have become so pervasive that there are few areas of the globe that have not been affected in some way by them. Mosaics of agriculture, urbanization, logged forest, and degraded land dominate much of the earth. This landscape conversion has not been random. The most accessible and productive habitats, such as floodplains, savannas, and riverine forests were the first to be developed. Conversely, forest remnants that exist today are often in steep, inaccessible, and less productive areas. For example, the remaining forested land on the island of Sumatra is typical of this selective conversion (Whitten et al., 1987). The consequences of habitat change are often chronicled in the changing geographic distributions of numerous species, and the effects of forest fragmentation are well documented for many vertebrate species (Malcolm, 1988; Rylands & Keuroghlian, 1988; Schwarzkopf & Rylands, 1989; Laurance, 1991; Lawton & Woodroffe, 1991; van Apeldoorn et al., 1992). In general, whether the effects are positive or negative depends on the species and the scale of change.

Many mammalian carnivores possess characteristics that may make them particularly vulnerable to landscape changes. As predators, carnivores tend to live at relatively low densities, occupy large home ranges, are long-lived, have low reproductive output, and often disperse long distances as juveniles or subadults. However, all species do not respond in the same way to landscape changes. To a wide-ranging species capable of traversing openings in a mix of habitat patches, the landscape may appear connected, whereas for another species the openings may function as barriers to movement and in this case the landscape appears to be unconnected. For both species, however, there is likely to be a threshold or critical scale where the landscape 'unravels' and movements become greatly restricted.

Similarly, habitat loss for some may be a gain for others. Habitat, as used in this chapter, is not simply vegetation but also includes the interactive components of food, cover, water and their spatial attributes (e.g. amount, dimension, juxtaposition). Interior species and habitat specialists, for example, may be impacted immediately by minor habitat changes; however, ecological generalists and edge-loving species are likely to prosper under intermediate levels of fragmentation.

In thinking about the issue of changing landscapes several questions come to mind: (1) are the landscape changes occurring today unprecedented?; (2) what types of habitat changes are occurring and what is the scale of change?; (3) how have these landscape changes affected carnivore populations and can we measure the impact?; (4) are some species predictably more sensitive to habitat change than others?; (5) what information is needed to assess the problem?; and (6) how can the problem be solved?

It is important to remember that natural landscapes are dynamic systems – habitats change continuously over time even in the absence of people. Humans do not create environmental change where none existed before, rather 'we are speeding up, magnifying and altering the nature of change' (Lawton, 1997). Whether the landscape is the Grand Canyon, the Scottish moors, or Kruger National Park, it was different 500 years ago and probably unrecognizable 5000 years ago. Landscape variation in space and time is a normal and essential part of evolution and the maintenance of biodiversity.

In this chapter we will explore these questions in an attempt to clarify the problems and issues; present selected case studies to illustrate specific responses to landscape changes; and then suggest what types of future research are needed to address the shortcomings in our understanding of the problems.

LANDSCAPE CHANGE
Prehistoric

Eighteen thousand years ago, ice covered the northern parts of Europe and Asia, and nearly all of Canada and northern United States (Pielou, 1991). By the end of the Pleistocene, some 10 000 years ago, rising global temperatures and altered patterns of precipitation triggered a dramatic transformation of the world's landscapes. These massive habitat changes were followed by a major loss of species as dozens of mammalian genera went extinct (Webb, 1984). In North America alone, between 35 and 40 species of large mammals disappeared, including carnivores like the dire wolf

(*Canis dirus*), giant short-faced bear (*Arctodus simus*), American cheetah (*Acinonyx trumani*), American lion (*Panthera leo atrox*), and sabre-toothed cats (*Smilodon* sp.). Other carnivores such as the gray wolf (*C. lupus*), coyote (*C. latrans*), grizzly bear (*Ursus arctos*), black bear (*U. americanus*), puma (*Puma concolor*) and wolverine (*Gulo gulo*) survived (Kurtén & Anderson, 1980; Pielou, 1991). Pollen records from many different sites in North America indicate that the habitat change was abrupt and extensive. Across a vast area, from Minnesota to Nova Scotia, spruce (*Picea* sp.) was replaced by a more diverse forest of pine (*Pinus* sp.), balsam fir (*Abies balsamea*) and hardwoods. At some sites the changes happened in less than a human lifetime (Watts, 1983).

Historical

Early humans may have given the North American Pleistocene fauna a nudge towards extinction but large-scale anthropogenic changes in the world's landscapes really began about 500 years ago (Thomas, 1956; Turner et al., 1990). Today there are massive levels of deforestation in the tropics, but we should be aware that similar levels of change occurred in the world's grasslands and temperate forests only a few hundred years ago. One of the most dramatic invasions of people and livestock occurred in the sixteenth Century when the colonists first loosed their domestic livestock upon the Americas. With the new ungulates came a new flora. The trampling and cropping of native grasses by millions of imported grazing animals combined with the practice of burning radically altered the species composition of the grasslands. Today, only a quarter of the plants growing wild in the Argentine pampas are native and in some areas the natural ground cover consists almost entirely of Old World grasses and clover (Crosby, 1994). Well-documented vegetation changes from that time period are rare, but historian Elinor Melville has shown that between 1560 and 1580, overgrazing by some four million sheep transformed the fertile oak and pine forests of the Valle del Mezquital, north of Mexico City to mesquite dominated scrubland (Melville, 1994).

Dramatic changes in landscapes were inevitable as some 50 million Europeans migrated to the New World between 1750 and 1930 (Crosby, 1994). In eastern North America colonists cleared forests for agriculture, lumber, and fuel wood. A typical New England household consumed 30 or 40 cords of firewood a year, which required cutting an acre of forest. When the effects of such burning is calculated for the whole colonial period, it is likely that New England colonists consumed more than 260 million cords of firewood between 1630 and 1800 (Cronon, 1983), and 6.5 to 8.7 million acres of forest was cut to meet this demand. As the forest disappeared so did the wolf, puma, black bear, and smaller carnivores. Similarly, the brown bear, wolf, and lynx (*Lynx lynx*) were extirpated from many of the forests of Europe by the seventeenth and eighteenth centuries. Deforestation and depletion of wild ungulate prey were largely responsible for the disappearance of these carnivores, but direct persecution was certainly a contributing factor (Breitenmoser 1998a).

Current

Human-caused habitat alteration has accelerated recently and the general consensus is that increasing rates of habitat degradation, loss, and fragmentation, coupled with the ecological effects of isolation and patch dynamics, are largely responsible for increasing the rate of species decline and endangerment (Laurence & Bierregaard, 1997). The world total of cultivated land is thought to have increased by 466% from 1700 to 1980, during which time an estimated 120 million km^2 were brought into cultivation (Richards, 1990). With few exceptions, most prime land for agriculture has been consumed, and agriculture is now expanding into more fragile, marginal areas (Meyer & Turner, 1992).

Raven (1988) reported, based on Food and Agriculture Organization (FAO) statistics, that about one-half (46%) of the world's tropical forests had been destroyed by the late 1970s, with 1.1% of the remaining forests being clear-cut annually. During the decade of the 1980s, FAO estimates indicate an average of 15.4 million hectares of tropical forest were destroyed each year and another 5.6 million hectares were selectively logged, which means 1.2% of the world's tropical forests were destroyed or degraded each year (Whitmore, 1997). These figures do not represent the full impact because no allowance is made for the effects of forest fragmentation. Almost half (48%) of the deforestation in the 1980s occurred in the Americas, with the remainder split evenly between Asia and Africa. Rates of forest loss were highest in Asia, followed by the Americas and Africa (Whitmore, 1997).

TYPES OF HABITAT CHANGE
Degradation

Landscapes across the globe are gradually degraded by a variety of activities such as logging, livestock grazing, firewood collection, and hunting and trapping of prey species. This largely unmeasured phenomenon affects carnivores because declines in habitat quality are associated with increases in home-range size, decreases in density, and increases in energy expenditure associated with rearing young (Gittleman & Harvey, 1982). Because some species may be more sensitive than others to humans and their activities, habitat degradation is relative. However, habitat degradation almost always results in habitat loss.

Loss

Most habitat loss is an absolute subtraction of functioning habitat that results in population reductions. Habitat loss has reduced the geographic distributions of many carnivore species (Schreiber *et al.*, 1989; Ginsberg & Macdonald, 1990; Nowell & Jackson, 1996). In addition, habitat loss has not been random, more productive habitats have been eliminated first, resulting in many species being forced into habitats of lower quality where survival probabilities are reduced. As more suitable habitat is lost population declines may exceed predictions because of the combined effects of isolation and loss of area (Gardner *et al.*, 1987; Pearson & Gardner, 1997). The threshold at which this occurs will vary because of species differences in willingness or ability to cross open or patchy landscapes (R. V. O'Neill *et al.*, 1988; With & Crist, 1995; Keitt *et al.*, 1997; Andrén, 1999; Mönkkönen & Reunanen, 1999). In this regard, those species that can use or traverse the matrix between the patches will be less affected. The matrix is the habitat type (e.g. agricultural field) in which a forest patch or patches are embedded.

Fragmentation

The effects of fragmentation are more subtle than those related to outright loss. Fragmentation of habitats not only results in a total loss of area and increased isolation of the remaining smaller patches, but it also introduces a suite of other potential problems. At the individual level these changes may limit movement between habitat fragments, alter home-range boundaries, modify habitat selection patterns, limit social interactions, and increase predation rates. These same changes may also isolate populations by reducing habitat connectivity, lowering dispersal success, alter reproductive and survival probabilities, increase the likelihood of inbreeding, and result in local extinction or population declines. Small isolated populations are particularly vulnerable since they have a high probability of extinction due to demographic and environmental stochasticity (Fahrig, 1997).

Habitat fragmentation may divide a population into spatially separated subpopulations, which together may form a metapopulation (Levins, 1970; Harrison, 1994). Such a model predicts a threshold in the proportion of suitable habitat below which the metapopulation cannot persist, although at what threshold of disturbance the habitat is no longer suitable will vary. In this regard, the dispersal capabilities of species and the nature of the matrix are particularly relevant.

Similarly, fragmentation may leave habitat patches that function as sources, where survivorship and natality are such that emigrants are produced, whereas other patches are sinks, where local populations cannot be maintained without continued immigration (Pulliam, 1988). The spatial arrangement of patches in the landscape, including their relative proximity to other patches and boundary effects can greatly affect whether a patch functions as a source or a sink (Dunning *et al.*, 1992).

Furthermore, fragmentation can change the contextual setting of the landscape as new habitat patches may be created in a matrix that is no longer familiar or suitable to the previous occupants. Because fragmentation increases the amount of edge, it may also degrade the quality of the remaining habitat for interior species (Yahner, 1988; Murcia, 1995; Fahrig, 1997) and at the same time improve foraging conditions for omnivores or carnivores that are habitat generalists (Andrén, 1995).

While there are numerous models available for analysis of spatially explicit data, theory has far outstripped empirical data. Furthermore, there is a problem teasing apart the effects of habitat loss and fragmentation. The generally negative effects of decreasing area and increasing isolation are often interpreted as fragmentation effects when more likely they are part of the negative effects of large scale habitat loss (Fahrig, 1997).

Unfortunately, there are few carnivore species for which the appropriate responses have been measured; the vast majority of studies that have examined the effects of habitat fragmentation and loss involve birds, small mammals, amphibians, butterflies, and plants (Andrén, 1994; Brown & Hutchings, 1997; Tocher *et al.*, 1997; Benitez-Malvido, 1998; Demaynadier & Hunter, 1998; Keyser *et al.*, 1998; Laurance *et al.*, 1998). A few studies do, however, illustrate some of the effects of habitat degradation, loss, and fragmentation on mammalian carnivores.

CASE STUDIES
Habitat degradation and tiger population viability
Habitat loss is responsible for population declines of tigers (*Panthera tigris*) (Schaller, 1967; Panwar, 1987; Thapar, 1992), however, recent assessments (Wikramanayake *et al.*, 1998) show that extensive areas of potential habitat exist. Surveys in some of these areas indicate that prey populations are severely depleted (Karanth, 1991; Redford, 1992; Rabinowitz, 1993). Be-

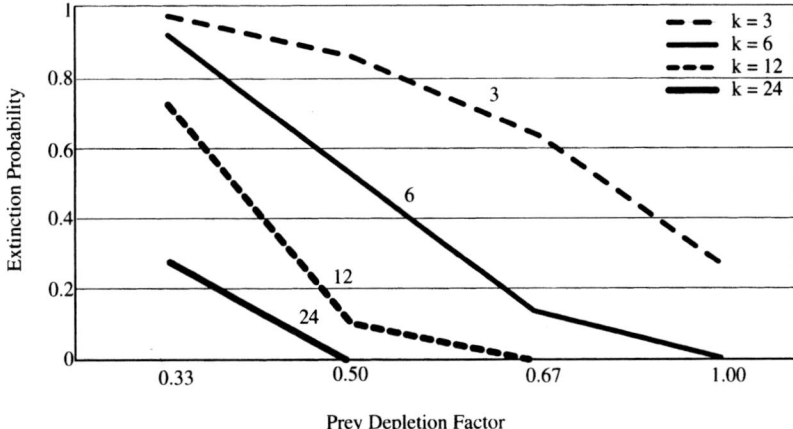

Figure 18.1. Probabilities of extinction for tiger populations at different carrying capacities for breeding females (k = 3 to k = 24), affected by different degrees of prey depletion (1.00 represents no prey depletion). Extinction probabilities increase with increasing degrees of prey depletion. (Redrawn from Karanth & Stith, 1999, by permission of Cambridge University Press.)

cause prey depletion lowers encounter rates with prey, tigers must travel farther to find prey, hunting success rates will be lower, and energy expenditure per kill will be higher. These ecological consequences will affect tiger condition, increase the possibility of intraspecific competition at kills, and reduce the time females can spend with their cubs. In effect, prey depletion is likely to affect survival of tigers in all demographic stages (e.g. breeder, transient, juvenile, cub).

Karanth & Stith (1999) developed a stochastic demographic (stage-based) model to evaluate the relative importance of both tiger poaching and prey depletion in reducing tiger populations. Their simulations show that prey depletion has a strong effect on tiger populations by reducing the carrying capacity for breeding females, decreasing cub survival, and decreasing population size. On the positive side, their results suggest that tigers may persist in relatively small reserves, even with low levels of poaching, provided their prey-base is protected and maintained at an adequate density (Figure 18.1).

Habitat loss and fragmentation: the importance of scale to martens
Throughout North America, marten (*Martes americana*) are generally associated with mature, closed-canopy conifer forest. However, in Maine, marten are found in broad-leaf/coniferous or mixed forest (Buskirk & Powell, 1994). Studies of habitat selection by marten in Utah, Maine, and

Newfoundland emphasize the importance of examining multiple scales (i.e. microhabitat, stand, home range, landscape) to understand the effects of fragmentation (Bissonette et al., 1989, 1997; Brainerd, 1990; Hargis & Bissonette, 1997; Chapin et al., 1998; Hargis et al., 1999).

At the *microhabitat scale*, marten in Utah, Newfoundland and Maine prefer habitats that are structurally complex, with trees providing escape or protective cover from predators (Hargis & McCullough, 1984), and coarse woody debris on the ground providing both thermal cover from cold (Buskirk et al., 1989), and cover for small mammals, the marten's main prey (Sherburne & Bissonette, 1994). Similar observations have been reported for other marten species (Brainerd, 1990; Bissonette & Broekhuizen, 1995; Zalewski et al., 1995; Kurki et al., 1998).

At the *stand scale*, marten in Utah and Newfoundland avoided second-growth forests and recent clearcuts, using residual coniferous stands regardless of size (Synder & Bissonette, 1987). In Maine, marten used coniferous, deciduous, mixed, mature, and regenerating forests in proportion to their availability; they did not select for specific cover types (Katnik, 1992).

At the *home-range scale*, marten in Utah and Newfoundland used mature forest, which was the only type available. While marten in Maine used all forest types, the proportion of mature forest within their home ranges was significantly greater than what was available, whereas the proportion of regenerating clearcuts and recent cuts was used less than what was available (Phillips, 1994; Katnik, 1992).

At the *landscape scale*, marten in both Utah and Maine show an avoidance of landscapes where the proportion of intact mature forest was less than 70% (Bissonette, 1997). At this level of fragmentation marten can still traverse the landscape and avoid openings (Hargis et al., 1999). Based on percolation theory (Stauffer & Aharony, 1992), which considers both the spatial aggregation and connectedness of habitat patches, marten numbers were predicted to decline with increasing loss of mature forest cover until about 55% to 65% of forest remained (Figure 18.2). According to the theory, at this point the forest matrix changes to disturbed habitat and will not be used by marten. That the critical proportion of mature forest for marten was 70% suggests they were responding not only to the loss of habitat but also to the spatio-temporal pattern of habitat patches remaining in the landscape.

While marten in both Utah and Maine responded similarly to the loss and fragmentation of habitat, the shape of the response curves (Figure 18.2) suggest different properties may be operating at each site (Bissonette et al., 1997). In Utah, marten density declined as soon as forestry operations be-

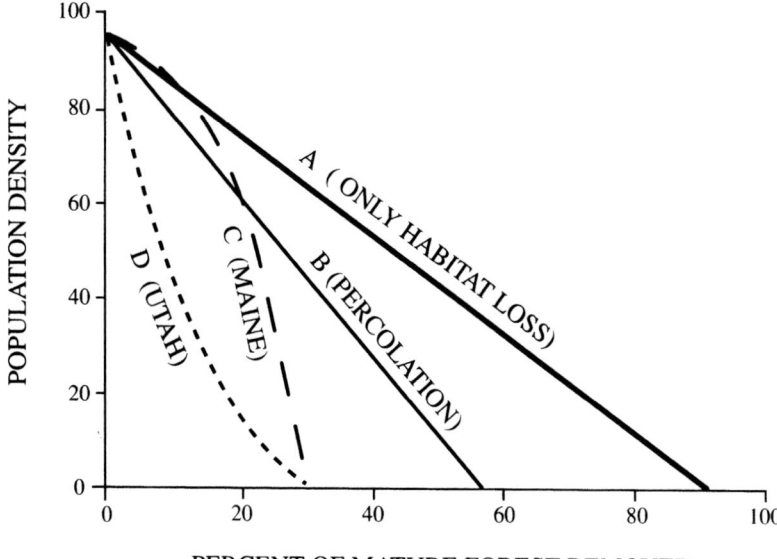

Figure 18.2. Reponses by American marten to habitat fragmentation. Line A: response if influenced only by habitat loss; line B: response to loss of habitat connectedness; lines C and D: actual responses to increasing habitat fragmentation in Maine and Utah. (Redrawn from Bissonette *et al.*, 1997, by permission of Springer-Verlag.)

gan because clearcuts and other openings generally function as nonhabitat, and thus any additional openings in the landscape represent a direct and immediate loss. In addition, as more habitat is fragmented the suitability of remaining tracts may also be reduced due to increased edge and area effects. In Maine, however, low levels of fragmentation (intermediate disturbance) may actually benefit marten by increasing access to alternate foods, but once a threshold of habitat fragmentation is reached the population declines exponentially (Bissonette, 1997).

At the landscape scale marten are clearly sensitive to changes caused by forestry practices, as forestry operations alter the size, isolation, and quality of remaining forest patches (Sturtevant *et al.*, 1996; Chapin *et al.*, 1998). Elimination of large tracts of mature forest will negatively impact marten populations far beyond what would have been predicted based simply on the animal's response to microhabitat or stand characteristics. The impact is magnified because marten respond not only to increasing levels of habitat loss but also to the spatio-temporal pattern of habitat patches remaining in the landscape (Bissonnette, 1997).

The marten study illustrates the difficulty of predicting critical thresholds for a particular species. In the marten's case the critical value of original remnant habitat was higher than that predicted from theoretical models. The bird and mammal data reviewed by Andrén (1994) suggest that the average threshold value was 10–30% of original habitat, but this assumes that species were responding only to habitat loss. The difficulty in determining the threshold is that the value is not simply a function of percentage of original habitat remaining, but that the species are also responding to fragmentation effects (fragment size, distance between fragments, matrix effects) which can compound the effects of pure habitat loss (Fahrig, 1997; Andrén, 1999; Mönkkönen & Reunanen, 1999).

Dispersal and pumas in southern California

In mammals, dispersal is the movement of independent young, predominately males, from a natal range to an unoccupied site where the animal settles and breeds (Greenwood, 1980; Shields, 1987). The process takes time and is fraught with risks; probably half of all dispersers will die before settling and reproducing. Dispersing young must be able to find enough food to sustain themselves during this phase of their lives, and in this regard the availability of suitable habitats in the landscape is critically important.

Carnivores often travel long distances during dispersal. Radio-collared and tagged puma, wolf, bear, lynx (*Lynx canadensis*), and red fox (*Vulpes vulpes*) have dispersed 100 to 500 km (Storm et al., 1976; Fritts & Mech, 1981; Anderson, 1983; Rogers, 1987; Koehler & Aubry, 1994), and snow leopards (*Uncia uncia*) have crossed 150 km of open steppe or desert while moving between isolated mountain ranges (McCarthy & Munkhtsog, 1997). Similarly, the appearance of young male jaguars (*Panthera onca*) and tigers at locations many kilometers from the nearest known population confirms their dispersal abilities (Heptner & Sludskii, 1992; Glen, 1996).

Beier (1995) monitored the dispersal movements of eight male and one female juvenile puma in the Santa Ana Mountains of southern California. The landscape was a mixture of national forest land, a military reservation, several smaller reserves, 17 cities, and seven unincorporated towns surrounding the mountain range. Three forested corridors (1.5 to 6 km long) connected the larger parcels of puma habitat.

Dispersal lasted for a few weeks to several months, and all cats encountered the urban–wildlife interface. All dispersers were located at least once within 100 m of urban areas and heavily used parklands, and some cats walked through residential areas at night. Puma crossed unlit openings that were 400 m wide, but only where native vegetation was present in the surrounding area. Such crossings were not made in urban areas.

Pumas used forested corridors, but only at night. Some corridors were only 400 m wide and narrowed to 3.3 m at road undercrossings. The cats regularly crossed under highway bridges built over watercourses but culverts under roads were usually avoided. Seven of nine dispersers died before establishing a permanent home range. Three deaths were due to collisions with vehicles, another was shot by police, one died of an intestinal diseases, and two deaths were probably natural causes. Nevertheless, the study shows that pumas will disperse via habitat corridors in an urbanized landscape and some will even use corridors with such unnatural features as golf courses and freeways.

Having documented that dispersing puma can find and use corridors, Beier (1993, 1996) then developed a metapopulation model to assess the viability of a subpopulation linked to adjacent populations via several levels of immigration. His model shows, as expected, that both size of area and the presence of an immigration corridor influenced the probability of a population going extinct. In the absence of a corridor, the critical size habitat lies between 1000 and 2200 km². Depending on carrying capacity, survival rates and other demographic parameters, habitat areas of 1000 to 2200 km² would hold 15 to 20 adult puma. These numbers are too small to preserve genetic variation over several centuries, although in the short-term (< 100 years) there is a low risk of extinction assuming demographic conditions remain unchanged.

The presence of an immigration corridor greatly improved the prognosis. Even at low levels of immigration (as low as one male per decade) the size of area needed was reduced by 20–27%. At higher levels of immigration, the probability of even small populations persisting was markedly improved. Thus maintaining or restoring corridors is a critical measure in areas where isolation or fragmentation of a population is imminent.

Isolation effects: large carnivores on habitat islands

Habitat fragmentation and loss can eventually isolate a population. The consequences will depend on the size of the population, the amount of genetic variation contained within the population, the length of isolation, the type of social system, the size of area and quality of the habitat, and other environmental variables.

The Florida panther is currently restricted to the Big Cypress Swamp and Everglades ecosystems, an area of about 8800 km² in southern Florida surrounded by water and human developments (Belden et al., 1997; Maehr, 1990). This relict population of 50–70 individuals is known to have low levels of genetic variation (O'Brien *et al.*, 1990), and produces animals with congenital abnormalities, which are presumably due to close inbreeding. A variety of physiological impairments such as spermatozoa defects, crypto-rchidism, and cardiac abnormalities are documented in the population (Roelke *et al.*, 1993a; Barone *et al.*, 1994; Cunningham *et al.*, 1999). Despite these problems, reproduction and survival are at levels seen in western pumas and demographically the population appears to be stable (Maehr & Caddick, 1995).

Some mortality in the panther population is human related; between 1986 and 1991 an average of five animals per year died or were killed, with 40% of the mortality associated with human activities such as road kills and illegal hunting (Roelke *et al.*, 1993a). However, since the early 1990s the greatest source of mortality has been intraspecific strife (Maehr *et al.*, 1991; Maehr, 1997). In this true island-like setting, young males cannot escape the confines of the local population and thus are repeatedly forced back into an established social setting, where they may be killed by resident males. 'Frustrated dispersal', as it is called (Gaines *et al.*, 1991; Lidicker, 1995), also forces dispersal-aged males into inhospitable and marginal fringe habitats where they are at high risk due to external factors.

While the evolutionary and environmental cards appear to be stacked against the panther, Maehr (1997) has long argued that the demographics of this population are in fact no different from other nonhunted puma populations in the western United States. Adult turnover is low, home ranges are stable, and reproduction in the panther's core range (source habitat) is sufficient to compensate for the loss of juveniles in marginal habitats. The only demographic dysfunction that he sees is the lack of dispersal, although two or possibly three males have recently crossed the Caloosahatchee River in apparent colonizing events. The problem of genetic impoverishment is currently being addressed by the introduction of female pumas from Texas into the south Florida population (but see Maehr & Caddick, 1995).

Wolves on Isle Royale in Lake Superior also show low levels of genetic variation (Wayne *et al.*, 1991). A single pair of wolves is thought to have crossed the ice from the Ontario mainland to the island in about 1949; wolf numbers peaked at 50 in 1980 and over the next three years plummeted to 12 individuals (Mech, 1966; Peterson, 1977; Peterson & Page, 1988). Molecular analyses indicate that about 50% of allozyme variability had been lost in the island population over 40 years of isolation. Low levels of genetic variability are also reported for brown bears on Kodiak Island (Paetkau *et al.*, 1998a). The thriving population, which is currently estimated at 2800 bears (Barnes *et al.*, 1995), has been isolated for about 10 000 years.

With carnivores, it appears that most small or declining populations are threatened primarily by habitat loss or human activities such as prey depletion, hunting, and highway mortality, rather than the consequences of inbreeding. Despite the fact that bear numbers on Kodiak Island are well below estimates required for long-term persistence, they have survived in isolation for more than 10 000 years, a time span that would satisfy most conservation plans. While loss of genetic diversity and inbreeding are potential problems for carnivores (Lacy, 1993a; L. Mills & Smouse, 1994), they pale in comparison to other factors. Absolute loss of habitat and general habitat degradation is likely to have a much greater impact (Fahrig, 1997). Limited available data suggest that carnivores such as pumas, bears, tigers, and lynx can survive in small (< 50 individuals) isolated populations provided that habitat and prey are protected and a low level of genetic exchange occurs. The level of exchange can be as low as one male per decade (Beier, 1996) or one per generation (Ralls *et al.*, 1985).

Matrix characteristics and opportunities for carnivore generalists

Island biogeography theory (MacArthur & Wilson, 1967) has been used to predict the effects of fragmentation on the colonization and extinction of species in forest patches and other island-like situations (Figure 18.3). The heuristic value of this approach is lessened by the assumption of a uniform and hostile environment between the patches and that individuals of each species use only one habitat patch. However, in landscapes where the matrix is not a vast expanse of concrete or an ocean, forest fragments are not truly isolated because many species can utilize and move through the surrounding matrix. Obviously, a matrix can function as a barrier to one species but be incorporated into the home range of another. Logged forest will be unusable to a margay (*Leopardus wiedii*), but provide new foraging opportunities for a crab-eating raccoon (*Procyon cancrivorus*).

Black bears are thought to have a high tolerance for natural habitat fragmentation (Hellgren & Maehr, 1992), and a study of black bears in northeastern Louisiana (Anderson, 1997) illustrates how an ecological generalist can survive and thrive in a highly fragmented habitat, if the matrix is 'friendly'. The 350 km² study area, originally bottomland hardwood forest, has been reduced to four major forest fragments totaling less than 35 km². These fragments are embedded in a matrix of intensively farmed corn and wheat fields. Wooded strips, 4–8 km long, link the forest tracts (Figure 18.4). Population density in the fragments is 1.9 bears/km², the highest density known for the species. Female home ranges are small and overlap

Figure 18.3. Island-like forest fragment embedded in an agricultural matrix.
(Photo courtesy of D. Maehr.)

extensively; females rarely move from one fragment to another. Males move between fragments through forested corridors and across fields. With low mortality and high reproduction, the population is thought to be stable or increasing. With little human activity in the area, the greatest threat to this population is that the remaining forest patches will be logged.

The landscape mosaics created by habitat fragmentation are well suited to the needs of habitat generalists such as fox, coyote (*Canis latrans*), raccoon (*Procyon lotor*), badger (*Taxidea taxus*) , and skunk (*Mephitis mephitis*). Fragmentation studies of birds and reptiles frequently refer to increased predation on edges (Patton, 1994; Yahner, 1988; Andrén, 1995) and there is little doubt that carnivore habitat generalists prosper in fragmented landscapes. There are, for example, an estimated 15 to 20 times more raccoons in the United States today than there were 60 years ago (Sanderson, 1987). Similarly, coyotes have recently expanded their range into the eastern and southeastern United States, reaching northwestern Florida in the 1970s. Over a 20-year period the coyote went from occupying a few counties in northern Florida to occupying nearly every county in the state (Brady & Campell, 1983; Wooding & Hardisky, 1990; Maehr *et al.*, 1996; Coates *et al.*, 1998). The numbers of several other carnivores, particularly the more omnivorous and smaller-bodied species, have also reportedly increased in Florida with increasing habitat fragmentation (Harris & Gallagher, 1989).

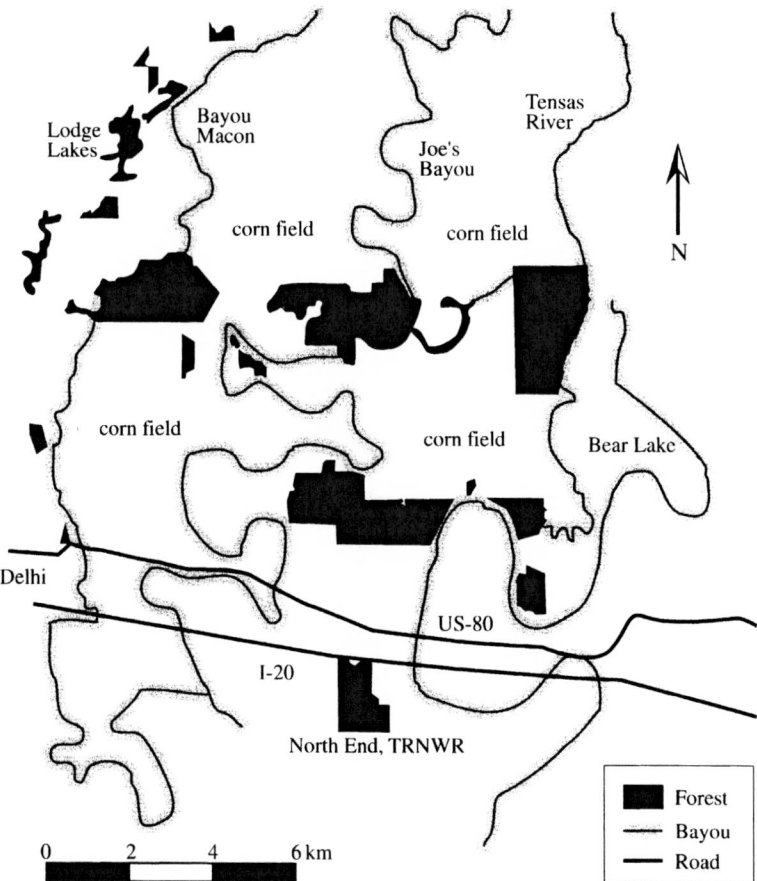

Figure 18.4. Black bear study site, Tensas River Basin, Louisiana. Bears live in remnant forest patches (black) in an agricultural matrix (corn fields) connected by wooded corridors along rivers (bayou). (Redrawn from Anderson 1997, by permission of author.)

In some cases, intermediate levels of disturbance can create favorable hunting conditions for larger carnivores, such as leopard (*Panthera pardus*), jaguar, and tiger, as edge habitats or habitat mosaics support higher densities of many ungulate prey species (Eisenberg & Seidensticker, 1976; Karanth & Sunquist, 1995). In many national parks in India, vegetation is cleared for 100 meters or so on either side of forest roads to enhance wildlife viewing for tourists. These cleared viewlines create more grazing habitat for ungulates like chital (*Axis axis*) and gaur (*Bos frontalis*). Karanth & Sunquist (1992) suggest that these man-made clearings enable the tropical moist and dry deciduous forest of Nagarahole National Park to support extremely high densities of large herbivores, which, in turn, support high carnivore densities.

In Laurence's (1991) study of rainforest mammals in Australia, a single trait, species' abundance in matrix habitats, was the best predictor of a species' vulnerability. Natural rarity was a misleading indicator of extinction likelihood, as was trophic level and body size. *K*-selected traits such as large size, low fecundity and high longevity were significant predictors of extinction disposition, but only for species that avoided matrix habitats (Laurance, 1991).

A matrix can be anything from degraded forest to urban development and as such can change quite rapidly. A population of puma may do quite well in an area fragmented by logging roads, but if the roads are opened to hunters, the result may be entirely different. A recently planted oil palm plantation may be unusable for both tigers and leopards, but a 30-year-old oil palm plantation may be quite suitable as a dispersal corridor for both cats.

Effects of roads and sources of carnivore mortality

In many ecosystems, roads and highways are one of the most obvious fragmenting forces; indeed their cumulative impact may exceed other types of fragmentation (Bennett, 1991; Reed *et al.*, 1996) and road densities continue to increase. In Florida, for example, the density of paved roads has increased from 0.11 km/km^2 in 1937 to 1.1 km/km^2 in 1995, and the volume of traffic on these roads also has increased substantially (D. Smith *et al.*, 1996). The increase in the number of black bears killed on the highways of Florida tracks the increase in vehicular traffic, from two recorded deaths in 1976 (Gilbert & Wooding, 1996) to 88 in 1998 (T. Eason, Florida Fish & Wildlife Conservation Commission, pers. comm.) (Figure 18.5). Heavily used roads and high-speed traffic can function as barriers to many wildlife species and collisions with vehicles are the major source of mortality for some carnivore populations. Because there are ways of mitigating the impact of roads on wildlife with fencing and underpasses, the topic has generated considerable interest (Evink *et al.*, 1996, 1998). Some carnivores will use corridors and underpasses (Beier, 1995; Foster & Humphrey, 1995) and the placement of underpasses at points where animals are known to cross highways has been shown to decrease road kills (Hellgren & Maehr, 1992; Foster & Humphrey, 1995).

Collisions with vehicles threaten to extirpate a small disjunct population of black bears in eastern North Carolina where vehicle-related mortality between 1988 and 1992 accounted for 71% of black bear deaths (Brandenburg, 1996). Roads were also the major source of mortality for

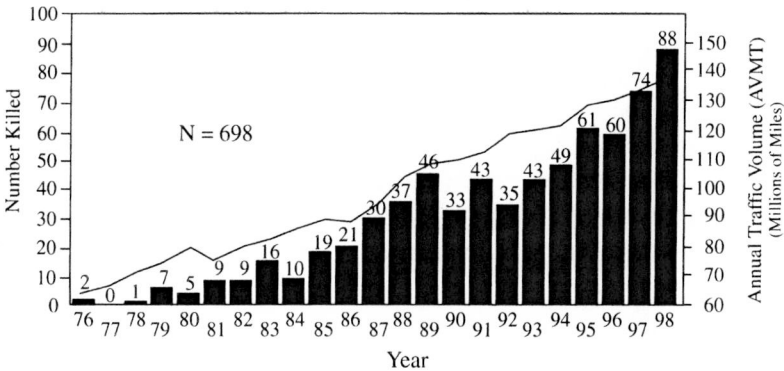

Figure 18.5. Black bear road kills in Florida from 1976 to 1998 (Source: Gilbert & Wooding, 1996, and T. Eason, Florida Fish & Wildlife Conservation Commission, pers. comm.). Traffic volume data (AVMT = annual vehicle miles traveled) from Florida Department of Transportation, Office of Statistics, Tallahassee, Florida.

Florida panthers, accounting for 49% of all documented panther deaths before 1990 (Maehr *et al.*, 1991). In southwestern Spain road traffic was the second most important cause of mortality for the endangered Iberian lynx (*Lynx pardinus*) (Ferreras *et al.*, 1992). A program to reintroduce lynx to New York's Adirondack Mountains was essentially foiled by high levels of highway mortality. Collisions with vehicles accounted for 37% of known deaths and the study concluded that road mortality was too high for reintroduction to be successful (Brocke *et al.*, 1990).

Badger (*Meles meles*) populations in Europe are also severely impacted by road mortality (Aaris-Sorensen, 1996; Wiertz, 1993; Lankester *et al.*, 1991; van der Zee *et al.*, 1992), and in the UK road traffic is the single largest cause of mortality for the species (Chesseman *et al.*, 1987; Davies *et al.*, 1987; Skinner *et al.*, 1991; Harris *et al.*, 1992; Clarke *et al.*, 1998). Highway mortality also affects wildlife in relatively undeveloped areas; road accidents are the largest known cause of mortality for wolves in a portion of the Canadian Rockies (Paquet, unpub. data in Noss *et al.*, 1996). Similarly, in parts of Africa where wild dogs (*Lycaon pictus*) occupy areas with good roads used by fast-moving traffic, road accidents may be the single most important cause of adult mortality (Woodroffe *et al.*, 1997).

Roads and human activities also disturb and displace wildlife populations. Bears tend to avoid heavily traveled roads, and may alter their movement patterns or shift home ranges (Mattson *et al.*, 1987; Brody & Pelton, 1989; McLellan & Schackleton, 1988; Mace *et al.*, 1996; Clevenger *et al.*, 1997). Other data suggest that highways displace wolves and grizzly bears and they generally avoid crossing them (Paquet & Hackman, 1995). Gibeau & Heuer (1996) have noted that wolverine and other carnivore ranges tend to be laid out along highways rather than straddling them.

Roads also bring development and provide access to hunters. Indeed, road density is often used as a measure of habitat suitability for large carnivores. Road densities exceeding 0.58 km/km^2 have been suggested as a threshold for survival of wolf populations in Wisconsin, Michigan, Ontario and Minnesota (Thiel, 1985; Jensen et al., 1986; Mech et al., 1988b). However, high road density may not in itself be a problem, rather the main threat from roads is the access they provide to hunters and trappers. In the Rocky Mountains, grizzly bear deaths are almost entirely human caused and bear mortality is positively correlated with human access (Mattson et al., 1995; Mattson et al., 1996a; McLellan, 1990). Mortality rates for puma in Montana and black bears in Arizona were also higher in areas where roads provided hunters with access to even remote areas (Murphy, 1983; Mollohan & LeCount, 1989).

Roads are also an important factor in the commercialization of wildlife hunting. In central and west Africa, faunal declines, primarily through bushmeat hunting, poses a greater immediate threat to wildlife conservation than deforestation (Wilkie et al., 1998). Throughout Africa, road access facilitates the export of bushmeat and allows remote forests to be trapped and hunted to provide meat for urban areas (Bowen-Jones & Pendry, 1999).

CONCLUSIONS

Though major landscape changes and species losses have occurred in the past, such changes have accelerated to unprecedented levels. The geographical distribution of many mammalian carnivores, especially the larger species, have declined drastically over the last 500 years due primarily to a combination of human-related activities. Agriculture, urbanization, and deforestation have fragmented carnivore populations and have influenced their dynamics by changing the extent, structure, and spatial arrangements of landscape components. The responses of wildlife species to such landscape changes are of great concern to conservationists and ecologists throughout the world (Schonewald-Cox et al., 1991; Fahrig & Merriam, 1994; Clark et al., 1996b; McCullough, 1996; Smallwood, 1999).

How carnivores respond to changing landscapes obviously depends on the timing and scale of alterations and the species' inherent ability to adapt. Because hypercarnivores (strictly meat eaters) are sensitive to the distribution and abundance of prey any change in these parameters, natural or anthropogenic, will potentially reduce carnivore populations via energetic constraints and altered spatial patterns. Habitat degradation will also change the timing and spatial arrangement of food patches for omnivorous carnivores such as bears.

Small to mid-sized species that feed on a variety of plant and animal foods, and that use both the matrix and the interior of patches may not be negatively impacted by habitat fragmentation. In fact, their populations may actually increase with increasing amounts of edge, as witnessed by the increases in raccoon, coyote, fox, and skunk populations in fragmented habitats. There is also a suggestion that where landscape changes have caused the disappearance of large carnivores, the populations of medium-sized predators have increased (ecological release) (Sargeant et al., 1993; Terborgh et al., 1997).

Carnivores that are associated with interior forests, avoid openings, and are area-sensitive will be impacted negatively by both habitat fragmentation and loss. An analysis of patch size effect (Bender et al., 1998) predicts that, for interior species, the effects of habitat fragmentation per se will be greater than that predicted from just habitat loss. The study of American marten is a good example of this effect. The fisher (Martes pennanti) is also an 'area-sensitive' species and their populations are expected to decline as forested landscapes become more fragmented (Powell & Zielinski, 1994). Similar responses may occur with obligate forest dwellers such as marbled cat (Pardofelis marmorata), clouded leopard (Neofelis nebulosa), margay, black-legged mongoose (Bdeogale nigripes), Nilgiri marten (Martes gwatkinsi), Malabar civet (Viverra civettina) and others.

Though many carnivores have traits that allow them to live with the effects of habitat fragmentation, the fact that they occupy large home ranges, travel widely to gain access to resources, and disperse long distances increases their vulnerability to these same effects. It has recently been shown that the combined effects of roads, hunters, and conflicts with humans along reserve boundaries contribute more to the extinction of populations of large carnivores isolated in small reserves than do stochastic processes. Simply put, wide-ranging carnivores are more likely to become extinct than those with smaller home ranges, regardless of population size (Woodroffe & Ginsberg, 1998). Nonetheless, it is important to consider all of the crucial issues when planning and assessing carnivore strategies for all carnivore species.

FUTURE CONSIDERATIONS

The empirical foundation for evaluating the effects of habitat change on carnivores is weak and there are few studies (e.g. Beier, 1993, 1996;

Bissonette *et al.*, 1997) that even come close to documenting the effects of habitat fragmentation on carnivore populations. Problems of small sample size, low densities, long generation times, cryptic, nocturnal study animals and difficult, time consuming data collection have combined to discourage biologists from using carnivores as study animals. Replication and experimental controls are rarely possible when working with species whose densities are $3-5/100$ km², have home ranges of 100 km² or more, and have lifespans of $15-20$ years. It will be all but impossible to experimentally modify landscapes on the spatial and temporal scale that is required for most carnivores. Some studies (e.g. American marten) have taken advantage of forestry harvesting schedules to look closely at the effects of fragmentation at different scales, and more studies of this kind are needed.

We particularly need information on how various species use their habitats relative to different kinds of edges, matrix characteristics, size of openings, and whether both sexes show the same responses. We also need information on responses to disturbances of various kinds (e.g. roads, logging, urban development). We need to know how generalist and specialist carnivores respond to the same landscape changes.

With only a few exceptions, the topic of dispersal remains almost completely unstudied. We need information on timing, frequency, distances, and pathways. We also need to understand how individual movements and landscape patterns interact. With increasing fragmentation, dispersal may only be accomplished via corridors or underpasses; thus identifying, characterizing, and siting the locations of these conduits will be an important research topic. At the very least, corridors are paths of least resistance that permit an occasional successful interchange between populations (Maehr *et al.*, 1988).

Management strategies should focus on spatial attributes and the behavioral aspects of movements across heterogeneous landscapes. Carnivores need large areas and maximizing space is often touted as the solution to reducing mortality factors along reserve boundaries. Unfortunately, expansion is rarely a management option. An alternative way to increase the effective size of an area is to increase connectedness. Short of physically transporting individuals between habitat patches, most carnivore populations will not survive without connections in the landscape. Increasing the probability of carnivore survival in a human-dominated landscape can best be accomplished by the creation of a system of interconnected protected areas.

19

Behaviour of carnivores in exploited and controlled populations

LAURENCE G. FRANK AND ROSIE WOODROFFE

INTRODUCTION

For as long as humans and our ancestors have existed, we have been in conflict with carnivores, especially the larger ones. The great cats, bears, wolves, and hyenas have preyed on primates since long before human beings evolved. As we tamed wildgame and became herders, conflict with predators took on a new form, as they found our increasingly helpless domestic livestock easy prey. Minimizing livestock losses to carnivores has always been one of the most serious problems confronting farmers and pastoralists.

Human beings have been killing large carnivores since their weapons were made of stone and bone (Cannon, 1995), but it took many hundreds of years to gradually eliminate large predators from western Europe. However, the technology of poison, traps, and firearms eradicated large carnivores from much of eastern North America in two centuries, and organized predator control programs eliminated them from most of the American West in a few decades. Today, large carnivores of most species face imminent extinction in Asia and have been drastically reduced in much of Africa.

Human exploitation, however, may have less dramatic effects than extinction. In this chapter, we outline the various forms that carnivore control and harvest may take, and review the impact of such disturbance upon the behavior of individuals that survive. In our discussions, we consider any deliberate large scale killing of carnivores to be exploitation. This includes predator control programs, persecution of 'nuisance' species, trapping for fur, sport hunting, and poaching. Thus, while our examples primarily involve legally hunted populations, our conclusions may also apply to illegal persecution or poaching of threatened species. By comparing species' responses to such exploitation, we identify the behavioral and ecological traits which allow some species to coexist successfully with humans, while others are more sensitive to disturbance.

Table 19.1. Summary of reasons for control of carnivore populations

Purpose	Examples
Protection of human life	'Problem' grizzly bears killed in Yellowstone National Park
Protection of livestock from depredation	Lions, hyenas, leopards, and cheetahs killed to reduce predation on cattle and sheep
	Jaguars killed to reduce predation on cattle
	Coyotes killed to reduce depredation on sheep
Protection of preferred or endangered wild populations	Gray wolves killed to protect moose and caribou in Canada and Alaska
	Lions and hyenas killed to promote antelope recovery in Kruger National Park
	Red foxes killed to increase bags of game birds
	Coyotes killed to protect endangered kit foxes
Disease control	Red foxes killed to limit spread of rabies
	European badgers killed to limit transmission of bovine tuberculosis
Sport	Grizzly bears shot for trophies
	Lions shot for trophies
	Red foxes hunted with hounds
Fur and other body parts	Canada lynx trapped for fur
	Bears killed for use in traditional Chinese medicine

WHY ARE CARNIVORE POPULATIONS CONTROLLED AND EXPLOITED?

Carnivores may be killed for a variety of reasons. In contrast with other large mammals, most carnivores are killed to remove a real or perceived nuisance, rather than for consumption. Thus, carnivores are controlled to prevent them from killing people, livestock, or other wildlife, or to avoid the transmission of infectious diseases (Table 19.1).

While there are often clear justifications for killing carnivores, human perception plays an important role in determining the effort with which persecution is pursued. For instance, on commercial ranches in Kenya, the spotted hyena (*Crocuta crocuta*) has the smallest *per capita* impact on livestock of any large predator (Figure 19.1), and damage by hyenas is largely preventable through livestock management (Frank, 1998b). The big cats, by comparison, kill more stock and those depredations are more difficult to prevent. However, most white ranchers admire the cats and avoid killing

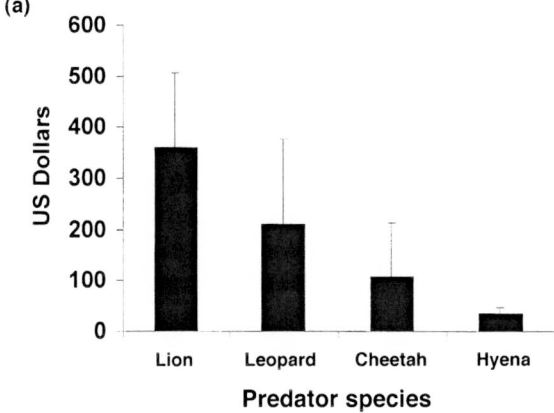

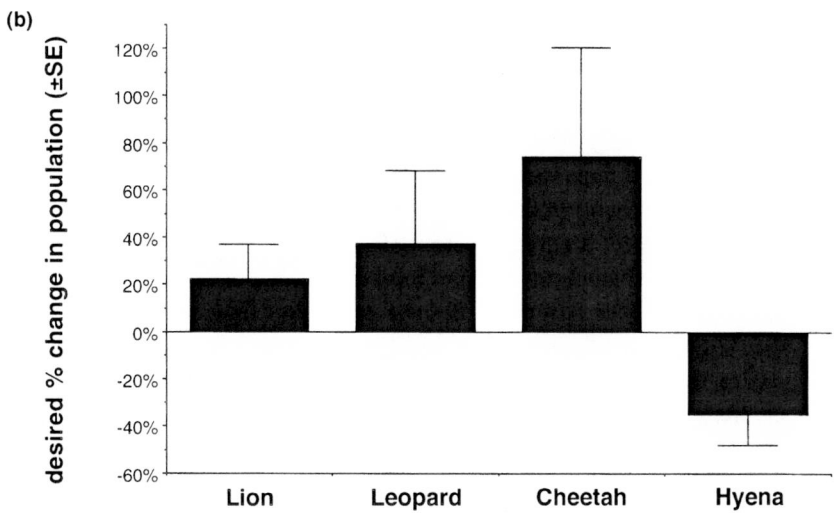

Figure 19.1. (a) Cost of maintaining large carnivores on commercial livestock ranches in Laikipia District Kenya, indicating that spotted hyenas do the least damage of all the larger predators. Data from Frank (1998a). (b) Human perceptions of the larger predators on commercial livestock ranches in Laikipia District, Kenya. White farmers (n = 12) were asked if they would like to have more or fewer of each species on their properties. Although the cats do more livestock damage than hyenas, people would like to have more cats and fewer hyenas. When asked why, most respondents mentioned negative images based on the perceived moral value of the animals. African pastoralists were generally much less positive in their attitudes toward predators. Data from Frank (1998a).

them when possible, while expressing antipathy for hyenas and sometimes shooting them on sight, even in the absence of serious depredation problems. These ranchers would like to see greater populations of lions, leopards, and cheetahs, but fewer or no hyenas (Frank, 1998b).

Predator control may also be instigated to increase populations of wild prey species that are more highly valued. In both Alaska and Canada, for instance, highly controversial large scale wolf (*Canis lupus*) culls are carried out in order to increase populations of moose (*Alces alces*) and caribou (*Rangifer tarandus*) for human hunters (Gasaway *et al.*, 1983; Boertje *et al.*, 1995; Stephenson *et al.*, 1995). In the belief that predators reduce populations of more 'desirable' (i.e. large ungulate) species, even national parks have been subject to intense predator control programs. For example, in the early part of the twentieth century Yellowstone National Park in the United States had a policy of exterminating predators (Phillips & Smith, 1996), while South Africa's Kruger National Park culled lions (*Panthera leo*) and spotted hyenas as recently as the mid-1970s (Smuts, 1978). In the same period, spotted hyenas were eradicated from Nairobi National Park in Kenya (J. Rudnai, pers. comm.). African wild dogs (*Lycaon pictus*) were shot in Niger's national parks until 1979 (Fanshawe *et al.*, 1991). Astonishingly, 'protected' wild dogs were shot by rangers in one of east Africa's premier game parks in 1999 (Anon, pers. comm.).

These control programs were undertaken in an effort to increase ungulates even where these were not hunted by humans, reflecting the notion that some species inherently deserve protection from others. The threat that large carnivores pose to human beings, livestock, and other wildlife makes them seem dangerous quarry; in many societies, killing large dangerous animals is highly prestigious. Perhaps for this reason trophy hunting of large predaotrs is a lucrative business, responsible for a high proportion of mortality in some populations (e.g. Miller, 1990; Hoogesteijn *et al.*, 1993; Creel & Creel, 1997). Sport hunting may also involve smaller carnivores such as red foxes (*Vulpes vulpes*) and coyotes (*Canis latrans*).

Other uses are more consumptive. People have trapped carnivores for their fur for thousands of years. More recently, however, demand has increased for other carnivore body parts (e.g. bear gall bladders and tiger bones) for use in traditional East Asian medicine. The resultant trade is an important cause of mortality for some species (e.g. Kumar & Wright, 1999; Servheen *et al.*, 1999). Finally, carnivores may be culled because they are reservoirs for zoonotic diseases (e.g. Macdonald, 1980; Krebs *et al.*, 1997).

THE SCALE OF CARNIVORE CONTROL PROGRAMS

Scenes on Egyptian tombs of pharaohs hunting lions, and medieval tapestries of European nobles hunting lions, bears (*Ursus arctos*) and wolves suggest that organized, large scale hunts probably have long been important in eliminating large carnivores from major geographic areas. Today, organized group sport hunting still goes on in the form of fox hunting in the UK, and large scale coyote 'roundups' in the American Southwest.

State-sponsored predator control is very old – laws promoting the persecution of carnivores date back over a thousand years. For example, the Emperor Charlemagne founded a dedicated wolf-hunting corps around 800 AD (Boitani, 1995), and laws have encouraged persecution of arctic foxes (*Alopex lagopus*) in Iceland since 1295 (Ginsberg & Macdonald, 1990).

In recent times, organized carnivore control programs by government agencies have been very important in many countries. These may be very intensive and highly effective: in 1965–66, over 10 000 red foxes were killed as part of organized control programs in Scotland (Hewson & Kolb, 1973). In the single year of 1964, the state of Kansas paid bounties on 52 000 coyotes (*Canis latrans*, Gier, 1968). In 1972, the Leopold Commission examined the history of federal predator control in the western United States over the period 1937 to 1970 (Cain *et al.*, 1972). These efforts, which continue today, are intended to protect the livestock industry from depredation losses. In that period, federal agents killed over 50 000 red wolves (*Canis rufus*) and nearly 24 000 grizzly and black bears (*Ursus arctos* and *U. americanus*), driving the wolf and grizzly bear to virtual extinction in the contiguous 48 states. An astonishing 2.8 million coyotes were killed by federal efforts alone (Figure 19.2). In spite of that unprecedented effort, coyotes not only thrived but dramatically expanded their range (Hilton, 1978) in response to the elimination of wolves, their primary predators and major competitors (Crabtree & Sheldon, 1999), and conversion of forest to open agricultural land. These numbers demonstrate the intensity at which predators can be killed through organized efforts (keeping in mind that they do not include the large but unknown number of animals killed by state agencies and private individuals), but also reflect the dramatic variability in species' response to large scale mortality.

Control has also occurred on a remarkably large scale within protected areas. Lions are the most sought-after wildlife among tourists in Africa (Western & Henry, 1979), but in the 1950s in South Africa's Kruger Park, 450 were killed in a five-year period, along with numerous other 'charismatic' predators (Figure 19.3).

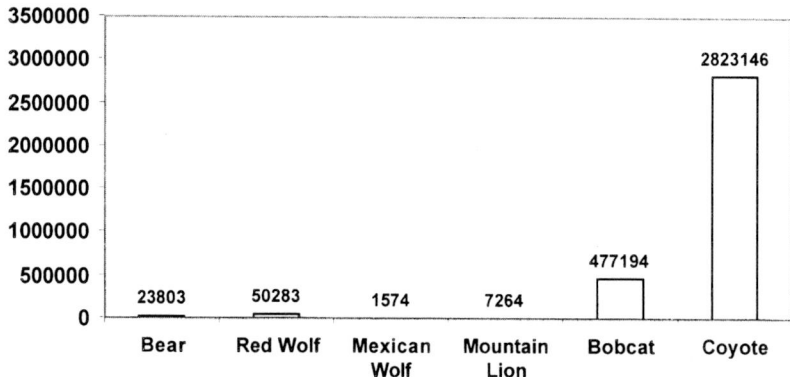

Figure 19.2. Number of large- and medium-sized carnivores killed by federal predator control programs in the western US, 1937–1970. These numbers do not reflect animals killed by state agencies or private individuals. Data from Cain *et al.* (1972).

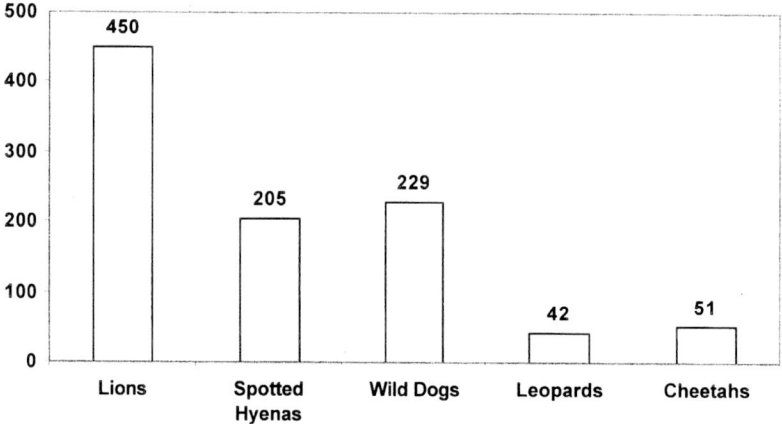

Figure 19.3. Numbers of large carnivores killed by rangers in Kruger National Park, South Africa, in the years 1954–58. Data from Smuts (1976).

BIOLOGICAL EFFECTS OF PREDATOR EXPLOITATION AND CONTROL

There have been many studies to determine whether control programs meet their objectives (reviewed in Harris & Smith, 1987; Ginsberg & Macdonald, 1990; Boertje *et al.*, 1995). In contrast, there has been surprisingly little research on less obvious effects on the behavior and ecology of target populations. Given the concern over dwindling populations of predators and other wildlife, it is important to examine aspects of the natural behavior of animals that make them more or less vulnerable to local or global extinction through control efforts (Woodroffe, Chapter 4).

Today, few control programs bring about local population extinction. Nevertheless, they may have marked effects upon behavior and population dynamics. Even solitary carnivores live in social systems comprising long-term stable neighborhoods, while group-living species are highly dependent upon the integrity and stability of those groups. Even when exploitation does not eliminate or drastically reduce populations, high mortality rates may disrupt social stability, inducing high rates of turnover. Such instability can have dramatic effects on animal movements, dispersal, reproductive rates, genetic relatedness, and offspring and adult sex ratios. These effects may even influence the efficacy of control itself. For example, stable populations of older individuals, that may have a low rate of killing livestock, may be replaced with younger inexperienced individuals that are more likely to kill stock (Frank, 1998b). Likewise, control aimed at managing disease may, through its effects on breeding and dispersal, promote disease persistence and spread (Swinton *et al.*, 1997).

How does exploitation affect populations?

Biologically, organized control programs, sport hunting, and trapping amount to increased rates of mortality. However, both the aims and methods of control will influence the effects on population density, structure, and distribution. For example, while trapping of coyotes to protect the sheep industry likely causes a simple reduction in density, control of European badgers (*Meles meles*) to limit tuberculosis spread to cattle leads to local eradication and 'holes' in the spatial structure of the population. Likewise, poisoning or trapping may cause preferential mortality of younger, less cautious animals, while digging litters out of dens may destroy entire cohorts. Trophy hunting has an entirely different effect, often concentrating on males, and so leading to female-biased populations. Each of these methods of exploitation and control target different sections of the population, and have different effects on behavior.

Where control leads to local reductions in population density, its effects are likely to resemble those of natural phenomena which place populations well below carrying capacity, such as disease outbreaks and sudden increases in food availability. Such events slacken density-dependent constraints on reproduction, ranging behavior, and dispersal, leading to marked changes in behavior and ecology (e.g. Lindström, 1992; Poole, 1994). Population reduction through exploitation is expected to have a similar effect. Exploited populations are *not* likely to resemble those which naturally occur at low densities such as striped hyenas *Hyaena hyaena,* or honey badgers *Meillvora capensis,* (Kruuk, 1976b; Kingdon, 1979); such populations will have a stable social structure without the flux generated by high mortality and superabundant food.

The great variation in carnivore social systems makes it difficult to generalize about the effects of persecution. Clearly, some species such as the red fox and coyote are able to persist in the face of chronic heavy mortality, while other species are more easily driven to local extinction (Woodroffe, Chapter 4).

Why do we know so little?

There has been very little research specifically on the behavioral effects of control programs or exlpoitation of carnivore populations. Social organization and the underlying behaviors are seen as largely of academic interest, without immediate bearing on the problem of reducing populations of problem species. Further, most work on behavioral ecology has been undertaken on protected populations, where interference by humans is minimal. Behavioral ecologists are interested in evolutionary explanations of behavior, and large scale killing programs disrupt normal behavior to such a degree that less drastic selective pressures are likely to be swamped. Further, it is simply harder to gather data on persecuted animals. Where controlled populations have elicited academic interest – for example, cyclic Canadian lynx (*Lynx canadensis*) populations tracked by export records of the Hudson Bay Company – control has been assumed to have no effect upon behavior or ecology (e.g. Stenseth *et al.*, 1998).

This reluctance to study disturbed populations has been justified. Early reports of coyote social systems were based on observations made on very heavily controlled populations, and suggested that coyotes were essentially solitary (Berg & Chesness, 1978). Later studies of protected populations (e.g. Camenzind, 1978; Crabtree & Sheldon, 1999) corrected this misinterpretation, showing a picture of a normal canid social system including territoriality, pair bonds and helpers at the den. Earlier studies were simply seeing the results of massive social disruption due to human-caused mortality.

Effects of exploitation on reproduction

Exploitation and control operations have the potential to influence carnivore reproduction through their effects on both population density and population structure. Lowering population density is likely to reduce density-dependent constraints on reproduction, allowing more rapid breeding. However, alterations to population age structure, social structure or sex ratios change the social environment in ways which might influence reproduction in more subtle ways.

Reproductive rates

There is comparatively little evidence to suggest that control influences carnivore litter sizes. Knowlton (1972) reported that coyotes had larger litter sizes under intensive control, but no such effects have been found in either red or arctic fox populations subject to control (Harris & Smith, 1987; Hersteinsson, 1992). Likewise, average litter size did not change in a grizzly bear population declining because of intense hunting (Miller, 1993). Another exploited grizzly bear population, however, appeared to show larger average litter sizes than nearby undisturbed populations (McLellan, 1989). This difference might also be caused by local variation in food availability (Miller, 1990).

While litter size changes relatively little under exploitation, individual females' probability of breeding is strongly influenced by control operations. In an urban red fox population subject to control, a higher proportion of females became pregnant, and fewer lost their litters around the time of birth, than in a similar, uncontrolled population (Harris & Smith, 1987). The chances of female wolves breeding also appear higher in heavily exploited populations. Packs usually produce a single litter in high-density populations, but Ballard *et al.* (1987) found that 7–10% of packs produced multiple litters in an exploited population (despite the fact that small pack size meant that relatively few packs contained more than one adult female). Van Ballenberghe *et al.* (1975) reported similar findings. In Kruger National Park, fecundity of female lions rose by 76% in the year after heavy control measures were instituted (Smuts, 1978).

In some species, however, control may have negative effects on females' reproductive rates. Miller (1993) found that the inter-birth intervals of female grizzly bears increased as population density declined due to hunting, with cubs being weaned markedly later. He attributed this change to alterations in population structure. Bear hunting selectively removes males, leading to increasingly female-biased populations. Miller (1993) suggested that females' lower reproductive rates – and greater investment in the cubs they did produce – might reflect a shortage of males in the population. As population density declined, Miller saw male–female consortships less often, and found that estrous females ranged more widely, suggesting that females were finding it more difficult to locate mates.

Low reproductive rates have also been reported from European badger populations subjected to intense control. While a small number of immigrant males and females rapidly re-occupied an area cleared by tuberculosis control operations, no breeding occurred in the first few years following the recolonization (Cheeseman *et al.*, 1993).

Infanticide

Carnivore control may also affect net reproductive rates through its impacts on infanticidal behavior by males. In many mammalian species, non-father males are a threat to juveniles (Hausfater & Hrdy, 1994). Because trophy hunting focuses on males, the resulting female-biased sex ratio may affect the behavior of males that remain. Depending on the species' social organization, both positive and negative effects have been postulated.

In North American black bears and grizzly bears, reductions in the proportion of males in the population have been linked to increased cub survival (Miller, 1990). Since males are occasionally infanticidal in both species, decreasing the number of males in the population might be expected to allow more cubs to survive; this logic has been used to suggest that hunting might be beneficial for bear populations (reviewed in Miller, 1990). However, removal of adult males from bear populations is known to cause immigration of younger males (Miller, 1990). Such disruption of male social relations might be expected to increase, rather than decrease, the threat to cubs fathered by the previous resident males. Such an effect has been proposed for brown bears in Sweden, where higher cub mortality was recorded in years when males were hunted than in years or areas where no hunting occurred (Swenson *et al.*, 1997). These authors argued that this effect was so strong that killing a single male had the same demographic impact as killing 0.5–1 females. However, the hunting in Sweden involved the killing of just five male bears over an 11-year period, in an area of 11 200 km² – an area larger than the home range of all but the most adventurous male bear (Blanchard & Knight, 1991). The fact that no similar effect has been documented in the many studies of bears in North America, where indeed, the opposite effects have been inferred (see above), suggests either that the social responses of bears to hunting are highly variable, or that Swenson *et al.*'s (1997) results might have some alternative explanation.

An effect of sport hunting on infanticide is also likely among lions. When new males take over a pride, they kill the existing cubs, which were sired by the previous pride males. In this way, they are able to increase the number of cubs they themselves father in the newly-acquired female groups (Packer & Pusey, 1983). While there are, as yet, no data to show that infanticide is more common in either controlled or hunted lion populations, the high male turnover likely to be associated with high male mortality may well create a social environment in which infanticide occurs frequently (Whitman & Packer, 1998).

Sex ratio effects

Under certain ecological or social circumstances, female mammals may produce more offspring of one sex than predicted by a 50:50 sex ratio (Clutton-Brock & Iason, 1986). Although the physiological or behavioral bases of this sex-ratio bias are not well understood, the evolutionary reasons are reasonably clear: females produce the sex that will maximize their own fitness in the social and ecological environment in which they find themselves (Frank, 1990). These evolutionary pressures mean that females should, where possible, produce offspring of the sex which is less common or more valuable in the population. In spotted hyenas, for instance, litter sex ratio responds rapidly to changes in the adult sex ratio (Holekamp & Smale, 1995; Frank, 1996)

Two studies suggest that hunting of male lions may cause females to produce more male cubs than would normally be expected (Creel & Creel, 1997; Whitman & Packer, 1998). In both studies, the overproduction of male cubs may be so great that the number of young females surviving to adulthood is reduced. This obviously produces lots of shootable males in the short term, but may cause severe population reduction in the longer term.

Home ranges and territoriality

Control operations can be expected to have profound effects upon ranging behavior and territoriality, through their effects on both population density and structure. In general, reduced population densities are expected to lead to larger home ranges and a less strict territorial structure (Sandell, 1989). This pattern has been seen among European badgers recolonizing an area cleared by tuberculosis control operations. Colonists ranged widely, using multiple den sites, and no territories were established until density had risen through breeding of the immigrant animals, several years after the initial recolonization (Cheeseman et al., 1993). Similarly overlapping ranges have been reported among lions and wolves recolonizing cleared areas (Smuts, 1978; Boyd et al., 1995).

Long-term lower-level control has similar effects. In a European badger population depressed by chronic persecution, home ranges were large and overlapping, and territorial scent marking reduced, (Sleeman & Mulcahy, 1993). Female bobcats (*Lynx rufus*) expand their home ranges into neighboring areas cleared by trapping, apparently to take advantage of high-quality habitat (Lovallo & Anderson, 1995); such behavior is likely to lead to larger home ranges in exploited populations. Female grizzly bears in populations made heavily female-biased by sport hunting ranged more widely, apparently in an attempt to locate mates (Miller, 1993).

While control is expected to allow expansion of home ranges as competition for food is reduced, it may occasionally trigger range contraction. Male lions typically have larger home ranges than females, overlapping the entire range of female groups (Schaller, 1972), but males in a sport-hunted population in the Luangwa Valley, Zambia, occupied smaller ranges than females (Yamazaki, 1996). While small, each of these ranges overlapped with those of two or more female groups, allowing males to mate with females from multiple kin groups. In undisturbed populations, male coalitions effectively exclude other males from their territories. Thus, in the female-biased Zambian population, there may also be greater opportunities for infanticide by non-father males.

Dispersal

In most mammal species, one or both sexes tend to disperse away from their natal home ranges, apparently to seek unoccupied habitat and unrelated mating partners (Greenwood, 1980). Such dispersal may allow populations to persist in the face of high mortality, or to recolonize cleared areas. Species in which at least some female juveniles disperse are likely to be more resilient than those in which females remain in their natal home range.

Dispersal within controlled and exploited populations

Increased mortality or reduced density may increase both the probability of dispersal and the distance traveled by dispersing individuals. For instance, Pils (1975) ear tagged red fox pups on a 65 km² area in Wisconsin (USA) in which there was heavy fox trapping. Only three out of 52 foxes killed in that area in subsequent years were tagged, while 27 tags were returned after dispersing an average of 23.2 km from the natal den. Virtually the entire cohort left the natal area, to be replaced by animals originating elsewhere. Effective control of such a population would require very wide scale population reduction, because local declines would rapidly be reversed by immigration of juveniles.

Allen & Sergeant (1993) reported on the results of a long-term study of

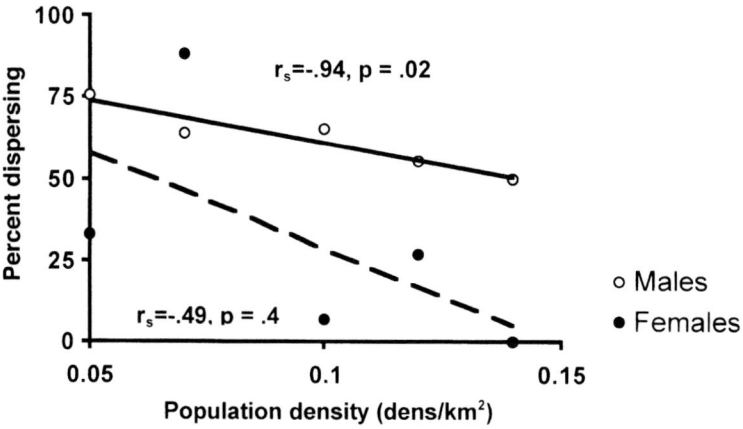

Figure 19.4. Increase of fox dispersal as a function of population density in Minnesota. As spring population decreases, more males disperse, and they travel longer distances, creating a tendency to equalize density over large areas. Data from Allan & Sargeant (1993).

red foxes in Minnesota (USA). Population density tripled over the course of the study, allowing them to examine dispersal as a function of density. For males, the probability of dispersing was inversely correlated with spring density – the lower the density, the more likely they were to disperse (Figure 19.4). Dispersal distance was as far as 302 km, with 8% moving over 80 km. The authors speculated that increased contact with conspecifics might inhibit dispersal at higher densities.

Woodroffe & Macdonald (1995) showed a similar relationship across undisturbed European badger populations, with males (but not females) more likely to disperse at low population densities. Dispersal distance was not affected – all males moved to neighboring territories. In a population highly disturbed by a tuberculosis control program, however, animals dispersed more frequently and moved farther than in natural populations (Sleeman, 1992).

On the other hand, social factors might serve to reduce dispersal in populations suffering heavy mortality. Gese et al. (1996a) showed that low ranking coyotes in an undisturbed population dispersed from the natal pack, while higher ranking siblings tended to remain with their parents. If human caused mortality is equally likely to affect high and lower ranking animals, the social gaps caused by control programs or trapping might allow lower ranking animals – which would otherwise disperse – to remain in the natal territory. Of course, if rank influences food availability, as in coyotes and spotted hyenas (Frank, 1986), different forms of mortality may tend to select for higher or lower ranking animals. Hungrier subordinate animals may be more likely to take risks with traps or poisoned baits than better fed dominants. In the absence of detailed information on a species' social system, it is rarely possible to predict the influence of excess mortality on dispersal.

Joining existing groups

As well as increasing the probability that individuals leave their natal groups, carnivore control may increase an animal's chance of joining an established group, as opposed to living alone or founding a new group. In unexploited populations, male lions may form single-sex coalitions with unrelated males but are extremely aggressive to incoming males once they have tenure of a group of females (Packer et al., 1988). In contrast, in a population made highly female-biased by sport hunting, males were able to join an established group with no resistance from the pride male (Yamazaki, 1996).

Similar results have been found among wolves. In unexploited populations, unrelated individuals are rarely accepted into wolf packs – and then only when a social role (such as breeding male or female) is vacant (Fritts & Mech, 1981). In contrast, in a harvested population, 22% of dispersers were able to join established packs (Ballard et al., 1987). Dispersers typically left large packs (mean: 7.6 wolves) and joined smaller ones (mean: 2.8 wolves). Since territory size correlates with pack size among wolves (Ballard et al., 1987), it is possible that small packs accepted unrelated immigrants to increase their access to resources.

Dispersal into cleared areas

Rather than causing an overall reduction in density, some forms of control create 'holes' in carnivores' spatial organization, leaving areas more or less completely unoccupied. Since such areas are likely to offer high food availability with few competitors, dispersing animals may be expected to reoccupy them rapidly. Such an effect was documented with lions in the Central District of Kruger National Park, where lions and hyenas were experimentally reduced in the 1970s (Smuts, 1978). Within 15 months, dispersing lions from unaffected areas quickly reestablished populations in the newly vacated habitat (Figure 19.5). Likewise, wolves rapidly re-colonized an area cleared by long-term control and persecution, once they were given legal protection (Fritts & Mech, 1981).

Curiously, some species respond relatively little to localized control in

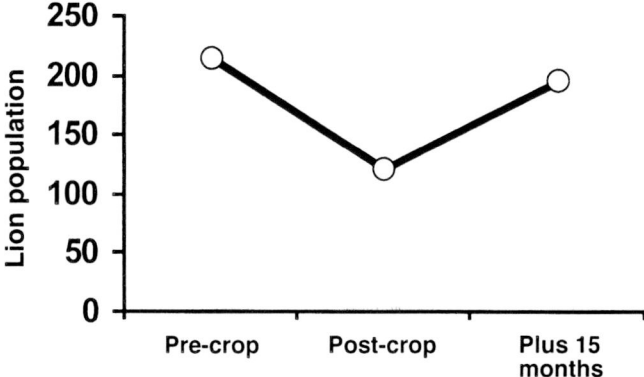

Figure 19.5. Lion population of the Central District of Kruger National Park before and after a culling program. Within 15 months, lion populations had recovered to pre-control levels, as a result of immigration and increased fecundity of survivors. Data from Smuts (1976).

neighboring areas. When part of a high-density badger population was cleared by tuberculosis control operations a few colonists moved into the cleared area, but neighboring social groups did not expand their territories into the vacant space (Cheeseman *et al.*, 1993). Likewise, spotted hyenas rarely alter their territory borders, even to exploit areas made vacant by predator control or persecution (Holekamp *et al.*, 1993; Smuts, 1978). It seems counterintuitive that highly opportunistic, highly vagile, generalist feeders should fail to occupy habitat vacuums. The distinctive social systems (Kruuk, 1972b; 1978) of these species is probably responsible for the inability of females to move around due to both social bonds at home and intolerance by other groups. Although this failure to disperse into newly available habitat may make such species highly vulnerable to local extinction, it also allows for effective local control.

Social organization

Control, in all its forms, may have profound effects upon social organization. Social structures evident in undisturbed populations may be poorly defined in populations drastically reduced in density through control (Berg & Chesness, 1978). For example, a high proportion of wolves recolonizing an area of the United States cleared by control could not be unequivocally assigned to a single pack; many were associated with more than one social unit, and some even contributed to pup care in two neighboring packs simultaneously (Boyd *et al.*, 1995). Social relations in controlled European badger populations are similarly confused, with group affiliations sometimes uncertain (Sleeman & Mulcahy, 1993). Preliminary results from our own study of lions, heavily persecuted for killing livestock, suggest that association patterns among females are more fluid than in either undisturbed or sport-hunted populations (Schaller, 1972; Yamazaki, 1996; L. Frank, unpubl. data).

While social structures may be strongly influenced by control, it is remarkable – and perhaps surprising – that sociality itself often remains. Despite the fact that badgers are believed to derive few benefits from group-living, young animals still remain in their natal ranges, forming groups in highly disturbed low-density populations where unoccupied areas and dens are available (Woodroffe & Macdonald, 1993).

In other species, however, sociality appears totally destroyed by control. While coyotes in undisturbed populations adopt a typical small canid social structure, in which territorial pairs sometimes recruit offspring to form small groups (Crabtree & Sheldon, 1999), those in a trapped population in Minnesota showed much more solitary behavior (Berg & Chesness, 1978). Females lived alone in small territories, with males occupying large, non-exclusive ranges overlapping those of several females. This spatial organization – more akin to that of cats and small mustelids than to canids (Sandell, 1989) – could have arisen because, in a situation of high mortality, males might encounter more breeding opportunities by ranging widely than by living as pairs (for example by mating with the widows of culled males, or by salvaging some reproductive success should their own mates be killed).

Even where animals remain social, mating systems may change in response to control or exploitation. For example, male lions in a trophy-hunted population in Zambia frequently mated with females from more than one pride (Yamazaki, 1996), which is extremely uncommon in undisturbed populations (Packer et al., 1991a).

Activity rhythms and communication

Persecuted animals become more secretive and often change their activity patterns to avoid humans. Thus, it is rare to see carnivores by day in areas where they are frequently hunted, as they tend to become more nocturnal. Africa's tourist industry depends heavily on the unafraid diurnal lions in national parks (Western & Henry, 1979). In livestock country, however, where predators are frequently killed to protect livestock, lions are extremely nocturnal, staying in heavy bush by day (L. Frank & R. Woodroffe, unpubl. data). Similarly, both wolves and European badgers are more nocturnal where persecuted (Vilá et al., 1995; Lindsay & Macdonald, 1985).

Similarly, poaching pressure in some national parks appears to make spotted hyenas exceptionally shy of human activity (Korb, 2000).

Disruption of local culture
Carnivore social groups may persist over long periods, representing many generations of animals that occupy the same home range (e.g. Frank et al., 1995), use traditional denning areas, and may have distinctive hunting or travel patterns (e.g. Malcolm & van Lawick, 1975). Such local tradition is passed from one generation to the next, and obviously cannot survive massive mortality, even if the former residents are eventually replaced by immigrants (Haber, 1996). In an extreme example, wild dogs released into Etosha National Park, Namibia, were unaware of the traditional migration routes of prey populations, and failed to follow prey migrating away from the release site (Scheepers & Venzke, 1995).

Through such effects, carnivore control programs may sometimes inadvertently increase rather than decrease depredation problems, by removing residents which do not prey on livestock with new ones that do. We have seen several instances where a ranch has resident lions but no depredation problems, while a neighbor may suffer heavy depredation in spite of regularly killing predators (Frank, 1998). A possible interpretation of these patterns is that when 'well-behaved' residents are removed, they are likely to be replaced by a series of naïve dispersers which have not learned to avoid livestock.

Genetic consequences of control
The studies presented above indicate that high rates of mortality frequently cause increased dispersal and break up of family groups and local neighborhoods. One would predict that these effects would be reflected in the genetic structure of the affected populations. There has been little research on these effects, but preliminary data are strongly suggestive. In a relatively undisturbed wolf population in Denali Park, where humans are responsible for only 4% of wolf mortality, there is a high degree of genetic similarity within packs (Mech et al., 1998). Fifty-three percent of individuals within a pack are related as siblings or parent–offspring, and a further 26% were related as cousins. Even higher levels of intra-pack relatedness were found in relatively undisturbed Minnesota packs. However, Lehman et al. (1992) described a much lower level of relatedness within packs in areas of the Northwest Territories, where heavy control measures annually kill a large percentage of the wolf population.

More severe evolutionary consequences of control have been postulated.

Genetic studies of the red wolf, once recognized as a distinct species of the southern United States, have suggested that it is actually a hybrid between coyotes and gray wolves (Reich *et al.*, 1999). Hybridization may have occurred when wolf persecution drove populations so low that male wolves started mating with female coyotes for lack of conspecific mates (Wayne & Jenks, 1991). Similarly, there may be elements of both domestic dog and wolf genomes in the large coyotes that have become established in the northeast United States in recent decades (Hilton, 1978), as well as introgression of coyote genes into wolf populations (Lehman *et al.*, 1991).

If a very small population becomes isolated in a protected area by intense exploitation or control in the surrounding region, inbreeding depression could become a serious concern. This effect has been suggested for both the Florida panther (Roelke *et al.*, 1993a) and the lions of the Gir forest (O'Brien *et al.*, 1987c). Possibly the best known cases showing loss of genetic diversity is that of the northern elephant seal (*Mirounga angustirostris*), which were reduced to a few individuals one hundred years ago. Although protection lead to population recovery, the species currently shows almost total loss of heterozygosity (Bonnell & Selander, 1974).

RATES OF RECOVERY FROM CONTROL

All of the behavioral and ecological effects of control outlined above influence the extent to which carnivore populations are able to recover when control is halted. Table 19.2 lists examples in which recovery times vary from as little as 5 months to periods in excess of a decade.

The data in Table 19.2 suggest that recovery rates depend, in part, on the species' propensity to disperse into vacant areas. While species such as wolves (Fritts & Mech, 1981; Figure 19.6a), lions (Smuts, 1978) and coyotes (Beasom, 1974b) readily move into cleared areas, and so are able to recolonize rapidly when control is halted, other species such as badgers (Cheeseman *et al.*, 1993; Figure 19.6b) and spotted hyenas (Holekamp *et al.*, 1993; Smuts, 1978) are more conservative. When an area was cleared in a highdensity badger population, only a small number of animals dispersed into the vacated territories. Gradual breeding among these few immigrants eventually led to population recovery, but the density did not equal that in the surrounding areas until 10 years following the initial removal (Cheeseman *et al.*, 1993; Figure 19.6b). Similarly, when 80% of the hyenas were culled in an area of Kruger National Park, recovery was extremely slow (Smuts, 1978). Since female hyenas are extraordinarily philopatric (females have never been found to join established groups), it seems likely that the

Table 19.2. Rates at which various carnivore species have recolonised areas cleared (more or less effectively) by control

Species	Area cleared	Time to recovery	Mode of recovery
Coyote[a]	23 km²	5 months	Immigration
Bobcat[a]	23 km²	5 months	Immigration
Lion[b]			
Kruger Area 1	c. 380 km²	17 months (90%)	Immigration and reproduction
Kruger Area 2	c. 600 km²	15 months (97%)	Immigration and reproduction
Kruger Area 2a	c. 95 km²	8 months (79%)	Immigration
Wolf[c]			Immigration and reproduction
African wild dog			
Save[d]		18 months	Immigration and reproduction
Laikipia[e]	4000 km²	≫ 10 years	Immigration
European badger			
Woodchester[f]	c. 3 km²	10 years	Reproduction of a small number of immigrants
Thornbury[g]	104 km²	> 11 years	Unknown
Spotted hyena[b,h]			
Kruger Area 1	c. 380 km²	≫ 17 months (≤ 10 years)	Reproduction (probably)

Data sources: [a]Beasom (1974b); [b]Smuts (1978); [c]Fritts & Mech (1981); [d]M. Mills et al. (1998); [e]R. Woodroffe, unpubl. data; [f]Cheeseman et al. (1993); [g]Clifton-Hadley et al. (1995); [h]M. Mills (1985).

population recovery which did eventually occur (M. Mills, 1985) must have been due to breeding of the females which remained, rather than to immigration.

Despite these clear differences, recovery rates are not solely dependent upon species' biology. For example, the dramatic difference in the recovery rates of African wild dogs in the Save Conservancy, Zimbabwe, and Laikipia District, Kenya, are almost certainly due to the proximity of source populations. Save is adjacent to Gona re Zhou National Park, which supports a substantial wild dog population (Fanshawe et al., 1997); under these circumstances, the remarkable dispersal abilities of wild dogs (e.g. Fuller et al., 1992b) and rapid reproduction allowed them to recolonize Save rapidly when persecution was halted (M. Mills et al., 1998). In contrast, Laikipia is distant from any high-density or protected wild dog populations (Fanshawe et al., 1997). Despite the fact that dispersal coalitions of both sexes pass through the district periodically, and appear to suffer little persecution

(a)

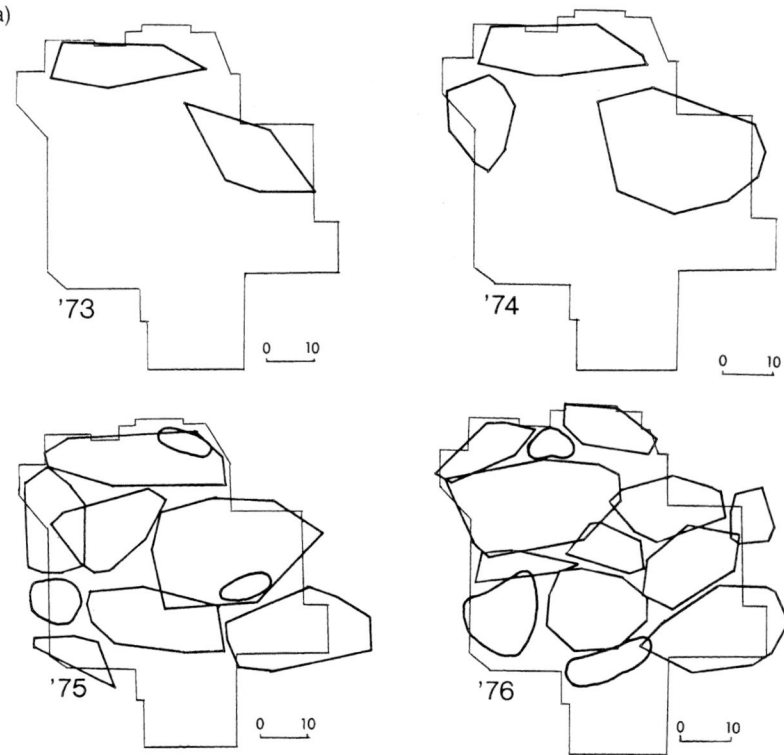

Figure 19.6. Contrasting patterns of recolonisation processes following the cessation of carnivore control. (a) In a newly protected wolf population, packs formed rapidly, and occupied territories which changed comparatively little as space was occupied. All space was filled within 4 years. Light line indicates the border of the study area, heavy lines give wolf home ranges. Scale bar is 10 km. (Redrawn from Fritts & Mech, 1981, reproduced by permission of The Wildlife Society); (b) European badgers recolonising an area following tuberculosis control operations ranged widely and did not form social groups for several years. Populations density did not reach that in surrounding areas until 10 years following the initial clearance. Circles indicate badger dens, lines indicate badger home ranges determined by bait-marking and radio tracking. Scale bar is 1 km. Redrawn from Cheeseman et al. (1993).

(Woodroffe, unpubl. data, Fanshawe et al., 1997), that prey density is high, and the density of competing predators low (Frank, 1998), no resident wild dog population has re-established. This appears to be a classic 'Allee effect' (Allee, 1931) – colonists occur at such low densities that male and female coalitions fail to meet and breeding cannot occur. A similar pattern

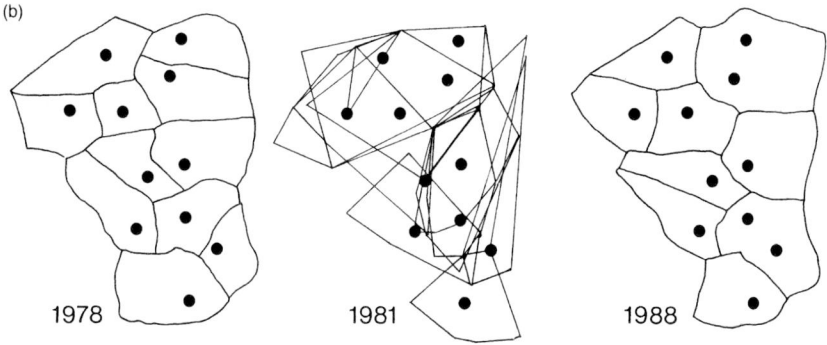

(b)

1978 1981 1988

Figure 19.6 (*cont.*)

has characterized wild dog recolonization of the Serengeti ecosystem, following extinction of the resident population due to disease (Alexander *et al.*, 1993; Gascoyne *et al.*, 1993a); while dispersers have continued to enter the area (Burrows *et al.*, 1994), no new packs have formed.

CONCLUSIONS: CHARACTERISTICS OF VULNERABLE VERSUS RESILIENT SPECIES

The various responses to control outlined above indicate substantial variation in species' susceptibility to control, and their ability to withstand high mortality. While some are able to compensate, others decline and disappear. In this section, we discuss the characteristics of resilient species.

Reproductive flexibility

Species' capacity to withstand the high mortality associated with control depends not only on their ability to breed rapidly, but also upon their ability to respond rapidly to changing conditions. For example, models of coyote populations suggest that they may continue to thrive with an annual offtake in excess of 50% (Connolly & Longhurst, 1975), while grizzly bears may decline with harvest rates an order of magnitude smaller. Coyotes not only multiply more quickly than grizzly bears in undisturbed populations – they may also modify both their litter sizes and their age at first breeding in response to control measures (Knowlton, 1972).

Such flexibility depends partly upon the physiological processes that govern breeding. In dingo (*Canis dingo*) populations, virtually all females become pregnant but a high proportion lose their litters around the time of birth (Thomson *et al.*, 1992). This delayed decision to breed or not allows females to take advantage of any breeding opportunities that may arise in dingoes' highly variable arid environment – and apparently pre-adapts them to resist intense control operations (Corbett, 1995). By contrast, female African wild dogs produce litters too large to raise without assistance from other group members, and subordinate group members do not become pregnant (Creel et al., 1997a). Thus, wild dogs have stringent social requirements for breeding opportunities, and can only respond comparatively slowly when such opportunities present themselves.

Social structure and vulnerability

The basic form of a carnivore's social system can have a major influence on the vulnerability of species to control measures. Species that travel and forage in packs are especially vulnerable – a poisoned carcass can kill off an entire pack of wolves or clan of hyenas in one night, and a pack of wild dogs can be eradicated in five minutes of shooting. Solitary foragers are less vulnerable to mass mortality, and those that rarely scavenge or return to kills are more difficult to poison. Colonially breeding seals are readily wiped out by intensive harvest for pelts or blubber (Bonnell & Selander, 1974).

Social structures also influence the ability of species to withstand high mortality. Species with highly flexible social systems – such as the smaller canids – are highly resilient. Coyotes, for example, can persist as group-living, pair-living, or even solitary animals (Berg & Chesness, 1978; Crabtree & Sheldon, 1999). By contrast, female African wild dogs cannot breed alone, and pairs or even small groups suffer low reproductive success (Creel et al., 1997a). Worse still, some species show behaviors, such as infanticide, which exacerbate the impact of exploitation, hindering, rather than promoting, population recovery. Finally, social structure influences a species' ability to recolonize cleared areas – a high degree of philopatry greatly inhibits recovery, perhaps making species more vulnerable to local extinction.

Ethological flexibility

Finally, resilient species are characterized by a high degree of flexibility in their day-to-day behavior. Species that modify their behavior in human-dominated landscapes – becoming less conspicuous in their activities, and switching their diets to use a variety of novel food sources – may also learn rapidly, making them more sensitive to risks in their environment. Thus, while coyotes suspiciously inspect bait stations that may conceal traps, wild dogs' bold investigations of well-armed people have often proven fatal.

Preadaptations and species resilience

Resilient species are characterized by flexibility in their reproduction, sociality, diet, and behavior. Several ecological circumstances may have equipped species with these preadaptations for persistence in the face of control and exploitation.

Competition and intraguild predation

In many ecosystems, carnivore species form 'guilds' of different sized predators taking different but overlapping size ranges of prey. Larger guild members typically do not tolerate smaller ones, such that dominant species often have a dramatic effect on numbers of subordinate ones; they may be responsible for up to 76% of mortality (Palomares & Caro, 1999). For example, wolves tend to exclude coyotes, which in turn exclude foxes (Cypher & Scrivner, 1992; Johnson et al., 1996; Crabtree & Sheldon, 1999). The most dominant member of a carnivore guild is typically a top predator, with no evolutionarily significant enemies, and consequently few antipredator adaptations such as increased wariness or nocturnality. In contrast, many subordinate guild members have evolved much the same antipredator defenses as prey animals. This gives them greater innate ability to cope with human-caused mortality pressures. Thus wolves, grizzly bears and lions are comparatively easy to kill and are relatively vulnerable to local extinction, while subordinate members of their guilds, coyotes, black bears and leopards, have greater wariness, behavioral and ecological flexibility and ability to cope with human pressure.

Life in fluctuating environments

Many of the reproductive and behavioral traits associated with resilience are those required to cope with fluctuating prey populations. For example, Canadian lynx exploit peaks in the arctic hare cycle by producing large litters, and then disperse very widely when prey crash and resources are scarce. The same behavioral and reproductive capabilities may pre-adapt them to tolerate high levels of exploitation for fur. Likewise, dingoes' reproductive and social flexibility – presumably shaped by evolution to withstand the extreme fluctuation in food availability within Australian desert regions – equip them to cope with heavy persecution.

A final word

Several behavioral traits characterize resilient species. Most have evolved in response to selection pressures that pre-date exploitation by humans, such as competition with larger predators, or a fluctuating prey base. However, it is worth stressing that resilience is conferred by a syndrome of traits – it is not sufficient to possess just one or two of the necessary characteristics. Spotted hyenas are behaviorally flexible, they learn rapidly, alter their behavior in human environments, and can exploit a wide variety of prey types – yet their extreme female philopatry makes them highly vulnerable to control operations. By contrast, African wild dogs disperse readily, and produce enormous litters – yet their social inflexibility and fearlessness of humans have made them incapable of withstanding persecution throughout the twentieth century. Few species have the necessary complement of features – and even the most resilient can be eradicated if control efforts are strong enough.

Carnivores' responses to control and exploitation are highly variable and unpredictable. Some breed more rapidly, others breed more slowly; some range more widely, others contract their home ranges. In the absence of detailed behavioral study of both disturbed and undisturbed populations, who would have predicted that sport hunting of male lions might have the potential to cripple the ability of populations to recover from exploitation, or that persecution of the adaptable hyena could have irreversible consequences? Similar phenomena might be occurring in little-known species subject to control, exploitation, persecution, or poaching. Further studies are urgently needed to assess human impact upon such carnivore species.

20

The role of disease in carnivore ecology and conservation

STEPHAN M. FUNK, CHRISTINE V. FIORELLO,
SARAH CLEAVELAND AND MATTHEW E. GOMPPER

INTRODUCTION

Although habitat destruction and fragmentation, direct persecution and over-exploitation of carnivores or their prey have been the most important endangering processes for carnivore populations, disease has now emerged as a central issue in carnivore conservation. Several dramatic population declines have occurred since the beginning of the 1990s, for example from rabies (African wild dogs, *Lycaon pictus* and Ethiopian wolves, *Canis simiensis*), from canine distemper virus (CDV) infections (lions, *Panthera leo*, and black-footed ferrets, *Mustela nigripes*), and from phocid distemper virus (PDV) in several seal species, which have highlighted the potential need for disease management (Table 20.1). Young's (1994) analysis of catastrophic population die-offs in large mammals indicates that disease impacts are influenced by tropic levels. Carnivore die-offs were significantly more often being attributed to disease, and herbivore die-offs more often attributed to other factors such as starvation. Therefore, disease is of particular concern for carnivore conservation.

The effect of disease and its dynamics in wild populations are complex. This is particularly true when pathogens that infect a wide range of species are involved and when a range of factors affect the occurrence and outcome of disease. Our understanding of disease mechanisms, dynamics and persistence is still very poor, but this complexity must at least be recognised and preferably understood for successful carnivore conservation. Ideally, a thorough understanding of the ecology and behaviour of both host and pathogen is a prerequisite for any attempt to manage disease. Unfortunately this understanding is a luxury that is not available to current conservationists, wildlife managers and scientists. As a result, where disease management has been attempted either for conservation or public health

Table 20.1. Examples of infectious diseases of carnivores that have resulted in population crashes of carnivores

Species	Locality	Cause	Reference
African wild dog	Serengeti*	CDV (suspected), rabies	Creel & Creel, 1998; Gascoyne et al., 1993a
	Madikwe Game Reserve, SA	rabies	Hofmeyr et al., 2000
	Etosha National Park, Namibia*	rabies	Scheepers & Venzke, 1995
	Masai Mara*	CDV	Alexander & Appel, 1994
Arctic fox	Medyni Island	otodectic mange	Goltsman et al., 1996
Black-footed ferret	Wyoming[1]*	CDV	Williams et al., 1988
Ethiopian wolf	Ethiopia	rabies	Sillero-Zubiri et al., 1996; Laurenson et al., 1997
African lions	Serengeti	CDV	Roelke-Parker et al., 1996
	Ngorongoro Crater	biting fly Stomoxys calcitrans	Fosbrooke, 1963; Packer et al., 1991b
Harbor seals	North Sea	PDV	Heide-Jørgenson & Härkönen, 1992; Dietz et al., 1989
		influenza A	Geraci et al., 1982
Lake Baikal seals	Lake Baikal	CDV	Grachev et al., 1989
Crab-eater seals	Antarctica	CDV	Bengston et al., 1991; Barrett, 1999
Mediterranean Monk seal	Mauritania	DMV suspected, but toxin most likely	*Osterhaus et al., 1998; Hernandez et al., 1998; van de Bilt et al., 1999; A. Dobson, pers. comm.

[1]The last black-footed ferrets from this population were taken into captivity.
*Indicates local extinction.
CDV: canine distemper virus; PDV: phocid distemper virus; DMV: dolphin morbillivirus.

reasons, controversy has frequently surrounded these well-intentioned efforts. The controversies surrounding the effect of handling and vaccination of African wild dogs (reviewed in Woodroffe *et al.*, 1997; Creel & Creel, 1998) and the role of badgers in tuberculosis transmission in the UK (Krebs *et al.*, 1998; Masood, 1998) both illustrate how a lack of knowledge can fuel debate. In addition, disease control strategies in wildlife have not always been successful or cost-effective when knowledge of the ecology and behaviour of both host and pathogen has been incomplete (Macdonald, 1980). For example, rabies control in Europe through fox culling, which was meant to reduce fox densities below a threshold level for rabies persistence (Bögel & Moegle, 1980) not only failed to eradicate rabies but may have even favoured rabies transmission. Culling, although sometimes reducing fox density, may have resulted in an increased number of new susceptible animals and in increasing contact rates, outcomes which were not predicted when the programme commenced (Macdonald, 1980; Funk, 1994). There is clearly, therefore, a need to improve our understanding of disease processes and evaluate the importance of disease for conservation.

REVIEW

Several recent reviews on the conservation implications of infectious disease have had a heavy emphasis on carnivores (Williams & Thorne, 1996; Creel & Creel, 1998; Murray *et al.*, 1999; Woodroffe, 1999c). We aim not to reiterate the points in these reviews or to discuss all diseases of relevance to carnivore conservation, but merely to provide a brief overview of the most important infectious and non-infectious diseases of carnivores and a synthesis of factors affecting the impact and dynamics of these diseases. We then examine how parasites and hosts may interact in populations of carnivores, taking epidemiological and ecological perspectives. Finally, we consider some management implications of disease in carnivore populations and highlight research priorities.

Infectious diseases

In this section we briefly review the pathogens that have had a serious impact on the persistence of carnivore populations (Table 20.1) and identify several common features that may be important as risk factors for wildlife epidemics.

Morbilliviruses are a group of viruses that includes rinderpest and measles. Two morbilliviruses are of importance to carnivores (reviewed by

Barrett, 1999): canine distemper virus (CDV); and phocid distemper virus (PDV). CDV is a cause of fatal disease in many carnivores and has resulted in large population fluctuations in species of canids, procyonids, felids, mustelids, and pinnipeds (Table 20.1). In canids, CDV-caused mortality can be substantial. For instance, CDV is an important contributing factor in African wild dog population declines (Creel & Creel, 1998), and 78% of mortality in gray foxes in the south-eastern United States was attributed to endemic CDV (Davidson *et al.*, 1992). Nevertheless, the impact of this virus on the persistence of endangered populations needs further exploration. CDV is best understood in the Canidae and was thought to not occur in felids outside of captivity until a large outbreak occurred in 1994 among lions of the Serengeti. The likely cause of this outbreak was spillover from domestic dogs (Roelke-Parker *et al.*, 1996). The outbreak reduced the Serengeti lion population by about one-third (Roelke-Parker *et al.*, 1996), and spread from the Serengeti into the Masai Mara where 55% of lions tested seropositive (Kock *et al.*, 1998). In hindsight, it appears that CDV may not have been new to felids; a retrospective study of lion and tiger deaths in Swiss zoos between 1972 and 1992 revealed that in 45% of the cases, CDV antigens were detected (Myers *et al.*, 1997). It is an enigma why CDV epidemics have not occurred earlier or have not been noticed (Harder & Osterhaus, 1997). Aggravating co-pathogens, predisposing genetic factors or strain variation of the virus may have contributed, but this requires more investigation (Harder & Osterhaus, 1997). CDV is also important in pinnipeds, and is the cause of massive die-offs of Lake Baikal seals, *Phoca sibirica*, in 1987–88 (Grachev *et al.*, 1989) and Caspian seals, *P. caspica*, in 1997 (Forsyth *et al.*, 1998). These CDV strains, which are more closely related to other strains found in terrestrial carnivores than they are to each other, may be a serious risk to the long-term survival of these species (Barrett, 1999). CDV was also implicated in a 1950's die-off of crab-eater seals, *Lobodon carcinophagus*, in Antarctica. The die-off occurred near a dog-sled base; these dogs were the likely cause of the outbreak, as the dogs were unvaccinated (Bengston *et al.*, 1991; Barrett, 1999).

PDV is closely related, and likely derived from, CDV (Barrett, 1999). Mortality is substantial, although once recovered, animals remain immune (Harder *et al.*, 1992). In 1988, a PDV epidemic killed an estimated 36% of the North Sea population of harbour seals, *P. vitulina*, and caused some mortality in grey seals, *Halichoerus grypus* (Dietz *et al.*, 1989; Heide-Jørgenson & Härkönen, 1992; Swinton *et al.*, 1998). The most likely cause of the outbreak is atypical contact with harp seals, *P. groenlandica*, usually found at more northern latitudes, which are believed to maintain endemic PDV due to extremely large population size (Goodhart, 1988).

The principal *parvovirus* of concern to conservation biologists is canine parvovirus (CPV). CPV is an emerging disease that was first detected in the 1970s as the cause of new diseases in canids which rapidly spread throughout domestic and wild dog populations world-wide (reviewed by Parrish, 1999; Truyen, 1999). Also referred to as CPV type-2 to distinguish it from a previously described, although unrelated canid parvovirus, CPV was soon shown to be a variant of the long recognised feline panleukopenia virus (FPV). By 1979, modified live virus vaccines were developed, and these have allowed much of the disease to be controlled. However, the virus remains widely distributed, and if pups are not vaccinated, or when maternal antibodies interfere with vaccination in pups, they generally become naturally infected. Soon after the discovery of CPV type-2, antigenic variants termed CPV type-2a and 2b were identified (Parrish, 1999; Truyen, 1999). Although those viruses only differed slightly from CPV type-2, in each case the viruses became globally distributed while CPV type-2 disappeared, indicating strong selective pressures. The genetic and antigenic variation in the virus strains was also correlated with changes in the host specificity of the virus, in particular in the ability to replicate in cats (Truyen *et al.*, 1996). CPV is enzootic in many canid populations and may cause increased pup mortality, which is higher than adult mortality (M. R. Johnson *et al.*, 1994; Windberg, 1995). Although pathogens such as CPV, that impact pup mortality but not adult mortality, are less significant in terms of population persistence than pathogens causing adult mortality (Woodroffe, 1999c), CPV is nevertheless of conservation concern. The ability of CPV to persist in the environment, for instance in faeces and carcasses, means that CPV is more likely to occur as a stable enzootic infection than pathogens such as rabies and CDV that cannot survive outside the host. Mech & Goyal (1995) found that gray wolf pup production and winter population size was inversely related to CPV prevalence, and predicted that winter wolf populations will decline once CPV prevalence in adults consistently exceeds 76%. Similarly, Creel *et al.* (1997b) note that in African wild dogs, litter sizes tend to be lower in regions where CPV titres are highest. If CPV is reducing recruitment in these species, it could threaten population persistence (Creel & Creel, 1998).

Rabies is caused by an RNA virus capable of infecting nearly all mammals (Carter *et al.*, 1995). Rabies epidemiology, zoonotic risk and threats to carnivore conservation have been reviewed extensively (Macdonald 1980; Bacon, 1985; Baer, 1991). Because of the zoonotic risk, massive immunisation programmes now target domestic and wild carnivores (Aubert, 1993;

Bingham *et al.*, 1999; Robbins *et al.*, 1998; Schubert *et al.*, 1998a,b; Smith, 1989; Tischendorf *et al.*, 1998). Although vaccination strategies vary greatly and are often controversial (e.g. type of vaccine, use of baits, role of hunting; e.g. Aubert, 1993; Funk, 1994; Woodroffe 1997, 1999c), some vaccination programmes have been effective (e.g. Tischendorf *et al.*, 1998; but see below). Where such programmes do not exist, however, rabies still kills thousands of people and untold numbers of animals every year (Smith, 1989). Although rabies is one of the best studied diseases of wildlife, many issues regarding its epidemiology remain unknown. Stress and physical condition has been suggested to influence the duration of the incubation period in red foxes (Winkler, 1975), but never demonstrated; how and which factors modify the largely varying incubation period in other species is uncertain.

Dealing with rabies in the context of conservation and wildlife management has led to widespread debate on the threat of disease for population persistence, appropriate methods of disease control and the risks and benefits of intervention. On the one hand, rabies has decimated packs of grey wolves (Weiler *et al.*, 1995) and wild dogs (Gascoyne *et al.*, 1993a; Kat *et al.*, 1995, Woodroffe *et al.*, 1997) and has caused decline in Ethiopian wolves (Laurenson *et al.*, 1997). Spill-over of rabies from domestic dogs is often the most parsimonious explanation (Ethiopian wolves, Laurenson *et al.*, 1997) or is indicated by modelling (side-striped jackals *Canis adustus*, Rhodes *et al.*, 1998). Transmission of disease from domestic carnivores to endangered populations may occur indirectly. Domestic dogs may have first transmitted rabies into the red fox population in Israel which in turn may have infected the endangered Blandford's fox, *Vulpes cana* (Macdonald, 1996). On the other hand, hypotheses have been raised that the stress of intervention (rabies vaccination) caused the local extinction of African wild dogs in the Serengeti. The debate surrounding the issue led to a moratorium on the handling of wild dogs (Creel, 1992; Macdonald *et al.*, 1992; Burrows *et al.*, 1994; De Villiers *et al.*, 1995; Ginsberg *et al.*, 1995; Kat *et al.*, 1995; East & Hofer, 1996) and approaches to rabies control now focus on domestic dog vaccination (Cleaveland, 1997; Laurenson *et al.*, 1998).

Bovine tuberculosis is a contagious disease caused by the bacterium *Mycobacterium bovis*. It is a serious disease of cattle and wild bovids, and diagnosis of the disease can have serious economic implications for affected farms. Transmission occurs between carnivore conspecifics, primarily by aerosol, and in some cases, carnivores may act as tuberculosis vectors to ungulates. Some evidence suggests that Eurasian badgers, *Meles meles*, may be a significant source of cattle infection in the United Kingdom, which has resulted in sustained efforts to manage badger populations in regions where bovine tuberculosis is enzootic (Krebs *et al.*, 1998; cf Masood, 1998). Tuberculosis is not just a disease of concern for badgers. For instance, recently an outbreak of *M. bovis* occurred among buffalo (> 70% prevalence) in Kruger National Park, South Africa, which then may have acted as a reservoir for spill-over into the lion population, causing a decline in lion numbers (Keet *et al.*, 1996; Weyer *et al.*, 1999).

The *mange* mites *Sarcoptes scabiei, Otodectes cynotis* and *Notoedres cati* cause sarcoptic, otodectic and notoedric mange respectively. Severe population declines in red foxes have been observed following epidemics of the canid variety of *Sarcoptes scabiei* (Danell & Hörnfield, 1987; Funk, 1995; Lindström & Mörner, 1985). In larger canids such as coyotes, sarcoptic mange may be enzootic or epizootic, but appears to cause only compensatory mortality (Pence & Windberg, 1994). The varieties of *Sarcoptes scabiei*, including *canis, suis* and others (although no felid varieties have been identified) are highly specialised to their hosts. However, spill-over from heavily infected canids to accidental hosts such as felids has been documented (Danell & Hörnfeldt, 1987) and may constitute a conservation threat for small carnivore populations. Infection of an endangered Arctic fox subspecies, *Alopex lagopus semenovi*, from domestic dogs, caused an epizootic of otoedric mange, a dramatic increase in cub mortality, and a population crash (Goltsman *et al.*, 1996). Treatment of infected foxes was attempted, but differences in cub mortality of treated and untreated animals were not significant, highlighting the difficulty of treating infected animals (Woodroffe, 1999c). Domestic cats have transmitted notoedric mange to coatis, *Nasua narica*, Florida panthers, *Felis concolor coryii*, and other Carnivora where it may constitute a regional conservation threat (Maehr *et al.*, 1995; Valenzuela *et al.*, 2000).

Disease resistance and susceptibility

While much attention has focused on the conservation importance of infectious diseases because of catastrophic die-offs, less attention has been given to non-infectious diseases that may act as factors modifying resistance and susceptibility to infectious and non-infectious diseases. This nonetheless deserves serious attention, as the often subtle effects of non-infectious diseases and disorders may sometimes be of great immediate or underlying importance. A variety of factors may be involved, such as genetics, environmental toxins, or chronic stress and their effects may also be apparent through altering the susceptibility of hosts or the pathogenicity of infectious agents.

Genetically inherited diseases are of little importance for the long-term survival of a species so long as the population remains relatively large, and indeed if a serious deleterious effect is apparent which affects host survival or reproduction, it is likely to be selected against. However, in small, inbred populations, where heterozygosity is lost, deleterious recessive alleles may occasionally become fixed due to random genetic drift (Lande, 1994) or be expressed more frequently (O'Brien *et al.*, 1985). For example, heritable blindness increased in a small, captive population of European wolves (Laikre & Ryman, 1991). Decreased genetic variation in harbour seals, *Phoca vitulina*, results in both decreased birth weight and neonatal survival (Coltman *et al.*, 1998). Similarly, the Florida panther population, which has been reduced to less than 50 individuals, exhibits a number of deleterious traits and appears unusually susceptible to microbial parasites compared to larger panther populations in the western United States (Roelke *et al.*, 1993a). Inbreeding may result in the loss of heterozygosity at particularly important loci such as the major histocompatibility complex (MHC) which are critical to the immune responses of vertebrates (Klein, 1986). Perhaps as a result, inbred animals in captivity may lack the variety of responses to pathogens that might be present in an outbred population. O'Brien *et al.* (1985) suggested that genetic impoverishment would make populations of cheetahs, *Acinonyx jubatus*, more susceptible to extinction from pathogens. A case study of mortality caused by feline infectious peritonitis in captive populations is consistent with but not necessarily the consequence of genetic uniformity at MHC and other loci (Caro & Laurenson, 1994; Caughley, 1994; O'Brien *et al.*, 1985). Similarly, Müller-Graf *et al.* (1999) found a significantly higher level of intensity of *Spirometra spp.* infection in an inbred lion population as compared to an outbred population, but it was not possible to distinguish whether differences in levels of parasite infection were due to genetic or ecological factors. Despite this unquestionable important theoretical concern, little observational or experimental evidence is available at present to demonstrate that this is currently a serious concern for carnivore conservation (Williams *et al.*, 1992) and some species, such as the northern elephant seal, *Mirounga angustirostris*, which have gone through severe bottlenecks, can either persist or even expand despite facing immunological challenges (Garza, 1998). In the face of the importance of human interference as the primary conservation threat for carnivora, genetic considerations appear to be relatively unimportant when assessing disease risk. However, the impact of loss of genetic variation on disease resistance and susceptibility will act on a slow time scale and is likely to be exaggerated by further habitat fragmentation and isolation of small populations. Once the impacts are large enough to be quantified, they may be irreversible.

The abiotic environment of an animal may result in mortality or reduced fitness when natural or man-made compounds are contacted or ingested at levels above which either the direct destruction of organs (liver, kidney) or indirect damage occurs. Changes in immunocompetence, thermoregulation, energy balance or other physiological processes may all lead indirectly to decreased life expectancy or reduced reproductive success. Direct toxicoses can occur when non-target carnivores succumb to accidental poisoning by consuming baits designed for other species (e.g. Berny et al., 1997; Schitoskey, 1975), or when carnivores contact toxic pollutants. For instance, the Exxon Valdez oil spill in 1989 resulted in declines in sea otter and river otter populations as a result of oil-saturated fur, which caused thermoregulatory difficulties as well as toxicosis due to direct ingestion of hydrocarbons (Bowyer et al., 1995; Duffy et al., 1994, 1996, 1999; Estes, 1991).

Unfortunately, the high trophic level of many Carnivora leaves them susceptible to the bioaccumulation of toxicants (Table 20.2) and this is pronounced when carnivores feed on other carnivores, such as PCB (polychlorinated biphenyl) and pesticide accumulation in walrus and polar bear feeding on ringed seals (Letcher et al., 1998; Muir et al., 1995). Because both predator and the prey can disperse long distances, toxicants may bioaccumulate in top carnivores inhabiting regions that are seemingly devoid of local contamination and individual differences in foraging preferences can lead to considerable variation in vulnerability to toxicosis. In addition, the age, sex and breeding status of an individual influence can affect toxicant burden. For instance, Born et al. (1981; see also Bernhoft et al., 1997) found a positive correlation between age and PCB concentrations for male but a negative relationship for female walrus. These intersexual differences may occur because of different foraging patterns and different metabolising capacities, but they also may occur because female toxicant load is reduced by lactation (Addison & Brodie, 1987a,b). Bernhoft et al. (1997) found that female polar bears could excrete considerable amounts of organochlorines into milk, and Nakata et al. (1995) estimated that a female Baikal seal transfers 20% and 14% of its DDTs and PCBs respectively to the infant during lactation. The impact of these phenomena from a conservation perspective needs further investigation.

Sublethal effects of pollutants such as decreased reproduction, decreased immune function, and endocrine disturbances, can be very important both at individual and populations levels, but are often difficult to

Table 20.2. Representative examples of bioaccumulated toxicants in carnivore species. The list is not inclusive of all species, all toxicants, or all studies for any given species

Species	Toxin	Reference
Phoca siberica (Baikal seal)	organochlorines	Kucklick *et al.*, 1996; Nakata *et al.*, 1995
P. vitulina (harbor seal)	PCBs and organochlorines	Hong *et al.*, 1996
P. groenlandica (harp seal)	organochlorines	G. Beck *et al.*, 1994
P. hispida (ringed seal)	PCBs and DDT	Addison *et al.*, 1986; Letcher *et al.*, 1998
Callorhinus ursinus (northern fur seal)	PCBs and DDT	Kurtz & Kim, 1976
Halichoerus grypus (grey seal)	PCBs	Addison & Brodie, 1987a,b
Mirounga angustirostris (northern elephant seal)	PCBs and DDE	Beckmen *et al.*, 1997
Eumetopis jubatus (Steller sea lion)	butyltin	Kim *et al.*, 1996
Odobenus rosmarus (Atlantic walrus)	PCBs, mercury, and organochlorines	Muir *et al.*, 1995; Born *et al.*, 1981
Lutra lutra (European otter)	PCBs and organochlorines	Mason & O'Sullivan, 1992; Leonards *et al.*, 1997
Mustela vision (mink)	hexachlorobenzen	Moore *et al.*, 1997
Martes martes (pine martin)	PCBs	Bremle *et al.*, 1997
Ursus maritimus (polar bear)	PCBs and organochlorines	Bernhoft *et al.*, 1997; Letcher *et al.*, 1998; Norstrom *et al.*, 1988
Procyon lotor (raccoon)	selenium	D. Clark *et al.*, 1989
Vulpes vulpes (red fox)	PCBs	Georgii *et al.*, 1994
Felis concolor coryi (Florida panther)	mercury	Roelke *et al.*, 1993b

PCB: polychlorinated biphenyl.

document. Harbour seals fed with fish contaminated with organochlorines exhibited depressed natural killer lymphocyte activity, indicating a direct impairment of the immune system, and lower reproductive success (Ross et al., 1996). Similarly, in female ringed seals levels of PCBs have been shown to be correlated with uterine lesions and decreased reproduction (Helle et al., 1976). Although the immediate cause of the mass mortality of Lake Baikal seals in 1987–1988 was a morbillivirus, the more distal cause was likely PCB toxicosis resulting in decreased immune system function (Grachev et al., 1989; Nakata et al., 1995). Toxicants may result in population declines without overt signs of disease. For example, using conservative simulations Moore et al. (1997) found a moderate to high probability that mink, Mustela vison, in the St. Clair River, Canada experience 5–20% declines in reproductive success due to fungicide bioaccumulation, although overt indications of toxicosis had not been reported. Conversely, populations inhabiting polluted regions may nonetheless thrive if pollutant related mortality is compensatory rather than additive to other causes of mortality. This has been shown for otter populations in Scotland, which are increasing in some areas despite high rates of PCB contamination (Kruuk & Conroy, 1996).

Finally, one should also consider that the parasite–host interaction does not occur in an ecological vacuum; independent health threats may be occurring simultaneously. Stress, caused by poor nutrition, subclinical disease, parasitic infection or human activities such as visitor pressure may alter susceptibility to some toxins or diseases such as sarcoptic mange in red foxes and wild dogs (Lloyd, 1975; M. Mills et al., 1998). Co-infections may substantially alter immune responses. Comparisons of feline immunodeficiency virus (FIV)-positive domestic cats and FIV-negative controls found more severe disease signs in FIV-positive animals after both groups were infected with feline calcivirus (Reubel et al., 1994). Similarly, replication of FIV is enhanced by the presence of feline leukemia virus (FeLV) (Pedersen et al., 1990). Well-documented cases of stress-induced diseases in wild carnivores are few. One example may be the finding of sarcocysts of Sarcocystis sp. in Florida panthers. This long-lasting stage of the life cycle is normally confined to the muscles of prey species, so the finding of these in adult panthers was unexpected. One explanation is that the presence of sarcocysts may be an indication of stress-induced immune compromise, enabling the atypical development of the sarcocysts. This immune suppression may have resulted from infections by FPV or FIV (Greiner et al., 1989). Overall, the role of stress in parasite–host interactions remains poorly understood and although stress may depress the immune response in some cases, in other cases immunity against parasites may actually be enhanced (Lloyd, 1995).

There is also some evidence that the number of new, 'emerging' diseases such as CPV have been increasing in general and in Carnivora in particular (Harvell *et al.*, 1999; Daszak *et al.*, 2000). In aquatic environments, environmental stress due to changes in climate and water temperature together with increasing numbers and concentrations of toxins may facilitate the emergence of new diseases (Harvell *et al.*, 1999). The infection of harbour seals with avian influenza viruses (Geraci, 1982; Webster, 1992) may be an example.

Which diseases are important for carnivore conservation?

Extinctions and near-extinctions of carnivore populations (Table 20.1) clearly identify three primary factors in pathogen-related extinction processes. Firstly, the majority of extinctions and near-extinctions were caused by generalist pathogens. Less generalist pathogens have either not caused any reported die-offs (e.g. the nematode *Skrjabingylus nasicola* in the stoat, *Mustela erminea;* King, 1991) or die-offs were rare (e.g. the stomoxys plague in African lions; Fosbrooke, 1963). Secondly, the world-wide distribution and high abundance of domestic Carnivora constitutes a particular risk for spillover of generalist pathogens to less abundant free-ranging Carnivora (Table 20.3). Thirdly, all extinctions and near-extinctions involved small populations – a trend generally observed in mammals (Woodroffe, 1999c). In the following section we highlight these and other aspects in the context of the ecology of parasite–host interactions that have relevance to conservation.

ECOLOGICAL INTERACTIONS

Effect of host population structure

Population size, density and distribution are fundamental factors affecting the impact and dynamics of diseases on carnivores (Dye *et al.*, 1995; Lyles & Dobson, 1993). Most examples of population extinction or near-extinction due to disease (see under 'Review') were in populations that were very small and isolated prior to the disease outbreak. As such, disease may be part of the 'extinction vortex' (Gilpin & Soulé, 1986). In small and declining populations, even diseases that cause only low additive mortality or decreased fecundity, may be enough to promote local extinction, particularly if re-colonisation is prevented by habitat fragmentation.

Some epidemics caused by morbilliviruses, rabies and mange have caused additive mortality. In other words, mortality due to the disease was

Table 20.3. Examples of spill-over of diseases from domestic carnivores into wild carnivore populations

Species	Disease	Domestic reservoir	Reference
Wild dogs	rabies	dog	Gascoyne et al., 1993a; Cleaveland & Dye, 1995; Kat et al., 1995
Ethopian wolves	rabies	dog	Sillero-Zubiri et al., 1996
Crabeater seals	CDV	dog	Bengston et al., 1991
Lake Baikal seals	CDV	dog	Grachev et al., 1989; Osterhaus et al., 1989
Wild dogs	CDV	dog	Alexander & Appel, 1994
Lions	CDV	dog?	Alexander et al., 1995; Roelke-Parker et al., 1996; Haas et al., 1996
Wolves	CPV	dog	Peterson et al., 1998
European wild cats	FeLV	cat?	Artois & Remond, 1994; McOrist et al., 1991
White-nosed coati	notoedric mange	cat	Valenzuela et al., 2000
Mednyi Arctic foxes	otodectic mange	dog	Goltsman et al., 1996

CDV: canine distemper virus; CPV: canine parvovirus; FeLV: feline leukemia virus.

in addition to mortality from other causes, resulting in a reduction in the population size. If disease-induced mortality is compensatory and infected animals would have died anyway for other reasons, it may have no effect on population size. Nevertheless, the underlying pathogens should not be ignored. Firstly, we rarely have information on whether disease-induced mortality is compensatory or additive. For example, anthrax (reviewed by Creel & Creel, 1998), caliciviruses (Barlough et al., 1986), herpesviruses (Gaskell & Willoughby, 1999), leptospirosis (Vedros et al., 1971), *Brucella* spp. (Nielsen et al., 1996) or FIV (Murray et al., 1999) cause clinical disease and mortality in the wild or in captivity, but their significance in wild Carnivora is generally unknown. Secondly, the effects of pathogens may switch from compensatory to additive, especially in small populations and when acting in combination with other factors such as toxins or stress. Thirdly, even if mortality is only compensatory, persistence still can be influenced if the disease influences reproductive success or the age structure of the population.

Examining parasite–host population dynamics without understanding the larger community structure can be problematic. Hypothetically, the small size of endangered populations could preclude them from maintaining species-specific pathogens because a critical number of hosts (the critical community size; Bartlett, 1957) are not available (McCallum & Dobson, 1995). In reality, however, most pathogens of conservation concern can infect multiple host species, resulting in reservoirs of infected hosts with the potential to spill-over into endangered species. This observation is supported by outbreaks of disease in both marine and terrestrial carnivores. Both domestic and wild carnivores, as well as non-carnivores, have appeared as reservoirs. For example, the 1988 PDV outbreak in harbour seals in Europe may have been caused by spill-over from the large population of harp seals, in which PDV is endemic. This occurred when harp seals migrated into the range of harbour seals following food shortages caused by overfishing around Greenland (Dietz *et al.*, 1989; Stuen *et al.*, 1994). Spillover infections from domestic animals may be even more common (Table 20.3). With domestic dog densities increasing in the developing world (Cleaveland, 1997, 1998) and carnivore populations becoming more fragmented, contact rates between wildlife and domestic carnivores can only increase.

Effect of disease on community structure

It should not be forgotten that when epidemics cause mass die-offs in large populations, but no extinction, there are implications and ramifications for the status and structure of carnivore populations and ecosystems as a whole. The best known example is that of rinderpest in the Serengeti ecosystem in Tanzania (reviewed by Dobson, 1995). Rinderpest was accidentally introduced into sub-Saharan Africa in the late nineteenth century and caused massive mortality in free-ranging ungulates. After wide-scale cattle vaccination in the 1950s, the disease effectively disappeared from wildlife, and in the Serengeti, wildebeest and buffalo numbers increased dramatically. As a result lion and spotted hyena densities also increased, a positive outcome for the conservation of these species. Unfortunately, predation and competition from these carnivores may be a limiting factor for cheetah and wild dog populations (Laurenson, 1994; Laurenson *et al.*, 1995; Creel & Creel, 1998; Vucetich & Creel, 1999). Another important consequence for carnivores is that cattle density has increased following the removal of a major mortality factor. During droughts, cattle are grazed and watered in some nature reserves (Dobson, 1995), potentially influencing ungulate and predator distributions and increasing human–carnivore conflicts.

An additional community-level issue, rarely considered by conservation biologists, is the effect of host loss on the structure of parasite communities. If host numbers decline, then host-specific parasites may also become extinct. For example, several parasites of black-footed ferrets were driven to extinction either as a result of the crash in ferret numbers or as a function of the quarantine programme the ferrets underwent when the last remaining individuals were captured to form a captive breeding programme (Gompper & Williams, 1998). Although this may appear to be an esoteric philosophical dilemma for conservation biologists, it is possible that parasites may compete with one another, with the loss of one species leaving the host vulnerable to a more pernicious parasite species which under normal circumstance would not be problematic.

Behavioural ecology

A behavioural perspective enhances the study of disease dynamics, by allowing insights into how individuals respond to disease, and providing a better understanding of factors underlying contact rates between individuals. Not host population density *per se*, but the contact rate between infected and susceptible animals determines the spread of contagious diseases. Contact rates can be frequency or density-dependent and are determined by the behaviour of the hosts. Because social aggregations of hosts represent a transmission opportunity for their parasites, selective pressures are exerted on host social structures. For instance, positive correlations of group size, number of parasite species per host and infection intensities have been shown (Keymer & Read, 1991; Møller *et al.*, 1993). Sociality can also result in decreased parasitism if it allows more efficient parasite or vector avoidance behaviours such as grooming, or if living with other potential hosts creates a dilution effect which decreases the chance of an individual being singled out by a vector or parasite (Mooring & Hart, 1992).

Carnivore social organisation can therefore influence parasite prevalence and population dynamics. For example, although the prevalence of FeLV increased with group size in domestic cats, FIV prevalence decreased with increasing group size (Fromont *et al.*, 1997). FeLV is transmitted horizontally between cats by grooming or sharing food, and thus transmission was enhanced in larger groups with non-aggressive contacts. FIV is transmitted by biting and occurred almost exclusively among adult males that fought to maintain dominance hierarchies. Similarly, the social structure of white-nosed coatis affects the prevalence and intensity of *Amblyomma* tick and *Eutrombicula* chigger infections (M. Gompper unpub. data). Although both parasites are acquired as free-living organisms that are attracted to a host by increased levels of CO_2, heat and other cues, and are not directly transmitted among individuals, they differ in how they relate to host sociality. Tick intensities are higher on solitary males than on band-living females, probably because ticks are removed through social grooming. In contrast, chigger parasitism is higher among band members than solitary coatis, and temporal chigger prevalence fluctuates greatly within bands and independently between bands, such that at any given time either all or none of the band members are infected. Thus social structure should be considered when examining seasonal fluctuations in parasitism.

Unfortunately, few researchers examining epizootics of carnivores have taken this perspective. We thus have little insight into how social structure may have influenced recent bovine tuberculosis or CDV outbreaks in lions (Weyer *et al.*, 1999; Roelke-Parker *et al.*, 1996). Similarly, we know that members of a wolf pack are doomed should one member contract rabies (Chapman, 1978), since intra-pack contact rate is very high, while inter-pack contact rate is much lower. However, serological surveys of disease in canids rarely take pack membership or social status into account when assessing prevalence. High prevalence of CPV, CDV, infectious canine hepatitis and *Yersina pestis* (plague) in coyotes in Yellowstone National Park may represent a spill-over threat to wolves (Gese *et al.*, 1997). If, however, prevalence is not uniform across coyote packs or between territorial and transient coyotes, then contact rates between infected coyotes and susceptible wolves might differ from what might be expected without knowledge of coyote social structure. Some management aims at stopping the spatial spread of infectious disease by reducing inter-group contact rates by culling entire groups. This was, for example, the aim of a fox culling scheme for rabies in Europe. Territory sizes are thought to be determined by the distribution and quality of resources, rather than population density (Kruuk & Macdonald, 1985), and thus empty territories may act as barriers for disease spread. However, foxes increase territory sizes when neighbouring territories become empty following rabies and mange epidemics, thus maintaining common territory boundaries and inter-territory contact rates (Funk, 1994, 1995; Baker *et al.*, 2000).

Comparison of the behaviour of healthy and diseased individuals, and how this affects disease transmission, has also received little attention in wildlife epidemiology (Kiesecker *et al.*, 1999). In some cases, parasites can change host behaviour to increase transmission (Gulland, 1995). In other cases, sick individuals may travel less and thus transmission may be less likely. In virtually all cases, contact rates and transmission probabilities are extrapolated from observations of healthy individuals and no attempt has been made to quantify differences in diseased individuals, although some anecdotal reports are available. For example Creel *et al.* (1995) noted that wild dog juveniles with anthrax were abandoned by the pack and suggested that this reduced the risk of disease transmission within the pack. There are, however, occasional reports of healthy individuals feeding and caring for sick individuals (Rasa, 1983), although these cases involved wounded not diseased individuals.

Individuals may also possess the ability to self medicate, although this is only occasionally hinted at among carnivores. In Panama, white-nosed coatis groom themselves and other band members with *Trattinnickia* terpene resin, perhaps for pharmacological reasons (Gompper & Hoylman, 1993). Pharmaceutical value has also been suggested for the use of *Ligusticum eallichii* root by black bears in North America (Rodriguez & Wrangham, 1993).

MANAGEMENT
Collecting and interpreting the data
Interpreting the impact of disease on populations and planning management strategies requires robust data sets; data that is often difficult to obtain even under the best of circumstances. Whether diseases regulate, limit or destabilise host populations, and whether diseases create compensatory or additive mortality depends on the extent to which the disease is density-dependent (Anderson & May, 1978; May & Anderson, 1978; Gulland, 1995; McCallum & Dobson, 1995). Assessment of the effects of a new epidemic on host populations or the effects of management is often based on studies of healthy animals. However, unexpected density-dependent effects can be observed, resulting in disagreement with the assessments. For example, by increasing hunting pressure to control rabies in red foxes in Europe did not account for compensatory reproduction resulting from changes in dispersal patterns, an increased proportion of females reproducing and increased litter sizes (Voigt & Macdonald, 1984; Artois *et al.*, 1990; Funk & Gürtler, 1991). Estimating population sizes in carnivores is inherently difficult and therefore the resulting estimates of disease-induced mortality are imprecise. Although the 1988 mass mortality in harbour seals and gray seals is perhaps the best documented epidemic in a marine carnivore, pre-epidemic population density estimates vary considerably (21×10^3 to 46×10^3 in British waters, Swinton *et al.*, 1998), and so mortality estimates differ dramatically among studies.

Even if correlations can be quantified between mortality, fecundity and disease prevalence on a population level, the causative link may remain uncertain, since additional factors may underlie these phenomena. For example, in 1997 more than half of the total population of Mediterranean monk seals, *Monachus monachus*, inhabiting the western Saharan coast of

Africa washed ashore in a mass die-off. Van de Bilt *et al.* (1999) isolated a morbilliviruses from this population that closely resembled previously identified dolphin morbilliviruses (DMV), indicating interspecies transmission from cetaceans to pinnipeds. The seal die-off was previously attributed to a morbillivirus (Harwood, 1998; Osterhaus *et al.*, 1998), but the epidemiological evidence in support of DMV remains weak and the ultimate cause of this mortality was more likely due to a toxic algae bloom (A. Dobson, pers. comm.). Thus, isolating a pathogen does not necessarily establish causality.

Wildlife diseases are still poorly understood because of the complexity of host–parasite systems and the general lack of quantitative data (Dobson & Grenfell, 1995; Grenfell & Gulland, 1995; Plowright, 1988). Traditional investigations concentrate on the examination of pathological manifestations in autopsies based on small sample sizes, while the impact of diseases at the population level is less commonly addressed (Grenfell & Gulland, 1995). Morbid animals are often difficult to find and examine in a timely fashion. From the 1997 mass mortality of Caspian seals, *Phoca caspica*, only one dead animal was analysed, indicating the presence of CDV (Forsyth *et al.*, 1998). This led the authors to conclude that CDV transmitted from terrestrial carnivores may have caused the mass die-off (Forsyth *et al.*, 1998; Barrett, 1999). We may never be able to confirm this. Many studies produce lists of parasites found in a population or infer parasite prevalence by using serology, but prevalence *per se* is a poor indicator of the impact of disease on populations. For example, Serengeti lions were found to be seropositive for CDV in the 1980s and 1990s, but only the 1994 epidemic caused a severe population decline (Roelke-Parker *et al.*, 1996).

Many studies are carried out retrospectively after mass mortality has been observed. Retrospective analysis of disease involvement in population declines in carnivores such as the European mink, *Mustela lutreola*, only started decades after the decline was observed (Maran & Henttonen, 1995). Health and disease screening in Hawaiian monk seal only started after an epizootic of unknown etiology killed a substantial number of animals (Aguirre, 1998). Similarly, the debate about the origin of the die-off of wild dogs in the Serengeti (see above) remains clouded and is unlikely to be resolved because of lack of data and samples. Unfortunately, the number of long-term studies of Carnivora is small, but long-term data are often the best method to achieve deeper insight into disease dynamics. Only the fact that serum samples were collected during long-term studies of Serengeti lions revealed the dramatic differences in mortality rates during the CDV epidemics in the 1980s and 1990s (Roelke-Parker *et al.*, 1996). More robust field-based experimental studies (e.g. Schubert *et al.*, 1998a,b) are urgently needed.

Management strategies

Diseases in both large and small populations are of concern for carnivore conservation, but require different management philosophies and strategies. In large populations, diseases may be of conservation concern for a number of reasons. Firstly, large populations may act as reservoirs for the transmission to small populations of endangered Carnivora. Secondly, there may be undesirable effects at the community level. Thirdly, they may profoundly increase human-carnivore conflicts where disease may be transmitted to humans or their livestock. Fourthly, if the disease is introduced by human activities, managers may want to limit its impact. In small populations, disease may be of more immediate concern. Disease may either result directly in population extinction or in a decrease in population size and thus increasing the likelihood of extinction due to other factors.

The complexity of disease dynamics requires an explicitly mathematical framework, for which a large theoretical literature exists (e.g. Grenfell & Dobson, 1995; Toft *et al.*, 1991). Using data from the 1988 PDV epidemic in harbour seals, for example, the development of the critical metapopulation distribution model has allowed determination of the threshold for disease persistence in fragmented or spatially structured populations more accurately than the traditionally used models of critical community size (Mollison & Levin, 1995; Swinton *et al.*, 1998). Moreover, techniques have recently been developed advancing the complex and difficult modelling of multihost, multipathogen interactions (Holt & Pickering, 1985; Begon *et al.*, 1992; Bowers & Turner, 1997; Greenman & Hudson, 1999). Models are important aids for predicting the impact of disease and alternative management actions and thus whether or not management action should be taken. Difficulties in classifying and quantifying disease risk, the production of models that perhaps oversimplify disease systems, communication problems between theoreticians, ecologists and decision makers, and the basic lack of data mean that disease risk has often been assessed subjectively or disregarded. For example, Gaona *et al.* (1998) did not model the impact of disease on Iberian lynx populations because it was assumed that any infectious disease introduction would lead to extinction. However, with recent advances in modelling techniques, stochastic, individual-based models incorporating disease risk into population viability assessments are becomingly increasingly useful for practical conservation (e.g. African wild dogs, Woodroffe *et al.*, 1997; Vucetich & Creel, 1999; Ethiopian wolves, Mace & Sillero-Zubiri, 1997). Therefore, modelling disease impacts should not be avoided.

In the past, management actions were generally reactive rather than proactive independent of models and available data. However, although preventative action is usually much harder to justify, it is fundamental to the long-term carnivore conservation. With relevant information often missing, management decisions are often based more on political consider-ations than ecological or epidemiological models and insights. For example, the German state of Saarland carried out rabies vaccination of red foxes for years without coordinating the efforts with neighboring regions. In reality, however, fox spatial structure and rabies epidemiology are de-pendent on patterns in adjacent regions (Funk, 1994). In addition, fox dis-persal patterns were not considered when target vaccination areas were chosen (Zimen, 1984).

Even when disease management programmes are based on the best available scientific knowledge, the occurrence of failed interventions (Woodroffe, 1999c) highlights inadequacies in our understanding of com-plex host–pathogen interactions. Therefore, we consider that any disease management strategy, even those where no action is taken, should be con-sidered experimental. Though rabies is amongst the best investigated wil-dlife diseases and rabies immunisation in central Europe has been spectacularly successful, there are still areas where rabies control failed (e.g. Funk, 1994). Therefore, the World Health Organization and European Union still regard oral vaccination of foxes in Europe as an ongoing experi-ment, allowing critical evaluation and subsequent feedback to correct the vaccination strategy (M. Artois, pers. comm.). Research and monitoring efforts should run concurrently with interventions so that the success of the programme can be critically evaluated. Indeed, failed attempts may occa-sionally be quite instructive by revealing limitations of reactive interven-tions. For example, attempted CDV control in black-footed ferrets and rabies control in African wild dogs identified difficulties with vaccines such as the unexpected susceptibility of ferrets to live CDV vaccine and the poss-ible ineffectiveness of killed rabies vaccines in wild dogs (Carpenter *et al.*, 1976; Woodroffe, 1997, 1999c).

At present, where disease has appeared to threaten small populations, three principal management strategies have been adopted. Firstly, attempts have been made to protect or treat individuals in the target population. For example, Medinyi arctic foxes were treated for mange, and successful vaccination campaigns were eventually carried out against CDV in black-footed ferrets (Woodroffe, 1999c). Although this approach has sometimes courted controversy, it may often be a cost-effective mechanism for achieving immediate conservation goals.

Secondly, where reservoir hosts are involved, managing disease in the reservoir and minimising contacts with the target population can prevent spillover infections (Laurenson et al., 1997; Woodroffe, 1999c). Where this reservoir host is a domestic carnivore, this may be a more manageable task than when the reservoir is a wild species. However, even in this latter situation, where the political will and economic resources are committed, as in the case of rabies control in coyotes in Texas and foxes in Europe, disease can be controlled in wildlife communities (e.g. Pastoret & Brochier, 1999). In domestic carnivores, possible indirect effects of disease control must be considered. For example, the vaccination of domestic dogs against rabies and CDV to protect wild carnivores may remove mortality factors and lead to increased dog populations and altered costs and efficacies of vaccination programmes. In both the Serengeti in Tanzania and the Bale mountains in Ethiopia where such experimental programmes have been set up, they have been designed to reveal secondary effects (Cleaveland, 1997; Laurenson et al., 1998). Preliminary results actually suggest that dog populations are limited by dog owners and that more stable dog populations may facilitate the implementation of such programmes (Cleaveland, 1997; Cleaveland & Laurenson, unpub. data).

Thirdly, the effect of metapopulation management strategies on small populations is unclear. A metapopulation approach, which involves movement of animals between populations, may help reduce the threat of extinction in the wild. Theoretical models, however, suggest that while such movement reduces the probability of extinction of the entire metapopulation, the spread of infectious diseases may be enhanced (Hanski, 1994). For example, highly contagious diseases causing moderate mortality have been suggested to increase extinction risk more than diseases causing high mortality (Hess, 1996). In addition, the critical community size for pathogen persistence may be reduced in a metapopulation (Swinton et al., 1998). Predictions, however, are greatly influenced by the model's assumptions (Hess, 1996) and the applicability of the metapopulation approach to host–parasite systems has never been tested empirically or experimentally.

Translocation and reintroduction of individuals hold great potential for augmenting the size, genetic diversity and persistence of small populations. However, the potential for transmission of pathogens from infected to non-infected populations has been repeatedly voiced (Simberloff & Cox,

1987; Griffith *et al.*, 1989; Woodford & Kock, 1991; Viggers *et al.*, 1993; Miller *et al.*, 1999). Managers have the opportunity to treat individuals and time reintroductions to minimise the possibility that the stress of reintroduction will increase the risk of disease. For instance, reintroduced captive-bred red wolves were treated to reduce the impacts of ectoparasites and endoparasites, and the timing of release of individuals was set to coincide with periods of low tick abundance (Phillips & Scheck, 1991). Although managers should have the opportunity to exclude or treat individuals whose translocation may decrease survival probabilities of the augmented population, such actions have not always been carried out. Raccoons, *Procyon lotor*, were translocated from the south-eastern to the mid-Atlantic United States in the 1970s by hunting clubs seeking to restock populations. Evidence exists that some translocated individuals were rabid, possibly leading to the spread of racoon rabies (Nettles *et al.*, 1979; Smith *et al.*, 1984).

Finally, because of the charismatic and flagship nature of carnivore conservation programmes, conservationists are often faced with philosophical and ethical issues that only rarely enter the scientific literature (e.g. the wild dog debate). In day-to-day management, confrontation may occur when issues are raised by the general public (e.g. mange in an urban environment) or by colleagues who voice their opinion in letters to editors (e.g. rabies control on domestic dogs and wildlife in the Serengeti system). It is thus important that where disease in natural populations requires active management, the design and goals of a management programme and the criteria for measuring success and failure are discussed and defined prior to initiation of any management strategy. Clearly, this requires the participation of various stakeholder groups and entails extensive co-ordination and information exchange. Differences of opinions and approach will be evident. For example, some groups may focus on individuals and welfare implications, whereas others may be most concerned with impacts at the population level.

RESEARCH PRIORITIES

Disease has only recently become a focus of research in conservation biology (Dobson & May, 1986; May, 1988; Dye *et al.*, 1995; Laurenson *et al.*, 1998; Woodroffe, 1999c). The increased awareness of the danger that diseases pose is mirrored in the relative importance IUCN Conservation Action Plans have placed on disease, ranging from the marginal discussion of disease in the volumes on procyonids and ailurids (1994) and cats (1996) to the very thorough risk assessment in more recent volumes on the African wild dog (1997) and Ethiopian wolf (1997). Most studies have focused on disease in large but neglected small carnivores (Murray et al., 1999; Williams & Thorne, 1996). The biological basis for the observed bias of catastrophic die-offs towards seals, felids and especially canids remains unclear. On the one hand, the bias may further emphasise the importance of domestic carnivores as reservoirs for generalist pathogens. Ursids or Viverrids may be unusually resistant to many of the pathogens affecting the other carnivore families. On the other hand, there are few in-depth studies on disease in large carnivore species and rarely do they address small carnivores.

This increased awareness of disease threats is aided by the large advances that have occurred in the statistical analysis and modelling of wildlife diseases (see under 'Management'). However, the gap is further widening between theory and empirical data. For instance, there are still very little data convincingly demonstrating the regulation of a Carnivora host by parasites and diseases. In addition, the circumstances under which mortality is compensatory or additive remains unclear. There is a need for basic studies of diseases that are not economically important. Because disease may kill study animals and thus reduce sample size, little data is available on how disease changes individual behaviour even for well-studied diseases such as rabies, although this may profoundly alter disease transmission (Artois & Aubert, 1985). Clearly, more long-term studies on the behaviour–epidemiology interface are required.

Effects of pathogens on host populations can often only be revealed by manipulative experiments, but those are rarely carried out (May, 1988; McCallum & Dobson, 1995; Schubert et al., 1998a). Much of the management debates, such as those surrounding rabies in wild dogs and tuberculosis in badgers, are fuelled by insufficient understanding of complex parasite–host interactions. The tuberculosis programme in Britain and the CDV and rabies vaccination programme of domestic dogs outside Serengeti National Park are examples of experiments that will shed light on the ongoing debates about the relative importance of badgers and domestic dogs as disease reservoirs in these two systems. Black-footed ferrets almost went extinct because early attempts at captive breeding were hindered by the death of several wild caught individuals following attempted vaccination with vaccines not tested in the species (Williams et al., 1992). Therefore, experiments are needed to test vaccines and effective vaccination protocols (Ginsberg, 1994; Woodroffe, 1999c).

To date, most research has been reactive and retrospective following outbreaks of catastrophic epidemics. A proactive approach of assessing the threat of disease before epidemics occur has rarely been followed. For some populations, epidemics have triggered comprehensive risk assessment such as in the Ethiopian wolf (Laurenson *et al.*, 1997, 1998), but other populations, including some 'flagship species', have not found similar attention despite concrete disease risks, possibly because catastrophic epidemics have not yet occurred. For example, CDV, canine coronavirus (CCV), and CPV may cause mortality in giant pandas and red pandas (Bush & Roberts, 1977; Qiu & Mainka, 1993; Mainka *et al.*, 1994). A serologic analysis of a small sample of domestic dogs, cats and captive and wild giant pandas in one Chinese Nature Reserve demonstrated antibodies against CDV, CCV and canine adenovirus, thus emphasising the disease risk for giant pandas and other Carnivora (Mainka *et al.*, 1994). Prevention and treatment of disease was implemented in the management of captive populations, but comprehensive disease screening or strategies of population and disease management of domestic dogs and cats within and adjacent to nature reserves have neither been devised nor implemented (F. Wei, pers. comm.). In this and many other cases, preventative health screening of potential disease is urgently needed.

Acknowledgements

We are most grateful for valuable comments by M. Artois, A. Dobson, J. Gittleman, K. Laurenson, and an anonymous referee.

Geographic priorities for carnivore conservation in Africa

M. GUS L. MILLS, STEFANIE FREITAG
AND ALBERT S. VAN JAARSVELD

INTRODUCTION

Being at the top of the food chain and given to often moving over large areas makes carnivores vulnerable to perturbations in ecosystems at almost any level. Therefore habitat deterioration and fragmentation may be expected to effect carnivores at an earlier stage of degradation than many other species. Furthermore, many carnivores frequently clash with human interests and are persecuted when they are considered a nuisance. For these reasons carnivores are often dependent on protected areas for their survival (M. Mills, 1991; Nowell & Jackson, 1996) and may remain vulnerable on reserve borders (Woodroffe & Ginsberg, 1998). Because of their wide ranging behaviour carnivores can serve as important umbrella species and their charismatic nature makes them suitable as important flagship species.

Protected areas remain limited. On average only 7% of the African continent is designated as protected area, ranging from 0% in equatorial Guinea to 14% in Tanzania (Singh *et al.*, 1998; WRI, 1998; Fig. 3). Because wild lands are becoming increasingly scarce and the economic resources allocated to biodiversity conservation fall far short of actual requirements (Reyers *et al.*, 1998; Balmford & Gaston, 1999), it is important to ensure that protected areas are selected optimally so that they provide protection for the greatest biodiversity. However, as pointed out by Pressey *et al.* (1993) and Pressey (1994), reserves have rarely been established for their representation of biodiversity features. Furthermore, the opportunism that has characterised the development of reserve systems may even have jeopardised the representation of biodiversity in reserves through the inefficient allocation of limited resources. Pressey *et al.* (1993) and Pressey (1994) called for a more systematic approach to reserve design if their role in protecting biodiversity is to be optimised.

In order to conserve biodiversity regional conservation actions should aim to sample the full spectrum of species found in that region (Pressey *et al.*, 1993; Freitag & van Jaarsveld, 1997). A number of strategies to achieve this goal have been developed. These include the identification of various types of biodiversity hotspots (e.g. Myers, 1988; 1990; Prendergast *et al.*, 1993; Mittermeier *et al.*, 1998) and the use of a number of iterative selection algorithms where representativeness is achieved using the principle of complementarity (e.g. Pressey *et al.*, 1993; Freitag *et al.*, 1996; Prendergast *et al.*, 1999), or a combination of both (e.g. Lombard, 1995; Freitag *et al.*, 1997). While much has been published on the relative merits of these different approaches, no single method used in isolation will provide the definitive outcome. Rarity and richness-based procedures have been shown to have limited overlap (Kershaw *et al.*, 1994), simple scoring procedures are less land-use efficient than iterative algorithms (Margules *et al.*, 1991), algorithm outputs are often rule-dependent (Underhill, 1994; Pressey *et al.*, 1997a,b) and have been criticised for not reaching mathematical optimality (Underhill, 1994).

In this chapter we address geographic priorities for terrestrial carnivore conservation on mainland Africa and the issue of reserve placement but not reserve design (Pressey *et al.*, 1993). Firstly, we attempt to prioritise each African carnivore species in order of regional conservation importance. Then we map the important areas (hotspots) of carnivore species richness, endemism and IUCN Red Data Book (RDB) vulnerability status, first individually and then collectively. Finally, we select grid cells (minimum area complementary sets) for representative carnivore conservation in Africa. We then discuss the value of these different conservation planning approaches for assessing the degree of protection currently afforded African carnivores, remaining aware of the dangers of basing conservation decisions on just one taxon (see P. Howard *et al.*, 1998; van Jaarsveld *et al.*, 1998).

MATERIALS AND METHODS

Carnivore database

The database used in the present study includes distribution data (presence only) for 70 carnivore species found on the African continent (Appendix 21.1). Due to numerous taxonomic uncertainties, the final species list was based on the scientific nomenclature taken from the Smithsonian Institution's 1993 'Mammal Species of the World' web site (Smithsonian Institute, 1993).

Carnivore distribution data for mainland Africa were digitised from published distribution maps presented in Skinner & Smithers (1990), Nowak & Paradiso (1983), Dorst & Dandelot (1972) and several IUCN Action Plans (Schreiber *et al.*, 1989; Forster-Turley *et al.*, 1990; Ginsberg & Macdonald, 1990; Nowell & Jackson, 1996; M. Mills & Hofer, 1998). Every attempt was made to use the most recent distribution maps, although data for the more obscure and little known species could be less accurate than those for the better known species. Digital maps were overlaid on a 2° × 2° grid (approximately 200 km × 200 km) and the presence of all species recorded within each grid cell ($n = 1482$). The size of the grids was of the same order of magnitude as the largest protected areas on the continent. The islands off the African coast and Madagascar were not included in the analysis for two reasons. First, because of problems of scale, 2° × 2° grids would offer far too little resolution on the islands or Madagascar for meaningful analyses, and secondly, because of the distinctiveness of their carnivore faunas, which would not affect conservation requirements on mainland Africa.

Data analyses

Continent-wide species priority ratings

The approach of prioritising species in order of regional conservation importance, proposed by Freitag & van Jaarsveld (1997), was used with minor modifications, including the incorporation of an additional body size category.

(a) *Taxonomic distinctiveness (TD)*: was calculated as

$$TD = \frac{1}{\sqrt{\text{no. subfamilies} \times \text{no. genera} \times \text{no. species}}} \qquad (1)$$

(b) *Degree of endemism (E)*: three categories of endemism were used, namely, true endemics (score = 1.0), partial endemics (score = 0.5) and non-endemics (score = 0). Partial endemics include those species that extend only marginally beyond Africa, or into Africa, or where the presence of viable populations outside Africa is unlikely.

(c) *Vulnerability status scores (V)*: were based on the IUCN's 1996 red list of threatened animals, or more recent IUCN action plans, and listings in Cites Appendix I and II (Smithsonian Institute, 1993; Baillie & Groombridge, 1996; Nowell & Jackson, 1996; M. Mills & Hofer, 1998). Categorisation and scoring was as follows: IUCN critically endangered (score = 1.0); IUCN endangered (score = 0.84); IUCN vulnerable (score = 0.7); IUCN lower risk: conservation dependent (score = 0.56);

IUCN lower risk: near threatened; IUCN data deficient; Cites Appendix I or II (score = 0.42); IUCN lower risk: least concern (score = 0.28); not evaluated or listed (score = 0).

(d) *Extent of occurrence (EO)*: was calculated as

$$EO = \frac{1}{\text{no. grid cells occupied in Africa}} \tag{2}$$

(e) *Body size category (BM)*: was included as an estimator of the potential for human conflict on the premise that the larger the carnivore the more likely it is to attack livestock. Four size class categories were used: very small (< 1 kg; score = 0.25); small (1–7 kg; score = 0.5); medium (7–25 kg; score = 0.75); and large (> 25 kg; score = 1.0).

(f) *Regional priority scores (RPS)*: were calculated by assigning equal weighting to each of the five categories:

$$RPS = \frac{TD + E + V + EO + BM}{5} \tag{3}$$

Richness, endemism and vulnerability hotspots

Species richness was calculated as the sum of species occurring within each grid cell. Similarly, endemism and vulnerability were determined by summing endemic and vulnerable species within grid cells. The degree of spatial overlap among the richness, endemism and vulnerability hotspots was assessed using the Jaccard coefficient (van Jaarsveld *et al.*, 1998):

$$\text{Spatial Overlap (Jaccard Coefficient)} = X / (A + B + X) \times 100 \tag{4}$$

where X is the number of grid cells shared, A is the number of additional grids selected using one criterion, and B is the number of additional grids selected for the other criterion.

Those grid cells that fell within the top 10% scoring sites for all three criteria – species, endemism and vulnerability – were ranked in order of importance by summing the RPSs for all species occurring within such grid cells and are referred to as combined hotspots.

African map of conservation areas

We developed a consolidated map of designated conservation areas in Africa. The map represents a refinement of the African Elephant Database (AED) conservation map. Additional conservation areas were mapped from localities from the World Conservation Monitoring Centre databases (WCMC, 1998). Where actual boundaries were not available, central localities were buffered to represent the reported sizes of these conservation areas.

Near-minimum complementarity sets

Minimum area requirements (in terms of the level of resolution of the data into 2° × 2° grid cells) for representative carnivore species conservation in Africa were selected using a multi-criteria iterative minimum set selection algorithm. The algorithm represents a slight modification of that described in Freitag *et al.* (1997); the modifications centre on the modified RPS allocations which included body mass.

RESULTS
Priority species for conservation

A breakdown of the individual category scores as well as the overall continent-wide RPSs of all species included in the study are provided in Appendix 21.1 (species are listed in decreasing order of conservation importance). According to this analysis the highest priority species requiring conservation action for the African continent is the Ethiopian wolf (*Canis simensis*) which was allocated maximum scores in three of the five evaluated categories (endemism, vulnerability and extent of occurrence). Next is the African wild dog (*Lycaon pictus*), followed by the spotted hyaena (*Crocuta crocuta*), the Liberian mongoose *(Liberiictis kuhni)* and the Congo clawless otter *(Aonyx congicus)*.

Hotspot assessment

Three grid cells, one each in Kenya, the Ugandan/Rwandan/Tanzanian/Democratic Republic of Congo border area and Botswana were the richest in species with 28 species each (Figure 21.1a). The same Ugandan/Rwandan/Tanzanian/Democratic Republic of Congo border area grid square was also richest in endemics, containing 22 species endemic to Africa (Figure 21.1b). Eleven grid squares were equally rich in species with vulnerability status scores. However, when the actual vulnerability scores per species within grid cells were summed, the highest ranking grid was in Ethiopia with three tied second place grids in Kenya and Tanzania (Figure 21.1c). East, south-central and southern Africa contain most hotspots (Figure 21.1a,b,c), although grids rich in vulnerable species are also concentrated in north-east and west Africa (Figure 21.1c).

The degree of overlap among the species richness, endemism and vulnerability hotspots, expressed through the Jaccard Coefficient, varied from

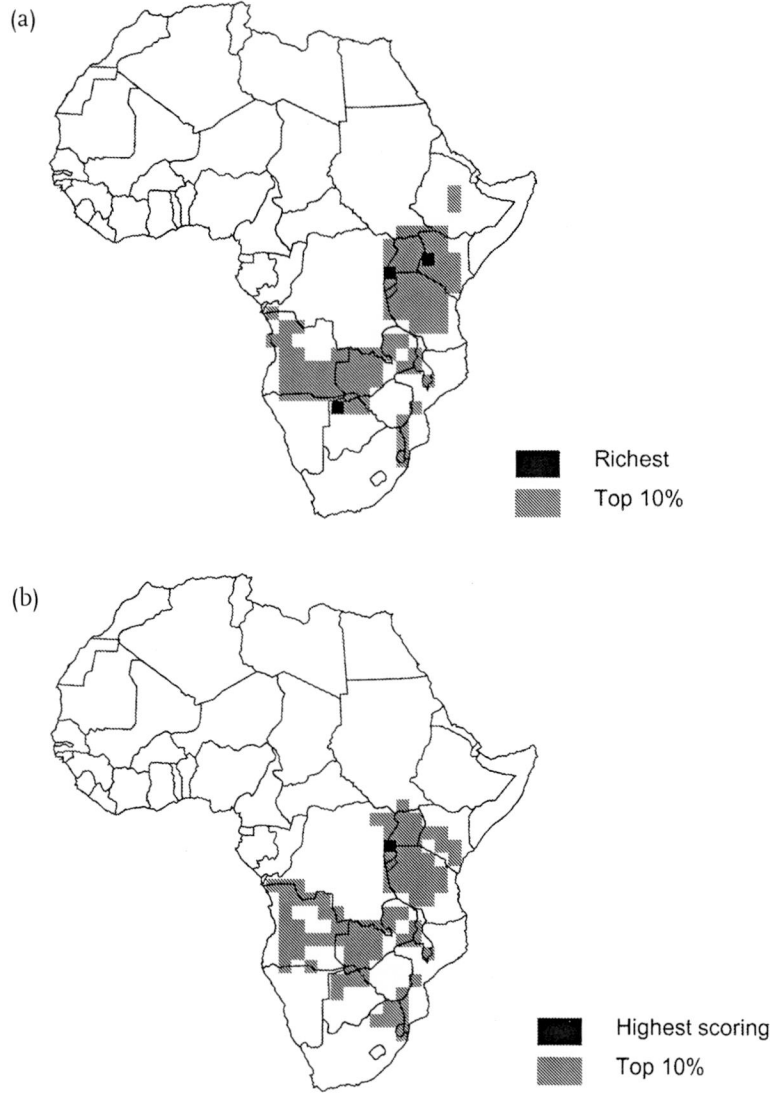

Figure 21.1. Hotspots (top 10% richest areas) in (a) species, (b) endemic species, and (c) vulnerable species of carnivores in Africa.

(c)

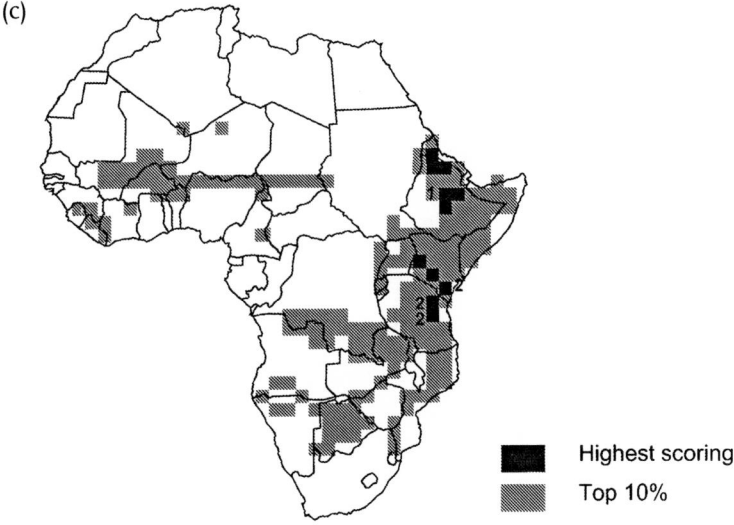

Highest scoring

Top 10%

Figure 21.1 (*cont.*)

23% to 72% (Table 21.1). This variable overlap led us to assess these categories collectively in order to adequately sample all three parameters. From this analysis 43 grid cells were identified as combined hotspots mainly concentrated in East Africa, with some in south-central, west-central and southern Africa. The highest scoring hotspot grid cell falls in northern Botswana with the next highest in western Kenya and the third in the Ugandan/Rwandan/Tanzanian/ Democratic Republic of Congo border area (Figure 21.2).

Complementarity network

The complementarity analysis suggests that at least 16 grid cells are required to represent the 70 carnivore species we identified to occur on the African continent at least once each (Figure 21.3). By employing RPS rankings to schedule the relative importance of these complementary conservation sites, we were able to identify the sequence in which these cells should be conserved. This sequence maximises the conservation gains in terms of species value added at each step. The highest priority cell is located in Ethiopia and contains the Ethiopian wolf plus a further four of the top 10 species from the RPS list (wild dog, spotted hyena, cheetah *Acinonyx jubatus*, lion *Panthera leo*). The second priority grid cell reflects the presence of the Liberian mongoose from the top 10 list while the third grid cell is selected for the presence of the Congo clawless otter from the Rwanda/

Table 21.1. The degree of spatial overlap (Jaccard Coefficient) among: (i) carnivore species richness hotspots, vulnerability hotspots and endemism hotspots; and (ii) combined species, endemism and vulnerability hotspots and complementarity sets

	Degree of spatial overlap (%)
10% hotspots	
(i) Species richness vs. endemism	72
Species richness vs. vulnerability	25
Endemism vs. vulnerability	23
(ii) Combined hotspots vs. complementarity	4

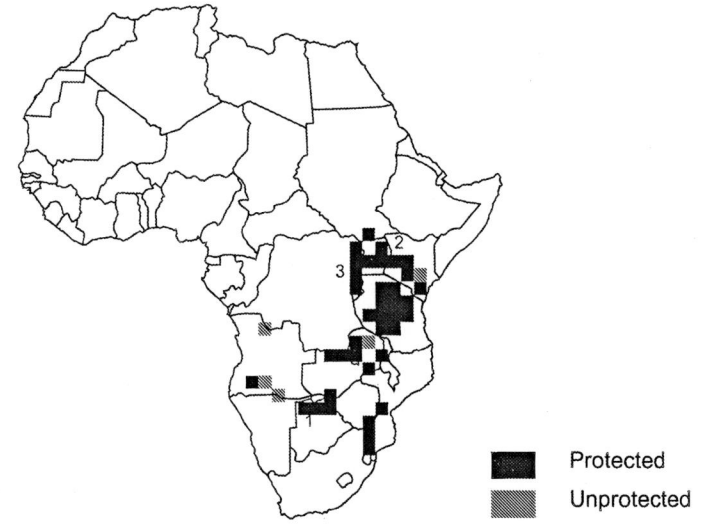

Protected
Unprotected

Figure 21.2. Combined hotspots (10%) rich in species, endemic species or vulnerable species in Africa and their protected status. The numbers refer to the top three ranking hotspots.

Burundi region as well as the spotted hyena and lion.

Efficacy of the hotspot and complementarity approaches to achieving full carnivore protection

Overlaying the conservation area map on the combined grid cell hotspots yielded the result that five (12%) of the 43 grid cells selected contain no protected areas. Three of these cells are in Angola and one each is in Zambia and Kenya (Figure 21.2). The hotspot approach falls far short of conserving all 70 carnivore species by failing to represent 21 species (Table 21.2), including such important species in terms of their RPS rank as the

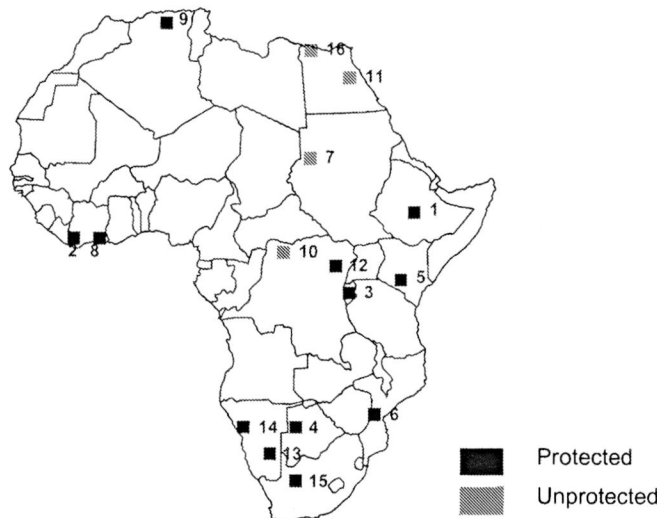

Figure 21.3. A complementary network that represents at least once every carnivore species in Africa. Rankings indicate the conservation importance of grid cells in terms of the regional conservation importance of the species contained in them. The protected status of grid cells is also indicated.

Ethiopian wolf and Liberian mongoose, as well as Johnston's genet (*Genetta johnstoni*), pale fox (*Vulpes pallida*) and Abyssinian genet (*Genetta abyssinica*) (Table 21.3). A further 11 grid cells (selected using the complementarity algorithm) are required to include these 21 species (Table 21.2).

Overlaying the conservation area map on the complementary set yielded the result that four (25%) of the 16 grid cells required for the complementary set contain no protected areas, the highest priority rank of these cells being 7th (Figure 21.3). Two cells are in Egypt, one in Sudan and one in the Democratic Republic of Congo (Figure 21.3). Seven species occur in these four grid cells with the pale fox with a RPS rank of 17th (Table 21.3), the most important. None of the other species appear to be seriously compromised by the fact that no protected areas occur in these particular grid cells as they are adequately protected in other areas, although little is known about the Congo cusimanse (*Crossarchus alexandri*) and the Angolan cusimanse (*Crossarchus ansorgei*).

Only 4% of grid cells were shared by the complementarity set and the combined hotspots network (Figure 21.4; Table 21.1) indicating that combined hotspots and complementary sets overlap little. This, together with

Table 21.2. *Number of grid cells required to fulfill species representation targets for hypothetical carnivore complementary versus 10% hotspot-based networks (richness, vulnerability and endemism), number of species represented in each network and number of additional grid cells required to complete full carnivore species representation for Africa*

	Number of grid cells required	Number of species represented	Number of species not represented	Number of additional grid cells required to complete species representation
Hotspots	43	49	21	11
Complementary	16	70	0	0

Table 21.3. *Species and their regional priority scales (RPS) rank not represented in (i) combined hotspot and (ii) complementarity selected grid cells*

Species	RPS Rank
(i) *Combined hotspots*	
Ethiopian wolf	1
Liberian mongoose	4
Johnston's genet	16
Pale fox	17
Abyssinian genet	18
Gambian mongoose	21
Villiers' genet	29
Eurasian otter	30
Cape fox	31
Forest genet	37
Fennec fox	40
Angolan cusimanse	48
Galerella swalius	50
Galerella flavescens	53
Suricate	56
Small grey mongoose	58
Jungle cat	63
Rueppel's fox	64
Sand cat	66
Libyan striped weasel	68
Red fox	69
(ii) *Complementary grids*	
Sand fox	17
Fennec fox	40
Congo cusimanse	42
Angolan cusimanse	48
Jungle cat	63
Sand cat	66
Red fox	69

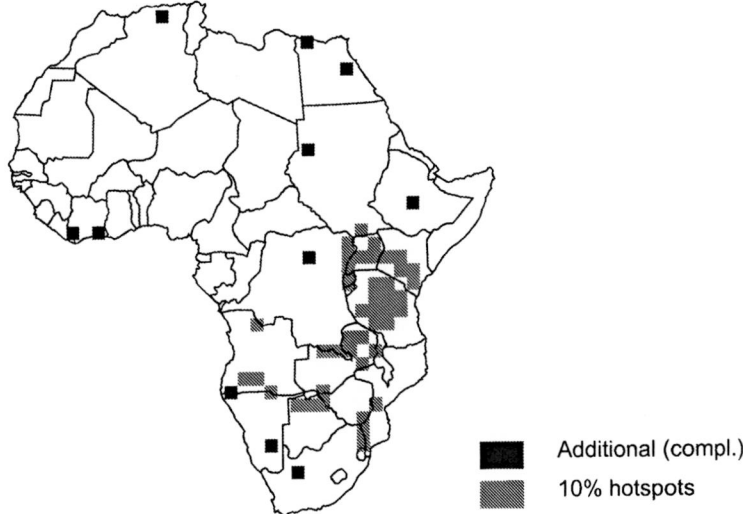

Figure 21.4. A combined hotspots and complementarity conservation map for African carnivores.

the fact that 21 species are not represented in the combined hotspots network, resulted in us adopting an approach integrating hotspots analysis with complementarity. This was achieved by using complementarity around the combined hotspots network to achieve complete species representation. Figure 21.4 depicts the hotspots network ($n = 43$ grid cells) together with the selection of an additional 11 grids, using complementarity, to achieve complete carnivore species representation in Africa.

DISCUSSION

By assigning five different but complementary 'rarity' criteria to all African carnivores, taking into account criteria not included in the IUCN's red list categories, we have attempted to strengthen the evaluation of regional species conservation priorities as each of the criteria applied have specific strengths and weaknesses. This is a relational approach where the relative conservation importance of a species is derived from a suite of criteria equally weighted, and rarity is treated as a continuous variable in a species-by-species assessment, thus circumventing problems of scale by standardisation (Freitag & van Jaarsveld, 1997). As far as we are aware this is the first time that this technique has been used for a multi-family, regional comparison on carnivores, although a similar technique was employed by

Nowell & Jackson (1996) on cats. We suggest that this approach is worthy of further use and development, although we do not presume that our criteria are necessarily optimal. For example, there is the possibility of partial overlap between the criteria vulnerability (V) and extent of occurrence (EO), but assigning a vulnerability category to a species using the 1994 IUCN red list categories may be done using any of three measures (declining population, shrinking range or extent of occurrence) so may not take extent of occurrence into consideration. There may also be merit in assigning unequal weightings to the criteria or to drop certain criteria should overlap become evident depending on the scale and objectives of the analysis.

Our analyses have produced some unexpected results. For example, the spotted hyena is given the third highest conservation priority while its IUCN RDB status is Lower Risk:Conservation dependent (M. Mills & Hofer, 1998), whereas the cheetah's RDB status is Vulnerable, but it is only placed eighth. This is mainly due to the fact that the spotted hyena is given extra importance because it is a true African endemic, whereas the cheetah is scored as a partial endemic. Consequently the spotted hyena's regional conservation importance elevates it above a species considered more vulnerable at a global scale.

The theoretical case for conserving areas rich in endemics or containing restricted range species and vulnerable taxa has been well argued (Myers 1988; 1990; Mittermeier *et al.*, 1998; Baillie & Groombridge, 1996). However, the case for conserving regions rich in species is usually based on the tenet that conserving species rich areas will conserve large numbers of species in small land areas (Myers, 1990; Prendergast *et al.*, 1993; Reid, 1998). This assumption has been repeatedly challenged in favour of more rational and land-use efficient approaches towards representing species efficiently. The use of heuristic algorithms employ the principles of complementarity to represent suites of species in a more land-use efficient manner by avoiding duplication (Pressey *et al.*, 1993; Camm *et al.*, 1996; P. Williams *et al.*, 1996; Freitag *et al.*, 1997). Consequently, it is now generally accepted that achieving conservation efficiency provides little space for including areas rich in species in regional conservation networks (Pressey *et al.*, 1993; P. Williams *et al.*, 1996; Pimm & Lawton, 1998; Figure 21.3).

Notwithstanding, a case for including areas rich in species in conservation networks can be made from an evolutionary–ecological perspective to conservation (Soulé & Wilcox, 1980; Frankel & Soulé, 1981; Soulé, 1986). This view essentially extends a strictly pragmatic conservation goal (Pressey *et al.*, 1993) to incorporate an added goal aimed at reducing the risks of extinction (Fjeldså, 1994). Here areas that are comparatively rich in species are viewed as regions: (1) containing larger numbers of restricted range species, or (2) where large numbers of species accumulate, or (3) that act as centres of diversification (see Fjeldså, 1994).

A related case can be made for conserving areas containing certain charismatic species as this might be the optimal form of land use through ecotourism and trophy hunting. This is particularly relevant for the larger carnivores. The emphasis in conservation, particularly in developing countries, should be on projects which seek to work with local people to make wild lands not only sustainable, but profitable (Nowell & Jackson, 1996). In this respect Thresher (1982) calculated the value of a male lion in Kenya's Amboseli National Park at US $128 570, while a leopard was calculated to be worth US $50 000 per year in a privately owned reserve in South Africa (Martin & de Meulenaer, 1988), and the price of a 21-day lion hunting safari in Tanzania was US $35 000 (Makombe, 1994).

These benefits derived from incorporating hotspots into conservation evaluation assessments led us to the combined hotspots and complementarity approach used here. The result that some categories of hotspots (richness, endemism and vulnerability) fail to overlap consistently with one another supports previous findings (Prendergast et al., 1993; Reid, 1998; van Jaarsveld et al., 1998) and resulted in the combination of the above categories into a single combined hotspot category to ensure sampling of all potentially important areas. This follows the Myers (1988, 1990) approach towards identifying hotspots which aims to identify areas where these variables are coincident. However, these combined hotspots were inadequately protected (12% unprotected, Figure 21.2) and only provide protection for 70% of species (Table 21.2).

The results presented here also reinforce the notion that the conservation goals of species representation and hotspot conservation may be spatially disjunct. This suggests that these conservation goals should be employed synergistically in order to minimise the probability of species extinction. We adopt just such a synergistic approach towards identifying priority conservation areas for carnivores in Africa which are summarised in Figure 21.4.

We emphasise that our aim here is to highlight the important regions in Africa in terms of reserve placement for carnivore conservation. It is important that these preliminary findings should be ground truthed to determine the current conservation status of the relevant species in the regions, the state of the habitats and threats of land transformation, as well as the integrity of any existing conservation areas. Once this has been achieved a reserve design phase should follow, whereby the size, shape and management framework of the protected areas are planned and negotiated (Pressey *et al.*, 1993) around these priority regions identified in the present analysis. The grid cells employed in the present analysis are of sufficient size ($\sim$ 200 × 200 km) to contain the bulk of any future reserves designed to accommodate viable populations of wide ranging carnivore species. This is particularly important for the larger species that need large areas to contain viable populations, or special management programs such as metapopulation management to ensure population viability.

It is striking and alarming that 21% of the combined hotspots and the third most important complementarity grid cell fall in countries where civil strife and war are continuing (Sudan, Democratic Republic of Congo, Burundi, Rwanda and Angola; Figures 21.2 & 21.3). These activities can have a major impact on biodiversity as conservation becomes a very low priority. There is little that conservationists can do in these situations except indirectly through encouraging and pursuing policies of sustainable utilisation, thereby improving the quality of life of local people and helping to stabilise volatile political conditions.

Acknowledgements

We would like to thank South African National Parks, the University of Pretoria, the Tony and Lisette Lewis Foundation and the Foundation for Research Development for financial assistance. Neil Burgess (Zoologisk Museum, Danemark) provided distribution maps for *Genetta angolensis*, *Crossarchus alexandri*, *Crossarchus ansorgei*, *Helogale hirtula*. Janet Webb (Scientific Services, KNP) is thanked for digitising certain distribution maps. Naledi Maré (Scientific Services, KNP), Heath Hull (UP) and Kinga Kuc (UP) provided GIS assistance and Willy Simmons (African Elephant Database, Nairobi) provided a digital version of the AED conservation map.

Appendix 21.1. Regional priority scores for African carnivores

Genus	Species	Common name	Taxonomic distinctivenness	Endemism	Extent occurrance	RDB status	Size	Priority score
Canis	simensis	Ethiopian wolf	0.25	1	1	1	0.75	0.8
Lycaon	pictus	wild dog	0.5	1	0.003	0.84	1	0.6686
Crocuta	crocuta	spotted hyena	0.5	1	0.003	0.56	1	0.6126
Liberiictis	kuhni	Liberian mongoose	0.154	1	0.5	0.84	0.5	0.5988
Aonyx	congicus	Congo clawless otter	0.204	1	0.5	0.42	0.75	0.5748
Hyena	brunnea	brown hyena	0.354	1	0.022	0.42	1	0.5592
Bdeogale	jacksoni	Jackson's mongoose	0.089	1	0.5	0.7	0.5	0.5578
Acinonyx	jubatus	cheetah	0.577	0.5	0.003	0.7	1	0.556
Proteles	cristatus	aardwolf	0.707	1	0.007	0.28	0.75	0.5488
Panthera	leo	lion	0.408	0.5	0.003	0.7	1	0.5222
Leptailurus	serval	serval	0.289	1	0.003	0.42	0.75	0.4924
Bdeogale	crassicauda	bushy-tailed mongoose	0.089	1	0.023	0.84	0.5	0.4904
Osbornictis	piscivora	aquatic civet	0.289	1	0.2	0.42	0.5	0.4818
Aonyx	capensis	Cape clawless otter	0.204	1	0.004	0.42	0.75	0.4756
Profelis	aurata	African golden cat	0.289	1	0.008	0.28	0.75	0.4654
Genetta	johnstoni	Johnston's genet	0.096	1	0.2	0.42	0.5	0.4432
Vulpes	pallida	pale/sand fox	0.224	1	0.011	0.42	0.5	0.431
Genetta	abyssinica	Abyssinian genet	0.096	1	0.125	0.42	0.5	0.4282
Felis	nigripes	black-footed cat	0.144	1	0.027	0.42	0.5	0.4182
Nandinia	binotata	two-spotted palm civet	0.577	1	0.009		0.5	0.4172
Mungos	gambianus	Gambian mongoose	0.109	1	0.043	0.42	0.5	0.4144
Civettictis	civetta	African civet	0.289	1	0.003		0.75	0.4084
Otocyon	megalotis	bat-eared fox	0.5	1	0.009		0.5	0.4018
Canis	mesomelas	black-backed jackal	0.25	1	0.007		0.75	0.4014

Appendix 21.1. (cont.)

Genus	Species	Common name	Taxonomic distinctiveness	Endemism	Extent occurrance	RDB status	Size	Priority score
Canis	adustus	side-striped jackal	0.25	1	0.005		0.75	0.401
Poiana	richardsonii	African linsang	0.289	1	0.014	0.42	0.25	0.3946
Panthera	pardus	leopard	0.408		0.002	0.42	1	0.366
Hyena	hyena	striped hyena	0.354		0.006	0.42	1	0.356
Genetta	thierryi	Villiers' genet	0.096	1	0.167		0.5	0.3526
Lutra	lutra	Eurasion otter	0.204		0.1	0.7	0.75	0.3508
Vulpes	chama	Cape fox	0.224	1	0.02		0.5	0.3488
Lutra	maculicollis	spotted-necked otter	0.204	1	0.005		0.5	0.3418
Rhynchogale	melleri	Meller's mongoose	0.154	1	0.053		0.5	0.3414
Crossarchus	ansorgei	Angolan cusimanse	0.089	1	0.1		0.5	0.3378
Paracynictis	selousi	Selous' mongoose	0.154	1	0.023		0.5	0.3354
Genetta	victoriae	giant genet	0.096	1	0.071		0.5	0.3334
Genetta	maculata	forest genet	0.096	1	0.063		0.5	0.3318
Atilax	paludinosus	water/marsh mongoose	0.154	1	0.003		0.5	0.3314
Ichneumia	albicauda	white-tailed mongoose	0.154	1	0.003		0.5	0.3314
Fennecus	zerda	fennec fox	0.224	0.5	0.009	0.42	0.5	0.3306
Herpestes	naso	long-snouted mongoose	0.109	1	0.024		0.5	0.3266
Crossarchus	alexandri	Congo cusimanse	0.089	1	0.04		0.5	0.3258
Genetta	tigrina	large-spotted genet	0.096	1	0.029		0.5	0.325
Mungos	mungo	banded mongoose	0.109	1	0.004		0.5	0.3226
Crossarchus	obscurus	dark mongoose	0.089	1	0.023		0.5	0.3224
Bdeogale	nigripes	black-legged mongoose	0.089	1	0.02		0.5	0.3218
Genetta	servalina	small-spotted genet	0.096	1	0.012		0.5	0.3216
Genetta	angolensis	Angolan genet	0.096	1	0.007		0.5	0.3206
Poecilogale	albinucha	African striped weasel	0.33	1	0.01		0.25	0.318

Galerella	*swalius*[a]		0.077	1	0.25		0.25	0.3154
Felis	*silvestris*	African wild cat	0.144	0.5	0.002	0.42	0.5	0.3132
Dologale	*dybowskii*	Pousargues' mongoose	0.154	1	0.143		0.25	0.3094
Galerella	*flavescens*[a]		0.077	1	0.2		0.25	0.3054
Ictonyx	*striatus*	striped polecat	0.236	1	0.003		0.25	0.2978
Caracal	*caracal*	caracal	0.289		0.002	0.42	0.75	0.2922
Suricata	*suricatta*	suricate/meerkat	0.154	1	0.024		0.25	0.2856
Cynictis	*penicillata*	yellow mongoose	0.154	1	0.018		0.25	0.2844
Galerella	*pulverulenta*	small gray mongoose	0.077	1	0.067		0.25	0.2788
Helogale	*hirtula*	East African dwarf mongoose	0.109	1	0.026		0.25	0.277
Helogale	*parvula*	dwarf mongoose	0.109	1	0.008		0.25	0.2734
Galerella	*sanguinea*	slender mongoose	0.077	1	0.003		0.25	0.266
Mellivora	*capensis*	honey badger	0.577		0.002		0.75	0.2658
Felis	*chaus*	jungle/swamp cat	0.144		0.2	0.42	0.5	0.2528
Vulpes	*rueppelli*	Rueppell's fox	0.224		0.006	0.42	0.5	0.23
Genetta	*genetta*	common genet	0.096	0.5	0.004		0.5	0.22
Felis	*margarita*	sand cat	0.144		0.016	0.42	0.5	0.216
Canis	*aureus*	common/golden jackal	0.25		0.002		0.75	0.2004
orctonyx	*libyca*	Libyan striped weasel	0.236	0.5	0.004		0.25	0.198
Vulpes	*vulpes*	red fox	0.224		0.067		0.5	0.1582
Herpestes	*ichneumon*	large gray/Egyptian mongoose	0.109		0.003		0.5	0.1224

[a]No common names could be found for these species.

22

Estimating interpopulation dispersal rates

PETER M. WASER, CURTIS STROBECK AND DAVID PAETKAU

In the mid-1960s, Gerry Storm and colleagues tagged more than 2000 red foxes in the midwestern United States, and compared initial capture locations with the locations at which these animals later died (Storm *et al.*, 1976). Beginning in the late 1960s, David Mech and coworkers radio-collared wolves in Minnesota (USA); many of these animals dispersed while carrying collars (Mech, 1987). Around the same time, George Schaller began to establish photographic identification files for individual East African lions. Years of intensive censusing of individually-recognizable animals allowed subsequent investigators to document lifetime movements (Hanby & Bygott, 1987; Pusey & Packer, 1987).

These and numerous similar studies have established the quantitative characteristics of short-distance dispersal, and the qualitative characteristics of long-distance dispersal, in many carnivore populations. Patterns of short-distance dispersal can have important consequences for conservation, and several of these may be better understood among carnivores than among any other mammals. Studies of within-population dispersal by lions as an inbreeding avoidance mechanism have become a model for our understanding of this issue in other species (Packer & Pusey, 1993). Philopatry, the absence of dispersal, is central to the complex and altruistic social interactions that give some carnivores their success as predators, and allow others to avoid becoming prey. Species like the dwarf mongoose are among the best-studied mammalian examples of this phenomenon (Rood, 1987; Creel & Waser, 1994).

But for carnivores, just as for other animals, long-distance dispersal rates remain largely a black box. Many of the most important conservation consequences of dispersal depend on rates of long-distance movement, and on dispersal between rather than within populations. Between-population dispersal can rescue declining populations, colonize new habitat or recolonize areas subject to past extinction, and prevent the loss of genetic variation that might otherwise seal the fate of isolated populations. But between-population dispersers represent the extreme tail of the dispersal distance distribution – they are a tiny proportion of most populations, and the most difficult of all dispersers to observe. Rarely do we know what habitat factors impede or assist dispersal between isolated populations, an issue of increasing conservation importance (van Vuren, 1998).

In this chapter, we briefly review and discuss the limitations of traditional methods available for estimating long-distance dispersal rates, and then focus on new approaches that tap the tremendous stores of information present in the genes of animals. These methods are developing rapidly, and hold the promise of allowing us to categorize each individual as a resident or an immigrant into the population within which it is sampled, and perhaps to identify its birth population.

DIRECT APPROACHES TO ESTIMATING DISPERSAL RATES

In principle, all we need to do to estimate interpopulation dispersal rates is to tag large numbers of young animals near their birthplaces, follow them throughout their lives, and see where they go. This is known traditionally as the 'direct' approach (Slatkin, 1985), and carnivore field studies provide excellent examples of most of its variants.

For example, very large numbers of Serengeti lions have been identified by individually-specific whisker patterns; the photographic record of such a pattern constitutes a tag (Pusey & Packer, 1987; Packer & Pusey, 1993). These animals constitute essentially a complete sample of all potential dispersers in the Serengeti shortgrass plains, and because nearly all adults in the same area have also been identified over a large number of years, we know more about lion dispersal within this study area than for all but a handful of other species. This approach of intensively tagging juveniles and then censusing adults within the same study area, nevertheless has two practical shortcomings. First, few carnivores are as accessible or visible as Serengeti lions. Rarely is it possible to tag adequate numbers of juveniles or to census adults completely enough to generate precise or reliable dispersal descriptors. Second, the approach tells us virtually nothing about long distance dispersal. Longer dispersal distances are necessarily undersampled, as an unknown fraction of juveniles born near the edge of the study area disperse out of it. Statistical corrections for this problem have been proposed (Barrowclough, 1978; Baker et al., 1995; Koenig et al., 1996) but they cannot solve the problem that, with this approach, distances greater than the maximum dimension of the study area are not sampled at all.

Studies of European badgers (Cheeseman *et al.*, 1988) exemplify another variant of the direct approach. Like Storm in his red fox study, Cheeseman and colleagues tagged individual badgers by a variety of means, and identified dispersal when tagged animals were radio-located, sighted, captured, killed in badger control operations, or found dead on roads. This approach detected some dispersal out of the initial study areas, providing extensive qualitative information on dispersal characteristics. But quantitative conclusions are more elusive. Are all juveniles equally likely to be sampled by roadkills or badger control operations? To what extent is dispersal complete when animals are opportunistically retrapped or resighted? Are these sampling mechanisms themselves spatially restricted, so that they undersample long dispersal distances?

As another approach to the limited-study-area problem, many carnivore studies use radio-collars as 'tags'. Juveniles are tagged in their natal population and followed wherever they go during the dispersal process. If radio-collars are attached to a large, unbiased sample of juveniles and continue to operate (and contact is maintained) for long enough, the study area size is effectively unlimited and even long-distance interpopulation dispersal rates can be measured. But in practice, rates of interpopulation movement are too low and the logistic difficulties of marking adequate numbers of juveniles are too severe to meet these conditions. The difficulties are well illustrated by numerous intensive studies of radio-collared wolves (Peterson *et al.*, 1984; Messier, 1985; Ballard *et al.*, 1987; Potvin, 1987; Gese & Mech, 1991). While these studies have told us a great deal about the process of wolf dispersal, only a small proportion of animals have completed the entire dispersal process while collared, and most of these have dispersed short distances, remaining within their population of origin.

Still another variant of the 'direct' approach involves elaboration of mark–recapture statistical techniques. If large enough random samples of individuals can be tagged, released, and repeatedly recaptured (or resighted), statistical models have been developed to distinguish immigration from recruitment and to estimate interpopulation dispersal rates (Nichols & Pollock, 1990; Spendelow *et al.*, 1995). To our knowledge, however, the large sample sizes required by such techniques have not yet allowed their application to carnivore populations.

Sometimes, repeated censuses are possible and most are individually recognizable so that immigrants into a study site and disappearances from it can be counted. Under these conditions, upper and lower bounds can be placed on rates of emigration (and on mortality rates associated with emigration). This approach has been used with dwarf mongooses and Eurasian badgers (Waser *et al.*, 1994b; Waser, 1996). Once again, however, few carnivores are easy enough to identify and census for this approach to be viable, and its assumption that the study area is neither a source nor a sink is not always tenable.

INDIRECT APPROACHES TO ESTIMATING DISPERSAL RATES

For almost as long as some biologists have tried to estimate interpopulation dispersal rates directly, others have tried to do it indirectly using the information present in the animals' genes (Slatkin, 1985, 1994). For 50 years, this approach has most often relied on some variant of Wright's (1951) F_{st} and its relationship to the expected number of migrants per generation Nm, $F_{st} = 1/(1 + 4Nm)$. In practice, however, this approach has suffered both practical and conceptual limitations. The overriding practical limitation has been a lack of allelic variation; prominent among the conceptual limitations has been the assumption inherent in Wright's approach that the populations analyzed are at genetic equilibrium (Neigel, 1997). Until recently, the most serious of these limitations for carnivores has been lack of variation. DNA variation was accessible only through protein sequences (allozymes), which are notoriously invariant in most carnivore species (Simonsen, 1982; Kennedy *et al.*, 1990; Wayne & Koepfli, 1996). Evans *et al.* (1989) convey the flavor of the usual results: only one of 23 allozyme loci was variable enough to estimate an F_{st} among four clusters of Eurasian badger groups; the estimate was consistent with the possibility of substantial adult transfer between clusters, but allowed no quantitative inferences about dispersal.

Recently, indirect studies of interpopulation dispersal have begun to tap the huge reservoirs of information in the DNA itself. Initial efforts used a combination of mitochondrial DNA (mtDNA) sequence information and multilocus minisatellite bandsharing to describe qualitative dispersal patterns. Some mtDNA sequences are much more variable than allozymes and mtDNA is transmitted clonally through the maternal line. As a result, distinctive mitochondrial haplotypes between populations imply the absence of dispersal, at least by females. Multilocus minisatellites are the basis of the now-familiar hypervariable DNA 'fingerprints'; bandsharing between individuals in different areas implies relatedness, and thus the presence of dispersal.

Combining these two types of information, Lehman *et al.* (1992) identified unrelated wolves as those with different mitochondrial haplotypes, and used these animals to calibrate the relationship of multilocus fingerprint bandsharing to relatedness. This approach identified certain wolves that appeared to be highly related, not to other wolves in the same group, but to members of other groups. Assuming that these individuals were dispersers, Lehman and coworkers argued that short-range dispersal among wolf packs in the same area is common and that it is more common in females than in males. Gompper *et al.* (1998) used multilocus fingerprint bandsharing to show that females in coati packs were closely related to each other and to males in the pack's vicinity, suggesting that both females and males are highly philopatric. Talbot & Shields (1996) and Paetkau *et al.* (1998b) showed that the distribution of mitochondrial, but not nuclear genotypes, in brown bears is structured by water gaps of 5–10 km. Because mtDNA is transmitted only through females, while dispersing bears of either sex can homogenize the distribution of nuclear genes, this result suggests that water gaps block dispersal of females, but not males.

The distribution of mitochondrial haplotypes has allowed important qualitative inferences about long-range dispersal in North American canids. Smaller canids – kit and swift foxes – show greater regional differentiation than larger canids, suggesting shorter dispersal distances, and mtDNA boundaries confirm several presumed ecological barriers to dispersal (Mercure *et al.*, 1993). Somewhat surprisingly, coyote mitochondrial genotypes show less geographical structure than those of wolves, implying higher levels of gene flow, even though coyotes are smaller than wolves (Lehman & Wayne, 1991; Wayne *et al.*, 1992).

MICROSATELLITES AND ASSIGNMENT INDICES

Promising as they are, approaches based on mtDNA and minisatellites are soon likely to be eclipsed in many applications by investigations based on the ubiquitous DNA sequences known as simple sequence repeats, short tandem repeats, or microsatellites. Microsatellites are hypervariable chunks of DNA that seem to be scattered through all eukaryotic genomes. For example, a common motif in mammals is the dinucleotide sequence 'CACACACAC...', with the polymorphism among animals resting in the number of CAs different individuals have at a particular genetic locus. Microsatellites offer many advantages to population biologists (Bruford & Wayne, 1993; Taberlet & Waits, 1998) but the most relevant to studies of dispersal is that all species examined to date have numerous microsatellite loci and numerous alleles at many loci, enough that every individual is likely to possess a unique microsatellite genotype. We can view these genotypes as tags that we don't have to apply – and because each individual's microsatellites are likely to be similar to those of its neighbors, but different from those of individuals born elsewhere, we can view an individual's microsatellite genotype as a tag containing information about its population of origin.

Microsatellite-based inferences about carnivore dispersal have already begun to appear. For example, Girman *et al.* (1997) used microsatellites to estimate relatedness among wild dog pack members. They found that members of neighboring packs were more closely related than random individuals, suggesting primarily short-distance dispersal. Microsatellites, like allozymes, can be used to derive F_{st} values, but they avoid allozymes' major limitation, lack of variation. Using seven microsatellite loci, Lade *et al.* (1996) found significant differentiation among Australian red fox populations separated by just a few tens of kilometers. But this example points up the conceptual limitation of F_{st}-based estimates of dispersal. Foxes on Phillip Island were differentiated slightly from those on the mainland but enough to suggest about 1.4 dispersers per generation across an approximately 1 km water gap if the populations are at genetic equilibrium. Phillip Island, however, was colonized by foxes, presumably from the adjacent mainland, only 90 years (~ 18 generations) ago. To what extent does this historical link obscure current isolation? Founding events and other aspects of genetic history are confounded with current gene flow in F_{st} values.

The same point arises when microsatellite-based F_{st} estimates are compared for wolves and coyotes (Roy *et al.*, 1994a). Microsatellite allelic variation is adequate enough to obtain such estimates, but F_{st} values reinforce the somewhat surprising result found with mtDNA, that interpopulation gene flow is higher for coyotes than for wolves. This result most likely reflects not higher current dispersal rates in coyotes, but rather the fact that coyotes have expanded their geographic range greatly in the past century. In other words, lack of microsatellite differentiation reflects past, not current, linkages between the sampled populations.

These and other problems will soon be avoidable as microsatellite-based dispersal analyses begin to take advantage of new statistical approaches that, unlike those based on Wright's F_{st}, are based on *individual* genotypes (Bowcock *et al.*, 1994; Paetkau *et al.*, 1995; Rannala & Mountain, 1997; Shriver *et al.*, 1997; Smouse & Chevillon, 1998). Several such approaches can trace their ancestry to the idea of genetic 'assignment', in which individual genotypes are assigned to populations in which their expected frequency is greatest (Waser & Strobeck, 1998).

Genetic 'assignment' is related to problems in management and forensics, when it may be important to determine what stock a harvested individual comes from, or whether a confiscated trophy comes from a protected population or not (e.g. Prager & Fabrizio, 1990; Shriver *et al.*, 1997). The idea depends intuitively on the fact that different populations tend to have different alleles, or perhaps the same alleles at different frequencies, and that once we have estimated allele frequencies, simple Hardy–Weinberg assumptions allow us to determine the expected frequency of any genotype in any population (a program to calculate such 'assignment indices' is on the web at www.biology.ualberta.ca/jbrzusto/Doh.php). An 'assignment test' assigns each individual to the population in which its assignment index is highest. The more variable loci the investigator has access to, and the more variation there is at each locus, the easier it is to differentiate members of different populations, and thus to determine where any given individual originated. The application of this approach to interpopulation dispersal was first recognized by Paetkau *et al.* (1995).

Paetkau and coworkers pointed out that when discrete populations are linked by dispersal, dispersers will often be 'misassigned'. In other words, the assignment test will (correctly) assign them to the population into which they were born, not the population within which they were sampled. By extension, the proportion of individuals in one population that are misassigned to another (P_{MA}) might index the rate of dispersal between those two populations.

Paetkau and coworkers developed this approach in a study of genetic structure in polar bears, another carnivore with a paucity of variation in allozymes or mitochondrial haplotypes. The study was motivated in part by concern about whether 'populations' historically managed as independent units were in fact demographically linked through dispersal. They sampled 124 bears in five Canadian populations; eight microsatellite loci proved adequate to distinguish the populations genetically. Paetkau and coworkers calculated the expected frequency of each bear's genotype in each population, and assigned it wherever that frequency was highest. Most bears captured in each population were correctly assigned to that population (Table 22.1).

However, some bears in each population were 'misassigned' to other populations. For example, 32% of South Beaufort (SB) bears were assigned to the North Beaufort (NB) population, while 37% of NB bears were assigned to SB. Paetkau and coworkers suggested that these misassigned bears included interpopulation dispersers. Of course, misassigned individuals include not only immigrants into the population, but also offspring or grandoffspring of immigrants, and if allele frequencies in the populations are very similar, some individuals will be misassigned by chance. Nevertheless, Paetkau *et al.* (1995) found that the proportions of genotypes misassigned

Table 22.1. Results of the assignment test for four Canadian polar bear populations, South Beaufort Sea (SB), North Beaufort Sea (NB), West Hudson Bay (WH) and Davis Strait (DS). The expected frequency of each individual's genotype was calculated and animals assigned to the population in which their genotype was most likely. Values are the proportions of animals from each source population assigned to each of the four sampled populations; sample sizes are given in parentheses. Data from Paetkau et al. (1995).

Source population	Assigned population			
	SB	NB	WH	DS
SB(22)	0.64	0.32	0.05	0.00
NB(30)	0.37	0.57	0.03	0.03
WH(30)	0.00	0.00	0.67	0.33
DS(26)	0.12	0.04	0.31	0.54

(P_{MA}) between pairs of polar bear populations reflected geographic proximity and suggested that they might also reflect movement patterns 'not obvious from geography'.

These inferences are reinforced by a second data set, from brown bears in seven populations in northwest Canada and Alaska (Paetkau et al., 1998b). The sampling sites include Kodiak Island, more than 50 km off the Alaskan mainland; a priori, we believe that bears have not moved between the mainland and Kodiak Island for a very long time. The sites also include the so-called ABC islands in the Alaskan panhandle, separated from the mainland by water gaps of up to 10 km. It would seem difficult, but possible, for brown bears to cross water gaps of this magnitude.

This time, the assignment test was performed with 261 bears, again with eight microsatellite loci; 92% of these bears were assigned to the population within which they were sampled. Misassignments indeed track factors expected to influence interpopulation dispersal in an intuitive way. In Figure 22.1 we plot the proportion of individuals misassigned between each pair of populations as a function of geographic distance and the presence and magnitude of water barriers between them. The following trends appear:

- No bears from Kodiak Island are ever assigned to any mainland population, and vice versa, regardless of the distance between Kodiak and the mainland population. This is consistent with the presumed impossibility of dispersal, even for brown bears, across 50 km of ocean.

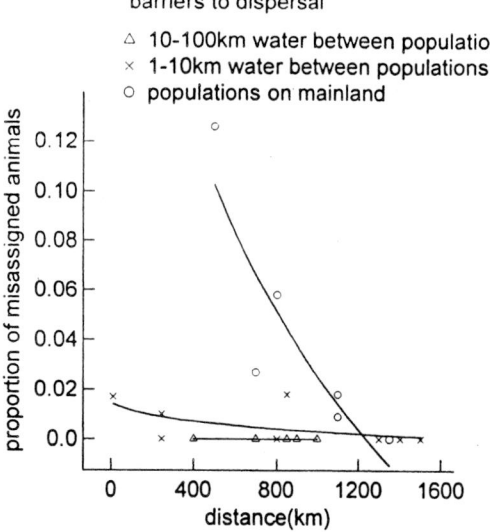

barriers to dispersal
△ 10-100km water between populations
× 1-10km water between populations
○ populations on mainland

Figure 22.1. The proportion of brown bears misassigned between populations (P_{MA}) behaves, intuitively, as dispersal rates between how those populations might be expected to behave. Each point represents a pair of the sampled populations; the x-axis is the geographic distance between those populations, and the y-axis is the proportion of animals sampled in either of the two populations that is 'misassigned' to the other. Substantial proportions of bears are misassigned between pairs of populations on the mainland, but P_{MA} declines with distance. The best-fit relationship is logarithmic ($r^2 = 0.89$), as expected if dispersers have an approximately constant probability of settling, as distance from their source population increases (Buechner, 1987). The relationship between P_{MA} and log distance is significant ($P = 0.04$, Mantel test). Fewer bears are misassigned between pairs of populations separated by a 1–10 km water barrier, but again P_{MA} declines with distance ($r^2 = 0.58$). No bears are misassigned between populations separated by 10–100 km of water.

- Among mainland populations, small numbers of bears are misassigned, but the proportion of bears misassigned between two populations depends on their separation. More bears are misassigned between populations that are close together, and the relationship appears to be logarithmic, as is common in dispersal distance distributions (Buechner, 1987).
- Small proportions of bears are misassigned between the ABC island populations and nearby mainland populations, but none between these populations and distant mainland populations. Far fewer bears are misassigned between populations separated by a short water barrier than between mainland populations, and there is more scatter in the relationship, which makes sense if the details of the water barriers

between populations, as well as the distance between them, influence dispersal rates.

ASSIGNMENT TEST SIMULATIONS

At least qualitatively, the proportion of animals misassigned between populations appears to track the rate of interpopulation dispersal. Limitations to the assignment test have been pointed out (e.g. Davies *et al.*, 1999); it is inappropriate when populations are not discrete and when some possible sources of immigrants are not sampled. When samples are small (less than 10) assignment becomes inaccurate, and when samples from different populations differ in size there may be a bias towards assigning individuals to the smaller population. However, for those carnivores for which the test is appropriate, does it have any precision? Could P_{MA} be used as a quantitative estimator of dispersal rates?

One way to explore the behavior of this index over a range of conditions is via computer simulation (the program is accessible on the web at http://bilbo.bio.purdue.edu/~pwaser/genetix). By simulating a set of populations that reproduce, die, and exchange individuals at defined rates, and asking the computer to keep track of individual genotypes, we can sample populations that have reached genetic equilibrium and perform assignment tests with the resulting 'virtual' carnivores. We can then ask if there is a simple, linear relationship between the proportion of animals misassigned between populations and the dispersal rate between them, and whether this relationship is robust to differing assumptions about mating and demography.

Figures 22.2 and 22.3 present sample simulation results, averaged over many replicate runs. P_{MA} indeed closely tracks interpopulation dispersal rate when the proportion of animals that move between populations is low. At high dispersal rates, dispersers carry so many alleles between populations that immigrants are no longer very distinctive; at the same time, residents are often misassigned simply because the two populations have similar allele frequencies. At the extreme, if dispersal were so high that the two 'populations' were actually a single panmictic group, then 50% of the animals in each population would be misassigned. In summary, at high interpopulation dispersal rates, P_{MA} saturates and cannot usefully track dispersal. But for carnivores, what's most intriguing is that the relationship is approximately linear over a broad range of biologically realistic dispersal rates.

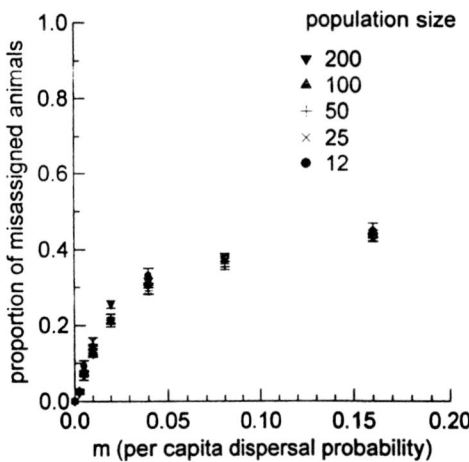

Figure 22.2. The proportion of animals misassigned between two simulated populations tracks the dispersal rate between them closely when interpopulation dispersal rates are low, although the relationship 'saturates' at higher dispersal rates. For example, between per capita dispersal rates of 0 and 0.02 (one in 50 animals in each population immigrated from the other), the value of P_{MA} is approximately 10 times the dispersal rate. This relationship is independent of the size of the populations. This graph plots the results of simulations with populations as small as 12 animals and as large as 200. For values of m between 0 and 0.02, the slope of the regression of P_{MA} on m is between nine and 10 for all population sizes. Each point plots the mean of 100 runs (± SE). These runs assumed: random mating; an equal sex ratio (and an equal sex ratio of dispersers); nonoverlapping generations; two populations exchanging dispersers; assignment indices calculated based on eight microsatellite loci; and stepwise mutation of microsatellites with 12 possible alleles and a mutation rate of 0.001.

The relationship between P_{MA} and dispersal is sensitive to the number of loci available for analysis: with data from more loci, the slope of relationship between P_{MA} and interpopulation dispersal is less steep and is informative over a broader range of dispersal rates. Figures 22.2 and 22.3 illustrate the relationship for a conservative set of assumptions – the analyses are based on just eight microsatellite loci.

Simulations indicate that the relationship between P_{MA} and dispersal rate is little influenced by a large number of factors that may differ among carnivore species, or that may be difficult to measure for particular populations. First and perhaps most surprisingly, it is virtually independent of population size (Figure 22.2). In other words, an investigator need not know the exact size of the populations under study to apply this approach. Secondly, the relationship between P_{MA} and dispersal rate is insensitive to the proportion of the population sampled. Thirdly, the relationship is unin-

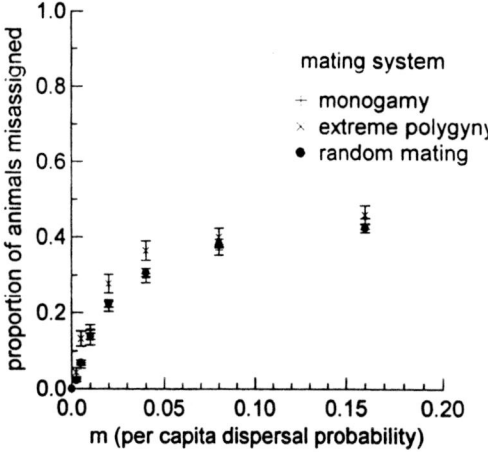

Figure 22.3. The relationship between P_{MA} and the per capita dispersal rate is surprisingly robust, and is uninfluenced by assumptions about the sex ratio of dispersers, the mortality rate of adults, the proportion of the populations that are sampled when estimating allele frequencies for the assignment indices, and the mating system. This figure plots the relationship for three assumptions about the mating system. 'Monogamy': each female is mated to a single, different male; each female produces at least one offspring and additional offspring are allocated randomly to these pairs. Polygyny: a single random male mates with all females, and female litter size is random (poisson-distributed). Random mating (promiscuity): a female is drawn randomly from the population (with replacement) and mated with a male drawn randomly from the population (with replacement) until all offspring are produced. In other words, both male and female 'litter sizes' are random.

fluenced by the number of populations that exchange dispersers; P_{MA} is the same for a given value of m, whether dispersers have only one other population to move into or many. Fourthly, P_{MA} indexes dispersal in the same way whether all adults die each year – the classic simplifying assumption of nonoverlapping generations – or whether generations are overlapping with annual mortality of only 25%. Finally, the relationship is little influenced by differences in mating system from strict monogamy through promiscuity (random mating) to extreme polygyny (Figure 22.3). Nor is the relationship influenced by the sex ratio of dispersers. Thus, P_{MA} should be equally useful as a dispersal index across carnivore species that differ greatly in demography and mating system.

Under all of these circumstances, P_{MA} tracks dispersal rate in a nearly linear fashion ranging from zero up to about 1 in 50 animals. Over this range, the proportion of animals misassigned between two populations appears to be approximately 10 times the proportion of animals dispersing between them (Figure 22.3; recall that this analysis is based on eight loci; with more loci, the slope of the relationship between P_{MA} and dispersal would be shallower).

The assignment test approach to estimating dispersal has a further advantage for conservation applications. Unlike the estimates of gene flow derived from F_{st}, P_{MA} values measure *current* dispersal rates and they do not assume genetic equilibrium. If some habitat manipulation – for example, the creation of a corridor – increases interpopulation dispersal, simulations show that P_{MA} increases immediately, in the first generation, because it is detecting the dispersers themselves. Conversely, if a barrier is created – a highway bisects a previously-continuous reserve – then P_{MA} decays almost as immediately. Because they contain some of their ancestor's alleles, offspring and more distant descendents of immigrants may be misassigned, but the chances decrease rapidly with each generation. Simulations confirm that the 'signature' of immigration decays to invisibility over just three to four generations.

FUTURE CONSERVATION APPLICATIONS

Beier (1993) has recently described how the range of Southern California cougars has become increasingly fragmented by roads and human 'development'. His population viability models indicate that the future of this population is crucially dependent on continued migration between parts of the formerly-continuous habitat. Clearly, to determine where corridors should be maintained and how effective they are, we need to monitor rates of interpopulation dispersal. The assignment test provides considerable promise for this kind of application, and variants of the method are rapidly developing (Estoup, 1996; Rannala & Mountain, 1997).

Many carnivore species of concern occur in discrete populations, which are becoming increasingly isolated through human activities. Yet we know virtually nothing about how many animals now move between protected populations, how movement rates are being influenced by various forms of development, what types of ecological factors constitute barriers to dispersal, or what types of corridors allow it (van Vuren, 1998). A combination of direct and indirect approaches will undoubtedly be necessary to address these questions, but the baseline information needed is a simple number – the rate of interpopulation dispersal – estimated for populations separated by different types of dispersal barrier.

The assignment test lends itself to this purpose where carnivores inhabit discrete habitat patches, and most or all available patches can be sampled. But genetic approaches are also on the horizon for situations in which populations are less discrete, or where one can sample only one population, but would like to know what proportion of its members have immigrated from elsewhere (Estoup *et al.* 1996). Favre *et al.* (1997), exploiting the fact that rare genotypes in a population often belong to immigrants, compared the distributions of male and female assignment indices and showed that sex-biased dispersal could be detected because the dispersing sex contains a disproportionate number of rare genotypes. Rannala & Mountain (1997) and Davies *et al.* (1999) have developed the Monte Carlo approach using a Bayesian estimator of microsatellite allele frequencies to distinguish immigrants into a population from residents. Individual-based approaches such as these may soon allow us to distinguish population sinks, because such areas will be 'immigrant-rich'. Methods to detect and monitor long-distance dispersal are, if not in our grasp, at least on the horizon, and can come none too soon as tools in our efforts to design reserve systems that maximize population viability.

Acknowledgements

We thank Andrew Waser for writing most of the assignment test simulation, and Brian Keane, Cathy Mossman, William Olupot, and Brad Swanson for discussion of earlier versions. Our work on these topics has been partially supported by the U.S. National Science Foundation.

23

Setting priorities for carnivore conservation: what makes carnivores different?

JOSHUA R. GINSBERG

Humanity badly needs things that are big and fearsome and homicidally wild. Counterintuitive as it may seem, we need to preserve those few remaining beasts, places, and forces of nature capable of murdering us with sublime indifference. We need the tiger, *Pathera tigris*, and the saltwater crocodile, *Crocodylus porosus*, and the grizzly bear, *Ursus arctos*, and the Komodo dragon, *Varanus komodoensis* ... to remind us that Homo sapiens isn't the unassailable zenith of all existence.

David Quammen, 1998, Deep Thoughts from Wild Places

Conservation is too often by default, and not design, a process of triage. Because of a series of insufficiencies – insufficient funds, insufficient expertise, insufficient time, and insufficient focus on priorities – populations and species will continue to decline, and in some cases go extinct, before strategies can be developed to save them. Establishing priorities for conservation is, therefore, of critical importance if we are to allocate intelligently scarce resources.

Clearly, we are not using our present resources in anything resembling an equitable, if not rational manner. Of the over 230 carnivore species, fewer than 15% have received serious scientific study: the small, nocturnal, solitary cats, mustelids, mongooses and civets have all but been ignored both by science, and by conservation (Schreiber *et al.*, 1989; Schaller, 1996). While the red fox *Vulpes vulpes* is, perhaps, the world's most studied medium sized carnivore, 18 species of fox in the genera *Vulpes* and *Dusicyon* remain a mystery to science (Ginsberg & Macdonald, 1990). We prefer to study the local, abundant, and problematic, the economically valuable fur bearers, or the large, wide ranging and mysterious, particularly if they occur in North America or Europe.

Legislative mandate, geographic accident, and aesthetic bias also skew our investments in carnivore conservation. Spending millions of dollars to successfully restore wolves *Canis lupus* to Yellowstone Park, while irrelevant for the global persistence of gray wolves, is arguably a good investment from both an economic and ecological perspective (Phillips & Smith, 1996). However, in either a global or national context, it is difficult to find a rationale to justify spending millions of dollars to prop up a fragmentary, in-bred subspecies of a relatively common, wide-ranging, species. Such actions are particularly perverse when the species is abundant and recovering throughout much of its range, but locally causes of decline have not been reversed (e.g. the Florida Panther Recovery Program, reviewed in Nowell & Jackson, 1996, pp. 135–7).

While setting priorities is critical, no priority setting exercise is without assumptions. Hence, in defining a hierarchical set of priorities, the methods we use, the assumptions we make, and the values we place on individual species, communities, or ecosystems must be transparent and openly enunciated. Too often, such clarity is missing. Even when present, however, most global priority setting exercises rarely give carnivores the attention they deserve, or the priority they require, if they are to persist. Carnivores are different, and these differences require new approaches both to the setting of conservation priorities and the implementation of conservation action.

WHAT MAKES CARNIVORE CONSERVATION DIFFERENT FROM BIODIVERSITY CONSERVATION?

Carnivore conservation is different from 'biodiversity' conservation (or conservation of other mammals, birds, amphibians, reptiles, insects, and plants) for several social and ecological reasons. Some of these factors share a common problem: the perception of threat to humans is greater than the real threat. As a result, people tend to persecute carnivores regardless of density, or number, or threat to their person or livelihood.

Large carnivores have the potential to kill people

While elephants, buffaloes, hippos, people with guns, and the average mid-sized car are all far more dangerous to humans than large carnivores, the mystique of man-eating lions and tigers and bears endlessly fascinates, repels and consumes us (see Quammen quote above). Yet predators are a statistically insignificant source of human mortality.

In the United States, in 1997, two children under the age of 14 were killed with guns every day (Hoyert *et al.*, 1999), yet gun control is a contentious issue. Approximately seven children under the age of 14 die every day in car accidents (Hoyert *et al.*, 1997), yet support for safer mass transportation is weak. Yet when a predator, for instance a mountain lion, attacks a child, or a jogger, cries are heard across the country for predator control (Associated Press, 1994).

How severe is the threat from mountain lions? In the 100 years from 1890 to 1990 there were 53 mountain lion attacks on humans in Canada and the United States. This resulted in 10 human deaths. In California, fatal attacks on humans were reported in 1890 and 1909. No other attacks were recorded until 1986 when a girl was injured in an Orange County park. Over a 10-year period between 1986 and 1995 the California Department of Fish and Game verified 10 attacks, or one per year, on humans (Torres, 1997). Only two of these were fatal to the humans, yet in 1994 alone 120 lions were killed under license, 10 because they threatened or attacked people (Mansfield, 1997). Mountain lion–human conflict is limited, but on the rise, perhaps reflecting both an increase in lion numbers and distribution, but also an increased inter-digitation of lion and human habitat with increased awareness of lions leading to increased sensitivity to their presence.

When predators live in fragmented populations with limited prey they appear more likely to attack humans. While tiger predation of humans is never common, it is certainly widespread (McDougal, 1987). Conflict with humans tends to be focused in areas where natural prey is limiting, and tigers are forced to hunt domestic animals to survive (McDougal, 1987). Recently, a tiger was reported to have attacked a human (Anon, 1999a) in an area where tigers are virtually extinct (Tilson *et al.*, 1997). In the twenty-first century, man-eating tigers may often be a better indication of prey loss and impending local extinction, than a problem that can be managed in the long-term (but see Rishi, 1988).

Whatever the actual level of threat posed by large predators, if the perception persists that predation constitutes a real threat to those living in communities in and around protected areas, conservation programs must address this threat if they are to reduce conflict, and thus be successful.

CARNIVORES LIKE TO EAT THE THINGS PEOPLE RAISE

In 1983, Schaller noted that predators, while often maligned, are rarely a significant factor in the survival of cattle and cattlemen:

> Indeed, cattle probably are the jaguar's main food, although kills account for only a small percentage of the animals dying annually. In the Pocone

district, which contains the largest remaining jaguar population in the Pantanal, cattle are said to have declined from 700,000 to 180,000 since the
1974 flood as a result of drowning, disease, and starvation ... during the
height of the calving season at Acurizal, I found 10 freshly dead young of
which only one had been killed by jaguar, the others having died of malnu-
trition and disease.

While frost, drought, bad husbandry, and a global economy may all have
more to do with the profitablity of ranching and farming of livestock, carnivores continue to be persecuted to local extinction for both the real and
perceived potential for livestock depredations.

Shooting wolves is common practice (Haber, 1996), even on the northern borders of Sweden and Norway where, perhaps, a few dozen wolves still
remain (Anon, 1989; Lüthcke, 1999; Plon, 1999), or in Arizona, where five
of 11 reintroduced Mexican wolves were shot in the first year following
release (Derr, 1999b; DiToro, 1999). Persecuting bears is equally good
sport, with a remnant population of perhaps two dozen bears in Spain
being under threat because of illegal hunting and persecution because of
perceived threats to livestock (Clevenger & Purroy, 1999). In reality the
economic policies of the EU, economic collapse in Asia, and New Zealand
overproducing lambs all probably have a far greater impact on the livelihoods of these farmers.

CARNIVORES CARRY RABIES AND OTHER ZOONOSES

Rabies is preventable and causes few deaths in developed countries (Macdonald, 1980). While rabies is, indeed, a horrible disease, and incurable
unless treated immediately after exposure, the number of deaths from rabies is rather small. In most developing countries, domestic dogs, not wild
carnivores, are a far more important resevoir for the disease, and a greater
source of human infection (Cleaveland & Dye, 1995; Rhodes et al., 1998).

Control of rabies is probably at least equally important to conserving
fragmented, threatened carnivore populations as it is to reducing human
suffering (Macdonald, 1996) – in the Serengeti, the last remnant population of African wild dogs was probably extirpated by rabies and/or distemper (Macdonald et al., 1992; Ginsberg et al., 1995; but see Burrows et al.,
1994). Rabies caused a population crash among critically endangered
Ethiopian wolves (Sillero-Zubiri et al., 1996), and has also thwarted African
wild dog reintroduction efforts in both Etosha and Madikwe (Scheepers &
Venzke, 1995; Hofmeyr et al., 2000).

For most carnivores, however, the threat of disease is linked to the indirect effects of land conversion (and closer proximity to humans, their livestock, companion animals, and the diseases they carry). For example, most research on epizootics and wildlife has focused on the role of wildlife as a potential source of infection for human and domestic animals, rather than as the victims of these diseases (Peterson, 1991; Barlow, 1996). Because animals that co-occur with humans (both wild and domestic) are found at relatively high densities, increased expansion of humans (and their associated fauna) into wild lands also increases the odds of disease epidemics obliterating wild carnivores as evidenced by outbreaks of distemper in species as diverse as pinnipeds (Harbor seals *Phoca vitulina*, Dietz *et al.*, 1989; Baikal seals *Phoca siberica*, Osterhaus *et al.*, 1989), black-footed ferrets (Williams *et al.*, 1988), and lions (Roelke-Parker *et al.*, 1996).

CARNIVORES EAT THINGS PEOPLE LIKE TO EAT: HUMAN–CARNIVORE COMPETITION FOR PREY

In many parts of the world where subsistence, or cultural perceptions of subsistence, are still tied to hunting, carnivores may be declining or disappearing because prey is becoming limited. There is broad overlap in the diets of humans and carnivores, and each share preferred prey with the other (e.g. Jorgenson & Redford, 1993; Cambell & Hofer, 1995; Ray, in press). Furthermore, incidental capture of carnivores in snares set for target prey species may also have a strong impact on carnivore populations (Hofer *et al.*, 1996). Carbone *et al.* (1999) show a transition from feeding on small prey (less than half of predator mass) to large prey (near predator mass), occurring at predator masses of 20–25 kg. Hence, those species which are more likely to compete most directly for prey with humans are these larger carnivores.

The reality of prey depletion as a factor driving declines in carnivore populations is probably worse than perception, although among ecologists and conservationists acknowledgement of the critical role of prey depletion as a threat is growing. In 1987, a review of tiger conservation (Tilson & Seal, 1987) makes no mention of prey depletion. Prey depletion is now thought to be the key immediate threat to persistence of tigers in many parts of their range (see papers in Seidensticker *et al.*, 1999). One could hypothesize that prey limitation is on a par with habitat loss as a threat to large and medium sized carnivores around the world. Unfortunately, few data exist to prove, or disprove, this hypothesis.

How severe is the competition between humans and carnivores? While data are few, indirect measures (such as off-take of bush meat) may give us some insight. In central Africa, for instance, it has been estimated that as much as a million metric tonnes of game meat is being exported annually – a billion kg of meat (Robinson et al., 1999). As carnivores are usually under-represented, numerically and by weight relative to their abundance in most surveys of human use of wildlife as food (see papers in Robinson & Bennett, 1999), very little game meat is probably derived from carnivores directly, with most of the biomass being represented by forest ungulates and primates.

How does harvesting of prey of this magnitude affect carnivores? One approach to answering this question is to estimate the relationship between densities of prey and their predators. To do this we have to assume first that a loss of prey has a direct, if not linear, impact on carnivore survival and reproduction, hence on carnivore densities. This appears to be the case for many carnivores, although factors such as interspecific competition, prey switching, disease, and human induced mortality may complicate this relationship (e.g. Fuller & Sievert, Chapter 8; Creel & Creel, 1996; Woodroffe & Ginsberg, 1998; see below). In reality, the loss of prey species would have a less direct, and therefore more difficult to measure impact mediated through a combination of reduced fecundity, reduced neonatal survival, reduced juvenile survival, and reduced survival of adults (see Fuller & Sievert Chapter 8). Extinction of the Javan tiger, in which loss of prey directly affected recruitment, however, had a smaller impact on adult survival (Seidenticker, 1986).

At each stage in a food web energy is lost. Classic studies of rodent–carnivore dynamics (Pearson, 1964) suggests an efficiency of transfer of about 10%. While such food web relationships are complex, if we make the simplifying and relatively conservative assumption that there is a loss of efficiency of approximately 20:1 as we move up a simplified food chain, a loss of 20 kg of prey would result in a loss of 1 kg of carnivore biomass. Hence, if a billion kg of bush meat are harvested, and this is all prey, then there would be a loss of approximately 50 million kg of carnivore biomass under these assumptions. If an average carnivore weighs 10 kg, five million individual carnivores will be lost each year from the indirect loss of prey. If only large carnivores weighing 25 kg are considered, the number of carnivores lost would be on the order of 2.5 million individuals. If one were to assume an average of two trophic levels separating top carnivores from the average human prey item the loss of carnivores would order of magnitude smaller, but nonetheless large. Throw out these assumptions, challenge them, but by any calculation present levels of extraction of bushmeat must be devastating to carnivores.

HARVESTING AND OVER-HARVESTING OF CARNIVORES

Controlled harvests of many carnivore species should not be a threat as off-take is balanced by relatively high rates of intrinsic growth (but see Haber, 1996; Weaver et al., 1996). As a result, few carnivore species, at present, are endangered in any way as a result of direct harvesting (Bailiee & Groombridge, 1996), although this has not always been the case.

When over-harvesting occurs the low densities common to many predators make loss of populations rapid and severe. In North America, over-exploitation of some furbearers in the nineteenth century resulted in the local extirpation of several species, and the near extinction of others (reviewed by Tapper & Reynolds, 1996). Marten were once threatened with extirpation in many parts of their range; their populations have recovered naturally (Tapper & Reynolds, 1996) and have been the successful focus of reintroduction (Slough, 1994). Because of low density, and relatively slow rates of intrinsic growth for carnivores, they remain vulnerable to a resurgence in the market for wild caught furs (Schneider & Yodzis, 1994).

Globalization of markets, combined with declining populations of predators, can certainly take its toll. Surging economies in Asia in the 1980s and 1990s led to increased exploitation of bears (J. Mills & Servheen, 1991; Servheen et al., 1999) and tigers (Hemley & Mills, 1999) for the traditional oriental medicine trade. These harvests have contributed to population declines, and in some areas local extirpation of a variety of bear species and tiger populations. As Schaller (1996) notes, from pandas to sea otters to spotted cats, there is a long and consistent history of over-exploitation of carnivores when their pelt values are high.

CARNIVORES, CARNIVORE GUILDS AND TOP DOWN ECOLOGICAL IMPACTS

The loss of top carnivores, and habitat modifying herbivores like elephants, results in top-down ecological impacts, many of which we are only now just beginning to study. The importance of top carnivores (or large herbivores for that matter) as agents of ecosystem change and the regulation of both herbivores and smaller carnivores are becoming increasingly evident (e.g. Orians et al., 1997; Terborgh, 1988; Crooks & Soulé, 1999; Terborgh et al., 1999). While the loss of a single carnivore species in an ecosystem may be compensated for by growth in the populations of other species, or eventual colonization by an allopatric guild competitor (e.g. spread of coyotes into the eastern United States; Harrison, 1986; Parker, 1995), both short- and

long-term consequences of removal of top predators are relatively unknown.

Competitive interactions are clearly responsible for determining relative densities within a carnivore guild (e.g. Sargeant *et al.*, 1987; Litvaitis &
Harrison 1989; Cypher & Spencer, 1998; Palomares & Caro, 1999). For
instance, study of the interaction between coyotes and swift fox provides
extensive, and convincing, evidence of the way in which the two species
interact, with coyotes responsible for up to 60% of fox mortality (Carbyn *et
al.*, 1994: Ralls & White, 1995; Cypher & Spencer, 1998; but see White &
Garrott, 1997; Kitchen *et al.*, 1999). A further set of examples comes from
shifts in the composition of carnivore guilds as a result of expansion of
human populations. Wild species that live, and often thrive, in human
dominated landscapes may not be a conservation priority. Yet, as human
associated invasives, these species are a potential threat to other more specialized carnivore species. For example, in West Africa, deforestation, and
even selective logging, allow savanna mongoose species, such as *Herpestes
ichneumon*, the Egyptian mogoose, to expand into forests (Ray, in press). In
the deserts of North Africa, expansion of human habitation, and human
use of wadis in the desert, has led to colonization of desert habitat by red
fox, and a concomitant reduction in the range of desert foxes (Ginsberg &
Macdonald, 1990). Other examples can be found in the literature of predator reintroductions which allow us a unique opportunity to study how
changes in carnivore guild composition affect both prey and competing
predatory species (Carbyn *et al.*, 1994; Phillips & Smith, 1996).

The secondary or indirect impacts of loss of top carnivores, such as
indirect effects on rodent and plant community dynamics (e.g. Wagner &
Stoddard, 1972), may be even greater ecologically but usually are ignored.
In a landscape of habitat fragments, loss of an apex predator may lead to an
increase in the abundance of smaller predators and local extinction of avian
prey (Crooks & Soulé, 1999). Few such examples of mesopredator release
are documented (Litvaitis & Villafuerte, 1996), perhaps because higher
level ecological interactions may have extremely long time lags (Balmford *et
al.*, 1998; Jackson, 1997) which makes them especially difficult to study,
and particularly easy to ignore.

LAND CONVERSION, FRAGMENTATION: DIRECT AND INDIRECT COMPETITION FOR SPACE

If persecution and loss of prey are the most immediate short-term threats to
carnivores, continued land conversion is probably the greatest long-term

threat to the persistence of carnivores, and probably affects carnivores geo-metrically more than herbivores. For the most part, because carnivores oc-cur at relatively low densities and have relatively large home ranges intrusion of humans into wildlands and wilderness has a host of direct and indirect effects on carnivore populations.

The direct impact of humans on large carnivores is striking. A majority of the mortality experienced by large carnivores living inside protected areas occurs outside the protected areas that are theoretically providing them with a safe haven (Woodroffe & Ginsberg, 1998, 2000). Hence large, sometimes very large, protected core areas will be central to all carnivore conservation. Yet, for many species, all but the largest protected areas, while necessary, will be insufficient to achieve long-term protection (Noss et al., 1996; Woodroffe & Ginsberg, 1998, 2000).

Perhaps more generally, expansion of human populations brings with it high densities of domestic livestock and carnivore companion animals. These animals can have multiple impacts, many of which are synergistic and can lead to the local extinction of native carnivores. These effects in-clude: livestock competing with wild ungulates for prey, thus reducing prey availability for wild carnivores (Seidensticker et al., 1990); reduction in wild prey leading carnivores to hunt domestic animals, thus increasing conflict (see above). Semi-feral and free-ranging cats and dogs may occur at den-sities higher than wild predators (Cleavland & Dye, 1995), sometimes as much as an order of magnitude higher (Sillero-Zubiri & Macdonald, 1997), resulting in both competition for prey and dirtect aggressive interactions.

A REVIEW OF SOME EXISTING PRIORITY SETTING EXERCISES

While existing priority setting exercises can be adapted to setting priorities for carnivore conservation, inevitably adaptation of these models will be difficult for the reasons reviewed above. In addition to some of the technical difficulties, however, each of the current methods of priority setting has its own problems.

Before reviewing other efforts at priority setting, a cautionary note is needed. If there are inherent biases in our knowledge of the population status and distribution of threatened species, such biases will have ramifi-cations for other types of priority setting. Hot spot models (Mittermeir et al., 1998), for instance, focus conservation efforts on areas of high species diversity. Other priority setting exercises focus on species representation and complementarity of sites chosen (Williams, 1998), or some combina-tion of representation, complementarity, hot spots and threat (Mills et al.,

Chapter 21; Williams, 1998). Yet, if we have a systematic lack of information on the distribution and status of smaller carnivores, such analyses will bias against establishing these species as conservation priorities. While it is simplistic to note that information is essential for analysis, such biases are rarely discussed.

Hot spot models

There is no lack of global priority setting exercises. The most widely used method for global priority setting is the hot spots model, first developed by (Myers, 1988) and more recently elaborated by Mittermeir *et al.* (1998). Hot spots models have been applied not just to areas of high species diversity, but to particular taxa (e.g. birds, Bibby *et al.*, 1992b). While hot spot models are appropriate as a means to designate priorities for conservation of a number of taxa, including small carnivores with relatively small home ranges and high densities, hot spot models will be, a priori, a poor surrogate for setting conservation priorities for larger carnivores.

Hotspot models focus on areas of high species diversity (Myers, 1988) or areas of high diversity of threatened taxa (Bibby *et al.*, 1992b). Often, hot spot models are combined with iterative selection algorithms (Williams, 1998), aimed to maximize species diversity in a minimum area, further reducing the overall area proposed to be set aside for conservation. Small areas and high diversity are not something one usually associates with conserving threatened large carnivores (Woodroffe & Ginsberg, 1998).

But other, more systematic problems exist with the application of hot spot models to carnivore conservation. Hot spots are identified by a calculus of spatial knowledge of species diversity, and spatial knowledge of species diversity, while generally poorly known, is particularly poorly known for most rare and threatened carnivores. As such, the probability of picking up threatened carnivores in hot spot models is limited to the congruence of carnivore distributions and that of other species used to determine hot spot distribution. For a host of civets, foxes, small cats, and mustelids, our knowledge of species distribution is next to zero (Schreiber *et al.*, 1989; Schaller, 1996; Nowell & Jackson, 1996; Ray, in press). Even for relatively well studied species, such as the lynx in North America, we are only now acquiring information about its current distribution (Weaver, 1993, pers. comm.)

Further survey effort is unlikely to give carnivores higher priority because their species diversity is almost always relatively low. In any one place at any one time, carnivores, particularly those in the tropics, tend to be relatively depauperate compared to other vertebrates, and wildly depauperate relative to invertebrates. Hot spot models can (and have) dealt with ameliorating this problem. Any calculus that gives stronger weight to taxonomic uniqueness, rather than just to species diversity, will favor carnivores. Furthermore, methods that include the role of a species in regulating (or at the very least, strongly influencing) other species in the ecosystem, will end up making a less random sample of carnivores.

In the end, the scale at which many larger carnivores live is often outside the structure of some hot spot models. As D. Miquelle *et al.* (unpub. data) note, if one were to protect all areas of high species diversity and endemism in the Russian Far East the resulting protected areas network would be too small, too highly fragmented, and in the wrong place to make much of a difference to tiger and leopard conservation.

Complementarity, minimum sets and iterative selection alogrithms

By focusing on carnviores, and combining hot spot models with the use of iterative selection algorithms where representation is achieved using the principle of complementarity, Mills *et al.* (Chapter 21) make an excellent first attempt at putting carnivores on the map of conservation priorities in Africa. As they note, their work is only a preliminary assessment, and is constrained by inadequacies of data, lack of field surveys, and a number of other factors. Nonetheless, the paper shows that large scale analyses which incorporate threat as a variable, but are not driven by it, can kick up new, and counter-intuitive priority species (e.g. spotted hyaenas: Mills *et al.*, Chapter 21).

As for hot spots (on which these models are based), there are some problems with the use of complementarity and representation in setting conservation priorities in general (Balmford *et al.*, 1998), and these problems are particularly notable for larger carnivores. Most species distributions used in hot spot and complementarity models are based on single 'snapshots' in time, while actual distributions can vary widely over time (Nicholls 1998). While effects may become more exaggerated on smaller scales, other processes (e.g. global warming) may strongly influence patterns on a larger scale in the near future (e.g. Graham & Grimm, 1990; Huntley, 1995). Changing patterns of distribution can result in priority setting exercises that reflect the 'ghosts' of former species distributions rather than the reality of present day distributions.

Another major weakness of such analyses is that the ecological requirements of a species are usually ignored by such algorithms (Nicholls, 1998) – the implicit assumption is that if one works on a large enough scale these requirements will be met. Hence, the value of hotspot models for any particular species will be inversely related to the home range area of the species we are trying to save. This is particularly problematic for medium and large carnivores where big is good, and bigger is better (Woodroffe & Ginsberg, 1998, 2000). To include larger areas, selection algorithms must include connectivity, and spatial requirements of space-limited species. They usually do not.

For smaller carnivores, what impact will a 'minimum set' approach have in the long-term? As stochastic threats, like disease, and deterministic edge effects, like the intrusion of human associated invasives, become increasingly important, will small, highly diverse, representational grids actually serve to protect these species?

Simply put, complementarity and hotspots (or combined approaches) do not take into account persistence of a species in any real way. For animals that require large areas, are the focus of persecution, and are susceptible to a host of other threats (including diseases carried by domestic companion animals) issues related to persistence, not representation, may be more important in determining the location, number and size, of protected areas. While these methods of priority setting are valuable, and may be critical to the conservation of many species, they will usually be inadequate for the conservation of larger carnivores.

Ecoregional models

These models, in which global conservation priorities are set by ensuring representation of habitat type uniqueness (e.g. Olson & Dinerstein, 1998), are potentially the best thing that has happened to carnivore conservation in a long time. By minimizing the need for precise data on species diversity and, instead, focusing on ensuring the conservation of representative blocks of habitat around the globe, we are probably more likely to save carnivores using these models than by opting for species diversity hotspots.

What we gain in terms of inclusiveness, space, and the potential to conserve carnivores is lost in generality, lack of clear goals and practicality of application. Ecoregional plans are a good first start, but in considering these plans, we need to keep in mind some of their implicit weaknesses.

Ecoregional models, such as the Global 200 (Olson & Dinerstein, 1997, 1998), cover a large proportion of the earth's surface (± 20%). We are currently inadequately managing less than 10% of the earth's surface which is formally protected, and doing a worse job on the remaining approximately 90% which lacks formal protection. Implementation of ecoregional priorities will require a massive increase in investment in both protection and management.

Most ecoregional models rely on an ecological approach in which humans and animals share the landscape. While a landscape approach is critical, zoning and large core areas in which human activity is minimal, or non-existent, will be required to conserve large carnivores (Noss *et al.*, 1996).

The great challenge in implementing ecoregional plans will be to ensure sufficient areas of contiguous habitat which are set aside for wildlife. By moving to a large scale, ecoregional models provide a broad justification for conservation of large areas, but no specific guidelines about how to focus effort on a scale relevant to management. They tell us little about how to effect the conservation of these priority areas and species. Unless we develop better ways of implementing priorities, and effecting conservation on the ground, we will debate priorities until our options are constrained by what options are left (Ginsberg, 1999).

ACTION PLANNING, RED DATA BOOKS AND CATEGORIES OF THREAT

The World Conservation Union's Species Survival Commission (IUCN) action plan and red data book (RDB) series provide a wealth of information for priority setting exercises. While neither is flawless, these initiatives provide tremendous assistance in developing species based strategies for carnivore conservation.

Red lists

The new IUCN red list categories use threats-based measures to estimate the probability of extinction of a species (Mace *et al.*, 1993; Baillie & Groombridge, 1996). Because categories of threat are linked to these probabilities they allow us to make cross taxa comparisons of the relative severity of threats faced by individual species within and among families (Baillie & Groombridge, 1996). In at least a superficial way, they establish priorities among families within a guild and they immediately allow us to determine the relative conservation threat facing species in different families.

While the Red List Categories are extremely valuable for assessment of threat, they are not always the best guide, for establishing conservation and research priorities. There are a minimum two good reasons why this is so.

Biases in the data

The first problem is related to inherent biases in classification. If we are going to set priorities among species, and we are using RDB categories of

Table 23.1. Distribution of species, by body weight class, Family, and category of threat

	Can	Fel	Herp	Hyn	Mus	Otr	Phoc	Proc	Urs	Viv	Total
Data dependent											
0–1	2	0	0	0	1	0	0	0	0	0	3
1–10	7	5	0	0	0	0	0	1	0	4	17
10–100	1	0	0	0	0	0	0	0	1	0	2
100–1000	0	0	0	0	0	0	0	0	0	0	10
> 1000	0	0	0	0	0	0	0	0	0	0	0
Least concern											
0–1	0	0	9	0	24	0	0	5	0	3	41
1–10	14	16	10	1	22	0	0	5	0	28	96
10–100	4	5	0	3	2	5	5	0	0	3	27
100–1000	0	0	0	0	0	4	9	0	1	0	14
> 1000	0	0	0	0	0	0	1	0	0	0	1
Threatened											
0–1	0	0	5	0	8	0	0	2	0	0	15
1–10	1	5	1	0	6	0	0	5	1	9	28
10–100	4	2	1	0	1	4	1	0	2	1	16
100–1000	0	2	0	0	0	2	2	0	3	0	9
> 1000	0	0	0	0	0	0	0	0	0	0	0

Can: Canidae; Fel: Felidae; Herp: Herpestidae; Hyn: Hyenidae; Must: Mustelidae; Otr: Otariidae; Phoc: Phocidae; Proc: Procyonidae; Urs: Ursidae; Viv: Viverridae.

Table 23.2. Results of generalized linear model analysis examining effect of Family and weight on category of threat

	Df	Deviance Resid.	Df	Resid. Deviance	F value	Pr(F)
Null		268	39.1057			
Weight factor	3	1.44283	265	37.6629	3.72437	0.01202
Family	9	3.40206	256	34.2608	2.92724	0.00259
Weight factor: Family	13	2.21094	243	32.0498	1.31702	0.20319

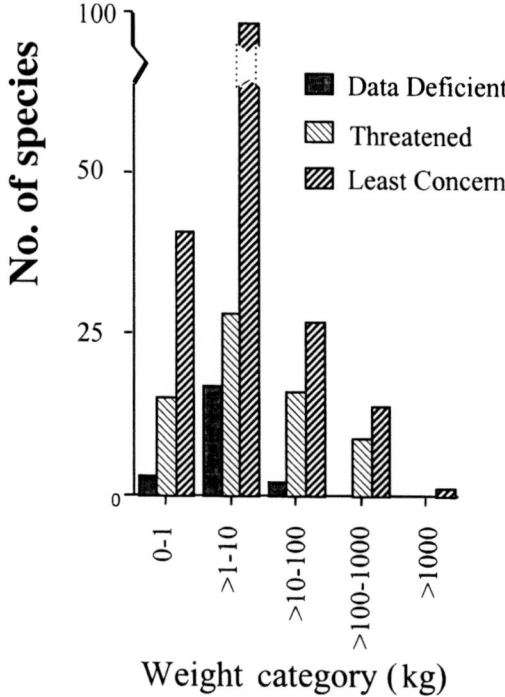

Figure 23.1. There is a significant effect of both body weight on threat status ($F = 3.7$, $P < 0.05$). Controlling for Family, larger species are more likely to be threatened. The preponderance of smaller animals in the Data Deficient category suggests a different systematic bias: we know less about smaller species than about larger ones.

threat as the mechanism by which to set priorities, then we would like data on the species to be without systematic biases if recommendations are to be comparable. In particular, while different taxonomic groups, or individual species, may be more prone to extinction (Baillie & Groombridge, 1996), if taxonomic history, or life history variables skew our ability to collect information on a species, then relative assessment of threat may be systematically biased by poor data.

To evaluate systematic biases in data information on category of threat (Baillie & Groombridge, 1996) and adult female body weight (database held by G. M. Mace, pers. com.) were compiled for 269 species in 10 Families of Carnivora. The effect of body size and Family on conservation status was determined using a generalized linear model (GLM) with a Possion model appropriate for multinomial count data (Venables & Ripley, 1997). The response variable was a three-level factor ('data dependent', 'low concern' and 'threatened'), where threatened is an amalgamation of critically endangered, endangered, vulnerable, and conservation dependent. The explanatory variables were Family and weight category (0–1 kg, 1–10 kg, 10–100 kg, 100–1000 kg, and > 1000 kg). A summary table of data is presented in Table 23.1. The significance of the reduction in deviance from adding each explanatory variable to the model was determined with an F-test. All analyses were conducted in S-plus for Windows (Mathsoft, Seattle, Washington). Results of the GLM analysis are presented in Table 23.2.

There is a highly significant effect of Family on threat status ($F = 2.92$, $P < 0.01$), suggesting that some groups (e.g. Ursidae; Procyonidae) are systematically more threatened than others (e.g. Mustelidae; Hyaenidae). Controlling for Family, there is also a strong effect of body weight on threat status ($F = 3.7$, $P < 0.05$). Larger species are more likely to be threatened (Figure 23.1), and larger species (or at least those with larger home ranges: Woodroffe & Ginsberg, 1998, 2000) are more likely to go extinct. The preponderance of smaller animals in the data deficient category (Table 23.1; Figure 23.1) suggests a different systematic bias – we know less about smaller species than about larger ones.

Ecological function versus extinction

The second major problem in using categories of threat to establish priorities relates to the difference between loss of ecological function and probability of extinction. The strength of the IUCN Red Lists is that they are an assessment of the global status of a species and an attempt to measure the global persistence of a species (Baillie & Groombridge, 1996). They measure evolutionary extinction, not loss of ecological function as a species dispappears, or is greatly reduced in numbers, across its range.

Hence, highly endemic species, no matter how healthy their populations, will nearly always be listed as threatened or conservation dependent (e.g. *Urocyon littoralis*, the Island Gray Fox, Ginsberg & Macdonald, 1990),

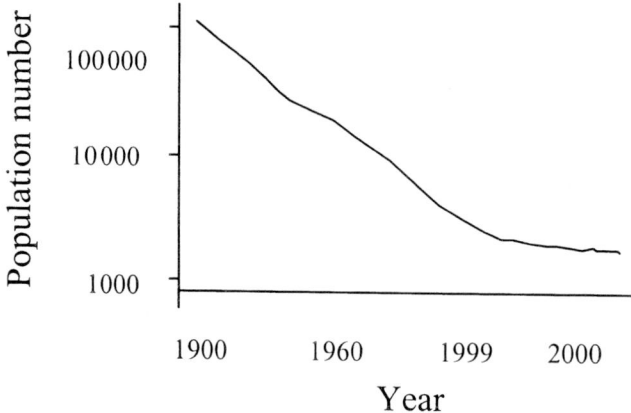

Figure 23.2. Approximate global population estimates for the African wild dog (*Lycaon pictus*: data from Woodroffe *et al.*, 1997). The species has been listed as endangered because of a rapid and persistent decline in range and population number. Having been extirpated across much of its range from 1950 to 1970, the Africa wild dog appears to have stabilized in numbers in the early 1990s. Having eliminated wild dogs from their most tenuous areas of occupation, a stable distribuion and absolute numbers of several thousand dogs spread over a large number of protected and unprotected areas suggest that the threat of extinction is declining, despite ecological extinction across much of the species' range.

while wide ranging, or widespread species (which includes a lot of carnivore species) may not be listed as endangered or even as vulnerable, even if they have been extirpated across much of their range. This problem is particularly salient for large carnivores.

Take, as an example, the African wild dog which has a continental distribution and absolute numbers of approximately 5000 individuals (Woodroffe *et al.*, 1997a; Figure 23.2). The species has been listed as endangered because of a rapid and persistent decline in range and population number, not because of its range-wide population size (> 2500), nor because of a small range, nor because their populations are highly fragmented (Baillie & Groombridge, 1996; Woodroffe *et al.*, 1997a; Woodroffe & Ginsberg, 1999).

Having been extirpated across much of its range from 1950 to 1970, the Africa wild dog appears to have become stable in number since the early 1990s. While declines continue in some parts of its range (e.g. parts of central and eastern Africa), numbers appear to be growing in other regions (e.g. parts of southern Africa). If present trends continue, wild dogs will probably go extinct in their last remnant populations in west and central Africa, and disappear across parts of eastern Africa, over the next decade or two. Having eliminated wild dogs from their most tenuous areas of occupation we will have a stable distribution and absolute numbers of several thousand dogs spread over a large number of protected and unprotected areas.

If this scenario is achieved, then wild dogs will need to be reclassified as vulnerable. Yet, the species will occupy a smaller range with fewer individuals than when it was listed as endangered. Rather than mocking the new Red List criteria, this classification validates the criteria because slowing rates of decline reflect a reduced probability of global (or 'evolutonary') extinction – the species is at a lower risk.

The same analysis probably would work for many carnivores, e.g. tigers, cheetah, a number of rare bear species, bush dogs and dhole. Historically, such an analysis would also apply to wolves (Ginsberg & Macdonald, 1990). More broadly, such an analysis clearly applies to both African (Cummings *et al.*, 1990) and Asian elephants (Santiapillai & Jackson, 1990), African rhinos (Cummings *et al.*, 1990) and is probably a general phenomenon for any large, wide-ranging species which is, or has been, extirpated across large parts of its range. While populations of many large, mobile, widespread species are locally threatened, the mere fact of a large range will increase the odds that somewhere, somehow, a few populations of the species will be given the space, and protection, they need to persist. Ecological extinction may occur across much of a species range with evolutionary extinction unlikely in any measurable time frame.

Action plans

The IUCN action plans series (Gimenez-Dixon & Stuart, 1993) represents the most comprehensive attempt to establish priorities for conservation of individual species, and taxa within families. To date, over 50 action plans have been completed, of which 10 address issues related to carnivore conservation. Some of these are Family-wide overviews (bears, Servheen *et al.*, 1999; hyenas, M. Mills & Hofer, 1998; felids, Nowell & Jackson, 1996; seals, fur seals, sea lions, and walrus, Reijnders *et al.*, 1993; procyonids and ailurids, Glatston, 1994; canids, Ginsberg & Macdonald, 1990; otters, Foster-Turley *et al.*, 1990; mustelids and viverrids, Schreiber *et al.*, 1989), while others take a more focused, species-specific approach (African wild dog, Woodroffe *et al.*, 1997a; Ethiopian wolf. Sillero-Zubiri & Macdonald, 1997).

Action plans, at their best, provide copious information and often set clear priorities for conservation of species within a Family. Some plans present priority ranking systems (e.g. Nowell & Jackson, 1996) which combine threat status, knowledge (or systematic lack thereof), and information about habitat preference. Almost all action plans outline information gaps, and set research priorities. Yet there remain significant, unimplemented recommendations in these plans, implementation of which would go a long way to effecting carnivore conservation.

TOWARDS A FEW GENERAL PRINCIPLES

Instead of trying to develop an integrated system which establishes priorities, I would like to establish a framework for analysis and implementation of carnivore conservation activities that attempts to find clusters of carnivore species with shared biological and conservation problems. The structure of this discussion reflects the IUCN Red List categories, but deviates from these categories in subtle but significant ways.

Unirradicable species

Species in the IUCN RDB category 'least concern' actually fall into (at least) two distinct categories, each with very different management needs. At one end of the carnivore spectrum lie the species that are essentially unirradicable – they thrive in human dominated landscapes. Coyotes, some jackal species, red fox, and Egyptian mongoose are but a few examples. While such species are unlikely to be conservation priorities, they do require conservation management because of their impact on rarer carnivores. Conservation is not just about recovering threatened species, but may increasingly involve manipulation of common species. While such activities have been discussed for introduced pests (Dobson, 1988; Courchamp & Sugihara, 1999), actions may be necessary in areas where human colonization of wilderness create continental islands.

Chronically data deficient species

There is a whole set of species classified as 'data deficient' in the IUCN RDB, but might better be called 'chronically data deficient species.' These are species about which we know almost nothing, and for which data on the simplest demographic and ecological variables are not only lacking, but may well never be collected (see Schreiber et al., 1989; Schaller, 1996). These species tend to be the cryptic, small, solitary carnivores of tropical forests, and inhospitable deserts – civets, mongooses, desert foxes and small cats. While numerically the most important group and perhaps the most threatened group of carnivores overall, they represent a real challenge for carnivore conservation. We know they are important, we suspect many of them are threatened, but what can we do about them?

There are several things that will probably not help us conserve these species:

- Increased effort in identifying distribution and abundance of these species across their entire potential range is probably not possible unless we develop quicker, cheaper, and better monitoring protocols (e.g. faecal genetics – Kohn & Wayne, 1997; Kohn *et al.*, 1999).
- Extensive autecological study of these animals is probably not worth the effort, and besides, no one will fund it on a time frame that will allow collection of data relevant to conservation.
- Detailed population biological studies are probably not necessary for conservation of these species.

Small carnivorous mammals tend to eat either rodents or insects (for the most part), and hence habitat loss and fragmentation are probably the critical factors affecting their persistence. Collection of a rather minimal data set on presence/absence, and a coarse-grain understanding of the ecological needs of these species should allow us to construct landscape analyses that predict ranges (e.g. Mills *et al.*, Chapter 21). Status of these species could then be confirmed by focused surveys to validate landscape analyses.

For small cryptic carnivores, therefore, the bedrock of conservation must be habitat preservation. By this I mean establishing and maintaining a landscape of inter-connected reserves that provide a reasonable sampling of habitat diversity, and in which human activity is severely restricted. Because space alone is not an adequate measure of persistence for many of these species, factors such as levels of disturbance, variation in patch size, and patch insularity will need to be included in such an analysis (e.g. for American martens, Hargis *et al.*, 1999). The only real difference between this approach, and prescriptions for priorities for larger carnivores, is that smaller, isolated patches of forest may be adequate, in the short-term, to conserve these species.

Wide-ranging, secure species

A second group of species that would be classified as 'least concern' by the IUCN RDB are those species that have relatively large populations across their range, and for which range areas are well secured. These species may have low conservation priority on a global scale, but may be important candidates for restoration of ecosystem function either by natural recovery (lynx, Weaver, 1993; Rocky Mountain carnivores, Weaver *et al.*, 1996;

wolves in the northeast United States, Harrison & Chapin, 1997) or reintroduction (e.g. kit fox in Canada, Carbyn *et al.*, 1994; wolves in the Lower
48 United States, Phillips & Smith, 1996). In Europe, and the United
States, extraordinary amounts of money, into the millions of dollars, have
been spent, and will continue to be spent, to recover populations of these
species. While on a global basis such expenditures are difficult to justify,
locally they can create a strong voice for conservation, and for wilderness
preservation (Kellert *et al.*, 1996).

SPECIES IN DECLINE: PAST, PRESENT, FUTURE

Species in persistent and continuing decline (IUCN RDB critical, endangered, and vulnerable) include a wide variety of taxa, but for the most part
are medium to larger sized carnivores. As discussed above, this size bias
may be real, or may be, in part, attributable to the paucity of data on the
status of smaller, unstudied carnivores. Because species which have declined across their range, and which are continuing to decline, are
threatened with extinction in a time frame recognizable to ecologists, if not
policy makers, they will always be among the highest priorities for conservation.

Understanding the spatial and temporal dynamics of declines, and the
threats which lead to these declines, is critical to priority setting for carnivores. At one extreme, the priorities for conservation of endemic species,
like the Ethiopian wolf, are relatively obvious, particularly when the entire
population of the species is within a single political jurisdiction (Sillero-
Zubiri & Macdonald, 1997): to secure the few remaining populations; establish a secure captive population; and work to reverse threats at a local
and national level

Species that are globally rare and, at best, occasionally locally abundant,
differ from wide-ranging secure species only in that they are, or have recently been, in persistent decline. On a global assessment of threat, these
species may or may not show up under highly threatened categories – some
populations may be stable, or expanding in range or number, while others
are highly threatened and/or extirpated (e.g. otter species, Foster-Turley *et
al.*, 1990; dhole, Ginsberg & Macdonald, 1990; Caro, 1994; Gros, 1998;
African wild dogs, Woodroffe *et al.*, 1997; cheetah). The IUCN Red List
category, however, masks the critical variable for many of these species,
many of which are unlikely to go extinct in the near future. Their extirpation across much of their range has made them functionally ecologically
extinct across most of their historical area of occupancy.

These species are candidates for recovery, hence, advance planning, and proactive conservation action similar to that developed for wide-ranging secure species may help avoid a conservation crisis. Management of these species can be complicated. Where the species is locally abundant, and populations stable or growing, the animal may be perceived as a low priority for conservation, a competitor for prey, or even a pest species as a threat to local domestic livestock (e.g. Cobb, 1981). Conversely, however, where the species is rare, it may be a conservation priority and be afforded high levels of protection. Such perverse investments can only be resolved with a combination of range-wide assessment, and regional level planning and implementation of conservation activities.

Tigers as a model

Conservation of species that are, or have been, in consistent and persistent decline across an entire continent, like the African wild dog or the tiger, require a nested set of approaches to setting of priorities. These approaches have been developed to a high degree for tigers.

The first step is to develop a distribution-wide priority setting exercise for these species. Early approaches for tigers focused on securing 'viable' populations of each 'subspecies' in each range state country (Seal et al., 1987), with similar sub-specific defined captive populations acting as a reserve for the metapopulation (Foose, 1987). Such an approach, while perhaps state of the art in 1987, is difficult to justify in 1999 for a number of reasons.

The first problem is that recent advances in molecular genetics of tigers suggest that sub-specific differences are small, justifying, perhaps, only the delineation of Sumatran and mainland 'subspecies' (Cracraft et al., 1998; but see Wentzel et al., 1999). Recent morphological analysis supports this conclusion (Kitchener, 1999).

A second issue is that tigers are found across a diversity of habitats, from near dry to rainforests to mangrove swamps to the near Arctic conditions of the Russian Far East. With small genetic differences among populations, preserving 'tigerness' is more ecological than genetic. Therefore, a more recent approach to setting priorities for tiger conservation is based on analyses that aim to secure ecological integrity and habitat representation across the range of the tiger (Noss, 1995; Wikramamayake et al., 1998). This process establishes priorities across the range of the tiger, which ensures both regional habitat representation of different habitats and gives priority to the largest intact blocks of potential tiger habitat or tiger conservation units (TCUs).

Working on a continent-wide scale has allowed a critical refocusing of tiger conservation priorities. At their best, TCUs offer a coarse-filter definition of priorities which has raised local and regional awareness of the importance of individual TCUs. The work has also assisted donors, such as the National Fish and Wildlife Foundation's 'Save the Tiger Fund,' and the U.S. Fish and Wildlife Service 'Tiger and Rhino Conservation Fund' to focus their investments, much as the hotspot analyses have assisted the MacArthur Foundation and the World Bank develop their portfolios.

But the TCU analysis is a coarse-filter approach in which the weighting of data used for setting priorities is similarly biased to data available at a larger scale – actual tiger distribution is only half as important as the estimated threat from poaching, and poaching is only half as important as habitat integrity (Wikramamayake et al., 1998). At this scale, the density, or even the presence or absence of tigers, is inconsequential compared to the existence of suitable habitat – if suitable habitat (and it is assumed, prey) exist, tigers will repopulate an area if poaching is controlled. But at many sites, the process speaks more about the potential to recover tigers than it does about saving them. This makes the approach an excellent first cut analyis, but demands a finer scale regional analysis in which tiger density, prey abundance, poaching pressures, and effective political structures are considered. Fortunately, progress is already being made at improving the resolution of both information and conservation action.

TURNING PRIORITIES INTO CONSERVATION ACTION

For widespread species in persistent decline across their range, the tools we use to effect local and regional conservation action continue with a tradition of landscape-based conservation planning already popular in the United States. These tools have been developed and refined in landscape management plans for species such as wolves and grizzly bears, which are locally rare, but globally abundant (Noss et al., 1996; Mladenoff & Sickley, 1998).

For tigers, there have been a diversity of approaches, each tailored to local conditions, but each of which views tiger conservation from a landscape approach to conservation. In the Russian Far East, a relative abundance of data on the ecology, distribution, and threats facing tigers has allowed for relatviely comprehensive planning to save the tiger (e.g. Miquelle et al., 1999; D. Miquelle et al., unpub. data). Here tiger density is very low, but land is abundant and human population densities are low, planning land-use with an eye to tiger conservation is still possible on a grand scale (e.g. Miquelle et al., 1999).

In Nepal or India, where a sea of humanity comes up against a hard-edge reserve boundary, options are often constrained by the available forest, and priorities can often be refined only with a better understanding the threats, both political and biological, facing each, isolated fragment of tiger range (Thapar, 1999). Here, a deep understanding of social and economic constraints is critical to developing conservation strategies which both limit access to the protected areas and offer benefits to the communities surrounding the tigers (e.g. Dinerstein *et al.*, 1999; Karanth *et al.*, 1999). Can these approaches be replicated? Perhaps. Although application of a model such as that developed in Chitwan is difficult for a number of reasons which can best be summarized by acknowledging that the model developed for Chitwan has met with some success because it has been carefully adapted to local conditions (Dinerstein *et al.*, 1999).

In some places, better data are needed at ground level to more finely define local and regional priorities. In Sumatra, both a strong interest from the government and a developing understanding of the factors that limit tigers and their prey (Franklin *et al.*, 1999; T. O'Brien/ M. Kinnaird pers. comm.) will allow for a better assessment of island-wide priorities for tigers. In Thailand, planning (J. Smith *et al.*, 1999) far outstrips current data on status of tigers and the threats they face (Rabinowitz, 1999). In Indochina, where forests are relatively vast, high priority 'level one' TCUs, encompass areas that represent half of Cambodia, or a third of Myanmar (Wikramanyake *et al.*, 1998). Simultaneously, there is a near total lack of data on tiger distribution, the status of the prey on which tigers rely, and the spatial dimension of threats (Rabinowitz, 1999).

CONCLUSIONS

There is a need to distinguish the difference between setting priorities and developing an operational strategy for doing conservation in priority areas. Global or regional priority setting exercises (for tigers, foxes, or beetles) are a coarse-filter that help us focus limited resources and manpower. These coarse-filters, however, tell us little about how to achieve conservation – at best, priority setting exercises tell us what to save first, but not how to save it (Ginsberg, 1999). Operational strategies are a fine-filter that do not tell us where, regionally or globally, to concentrate our efforts, or spend limited funds, they merely set priorities for action at a local or landscape level. Both coarse-filter priority setting, and fine-filter operational strategies adapted to local political, ecological and social needs are required to effect conservation.

Global scale priority setting exercises do not work particularly well for carnivores. Carnivores are different from other species, both ecologically, and in our cultural relationships to them. Models that look for small areas of high diversity are unsuitable for large carnivores, and problematic for the rarer, cryptic small to medium sized carnivores. Complementarity models must evolve to include variables that affect spatial and temporal persistence of a species if they are to be of value to carnivore conservation.

Strategies that advocate conservation of larger areas inform us on global priorities for conservation at a scale appropriate for large carnivores. We need to adapt these priority setting exercises, however, to focus directly on carnivores as it is entirely possible to save priority areas defined by other measures of biodiversity within an ecoregion, but nonetheless fail to protect adequate space for large carnivores (D. Miquelle et al., unpub. data; Noss et al., 1996).

Data provided by the IUCN, both in Red Lists and in Action Plans, go a long way towards achieving a framework for carnivore conservation. These data, while necessary, are insufficient to effect conservation or set priorities for carnivores. Because the IUCN has no funds to implement action plan recommendations, and because the conservation community has not adequately enunciated an integrated strategy for carnivore conservation, implementation remains idiosyncratic.

Synthesis of action plans with a focus on regional, or local conservation priorities is clearly needed (see Mills et al., Chapter 2) and would help raise both funds and awareness of the problems facing carnivores. An integrative approach to carnivore conservation would also promote desperately needed guild-level studies, and further elucidate the conservation importance of guild shifts and mesopredator release.

The Red List categories inform us only of the probability of global persistence of a species – they discuss evolutionary extinction but do not grapple with the thorny issue of ecological extinction, or the role a species plays in generating or maintaining species diversity within an ecosytem. Many priority setting exercises seek to maximize diversity (often in the smallest area possible) but do not seek to conserve the processes that generate and maintain this diversity (Balmford et al., 1998). Yet, as our understanding of community level processes improves, the cascading effects of the loss of top predators suggests that such top–down processes are critical to maintaining diversity in complex mammalian and avian communities (e.g. Terborgh et al., 1999; Crooks & Soulé, 1999). In addition to assessing the global persistence of species, as is ably done by the IUCN, the conservation community must develop algorithms and rankings that measure the importance of a species or guild in regulating ecosystem processes and that, sets value on conserving those species.

Acknowledgements

This work was supported by the Wildlife Conservation Society. I would like to thank the editors for their patience, input and support, and for bringing together a remarkable group of carnviore conservationists. Dr. Georgina Mace provided body weight data for the analysis in Figure 23.1, and Dr. Beth Babcock assisted in the GLIM analysis. I thank them both, but take all responsibility for errors in data or analysis. Dr. Justina Ray provided comments on an earlier draft of the manuscript.

24

Postscript – carnivore conservation: science, compromise and tough choices

DAVID W. MACDONALD

A DIFFICULT DOZEN ISSUES FOR THE FUTURE

At the end of the conference that spawned this book, I was asked to offer some concluding thoughts, as I have now been asked to provide this postscript. Then, as now, one thing is clear in these flattering invitations: the least useful thing would be to plod dutifully through a précis of each chapter – the skill with which our authors have written would doom any such review to repetitive irrelevance. Our authors have produced a veritable electrical storm of ideas within these covers, celebrating what are, after all, amongst the world's most glorious animals. Having spent close to 30 years in the thrall of carnivores, this compendium strikes me as a major punctuation mark in the story of efforts by scientists to understand and, thereby, to conserve them. From this editorial vantage point, I have wondered whether there are any patterns that shimmer, as unwritten meta-conclusions, through the tapestry woven by the chapters herein, or other topics that have escaped our net. A dozen loosely related, difficult issues come to mind. This 'difficult dozen' springs from the foregoing chapters, but I will draw on examples from my own experience.

HOW TO RETAIN SIGHT OF THE TREES WHILE SEEING THE WOOD?

One hallmark of penetrating science is the stripping away of distracting detail to expose an essential principle – conversely, the inability to untangle the untidiness of reality is captured in the English idiom of failure to see the wood for the trees. Watching the travails of careers-worth of fieldwork ingeniously distilled down to a reassuring generalization, reminds me that fully understanding the wood sometimes requires a close look at the trees. Amongst my own encounters with carnivores some were of episodes so intricate and so rarely repeated that they defy neat packaging. Sometimes, an opportunity arises to publish raw field notes (Macdonald, 2000a) – for example, vignette accounts such as the time I watched one male meerkat, *Suricata suricatta*, appear to foment a clash between rival groups which led him to commandeer a harem of females, or the time I saw a mother red fox, *Vulpes vulpes*, appear to teach her cub the technique of worm-hunting. Inspirational glimpses are a reward of long hours of fieldwork, but the business of science properly requires that this pastime be conducted methodically, revealing generalizations based on repeated observations of a particular phenomenon – indeed the several days of enthralling observation during which that meerkat kidnapped the females are hardly done justice by the datum to which they distil somewhere in Doolan & Macdonald (1996).The requirement for replication fosters the sometimes-brutal language that demotes a single observation to the disparaged status of anecdote whereas, had it been repeated, it would have achieved the lofty designation of an insight. Nonetheless, the lonely anecdote can be inspirational. Of course, such observations are likely to remain so rare as always to elude the statistician's rapier, and, of course, I see the demon anthropomorphism skulking over my shoulder. But things you see only rarely may nonetheless be important scientifically, as may be events that remind us of the individuality – perhaps personality – of our subjects. Too much time spent staring at a computer screen can dim our memories of the enthralling complexity of animals in the wild, increasing the risk that we are not only guided and inspired by theory, as we should be, but blinkered by it.

HOW CAN WE GET THE INTERPRETATIVE TIME-SCALE RIGHT?

Many important questions about carnivore ecology can only be answered by studies that take the long view (Chapters 12 and 13). This is for at least three types of reason. First, many carnivores are sufficiently long-lived that the course of their lives is only revealed over several years. Second, the answers to many questions lie in the relationships – social and genetic – between individuals of a population, and uncovering these networks of interaction may take the lifetime of at least one cohort. Third, many ecological processes, including cyclical ones and those currently driven by human intervention, proceed at a pace slower than can be encapsulated in short-term studies. Of course, analytical and experimental ingenuity can allow the long-term process to be unveiled in rapid, efficient cross-section, but I am impressed by both the power and, yet, the scarcity of long-term studies. Perhaps the most striking mammalian example comes from the red deer,

Cervus elephas, of the Isle of Rhum (most recently Kruuk *et al.*, 1999), but amongst carnivores the wolves, *Canis lupus* of Isle Royale (Post *et al.*, 1999), foxes, *Vulpes vulpes*, in the city of Bristol (P. Baker *et al.*, 2000), lions, *Panthera leo*, cheetah, *Aconyx jubatus*, and spotted hyenas, *Crocuta crocuta*, and dwarf mongooses, *Helogale parvula*, of the Serengeti (Rood, 1987; Packer et al 1991b; Hofer & East, 1995; Durant, 1998) all illustrate the point, as does McNutt's (1996) study of African wild dogs in Botswana and Holekamp and co-workers' study of hyenas in Kenya (Frank *et al.*, 1995). My own team's work with the badgers, *Meles meles*, of Wytham Woods has involved marking almost 750 individuals during more than 4500 captures since 1987, and logging something close to quadrupling of population density from 10 adults/km^2 to 38 adults/km^2. This passage of time has revealed that explanations, which appeared sufficient for the early stages of this population increase (changing land use, Da Silva *et al.*, 1993), were interwoven with many other factors (summer rainfall Woodroffe & Macdonald, 2000) that ultimately pointed to the involvement of larger processes (climate change, D. Macdonald & Newman, in press). Indeed, a spatial system of neatly tessellating territories each centred on one communal sett or den in which most individuals spent most days through the 1970s and 1980s (e.g. Kruuk, 1978; Woodroffe & Macdonald, 1995) was largely unrecognizable by the late 1990s when members of social groups split between twice as many dens and appeared to move freely between several former territories linked as super-groups (D. Macdonald & Newman, in press).

In summary, the value of long-term studies, and the dedication to monitoring that these involve, should be amongst the priorities of research funders.

ARE CARNIVORES DIFFERENT?

My next point is that while the fundamental principles of modern biology are equally applicable to any order of mammals, and while the conservation of any taxon can raise tough choices, there are senses in which, both biologically and managerially, the carnivores are special. There are good reasons to devote a book entirely to their conservation. Biologically, perched at the top of the food pyramid, their energy requirements place special constraints on the life histories especially of larger carnivores (Carbone *et al.*, 1999), exposing them to the perils of low population densities and wideranging movements (Woodroffe & Ginsberg, 1998). These linked problems are vividly illustrated by Gorman *et al.*'s (1998) analysis of the costs of living facing African wild dogs.

The list of qualities that distinguish carnivores each includes interesting areas of both theory and practice. As targets of trophy hunting they have a special place, and with mounting debate about sustainable use the biological impact of this activity merits special attention (Chapters 9, 10 and 19). A linked issue is the widespread conflict between carnivores and game, another global aspect of commercializing wildlife. Also there is their possible role as keystone species and indisputable role as flagship species (Chapter 23). All these factors mean that while we should be sensitive to unfairness, and illogical speciesism that draws more attention to clouded leopards than to depressed mussels, we should also not be blind to the reality that carnivores are both biologically species and potent symbols for conservation.

HOW CAN WE RECONCILE CONTRADICTORY PERCEPTIONS OF CARNIVORES?

Carnivores also pose special problems to conservationists because, more than any other group, they are simultaneously different things to different people (and sometimes also to the same people). The red fox glimpsed as it trots through the frost-crisp dawn may be watched by the naturalist thrilled by its beauty, by the hunter challenged by its ingenuity as quarry, by the trapper to whom its pelt is a resource, by the health officer to whom it is a disease vector, to the chicken farmer and a pheasant-shooter who see it as a competing vermin and the arable farmer who should see it as a pest-control agent (D. Macdonald et al., in press). The list of contrasts goes on. We must, therefore, become familiar with finding solutions that acknowledge, even if they cannot fully accommodate, the multiple personas ascribed to many carnivores. We should also remember, I think, that carnivore conservation is not just about saving rare species, but also prudently managing abundant ones. The force of this point, and indeed the fox example, is especially plain to British readers in view of the published results of the Government's Inquiry into Hunting with Dogs (see also D. Macdonald et al., 2000). The very existence of this inquiry illustrates how exacting scientific questions about managing a carnivore can challenge society's values (e.g. D. Macdonald & Johnson, 1996, 2000; Baker & Macdonald, 2000) and even threaten governmental and prime ministerial reputations. Around the world carnivores are no strangers to bio-political turmoil, which has embroiled the wolves of Yellowstone to the tigers of Nepal (McDougal, 1987; McNamee, 1997; Phillips & Smith, 1996).

HOW ARE WE TO RESOLVE CONFLICT BETWEEN CARNIVORES AND PEOPLE?

One special thing about carnivores is that they get into trouble with people – sometimes that trouble is imagined, but sometimes it is real. In Britain, society greets with almost unanimous delight the hesitant recovery of the otters, *Lutra lutra*, that were largely obliterated during the 1960s by profligate use of pesticides (Chanin & Jefferies, 1978; Mason & Macdonald, 1986). Indeed, otters are a prized rarity throughout much of continental Europe. In stark contrast, fish farmers of the Czech Republic, many of them operating in subsistence economies, take a different view as otters are perceived to drive them to destitution (Kranz, 2000; Kranz & Toman, 2000). Indeed, in a tangle of paradoxes, it may only be because of the bountiful food offered by carp ponds that otters abound in that Czech landscape, and it is even possible that the otter densities sustained by piscicultural productivity threaten native fish populations in local streams. Claims against the badgers perceived to spread the tuberculosis that bankrupts English dairy farmers (Krebs *et al.*, 1997), or against the tigers that disenfranchise Indian villagers (McDougal, 1987) or against the cheetah killing Namibian calves (Marker-Kraus *et al.*, 1996b) may or may not turn out to be exaggerated, but as the twenty-first century advances, we are unlikely to be able always to get away with debunking the criticisms of carnivores, falsely accused as they often are. We will have to face up to some uncomfortable facts that there can be real costs to living alongside them, and we'll have to decide who is going to pay.

HOW ARE WE TO RESOLVE CONFLICT BETWEEN CONSERVATION GOALS FOR CARNIVORES?

As if the tough choices provoked by conflict between people and carnivores were not bad enough, there are increasing instances where conservation goals lie together uneasily. In natural communities carnivores may impact on their prey and on their competitors. Indeed, the generalization emerges that within guilds larger carnivores brutalize smaller ones; thus more lions or spotted hyenas may mean fewer African wild dogs or cheetah (Creel & Creel, 1996). Hunters have long maintained that populous predators were inimical to their prey, but conservationists now face a similar dilemma. Consider the agonising example of the Asian wild dogs, *Cuon alpinus*, or dholes, known locally as ajag, in Java. Observations of these locally endangered canids in Alas Purwow National Park in Java suggested that they were obliterating the calf production of a rare wild cow, the banteng (*Bos javanicus*) – one endangered species seemed set to exterminate another. The Park authorities were attracted by the idea of shooting the one, or perhaps translocating it, in the attempt to preserve the other. Clearly an ethical and practical morass lay ahead. As it happens, the ajag seem to have shifted their focus and the banteng calves are flourishing, but the point remains. Conservation of the diminutive swift fox, *Vulpes velox*, reintroduced into the Canadian prairies may hinge on protecting them from coyotes, *Canis latrans*, or red foxes. Indeed, the same point is vividly illustrated by the Island fox, *Urocyron littoralis*. These foxes provide marvellous examples of evolution in action. On all six Channel Islands off the coast of California, there are miniature replicas of the grey fox found on the mainland, most of them distinguished as subspecies. They weigh between a mere 1.1 and 2.7 kg, whereas their larger Californian counterpart weighs 5 kg. The three northern subspecies of the Island fox on the northern three islands have recently declined dramatically. Less than six individuals are now known on San Miguel, for example, from a population of several hundred in 1993. Golden eagle predation and disease have been implicated in the decline but the reasons are really unknown. A high priority plan to save the species on the northern islands, including captive breeding, has been initiated by the National Park Service. However, less than 200 km to the south, on San Clemente island another subspecies of the island fox (found only on San Clemente island) is being killed by the US Navy in an attempt to eliminate predators of the San Clemente island shrike, itself an endangered subspecies of the widely distributed loggerhead shrike. Clearly, there is a risk of inconsistency here, and it highlights one of the major issues facing conservation – how to reconcile conflicting priorities. We have so badly driven so much wildlife into tight corners that we may expect to face ever-more gruelling trade-offs in deciding what we want.

Superficially, the dilemma seems less uncomfortable when people have precipitated the imbalance (Chapters 5, 6 and 7) – American mink, *Mustela vison*, are only able to drive out European mink, *M. lutreola*, because people foolishly introduced them (Sidorovich *et al.*, 1998). The trade-off is straightforward if lamentable – basically, kill American mink or accept the extinction of European mink. However, even the distinction between alien and native is not so convenient as it at first appears. The American mink, currently making their way through the Iberian peninsula, are widely considered as legitimate targets for extermination because they were introduced during the twentieth century, but the Egyptian mongooses, *Herpestes ichneumon*, and perhaps even the genets, *Genetta genetta*, that are now welcome components of that fauna were probably introduced by the Romans. How long do you have to be resident before being naturalized? We may be eager to dodge such questions, but all these fascinating discoveries threaten to face us with awkward consumer choices about what sort of balance we want from the conservation of carnivores. We may lament that the actions of our forefathers, and indeed ourselves, compel us to address such issues, but the likelihood is that tough choices are likely to get worse rather than better. What can we do about this? It may be delusion to believe that every problem has a solution, but there are at least three general courses of action that may be helpful in the context of the foregoing examples. First, as the public's sophistication increases, it may be advantageous to move away from the easily marketable, but often untrue, caricature that conservation issues boil down to stand-offs between the good guys and the bad guys. Reality is more often a difficult and painfully utilitarian balance of considerations that have the annoying qualities of being simultaneously valid yet contradictory! A second goal might be to foster greater transparency and consistency in the presentation of dilemmas. A third may be to acknowledge more fully that animal welfare is an important element of conservation biology's interdisciplinary remit (Chapter 9). On this latter point, my own prediction is that both public perception and regulatory action will make the science (alongside the ethics) of welfare an increasingly important ingredient of conservation in the future (see, for example, Bradshaw & Bateson, 2000). While each of these three suggestions may have virtue, the raw fact is that tough choices are going to get tougher.

HOW ARE WE TO EMBRACE INTERDISCIPLINARITY?

To tackle such problems, of course, will require the ingenuity and wisdom that has been exhibited in so many of the chapters of this book. This leads me to interdisciplinarity, in two senses. First, within our professional circle it is thrilling to read in this book of the different ways each of the authors has wracked the microscope lens up or down to tackle problems at hugely contrasting scales. The molecular experts perform their wizardry to unravel the past (Chapters 14, 15 and 22), those with a geological perspective use an analytical sophistication that boggles the mind in unravelling millions of years of phylogeny (Chapter 2), the field-workers probe ever deeper into the intricacies of individual lives, and all these people benefit hugely form uniting their efforts. Carnivore biologists have always coveted droppings with a fervor that has perplexed many of those not alert to scatalogical delights, but the thought that today the DNA extracted from the treasured scat can reveal the depositor's identity, parentage, hormonal and social state is truly wonderful.

This is one sort of interdisciplinarity in which we must surely rejoice (Chapter 11). But I have another sort in mind, which I think will be even more important for us in the future. However elegant the science, and there are some masterpieces within these covers, time and again we discover that science is not enough. Take the question: what causes extinction of carnivores (Chapters 3 and 4)? If extinction is non-random, it makes sense to try and explain it. First, the evidence is that it is indeed non-random. Although Nee & May (1997) showed that a surprisingly high proportion of evolutionary history would survive a random extinction episode (because most species have a close relative), Purvis *et al.* (2000a) argue that clusters of relatives (clades) are likely to sink together (being prone to the same factors). Measuring extinction in terms of lost phylogenetic diversity (total phylogenetic branch length) suggests that the likely degree of non-randomness evidenced by the IUCN Red Book threat categories threatens the loss of 85 more mammalian general than would random extinctions. This loss would amount to 850 million years of additional lost evolutionary history. Of course, although it has undeniable gravity, lost phylogenetic diversity is not the only measure of loss that may disturb a wide public. One can think of many circumstances in which survival of a related species will provide scant solace for the extinction of its congener.

Returning to the question of what causes extinction of carnivores, all too often the answers point towards the activities of people, and have done for a long time – 40–70% of fossil genera of Carnivora went extinct at roughly the time of human population expansion and global colonization. Nowadays, prevailing extinction rates are 100 to 1000-fold expected background rates (Lawton & May, 1995). The quest for solutions must be interdisciplinary. Biological knowledge is good stuff, but it isn't enough. So many of the problems we have discussed in this book have raised questions of human behaviour, of community development and, ultimately, of economics, ethics and on to politics. We need large protected areas, but how are they to be paid for? How is the hardship caused by their creation to be mitigated? Try though I might to wriggle back into the snug shell of biological familiarities, I know that it just won't do. If we are hermit crabs who wish to grow our contribution to conservation, it's time we changed our shells. Of course, we mustn't waste one droplet of the biological wisdom that has been so hard won, but my hunch is that we must move swiftly to blending it much more with the wisdom of planners, sociologists, development experts and economists. We must expect this interdisciplinarity to face us with new and perplexing trade-offs.

But beware motherhood-and-apple pie platitudes. In practice, how are we to do it? An exhilarating but unnerving question. It seems to me that this new orientation of our focus does not resonate well with the career structure with which many biologists are enmeshed. That which results in papers in top journals and high citation indices, or even high ranking in competitive grant proposals, may not always be that which is most useful for conservation. Indeed, it's a lot easier to be interesting than it is to be useful.

HOW CAN WE BE USEFUL?

Empowered by more and better research, how can biologists contribute to conservation (Chapters 11 and 23)? Elsewhere (Macdonald *et al.*, 2000), I have suggested an answer at two levels. At a strategic level, applied biological research – that is, problem solving – is just one component of what I term the conservation quartet. These are the four interdependent ingredients of successful conservation: research (to crack the problem and identify the solution); education (to inform society, and influence opinion); community (to involve the stake-holders; Chapter 13); and implementation (to get the job done). At an operational level, at least four courses of action are embraced by the acronym FREE: standing for Foster, Record, Enhance and Enrich. Obviously we must foster what remains. We must also record what we've got, and how it's changing (Chapters 8, 17 and 18). Monitoring, in one guise or another, thus contributes to each part of the conservation quartet. Counting what we have and how it is changing is essential to research, provides grist for the educational mill, involves the active participation of community and will be the yardstick against which the success of implementation will be judged. Enhancement revolves especially around habitat restoration, and enrichment involves putting creatures back into faunal communities that we have removed them from in the past. The swift fox in Canada springs to mind, along with the black-footed ferret, *Mustela nigripes*.

To make all these things happen requires changing policy and changing economies. Conserving carnivores (and most other things) is going to require money, and re-evaluation of priorities. The day of the conference which spawned this book saw the launch of Zarya, part of the 450 tonne international manned space station at a cost of the better part of US $100 bn. With the majesty of the world's fauna poised to go down the drain, and brewing environmental disasters threatening to take us and our children down the same drain, I wonder whether humanity is spending its pocket money prudently. I make this point cautiously, and without being blind to the fact that there have been numerous breakthroughs in the prediction and monitoring of biodiversity, climate change and human induced change as well as applied pest management derived from Earth-orbiting sensors. These were not possible before, and yet, they are vital precisely because of rapidity of biodiversity loss especially in inaccessible places. Even the US military satellite programme has spawned GPS technology, a great tool for biologists. However, to acknowledge, as I do, that the prioritization argument must be wielded carefully, is not to accept that the balance of world spending is sensible. It is hard to be at ease with the estimate (see below) that US spending on perfumery is roughly the same as world spending on biodiversity conservation.

HOW TO VALUE CARNIVORES?

All too often, conservation is about tough choices between heartbreaking, or at least incommensurable, options (Chapters 21 and 23). If Carnivores, or other elements of biodiversity, are to be conserved then, sooner or later, the question arises of their value (against some alternative, such as keeping the game park or converting it to an industrial site). Sometimes it is possible to monetize elements of the environment. For example, one might calculate how much, in theory, it would cost to generate the so-called ecosystem services on which we depend for energy, nutrients, water, etc., if they were not provided 'free' by Nature (a first, albeit controversial, attempt to value these services estimated their annual worth to humanity at US $33 trillion per annum; Costanza et al., 1997). Biodiversity contributes to these services, and James et al. (1999) estimate the cost of conserving biodiversity across 15% of the global land area (which might be enough to sustain ecosystem services) would be US $22.6 billion annually. This sum is small in comparison to the estimated US $1 trillion currently spent annually on environmentally harmful, or 'perverse' subsidies (those used to promote over-exploitative agricultural production, energy use, road transportation, water consumption and commercial fishing, by keeping resource prices below market values) (Myers 1998). James et al. (1999) estimate that US $6bn is spent globally on biodiversity conservation – a figure put in perspective by the revelation that in North America alone this sum is spent on perfumes annually ($1bn on nail polish; Newman, 2000). But Carnivores have a worth beyond their contribution to ecosystem services – for example, through tourism they may generate revenue (Thresher, 1982, estimated the revenue generation by a male lion in Kenya's Amboseli National Park at US

$128,750; see also Creel & Creel, 1997; a lion hunt in Zimbabwe is currently auctioned at c. £10,000). Furthermore, carnivores have an existence value that has no market price. An existence value is that which an individual attaches to something that he/she does not visit, consume or use. Well-known examples are the existence values attached to the preservation of Antarctica and whales by individuals who come into contact with neither nor derive any services from them. There are techniques that purport to arrive at such values. For example, expressed preference techniques such as contingent ranking (people choose between baskets of options, thereby establishing their relative preferences for particular species relative to goods that have an easily-measured market price). These valuations are woefully subjective, and generally rarefied (Hanley & Milne, 1996 note that while 99% of people questioned thought wildlife had the right to exist, only 49% were at ease with the idea of this right translating into a cost in money or jobs). If the fate of carnivores is to hinge on expressed or indirect preference techniques, then one thing is clear: education is essential – willingness to pay will be directly driven by understanding and appreciation that can only be fostered by education. Insofar as education is never complete, these valuations will be underestimates and dynamic, and a nerve-wrackingly labile – not to say whimsical – basis for globally important decisions. A great difficulty for conservation is how to incorporate an economic definition of sustainability (e.g the Hardwick–Solow rule) when equations cannot be solved in the absence of a method to trade-off the basic components. Although there are few contingent valuations available with which to undertake these trade-offs, we could tackle the question of what the trade-offs would have to be in order to meet the sustainability criterion. Asking this question is no more than recognition that there is only limited, scarce money to spend, and it makes sense to get the maximum environmental benefit from it. Because conservation decisions can only be made through cost-benefit analysis of alternative actions, and because the only currency available for this analysis is money (Pearce, 2000), the question of how to value carnivores in particular and Nature in general is amongst the most daunting challenges ahead. A linked question, is how to raise the money to pay for their conservation.

HOW TO PAY FOR CONSERVATION?

Clearly, governments and non-government organisations are substantial sponsors of conservation, but to judge by results, current funding is inadequate. Conserving carnivores is expensive, so where else might we turn?

Perhaps inspiration lurks in an extension of the idea of carbon emissions offset. The notion is this: that where the functioning of society necessitates an environmental 'bad' (such as carbon emissions), we might demand restitution, in the form of environmental 'goods', and what could be more good than conservation. Conservation need not be limited to scrabbling for grants from charities or research councils, when much bigger roles await industry and its regulators. My own proposal (D. Macdonald, 2000b) is based on the premise that the ultimate goal of impact management is to cause no damage to the environment. However, since some commercial activities clearly involve unavoidable environmental impacts, net environmental neutrality necessarily involves compensatory investment in environmental improvement. An iterative process of improvement is suggested by the obvious fact that current impacts of business on biodiversity can be categorized as either irreducible (consequences of essential business activities the impact of which can be neither avoided nor mitigated) or reducible (by current best practice). First, the mitigation of reducible impacts should be mandatory, and as know-how and technology improve, continual reassessment should result in the transfer of formerly irreducible impacts into the reducible category. My suggestion is that companies causing unavoidable environmental damage could make compensatory payments into an environmental pool, which then funded biodiversity conservation worldwide (D. Macdonald, 2000b). This system – the Biodiversity Impacts Compensation Scheme – would hinge on: (a) the calculation of the compensatory levy imposed on companies; (b) the prioritization of conservation activities to which payments were made; and (c) development of robust performance indicators, which could be applied to monitoring both the company's (hopefully diminishing) impact on biodiversity and to the (hopefully enhanced) biodiversity conservation achieved by compensatory projects.

This compensatory mechanism for generating conservation revenue is just one proposal, but the point is clear that no account of significant questions facing carnivore conservation can avoid the issue of how to pay for it!

HOW CAN CARNIVORE CONSERVATION SQUEEZE INTO A DEVELOPED LANDSCAPE, AND AVOID DOUBLE STANDARDS?

There are occasions when the comforts of the developed world tempt the armchair conservationist to double standards, urging other countries to noble, self-sacrificing conservation efforts on a scale that we would scarcely dream of implementing at home. It may be a hackneyed thought, but the ethical certainty that tigers should prowl unfettered in India might be less straightforward if the prowling were in our own backyard! Of course, writing this from an industrialized, over-populated island I should not push the argument of practicing what we preach to fatuous extremes – but how far should we push it? In proposing carnivore conservation in far-off lands, one should have in mind the realities of living alongside mega-predators, and consider what radical options for carnivore conservation exist in our own back yards (Macdonald *et al.*, 2000).

Conservation of large expanses of pristine habitat may be the utopia of carnivore conservationists, but sadly this is a possibility already lost to history. However, if the only option is intervention, that may yet have merit. One example stems from an IUCN workshop to explore the conservation of African wild dogs (Mills *et al.*, 1998). It transpired that there is the possibility, by a combination of small national parks and by coalescing areas of game ranching, to create a constellation of wild dog refuges. Our eggs could therefore be spread between a larger number of baskets. The problem is that the possibilities for dispersal between these refuges are almost nil and, indeed, many of them would have to be fenced. Inbreeding and over-population would be inevitable in these enclaves. The concept which emerged from this workshop was that of metapopulation management. As necessary, wild dogs would be captured, anaesthetized and artificially dispersed between the elements of this burgeoning metapopulation. This is a deeply pragmatic idea. If we want more wild dog populations in South Africa, and I think we do, then the only way to have them for the foreseeable future seems to be within huge fenced reserves, and if they are to be fenced, then intervention is required to manage their numbers and demography.

Greatly though we should rejoice in naturalness, the fulfillment of naturalness is often somewhere just beyond the tip of the rainbow. Some wilderness remains, and clearly it must be cherished. But for many of the arenas in which carnivores require conservation, withdrawal of human influence is simply not an option. All too often, intervention is unavoidable. The metapopulation of South African wild dogs may be an extreme case, but it forces us to think about the tolerable limits between naturalness and intervention. Nature as a theme park is an unattractive prospect, but there are parts of the world where the option of laissez-faire conservation is long past.

WHY BOTHER?

And so to my final point. In drafting this postscript, I asked myself, what points stick in my mind about carnivores and their conservation. Of course, the key motivators are clear – on the one hand the urgency of conservation and the probable irretrievability of much biodiversity loss, on the other the beauty of ingenious science. But while one half of my brain was answering this question with the words you have just read, the other half was off on a different track. I thought about the humour in carnivores, and the outrageously funny antics of the foxes I watched beside the Dead Sea, dancing and whirling as they teased and tormented a lumbering striped hyena – and the shock I felt, as I watched this theatre, of feeling the hot breath of the second hyena blasting into my ear as it crept up to watch me watching its companion. I thought how you simply have to feel happy when you see the Disney-esque honky-tonk gait of a honey badger jauntily going about its business, and I thought about the occasion late at night in the Kalahari, when leaning out of the window of the jeep to take a photo in the flurry of a spotted hyena kill, I couldn't understand why my elbow kept slipping, until I looked down to see the young spottie gently tugging on the sleeve of my pullover grasped delicately in his incisors. Thinking about carnivores, that other side of my brain also thought about their sensitivity, and remembered the moment when Toothypeg, a Methuselah among red foxes, delicately grasped earthworms in her teeth, and by patting them slowly with her forepaw, mesmerized the earthworms out of the ground til her cub could grasp them as he served his apprenticeship as a worm-catcher. And I remember just how great it felt to be so much a part of the scenery that a sentinel meerkat clambered aloft my shoulder to do his stint of guard duty or smeared his anal pouches endearingly onto my leg. I recall the knee-jellying majesty of lions pulling down a wildebeest within a few feet of me, and the stunning superiority in the eyes of a tiger just one elephant's haunch below me. Finally, I remember the startling, stunningly aching beauty of the evening sunlight on African wild dogs as they scythed through a sea of parched grass.

Everyone who has written in this book has a similar treasure trove of such memories and, personally, I'm not embarrassed to own up to sentiment. Of course, sentiment is certainly not the strongest argument for conservation or research – it may even be the weakest – but, for me at least, amidst the flurry of computer screens, statistics and citation indices, it's a good idea to get such memories out once in a while and dust them off, lest we forget why we got into all this, and why it is worth persevering.

Acknowledgements

I am grateful for helpful comments on an earlier draft from Scott Creel, Dominic Johnson, Tico McNutt, Gus Mills, Philip Riordan and Michael Thom, and from my co-editors, and for many stimulating conversations with members of Oxford University's Wildlife Conservation Research Unit.

References

Aaris-Sorensen, J. (1996). Road-kills of badgers (*Meles meles*) in Denmark. *Annales Zoologici Fennici* **32**: 31–36.

Abrams, P. A. (1996). Evolution and the consequences of species introductions and deletions. *Ecology* **77**: 1321–1328.

Abramsky, Z., Bowers, M. A. & Rosenzweig, M. L. (1986). Detecting interspecific competition in the field: testing the regression method. *Oikos* **47**: 199–204.

Ackerman, B. B., Hemker, T. P., Lindzey, F. G. & Button, A. J. (1981). Cougar numbers in the Henry Mountains, Utah. *Encyclia* **58**: 57–62.

Addin Enbbu, Baillie, J. & Groombridge, B. (1996). *1996 IUCN Red List of Threatened Animals*. Gland: IUCN.

Addin Enbbu, Balmford, A., Mace, G. M. & Leader-Williams, N. (1996). Redesigning the ark: setting priorities for captive breeding. *Conservation Biology* **10**: 719–727.

Addison, R. F. & Brodie, P. F. (1987a). Organochlorine residues in maternal blubber, milk, and pup blubber from grey seals (*Halichoerus grypus*) from Sable Island, Nova Scotia. *Journal of the Fisheries Research Board Canada* **34**: 937–941.

Addison, R. F. & Brodie, P. F. (1987b). Transfer of organochlorine residues from blubber through the circulatory system to milk in the lactating grey seal *Halichoerus grypus*. *Canadian Journal of Fisheries and Aquatic Sciences* **44**: 782–786.

Addison, R. F., Zinck, M. E. & Smith, T. G. (1986). PCBs have declined more than DDT-group residues in Arctic ringed seals (*Phoca hispida*) between 1972 and 1981. *Environmental Science and Technology* **20**: 253–256.

Adorjan, A. S. & Kolenosky, G. B. (1969). A manual for the identification of hairs of selected Ontario mammals. *Ontario Department of Lands and Forests Research Report* No. 90.

Aguirre, A. A. (1998). Biomedical management of the most endangered pinniped in US waters: the Hawaiian monk seal (*Monarchus schauinslandii*). *Proceedings of the AAZV and AAWV Joint Conference* p. 374.

Aitken, G. (1997). Conservation and individual worth. *Environmental Values* **6**: 439–454.

Aitken, G. (1998). Extinction. *Biology and Philosophy* **13**: 393–411.

Albrecht, G. (1998). Thinking like an ecosystem: the ethics of the relocation, rehabilitation and release of wildlife. *Animal Issues* **2**: 21–46.

Aldrich, P. & Hamrick, J. (1998). Reproductive dominance of pasture trees in a fragmented tropical forest mosaic. *Science* **281**: 103–105.

Alexander, K. A. & Appel, M. J. G. (1994). African wild dogs (*Lycaon pictus*) endangered by a canine distemper epidemic among domestic dogs near the Masai Mara National reserve Kenya. *Journal of Wildlife Diseases* **30**: 481–485.

Alexander, K. A., Kat, P. W., Frank, L. G., Holekamp, K. E., Smale, L., House, C. & Appel, M.J.G. (1995). Evidence of canine distemper virus infection among free-ranging spotted hyenas (*Crocuta crocuta*) in the Masai Mara, Kenya. *Journal of Zoo and Wildlife Medicine* **26**: 201–206.

Alexander, K. A., Maclachlan, N. J., Kat, P. W., House, C., O'Brien, S. J., Lerche, N. W., Sawyer, M., Frank, L. G., Holekamp, K., Smale, L., McNutt, J. W., Laurenson, M. K., Mills, M. G. L. & Osburn, B. I. (1994). Evidence of natural bluetongue virus infection among African carnivores. *American Journal of Tropical Medicine and Hygiene* **51**: 568–576.

Alexander, K. A., Smith, J. S., Macharia, M. J. & King, A. A. (1993). Rabies in the Masai Mara, Kenya: a preliminary report. *Onderstepoort Journal of Veterinary Research* **60**: 411–414.

Allee, W. C. (1931). *Animal Aggregations. A Study in General Sociology.* Chicago: University of Chicago Press.

Allen, L. & Engeman, R. (1995). Assessing the impact of dingo predation on wildlife using an activity index. *Australian Vertebrate Pest Control Conference* **10**: 72–79.

Allen, L., Engeman, R. M. & Krupa, H. W. (1996). Evaluation of three relative abundance indices for assessing dingo populations. *Wildlife Research* **23**: 197–206.

Allen, P. J., Amos, W., Pomeroy, P. P. & Twiss, S. D. (1995). Microsatellite variation in grey seals (*Halichoerus grypus*) shows evidence of genetic differentiation between two British breeding colonies. *Molecular Ecology* **4**: 653–662.

Allen, S. H. (1984). Some aspects of reproductive performance in female red fox in North Dakota. *Journal of Mammalogy* **65**: 236–255.

Allen, S. H. & Sargeant, A. B. (1975). A rural mail-carrier index of North Dakota red foxes. *Wildlife Society Bulletin* **3**: 74–77.

Allen, S. H. & Sargeant, A. B. (1993). Dispersal patterns of red foxes relative to population density. *Journal of Wildlife Management* **57**: 526–533.

Alterio, N., Brown, K. & Moller, H. (1997). Secondary poisoning of mustelids in a New Zealand nothofagus forest. *Journal of Zoology* **243**: 863–869.

Amarasekare, P. (1994). Ecology of introduced small mammals on western Mauna Kea, Hawaii. *Journal of Mammalogy* **75**: 24–38.

Amos, W., Twiss, S. D., Pomeroy, P. P. & Anderson, S. S. (1993). Male mating success and paternity in the grey seal, *Halichoerus grypus:* a study using DNA fingerprinting. *Proceedings of the Royal Society of London, Series B, Biological Sciences* **252**: 199–207.

Andelt, W. F. & Andelt, S. H. (1984). Diet bias in scat deposition-rate surveys of coyote density. *Wildlife Society Bulletin* **12**: 74–77.

Andersen, L. W., Born, E. W., Gjertz, I., Wiig, O., Holm, L.-E. & Bendixen, C.(1998). Population structure and gene flow of the Atlantic walrus (*Odobenus rosmarus rosmarus*) in the eastern Atlantic Arctic based on mitochondrial DNA and microsatellite variation. *Molecular Ecology* **7**: 1323–1336.

Anderson, A. E. (1983); A critical review of literature on puma (*Felis concolor*). *Colorado Division of Wildlife, Special Report* **54**: 1–91.

Anderson, D. R. (1997). *Corridor Use, Feeding Ecology, and Habitat Relationships of*

Black Bears in a Fragmented Landscape in Louisiana. Unpublished M.S. thesis, University of Tennessee, Knoxville.

Anderson, J. L. (1981). The re-establishment and management of a lion population in Zululand, South Africa. *Biological Conservation* **19**: 107–117.

Anderson, R. M. & May, R. M. (1978). Regulation and stability of host–parasite population interactions. I. Regulatory processes. *Journal of Animal Ecology* **47**: 219–247.

Anderson, S. S. & Hawkins, A. D. (1978). Scaring seals by sound. *Mammal Review* **8**: 19–24.

Andrén, H. (1994). Effects of habitat fragmentation on birds and mammals in landscapes with different proportions of suitable habitat: a review. *Oikos* **71**: 355–366.

Andrén, H. (1995). Effects of landscape composition on predation rates at habitat edges. In *Mosaic Landscapes and Ecological Processes*, eds L. Hansson, L. Fahrig & G. Merriam, pp. 225–255. London: Chapman & Hall.

Andrén, H. (1999). Habitat fragmentation, the random sample hypothesis and critical thresholds. *Oikos* **84**: 306–308.

Andrews, R. D. (1979). Furbearer population surveys and techniques: their problems and uses in Iowa. In *Proceedings of the Midwest Furbearer Conference*. Kansas State University, Coop. Ext. Serv., Manhattan, pp. 45–55.

Angelstam, P., Lindstrom, E. & Widen, P. (1984). Role of predation in short-term population fluctuations of some birds and mammals in Fennoscandia. *Oecologia* **62**: 199–208.

Angerbjörn, A., Tannerfeldt, M. & Erlinge, S. (1999). Predator–prey relationships: arctic foxes and lemmings. *Journal of Animal Ecology* **68**: 34–49.

Angerbjörn, A., Arvidson, B., Noren, E. & Stromgren, L. (1991). The effect of winter food on reproduction in the arctic fox, *Alopex lagopus*: a field experiment. *Journal of Animal Ecology* **60**: 705–714.

Anon. (1989). Wolf persecution continues in Norway. *Oryx* **23**: 106.

Anon. (1999a). South China tiger killed one person – five tigers believed to survive in Jiangxi Province. *People's Daily (Beijing)* July 22.

Anon. (1999b). Colorado lynx re-introduction sparks bitter controversy. *Cat News* **30**: 15.

Anthony, R. M. (1996). Den use by arctic foxes (*Alopex lagopus*) in a subarctic region of western Alaska. *Canadian Journal of Zoology* **74**: 627–631.

Anzalone, C. R., Bench, G. S., Balhorn, R. & Wildt, D. E. (1998). Chromatin alterations in sperm of teratospermic domestic cats. *Proceedings of the Society for the Study of Reproduction, Biology of Reproduction Supplement*, abstract 364.

Apps, P. J. (1984). Cats on Dassen Island. *Acta Zoologica Fennica* **172**: 115–116.

Arcese, P., Hando, J. & Campbell, K. (1995). Historical and present-day anti-poaching efforts in Serengeti. In *Serengeti II: Dynamics, Conservation and Management of an Ecosystem*, eds A. R. E. Sinclair & P. Arcese, pp. 506–533. Chicago: University of Chicago Press.

Arnason, A. N., Schwarz, C. J. & Gerrard, J. M. (1991). Estimating closed population size and number of marked animals from sighting data. *Journal of Wildlife Management* **55**: 716–730.

Arnason, U., Bodin, K., Gullberg, A., Ledje, C. & Mouchaty, S. (1995). A molecular view of pinniped relationships with particular emphasis on the true seals. *Journal of Molecular Evolution* **40**: 78–85.

Arnold, M. L. (1997). *Natural Hybridization and Evolution*. Oxford: Oxford University Press.

Artois, M. & Aubert, M. F. A. (1985). Behaviour of rabid foxes. *Revue d'Écologie (La Terre et la Vie)* **40**: 171–176.

Artois, M., Aubert, M. & Stahl, P. (1990). Organization spatiale du renard roux (*Vulpes vulpes* L., 1758) en zone d'enzootie de rage en Lorraine. *Revue d'Écologie (La Terre et la Vie)* **45**: 113–134.

Artois, M. & Redmond, M. (1994). Viral diseases as a threat to free-living wild cats (*Felis silvestris*) in continental Europe. *Veterinary Record* **134**: 651–652.

Ashmole, N. P., Ashmole, M. J. & Simmons, K. E. L. (1994). Seabird conservation and feral cats on Ascension Island, South Atlantic. In *Seabirds on Islands, Threats, Case Studies and Action Plans*, eds D. N. Nettleship, J. Burger & M. Gochfeld, pp. 94–121. Cambridge: Birdlife International.

Asquith, P. J. (1996). Japanese science and western hegemonies: primatology and the limits set to questions. In *Naked Science: Anthropological Inquiry into Boundaries, Power, and Knowledge*, ed. L. Nader, pp. 239–256. New York: Routledge.

Associated Press (1994). Cougar is killed for jogger's death. *New York Times*, May 2, p. 1, A23.

Atkinson, I. (1989). Introduced animals and extinctions. In *Conservation for the Twenty-first Century*, eds D. Western & M. Pearl, pp. 54–75. Oxford: Oxford University Press.

Atkinson, I. A. E. (1996). Introductions of wildlife as a cause of species extinctions. *Wildlife Biology* **2**: 135–141.

Atkinson, I. A. E. & Cameron, E. K. (1993). Human influence on the terrestrial biota and biotic communities of New Zealand. *Trends in Ecology and Evolution* **8**: 447–451.

Atkinson, M. W. & Wood, P. (1995). The re-introduction of the cheetah into the Matusadona National Park, Zimbabwe. *Re-Introduction News* **10**: 7–8.

Aubert, M. F. A. (1993). Control of rabies in wildlife by depopulation. In *Proceedings of International Conference on Epidemiology, Control and Prevention of Rabies in Eastern and Southern Africa 1992*, ed. A. A. King, pp. 141–145. Lyon: Editions Foundation Marcel Merieux.

Aune, K. E. (1991). Increasing mountain lion populations and human–mountain lion interactions in Montana. In *Mountain lion – human Interactions*, ed. C. L. Braun, pp. 86–94. Denver: Symposium of Colorado Division for Wildlife.

Avise, J. C. (1989). Role of molecular genetics in recognition and conservation of endangered species. *Trends in Ecology and Evolution* **4**: 279–281.

Avise, J. C. (1994). *Molecular Markers, Natural History, and Evolution*. New York: Chapman and Hall.

Avise, J. C. (1996). Toward a regional conservation genetics perspective: phylogeography of faunas in the Southeastern United States. In *Conservation Genetics: case histories from nature*, eds J. C. Avise & J. L. Hamrick, pp. 431–467. New York: Chapman and Hall.

Avise, J. C. & Ball, R. M. (1990). Principles of genealogical concordance in species concepts and biological taxonomy. In *Oxford Surveys in Evolutionary Biology*, vol. 7, eds D. Futuyma & J. Antonovics, pp. 54–67. Oxford: Oxford University Press.

Avise, J. C. & Nelson, W. S. (1989). Molecular genetic relationships of the extinct Dusky seaside sparrow. *Science* **243**: 646–648.

Ayres, J. M., de Magalhaes Lima, D., de Souza Martins, E. & Barreiros, J. L. K. (1991). On the track of the road: changes in subsistence hunting in a Brazilian Amazonian Village. In *Neotropical Wildlife Use and Conservation*, eds. J. G. Robinson & K. H. Redford, pp. 82–91. Chicago: Chicago University Press.

Bacon, P. J. (1985). *Population Dynamics of Rabies in Wildlife*. London: Academic Press.

Badridze, J. (1999). Preparing captive-raised wolves for re-introduction, Georgia, Commonwealth of Independent States (C.I.S.). *Re-Introduction News* 18: 5–6.

Bailey, E. P. (1992). Red foxes, *Vulpes vulpes*, as biological control agents for introduced arctic foxes, *Alopex lagopus*, on Alaskan islands. *Canadian Field-Naturalist* 106: 200–205.

Bailey, E. P. (1993). Introduction of foxes to Alaskan islands: history, effects on avifauna, and eradication. *Resource Publication* No. 198. Washington, D.C.: U.S. Fish and Wildlife Service.

Bailey, E. P. & Kaiser, G. W. (1993). Impacts of introduced predators on nesting seabirds in the northeast Pacific. In *The Status, Ecology, and Conservation of Marine Birds of the North Pacific*, eds K. Vermeer, K. T. Briggs, K. H. Morgan & D. Siegel-Causey, pp. 218–226. Ottawa, Ontario: Canadian Wildlife Service Special Publication.

Bailey, J. L., Snow, N. B., Carpenter, A. & Carpenter, L. (1995). Cooperative wildlife management under the Western Arctic Inuvialuit. In *Integrating People and Wildlife for a Sustainable Future*, eds J. A. Bissonette & P. R. Krausmann, pp. 11–15. Bethesda, MA: The Wildlife Society.

Bailey, T. N. (1971). Biology of striped skunks on a southwestern Lake Erie marsh. *American Midland Naturalist* 85: 196–207.

Bailey, T. N. (1993). *The African Leopard: Ecology and Behavior of a Solitary Felid*. New York: Columbia University Press.

Bailey, T. N., Bangs, E. E., Portner, M. F., Malloy, J. C. & McAvinchey, R. M. (1986). An apparent overexploited lynx population on the Kenai peninsula Alaska. *Journal of Wildlife Management* 50: 279–290.

Baillie, J. & Groombridge, B. (Eds) (1996). *The 1996 IUCN Red List of Threatened Animals*. Gland: IUCN.

Baker, C. S. & Palumbi, S. R. (1994). Which whales are hunted? A molecular genetic approach to monitoring whaling. *Science* 265: 1538–1539.

Baker, M., Nur, N. & Geupel, G. R. (1995). Correcting biased estimates of dispersal and survival due to limited study area: theory and an application using wrentits. *Condor* 97: 663–674.

Baker, P. J., Funk, S. M., Harris, S. & White, P. C. L. (2000). Flexible spatial organisation of urban foxes (*Vulpes vulpes*) before and during an outbreak of sarcoptic mange. *Animal Behaviour* 59: 127–146.

Baker, S. E. & Macdonald, D. W. (2000). Foxes and foxhunting on farms in Wiltshire: a case study. *Journal of Rural Studies* 16: 185–201.

Baker, S. J. (1986). Irresponsible introductions and reintroductions of animals into Europe with particular reference to Britain. *International Zoo Yearbook* 24/25: 200–205.

Balding, D. & Nichols, R. (1994). DNA profile match probability calculation: how to allow for population stratification, relatedness, database selection and single bands. *Forensic Science International* 64: 125–140.

Baldwin, P. H., Schwartz, C. W. & Schwartz, E. R. (1952). Life history and economic status of the mongoose in Hawaii. *Journal of Mammalogy* **33**: 335–356.

Balharry, D. & Daniels, M. J. (1997). Wild living cats in Scotland. *Scottish National Hert. Research, Survey and Monitoring Rep.* No. 23.

Ballard, W. B., Edwards, M., Fancy, S. G., Boe, S. & Krausman, P. R. (1998). Comparison of VHF and satellite telemetry for estimating sizes of wolf territories in northwest Alaska. *Wildlife Society Bulletin* **26**: 823–829.

Ballard, W. B., McNay, M. E., Gardner, C. L. & Reed, D. J. (1995). Use of line-intercept track sampling for estimating wolf densities. In *Ecology and Conservation of Wolves in a Changing World*, L. N. Carbyn,, S. H. Fritts & D. R. Seip, pp. 469–480. Canadian Circumpolar Institute Occasional Publication No. 35. Edmonton, Alberta.

Ballard, W. B., Whitman, J. S. & Gardner, C. L. (1987). Ecology of an exploited wolf population in south-central Alaska. *Wildlife Monographs* **98**: 1–54.

Balmford, A. (1996). Extinction filters and current resilience: the significance of past selection pressures for conservation biology. *Trends in Ecology and Evolution* **11**: 193–196.

Balmford, A. & Gaston, K. J. (1999). Why biodiversity surveys are good value. *Nature* **398**: 204–205.

Balmford, A., Mace, G. M. & Ginsberg, J. R. (1998). The challenges to conservation in a changing world: putting processes on the map. In *Conservation in a Changing World*, eds G. M. Mace, A. Balmford & J. R. Ginsberg, pp. 1–28. Cambridge: Cambridge University Press.

Ballou, J., Gilpin, M. & Foose, T. (1994). *Population Management for Survival and Recovery*. Cambridge: Cambridge University Press.

Banfield, A. W. F. (1974). *The Mammals of Canada*. Toronto: University of Toronto Press.

Bangs, E. E., Fritts, S. H., Fontaine, J. A., Smith, D. W., Murphy, K. M., Mack, C. M. & Niemeyer, C. C. (1998). Status of gray wolf restoration in Montana, Idaho, and Wyoming. *Wildlife Society Bulletin* **26**: 785–798.

Barbour, R. W. & Davis, W. H. (1974). *Mammals of Kentucky*. Lexington: University of Kentucky Press.

Barlough, J. E., Berry, A. S., Skilling, D. E., Smith, A. W. & Fay, F. H. (1986). Antibodies to marine calciviruses in the Pacific walrus (*Odobenus rosmarus divergens* Illiger). *Journal of Wildlife Diseases* **22**: 165–168.

Barlow, N. D. (1996). The ecology of wildlife disease control: simple models revisited. *Journal of Applied Ecology* **33**: 303–314.

Barnes, R. F. W. & Tapper, S. C. (1985). A method for counting hares by spotlight. *Journal of Zoology* (London) **206**: 273–276.

Barnes, V. G., Smith, R. B., Udevitz, M. S. & Bellinger, J. R. (1995). Kodiak brown bears. In *Our Living Resources: A Report to the Nation on the Distribution, Abundance, and Health of U.S. Plants, Animals, and Ecosystems*, eds E. T. LaRoe, G. S. Farris, C. E. Puckett, P. D. Doran & M. J. Mac, pp. 349–350. Washington, DC: National Biological Service.

Barone, M. A., Roelke, M. E., Howard, J., Brown, J. L., Anderson, A. E. & Wildt, D. E. (1994). Reproductive characteristics of male Florida panthers: comparative studies of *Felis concolor* from Florida, Texas, Colorado, Chile, and North American Zoos. *Journal of Mammalogy* **75**: 150–162.

Barrett, C. B. & Arcese, P. (1995). Are integrated conservation-development projects (ICDPs) sustainable? *World Development* **23**: 1073–1084.

Barreto, G. R., Rushton, S. P., Strachan, R. & Macdonald, D. W. (1998). The role of habitat and mink predation in determining the status and distribution of declining populations of water voles in England. *Animal Conservation* **1**: 129–137.

Barrett, R. H. (1983). Smoked aluminum track plots for determining furbearer distribution and relative abundance. *California Department of Fish and Game* **69**: 188–190.

Barrett, T. (1999). Morbillivirus infections, with special emphasis on morbilliviruses of carnivores. *Veterinary Microbiology* **69**: 3–13.

Barrow, E., Gichochi, H. & Infield, M. (2000). Community conservation in East Africa: a comparative analysis. In *African Wildlife and African Livelihoods: The Promise and Performance of Community Conservation*, eds D. Hulme & M. Murphree. Oxford: James Currey.

Barrowclough, G. F. (1978). Sampling bias in dispersal studies based on finite area. *Bird Banding* **49**: 333–341.

Bartlett, M. S. (1957). Measels periodicity and community size. *Journal of the Royal Statistical Society* A **120**: 48–70.

Bartlett, P. C. & Martin, R. J. (1982). Skunk rabies surveillance in Illinois. *Journal of American Veterinary Medical Association* **180**: 1448–1450.

Bauer, E. A. & Bauer, P. (1996). *Bear Behaviour, Ecology and Conservation*. Stillwater: Voyageur Press.

Beasom, S. L. (1974a). Relationships between predator removal and white-tailed deer net productivity. *Journal of Wildlife Management* **38**: 854–859.

Beasom, S. L. (1974b). Intensive short-term predator removal as a game management tool. *Transactions of the North American Wildlife and Natural Resources Conference* **39**: 230–240.

Beck, B. B. (1996). Reintroduction of captive-bred animals. In *The Well-being of Animals in Zoo and Aquarium Sponsored Research*, eds G. M. Burghardt, J. T. Bielitzki, R. Boyce & D. O. Schaeffer, pp. 61–65. Greenbelt, MD: Scientists Center for Animal Welfare.

Beck, B. B., Rapaport, L. G., Stanley Price, M. R. & Wilson, A. C. (1994). Reintroduction of captive-born animals. In *Creative Conservation. Interactive Management of Wild and Captive Animals*, eds P. J. S. Olney, G. M. Mace & A. T. C. Feistner, pp. 265–286. London: Chapman & Hall.

Beck, G. G., Smith, T. G. & Addison, R. F. (1994). Organochlorine residues in harp seals, *Phoca groenlandica*, from the Gulf of St. Lawrence and Hudson Strait: an evaluation of contaminant concentrations and burdens. *Canadian Journal of Zoology* **72**: 174–182.

Beck, M. L. & Kennedy, M. L. (1980). Biochemical genetics of the raccoon, *Procyon lotor*. *Genetica* **54**: 127–132.

Beck, T. (1990). River otter recovery program. Job progress report Colorado Division of Wildlife.

Beckmen, K. B., Lowenstine, L. J., Newman, J. Hill, J., Hanni, K. & Gerber, J. (1997). Clinical and pathological characterization of northern elephant seal skin disease. *Journal of Wildlife Diseases* **33**: 438–449.

Beerli P. & Felsenstein, J. (1999). Maximum-likelihood estimation of migration rates and effective population numbers in two populations using a coalescent approach. *Genetics* **152**: 763–773.

Begon, M., Harper, J. L. & Townsend, C. R. (1990). *Ecology: Individuals, Populations and Communities*. 2nd edn. Boston: Blackwell Scientific Publications.

Begon, M., Bowers, R. G., Kadianakis, N. & Hodgkinson, D. E. (1992). Disease and community structure – the importance of host self-regulation in a host–host-pathogen model. *American Naturalist* **139**: 1131–1150.

Beier, P. (1991). Cougar attacks on humans in the United States and Canada. *Wildlife Society Bulletin* **19**: 403–412.

Beier, P. (1993). Determining minimum habitat areas and habitat corridors for cougars. *Conservation Biology* **7**: 94–108.

Beier, P. (1995). Dispersal of juvenile cougars in fragmented habitat. *Journal of Wildlife Management* **59**: 228–237.

Beier, P. (1996). Metapopulation models, tenacious tracking and cougar conservation. In *Metapopulations and Wildlife Conservation*, ed. D. R. McCullough, pp. 293–323. Washington, DC: Island Press.

Beier, P. & Cunningham, S. C. (1996). Power of track surveys to detect changes in cougar populations. *Wildlife Society Bulletin* **24**: 540–546.

Bekoff, M. (1995). Naturalizing and individualizing animal well-being and animal minds: an ethologist's naiveté exposed? In *Wildlife Conservation, Zoos, and Animal Protection. A Strategic Analysis*, ed. A. Rowan, pp. 63–129. Grafton, MA: Tufts Center for Animals and Public Policy.

Bekoff, M. (Ed.) (1998a). *Encyclopedia of Animal Rights and Animal Welfare*. Westport, CT: Greenwood Publishing Group.

Bekoff, M. (1998b). Resisting speciesism and expanding the community of equals. *BioScience* **48**: 638–641.

Bekoff, M. (1998c). Deep ethology, animal rights, and the Great Ape/Animal Project: Resisting speciesism and expanding the community of equals. *Journal of Agricultural and Environmental Ethics* **10**: 269–296.

Bekoff, M. (1999). Lynx and academic freedom. *Boulder Camera*, July 22, 7A. (http://www.bouldernews.com/opinion/columnists/bekmarc.html)

Bekoff, M. & Elzanowski A. (1997). Collecting birds: The importance of moral debate. *Bird Conservation International* **7**: 357–361.

Bekoff, M. & Jamieson, D. (1996). Ethics and the study of carnivores. In *Carnivore Behavior, Ecology, and Evolution*, ed. J. L. Gittleman, pp. 16–45. Ithaca: Cornell University Press.

Belanger, D. O. (1988). *Managing American Wildlife, A History of the International Association of Fish and Wildlife Agencies*. Amherst: The University of Massachusettes Press.

Beldon, R. C., Frankenberger, W. B. & Roof, J. C. (1997). Florida panther distribution. In *Proceedings of the Florida Panther Conference*, ed. D. B. Jordan, pp. 47–70. Washington, DC: U.S. Fish and Wildlife Service.

Beldon, R. C., Hagdorn, B. W. & Frankenberger, W. B. (1990). Panther captive breeding/reintroduction feasibility. *Final Performance Report, Florida Panther Research, 1 July 1985–30 July 1990*.

Beldon, R. C. & McCown, J. W. (1996). Florida panther reintroduction feasibility study. *Final Report, Study No 7507*. Tallahassee, FL: Florida Game and Fresh Water Fish Commission.

Ben-David, M., Bowyer, R. T. & Faro, J. B. (1995). Niche separation by mink and river otters: coexistence in a marine environment. *Oikos* **75**: 41–48.

Ben-David, M., Flynn, R. W. & Schell, D. M. (1997). Annual and seasonal changes in diets of martens: evidence from stable isotope analysis. *Oecologia* **111**: 280–291.

Bender, D. J., Contreras, T. A. & Fahrig, L. (1998). Habitat loss and population decline: a meta-analysis of the patch size effect. *Ecology* **79**: 517–533.

Bengtson J. L., Boveng, P., Franzen, U., Have, P., Heide-Jørgensen, M. P. & Härkönen, T. J. (1991). Antibodies to canine distemper virus in Antarctic seals. *Marine Mammal Science* **71**: 85–87.

Benitez-Malvido, J. (1998). Impact of forest fragmentation on seedling abundance in a tropical rain forest. *Conservation Biology* **12**: 380–389.

Bennett, A. F. (1991). Roads, roadsides and wildlife conservation: a review. In *Nature Conservation 2: The Role of Corridors*, eds D. A. Saunders & R. J. Hobbs, pp. 99–117. Sydney: Surrey Beatty.

Bennett, P. M. & Owens, I. P. F. (1997). Variation in extinction risk among birds: chance or evolutionary predisposition? *Proceedings of the Royal Society of London, Series B* **264**: 401–408.

Bere, R. M. (1955). The African wild dog. *Oryx* **3**: 180–182.

Berg, W. E. (1982). Reintroduction of fisher, pine marten and river otter. In *Midwest Furbearer Management*, ed. G. C. Sanderson, pp. 159–73. Wichita, KS: Chapter of The Wildlife Society.

Berg, W. E. & Chesness, R. A. (1978). Ecology of coyotes in northern Minnesota. In *Coyotes: Biology, behavior and management*, ed. M. Bekoff, pp. 229–247. New York: Academic Press.

Berg, W. E. & Kuehn, D. W. (1982). Ecology of wolves in north-central Minnesota. In *Wolves of the World: Perspectives of Behaviour, Ecology and Conservation*, eds F. H. Harrington & P. L. Paquet, pp. 4–11. Park Ridge, NJ: Noyes Publications.

Berg, W. E. & Kuehn, D. W. (1994). Demography and range of fishers and American martens in a changing Minnesota landscape. In *Martens, Sables, and Fishers. Biology and Conservation*, eds S. W. Buskirk, A. S. Harestad, M. G. Raphaël & R. A. Powell, pp. 262–271. Ithaca: Cornell University Press.

Berger, J. (1994). Science, conservation, and black rhinos. *Journal of Mammalogy* **75**: 298–308.

Berger, J. (1998). Future prey: consequences of the loss and restoration of large carnivores. In *Behavioral Ecology and Conservation Biology*, ed. T. Caro, pp. 80–100. New York: Oxford University Press.

Berger, J. (1999). Anthropogenic extinction of top carnivores and interspecific animal behaviour: implications of the rapid decoupling of a web involving wolves, bears, moose, and ravens. *Proceedings of the Royal Society of London, Series B* **266**: 2261–2267.

Berger, J. & Cunningham, C. (1994). Horns, hyenas, and black rhinos. *Research and Exploration* **10**: 241–244.

Bernhoft, A., Wiig, A. & Skaare, J. U. (1997). Organochlorines in polar bears (*Ursus maritimus*) at Svalbard. *Environmental Pollution* **95**: 159–175.

Bertorelle, G. & Excoffier, L. (1998). Inferring admixture proportions from molecular data. *Molecular Biology and Evolution* **15**: 1298–1311.

Berny, P. J., Buronfosse, T., Buronfosse, F., Lamarque, F. & Lorgue, G. (1997). Field evidence of secondary poisoning of foxes (*Vulpes vulpes*) and buzzards (*Buteo*

buteo) by bromadiolone, a 4-year study. *Chemosphere* **35**: 1817–1829.

Bibby, C. J., Burgess, N. D. & Hill, D. A. (1992a). *Bird Census Techniques*. London: Academic Press.

Bibby, C. J., Collar, N. J., Crosby, M. J., Heath, M. F., Imboden, C., Johnson, T. H., Long, A. J., Stattersfield, A. J. & Thirgood, S. J. (1992b). *Putting Biodiversity on the Map: Priority Areas for Global Conservation*. Cambridge: International Council for Bird Preservation.

Bibikov, D. I. (1988). *Der Wolf. Die Neue Brehm-Bucherei*. Wittenberg Lutherstadt: A. Ziemsen.

Bickham, J. W., Patton, J. C. & Loughlin, T. R. (1996). High variability for control-region sequences in a marine mammal: implications for conservation and biogeography of Steller sea lions (*Eumetopias jubatus*). *Journal of Mammalogy* **77**: 95–108.

Bignal, E. (1978). Mink predation on shelduck and other wildfowl at Loch Lomond. *The Western Naturalist* **7**: 47–53.

Bingham, J., Foggin, C. M., Wandeler, A. I. & Hill, F. W. G. (1999). The epidemiology of rabies in Zimbabwe. 2. Rabies in jackals (*Canis adustus* and *Canis mesomelas*). *Onderstepoort Journal of Veterinary Research* **66**: 11–23.

Bininda-Emonds, O. R. P., Gittleman, J. L. & Purvis, A. (1999). Building large trees by combining phylogenetic information: a complete phylogeny of the extant Carnivora (*Mammalia*). *Biological Reviews of the Cambridge Philosophical Society* **74**: 143–175.

Birks, J. D. S. (1990). Feral mink and nature conservation. *British Wildlife* **1**: 313–323.

Birks, J. D. S. (1997). A volunteer based system for sampling variations in the abundance of the polecat (*Mustela putorius*). *Journal of Zoology* **243**: 857–863.

Birks, J. D. S. & Dunstone, N. (1991). Mink. In *The Handbook of British Mammals*. 3rd edn. eds G. Corbet & S. Harris, pp. 406–415. Oxford: Blackwell Scientific Publications.

Birks, J. D. S. & Kitchener, A. C. (1999). *The distribution and status of the polecat, Mustela putorius, in Britain in the 1990s*. London: Vincent Wildlife Trust.

Bissonette, J. A. (1997). Scale-sensitive ecological properties: historical context, current meaning. In *Wildlife and Landscape Ecology: Effects of Pattern and Scale*, ed. J. A. Bissonette, pp. 3–31. New York: Springer-Verlag.

Bissonette, J. A. & Broekhuizen, S. (1995). *Martes* populations as indicators of habitat scale patterns: the need for a multiscale approach. In *Landscape Approaches in Mammalian Ecology and Conservation*, ed. W. Z. Lidicker Jr., pp. 95–121. Minneapolis: University of Minnesota Press.

Bissonette, J. A., Frederickson, R. J. & Tucker, B. J. (1989). American marten: a case for landscape-level management. *Transactions of the North American Wildlife and Natural Resources Conference* **54**: 89–101.

Bissonette, J. A., Harrison, D. J., Hargis, C. D. & Chapin, T. G. (1997). The influence of spatial scale and scale-sensitive properties on habitat selection by American marten. In *Wildlife and Landscape Ecology: Effects of Pattern and Scale*, ed. J. A. Bissonette, pp. 368–385. New York: Springer-Verlag.

Bjorge, R. R., Gunson, J. R. & Samuel, W. M. (1981). Population characteristics and movements of striped skunks (*Mephitis mephitis*) in central Alberta. *Canadian Field-Naturalist* **95**: 149–155.

Blanchard, B. M. & Knight, R. R. (1991). Movements of Yellowstone grizzly bears.

Biological Conservation **58**: 41–67.

Blanco, J., Reig, S. & Cuesta, L. (1992). Distribution, status and conservation problems of the wolf *Canis lupus* in Spain. *Biological Conservation* **60**: 73–80.

Böer, M., Smielowski, J. & Tyrala, P. (1995). Reintroduction of the European lynx (*Lynx lynx*) to the Kampinoski Nationalpark/Poland – a field experiment with zooborn individuals. Part II. Release phase: procedures, and activities of the lynxes during the first year after. *Zoologischer Garten NF* **65**: 333–342.

Boertje, R. D., Kelleyhouse, D. G., & Hayes, R. D. (1995). Methods for reducing natural predation on moose in Alaska and Yukon: an evaluation. In *Ecology and Conservation of Wolves in a Changing World*, eds L. N. Carbyn, S. H. Fritts & D. R. Seip, pp. 505–514. Canadian Circumpolar Institute, Occasional Publication No. 35.

Boertje, R. D. & Stephenson, R. O. (1992). Effects of ungulate availability on wolf reproductive potential in Alaska. *Canadian Journal of Zoology* **70**: 2441–2443.

Bögel, K. & Moegle, H. (1980). Characteristics of the spread of a wildlife rabies epidemic in Europe. In *The Red Fox*, ed. E. Zimen, pp. 251–258. The Hague: Dr. W. Junk Publishers.

Boitani, L. (1982). Wolf management in intensively used areas of Italy. In *Wolves of the World*, ed. F. H. Herrington & P. C. Paquet, pp. 158–172. New Jersey: Hayes Publications.

Boitani, L. (1983). Wolf and dog competition in Italy. *Acta Zoologica Fennica* **174**: 259–264.

Boitani, L. (1992). Wolf research and conservation in Italy. *Biological Conservation* **61**: 125–132.

Boitani, L. (1995). Ecological and cultural diversities in the evolution of wolf–human relationships. In *Ecology and Conservation of Wolves in a Changing World*, eds L. N. Carbyn, S. H. Fritts & D. R. Seip, pp. 3–11. Edmonton: Canadian Circumpolar Institute.

Boitani, L. & Fabbri, M. L. (1984). National strategy for wolf (*Canis lupus*). *Ricerche di Biologica della Selvaggina* **72**: 1–30.

Boitani, L. & Francisci, F. (1978). Ein Versuch der Wiedereinbürgerung des Luchses im Nationalpark Gran Paradiso, Italien. In *Der Luchs – Erhaltung und Wiedereinbürgerung in Europa*, ed. U. Wotschikowsky, pp. 63–65. Mammendorf: Druckerei Bernhard.

Boitani, L., Francisci, F., Ciucci, P. & Andreoli, G. (1995). Population biology and ecology of feral dogs in central Italy. In *The Dog, its Ecology, Behaviour and Evolution*, ed. J. Serpell, pp. 217–244. Cambridge: Cambridge University Press.

Bolger, D. T., Alberts, A. C. & Soulé, M. E. (1991). Occurrence patterns of bird species in habitat fragments: sampling, extinction and nested species subsets. *American Naturalist* **137**: 155–166.

Bomford, M. & O'Brien, P. (1995). Eradication or control for vertebrate species? *Wildlife Society Bulletin* **23**: 249–255.

Bonnell, M. L. & Selander, R. K. (1974). Elephant seals: genetic variation and near extinction. *Science* **134**: 908–909.

Bonner, W. N. (1982). *Seals and Man: A Study of Interactions*. Washington Sea Grant Publication. Seattle: University of Washington Press.

Bonner, W. N. (1984). Introduced mammals. In *Antarctic Ecology*, vol. 1, ed. R. M. Laws, pp. 237–278. London: Academic Press.

Boo, E. (1990). *Ecotourism: The Potentials and Pitfalls*, vol. 2. WWF, USA.

Born, E. W., Kraul, I. & Kristensen, T. (1981). Mercury, DDT, and PCB in Atlantic walrus (*Odobenus rosmarus*) from the Thule District, north Greenland. *Artic* **34**: 255–260.

Bowcock, A. M., Ruiz-Linares, A., Tomfohrde, J., Minch, E., Kidd, J. R. & Cavalli-Sforza, L. (1994). High resolution of human evolutionary trees with polymorphic microsatellites. *Nature* **368**: 455–457.

Bowen-Jones, E. & Pendry, S. (1999). The threat to primates and other mammals from the bushmeat trade in Africa, and how this threat could be diminished. *Oryx* **33**: 233–246.

Bower, W. T. & Aller, H. D. (1917). Alaska fisheries and fur industries in 1916. Document No. 838. *Department of Commerce, Bureau of Fisheries (USA)*.

Bowers, R. G. & Turner, J. (1997). Community structure and the interplay between interspecific infection and competition. *Journal of Theoretical Biology* **187**: 95–109.

Bowles, D. (1996). Wildlife trade – a conserver or exploiter? In *The Exploitation of Mammal Populations*, eds V. J. Taylor & N. Dunstone, pp. 266–291. London: Chapman and Hall.

Bowyer, R. T., Testa J. W. & Faro, J. B. (1995). Habitat selection and home ranges of river otters in a marine environment: effects of the Exxon Valdez oil spill. *Journal of Mammalogy* **76**: 1–11.

Boyce, M. S. (1992). Population viability analysis. *Annual Review of Ecology and Systematics* **23**: 481–506.

Boyd, D. K., Paquet, P. C., Donelon, S., Ream, R. R., Pletscher, D. H. & White, C. C. (1995). Transboundary movements of a recolonizing wolf population in the Rocky Mountains. In *Ecology and Conservation of Wolves in a Changing World*, eds L. N. Carbyn, S. H. Fritts & D. R. Seip, pp. 135–140. Edmonton: Canadian Circumpolar Institute.

Boyd, R. & Silk, J. B. (1983). A method for assigning cardinal dominance ranks. *Animal Behaviour* **31**: 45–58.

Bradshaw, E. L. & Bateson, P. (2000). Animal welfare and wildlife conservation. In *Behaviour and Conservation*, eds L. M. Gosling & W. J. Sutherland, pp. 330–348. Cambridge: Cambridge University Press.

Brady, J. R. & Campbell, H. (1983). Distribution of coyotes in Florida. *Florida Field Naturalist* **11**: 40–1.

Brainerd, S. M. (1990). The pine marten and forest fragmentation: a review and general hypothesis. *Transactions of the International Congress of Game Biologists* **19**: 421–434.

Brand, C. J. & Keith, L. B. (1979). Lynx demography during a snowshoe hare decline in Alberta. *Journal of Wildlife Management* **43**: 827–849.

Brandenburg, D. M. (1996). *Effects of Roads on Behavior and Survival of Black Bears in Coastal North Carolina*. Unpublished M.S. thesis, University of Tennessee, Knoxville.

Brandon, K. E. & Wells, M. (1992). Planning for people and parks: design dilemmas. *World Development* **20**: 557–570.

Brechtel, S., Carbyn, L.N., Hjertaas, D. & Mamo, C. (1993). Canadian swift fox feasibility study: 1989 to 1992. *Swift Fox Recovery Team Report*.

Breitenmoser, U. (1998a). Large predators in the Alps: the fall and rise of man's competitors. *Biological Conservation* **83**: 279–89.

Breitenmoser, U. (1998b). Recovery of the Alpine lynx population: conclusions from the first SCALP report. In *The re-introduction of the lynx into the Alps*, eds C. Breitenmoser-Würsten, C. Rohner & U. Breitenmoser, pp. 135–144. Strasbourg: Council of Europe Publishing, Environmental Encounters, No. 38.

Breitenmoser, U. (2001). Die ökologischen und anthropogenen Voraussetzungen für die Existenz grosser Beutegreifer in der Kulturlandschaft. In *Raubtierakzeptanz*, eds M. Hunziker & R. Landolt, Forest Snow and Landscape Research. (In press.)

Breitenmoser, U. & Baettig, M. (1992). Wiederansiedlung und Ausbreitung des Luchses *Lynx lynx* im Schweizer Jura. *Revue suisse de Zoologie* **99**: 163–76.

Breitenmoser, U. & Breitenmoser-Würsten, Ch. (1990). *Status, Conservation Needs and Re-introduction of the Lynx* (Lynx lynx) *in Europe*. Strasbourg: Council of Europe Publishing, Nature and Environment Series, No. 45.

Breitenmoser, U., Breitenmoser-Würsten, Ch. & Capt, S. (1998). Re-introduction and present status of the lynx in Switzerland. *Hystrix* **10**: 17–30.

Breitenmoser, U., Breitenmoser-Würsten, Ch., Capt, S., Ryser, A., Zimmermann, F., Angst, C., Olsson, P., Baumgartner, H. J., Siegenthaler, A., Molinari, P., Laass, J., Burri, A., Jobin, A. & Weber, J. M. (1999). Lynx Management problems in the Swiss Alps. *Cat News* **30**: 16–8.

Breitenmoser, U., Breitenmoser-Würsten, Ch., Okarma, H., Kapeghyi, T., Kaphegyi-Wallmann, U. & Müller, U. M. (2000). *Final Draft Action Plan for Conservation of the Eurasian Lynx* (Lynx lynx) *in Europe*. Strasbourg: Council of Europe Publishing.

Breitenmoser, U., Kavczensky, P., Doetterer, M., Breitensmoser-Würster, Ch., Capt, S., Berhart, F. & Liberek, M. (1993b). Spatial organisation and recruitment of lynx (*Lynx lynx*) in a reintroduced population in the Swiss Jura Moutains. *Journal of Zoology* **231**: 449–464.

Breitenmoser, U., Slough, B. G. & Breitenmose-Wursten, Ch. (1993a). Predators of cyclic prey: is the Canada lynx victim or profiteer of the snowshoe hare cycle? *Oikos* **66**: 551–554.

Breitenmoser-Würsten, Ch., Robin, K., Landry, J. M., Gloor, S., Olsson, P. & Breitenmoser, U. (2001). Die geschichle von Fuchs, Luchs, Bartgeier, Wolf und Braunbär in der Schweiz – ein kurzer Überblick. In *Raubtierakzeptanz*, ed. M. Hunziker & R. Landolt. Forest Snow and Landscape Research. (In Press.)

Breitenmoser-Würsten, Ch., Rohner, C. & Breitenmoser, U. (Eds). (1998). *The Reintroduction of the Lynx into the Alps*. Strasbourg: Council of Europe Publishing, Environmental Encounters, No. 38.

Bremle, G., Larsson, P. & Helldin, J. O. (1997). Polychlorinated biphenyls in a terrestrial predator, the pine marten (*Martes martes* L.). *Environmental Toxicology and Chemistry* **16**: 1779–1784.

Bretram, B. C. R. (1975). Social factors influencing reproduction in wild lions. *Journal of Zoology (London)* **177**: 463–482.

Brewer, G. D. & de Leon, P. (1983). *The Foundations of Policy Analysis*. Homewood, IL: The Dorsey Press.

Brewer, G. D. & Clark, T. W. (1994). A policy sciences perspective: improving implementation. In *Endangered Species Recovery: Finding the Lessons, Improving the Process*, eds T. W. Clark, R. P. Reading & A. L. Clarke, pp. 391–413. Washington:

Island Press.

Bright, C. (1999). *Life Out of Bounds*. London: Earthscan.

Broad, S., Luxmore, R. & Jenkins, M. (1988). *Significant trade in wildlife: A review of selected species in CITES Appendix II*, vol. 1. *Mammals*. Gland: IUCN.

Brocke, R. H. & Gustafson, K. A. (1992). Lynx in New York State. *Re-Introduction News* 4: 6–7.

Brocke, R. H., Gustafson, K. A. & Major, A. R. (1990). Restoration of lynx in New York: biopolitical lessons. *Transactions of the North American Wildlife and Natural Resources Council* 55: 590–598.

Brody, A. J. & Pelton, M. R. (1989). Effects of roads on black bear movements in western North Carolina. *Wildlife Society Bulletin* 17: 5–10.

Brook, B. W., Cannon, J. R., Lacy, R. C., Mirande, C. & Frankham, R. (1999). Comparison of the population viability analysis packages GAPPS, INMAT, RAMAS and VORTEX for the whooping crane (*Grus americana*). *Animal Conservation* 2: 23–31.

Brookfield, J. (1996). A simple new method for estimating null alleles frequency from heterozygote deficiency. *Molecular Ecology* 5: 453–455.

Brothers, N. P., Skira, I. J. & Copson, G. R. (1985). Biology of the feral cat, *Felis catus*, on Macquarie Island. *Australian Wildlife Reserach* 12: 425–436.

Brown, E. W., Yuhki, N., Packer, C. & O'Brien, S. J. (1994). A lion lentivirus related to feline immunodeficiency virus: epidemiologic and phylogenetic aspects. *Journal of Virology* 68: 5953–5968.

Brown, J. H. (1971). Mammals on mountaintops: nonequilibrium insular biogeography. *American Naturalist* 105: 467–478.

Brown, J. L. & Wildt, D. E. (1997). Assessing reproductive status in wild felids by non-invasive faecal steroid monitoring. *International Zoo Yearbook* 35: 173–191.

Brown, J. L., Wildt, D. E., Wielebnowski, N., Goodrowe, K. L., Wells, S., Graham, L. H. & Howard, J. G. (1996). Reproductive activity in captive female cheetahs (*Acinonyx jubatus*) as assessed by faecal steroids. *Journal of Reproduction and Fertility* 106: 337–46.

Brown, K. S., Jr. & Hutchings, R. W. (1997). Disturbance, fragmentation, and the dynamics of diversity in Amazonian forest butterflies. In *Tropical Forest Remnants: Ecology, Management, and Conservation of Fragmented Communities*, eds W. F. Laurance & R. O. Bierregaard, Jr., pp. 91–110. Chicago: University of Chicago Press.

Brown, L. E. (1969). Field experiments on the movements of *Apodemus sylvaticus* using trapping and tracking techniques. *Oecologia* 2: 198–222.

Bruford, M. W. & Wayne, R. K. (1993). Microsatellites and their application to population genetic studies. *Current Opinion in Genetics and Development* 3: 939–943.

Bruggers, R. A. & Zaccagnini, M. E. (1994). Vertebrate pest problems related to agricultural production and applied research in Argentina. *Vida Silvestre Neotropical* 3: 71–83.

Brunner, S. (1998). Cranial morphometrics of the southern fur seals *Arctocephalus fosteri* and *A. pusillus* (Carnivora: Otarridae). *Australian Journal of Zoology* 46: 67–108.

Buckland, S. T., Anderson, D. R., Burnham, K. P. & Laake, J. L. (1993). *Distance Sampling: Estimating Abundance of Biological Populations*. London: Chapman and Hall.

Buechner, M. (1987). A geometric model of vertebrate dispersal: tests and implications. *Ecology* **68**: 310–318.

Bueno, F. (1996). Competition between American mink *Mustela vison* and otter *Lutra lutra* during winter. *Acta Theriologica* **41**: 149–154.

Bunin, J. S. & Jamieson, I. G. (1994). New approaches toward a better understanding of the decline of takahe (*Porphyrio mantelli*) in New Zealand. *Conservation Biology* **9**: 100–106.

Bunnell, F. L. & Tait, D. E. N. (1981). Population dynamics of bears – implications. In *Dynamics of Large Mammal Populations*, eds C. W. Fowler & T. D. Smith, pp. 75–98. New York: John Wiley & Sons.

Burbidge, A. A. & Jenkins, R. W. G. (1984). Endangered vertebrates of Australia and its island territories. *Report of the Working Group on Endangered Fauna of the Standing Committee of the Council Nature Conservation*. Canberra: Australian National Parks and Wildlife Service.

Burbidge, A. A. & McKenzie, N. L. (1989). Patterns in the modern decline of Western Australia's vertebrate fauna: causes and conservation implications. *Biological Conservation* **50**: 143–198.

Burgess, S. A. & Bider, J. R. (1980). Effects of stream habitat improvements on invertebrates, trout populations, and mink activity. *Journal of Wildlife Management* **44**: 871–880.

Burnham, K. P., Anderson, D. R. & Laake, J. L. (1980). Estimation of density from line-transect sampling of biological populations. *Wildlife Monographs* **72**: 1–202.

Burrows, R. (1995). Demographic changes and social consequence in wild dogs, 1964–1992. In *Serengeti II: Dynamics, Conservation and Management of an Ecosystem*, eds A. R. E. Sinclair & P. Arcese, pp. 400–420. Chicago: University of Chicago Press.

Burrows, R., Hofer, H. & East, M. L. (1994). Demography, extinction and intervention in a small population: the case of the Serengeti wild dogs. *Proceedings of the Royal Society of London, Series B* **256**: 281–292.

Burton, R. G. (1933). *The Book of the Tiger*. Boston: Houghton Mifflin.

Bush, M. & Roberts, M. (1977). Distemper in captive red pandas. *International Zoo Yearbook* **17**: 194–196.

Buskirk, S. W., Forrest, S. C., Raphael, M. G. & Harlow, H. J. (1989). Winter resting ecology of marten in the central Rocky Mountains. *Journal of Wildlife Management* **53**: 191–196.

Buskirk, S. W., Harestad, A. A., Raphael, M. G. & Powell, R. A. (Eds) (1994). *Martens, Sables, and Fishers. Biology and Conservation*. Ithaca: Cornell University Press.

Buskirk, S. W. & Powell, R. A. (1994). Habitat ecology of fishers and American marten. In *Martens, Sables, and Fishers: Biology and Conservation*, eds S. W. Buskirk, A. S. Harestad, M. G. Raphael & R. A. Powell, pp. 283–296. Ithaca: Cornell University Press.

Butler, D. (1994). Bid to protect wolves from genetic pollution. *Nature* **370**: 497.

Byrne, G. & Shenk, T. (1999). Lynx recovery project background and post release monitoring of lynx re-introduced to the southern Rocky Mountains of southwestern Colorado, USA. *Re-introduction News* **18**: 15–17.

Byun, S. A., Koop, B. F. & Reimchen, T. E. (1997). North American black bear mtDNA phylogeography: implications for morphology and the Haida Gwaii glacial refugium controversy. *Evolution* **51**: 1647–1653.

Cain, S. A., Kadlec, J. A., Allen, D. L., Cooley, R. A., Hornocker, M. G., Leopold, A. S. & Wagner, F. H. (1972). *Predator Control–1971, Report to the Council on Environmental Quality and the Department of the Interior by the Advisory Committee on Predator Control.* Ann Arbor: Institute for Environmental Quality, University of Michigan.

Callicott, J. B. (1998). 'Back together again' again. *Environmental Values* **7**: 461.

Cambell, K. & Hofer, H. (1995). People and wildlife: spatial dynamics and zones of interaction. In *Serengeti II: Dynamics, Management and Conservation of an Ecosystem*, eds A. R. E. Sinclair & P. Arcese, pp. 534–570. Chicago: Chicago Unviersity Press.

Camenzind, F. J. B. (1978). Behavioral ecology of coyotes on the National Elk Refuge, Jackson, Wyoming. In *Coyotes: Biology, Behavior and Management*, ed. M. Bekoff, pp. 267–296. New York: Academic Press.

Camm, J. D., Polasky, S., Solow, A. & Csuti, B. (1996). A note on optimal algorithms for reserve site selection. *Biological Conservation* **78**: 353–355.

Campbell, T. M., III, Biggins, D., Forrest, S. & Clark, T. W. (1985). Spotlighting as a method to locate and study black-footed ferrets. In *Proceedings of the Black-footed Ferret Workshop*, eds S. H. Anderson & D. B. Inkley, pp. 24.1–24.7. Laramie: University of Wyoming.

Cannon, K. P. (1995). Blood residue analysis of ancient stone tools reveal clues to prehistoric subsistence patterns. *Culutral Resource Management* **18**: 14–16.

Carbone, C., Mace, G. M., Roberts S. C. & Macdonald, D. W. (1999). Energetic constraints on the diet of terrestrial carnivores. *Nature* **402**: 286–288.

Carbyn, L. N. (1982). Coyote population fluctuations and spatial distribution in relation to wolf territories in Riding Mountain National Park, Manitoba. *Canadian Field-Naturalist* **96**: 176–183.

Carbyn, L. N. (1998). Swift fox status in Canada. *Committee on the Status of Endangered Species Report.*

Carbyn, L. N., Armbruster, H. J. & Mamo, C. (1994). The swift fox re-introduction program in Canada from 1983 to 1992. In *Restoration of Endangered Species: Conceptual Issues, Planning and Implementaion*, eds M. L. Bowles & C. J. Whelan, pp. 247–271. Cambridge: Cambridge University Press.

Carey, P. W. (1992). Fish prey species of the New Zealand fur seal (*Arctocephalus forsteri*, Lesson). *New Zealand Journal of Ecology* **16**: 41–46.

Carlington, B. (1978). Re-introduction of the Swift Fox to southern Alberta. Feasibility study.

Carlington, B. (1980). *Re-introduction of the Swift Fox* (Vulpes velox) *to the Canadian Prairies.* MSc University of Calgary.

Caro, T. M. (1994). *Cheetahs of the Serengeti Plains: Group Living in an Asocial Species.* Chicago: University of Chicago Press.

Caro, T. M. (Ed.) (1998). *Behavioral Ecology and Conservation Biology.* New York: Oxford University.

Caro, T. M. (1999a). The behaviour-conservation interface. *Trends in Ecology and Evolution* **14**: 366–369.

Caro, T. M. (1999b). Demography and behaviour of African mammals subject to exploitation. *Biological Conservation* **91**: 91–97.

Caro, T. M. & Durant, S. M. (1995). The importance of behavioural ecology for conservation biology: examples from Serengeti carnivores. In *Serengeti II: Dynamics, Conservation and Management of an Ecosystem*, eds A. R. E. Sinclair & P. Arcese, pp. 451–472. Chicago: University of Chicago Press.

Caro T. M. & Laurenson, M. K. (1994). Ecological and genetic factors in conservation: A cautionary tale. *Science* 263: 485–486.

Carpaneto G. M. (1990). The Indian grey mongoose (*Herpestes edwardsi*) in the Circeo National Park: a case of incidental introduction. *Mustelid and Viverrid Conservation* (The Newsletter of the IUCN/SSC Mustelid & Viverrid Specialist Group), 2: 10.

Carpenter, J. W., Appel, M. J. G., Erickson, R. C. & Novilla, M. N. (1976). Fatal vaccine-induced canine distemper virus infection in black-footed ferrets. *Journal of American Veterinary Medical Association* 169: 961–964.

Carpenter, M., Brown, E. W., Culver, M., Johnson, W. E., Pecon-Slattery, J., Brousset, D. & O'Brien, S. J. (1996). Genetic and phylogenetic divergence of feline immunodeficiency virus in the puma (*Puma concolor*). *Journal of Virology* 70: 6682–6693.

Carter, G. R., Chengappa, M. M. & Roberts, A. W. (1995). *Essentials of Veterinary Microbiology*, 5th edn. Philadelphia: Williams and Wilkins.

Carvell, C., Inglis, N. F. J., Mace, G. M. & Purvis, A. (1998). How Diana climbed the ratings at the zoo. *Nature* 395: 213.

Case, T. J. (1996). Global patterns in the establishment and distribution of exotic birds. *Biological Conservation* 78: 69–96.

Case, T. J. & Gilpin, M. E. (1974). Interference competition and niche theory. *Proceedings of the National Academy of Sciences of the United States of America* 71: 3073–3077.

Caughley, G. (1977). *Analysis of Vertebrate Populations*. New York: John Wiley and Sons.

Caughley, G. (1994). Directions in conservation biology. *Journal of Animal Ecology* 63: 215–244.

Caughley, G. & Gunn, A. (1996). *Conservation Biology in Theory and Practice*. Oxford: Blackwell Science.

Central Statistics Office (1992). *Population of Towns, Villages and Associated Localities in August 1991*. Gaborone: Ministry of Finance & Development Planning, Republic of Botswana.

Central Statistics Office, Zimbabwe (1992). *Census 1992*. Harare: Central Statistics Office.

Cerveny, J. & Bufka, L. (1996). Lynx (*Lynx lynx*) in south-western Bohemia. In *Lynx in the Czech and Slovak Republics*, eds P. Koubek & J. Cerveny, pp. 16–33. Brno: Institute of Landscape Ecology.

Chakrabarty, K. (1992). *Man-eating Tigers*. Calcutta: Darbaru Prokashan.

Chakraborty, R., Meagher, T. & Smouse, P. (1988). Parentage analysis with genetic markers in natural populations. I. The expected proportion of offspring with unambiguous paternity. *Genetics* 118: 527–536.

Chakravarti, A. (1999). Populations genetics – making sense out of sequence. *Nature Genetics Supplement* 21: 56–60.

Chanin, P. R. F. & Jeffries, D. J. (1978). The decline of the otter, *Lutra lutra* L., in Britain: an analysis of hunting records and discussion of causes. *Biological Journal of the Linnean Society* 10: 305–328.

Chanin, P. R. F. & Linn, I. (1980). The diet of the feral mink (*Mustela vison*) in southwest Britain. *Journal of Zoology (London)* 192: 205–223.

Chaparro, F. & Esterhuysen, J. J. (1993). The role of yellow mongooses (*Cynictis penicillata*) in the epidemiology of rabies in South Africa – preliminary results. *Onderstepoort Journal of Veterinary Research* 60: 373–377.

Chapin, T. G., Harrison, D. J. & Katnik, D. D. (1998). Influence of landscape pattern on habitat use by American marten in an industrial forest. *Conservation Biology* 12: 1327–1337.

Chapman, R. C. (1978). Rabies: decimation of a wolf pack in Arctic Alaska. *Science* 201: 365–367.

Chapuis, J. L., Boussès, P. & Barnaud, G. (1994). Alien mammals, impact and management in the French subantarctic islands. *Biological Conservation* 67: 97–104.

Charlton, K. M., Webster, W. A., Casey, G. A. & Rupprecht, C. E. (1988). Skunk rabies. *Reviews of Infectious Diseases* 10: 626–628.

Cheeseman, C. L., Cresswell, W. J., Harris, S. & Mallinson, P. J. (1988). Comparison of dispersal and other movements in two badger (*Meles meles*) populations. *Mammal Review* 18: 51–59.

Cheeseman, C. L., Mallinson, P. J., Ryan, J. & Wilesmith, J. W. (1993). Recolonization by badgers in Gloucestershire. In *The Badger:* 78–93. Hayden, T. J. (Ed.) Dublin: Royal Irish Academy.

Cheeseman, C. L., Wilesmith, J. W., Ryan, J. & Mallinson, P. J. (1987). Badger population dynamics in a high density area. *Symposium Zoological Society London* 58: 279–294.

Child, B. (1996). The practice and principles of community-based wildlife management in Zimbabwe: The CAMPFIRE programme. *Biodiversity and Conservation* 5: 369–398.

Childes, S. L. (1988). The past history, present status and distribution of the hunting dog *Lycaon pictus* in Zimbabwe. *Biological Conservation* 44: 301–316.

Chitty, H. (1950). Canadian arctic wildlife enquiry, 1943–49: with a summary of results since 1933. *Journal of Animal Ecology* 19: 180–193.

Ciucci, P., Boitani, L., Francisci, F. & Andreoli, G. (1997). Home range, activity and movements of a wolf pack in central Italy. *Journal of Zoology* 243: 803–819.

Clark, D. R. Jr., Ogasawara, P. A., Smith, G. J. & Ohlendorf, H. M. (1989). Selenium accumulation by raccoons exposed to irrigation drainwater at Keterson National Wildlife Refuge, California, 1986. *Archive for Environmental Contamination and Toxicology* 18: 787–794.

Clark, F. W. (1972). Influence of jackrabbit density on coyote population change. *Journal of Wildlife Management* 36: 343–356.

Clark, H. (1991). Mink hunting on the Nene. *International Fieldsports and Conservation* Sep–Oct: 76–78.

Clark, J. D. (1998). Black bear repatriation to Cumberland Plateau. *Re-Introduction News* 12: 12–13.

Clark, J. D. & Smith, K. G. (1994). A demographic comparison of two black bear populations in the interior highlands of Arkansas. *Wildlife Society Bulletin* 22: 593–603.

Clark, T. W. (1994). Restoration of the endangered black-footed ferret: a 20-year overview. In *Restoration and recovery of endangered species*, eds M. L. Bowles & C. J. Whelan, pp. 272–297. Cambridge: Cambridge University Press.

Clark, T. W. (1997). *Averting Extinction: Reconstructing Endangered Species Recovery.* New Haven: Yale University Press.

Clark, T. W. (1999). Interdisciplinary problem-solving: next steps in the Greater Yellowstone ecosystem. *Policy Sciences* **32**: 393–414.

Clark, T. W. & Brunner, R. D. (1996). Making partnerships work: an introduction to decision process. *Endangered Species Update* **13**(1): 1–4.

Clark, T. W. & Campbell, T. M., III. (1983). A small carnivore survey technique. *Great Basin Naturalist* **43**: 438–440.

Clark, T. W., Curlee, A. P., Minta, S. C. & Kareiva, P. M. (Eds) (1999). *Carnivores in Ecosystems: The Yellowstone Experience.* New Haven: Yale University Press.

Clark, T. W., Curlee, A. P. & Reading, R. P. (1996a). Crafting effective solutions to the large carnivore conservation problem. *Conservation Biology* **10**: 940–948.

Clark, T. W., Paquet, P. C. & Curlee, A. P. (1996b). Special section: Large carnivore conservation in the Rocky Mountains of the United States and Canada. *Conservation Biology* **10**: 936–1058.

Clark, T. W., Reading, R. P. & Backhouse, G. N. (1995a). Prototyping for successful conservation: the eastern barred bandicoot program. *Endangered Species Update* **12**(10&11): 5–8.

Clark, T. W., Reading, R. P. & Clarke, A. L. (Eds) (1995b). *Endangered Species Recovery: Finding the Lessons, Improving the Process.* Washington, DC: Island Press.

Clark, T. W., Reading, R. P. & Wallace, R. L. (2000b). Research in endangered species conservation: An introduction to multiple methods. *Endangered Species Update* **16**: 96–101.

Clark, T. W. & Wallace, R. L. (1998). Understanding the human factor in endangered species recovery: an introduction to the human social process. *Endangered Species Update* **15**(1): 2–9.

Clark, T. W. & Wallace, R. L. (1999). The professional in endangered species conservation: An introduction to standpoint clarification. *Endangered Species Update* **16**(1): 9–13.

Clark, T. W., Willard, A. R. & Cromley, C. R. (Eds) (2000a). *Foundations of Natural Resources Policy and Management.* New Haven: Yale University Press.

Clark, W. R. & Andrews, R. D. (1982). Review of population indices applied in furbearer management. In *Midwest Furbearer Management*, ed. G. C. Sanderson, pp. 11–22. Wichita, KS: Proceedings of Symposium of 43rd Midwest Fish and Wildlife Conference.

Clark, W. R., Hasbrouck, J. J., Kienzler, J. M. & Glueck, T. F. (1989). Vital Statistics and harvest of an Iowa raccoon population. *Journal of Wildlife Management* **53**: 982–990.

Clarke, G. P., White, P. C. L. & Harris, S. (1998). Effects of roads on badger *Meles meles* populations in south-west England. *Biological Conservation* **86**: 117–124.

Cleaveland S. (1997). Dog vaccination around the Serengeti – Reply. *Oryx* **31**: 13–14.

Cleaveland S. (1998). Epidemiology and control of rabies: the growing problem of rabies in Africa. *Transactions of the Royal Society of Tropical Medicine and Hygiene* **92**: 131–134.

Cleaveland, S. & Dye, C. (1995). Maintenance of a microparasite infecting several host species: rabies in the Serengeti. *Parasitology* **111**: S33–S47.

Clevenger, A. P. & Purroy, F. J. (1999). Status and management of the brown bear in eastern Cantabira, Spain. In *Bears: Status Survey and Action Plan*, eds

C. Servheen, S. Herrero & B. Peyton, pp. 100–104. Gland, Switzerland: IUCN.

Clevenger, A. P., Purroy, F. J. & Campos, M. A. (1997). Habitat assessment of a relict brown bear *Ursus arctos* population in northern Spain. *Biological Conservation* **80**: 17–22.

Clifton-Hadley, R. S., Wilesmith, J. W., Richards, M. S., Upton, P. & Johnston, S. (1995). The occurrence of *Mycobacterium bovis* infection in and around an area subject to extensive badger (*Meles meles*) control. *Epidemiology and Infection* **114**: 179–193.

Clode, D., Birks, J. D. S. & Macdonald, D. W. (2000). The influence of risk and vulnerability on predator mobbing by terns (*Sterna* spp.) and gulls (*Larus* spp.). *Journal of Zoology (London)* **252**: 53–59.

Clode, D. & Macdonald, D. W. (1995). Evidence for food competition between mink (*Mustela vison*) and otter (*Lutra lutra*) on Scottish islands. *Journal of Zoology, (London)* **237**: 435–444.

Clout, M. & Lowe, S. (1997). Biodiversity loss due to biological invasion: prevention and cure. In *Conserving vitality and diversity: proceedings of the World Conservation Congress workshop on alien invasive species*, eds C. Rubec & G. Lee, pp. 29–40. IUCN Species Survival Commission and the North American Wetlands Conservation Council (Canada). Canada: Ottowan.

Clutton-Brock, J. (1996). Competitors, companions, status symbols, or pests? A review of human associations with other carnivores. In *Carnivore Behavior, Ecology, and Evolution*, ed. J. L. Gittleman, pp. 375–392. Ithaca: Cornell University Press.

Clutton-Brock, J., Kitchener, A. C. & Lynch, J. M. (1994). Changes in the skull morphology of the Arctic wolf, *Canis lupus arctos*, during the twentieth century. *Journal of Zoology (London)* **233**: 19–36.

Clutton-Brock, T. H., Albon, S. D. & Guinness, F. E. (1985). Parental investment and sex differences in juvenile mortality in birds and mammals. *Nature* **313**: 131–133.

Clutton-Brock, T. H. & Iason, G. R. (1986). Sex ratio variation in mammals. *Quarterly Review of Biology* **61**: 339–374.

Coates, S. F., Main, M. B., Mullahey, J. J., Schaefer, J. M., Tanner, G. W., Sunquist, M. E. & Fanning, M. D. (1998). The coyote (*Canis latrans*). Florida's newest predator. *Florida Cooperative Extension Service* WEC-**124**: 1–7.

Cobb, S. (1981). The leopard – problems of an overabundant, threatened terrestrial carnivore. In *Problems in Management of Locally Abundant Wild Mammals*, ed. P. A. Jewell, pp. 66–77. New York: Academic Press.

Coblentz, B. E. (1990). Exotic organisms: a dilemma for conservation biology. *Conservation Biology* **4**: 261–265.

Cochran, W. W. & Lord, R. D. Jr. (1963). A radio-tracking system for wild animals. *Journal of Wildlife Management* **27**: 9–24.

Coltman, D. W., Bowen, W. D. & Wright, J. M. (1998). Birth weight and neonatal survival of harbour seal pups are positively correlated with genetic variation measured by microsatellites. *Proceedings of the Royal Society of London, Series B* **265**: 803–809.

Connell, J. H. (1980). Diversity and the coevolution of competitors, or the ghost of competition past. *Oikos* **35**: 131–138.

Connell, J. H. (1983). On the prevalence and relative importance of interspecific competition: evidence from field experiments. *The American Naturalist* **122**: 661–696.

Conner, M. C., Labisky, R. F. & Progulske, D. R. (1983). Scent-station indices as measures of population abundance for bobcats, raccoons, gray foxes, and opossums. *Wildlife Society Bulletin* **11**: 146–152.

Conner, M. C. & Labisky, R. F. (1985). Evaluation of radioisotope tagging for estimating abundance of raccoon populations. *Journal of Wildlife Management* **49**: 326–332.

Connolly, G. E. & Longhurst, W. M. (1975). The effects of control on coyote populations: a simulation model. *Bulletin of the Division of Agricultural Sciences, University of California* **1872**: 1–37.

Connolly, G. E. (1978). Predator control and coyote populations: a review of simulation models. In *Coyotes: Biology, Behavior, and Management*, ed. M. Bekoff, pp. 327–345. New York: Academic Press

Conroy, J. & French, D. D. (1987). The use of spraints to monitor populations of otters (*Lutra lutra*). *Symposium of the Zoological Society of London* **58**: 247–262.

Constantine, R. (1999). Effects of tourism on marine mammals in New Zealand. *New Zealand Department of Conservation report, Science for Conservation No. 104. NZ Department of Conservation PO Box 10–420, Wellington, New Zealand.*

Cooper, N. S. & Carling, R. C. J. (Eds) (1996). *Ecologists and Ethical Judgements.* London: Chapman & Hall.

Cooper, S. M. (1991). Optimal hunting group size: the need for lions to defend their kills against loss to spotted hyaenas. *African Journal of Ecology* **29**: 130–136.

Cop, J. & Frkovic, A. (1998). The re-introduction of the lynx in Slovenia and its present status in Slovenia and Croatia. *Hystrix* **10**: 65–76.

Copeland, C. & Lewis, D. (1997). *Saving our Natural Heritage? The Role of Science in Managing Australia's Ecosystems.* Ruscutters Bay, New South Wales: Halstead Press.

Corbett, J. (1957). *Man-eaters of India.* Oxford: Oxford University Press.

Corbett, L. K. (1995). *The Dingo in Australia and Asia.* Sydney: University of New South Wales Press.

Corbett, L. K. & Newsome, A. E. (1987). The feeding ecology of the dingo. 3. Dietary relationships with widely fluctuating prey populations in arid Australia: an hypothesis of alternation of predation. *Oecologia* **74**: 215–227.

Corn, J. L. & Conroy, M. J. (1998). Estimation of density of mongooses with capture-recapture and distance sampling. *Journal of Mammalogy* **79**: 1009–1015.

Cornuet, J.-M. & Luikart, G. (1996). Description and power analysis of two tests for detecting recent population bottlenecks from allele frequency data. *Genetics* **144**: 2002–2014.

Cornuet, J.-M., Piry, S., Luikart, G., Estoup, A. & Solignac, M. (1999). New methods employing multilocus genotypes for selecting or excluding populations as origins of individuals. *Genetics* **153**: 1989–2000.

COSEWIC (1978). *Committee on the Status of Endangered Wildlife in Canada 1978. Status report and Evaluations*, vol. 1, *Official Classification of the Swift Fox.*

Costanza, R., D'arge, R., De Groot, R., Farber, S., Grasso, M., Hannon, B., Limburg, K., Naeem, S., O'Neill, R. V. O., Paruelo, J., Raskin, R. G., Sutton, P., & Van Den Belt, M. (1997). The value of the world's ecosystem services and natural capital. *Nature* **387**: 253–260.

Cotterill, S. (1997). *Population census of swift fox* (Vulpes velox) *in Canada: Winter 1996–1997*. Prepared for the Swift Fox National Recovery Team Alberta Environmental Protection. Wildlife Management Division.

Coulson, T., Pemberton, J., Albon S. *et al.* (1998). Microsatellites reveal heterosis in red deer. *Proceedings of the Royal Society of London, Series B* **256**: 489–495.

Coulter, M. W. (1966). *Ecology and Management of Fishers in Maine*. PhD thesis, Syracuse University, Syracuse, New York.

Council of Europe (1989). *Workshop on the Situation and Protection of the Brown Bear* (Ursus arctos) *in Europe*. Strasbourg: Council of Europe.

Courchamp, F. & Suigihara, G. (1999). Modeling the biological control of an alien predator to protect island species from extinction. *Ecological Application* **9**: 112–123.

Couturier, M. A. J. (1954). *L'Ours brun*. Grenoble.

Cox, D. R. (1970). *The Analysis of Binary Data*. London: Methuen.

Cozza, K., Fico, R., Battistini, M. L. & Rogers, E. (1996). The damage-conservation interface illustrated by predation on domestic livestock in central Italy. *Biological Conservation* **78**: 329–336.

Crabtree, R. (1998). Total impact. *The Tracker* **5**: 12.

Crabtree, R. L. & Sheldon, J. W. (1999). Coyotes and canid coexistence in Yellowstone. In *Carnivores in Ecosystems*, eds T. W. Clark, A. P. Curlee, S. C. Minta & P. M. Kareiva, pp. 127–163. New Haven, Connecticut: Yale University Press.

Craighead, F. L. & Vyse, E. R. (1996). Brown/grizzly bear metapopulations. In *Metapopulations and Wildlife Conservation*, ed. D. R. McCullough, pp. 325–351. Washington, DC: Island Press.

Craik, J. C. A. (1995). Effects of North American mink on the breeding success of terns and smaller gulls in west Scotland. *Seabird* **17**: 3–11.

Craik, J. C. A. (1997). Long-term effects of North American mink *Mustela vison* on seabirds in western Scotland. *Bird Study* **44**: 303–309.

Craik, J. C. A. (1998). Recent mink-related declines of gulls and terns in west Scotland and the mitigating effects of mink control. *Argyll Bird Report* **14**: 98–110.

Crandall, K. A., Bininda-Emonds, O. R. P, Mace, G. M. & Wayne, R. K. (2000). Considering evolutionary processes in conservation biology. *Trends in Ecology and Evolution* **15**: 290–295.

Crawley, M. J. (1986). The population biology of invaders. *Philosophical Transactions of the Royal Society* (B) **314**: 711–731.

Crawshaw Jr., P. G. (1995). *Comparative Ecology of Ocelot* (Felis pardalis) *and Jaguar* (Panthera onca) *in a Protected Subtropical Forest in Brazil and Argentina*. PhD dissertation, University of Florida, Gainesville.

Crawshaw, Jr., P. G. & Quigley, H. B. (1991). Jaguar spacing, activity and habitat use in a seasonally flooded environment in Brazil. *Journal of Zoology* **223**: 357–370.

Cracraft, J. (1983). Species concepts and speciation analysis. In *Current Ornithology*, ed. R. F. Johnson, pp. 159–187. New York: Plenum Press

Cracraft, J., Feinstein, J., Vaughan, J. & Helm-Bychowski, K. (1998). Sorting out tigers (*Panthera tigris*): mitochondrial sequences, nuclear inserts, systematics, and conservation genetics. *Animal Conservation* **1**: 139–150.

Creagh, C. (1992). New approaches to rabbit and fox control. *Ecosystems* **71**: 18–24.

Cree, A., Daugherty, C. H. & Hay, J. M. (1995). Reproduction of a rare New Zealand reptile, the tuatara *Sphenodon punctatus*, on rat-free and rat-inhabited islands.

Conservation Biology **9**: 373–383.

Creel, S. (1992). Cause of wild dog deaths (letter to the editor). *Nature* **360**: 633.

Creel, S. R. (1996). Behavioral endocrinology and social organisation in dwarf mongooses. In *Carnivore Behaviour, Ecology, and Evolution*, ed. J. L. Gittleman, pp. 46–77. Ithaca: Cornell University press.

Creel, S. & Creel, N. M. (1995). Communal hunting and pack size in African wild dogs, *Lycaon pictus*. *Animal Behaviour* 50: 1325–1339.

Creel, S. & Creel, N. M. (1996). Limitation of African wild dogs by competition with larger carnivores. *Conservation Biology* 10: 526–538.

Creel, S. & Creel, N. (1997). Lion density and population structure in the Selous Game Reserve: evaluation of hunting quotas and offtake. *African Journal of Ecology* 35: 83–93.

Creel, S. & Creel, N. M. (1998). Six ecological factors that may limit African wild dogs, *Lycaon pictus*. *Animal Conservation* 1: 1–9.

Creel S., Creel, N. M., Matovelo, J. A., Mtambo, M. M. A., Batamuzi, E. K. & Cooper, J. E. (1995). The effect of anthrax on endangered African wild dogs (*Lycaon pictus*). *Journal of Zoology (London)* 236: 199–209.

Creel, S., Creel, N. M., Mills, M. G. L. & Monfort, S. L. (1997a). Rank and reproduction in cooperatively breeding African wild dogs: behavioural and endocrine correlates. *Behavioural Ecology* 8: 298–306.

Creel, S., Creel, N. M., Munson, L., Sanderlin, D. & Appel, M. J. G. (1997b). Serosurvey for selected viral diseases and demography of African wild dogs in Tanzania. *Journal of Wildlife Diseases* 33: 823–832.

Creel, S., Creel, N., Wildt, D. E. & Monfort, S. L. (1992). Behavioural and endocrine mechanisms of reproductive suppression in Serengeti dwarf mongooses. *Animal Behaviour* 43: 231–45.

Creel, S. R. & Waser, P. M. (1994). Inclusive fitness and reproductive strategies in dwarf mongooses. *Behavioral Ecology* 5: 339–348.

Crête, M. & Messier, F. (1987). Evaluation of indices of gray wolf, *Canis lupus*, density in hardwood-conifer forests of southwestern Quebec. *Canadian Field-Naturalist* 101: 147–152.

Crisp, R. (1998). Animal liberation is not an environmental ethic. *Environmental Values* 7: 476–478.

Cromley, C. M. (2000). The killing of grizzly bear #209 in Grand Teton National Park, Wyoming: Identifying norms for grizzly bear management. In *Foundations of Natural Resources Policy and Management*, eds T. W. Clark, A. R. Willard & C. M. Cromley, pp. 173–220. New Haven: Yale University Press.

Cronin, M. A., Hills, S., Born, E. W. & Patton, J. C. (1994). Mitochondrial DNA variation in Atlantic and Pacific walruses. *Canadian Journal of Zoology* 72: 1035–1043.

Cronon, W. (1983). *Changes in the Land: Indians, Colonists, and the Ecology of New England*. New York: Hill & Wang.

Crooks, K. (1994). Demography and status of the island fox and the island spotted skunk on Santa Cruz Island, California. *Southwestern Naturalist* 39: 257–262.

Crooks, K. R. & Soulé, M. E. (1999). Mesopredator release and avifaunal extinctions in a fragmented system. *Nature* 400: 563–566.

Crosby, A. W. (1994). *Germs, Seeds and Animals: Studies in Ecological History*. Armonk, NY: M. E. Sharpe, Inc.

Culver, M. (1999). *Molecular Genetic Variation, Phylogeography and Natural History of the Puma* (Puma concolor). PhD Dissertation, University of Maryland, College Park.

Culver, M., Johnson, W. E., Pecon-Slattery, J. & O'Brien, S. J. (2000). Genomic ancestry of the American puma (*Puma concolor*). *Journal of Heredity* **91**: 186–197.

Cumming, D. H. M. (1982). A case history of the spread of rabies in an African country. *South African Journal of Science* **78**: 443–447.

Cumming, D. H. M., du Toit, R. F. & Stuart, S. N. (1990). *African Elephants and Rhinos. Status Survey and Conservation Action Plan*. Gland: IUCN.

Cunningham, M. W., Dunbar, M. R., Buergelt, C. D., Homer, B. L., Roelke-Parker, M. E., Taylor, S. K., King, R., Citino, S. B. & Glass, C. (1999). Atrial septal defects in Florida panthers. *Journal of Wildlife Discovery* **35**: 519–530.

Cunningham, C. & Berger, J. (1998). *Horn of darkness*. USA: Oxford University Press Inc.

Cunningham, S., Haynes, L. A., Gustavson, C. & Haywood, D. D. (1995). Evaluation of the interaction between mountain lions and cattle in the Aravaipa-Klondyke area of southeast Arizona. *Arizona Game and Fish Department Technical Report* No. 17, Phoenix.

Currier, J. P. (1983). *Felis concolor*. Mammalian Species **200**: 1–7.

Currier, M. J. P., Sheriff, S. L. & Russell, K. R. (1977). Mountain lion population and harvest near Canon City, Colorado, 1974–1977. *Colorado Division of Wildlife Special Report* **42**: 1–12.

Curry-Lindahl, K. (1972). The brown bear (*Ursus arctos*) in Europe: decline, present distribution, biology and ecology. In *Bears – Their Biology and Management*, ed. S. Herrero. Morges: IUCN.

Cypher, B. L. (1997). Effects of radiocollars on San Joaquin kit foxes. *Journal of Wildlife Management*, **61**: 1412–1423.

Cypher, B. L. & Scrivener, J. H. (1992). Coyote control to protect the endangered San Joaquin kit foxes in the then Naval Petroleum Reserves, California. In *Proceedings of 15th Vertebrate Pest Conference*, eds J. E. Borecco & R. E. Marsh, pp. 42–47. Davis: University of California, Davis.

Cypher, B. L. & Spencer, K. A. (1998). Competitive interactions between coyotes and San Joaquin kit foxes. *Journal of Mammalogy* **79**: 204–214.

Danell, K. & Hörnfeldt, B. (1987). Numerical responses by populations of red fox and mountain hare during an outbreak of sarcoptic mange. *Oecologia* **73**: 533–536.

Daniels, M. J., Balharry, D., Hirst, D., Kitchener, A. C. & Aspinall, R. J. (1998). Morphological and pelage characteristics of wild living cats in Scotland: implications for defining the 'wildcat'. *Journal of Zoology* **244**: 231–247.

Daniels, M. J., Golder, M. C., Jarrett, O. & Macdonald, D. W. (1999). Feline viruses in wildcats from Scotland. *Journal of Wildlife Diseases* **35**: 121–124.

Daniels, M. J., Macdonald, D. W., Johnson, P. J. & Barratt, E. M. (In press). Wild living cats in the north east of Scotland: when is a wildcat not a wild cat? *Journal of Applied Ecology*.

Danielson, B. J. (1991). Communities in a landscape: the influence of habitat heterogeneity on the interactions between species. *American Naturalist* **138**: 1105–1120.

Da Silva, J., Woodroffe, R. & Macdonald, D.W. (1993). Habitat, food availability and group territoriality in the European badger, *Meles meles*. *Oecologia* **95**: 558–564.

Daszak, P., Cunningham, A. A. & Hyatt, A. D. (2000). Emerging infectious diseases of wildlife. Threats to biodiversity and human health. *Science.* **287**: 443–449.

Davidson W. R., Nettles, V. F., Hayes, L. E., Howerth, E. W. & Couvillon, C. W. (1992). Diseases diagnosed in gray foxes (*Urocyon cinereoargenteus*) from the southeastern United States. *Journal of Wildlife Diseases* **28**: 28–33.

Davies, J. M., Roper, T. J. & Shepherdson, D. J. (1987). Seasonal distribution of road kills in the European badger (*Meles meles*). *Journal of Zoology (London)* **211**: 525–9.

Davies, N., Villablanca, F. X. & Roderick, G. K. (1999). Determining the source of individuals: multilocus genotyping in nonequilibrium population genetics. *Trends in Ecology and Evolution* **14**: 17–21.

Davison, A., Birks, J. D. S., Griffiths, H. I., Kitchener, A. C., Biggins, D. & Butlin, R. K. (1999). Hybridization and the phylogenetic relationship between polecats and domestic ferrets in Britain. *Biological Conservation* **87**: 155–161.

Davison, R. P. (1980). *The Effect of Exploitation on Some Parameters of Coyote Populations*. PhD thesis, Utah State University, Logan.

Dawkins, M. S. (1998). Evolution and animal welfare. *Quarterly Review of Biology* **73**: 305–328.

Dayan, T., Simberloff, D., Tchernov, E. & Yom-Tov, Y. (1989). Inter- and intraspecific character displacement in mustelids. *Ecology* **70**: 1526–1539.

Dean, E. E. (1979). *Training of Dogs to Detect Black-footed Ferrets*. Santa Fe: New Mexico Department of Game and Fish, Southwestern Research Institute.

Deems, E. F. & Pursley, D. (Eds) (1983). North American Furbearers, a Contemporary Reference. Worldwide Furbearer Conference 1983. International Association of Fish and Wildlife Agencies, and Maryland Department of Natural Resources.

Delahay, R. J., Daniels, M. J., Macdonald, D. W., McGuire, K. & Balharry, D. (1998). Do patterns of helminth parasitism differ between groups of wild living cats in Scotland? *Journal of Zoology (London)* **245**: 175–183.

Delibes, M. (1990). *Status and Conservation Needs of the Wolf* (Canis lupus) *in the Council of Europe Member States*. Strasbourg: Council of Europe.

DeMaster, D. P., Kingsley, M. C. S. & Stirling, I. (1980). A multiple mark and recapture estimate applied to polar bears. *Canadian Journal of Zoology* **58**: 633–638.

Demaynadier, P. G. & Hunter, M. L. Jr. (1998). Effects of silvicultural edges on the distribution and abundance of amphibians in Maine. *Conservation Biology* **12**: 340–352.

Derr, M. (1999a). Texas rescue squad comes to aid of Florida panther. (*SciencePages*) *New York Times* (November 2, 1999).

Derr, M. (1999b). A Rescue Plan for Threatened Species, Breeding Programs Falter. *New York Times* (January 19, 1999).

De Ruiter, J. & Geffen, E. (1998). Relatedness of matrilines, dispersing males and social groups in long-tailed macaques (*Macaca fascicularis*). *Proceedings of the Royal Society of London, Series B* **265**: 79–87.

DeVilliers, M.S., Meltzer, D. G. A., Van Heerden, J., Mills, M. G. L., Richardson, P. R. K., & Van Jaarsveld, A. S. (1995). Handling-induced stress and mortalities in African wild dogs (*Lycaon pictus*). *Proceedings of the Royal Society of London,*

Series B. **262**: 215–220.

deVos, A. (1952). *Ecology and management of fisher and marten in Ontario.* Ontario Department of Lands & Forests, Technical Bulletin, Wildlife Series **1**: 1–90.

deVos, A. (1977). Biological effects of terrestrial vertebrates introduced into non-native environments. *Tigerpaper* **4**: 2–5.

deVos, A., Manville, R. H. & Velder, R. G. (1956). Introduced mammals and their influence on native biota. *Zoologica* (New York Zoological Society) **41**: 163–194.

deVos, A. & Petrides, G. A. (1967). Biological effects caused by terrestrial vertebrates introduced into non-native environments. *Proceedings and Papers Tenth Technical Meeting, IUCN Publications, New Series* **9**: 113–119.

Dew, R. D. & Kennedy, M. L. (1980). Genic variation in raccoons, *Procyon lotor. Journal of Mammalogy* **61**: 697–702.

Dexter, N. & Meek, P. (1998). An analysis of bait-uptake and non-target impacts during a fox-control exercise. *Wildlife Research* **25**: 147–155.

Diamond, J. M. (1984a). Historic extinctions: a Rosetta Stone for understanding prehistoric extinctions. In *Quaternary Extinctions,* eds P. S. Martin & R. G. Klein, pp. 824–862. Tucson: University of Arizona Press.

Diamond, J. M. (1984b). 'Normal' extinctions of isolated populations. In *Extinctions,* ed. M. H. Nitecki, pp. 191–246. Chicago: University of Chicago Press.

Diamond, J. M. (1989). The present, past and future of human-caused extinctions. *Philosophical Transactions of the Royal Society of London, Series B* **325**: 469–477.

Diamond, J. M. & Case, T. J. (1986). Introductions, extinctions, exterminations, and invasions. In *Community Ecology,* eds J. M. Diamond & T. J. Case, pp. 65–79. New York: Harper & Row.

Dickman, C. R. (1996a). *Overview of the Impact of Feral Cats on Australian Native Fauna.* Canberra: Australian Nature Conservation Agency.

Dickman, C. R. (1996b). Impact of exotic generalist predators on the native fauna of Australia. *Wildlife Biology* **2**: 185–195.

Dietz, R., Heide-Jørgensen, M. P. & Härkönen, T. (1989). Mass death of harbour seals *Phoca vitulina* in Europe. *Ambio* **18**: 258–264.

Dietz, L. A. H. & Nagagata, E. H. (1995). Golden lion tamarin conservation program: a community educational effort for forest conservation in Rio de Janeiro State, Brazil. In *Conserving Wildlife. International Education and Communication Approaches,* ed. S. K. Jackobson, pp. 64–86. New York: Columbia University Press.

Dietz, T. & Stern, P. C. (1998). Science, values, and biodiversity. *BioScience* **48**: 441–444.

Dilks, P. J. (1979). Observations on the food of the feral cats on Campbell Island. *New Zealand Journal of Ecology* **2**: 64–66.

Dinerstein, E., Rijal, A., Bookbinder, M., Kattel, B. & Rajuria, A. (1999). Tigers as neighbours: efforts to promote local guardianship of endangered species in lowland Nepal. In *Riding the Tiger – Tiger Conservation in Human-Dominated Landscapes,* eds J. Seidensticker, S. Christie & P. Jackson, pp. 316–333. Cambridge: Cambridge University Press.

Dinerstein, E., Wikramanayake, E., Robinson, J., Karanth, U., Rabinowitz, A., Olson, D., Mathew, T., Hedao, P., Connor, M., Hemley, G. & Bolze, D. (1997). A framework for identifying high priority areas and actions for the conservation of tigers in the wild. Washington/New York: WWF/WCS.

DiToro, D. (1999). *Mexican Wolf Recovery Program Project Update: July 1998 – March 1999*. Arizona: Arizona Game and Fish Department and New Mexico Department of Game and Fish.

Divyabhanusinh (1995). *The End of a Trail – The Cheetah in India*. New Delhi: Banyan Books.

Dix, B. (1993). *Population Changes and Diet Preferences of the New Zealand Fur Seal* (Arctocephalus forsteri) *in eastern Cook Strait*. MSc thesis, Victoria University of Wellington, Wellington, New Zealand.

Doak, D. F. (1995). Source-sink models and the problem of habitat degradation: general models and applications to the Yellowstone grizzly. *Conservation Biology* 9: 1370–1379.

Dobson, A. (1995). The ecology and epidemiology of rinderpest virus in Serengeti and Ngorongoro Conservation area. In *Serengeti II: Dynamics, Management, and Conservation of an Ecosystem*, eds A. R. E. Sinclair & P. Arcese, pp. 485–505. Chicago: University of Chicago Press.

Dobson, A. P. (1988). Restoring island ecosystems: the potential of parasites to control introduced mammals. *Conservation Biology* 2: 31–39.

Dobson, A. P. & Grenfell, B. T. (1995). Introduction. In *Ecology of Infectious Diseases in Natural Populations*, eds B. T. Grenfell & A. P.Dobson, pp. 1–19. Cambridge: Cambridge University Press.

Dobson A. P. & May, R. M. (1986). Disease and conservation. In *Conservation Biology: the Science of Scarcity and Diversity*, ed. M. E. Soulé, pp. 345–365. Sunderland, MA: Sinauer Associates.

Dobson, M. (1994). Patterns of land distribution in Japanese land mammals. *Mammal Review* 24: 91–111.

Donoghue, M. (1997). Seal/fisheries interactions in New Zealand. In *Pinniped Populations, Eastern North Pacific: status, trends and issues*, eds G. Stone, J. Goebel & S. Webster, pp. 56–62. A symposium of the 127th Annual Meeting of the American Fisheries Society. New England Aquarium and Monterey Bay Aquarium, Conservation Department of the New England Aquarium, Central Wharf, Boston, MA.

Doolan, S. P. & Macdonald, D. W. (1996). Dispersal and extra-territorial prospecting by slender-tailed meerkats (*Suricata suricata*) in the southwestern Kalahari. *Journal of Zoology (London)* 240: 59–73.

Dorst, J. & Dandelot, P. (1972). *Collins field guide – Larger Mammals of Africa*. London: Harper Collins.

Douglas, C. W. & Strickland, M. A. (1987). Fisher. In *Wild Furbearer Management and Conservation in North America*, eds M. Novak, J. A. Baker, M. E. Obbard & B. Malloch, pp. 511–529. Ontario Ministry of Natural Resources.

Dowling, T. E. & Demarais, B.D. (1993). Evolutionary significance of introgressive hybridization in cyprinid fishes. *Nature* 362: 444–446.

Dowling, T. E., Mminckley, W. L., Douglas, M. E., Marsh, P. C. & Demarais, B. D. (1992). The use of molecular characters in conservation biology – response. *Conservation Biology* 6: 600–603.

Dragoo, J. W., Bradley, R. D., Honeycutt, R. L. & Templeton, J.W. (1993). Phylogenetic relationships among the skunks: a molecular perspective. *Journal of Mammalian Evolution* 1: 255–267.

Dragoo, J. W., Choate, J. R., Yates, T. L. & O'Farrell, T. P. (1990). Evolutionary and taxonomic relationships among North American arid land foxes. *Journal of Mammalogy* **71**: 318–332.

Dragoo, J. W. & Honeycutt, R. L. (1997). Systematics of the mustelid-like carnivores. *Journal of Mammalogy* **78**: 426–443.

Drake, J. A., Mooney, H. A., di Castri, F., Groves, R. H., Kruger, F. J., Rejmanek, M. & Williamson, M. (Eds) (1989). *Biological Invasions, a Global Perspective.* Chichester: John Wiley & Sons.

Driscoll, C. (1998). *A characterization of microsatellite loci variation in* Panthera. MS Thesis, Hood College, Frederick, MD.

Duckler, G. & Van Valkenburgh, B. (1998). Osteological corroboration of pathological stress in a population of endangered Florida pumas (*Puma concolor coryi*), *Animal Conservation* **1**: 39–46.

Duffy, L. K., Bowyer, R. T., Testa, J. W. & Faro, J. B. (1994). Chronic effects of the Exxon Valdez oil spill on blood and enzyme chemistry of river otters. *Environmental Toxicology and Chemistry* **13**: 643–647.

Duffy, L. K., Bowyer, R. T., Testa, J. W. & Faro, J. B. (1996). Acute phase proteins and cytokines in Alaskan mammals as markers of chronic exposure to environmental pollutants. *American Fisheries Society Symposium* **18**: 809–813.

Duffy, L. K., Hecker, M. K., Blundell, G. M. & Bowyer, R. T. (1999). An analysis of the fur of river otters in Prince William Sound, Alaska: oil related hydrocarbons 8 years after the Exxon Valdez oil spill. *Polar Biology* **21**: 56–58.

Dulamtseren, S. (1970). Guidebook to the mammals of Mongolia. Ulaanbaatar: State Publishing House. (In Mongolian.)

Dunlap, T. R. (1988). *Saving America's Wildlife: Ecology and the American Mind, 1850–1990.* New Jersey: Princeton University Press.

Dunning, J. B., Danielson, B. J. & Pulliam, H. R. (1992). Ecological processes that affect populations in complex landscapes. *Oikos* **65**: 169–75.

Dunstone, N. & Ireland, M. (1989). The mink menace? A reappraisal. In *Mammals as Pests*, ed. R. J. Putman, pp. 225–241. London: Chapman and Hall.

Durant, S. M. (1998). Competition refuges and coexistence: an example from Serengeti carnivores. *Journal of Animal Ecology* **67**: 370–386.

Dye C., Barlow, N. D., Begon, M. *et al.* (1995). Microparasite group report: Persistence of microparasites in natural populations. In *Ecology of Infectious Diseases in Natural Populations*, eds B. T. Grenfell & A. P. Dobson, pp. 123–143. Cambridge: Cambridge University Press.

East, M. L. & Hofer, H. (1991). Loud calling in a female-dominated mammalian society: I. Structure and composition of whooping bouts of spotted hyaenas, *Crocuta crocuta. Animal Behaviour* **42**: 637–649.

East, M. L. & Hofer, H. (1996). Wild dogs in the Serengeti ecosystem: what really happened? *Trends in Ecology and Evolution* **11**: 509.

Easteal, S. (1981). The history of introductions of *Bufo marinus* (Amphibia: Anura); a natural experiment in evolution. *Biological Journal of the Linnean Society* **16**: 93–113.

Easterbee, N. (1991). The wildcat. In *The Handbook of British Mammals*, eds G. Corbet & S. Harris, pp. 431–437. 3rd edn. Oxford: Blackwell Scientific Publications.

Eaton, R. L. (1974). *The Cheetah: The Biology, Ecology and Behavior of an Endangered Species.* New York: Van Nostrand Reinhold.

Ebenhard, T. (1988). Introduced birds and mammals and their ecological effects.

Swedish Wildlife Research **13**: 1–107.

Eberhardt, L. L. & Simmons, M. A. (1992). Assessing rates of increase from trend data. *Journal of Wildlife Management* **56**: 603–610.

Edwards, S. R. & Allen, C. M. (1992). Sport hunting as a sustainable use of wildlife. IUCN Sustainable Use of Wildlife Programme. Washington DC: IUCN.

Effenberger, S. & Suchentrunk, F. (1999). RFLP analysis of the mitochondrial DNA of otters (*Lutra lutra*) from Europe – implications for conservation of a flagship species. *Biological Conservation* **90**: 229–234.

Ehrenfeld, D. W. (1970). *Biological Conservation*. Toronto: Holt, Rinehart, and Winston of Canada.

Eisenberg, J.F. (1981). *The Mammalian Radiations*. Chicago: University of Chicago Press.

Eisenberg, J. F. & Seidensticker, J. (1976). Ungulates in southern Asia: a consideration of biomass estimates for selected habitats. *Biological Conservation* **10**: 293–308.

Eizirik, E., Bonatto, S. L., Johnson, W. E., Crawshaw, P., Vie, C., Brousset, D., O'Brien, S. J. & Salzano, F. M. (1998). Phylogeographic patterns and mitochondrial DNA control region evolution in two Neotropical cats (*Mammalia, Felidae*). *Journal of Molecular Evolution* **47**: 613–624.

Eizirik, E., Indrusiak, C. B. & Johnson, W. E. (2000). Jaguar population viability analyses: evaluation of parameters and a case study of a remnant population in a South American subtropical rain forest. In *Jaguars in the New Millennium. A Status Assessment, Priority Detection, and Recommendations for the Conservation of Jaguars in the Americas*, eds R. A. Medellin, C. Chetkiewicz, A. Rabinowitz, K. H. Redford, J. G. Robinson, E.Sanderson & A. Taber. Universidad Nacional Autonoma de Mexico/Wildlife Conservation Society.

Elgmork, K. (1994). The decline of a brown bear (*Ursus arctos L.*) population in central south Norway. *Biological Conservation* **69**: 123–129.

Elliot, R. (1997). *Faking Nature: The Ethics of Environmental Restoration*. New York: Routledge.

Elowe, K. D. & Dodge, W. E. (1989). Factors affecting black bear reproductive success and cub survival. *Journal of Wildlife Management* **53**: 962–968.

Elton, C. (1942). *Voles, Mice and Lemmings: Problems in Population Dynamics*. Oxford, UK: Oxford University Press.

Elton, C. & Nicholson, M. (1942). The ten-year cycle in numbers of the lynx in Canada. *Journal of Animal Ecology* **11**: 215–244.

Elton, C. S. (1958). *The Ecology of Invasions by Animals and Plants*. London: Methuen and Co.

Engeman, R. M., Allen, L. & Zerbe, G. O. (1998). Variance estimate for the Allen activity index. *Wildlife Research* **25**: 643–648.

Englund, J. (1970). Some aspects of reproduction and mortality rates in Swedish foxes (*Vulpes vulpes*), 1961–63 and 1966–69. *Swedish Wildlife* **81**: 1–82.

Erickson, A. W. & Siniff, D. B. (1963). A statistical evaluation of factors influencing aerial survey results for brown bears. *Transactions of the North American Wildlife and Natural Resources Conference* **28**: 391–409.

Erickson, D. W. (1982). Estimating and using furbearer harvest information. In *Midwest Furbearer Management*, ed. G. C. Sanderson, pp. 53–65. Proceedings of the Symposium of the 43rd Midwest Fish and Wildlife Conference, Wichita,

Kansas.

Erlich, P. R. (1986). Which animal will invade ? In *Ecology of Biological Invasions of North America and Hawaii*, eds H. A. Mooney & J. A. Drake, pp. 79–95. New York: Springer Verlag.

Erlinge, S. (1969). Food habits of the otter (*Lutra lutra*) and mink (*Mustela vison*) in a trout water in southern Sweden. *Oikos* **20**: 1–7.

Erlinge, S. (1972). Interspecific relations between otter *Lutra lutra* and mink *Mustella* [sic] *vison* in Sweden. *Oikos* **23**: 327–335.

Erlinge, S. (1974). Distribution, territoriality and numbers of the weasel *Mustela nivalis* in relation to prey abundance. *Oikos* **25**: 308–314.

Erlinge, S. (1983). Demography and dynamics of a stoat, *Mustela erminea*, population in a diverse community of vertebrates. *Journal of Animal Ecology* **52**: 507–526.

Errington, P. L. (1956). Factors limiting higher vertebrate populations. *Science* **124**: 304–307.

Estes, J. A. (1979). Exploitation of marine mammals: r-selection or K-strategists? *Journal of the Fisheries Research Board of Canada* **36**: 1009–1017.

Estes, J. A. (1991). Catastrophes and conservation: lessons from sea otters and the Exxon Valdez. *Science* **254**: 1596.

Estes, J. A. (1998). Concerns about rehabilitation of oiled wildlife. *Conservation Biology* **12**: 1156–1157.

Estes, R. D. & Goddard, J. (1967). Prey selection and hunting behavior of the African wild dogs. *Journal of Wildlife Management* **31**: 52–70.

Estoup, A., Solignac, M., Cornuet, J-M., Gudet, J. & Scholl, A. (1996). Genetic differentiation of continental and island populations of *Bombus terrestris* (Hymenoptera, Apidae) in Europe. *Molecular Ecology* **5**: 19–32.

Evans, P. G. H., Macdonald, D. W. & Cheeseman, C. L. (1989). Social structure of the Eurasian badger (*Meles meles*). Genetic evidence. *Journal of Zoology (London)* **218**: 587–595.

Everard, C. O. R., Baer, G. M. & James, A. (1974). Epidemiology of mongoose rabies in Grenada. *Journal of Wildlife Diseases* **10**: 190–196.

Evermann, B. W. (1914). *Alaska Fisheries and Fur Industries in 1913*. Department of Commerce, Bureau of Fisheries (USA), Document No. 797.

Evermann, J., Burns, F. G., Roelke, M. E., McKeirnan, A. J., Greelee, A., Ward, A. C. & Pfeifer, M. L. (1984). Diagnostic features of an epizootic of feline infectious peritonitis in captive cheetahs. *American Association of Veterinary Lab Diagnosis* **26**: 265–282.

Evink, G. L., Garrett, P., Zeigler, D. & Berry, J. (Ed.) (1996). *Proceedings of the Transportation Related Wildlife Mortality Seminar*, FL-ER-58-96. Tallahassee: Florida Department of Transportation.

Evink, G. L., Garrett, P., Zeigler, D. & Berry, J. (Ed.) (1998). *Proceedings of the International Conference on Wildlife Ecology and Transportation*, FL-ER-69-98. Tallahassee: Florida Department of Transportation.

Fahrig, L. (1997). Relative effects of habitat loss and fragmentation on population extinction. *Journal of Wildlife Management* **61**: 603–10.

Fahrig, L. & Merriam, G. (1994). Conservation of fragmented populations. *Conservation Biology* **8**: 50–9.

Fanshawe, J. H. & Fitzgibbon, C. D. (1993). Factors influencing the hunting success of an African wild dog pack. *Animal Behaviour* **45**: 479–490.

Fanshawe, J. H., Frame, L. H. & Ginsberg, J. R. (1991). The wild dog – Africa's vanishing carnivore. *Oryx* **25**: 137–146.

Fanshawe, J. H., Ginsberg, J. R., Sillero-Zubiri, C. & Woodroffe, R. (1997). The status and distribution of remaining wild dog populations. In *The African Wild Dog: Status Survey and Conservation Action Plan*, eds R. Woodroffe, J. Ginsberg & D. Macdonald, pp. 11–57. Gland: IUCN.

Favre, L., Balloux, F., Goudet, J. & Perrin, N. (1997). Female-biased dispersal in the monogamous mammal *Crocidura russula:* evidence from field data and microsatellite patterns. *Proceedings of the Royal Society of London, Series B*, **264**: 127–132.

Fay, F. H. (1982). Ecology and biology of the Pacific walrus, *Odobenus rosmarus divergens* Illiger. North American Fauna, No. 74. Washington, DC: US Dept of the Interior, Fish and Wildlife Service.

Feiler, A. (1984). Uber die Saugetiere der Insel Sao Tomé. Zoologische Abh. *Museum Tierkunde Dresden* **40**: 75–78.

Felsenstein, J. (1985). Phylogenies and the comparative method. *American Naturalist* **125**: 1–15.

Ferreras, P., Aldama, J.J., Beltrán, J.F. & Delibes, M. (1992). Rates and causes of mortality in a fragmented population of Iberian lynx *Felis pardina* Temminck, 1824. *Biological Conservation* **61**: 197–202.

Ferreras, P. & Macdonald, D. W. (1999). The impact of American mink *Mustela vison* on water birds in the upper Thames. *Journal of Applied Ecology* **36**: 701–708.

Festetics, A. (1980). Die Wiedereinbürgerung des Luchses in Europa. In *Der Luchs in Europa*, ed. A. L. Festetics, pp. 224–254. Greven: Kilda Verlag.

Field, J., Solis, C., Queller, D. & Srassmann, J. (1998). Social and genetic structure of paper wasp cofoundress associations: test of reproductive skew models. *American Naturalist* **151**: 545–563.

Finlayson, B. & McMahon, T. (1994). Funding and conduct of environmental research. In *Restoring the Land: Environmental Values, Knowledge, and Action*, eds D. Evans & D. Yenchken, pp. 44–62. Melbourne: Melbourne University Press.

Fisher, H. I. (1948). The question of avian introductions in Hawaii. *Pacific Science* **2**: 58–64.

Fitzgerald, B. M. (1977). Weasel predation on a cyclic population of the montane vole (*Microtus montanus*) in California. *Journal of Animal Ecology* **46**: 367–397.

Fitzgerald, B. M. (1988). Diet of domestic cats and their impact on prey populations. In *The Domestic Cat: The Biology of its Behaviour*, eds D. Turner & P. Bateson, pp. 123–145. Cambridge: Cambridge University Press.

Fjeldså, J. (1994). Geographical patterns for relic and young species of birds in Africa and South America and implications for conservation priorities. *Biodiversity and Conservation* **3**: 207–226.

Fleming, P. J. S., Allen, L. R., Berghout, M. J., Meek, P. D., Pavlov, P. M., Stevens, P., Strong, K. Thompson, J. A. & Thomson, P. S. (1998). The performance of wild-canid traps in Australia: Efficiency, selectivity and trap-related injuries. *Wildlife Research* **25**: 327–338.

Flynn, J. J. (1996). Carnivoran phylogeny and rates of evolution: morphological, taxic, and molecular. In *Carnivore Behavior, Ecology, and Evolution*, vol. 2, ed. J. L.

Gittleman, pp. 542–581. Ithaca: Cornell University Press.

Flynn, J. J. & Galiano, H (1982). Phylogeny of early Tertirary Carnivora, with a description of a new species *Protictis* from the middle Eocene of northwestern Wyoming. *American Museum Novitates* **2725**: 1–64.

Flynn, J. J. & Nedbal, M. A. (1998). Phylogeny of the Carnivora (Mammalia). congruence vs. incompatibility among multiple data sets. *Molecular Phylogenetics and Evolution* **9**: 414–426.

Follmann, E. H. & Hachtel, J. L. (1990). Bears and pipeline construction in Alaska (USA). *Arctic* **43**: 103–109.

Foose, T. J. (1987). Species Survival Plans and Overall Management Strategies. In *Tigers of the World*, eds R. L. Tilson & U. S. Seal, pp. 304–316. Park Ridge, NJ: Noyes Publications.

Foose, T. J. & Seal, U. S. (1992). Conservation assessment and management plans, global captive action plans summary reports. Minnesota: IUCN-CBSG.

Foote, M. (1997). Estimating taxonomic durations and preservation probability. *Paleobiology* **23**: 278–300.

Foran, D. R., Crooks, K. R. & Minta, S. C. (1997a). Species identification from scat: an unambiguous genetic method. *Wildlife Society Bulletin* **25**: 835–839.

Foran, D. R., Minta, S. C. & Heinemeyer, K. S. (1997b). DNA-based analysis of hair to identify species and individuals for population research and monitoring. *Wildlife Society Bulletin* **25**: 840–847.

Forbes, S. & Boyd, D. (1997). Genetic structure and migration in native and reintroduced Rocky Mountain wolf populations. *Conservation Biology* **11**: 1226–1234.

Forbes, G. J. & Theberge, J. B. (1996). Response by wolves to prey variation in central Ontario. *Canadian Journal of Zoology* **74**: 1511–1520.

Foreman, G. E. (1992). Pumas and people. *Cat News* **16**: 11–12.

Foresman, K. R. & Pearson, D. E. (1998). Comparison of proposed survey procedures for detection of forest carnivores. *Journal of Wildlife Management* **62**: 1217–1226.

Forsyth, M. A., Kennedy, S., Wilson, S., Eybatov, T. & Barrett, T. (1998). Canine distemper virus in a Caspian seal. *Veterinary Record* **143**: 662–664.

Fortenbery, D. K. (1970). Ecology and management of the black-footed ferret. In *Annual Progress Report of the Patuxent Wildlife Research Center*, pp. 216–218. Laurel, MD: U.S. Fish and Wildlife Service.

Fosbrooke H. H. (1963). The stomoxys plague in Ngorongoro, 1962. *East African Wildlife Journal* **1**: 124–126.

Foster, M. L. & Humphrey, S. R. (1995). Use of highways and underpasses by Florida panthers and other wildlife. *Wildlife Society Bulletin* **23**: 95–100.

Foster-Turley, P., Macdonald, S. & Mason, C. (Eds) (1990). *Otters. An Action Plan for their Conservation*. Gland: IUCN. 126pp.

Foucault, F., Praz, F., Jaulin, C. & Amor-Gueret, M. (1996). Experimental limits of PCR analysis of (CA)n repeat alterations. *Trends in Genetics* **12**: 450–452.

Fountain, M. (1975). Censusing and collecting marsh raccoons via an air-boat. *Proceedings of the Annual Conference of Southeast Assoc. Game and Fish Comm.* **29**: 680–681.

Fox, M. W. (1984). *The Whistling Hunters: Field Studies of the Asiatic Wild Dog* (Cuon alpinus). Albany: SUNY Press.

Frafjord, K., Becker, D. & Angerbjörn, A. (1989). Interactions between arctic and red foxes in Scandinavia – predation and aggression. *Arctic* **42**: 354–356.

Frame, L. H., Malcolm, J. R., Frame, G. W. & van Lawick, H. (1979). Social organization of African wild dogs (*Lycaon pictus*) on the Serengeti Plains, Tanzania, 1967–1978. *Zeitschrift fur Tierpsychologie* **50**: 225–249.

Framstad, E., Stenseth, N. C., Bjornstad, O. N. & Falk, W. (1997). Limit cycles in Norwegian lemmings: tensions between phase-dependence and density-dependence. *Proceedings of the Royal Society of London, Series B* **264**: 31–38.

Frank, L. G. (1986). Social organization of the spotted hyena (*Crocuta crocuta*). II. Dominance and reproduction. *Animal Behaviour* **34**: 1510–1527.

Frank, L.G. (1996). Female masculinization in the spotted hyena: endocrinology, behavioral ecology and evolution. In *Carnivore Behavior, Ecology and Evolution*, vol. II, ed. J. L. Gittleman, pp. 78–131. Ithaca: Cornell University Press.

Frank, L. G. (1998a). The Laikipia Predator Project – First biannual progress report. Unpublished Report, Mpala Research Centre.

Frank, L. G. (1998b). Living with lions: Carnivore conservation and livestock in Laikipia District, Kenya. Unpublished Report, Development Alternatives, Inc.

Frank, L. G., Holekamp, K. E. & Smale, L. (1995). Dominance, demography and reproductive success of female spotted hyenas. In *Serengeti II: Research, Management, and Conservation of an Ecosystem*, eds A. R. E. Sinclair & P. Arcese, pp. 364–384. Chicago: Chicago University Press.

Frank, S. A. (1990). Sex allocation theory for birds and mammals. *Annual Review of Ecology and Systematics* **21**: 13–55.

Frankel, O. & Soulé, M. (1981). *Conservation and Evolution*. Cambridge: Cambridge University Press.

Frankham, R. (1995). Effective population size/adult population size ratios in wildlife: a review. *Genetic Research* **66**: 95–107.

Frankham, R., Hemmer, H., Ryder, O. A., Cothran, E. G., Soulé, M. E., Snyder, M. & Murray, N. D. (1986). Selection in captive populations. *Zoo Biology* **5**: 127–138.

Franklin, I. R. (1980). Evolutionary changes in small populations. In *Conservation Biology. An Evolutionary Ecological Perspective*, eds M. E. Soulé & B. A. Wilcox-Sinauer, Associates, pp. 135–139. Sunderland, MA: Sinauer.

Franklin, I. R. & Frankham, R. (1998). How large must populations be to retain evolutionary potential? *Animal Conservation*, **1**: 69–70.

Franklin, N., Sriyanto, B., Siswormartono, D., Manansang, J. & Tilson, R. (1999). Last of the Indonesian Tigers: A Cause For Optimism. In *Riding the Tiger: Tiger Conservation in Human-dominated Landscape*, eds J. Seidensticker, S. Christie & P. Jackson, pp. 114–122. Cambridge: Cambridge University Press.

Franklin, W. L., Johnson, W. E., Sarno, R. J. & Iriarte, J. A. (1999). Ecology of the Patagonia puma (*Felis concolor patagonica*) in southern Chile. *Biological Conservation* **90**: 33–40.

Frati, F., Hartl, G. B., Lovari, S., Delibes, M. & Markov, G. (1998). Quaternary radiation and genetic structure of the red fox *Vulpes vulpes* in the Mediterranean Basin, as revealed by allozymes and mitochondrial DNA. *Journal of Zoology (London)* **245**: 43–51.

Frederickson, L. (1979). Furbearer population surveys: techniques, problems and uses in South Dakota. In *Proceedings of the Midwest Furbearer Conference*, pp. 62–70. Kansas State University, Coop. Ext. Serv., Manhattan.

Freitag, S., Nicholls, A. O. & van Jaarsveld, A. S. (1996). Nature reserve selection in the Transvaal, South Africa: what data should we be using? *Biodiversity and Conservation* 5: 285–298.

Freitag, S. & van Jaarsveld, A. S. (1997). Relative occupancy, endemism, taxonomic distinctiveness and vulnerability: prioritising regional conservation actions. *Biodiversity and Conservation* 6: 211–32.

Freitag, S., van Jaarsveld, A. S. & Biggs, H. C. (1997). Ranking priority biodiversity areas: an iterative conservation value-based approach. *Biological Conservation* 82: 263–72.

French, D. D., Corbett, L. K. & Easterbee, N. (1988). Morphological discriminants of Scottish wildcats (*Felis silvestris*), domestic cats (*F. catus*) and their hybrids. *Journal of Zoology (London)* 161: 75–123.

French, W. C. (1995). Against biospherical egalitarianism. *Environmental Ethics* 17: 39–57.

Fritts, S. H. (1982). Wolf depredation on livestock in Minnesota. *U.S. Fish and Wildlife Service Resource Publication* 145: 1–11.

Fritts, S. H. (1983). Record dispersal by a wolf from Minnesota. *Journal of Mammalogy* 64: 166–167.

Fritts, S. H., Bangs, E. E., Fontaine, J. A., Brewster, W. G. & Gore, J. F. (1995). Restoring wolves to the northern Rocky Mountains of the United States. In *Ecology and Conservation of Wolves in a Changing World*, eds L. N. Carbyn, S. H. Fritts & D. R. Seip, pp. 107–126. Edmonton: Canadian Circumpolar Institute.

Fritts, S. H. & Mech, L. D. (1981). Dynamics, movements, and feeding ecology of a newly protected wolf population in northwestern Minnesota. *Wildlife Monographs* 80: 1–79.

Fritzell, E. K., Hubert, G. F., Meyen, B. E. & Sanderson, G. C. (1985). Age-specific reproduction in Illinois and Missouri raccoons. *Journal of Wildlife Management* 49: 901–905.

Fromont, E., Courchamp, F., Artois, M. & Pontier, D. (1997). Infection strategies of retroviruses and social grouping of domestic cats. *Canadian Journal of Zoology* 75: 1994–2002.

Fuller, T. K. (1989). Population dynamics of wolves in north-central Minnesota. *Wildlife Monographs* 105: 1–41.

Fuller, T. K. (1990). Dynamics of a declining population of white-tailed deer population in north-central Minnesota. *Wildlife Monographs* 110: 1–39.

Fuller, T. K. (1995). An international review of large carnivore conservation status. In *Integrating People and Wildlife for a Sustainable Future*, eds J. A. Bissonette & P. R. Krausmann, pp. 410–412. Bethesda, MA: The Wildlife Society.

Fuller, T. K. (1997). Guidelines for gray wolf management in the northern Great Lakes region. *Educucational Publication No. IWC97-271*, 2nd edn, p. 20. Ely, MN: International Wolf Center.

Fuller, T. K., Berg, W. E., Radde, G. L., Lenarz, M. S. & Joselyn, G. B. (1992a). A history and current estimate of wolf distribution and numbers in Minnesota. *Wildlife Society Bulletin* 20: 42–55.

Fuller, T. K. & Kat, P. W. (1993). Hunting success of African wild dogs in southwestern Kenya. *Journal of Mammalogy* 224: 464–467.

Fuller, T. K. & Keith, L. B. (1981). Non-overlapping ranges of coyotes and wolves in Northeastern Alberta. *Journal of Mammalogy* 62: 403–405.

Fuller, T. K., Mills, M. G. L., Borner, M., Laurenson, K. & Kat, P. W. (1992b). Long distance dispersal by African wild dogs in East and South Africa. *Journal of African Zoology* **106**: 535–537.

Fuller, T. K. & Murray, D. L. (1998). Biological and logistical explanations of variation in wolf population density. *Animal Conservation* **1**: 153–157.

Fuller, T. K. & Sampson, B. A. (1988). Evaluation of a simulated howling survey for wolves. *Journal of Wildlife Management* **52**: 60–63.

Funk, S. M. (1994). *Zur Dichteabhängigkeit der räumlichen und sozialen Organsiation und der Reproduktion beim Rotfuchs* (Vulpes vulpes L.): *eine Studie bei zeitlich und räumlich durch Jagd und Tollwut variierenden Populationsdichten in Südwest-Deutschland und Ost-Frankreich.* PhD Thesis, Universität des Saarlandes, Saarbrücken.

Funk, S. M. (1995). *Epidemiology of an outbreak of sarcoptic mange and consequences for spatial and social organisation in an urban fox* (Vulpes vulpes) *population.* MSc Thesis, University of Bristol.

Funk, S. M. & W.-D. Gürtler (1991). Uber den Zusammenhang zwischen Reproduktionserfolg und Populationsdichte beim Rotfuchs, *Vulpes vulpes* L. *Schriften des Arbeitskreises Wildbiologie an der Justus-Liebig-Universität Gieen e.V* **20**: 39–48.

Gagneux, P., Boesch, C. & Woodruff, D.S. (1997). Microsatellite scoring errors associated with noninvasive genotyping based on nuclear DNA amplified from shed hair. *Molecular Ecology* **6**: 861–868.

Gaines, M. S., Stenseth, N. C., Johnson, M. L., Ims, R. A. & Bondruop-Nielsen, S. (1991). A response to solving the enigma of population cycles with a multifactorial perspective. *Journal of Mammalogy* **72**: 627–631.

Gales, N. J. & Fletcher., D. J. (1996). Abundance, distribution and status of the New Zealand sea lion *Phocarctos hookeri.* New Zealand sea lion population management plan technical meeting. Wellington, NZ 8/9 October 1996.

Gales, N. & Mattlin, R. (1997). Summer diving behaviour of lactating New Zealand sea lions (*Phocarctos hookeri*). *Canadian Journal of Zoology* **75**: 1695–1706.

Galhano, A. (1996). 'Down to Earth' **5 (11)**: 32–34.

Gaona P., Ferreras, P. & Delibes, M. (1998). Dynamics and viability of a metapopulation of the endangered Iberian lynx (*Lynx pardinus*). *Ecological Monographs* **68**: 349–370.

García-Moreno, J., Matocq, M. D., Roy, M. S., Geffen, E. & Wayne, R. K. (1996). Relationships and genetic purity of the endangered Mexican wolf based on analysis of microsatellite loci. *Conservation Biology* **10**: 376–389.

Gardner, R. H., Milne, B. T, Turner, M. G. & O'Neill, R.V. (1987). Neutral models for the analysis of broad-scale landscape pattern. *Landscape Ecology* **1**: 19–28.

Garland, T. J., Harvey, P. H. & Ives, A. R. (1992). Procedures for the analysis of comparative data using phylogenetically independent contrasts. *Systematic Biology* **41**: 18–32.

Garrott, R. A., Eberhardt, L. E. & Hanson, W. C. (1983). Arctic fox identification and characteristics in northern Alaska. *Canadian Journal of Zoology* **61**: 423–426.

Garshelis, D. L. (1987). Sea otter. In *Wild Furbearer Management and Conservation in North America*, eds M. Novak, J. A. Baker, M. E. Obbard & B. Malloch, pp. 634–655. North Bay: Ontario Trappers Association.

Garshelis, D. L. (1994). Density-dependent population regulation of black bears. In *Density Dependent Population Regulation in Black, Brown, and Polar Bears*, ed.

M. Taylor, pp. 3–13. International Conference on Bear Research and Management Monographs Series No. 3.

Garza, J. C. (1998). *Population Genetics of the Northern Elephant Seal*. PhD dissertation, University of California, Berkeley.

Gasaway, W. C, Boertje, R. D., Grangaard, D. V., Kelleyhouse, D. G., Stephenson, R. O. & Larsen, D. G. (1992). The role of predation in limiting moose at low densities in Alaska and Yukon and implications for conservation. *Wildlife Monographs* 120: 1–59.

Gasaway, W. C., Mossestad, K. T. & Stander, P. E. (1991). Food acquisition by spotted hyaenas in Etosha National Park, Namibia: predation versus scavenging. *African Journal of Ecology* 29: 64–75.

Gasaway. W. C., Stephenson, R. O., Davis, J. L., Shepherd, P. E. K. & Burris, O. E. (1983). Interrelationships of wolves, prey, and man in interior Alaska. *Wildlife Monographs* 84. 1–50.

Gascoyne S. C., King, A. A., Laurenson, M. K., Borner, M., Schildger, B. & Barrat, J. (1993a). Aspects of rabies infection and control in the conservation of the African wild dog (*Lycaon pictus*) in the Serengeti region, Tanzania. *Onderstepoort Journal of Veterinary Research* 60: 415–420.

Gascoyne, S. C., Laurenson, M. K., Lelo, S. & Borner, M. (1993b). Rabies in African wild dogs *Lycaon pictus* in the Serengeti region, Tanzania. *Journal of Wildlife Diseases* 29: 396–402.

Gaskell R. & Willoughby, K. (1999). Herpes viruses of carnivores. *Veterinary Microbiology* 69: 73–88.

Gaston, K. J. (1996). Species richness: measure and measurement. In *Biodiversity: a Biology of Numbers and Difference*, ed. K. J. Gaston, pp. 77–113. Oxford: Blackwell Science.

Gaston, K. J. & Blackburn, T. M. (1995). Birds, body size and the threat of extinction. *Philosophical Transactions of the Royal Society Of London, Series B* 347: 205–212.

Gates, C. E. (1979). Line transects and related issues. In *Sampling Biological Populations*, eds R. M. McCormick, P. Patil & D. S. Robson, pp. 71–154. *Statistics and Ecology Series*, vol. 5. Fairland, MD: Int. Coop. Publishing House.

Geiger, A. C. (1985). Evaluation of seal harassment devices to protect salmon in gillnet fisheries. In *Proceedings of the Sixth Biennial Conference on the Biology of Marine Mammals November 22–26, Vancouver, British Columbia*.

Geist, V. (1988). How markets in wildlife meat and parts, and the sale of hunting privileges jeopardize wildlife conservation. *Conservation Biology* 2: 1–12.

Geist, V. (1994). Wildlife conservation as wealth. *Nature* 368: 491–492.

Genovesi, P. (2000). Brown bear reintroduction in the Italian central Alps. *International Bear News* 9: 13.

Georgii, S., Bachour, G., Failing, K., Eskens, U., Elmadfa, I. & Brunn, H. (1994). Polychlorinated biphenyl congeners in foxes in Germany from 1983 to 1991. *Archives of Environmental Contamination and Toxicology* 26: 1–6.

Geraci, J. R., Staubin, D. J., Barker, I. K., Webster, R. G., Hinshaw, V. S., Bean, W. J., Ruhnke, H. L., Prescott, J. H., Early, G., Baker, A. S., Madoff, S. & Schooley, R. T. (1982). Mass mortality of harbor seals – pneumonia associated with influenza-a virus. *Science* 215: 1129–1131.

Gerell, R. (1967). Dispersal and acclimatization of the mink (*Mustela vison*) in Sweden. *Viltrevy* 4: 1–38.

Gerell, R. (1985). Habitat selection and nest predation in a common eider population in southern Sweden. *Ornis Scandinava* **16**: 129–139.

Gerloff, U., Schlötterer, C., Rassmann, K. *et al.* (1995). Amplification of hypervariable simple sequence repeats (microsatellites) from excremental DNA of wild living bonobos (Pan paniscus). *Molecular Ecology* **4**: 515–518.

Gerrodette, T. (1987). A power analysis for detecting trends. *Ecology* **68**: 1364–1372.

Gese, E. M. & Mech, L. D. (1991). Dispersal of wolves (*Canis lupus*) in northeastern Minnesota 1969–1989. *Canadian Journal of Zoology* **69**: 2946–2955.

Gese, E. M., Rongstad, O. J. & Mytton, W. R. (1989). Population dynamics of coyotes in southeastern Colorado. *Journal of Wildlife Management* **53**: 174–181.

Gese, E. M. & Ruff, R. L. (1998). Howling by coyotes (*Canis latrans*): variation among social classes, seasons, and pack sizes. *Canadian Journal of Zoology* **76**: 1037–1043.

Gese, E. M., Ruff, R. L. & Crabtree, R. L. (1996a). Social and nutritional factors influencing the dispersal of resident coyotes. *Animal Behaviour* **52**: 1025–1043.

Gese, E. M., Ruff, R. L. & Crabtree, R. L. (1996b). Foraging ecology of coyotes (*Canis latrans*): the influence of extrinsic factors and a dominance hierarchy. *Canadian Journal of Zoology* **74**: 769–783.

Gese, E. M., Schultz, R. D., Johnson, M. R., Williams, E. S., Crabtree, R. L. & Ruff, R. L. (1997). Serological survey for diseases in free-ranging coyotes (*Canis latrans*) in Yellowstone National Park, Wyoming. *Journal of Wildlife Diseases* **33**: 47–56.

Gese, E. M., Stotts, T. E. & Grothe, S. (1996c). Interactions between coyotes and red foxes in Yellowstone National Park, Wyoming. *Journal of Mammalogy* **77**: 377–382.

Gibeau, M. L. & Heuer, K. (1996). Effects of transportation corridors on large carnivores in the Bow River Valley, Alberta. In *Proceedings of the Transportation Related Wildlife Mortality Seminar*, FL-ER-58-96. eds G. L. Evink, P. Garrett, D. Zeigler & J. Berry, Tallahassee: Florida Department of Transportation.

Gibson, D. & Isakseen, B. (1998). *Functionality of a Full-sized Marine Mammal Exclusion Device*. New Zealand Department of Conservation report, Science for Conservation No. 81. Wellington: NZ Department of Conservation.

Gier, H. T. (1968). *Coyotes in Kansas*. Manhattan, KS: Kansas State University of Agriculture and Applied Science.

Gilbert, D. A., Packer, C., Pusey, A. E., Stephens, J. C. & O'Brien, S. J. (1991). Analytical DNA fingerprinting in lions: parentage, genetic diversity, and kinship. *Journal of Heredity* **82**: 378–386.

Gilbert, J. R., Kordek, W. S., Collins, J. & Conley, R. (1978). Interpreting sex and age data from legal kills of bears. In *Proceedings of 4th East. Black Bear Workshop*, ed. R. D. Hugie, pp. 253–262. Maine: Greenville.

Gilbert, T. & Wooding, J. (1996). An overview of black bear roadkills in Florida 1976–1995. In *Proceedings of the Transportation Related Wildlife Mortality Seminar*, FL-ER-58-96. eds G. L. Evink, P. Garrett, D. Zeigler & J. Berry, Tallahassee: Florida Department of Transportation.

Gilinsky, N. L. (1998). Evolutionary turnover and volatility in higher taxa. In *Biodiversity Dynamics*, eds M. L. McKinney & J. A. Drake, pp. 162–184. New York: Columbia University Press.

Gilpin, M. & Soulé, M. (1986). Minimum viable populations: processes of species extinction. In *Conservation Biology: The Science of Scarcity and Diversity*, ed. M. E. Soulé, pp. 13–34. Sunderland, MA: Sinauer.

Gimenez-Dixon, M. & Stuart, S. (1993). Action plans for species conservation, an evaluation of their effectiveness. *Species* **20**: 6–10.

Gingerich, P. D. (1984). Pleistocene extinction in the context of origination-extinction equilibria in Cenozoic mammals. In *Quaternary Extinctions*, eds P. S. Martin & R. G. Klein, pp. 211–222. Tucson: University of Arizona Press.

Ginsberg, J. R. (1994). Captive breeding, reintroduction and the conservation of canids. In *Creative conservation. Interactive Management of Wild and Captive Animals*, eds P. J. S. Olney, G. M. Mace & A. T. C. Feistner, pp. 365–383. London: Chapman & Hall.

Ginsberg, J. R. (1999). Global conservation priorities. *Conservation Biology* **13**: 5.

Ginsberg, J. R., Alexander, K. A., Creel, S., Kat, P. W., McNutt, J. W. & Mills, G. L. (1995). Handling and survivorship of African wild dog (*Lycaon pictus*) in five ecosystems. *Conservation Biology* **9**: 665–674.

Ginsberg, J. R. & Macdonald, D. W. (1990). *Foxes, Wolves, Jackals and Dogs: An Action Plan for the Conservation of Canids*. Gland: IUCN.

Ginsberg, J. R. & Woodroffe, R. (1997). Extinction risk faced by remaining wild dog populations. In *The African Wild Dog. Status and Conservation Action Plan*, eds R. Woodroffe, J. R. Ginsberg & D. Macdonald, pp. 75–87. Gland: IUCN/SSC Canid Specialist Group.

Gipps, J. (1991). Preface. In *Beyond Captive Breeding: Reintroducing Endangered Mammals to the Wild*, ed. J. Gipps, pp. v–vii. Oxford: Oxford Scientific Publishing.

Girman, D. J., Kat, P. W., Mills, M. G. L., Ginsberg, J. R., Borner, M., Wilson, V., Fanshaw, J. H., Fitzgibbon, C., Lau, L. M. & Wayne, R. K. (1993). Molecular genetic and morphological analyses of the African wild dog (*Lycaon pictus*). *Journal of Heredity* **84**: 450–459.

Girman, D., Miles, G., Geffen, E. & Wayne, R. (1997). A molecular genetic analysis of social structure, dispersal, and pack interactions in the African wild dog (*Lycaon pictus*). *Behavioral Ecology and Sociobiology* **40**: 187–198.

Girman, D. J. & Wayne, R. K. (1997). Genetic perspectives on wild dog conservation. In *The African Wild Dog, Status Survey and Conservation Action Plan*, eds R. Woodroffe, J. Ginsberg & D. Macdonald, pp. 7–10. Gland: IUCN.

Gittleman, J. L. (Ed.) (1989). *Carnivore Behavior, Ecology, and Evolution*. Ithaca: Cornell University Press.

Gittleman, J. L. (Ed.) (1996). *Carnivore Behavior, Ecology, and Evolution*, vol. 2. Ithaca: Cornell University.

Gittleman, J. L. (1994). Are the pandas successful specialists or evolutionary failures? *BioScience* **44**: 456–464.

Gittleman, J. L., Anderson, C. G., Cates, S. E., Luh, H.-K. & Smith, J. D. (1998). Detecting ecological pattern in phylogenies. In *Biodiversity Dynamics – Turnover of Populations, Taxa and Communities*, eds M. L. McKinney & J. A. Drake, pp. 51–69. New York: Columbia University Press.

Gittleman, J. L. & Harvey, P. H. (1982). Carnivore home-range size, metabolic needs and ecology. *Behavioural Ecology & Sociobiology* **10**: 57–63.

Gittleman, J. L. & Purvis, A. (1998). Body size and species richness in primates and carnivores. *Proceedings of the Royal Society of London, Series B* **265**: 113–119.

Giulianelli, S. & Laikre, L. (1994). Mitochondrial DNA variability in otters (*Lutra lutra*) from northern Europe. In *Directorate of Environment and Local Authorities. Seminar on the Conservation of the European Otter (Lutra lutra)*, pp. 205–208. The Netherlands: Leeuwarden. 7–11 June 1994. Strasbourg: Council of Europe.

Gjertz, I. & Persen, E. (1987). Confrontations between humans and polar bears in Svalbard (Norway). *Polar Research* 5: 253–256.

Glatston, A. R. (1994). *The Red Panda, Olingos, Coatis, Raccoons, and their Relatives. Status Survey and Conservation Action Plan for Procyonids and Ailurids.* Gland: IUCN. (In English and Spanish.)

Glen, W. (1996). *Eyes of Fire.* El Paso, TX: Printing Corner Press.

Glenn, L. P. & Miller, L. H. (1980). Seasonal movements of an Alaska peninsula brown bear population. *International Conference on Bear Research & Management* 4: 307–312.

Goldman, D. P., Giri, R. & O'Brien, S. J. (1989). Molecular genetic distance estimates among the Ursidae as indicated by one- and two-dimensional protein electrophoresis. *Evolution* 43: 282–295.

Golden, H. N. (1986). Survey of furbearer populations on the Yukon Flats National Wildlife Refuge. Fairbanks: Alaska Department of Fish and Game, Progress Report.

Goldstein, D. B., Roemer, G. W., Smith, D. A., Reich, D. E., Bergman, A. & Wayne, R. K. (1999). The use of microsatellite variation to infer population structure and demographic history in a natural model system. *Genetics* 151: 797–801.

Goltsman, M., Kruchenkova, E. P. & Macdonald, D. W. (1996). The Mednyi arctic foxes: treating a population imperilled by disease. *Oryx* 30: 251–258.

Gompper, M. E. & Gittleman, J. L. (1991). Home range scaling: intraspecific and comparative trends. *Oecologia* 87: 343–348.

Gompper, M. E., Gittleman, J. L. & Wayne, R. K. (1997). Genetic relatedness, coalitions and social behaviour of white-nosed coatis, *Nasua narica. Animal Behaviour* 53: 781–797.

Gompper, M. E., Gittleman, J. L. & Wayne, R. K. (1998). Dispersal, phylopatry, and genetic relatedness in a social carnivore: comparing males and females. *Molecular Ecology* 7: 157–163.

Gompper M. E. & Hoylman, A. M. (1993). Grooming with trattinnickia resin – possible pharmaceutical plant use by coatis in Panama. *Journal of Tropical Ecology* 9: 533–540.

Gompper, M. E., Stacey, P. B. & Berger, J. (1997). Conservation implications of the natural loss of lineages in wild mammals and birds. *Conservation Biology* 11: 857–867.

Gompper, M. E. & Wayne, R. K. (1996). Genetic relatedness among individuals within carnivore societies. In *Carnivore Behaviour, Ecology and Evolution*, vol. 2. ed. J. L. Gittleman, pp. 429–452. Ithaca: Cornell University Press.

Gompper, M. E. & Williams, E. S. (1998). Parasites and conservation: insights from the black-footed ferret recovery program. *Conservation Biology* 12: 730–732.

Goodhart C. B. (1988). Did virus transfer from harp seals to common seals? *Nature* 336: 21.

Goodman, S. J. (1998). Patterns of extensive genetic differentiation and variation among European harbor seals (*Phoca vitulina vitulina*) revealed using microsatellite DNA polymorphisms. *Molecular Biology and Evolution* 15: 104–118.

Goodnight, K. F. & Queller, D. C. (1999). Computer software for performing likelihood tests of pedigree relationship using genetic markers. *Molecular Ecology* **8**: 1231–1234.

Gosselin, L. (1999). Too cute to be killed? *Audubon* (May–June), No. 6.

Goossens, B., Waits, L. P. & Taberlet, P. (1998). Plucked hair samples as a source of DNA: reliability of dinucleotide microsatellite genotyping. *Molecular Ecology* **7**: 1237–1241.

Gorman, M. L., Mills, M. G. L., Raath, J. P. & Speakman, J. R. (1998). High hunting costs make African wild dogs vulnerable to kleptoparasitism by hyaenas. *Nature* **391**: 479–481.

Gottelli, D. & Sillero-Zubiri, C. (1992). The Ethiopian wolf – an endangered endemic canid. *Oryx* **26**: 205–214.

Gottelli, D., Sillero-Zubiri, C., Applebaum, G. D., Roy, M. S., Girman, D. J., Garcia-Moreno, J., Ostrander, E. A. & Wayne, R. K. (1994). Molecular genetics of the most endangered canid: the Ethiopian wolf *Canis simensis*. *Molecular Ecology* **3**: 301–312.

Goude, A. (Ed.) (1994). *The Human Impact on the Natural Environment*. Cambridge, MA: MIT Press.

Government of Brazil (1998). Population statistics. World Wide Web: http://www.ibge.gov.br/english/e-home.htm.

Grachev, M. A., Kumarev, V. P., Mamaev, L. V., Zorin, V. L., Baranova, L. V., Denikina, N. N., Belikov, S. I., Petrov, E. A., Kolesnik, V. S., Kolesnik, R. S., Dorofeev, V. M., Beim, A. M., Kudelin, V. N., Nagieva, F. G. & Sidorov, V. N. (1989). Distemper virus in Baikal seals. *Nature* **338**: 209.

Graham, L. H. & Brown, J. L. (1996). Cortisol metabolism in the domestic cat and implications for developing a non-invasive measure of adrenocortical activity in nondomestic felids. *Zoo Biology* **15**: 71–82.

Graham, R. W. &. Grimm, E. C (1990). Effects of global climate change on the patterns of terrestrial biological communities. *Trends in Ecology and Evolution* **5**: 289–292.

Grakov, N. N. (1994). Kidus – a hybrid of the sable and the pine marten. *Lutreola* **3**: 1–4.

Grant, P. R. & Grant, B. R. (1992). Hybridization of bird species. *Science* **256**: 193–197.

Green, K. A. (1991). Development of a database for analysis of information about lion sightings and lion-human interactions. In *Mountain Lion–Human Interactions*, ed. C. L. Braun, pp. 18–19. Symposium of the Colorado Division for Wildlife Denver 114pp.

Greenman, J. V. & Hudson, P. J. (1999). Host exclusion and coexistence in apparent and direct competition: an application of bifurcation theory. *Theoretical Population Biology* **56**: 48–64.

Greenway, J. C. (1967). *Extinct and Vanishing Birds of the World*. New York: Dover Publications.

Greenwood, P. J. (1980). Mating systems, philopatry, and dispersal in birds and mammals. *Animal Behaviour* **28**: 1140–62.

Greiner, E. C., Roelke, M. E., Atkinson, C. T., Dubey, J. P. & Wright, S. D. (1989). *Sarcocystis* sp. in muscles of free-ranging Florida panthers and cougars (*Felis concolor*). *Journal of Wildlife Diseases* **25**: 623–628.

Grenfell, B. T. & Dobson, A. P. (Eds) (1995). *Ecology of Infectious Diseases in Natural Populations.* Cambridge: Cambridge University Press.

Grenfell, B. T. & Gulland, F. M. D. (1995). Introduction: ecological impact of parasitism on wildlife host populations. *Parasitology* 111: 3–14.

Griffith, B., Scott, J. M., Carpenter, J. W. & Reed, C. (1989). Translocation as a species conservation tool: status and strategy. *Science* 245: 477–80.

Groombridge, B. (1992). *Global Biodiversity: Status of the Earth's Living Resources.* London: Chapman & Hall.

Gros, P. M. (1998). Status of the cheetah Acinonyx jubatus in Kenya: a field-interview assessment. *Biological Conservation* 85: 137–149.

Gros, P. M., Kelly, M. & Caro, T. M. (1996). Estimating carnivore densities for conservation purposes: indirect methods compared to baseline demographic data. *Oikos* 77: 197–206.

Groves, C. P. (1976). The origin of the mammalian fauna of Sulawesi (Celebes). *Zeitschrift fur Saugetierkunde* 41: 201–216.

Guggisberg, C. A. W. (1975). *Wild Cats of the World.* New York: Taplinger Publishing Company.

Gulland, F. M. D. (1995). The impact of infectious diseases on wild animal populations – a review. In *Ecology of Infectious Diseases in Natural Populations,* eds B. T. Grenfell & A. P. Dobson, pp. 20–51. Cambridge: Cambridge University Press.

Gustafson, K. A. & Brocke, R. H. (1998). *Lynx translocation in New York: Ecology, Biopolitics and Human Impacts.* Buffalo: Wildlife Society Meetings.

Gutleb, B. (1994). Verbreitung, Situation und Schadensproblematik beim Braunbären sowie der aktuelle Stand des Bärenprojektes in Kärnten. In *Braunbären in den Ländern Alpen-Adria,* ed. M. Adamic, pp. 1233–130. Ljubliana: Ministrstvo za kmetijstvo in gozdarstvo.

Haas, L., Hofer, H., East, M., Wohlsein, P., Liess, B. & Barrett, T. (1996). Canine distemper virus infection in Serengeti spotted hyaenas. *Veterinary Microbiology* 49: 147–152.

Haber, G. C. (1996). Biological, conservation and ethical implications of exploiting and controlling wolves. *Conservation Biology* 10: 1068–1081.

Hackel, J. D. (1999). Community conservation and the future of African wildlife. *Conservation Biology* 13: 726–734.

Halfpenny, J. C., Saunders, M. R. & McGrath, K. A. (1991). Human–lion interactions. In *Mountain Lion–Human Interactions. Boulder County, Colorado: Past, Present, and Future,* ed. C. L. Braun, pp. 10–16. Denver: Symposium of the Colorado Division for Wildlife.

Hall, E. R. (1984). *Geographic Variation Among Brown and Grizzly Bears* (Ursus arctos) *in North America.* Special publication 13. Lawrence: Museum of Natural History, University of Kansas.

Hamilton, D. (1998). *Missouri River Otter Population Assessment. Final Report: 1996–97 and 1997–98 Trapping Seasons and Petition for Multi-Year Export Authority.* Missouri: Missouri Department of Conservation.

Hamilton, M. J. & Kennedy, M. L. (1987). Genic variability in the raccoon, *Procyon lotor. American Midland Naturalist* 118: 266–274.

Hamilton, P. H. (1986a). Status of the cheetah in Kenya, with reference to sub-Saharan Africa. In *Cats of the World: Biology, Conservation and Management,* eds S. D. Miller & D. D. Everett, pp. 65–76. Washington DC: National Wildlife Federation.

Hamilton, P. H. (1986b). Status of the leopard in Kenya, with reference to sub-Saharan Africa. In *Cats of the World: Biology, Conservation and Management*, eds S. D. Miller & D. D. Everett, pp. 447–457. Washington DC: National Wildlife Federation.

Hamilton, W. J., Jr. (1933). The weasels of New York. *American Midland Naturalist* 14: 289–344.

Hammond, K. A. & Diamond, J. (1997). Maximal sustained energy budgets in humans and animals. *Nature* 386: 457–462.

Hanby, J. P. & Bygott, J. D. (1979). Population changes in lions and other predators. In *Serengeti: Dynamics of an Ecosystem*, eds A. R. E. Sinclair & M. Norton-Griffiths, pp. 249–262. Chicago: University of Chicago Press.

Hanby, J. P. & Bygott, J. D. (1987). Emigration of subadult lions. *Animal Behaviour* 35: 161–169.

Hanby, J. P., Bygott, J. D. & Packer, C. (1995). Ecology, demography and behavior in lions in two contrasting habitats: Ngorongoro Crater and the Serengeti plains. In *Serengeti II: Dynamics, Conservation and Management of an Ecosystem*, eds A. R. E. Sinclair & P. Arcese, pp. 315–331. Chicago: University of Chicago Press.

Hanley, N. & Milne, J. (1996). *Ethical Beliefs and Behaviour in Contingent Valuation*. Discussion Papers in Ecological Economics, Department of Economics, University of Stirling, 96/1.

Hanni, C., Laudet, V., Stehelin, D. & Taberlet, P. (1994). Tracking the origins of the cave bear (*Ursus spelaeus*) by mitochondrial DNA sequencing. *Proceedings of the National Academy of Sciences* USA 91: 12336–12340.

Hannon, B. (1996). Adding to the Nee-May model. *Journal of Animal Ecology* 65: 850.

Hansen, B. (1999). Cat III controversy. *Colorado Daily*, May 10th 1,3.

Hanski I. (1994). A practical model of metapopulation dynamics. *Journal of Animal Ecology* 63: 151–162.

Hanski, I. A. (1997). Metapopulation dynamics: from concepts and observations to predictive models. In *Metapopulation Biology: Ecology, Genetics and Evolution*, eds I. A. Hanski & M. E. Gilpin, pp. 69–91. London: Academic Press.

Harcourt, A. H. (1998). In *Behavioural Ecology and Conservation Biology*, ed. T. M. Caro. New York: Oxford University Press.

Harcourt, A. H., Parks, S. A. & Woodroffe, R. (in press). Small reserves face a double jeopardy: small size and high surrounding human density. *Biodiversity and Conservation*.

Harder T. C. & Osterhaus, A. D. M. E. (1997). Canine distemper virus – a morbillivirus in search of new hosts? *Trends in Microbiology* 5: 120–124.

Harder T. C., Willhaus, T., Leibold, W. & Liess, B. (1992). Investigations on course and outcome of phocine distemper virus infection in harbor seals (*Phoca vitulina*) exposed to polychlorinated biphenyls: virological and serological investigations. *Journal of Veterinary Medicine Series B-Zentralblatt Fur Veterinarmedizin Reihe B-Infectious Diseases and Veterinary Public Health* 39: 19–31.

Harfenist, A. & Kaiser, G. W. (1997). Effects of introduced predators on the nesting seabirds of the Queen Charlotte Islands. In *The Ecology, Status, and Conservation of Marine and Shoreline Birds of the Queen Charlotte Islands*, eds K. Vermeer & K. H. Morgan, pp. 132–137. Occasional Paper No. 93. Ottawa: Canadian Wildlife Service.

Hargis, C. D. & Bissonette, J. A. (1997). Effects of forest fragmentation on populations of American marten in the inter-mountain west. In *Martes: Taxonomy, Ecology, Techniques and Management*, eds G. Proulx, H. Bryant & P. M. Woodard, pp. 437–451. Edmonton: Provincial Museum of Alberta.

Hargis, C. D., Bissonette, J. A. & Turner, D. L. (1999). The influence of forest fragmentation and landscape pattern on American martens. *Journal of Applied Ecology* **36**: 157–172.

Hargis, C. D. & McCullough, D. R. (1984). Winter diet and habitat selection of marten in Yosemite National Park. *Journal of Wildlife Management* **48**: 140–46.

Harrington, F. H. & Mech, L. D. (1982). An analysis of howling response parameters useful for wolf pack censusing. *Journal of Wildlife Management* **46**: 686–693.

Harris, L. D. & Gallagher, P. B. (1989). New initiatives for wildlife conservation: the need for movement corridors. In *Defense of Wildlife: Preserving Communities and Corridors*, ed. G. Mackintosh, pp. 11–34. Washington, DC: Defenders of Wildlife.

Harris, R. & Allendorf, F. (1989). Genetically effective population size of large mammals: an assessment of estimators. *Conservation Biology* **3**: 181–191.

Harris, S. (1979). Age-related fertility and productivity in red foxes, *Vulpes vulpes*, in suburban London. *Journal of Zoology, London* **187**: 195–199.

Harris, S. (1981). An estimation of the number of foxes (*Vulpes vulpes*) in the city of Bristol, and some possible factors affecting their distribution. *Journal of Applied Ecology* **18**: 455–465.

Harris, S., Cresswell, W., Reason, P. & Cresswell, P. (1992). An integrated approach to monitoring badger (*Meles meles*) population changes in Britain. In *Wildlife 2001*, eds D. R. McCullough & R. H. Barnett, pp. 945–953. New York: Elsevier Applied Science.

Harris, S. & Saunders, G. (1993). The control of canid populations. *Symposium of the Zoological Society of London* **65**: 441–464.

Harris, S. & Smith, G. C. (1987). Demography of two urban fox (*Vulpes vulpes*) populations. *Journal of Applied Ecology* **24**: 75–86.

Harris, S. & Trewhella, W. J. (1988). An analysis of some of the factors affecting dispersal in an urban fox (*Vulpes vulpes*) populations. *Journal of Applied Ecology* **25**: 409–422.

Harrison, D. J. (1986). Coyotes in the Northeast: their distribution, origin, and ecology. *Appalachia* **182**: 30–39.

Harrison, D. J. & Chapin, T. G. (1997). An assessment of potential habitat for eastern timber wolves in the northeastern United States and connectivity with occupied habitat in southeastern Canada. *Wildlife Conservation Society Working Paper* **7**: 11.

Harrison, R. L. (1993). A survey of anthropogenic ecological factors potentially affecting gray foxes (*Urocyon cinereoargenteus*) in a rural residential area. *The Southwestern Naturalist* **38**: 352–356.

Harrison, R. L. (1997). A comparison of gray fox ecology between residential and undeveloped rural landscapes. *Journal of Wildlife Management* **61**: 112–122.

Harrison, S. (1994). Metapopulations and conservation. In *Large-scale Ecology and Conservation Biology*, eds P. J. Edwards, R. M. May & N. R. Webb, pp. 111–128. Oxford: Blackwell Scientific Press.

Hart, J. A., Katembo, M. & Punga, K. (1996). Diet, prey selection and ecological relations of leopard and golden cat in the Ituri Forest, Zaire. *African Journal of Ecology* **34**: 364–379.

Harting, J. E. (1880). *A Short History of the Wolf in Britain*. Whitstable: Pryor Publications.

Hartl, D. L. & Clark, A. G. (1997). *Principles of Population Genetics*. Sunderland, MA: Sinauer Associates.

Hartwig, S. (1998). Observations of an introduction of an African wild dog (*Lycaon pictus*) from captivity into a free-ranging pack. *Advances in Ethology (Supplement to Ethology)* **33**: 81.

Harvell, C. D., Kim, K., Burkholder, J. M., Colwell, R. R., Epstein, P. R., Grimes, D. J., Hofmann, E. E., Lipp, E. K., Osterhaus, A. D. M. E., Overstreet, R. M., Porter, J. W., Smith, G. W. & Vasta, G. R. (1999). Review: marine ecology – emerging marine diseases – climate links and anthropogenic factors. *Science* **285**: 1505–1510.

Harvey, P. H., May, R. M. & Nee, S. (1994). Phylogenies without fossils. *Evolution* **48**: 523–529.

Harvey, P. H. & Pagel, M. D. (1991). *The Comparative Method in Evolutionary Biology*. Oxford: Oxford University Press.

Harwood, J. (1998). Conservation biology – what killed the monk seals? *Nature* **393**: 17–18.

Hash, H. (1987). Wolverine. In *Wild Furbearer Management and Conservation in North America*, eds M. Novak, J. A. Baker, M. E. Obbard & B. Malloch, pp. 575–585. North Bay: Ontario Trappers Association.

Hausfater, G. & Hrdy, S.B. (Eds) (1994). *Infanticide: Comparative and Evolutionary Perspectives*. New York: Aldine.

Havens, K. J. & Sharp, E. J. (1998). Using thermal imagery in the aerial survey of animals. *Wildlife Society Bulletin* **26**: 17–23.

Hedges, S. & Tyson, M. (1996). *Is Predation by Ajag (Asiatic Wild Dog, Cuon alpinus) A Threat to the Banteng* (Bos javanicus) *Population in Alas Purwo?* Review of the evidence and discussion of management solutions. Report to the Directorate General of Forest Protection and Nature Conservation, Indonesia.

Hedges, S. B., Bogart, J P. & Maxson, L R. (1992). Ancestry of unisexual salamanders. *Nature* **356**: 708–710.

Hedrick, P. W. (1995). Gene flow and genetic restoration: the Florida panther as a case study. *Conservation Genetics* **9**: 996–1007.

Hedrick, P. W. (1996). Genetics of metapopulations: aspects of a comprehensive perspective. In *Metapopulations and Wildlife Conservation*, ed. D. R. McCullough *et al.*, pp. 29–51. Washington, DC: Island Press.

Hedrick, P. W., Miller, P. S., Geffen, E. & Wayne, R. (1997). Genetic evaluation of the three captive Mexican wolf lineages. *Zoo Biology* **16**: 47–69.

Heeney, J. L., Evermann, J. F., McKeirnan, A. J., Marker-Kraus, L., Roelke, M. E., Bush, M., Wildt, D. E., Meltzer, D. G., Colly, L, Lucas, J, Manton, V. J., Caro, T. & O'Brien, S. J. (1990). Prevalence and implications of feline coronavirus infections of captive and free-ranging cheetahs (*Acinonyx jubatus*). *Journal of Virology* **64**: 1964–1972.

Heide-Jørgensen, M. P. & Härkönen, R. (1992). Epizootiology of the seal disease in the eastern North Sea. *Journal of Applied Ecology* **29**: 99–107.

Hein, E.W. (1997). Improving translocation programs. *Conservation Biology* 11: 1270–1271.

Hein, E. W. & Andelt, W. F. (1995). Evaluation of indices of abundance for an unexploited badger population. *Southwestern Naturalist* 40: 288–292.

Heisey, D. M. & Fuller, T. K. (1985). Evaluation of survival and cause-specific mortality rates using telemetry data. *Journal of Wildlife Management* 49: 668–674.

Helle, E., Olsson, M. & Jensen, S. (1976). DDT and PCB levels and reproduction in ringed seal from Bothnian Bay. *Ambio* 5: 188–189.

Hellgren, E. C. & Maehr, D. S. (1992). Habitat fragmentation and black bears in the eastern United States. *Eastern Workshop Black Bear Management and Research* 11: 154–166.

Hellgren, E. C. & Vaughan, M. R. (1989). Demographic analysis of a black bear population in the Great Dismal Swamp. *Journal of Wildlife Management* 53: 969–977.

Hemker, T. P. (1982). *Population characteristics and movement patterns of cougars in southern Utah*. MS thesis, Utah State University, Logan.

Hemker, T. P., Lindzey, F. G. & Ackerman, B. B. (1984). Population characteristics and movement patterns of cougars in southern Utah. *Journal of Wildlife Management* 48: 1275–1284.

Hemley, G. & Mills, J. A. (1999). The beginning of the end of tigers in trade? In *Riding the Tiger: Tiger Conservation in Human-dominated Landscape*, eds J. Seidensticker, S. Christie & P. Jackson, pp. 215–252. Cambridge: Cambridge University Press.

Hemmer, H., Grubb, P. & Groves, C. P. (1976). Notes on the sand cat, *Felis margarita* Loche, 1858. *Zeitschriftfuer Saugetierkunde* 41: 286–303.

Henderson, R. W. (1992). Consequences of predator introductions and habitat destruction on amphibians and reptiles in the post-Columbus West Indies. *Caribbean Journal of Science* 28: 1–10.

Hengeveld, R. & van den Bosch, F. (1996). Predicting the rate of spread of introduced animals and plants. *Wildlife Biology* 2: 151–158.

Henry, G.V. & Lucash, C.F. (1998). Red wolf status report. http://members.xoom.com/mthor/dogs/rwolfstatus.htm (21 Oct 1998).

Heptner, V. G. & Sludskii, A. A. (1992). *Mammals of the Soviet Union*, vol. II, part 2, *Carnivora (Hyaenas and Cats)*, ed. R. S. Hoffmann. (English translation). Washington, DC: Smithsonian Institution Libraries and The National Science Foundation.

Herbold, B. & Moyle, P. B. (1986). Introduced species and vacant niches. *The American Naturalist* 128: 751–760.

Hernandez M., Robinson, I., Aguilar, A., Gonzalez, L. M., LopezJurado, L. F., Reyero, M. I., Cacho, E., Franco, J., LopezRodas, V. & Costas, E. (1998). Did algal toxins cause monk seal mortality? *Nature* 393: 28–29.

Herrero, S. (Ed.) (1972). *Bears – Their Biology and Management*. Morges: IUCN.

Herrero, S. (1985). *Bear Attacks: Their Causes and Avoidance*. New York: Nick Lyons Books.

Hersteinsson, P. (1992). Demography of the arctic fox (*Alopex lagopus*) population in Iceland. In: *Wildlife 2001: Populations*, eds D. R. McCullough & R. H. Barrett. London: Elsevier Science.

Hersteinsson, P., Angerbjörn, A., Frafjord, K. & Kaikusalo, A. (1989). The arctic fox in Fennoscandia and Iceland: management problems. *Biological Conservation* **49**: 67–81.

Hersteinsson, P. & Macdonald, D. W. (1992). Interspecific competition and the geographical distribution of red and arctic foxes *Vulpes vulpes* and *Alopex lagopus*. *Oikos* **64**: 505–515.

Hess G. (1996). Disease in metapopulation models: implications for conservation. *Ecology* **77**: 1617–1632.

Hettinger, N. (1996). Enhancing natural value? *Human Ecology Review* **3**: 8–11.

Hettinger, N. (1998). The problem of finding a positive role for humans in the natural world: comments on Eric Katz's *Nature as Subject*. Meetings of the International Society for Environmental Ethics. Chicago.

Hewitt, C. G. (1921). *The Conservation of the Wild Life of Canada*. New York: Charles Scribner's Sons.

Hewitt, G. M. (1988). Hybrid zones – natural laboratories for evolutionary studies. *Trends in Ecology and Evolution* **3**: 158–166.

Hewson, R. & Kolb, H. H. (1973). Changes in the numbers and distribution of foxes (*Vulpes vulpes*) killed in Scotland from 1948–1970. *Journal of Zoology (London)* **171**: 345–365.

Hewson, R. (1984). Scavenging and predation upon sheep and lambs in west Scotland (UK). *Journal of Applied Ecology* **21**: 843–868.

Hillman, C. N. & Linder, R. L. (1973). The black-footed ferret. In *Proceedings of the Black-footed Ferret and Prairie Dog Workshop*, eds R. L. Linder & C. N. Hillman, pp. 10–23. South Dakota State University, Brookings.

Hilton, H. (1978). Systematics and ecology of the eastern coyote. In *Coyotes: Biology, Behavior and Management*, ed. M. Bekoff, pp. 209–228. New York: Academic Press.

Hoelzel, A. R. (Ed.) (1998). *Molecular Genetic Analysis of Populations: A Practical Approach*, 2nd edn. New York: Oxford University Press.

Hoelzel, A. R., Halley, J., O'Brien, S. J., Campagna, C., Arnbom, T., le Boeuf, B., Ralls, K. & Dover, G. A. (1993). Elephant seal genetic variation and the use of simulation models to investigate historical population bottlenecks. *Journal of Heredity* **84**: 443–449.

Hofer, H., Campbell, K. L. I., East, M. L. & Huish, S. (1996). The impact of game meat hunting on target and non-target species in the Serengeti. In *The Exploitation of Mammal Populations*, eds V. Taylor & N. Dunstone, pp. 117–146. London: Chapman and Hall.

Hofer, H & East, M. (1995). Population dynamics, population size and the commuting system of Serengeti spotted hyenas. In *Serengeti II: Dynamics, Management and Conservation of an Ecosystem*, eds A. R. E. Sinclair & P. Arcese, pp. 332–363. Chicago: University of Chicago Press.

Hofer, H., East, M. L. & Campbell, K. L. I. (1993). Snaring, commuting hyaenas and migratory herbivores: humans as predators in the Serengeti. *Symposia of the Zoological Society of London* **65**: 347–366.

Hofer, H., James, A. N., Gaston, K. J. & Balmford, A. (1999). Balancing the Earth's accounts. *Nature* **401**: 323–324.

Hofmeyr, G. J. G., Bester, M. N. & Jonker, F. C. (1997). Changes in the population sizes and distribution of fur seals at Marion Island. *Polar Biology* **17**: 150–158.

Hofmeyr, M., Bingham, J., Lane, E. P., Ide, A. & Nel, L. (2000). Rabies in African wild dogs (*Lycaon pictus*) in the Madikwe game reserve, South Africa. *Veterinary Record* **146**: 50–52.

Holdgate, M.W. (1986). Summary and conclusions: characteristics and consequences of biological invasions. *Philosophical Transactions of the Royal Society. Series B* **314**: 733–742.

Holekamp, K. E., Ogutu, J. O., Frank, L. G., Dublin, H. T. & Smale, L. (1993). Fission of a spotted hyena clan: consequences of prolonged female absenteeism and causes of female emigration. *Ethology* **93**: 285–299.

Holekamp, K. E. & Smale, L. (1995). Rapid change in offspring sex ratio after clan fission in the spotted hyena. *American Naturalist* **145**: 261–278.

Holt, R. D. & Pickering, J. (1985). Infectious-disease and species coexistence – a model of Lotka-Volterra form. *American Naturalist* **126**: 196–211.

Holt, R. D. & Polis, G. A. (1997). A theoretical framework for intraguild predation. *American Naturalist* **149**: 745–764.

Holt, W. V., Bennett, P. M., Volobouev, V. & Watson, P. F. (1996). Genetic resource banks in wildlife conservation. *Journal of Zoology (London)* **238**: 531–544.

Hong, C.-S., Calambokidis, J., Bush, B., Steiger, G. H. & Shaw, S. (1996). Polychlorinated biphenyls and organochlorine pesticides in harbor seal pups from the inland waters of Washington state. *Environmental Science and Technology* **30**: 837–844.

Hoogesteijn, R., Hoogesteijn, A. & Mondolfi, E. (1993). Jaguar predation and conservation: cattle mortality caused by felines on three ranches in the Venezuelan llanos. *Symposia of the Zoological Society of London* **65**: 391–407.

Hook, J. C. & Robinson, W. L. (1982). Attitudes of Michigan citizens toward predators. In *Wolves of the World: Perspective on Behavior, Ecology, and Conservation*, eds H. Harrington & P. C. Paquet, pp. 382–394. New Jersey: Park Ridge.

Hope, J. H. (1973). Mammals of the Bass Straights Islands. *Proceedings of the Royal Society, Victoria* **85**: 163–196.

Hornocker, M. G. (1970). An analysis of mountain lion predation upon mule deer and elk in the Idaho Primitive Area. *Wildlife Monographs* **21**: 1–39.

Hornocker, M. G. & Hash, H. S. (1981). Ecology of the wolverine in northwestern Montana. *Canadian Journal of Zoology* **59**: 1286–1301.

Höss, M., Kohn, M., Pääbo, S., Knauer, F. & Schröder, W. (1992). Excrement analysis by PCR. *Nature* **359**: 199.

Houji, L. & Heli, S. (1986). The status and population fluctuation of the leopard cat in China. In *Cats of the World*, eds S. D. Miller & E. D. Everett, pp. 59–62. Washington DC: National Wildlife Federation.

Howard, J. G. (1998). Assisted reproductive techniques in nondomestic carnivores. In *Zoo & Wild Animal Medicine, Current Therapy*, eds M. E. Fowler & R. E. Miller, pp. 449–457. Philadelphia: W. B. Saunders.

Howard, J. G., Brown, J. L., Bush, M. & Wildt, D. E. (1990). Teratospermic and normospermic domestic cats: ejaculate traits, pituitary-gonadal hormones, and improvement of spermatozooal motility and morphology after swim-up processing. *Journal of Andrology* **11**: 204–215.

Howard, J. G., Bush, M. & Wildt, D. E. (1991). Teratospermia in domestic cats compromises penetration of zona-free hamster ova and cat zonae pellucidae. *Journal of Andrology* **12**: 36–45.

Howard, J. G., Donoghue, A. M., Johnston, L. A. & Wildt, D. E. (1993). Zona pellucida filtration of structurally abnormal spermatozoa and reduced fertilization in teratospermic cats. *Biology of Reproduction* **49**: 131–139.

Howard, J. G., Wolf, K. N., Marinari, P. E., Kreeger, J. S., Anderson, T. R., Vargas, A. & Wildt, D. E. (1999). Delayed onset of sperm production in 1-year old male black-footed ferrets. *Proceedings of the Society for the Study of Reproduction, Biology of Reproduction Supplement*, abstract 170.

Howard, J. G., Wolf, K., Vargas, A., Marinari, P., Kreeger, J., Williamson, L. & Wildt, D. E. (1997). Enhanced reproductive efficiency and pregnancies after artificial insemination in black-footed ferrets. *Proceedings of the American Association of Zoo Veterinarians*, pp. 351–352.

Howard, P. C., Viskanic, P., Davenport, T. R. B., Kigenyi, F. W., Baltzer, M., Dickinson, C. J., Lwanga, J. S., Matthews, R. A. & Balmford, A. (1998). Complementarity and the use of indicator groups for reserve selection in Uganda. *Nature* **394**: 472–475.

Hoyert, D. L., Kochanek, K. D., Murphy, S. L. (1999). *Deaths: Final Data for 1997. National Vital Statistics Reports* 47 (19). Atlanta: Center for Disease Control and Prevention.

Hubbard, A. L., McOrist, S., Jones, T. W., Boid, R., Scott, R. & Easterbee, N. (1992). Is survival of European wildcats (*Felis sylvestris*) in Britain threatened by interbreeding with domestic cats? *Biological Conservation* **61**: 203–208.

Huber, T. & Kaczensky, P. (1998). The situation of the lynx in Austria. *Hystrix* **10**: 43–54.

Hudson, W. E. (Ed.) (1993). *Buildling economic incentives into the Endangered Species Act.* Washington DC: Defenders of Wildlife.

Hudson, W. H. (1895). *The Naturalist in La Plata.* London: Chapman and Hall.

Hulme, D. & Infield, M. (2000) From criminals to citizens? Community conservation at Lake Mburo National Park, Uganda. In *African Wildlife and African Livelihoods: The Promise and Performance of Community Conservation*, eds D. Hulme & M. Murphree, Oxford: James Currey.

Human, K. (1999). Return of the missing lynx. *Boulder Camera*, 10 January, 1A, 10A.

Humphrey, S. R. & Zinn, T. L. (1982). Seasonal habitat use by river otters and Everglades mink in Florida. *Journal of Wildlife Management* **46**: 375–381.

Hunt, R. M., Jr. (1996). Biogeography of the order Carnivora. In *Carnivore Behavior, Ecology, and Evolution*, vol. 2, ed. J. L. Gittleman, pp. 485–541. Ithaca: Cornell University Press.

Hunter, L. (1998). *The Behavioural Ecology of Reintroduced Lions and Cheetahs in the Phinda Reserve, Northern KwaZulu-Natal, South Africa.* Doctoral thesis, University of Pretoria.

Huntley, B. (1995). Plant species' response to climate change: implications for the conservation of European birds. *Ibis* 137 (Suppl. 1). S127–138.

India Census Commissioner (1901). *Census of India.* Government of India.

India Census Commissioner (1941). *Census of India.* Government of India.

India Network Foundation (1999). States of India. World Wide Web: http://www.indianetwork.org/res/states.html

Infield, M. & Adams, W. M. (1999). Institutional sustainability and community conservation: a case study from Uganda. *Journal of International Development* **11**: 305–315.

Inoue, T. (1994). [Searching the Japanese sea lions.] *Seton* 3: 8–13. (In Japanese.)

Iriarte, J. A. & Jaksic, F. M. (1986). The fur trade in Chile: an overview of seventy-five years of export data (1910–1984). *Biological Conservation* 38: 243–253.

Istat (1999). Italy in figures. World Wide Web: http://www.istat.it.

Istituto Nacional de Estadística (1999). España en Cifras 1998. World Wide Web: http://www.ine.es

IUCN (1988). *Significant Trade in Wildlife: A Review of Selected Species in CITES Appendix II*, vol. 1. *Mammals*. Gland: IUCN.

IUCN / UNEP / WWF (1991). *Caring for the Earth. A Strategy for Sustainable Development*. Gland: IUCN.

IUCN Species Survival Commission (1994). *IUCN Red List Categories*. Gland: IUCN.

IUCN/SSC Re-introduction Specialist Group (1998). *IUCN Guidelines for Re-introductions*. Gland: IUCN.

Iverson, J. B. (1978). The impact of feral cats and dogs on populations of the West Indian rock iguana, *Cyclura carinata*. *Biological Conservation* 14: 63–73.

Jablonski, D. (1995). Extinctions in the fossil record. In *Extinction Rates*, eds J. H. Lawton & R. M. May, pp. 25–44. Oxford: Oxford University Press.

Jackson, J. (1997). Reefs since Columbus. *Coral Reefs* 16: 23–32.

Jackson, J.A. (1977). Alleviating problems of competition, predation, parasitism, and disease in endangered birds. In *Endangered birds*, ed. S. A. Temple, pp. 75–84. Madison: University of Wisconsin Press and Croom Helm.

Jackson, P. (1996). *The Tiger Re-introduction Project of the Heilongjiang Felid Breeding Centre and Harbin Siberian Tiger Park. An Assessment and Recommendations*. Gland: IUCN.

Jackson, R. & Ahmad, A. (Eds) (1997). *Proceedings of the Eighth International Snow Leopard Symposium*. Islamabad: International Snow Leopard Trust and World Wide Fund for Nature.

Jalkotzy, M. G., Ross, P. I. & Gunson, R. R. (1992). *Management Plan for Cougars in Alberta*. *Wildlife Management Planning Series* No. 5. Edmonton: Alberta Forestry, Lands, and Wildlife.

James, A. N., Gaston, K. J. & Balmford, A. (1999). Balancing the Earth's accounts, *Nature*, 401, 323–324.

Jamieson, D. (1998). Animal liberation is an environmental ethic. *Environmental Values* 7: 41–57.

Jamieson, D. & Bekoff, M. (1996). Ethics and the study of animal cognition. In *Readings in Animal Cognition*, eds M. Bekoff & D. Jamieson, pp. 359–371. Cambridge MA: MIT Press.

Janis, C. M., Baskin, J. A., Berta, A., Flynn, J. J., Gunnell, G. F., Hunt, R. M. Jr., Martin, L. D. & Munthe, K. (1998). Carnivorous mammals. In *Evolution of Tertiary Mammals of North America*, eds C. M. Janis, K. M. Scott & L. L. Jacobs, pp. 73–90. Cambridge: Cambridge University Press.

Jedrzejewski, B. Jedrzejewski, W. (1998). *Predation in Vertebrate Communities: The Bialowieza Primeval Forest as a Case Study*. Berlin: Springer-Verlag.

Jedrzejewski, W., Jedrzejewski, B. & Szymura, L. (1995). Weasel population response, home range, and predation on rodents in a deciduous forest in Poland. *Ecology* 76: 179–195.

Jenks, S. M. & Wayne, R. K. (1992). Problems and policy for species threatened by hybridization: the red wolf as a case study. In *Wildlife 2001: Populations*, eds D. R. McCullough & R. H. Barrett, pp. 237–251. London: Elsevier Science Publishers.

Jenkins, S.R., Perry, B. D. & Winkler, W. G. (1998). Ecology and epidemiology of raccoon rabies. *Reviews of Infectious Diseases* **10**: 620–625.

Jensen, W. F., Fuller, T. K. & Robinson, W. L. (1986). Wolf (*Canis lupus*) distribution on the Ontario–Michigan border near Sault Ste. Marie. *Canadian Field-Naturalist* **100**: 363–366.

Jhala, Y. V. & Giles, R. H. (1991). The status and conservation of the wolf in Gujarat and Rajasthan, India. *Conservation Biology* **5**: 476–483.

Jimenez, J. A., Hughes, K. A., Alaks, G., Graham, L. & Lacy, R. C. (1994). An experimental study of inbreeding depression in a natural habitat. *Science* **266**: 271–273.

Johnsingh, A.J.T. (1985). Distribution and status of dhole *Cuon alpinus* Pallas 1811 in South Asia. *Mammalia* **49**: 203–208.

Johnson, K. G. & Pelton, M. R. (1981). A survey of procedures to determine relative abundance of furbearers in the southeastern United States. *Proceedings of the Annual Conference of the Southeastern Association of Fish and Wildlife Agencies* **35**: 261–272.

Johnson, M. K., Aldred, D. R. & Martin, T. E. (1981). Feces, bile acids and furbearers. In *Proceedings of the Worldwide Furbearer Conference*, Frostburg, Maryland, eds J. A. Chapman & D. Pursley, pp. 1143–1150.

Johnson, M. R., Boyd, D. K. & Pletscher, D. H. (1994). Serologic investigations of canine parvovirus and canine-distemper in relation to wolf (*Canis lupus*) pup mortalities. *Journal of Wildlife Diseases* **30**: 270–273.

Johnson, N. F. & Holloran, D. F. (1985). Reproductive activity of Kansas bobcats. *Journal of Wildlife Management* **49**: 42–46.

Johnson, S. A. (1984). *Home range, Movements, and Habitat use of Fishers in Wisconsin*. MS thesis, University of Wisconsin, Stevens Points. 78 pp.

Johnson, W. E., Culver, M., Iriarte, J. A., Eizirik, E., Seymour, K. & O'Brien, S. J. (1998). Tracking the elusive Andean Mountain Cat (*Oreailurus jacobita*) from mitochondrial DNA. *Journal of Heredity* **89**: 227–232.

Johnson, W. E. & Franklin, W. L. (1994a). Conservation implications of South American Grey fox (*Dusicyon griseus*) socioecology in the Patagonia of southern Chile. *Vida Silvestre Neotropical* **3**: 16–23.

Johnson, W. E. & Franklin, W. L. (1994b). Spatial resource partitioning by sympatric grey fox (*Dusicyon griseus*) and culpeo fox (*Dusicyon culpaeus*) in southern Chile. *Canadian Journal of Zoology* **72**: 1788–1793.

Johnson, W. E., Fuller, T. K. & Franklin. W. L. (1996). Sympatry in canids: a review and assessment. In *Carnivore Behavior, Ecology, and Evolution*, ed. J. L. Gittleman, pp. 189–218. Ithaca: Cornell University Press.

Johnson, W. E. & O'Brien, S. J. (1997). Phylogenetic reconstruction of the Felidae using 16S rRNA and NADH-5 mitochondrial genes. *Journal of Molecular Evolution* **44**: S98–S116.

Johnson, W. E., Pecon Slattery, J., Eizirik, E., Kim, J., Menotti Raymond, M., Bonacic, C., Cambre, R., Crawshaw, P., Nunes, A., Seuanez, H., Moreira, M. A., Seymour, K. L., Simon, F., Swanson, W. & O'Brien, S. J. (1999a). Disparate phylogeographic patterns of mitochrondrial DNA variation in four closely related South American small cat species. *Molecular Ecology* **8**: S79–S94.

Johnson, W. E., Shinyashiku, F., Menoti, M., Raymond, M., Driscoll, C., Leh, C., Sunquist, M., Johnson, L, Bush, M., Wildt, D., Yuhki, N. & O'Brien, S. J. (1999b). Molecular genetic characterization of two insular Asian cat species, Bornean bay cat and Iriomote cat. In *Evolutionary Theory and Processes: Modern Perspecitives*, ed. S. P. Wasser, pp. 223–248. Klewar Academic Publishers.

Johnston, J. J., Windberg, L. A., Furcolow, C. A., Engeman, R. M. & Roetto, M. (1998). Chlorinated benzenes as physiological markers for coyotes. *Journal of Wildlife Management* 62: 410–421.

Johnstone, G. W. (1985). Threats to birds on subantarctic islands. In *Conservation of Island Birds: Case Studies for the Management of Threatened Island Species*, ed. P. Moors, pp. 101–121. Technical Publication 3. Cambridge: International Council for Bird Preservation.

Jolly, G. M. (1982). Mark-recapture models with parameters constant in time. *Biometrics* 38: 301–321.

Jones, E. (1977). Ecology of the feral cat *Felis catus* (L.), (Carnivora, Felidae) on Macquarie Island. *Australian Wildlife Research* 4: 249–262.

Jones, E. (1990). Physical characteristics and taxonomic status of wild canids, *Canis familiaris*, from the eastern highlands of Victoria. *Australian Wildlife Research* 17: 69–81.

Jordan, P. A., Shelton, P. C. & Allen, D. L. (1967). Numbers, turnover, and social structure of the Isle Royale wolf population. *American Zoologist* 7: 233–252.

Jorgensen, J. P. & Redford, K. H. (1993). Humans and big cats as predators in the Neotropics. *Symposium Zoological Society of London* 65: 367–390.

Joshi, M. N. V. & Gadgil, M. (1991). On the role of refugia in promoting prudent use of biological resources. *Theoretical Population Biology* 40: 211–229.

Joslin, P. (1982). Status of the Caspian tiger in Iran. In *Cats of the World: Biology, Conservation, and Management*, eds S. D. Miller & D. D. Everett, p. 63. Washington, DC: National Wildlife Federation.

Junge, R. E., Miller, R. E., Boever, W. J., Scherba, G. & Sunberg, J. (1991). Persistant cutaneous ulcers associated with feline herpesvirus type I infection in a cheetah. *Journal of American Veterinary Medical Association* 198: 1057–1058.

Kaczensky, P. (1998). Status and distribution of the lynx in the German Alps. *Hystrix* 10: 39–42.

Kaikusalo, A. & Angerbjörn, A. (1995). The arctic fox population in Finnish Lapland during 30 years, 1964–1993. *Annales Zooloogici Fennici* 32: 69–77.

Kajimura, H. (1984). *Opportunistic Feeding of the Northern Fur Seal*, Callorhinus ursinus, *in the Eastern North Pacific Ocean and Eastern Bering Sea*. U.S. Dept. of Commerce, NOAA Technical Report NMFS SSRF-779.

Kangwana, K. & Ole Mako, R. (2000). The impact of community conservation initiatives around Tarangire National Park, Tanzania (1992–1997). In *African Wildlife and African Livelihoods: The Promise and Performance of Community Conservation*, eds D. Hulme & M. Murphree, pp. Oxford: James Currey.

Kappe, A. L., Bijlsma, R., Osterhaus, A., Van-Delden, W. & Van De Zande, L. (1997). Structure and amount of genetic variation at minisatellite loci within the subspecies complex of *Phoca vitulina* (the harbour seal). *Heredity*. 78: 457–463.

Karanth, K. U. (1987). Status of wildlife and habitat conservation in Karnataka. *Journal of the Bombay Natural History Society Supplement* 83: 166–179.

Karanth, K. U. (1991). Ecology and management of the tiger in tropical Asia. In

Wildlife Conservation: Present Trends and Perspectives for the 21st Century, eds N. Maruyama, B. Bobek, Y. Ono, W. Reglin, L. Bartos & R. Ratcliffe, pp. 156–159. Tokyo: Japan Wildlife Research Centre.

Karanth, K. U. & Stith, B. M. (1999). Prey depletion as a critical determinant of tiger population viability. In *Riding the Tiger: Tiger Conservation in Human-Dominated Landscapes*, eds J. Seidensticker, S. Christie & P. Jackson, pp. 100–113. Cambridge: Cambridge Univesity Press.

Karanth, K. U. & Sunquist, M. E. (1992). Population structure, density and biomass of large herbivores in the tropical forests of Nagarahole, India. *Journal of Tropical Ecology* **8**: 21–35.

Karanth, K. U. & Sunquist, M. E. (1995). Prey selection by tiger, leopard and dhole in tropical forests. *Journal of Animal Ecology* **64**: 439–450.

Karanth, K. U., Sunquist, M. & Chinnappa, K. M. (1999). Long-term monitoring of tigers: lessons from Nagarahole. In *Riding the Tiger: Tiger Conservation in Human-Dominated Landscape*, eds J. Seidensticker, S. Christie & P. Jackson, pp. 114–122. Cambridge: Cambridge University Press.

Kareiva, P. (1996). Developing a predictive ecology for non-indigenous species and ecological invasions. *Ecology* **77**: 651–1652.

Kat, P.W., Alexander, K. A., Smith, J. S. & Munson, L. (1995). Rabies and African wild dogs in Kenya. *Proceedings of the Royal Society of London, Series B* **262**: 229–233.

Katnik, D. D. (1992). *Spatial Use, Territoriality, and Summer–Autumn Selection of Habitat in an Intensively Harvested Population of Martens on Commercial Forestland in Maine*. Unpublished M.S. thesis, University of Maine, Orono.

Katz, E. (1996). *Nature as Subject: Human Obligation and Natural Community*. Lanham, MD: Rowman and Littlefield.

Kaufman, J. H. (1987). Ringtail and coati. In *Wild Furbearer Management and Conservation in North America*, eds M. Novak, J. A. Baker, M. E. Obbard & B. Malloch, pp. 501–508. North Bay: Ontario Trappers Association.

Kaufman, J. H., Lanning, D. V. & Poole, S. E. (1976). Current status and distribution of the coati in the United States. *Journal of Mammalogy* **57**: 621–637.

Kauhala, K. (1995). Changes in distribution of the European badger *Meles meles* in Finland during the rapid colonization of the raccoon dog. *Annales Zoologici Fennici* **32**: 183–191.

Kauhala, K. (1996). Introduced carnivores in Europe with special reference to central and northern Europe. *Wildlife Biology* **2**: 197–204.

Kauhala, K., Kaunisto, M. & Helle, E. (1993). Diet of the raccoon dog, *Nyctereutes procyonoides*, in Finland. *Zeitschrift fur Saugetierkunde* **58**: 129–136.

Kauhala, K., Laukkanen, P. & von Rége, I. (1998). Summer food composition and food niche overlap of the raccoon dog, red fox and badger in Finland. *Ecography* **21**: 457–463.

Keet D. F., Kriek, N. P. J., Penrith, M. L., Michel, A. & Huchzermeyer, H. (1996). Tuberculosis in buffaloes (*Syncerus caffer*) in the Kruger National Park: Spread of the disease to other species. *Onderstepoort Journal of Veterinary Research* **63**: 239–244.

Keith, L. B. (1983). Population dynamics of wolves. In *Wolves in Canada and Alaska*, ed. L. N. Carbyn, pp. 66–77. *Canadian Wildlife Services Report Series* 45.

Keitt, T. H., Urban, D. L. & Milne, B. T. (1997). Detecting critical scales in fragmented landscapes. *Conservation Ecology* 1(1): 4.

Keller, L. F., Arcese, P., Smith, J. N., Hochachka, W. M. & Stearns, S. C. (1994). Selection against inbred song sparrows during a natural population bottleneck. *Nature* 372: 356–357.

Kellert, S. R. (1985). Public perceptions of predators, particularly the wolf and coyote. *Biological Conservation* 31: 167–189.

Kellert, S. R. (1996). *The Value of Life: Biological Diversity and Human Society.* Washington, DC: Island Press.

Kellert, S. R., Black, M., Rush, C. R. & Bath, A. J. (1996). Human culture and large carnivore conservation in North America. *Conservation Biology* 10: 977–990.

Kellert, S. R. & Clark, T. W. (1991). The theory and application of a wildlife policy framework. In *Public Policy Issues in Wildlife Management*, ed. W. R. Mangun, pp. 17–36. New York: Greenwood Press.

Kelly, B. & Phillips, M. (2000). Red wolf. In *Endangered Species: Conflict and Context.* eds R. P. Reading & B. J. Miller, Westport, CT: Greenwood Press.

Kelly, B.T. (1999). Red wolves 12 years after re-introduction: managing hybridization, North Carolina, USA. *Re-introduction News* 18: 6–7.

Kendall, K. C., Metzgar, L. H., Patterson, D. A. & Steele, B. M. (1992). Power of sign surveys to monitor population trends. *Ecological Applications* 2: 422–430.

Kendall, M. & Stuart, A. (1977). *The Advanced Theory of Statistics.* New York: Macmillan.

Kennedy, P. K., Kennedy, M. L., Clarkson, P. L. & Liepins, I. S. (1990). Genetic variability in wild populations of the gray wolf, Canis lupus. *Canadian Journal of Zoology* 69: 1183–1188.

Kenney, J. S., Starfield, A. M. & McDougal, C. W. (1995). The long-term effects of tiger poaching on population viability. *Conservation Biology* 9: 1127–1133.

Kenya Game Department (1958). *Colony and Protectorate of Kenya – Game Department Annual Report.* Nairobi: Game Department.

Kershaw, M., Williams, P. H. & Mace, G. M. (1994). Conservation of Afrotropical antelopes: consequences and efficiency of using different site selection methods and diversity criteria. *Biological Conservation* 3: 354–72.

Kessler, W. B. & Eastland, W. G. (1995). Strategies to sustain human and wildlife communities. In *Integrating People and Wildlife for a Sustainable Future*, eds J. A. Bissonette & P. R. Krausmann, pp. 1–3. Bethesda, MA: The Wildlife Society.

Keymer A. E. & Read, A. F. (1991). Behavioural ecology: the impact of parasitism. In *Parasite–Host Associations*, eds C. A. Toft, A. Aeschlimann & L. Bolis, pp. 37–61. Oxford: Oxford University Press.

Keyser, A. J., Hill, G. E. & Soehren, E.C. (1998). Effects of forest fragment size, nest density, and proximity to edge on the risk of predation to ground-nesting passerine birds. *Conservation Biology* 12: 986–94.

Kholkute, S. D., Meherji, P. & Puri, C. P. (1992). Capacitation and the acrosome reaction in sperm from men with various semen profiles monitored by a chlortetracycline fluorescence assay. *International Journal of Andrology* 15: 43–53.

Kiesecker J. M., Skelly, D. K., Beard, K. H. & Preisser, E. (1999). Behavioral reduction of infection risk. *Proceedings of the National Academy of Science of the United States of America* 96: 9165–9168.

Kim, G. B., Tanabe, S., Tatsukawa, R., Loughlin, T. R. & Shimazaki, K. (1996). Characteristics of butyltin accumulation and its biomagnification in Steller sea lions (*Eumetopias jubatus*). *Environmental Toxicology and Chemistry* 15: 2043–2048.

King, C. (1984). *Immigrant Killers: Introduced Predators and the Conservation of Birds in New Zealand*. Auckland: Oxford University Press.

King, C. M. (1981). The reproductive tactics of the stoat (*Mustela erminea*) in New Zealand forests. In *Worldwide Furbearer Conference Proceedings*, eds J. A. Chapman & D. Pursley, pp. 443–468. Frostburg, Maryland.

King, C. M. (1983a). Factors regulating mustelid populations. *Acta Zoologica Fennica* 174: 217–220.

King, C. M. (1983b). The relationships between beech (*Nothofagus* sp.) seedfall and populations of mice (*Mus musculus*), and the demographic and dietary responses of stoats (*Mustela erminea*) in three New Zealand forests. *Journal of Animal Ecology* 52: 141–166.

King, C. M. (1989). *The Natural History of Weasels and Stoats*. Ithaca: Cornell University Press.

King, C. M. (1991). Age-specific prevalence and a possible transmission route for skrjabingylosis in New Zealand stoats, *Mustela erminea*. *New Zealand Journal of Ecology* 15: 23–30.

King, C. M. & Edgar, R. L. (1977). Techniques for trapping and tracking stoats (*Mustela erminea*): a review and a new system. *New Zealand Journal of Zoology* 4: 193–212.

King, W. B. (1985). Island birds: will the future repeat the past? In *Conservation of Island Birds: Case Studies for the Management of Threatened Island Species*, ed. P. Moors, pp. 3–15. Technical Publication 3. Cambridge: International Council for Bird Preservation.

Kingdon, J. (1979). *East African Mammals: An Atlas of Evolution in Africa*. London: Academic Press.

Kingdon, J. (1997). *The Kingdon Field Guide to African Mammals*. London: Academic Press.

Kinnear, J. E., Onus, M. L. & Bromilow, R. N. (1988). Fox control and rock-wallaby population dynamics. *Australian Wildlife Research* 15: 435–450.

Kirkwood, J. (1992). Wild animal welfare. In *Animal Welfare and the Environment*, eds D. Ryder & P. Singer, pp. 139–154. London: Duckworth.

Kitchen, A. M., Gese, E. M. & Schauster, E. R. (1999). Resource partitioning between coyotes and swift foxes: space, time, and diet. *Canadian Journal of Zoology* 77: 1645–1656.

Kitchener, A. C. (1999). Tiger distribution, phenotypic variation and conservation issues. In *Riding the Tiger: Tiger Conservation in Human-Dominated Landscape*, eds J. Seidensticker, S. Christie & P. Jackson, pp. 19–39. Cambridge: Cambridge University Press.

Kleiman, D. G. (1989). Reintroduction of captive mammals for conservation. *BioScience* 39: 152–61.

Kleiman, D. G., Reading, R. P., Miller, B. J., Clark, T. W., Scott, J. M., Robinson, J., Wallace, R. L., Cabin, R. & Felleman, F. (in press). Improving the evaluation of conservation programs. *Conservation Biology*.

Kleiman, D. G., Stanley Price, M. R. & Beck, B. B. (1994). Criteria for reintroductions. In *Creative Conservation. Interactive Management of Wild and Captive Animals*, eds P. J. S. Olney, G. M. Mace & A. T. C. Feistner, pp. 287–303. London:

Chapman & Hall.

Klein, D. R. (1968). The introduction, increase, and crash of reindeer on St. Matthew Island. *Journal of Wildlife Management* **32**: 350–367.

Klein, J. (1986). Natural history of the major histocompatibility complex. New York: Wiley.

Kloor, K. (1999). Lynx and biologists try to recover after disastrous start. *Science* **285**: 320–321.

Knaus, R. M., Kinler, N. & Linscombe, R. G. (1983). Estimating river otter populations: the feasibility of 65Zn to label feces. *Wildlife Society Bulletin* **11**: 375–377.

Knick, S. T. (1990). Ecology of bobcats relative to exploitation and a prey decline in southeastern Idaho. *Wildlife Monographs* **108**: 1–42.

Knight, R. R., Blanchard, B. M. & Eberhardt, L. L. (1988). Mortality patterns and population sinks for Yellowstone grizzly bears, 1973–1985. *Wildlife Society Bulletin* **16**: 121–125.

Knowlton, F. F. (1972). Preliminary interpretations of coyote population mechanics with some management implications. *Journal of Wildlife Management* **36**: 369–382.

Knowlton, F. F. (1984). *Feasibility of Assessing Coyote Abundance on Small Areas. Final Report*, Work Unit 909: 01. Denver: Denver Wildlife Research Center.

Knowlton, F. F., Savarie, P. J., Wahlgren, C. E. & Hayes, D. H. (1988). Retention of physiological marks by coyotes ingesting baits containing iophenoxic acid, mirex, and rhodamine B. *Vertebrate Pest Control and Management Materials* **5**: 141–147.

Kock, R., Chalmers, W. S. K., Mwanzia, J., Chillingworth, C., Wambua, J., Coleman, P. G. & Baxendale, W. (1998). Canine distemper antibodies in lions of the Masai Mara. *Veterinary Record* **142**: 662–665.

Koehler, G. M. & Hornocker, M. G. (1991). Seasonal resource use among mountain lions, bobcats, and coyotes. *Journal of Mammalogy* **72**: 391–396.

Koehler, G. M. & Aubry, K. B. (1994). Lynx. In *The Scientific Basis for Conserving Forest Carnivores: American Marten, Fisher, Lynx, and Wolverine in the Western United States*, eds L. F. Ruggiero, K. B. Aubry, S. W. Buskirk, L. J. Lyon & W. J. Zielinski, pp. 74–98. General Technical Report RM-254. Fort Collins, CO: US Department of Agriculture, Forest Service, & Rocky Mountain Forest and Range Experimental Station.

Koenig, W. D., Van Vuren, D. & Hooge, P. N. (1996). Detectability, philopatry and the distribution of dispersal distances in vertebrates. *Trends in Ecology and Evolution* **11**: 514–517.

Koepfli, K.-P. & Wayne, R. K. (1998). Phylogenetic relationships of otters (*Carnivora: Mustelidae*) based on mitochondrial cytochrome b sequences. *Journal of Zoology (London)* **246**: 401–416.

Koford, C. B. (1976). Latin American cats: economic values and future prospects. In *The World's Cats: Contributions to Status, Management and Conservation*, ed. R. L. Eaton, pp. 79–88. Seattle: Carnivore Research Institute, University of Washington.

Kohn, B. E. (1982). Status and management of black bears in Wisconsin. *Wisconsin Department of Natural Resources Technical Bulletin* No. 129.

Kohn, M., Knauer, F., Stoffella, A., Schröder, W. & Pääbo, S. (1995). Conservation genetics of the European brown bear – a study using excremental PCR of nuclear and mitochondrial sequences. *Molecular Ecology* 4: 95–103.

Kohn, M. H. & Wayne, R. K. (1997). Facts from feces revisited. *Trends in Ecology and Evolution* 12: 223–227.

Kohn, M. H., York, E. C., Kamradt, D. A., Haught, G., Sauvajot, R. M. & Wayne, R. K. (1999). Estimating population size by genotyping faeces. *Proceedings of the Royal Society of London, Series B* 266: 657–663.

Kolenosky, G. B. (1987). Polar bear. In *Wild Furbearer Management and Conservation in North America*, eds M. Novak, J. A. Baker, M. E. Obbard & B. Malloch, pp. 475–485. North Bay: Ontario Trappers Association.

Kolenosky, G. B. & Strathearn, S. M. (1987). Black bear. In *Wild Furbearer Management and Conservation in North America*, eds M. Novak, J. A. Baker, M. E. Obbard & B. Malloch, pp. 443–454. North Bay: Ontario Trappers Association.

Korb, J. (2000). Methods to study elusive spotted hyenas in the Comoe National Park, Cote d'Ivoire. *Hyena Specialist Group Newsletter* No. 7. Gland: IUCN.

Korpimaki, E., Norrdahl, K. & Rinta-Jaskari, T. (1991). Response of stoats and least weasels to fluctuating food abundances: is the low phase of the vole cycle due to mustelid predation? *Oecologia* 88: 552–561.

Kovacs, K. M., Lydersen, L., Hammill, M. O., White, B. N., Wilson, P. J. & Malik, S. (1997). A harp seal x hooded seal hybrid. *Marine Mammal Science* 13: 460–468.

Kranz, A. (2000). Otters, *Lutra lutra* increasing in Central Europe: from the threat of extinction to locally perceived over-population. *Mammalia*, 64: 357–368.

Kranz, A. & Toman, A. (2000). Otters recovering in man-made habitats in central Europe. In *Mustelids in a Modern World: Management and Conservation Aspects of Small Carnivore–Human Interactions*, ed. H. I. Griffiths, pp. 163–183. Backhuys Publishing.

Krebs, C. J. (1985). *Ecology: The Experimental Analysis of Distribution and Abundance*. New York: Harper & Row.

Krebs J. R., Anderson, R. M., Clutton-Brock, T., Donnelly, C. A., Frost, S., Morrison, W. I., Woodroffe, R. & Young, D. (1998). Policy: biomedicine – badgers and bovine TB. Conflicts between conservation and health. *Science* 279: 817–818.

Krebs, J. R., Anderson, R., Clutton-Brock, T., Morrison, I., Young, D., Donelly, C., Frost, S. & Woodroffe, R. (1997). *Bovine Tuberculosis in Cattle and Badgers*. London: H.M.S.O.

Kretzmann, M. B., Gilmartin, W. G., Mayer, A., Zegers, G. P., Fain, S. R.,Taylor, B. F. & Costa, O. P. (1997). Low genetic variability in the Hawaiian Monk Seal. *Conservation Biology* 11: 482–490.

Kruuk, H. (1972a). Surplus-killing by carnivores. *Journal of Zoology (London)* 166: 233–244.

Kruuk, H. (1972b). *The Spotted Hyena*. Chicago: University of Chicago Press.

Kruuk, H. (1976a). Carnivores and conservation. In *Proceedings of a Symposium on Endangered Wildlife on Southern Africa*, pp. 1–13. Pretoria: The Endangered Wildlife Trust and University of Pretoria.

Kruuk, H. (1976b). Feeding and social behavior of the striped hyena (*Hyena vulgaris* Desmarest). *East African Wildlife Journal* 14: 91–111.

Kruuk H. (1978). Spacial organisation and territorial behaviour of the European badger, *Meles meles*. *Journal of Zoology* 184: 1–19.

Kruuk, H. (1980). *The Effect of Large Carnivores on Livestock and Animal Husbandry in*

Marsabit District, Kenya. IPAL Technical Report E – 4. Integrated Project on Arid Lands. UNEP & MAB (Man and the Biosphere Programme).

Kruuk, H. (1995). *Wild otters: Predation and Populations*. New York: Oxford University Press.

Kruuk, H., Carss, D. N., Conroy, J. W. H., & Durbin, L. (1993). Otter (*Lutra lutra L.*) numbers and fish productivity in north-east Scotland. *Symposium of the Zoological Society of London* **65**: 171–191.

Kruuk, L., Clutton-Brock, T. H., Albon, S. D., Pemberton, J. M. & Guinness, F. E. (1999). Population density affects sex ratio variation in red deer. *Nature* **399**: 459–461.

Kruuk, H. & Conroy, J. W. H. (1996). Concentrations of some organochlorines in otters (*Lutra lutra L.*) in Scotland: implications for populations. *Environmental Pollution* **92**: 165–171.

Kruuk, H., Conroy, J. W. H., Glimmerveen, U. & Ouwerkerk, E. J. (1986). The use of spraints to survey populations of otters *Lutra lutra*. *Biological Conservation* **35**: 187–194.

Kruuk, H., Gorman, M. & Parrish, T. (1980). The use of 65Zn for estimating populations of carnivores. *Oikos* **34**: 206–208.

Kruuk H. & Macdonald, D. W. (1985). Group territories of carnivores: empires and enclaves. In *Behavioural Ecology: Ecological Consequence of Adaptive Behaviour*, eds R. M. Sibly & R. H. Smith, pp. 521–536. Oxford: Blackwell Scientific Publications.

Kruuk, H. & Parrish, T. (1982). Factors affecting the population density, group size and territory size of the European badger, *Meles meles*, in relation to earthworm population, Scotland. *Journal of Zoology (London)* **196**: 31–39.

Kruuk, H. & Parrish, T. (1987). Changes in the size of groups and ranges of the European badger (*Meles meles L.*) in an area in Scotland. *Journal of Animal Ecology* **56**: 351–364.

Kubo, T. & Iwasa, Y. (1995). Inferring the rates of branching and extinction from molecular phylogenies. *Evolution* **49**: 694–704.

Kucera, T. E., Soukkala, A. M. & Zielinski W. J. (1995). Photographic bait stations. In *American Marten, Fisher, Lynx, and Wolverine: Survey Methods for their Detection*, eds W. J.Zielinski & T. E. Kucera, pp. 25–65. USDA For. Serv. Gen. Tech. Rep. PSW-GTR-157.

Kucklick, J. R., Harvey, H. R., Ostrom, P. H., Ostrom, N. E. & Baker, J. E. (1996). Organochlorine dynamics in the pelagic food web of Lake Baikal. *Environmental Toxicology and Chemistry* **15**: 1388–1400.

Kumar, A. & Wright, B. (1999). Combating tiger poaching and illegal wildlife trade in India. In *Riding the Tiger – Tiger Conservation in Human-Dominated Landscapes*, eds J. Seidensticker, S. Christie & P. Jackson, pp. 243–251. Cambridge: Cambridge University Press.

Kurki, S., Nikula, A., Helle, P. & Lindén, H. (1998). Abundances of red fox and pine marten in relation to the composition of boreal forest landscapes. *Journal of Animal Ecology* **67**: 874–86.

Kurtén, B. & Anderson, E. (1980). *Pleistocene Mammals of North America*. New York: Columbia University Press.

Kurtz, D. A. & Kim, K. C. (1976). Chlorinated hydrocarbon and PCB residues in tissues and lice of northern fur seals, 1972. *Pesticides Monitoring* **10**: 79–83.

Laake, J. L., Buckland, S. T., Anderson, D. R. & Burnham, K. P. (1993). Distance user's guide, v2.0. *Colorado Cooperative Fish and Wildlife Research Unit, Colorado State University, Fort Collins.*

Lack, D. (1954). *The Natural Regulation of Animal Numbers.* London: Oxford University Press.

Lack, D. (1971). *Ecological Isolation in Birds.* Oxford: Blackwell Scientific Publications.

Lacy, M. K. (1983). *Home Range Size, Intraspecific Spacing, and Habitat Preference of Ringtails* (Bassariscus astutus) *in a Riparian Forest in California.* M.A. thesis, California State University, Sacramento.

Lacy, R. C. (1987). Loss of genetic diversity from managed populations: interacting effects of drift, mutation, immigration, selection, and population subdivision. *Conservation Biology* 1: 143–158.

Lacy, R. C. (1993a). Impacts of inbreeding in natural and captive populations of vertebrates: implications for conservation. *Perspectives in Biology and Medicine.* 36: 480–496.

Lacy, R. C. (1993b). VORTEX: a computer simulation model for population viability analysis. *Wildlife Research* 20: 45–65.

Lacy, R. C. (1997). Importance of genetic variation to the viability of mammalian populations. *Journal of Mammalogy* 18: 320–335.

Lacy, R. C., Hughes, K. A. & Miller, P. S. (1995). *VORTEX: A Stochastic Simulation of the Extinction Process.* Version 5.1 users's manual. Chicago: Brookfield Zoo, Chicago Zoological Society.

Lacy, R. C., Petric, A. M. & Warneke, M. (1993). Inbreeding and outbreeding depression in captive populations of wild animal species. In *The Natural History of Inbreeding and Outbreeding*, ed. N. W. Thronhill, pp. 352–374. Chicago: The University of Chicago Press.

Lade, J. A., Murray, N. D., Marks, C. A. & Robinson, N. A. (1996). Microsatellite differentiation between Phillip Island and mainland Australian populations of red fox *Vulpes vulpes. Molecular Ecology* 5: 81–88.

Laikre, L., Andier, R., Larsson, H. O. & Ryman, N. (1996). Inbreeding depression in Brown bear *Ursus arctos. Biology Conservation* 76: 69–72

Laikre, L. A. & Ryman, N. (1991). Inbreeding depression in a captive wolf (*Canis lupus*) population. *Conservation Biology* 5: 33–40.

Laikre, L., Ryman, N. & Thompson, E.A. (1993). Hereditary blindness in a captive wolf (*Canis lupus*) population: frequency reduction of a deleterious allele in relation to gene conservation. *Conservation Biology* 7: 592–601.

Lambretchs, A. V. W. (1995). Meeting wildlife and human needs by establishing collaborative nature reserves: the Transvaal system. In *Integrating People and Wildlife for a Sustainable Future*, eds J. A. Bissonette & P. R. Krausmann, pp. 37–43. Bethesda: The Wildlife Society.

Lampio, T. (1982). [Occurrence of the Samson fox in Finland, 1960–1980]. *Suomen Riista* 29: 21–28. (In Finnish with English summary).

Lancia, R. A., Nichols, J. D. & Pollock, K. H. (1994). Estimating the number of animals in wildlife populations. In *Research and Management Techniques for Wildlife and Habitats*, 5th edn, ed. T. A. Bookhout, pp. 215–253. Bethesda: The Wildlife Society.

Land, D. & Taylor, S. K. (1998). Florida panther genetic restoration and management annual performance report 1997–1998. Naples: Florida Game and Fresh Water Fish Commission.

Landa, A., Strand, O., Swenson, J. E. & Skogland, T. (1997). Wolverines and their prey in southern Norway. *Canadian Journal of Zoology* **75**: 1292–1299.

Landa, A. & Tommeras, B. A. (1997). A test of aversive agents on wolverines. *Journal of Wildlife Management* **61**: 510–516.

Lande, R. (1988). Genetics and demography in biological conservation. *Science* **241**: 1455–1460.

Lande, R. (1994). Risk of population extinction from fixation of new deleterious mutations. *Evolution* **48**: 1460–1469.

Lande, R. & Barrowclough, G. (1987). Effective population size, genetic variation, and their use in population management. In *Viable Populations for Conservation*, ed. M. Soulé, pp. 87–123. Cambridge: Cambridge University Press.

Langley, P. J. W. & Yalden, D. W. (1977). The decline of the rare carnivores in Great Britain during the nineteenth century. *Mammal Review* **7**: 95–116.

Lankester, K., van Apeldoorn, R., Meelis, E. & Verbom, J. (1991). Management perspectives for populations of the Eurasian badger (*Meles meles*) in a fragmented landscape. *Journal of Applied Ecology* **28**: 561–73.

Lanning, D. V. (1976). Density and movements of the coati in Arizona. *Journal of Mammalogy* **57**: 609–611.

Larsen, T. (1972). Air and ship census of polar bears in Svalbard (Spitsbergen). *Journal of Wildlife Management* **36**: 562–570.

Lasswell, H. D. (1971). *A Pre-view of Policy Sciences*. New York: American Elsevier.

Laundré, J. W. (1981). Temporal variation in coyote vocalization rates. *Journal of Wildlife Management* **45**: 767–769.

Laurance, W. F. (1991). Ecological correlates of extinction proneness in Australian tropical rain forest mammals. *Conservation Biology* **5**: 80–89.

Laurance, W. F. & Bierregaard, R. O., Jr. (Ed.) (1997). *Tropical Forest Remnants: Ecology, Management, and Conservation of Fragmented Communities*. Chicago: University of Chicago Press.

Laurance, W. F., Ferreira, L. V., Rankin-de Merona, J. M., Laurance, S. G., Hutchings, R. W. & Lovejoy, T. E. (1998). Effects of forest fragmentation on recruitment patterns in Amazonian tree communities. *Conservation Biology* **12**: 460–464.

Laurenson, M. K. (1994). High juvenile mortality in cheetahs (*Acinonyx jubatus*) and its consequences for maternal care. *Journal of Zoology* **234**: 387–408.

Laurenson, M. K. (1995a). Implications of high offspring mortality for cheetah population dynamics. In *Serengeti II: Dynamics, Conservation and Management of an Ecosystem*, eds A. R. E. Sinclair & P. Arcese, pp. 385–399. Chicago: University of Chicago Press.

Laurenson, M. K. (1995b). Cub growth and maternal care in cheetahs. *Behavioral Ecology* **6**: 405–409.

Laurenson, M. K. & Caro, T. M. (1994). Monitoring the effects of non-trivial handling in free-living cheetahs. *Animal Behaviour* **47**: 547–557.

Laurenson, K. M., Shiferaw, F. & Sillero-Zubiri, C. (1997). Disease, domestic dogs and the Ethiopian wolf: the current situation. In *The Ethiopian Wolf. Status and Conservation Action Plan*, eds C. Sillero-Zubiri & D. Macdonald, pp. 32–40. Gland: IUCN/SSC Canid Specialist Group.

Laurenson, K. M., Sillero-Zubiri, C., Thompson, H., Shiferaw, F., Thirgood, S. & Malcolm, J. (1998). Disease as a threat to endangered species: ethiopian wolves, domestic dogs and canine pathogens. *Animal Conservation* **1**, 273–280

Laurenson K. M., Wielebnowski, N. & Caro, T. M. (1995). Extrinsic factors and juvenile mortality in cheetahs. *Conservation Biology* **9**: 1329–1331.

Lavigne, D. M., Callaghan, C. J., & Smith, R. J. (1996). Sustainable utilization: the lessons of history. In *The Exploitation of Mammal Populations*, eds V. J. Taylor & N. Dunstone, pp. 250–265. London: Chapman and Hall.

Lavigne, D. M. & Øritsland, N. A. (1974). Black polar bears. *Nature* **251**: 218–219.

Lavrov, N. P. (1970). Increasing the productivity of hunting grounds in the USSR by introducing new species of carnivores. *Transactions IX International Congress Game Biologists (Moscow, USSR)*, pp. 199–200.

Lawton, J. H. (1995). Population dynamic principles. In *Extinction Rates*, eds J. H. Lawton & R. M. May, pp. 147–163. Oxford: Oxford University Press.

Lawton, J. H. (1997). The science and non-science of conservation biology. *Oikos* **79**: 3–5.

Lawton, J. H. & Brown, K.C. (1986). The population and community ecology of invading insects. *Philosophical Transactions of the Royal Society, Series B* **314**: 606–617.

Lawton, J. H. & Hassell, M. P. (1981). Asymmetrical competition in insects. *Nature* **289**: 793–795.

Lawton, J. H. & May, R. M. (Eds) (1995). *Extinction Rates*. Oxford: Oxford University Press.

Lawton, J. H. & Woodroffe, G. L. (1991). Habitat and the distribution of water voles: why are there gaps in a species' range? *Journal of Animal Ecology* **60**: 79–91.

Leader-Williams, N. & Albon, S. D. (1988). Allocation of resources for conservation. *Nature* **336**: 533–535.

Leader-Williams, N., Albon, S. D. & Berry, P. S. M. (1990). Illegal exploitation of black rhinoceros and elephant populations: patterns of decline, law enforcement and patrol effort in Luangwa Valley, Zambia. *Journal of Applied Ecology* **27**: 1055–1087.

Leader-Williams, N., Kayera, J. A. & Overton, G. L. (1996). *Tourist Hunting in Tanzania*. Gland: IUCNNR.

Lehman, N., Clarkson, P., Mech, L. D., Meier, T. J. & Wayne, R. K. (1992). A study of the genetic relationships within and among wolf packs using DNA fingerprinting and mitochondrial DNA. *Behavioral Ecology Sociobiology* **30**: 83–94.

Lehman, N., Eisenhawer, A., Hansen, K., Mech, L. D., Peterson, R. O., Gogan P. J. P. & Wayne, R. K. (1991). Introgression of coyote mitochondrial DNA into sympatric North American gray wolf populations. *Evolution* **45**: 104–119.

Lehman, N. & Wayne, R. K. (1991). Analysis of coyote mitochondrial DNA genotype frequencies: estimation of the effective number of alleles. *Genetics* **128**: 405–416.

Lemke, C. W. & Thompson, D. R. (1960). Evaluation of a fox population index. *Journal of Wildlife Management* **24**: 406–412.

Lento, G. M. (1995). *Molecular Systematic and Population Genetic Studies of Pinnipeds: Phylogenetics of our Fin-Footed Friends and their Surreptitious Species Status*. PhD Thesis. Victoria University of Wellington, Wellington, New Zealand.

Lento, G. M., Haddon, M., Chambers, G. K. & Baker, C. S. (1997). Genetic variation, population structure, and species identity of southern hemisphere fur seals, *Arctocephalus* spp. *Journal of Heredity* **88**: 202–208.

Lento, G. M., Hickson, R. E., Chambers, G. K. & Penny, D. (1995). Use of spectral analysis to test hypotheses on the origin of pinnipeds. *Molecular Biology and Evolution* **12**: 28–52.

Lento, G. M., Mattlin, R. H., Chambers, G. K. & Baker, C. S. (1994). Geographic distribution of mitochondrial cytochrome b DNA haplotypes in New Zealand fur seals (*Arctocephalus forsteri*). *Canadian Journal of Zoology* **72**: 293–299.

Leonard, J. A., Wayne, R. K. & Cooper, A. (2000). Population genetics of Ice Age brown bears. *Proceedings of the National Academy of Sciences USA* **97**: 1651–1654.

Leonards, P. E. G., Zierikzee, Y., Brinkman, U. A. T., Cofino, W. P., van Straalen, N. M. & van Hattum, B. (1997). The selective dietary accumulation of planar polychlorinated biphenyls in the otter (*Lutra lutra*). *Environmental Toxicology and Chemistry* **16**: 1807–1815.

Leopold, A. S., Cain, S. A., Cottam, C. M., Gabrielson, I. N. & Kimball, T. L. (1964). Predator and rodent control in the United States. *Transactions of the North American Wildlife and Natural Resources Conference* **29**: 27–49.

Letcher, R. J., Norstrom, R. J. & Muir, D. C. G. (1998). Biotransformation versus bioaccumulation: sources of methyl sulfone PCB and 4,4í-DDE metabolites in the polar bear food chain. *Environmental Science and Technology* **32**: 1656–1661.

Lever, C. (1978). The not so innocuous mink? *New Scientist* **78**: 812–814.

Lever, C. (1985). *Naturalized Mammals of the World*. London: Longmans.

Lever, C. (1994). *Naturalized Animals: the Ecology of Successfully Introduced Species*. London: T. & A.D. Poyser Ltd.

Levins, R. (1970). Extinction. In *Some Mathematical Questions in Biology. Lectures on Mathematics in the Life Sciences*, vol. 2, ed. M. Gerstenhaber, pp. 77–107. Province, Rhode Island: American Mathematical Society.

Li, W. H., Gouy, M., Sharp, P. M., Huigin, O. C. & Yang, Y. W. (1990). Molecular phylogeny of Rodentia, Lagomorpha, Primates, Artiodactyla, and Carnivora and molecular clocks. *Proceedings of the National Academy of Sciences* **87**: 6703–6707.

Lidgard, D. C. (1997). *The Effects of Human Disturbance on the Maternal Behaviour and Performance of Gray Seals* (Halichoerus grypus) *at Donna Nook, Lincolnshire, UK*. Preliminary report to the British Ecological Society.

Lidicker, W. Z., Jr. (1995). The landscape concept: something old, something new. In *Landscape Approaches in Mammalian Ecology*, ed. W. Z. Lidicker, Jr., pp. 3–19. Minneapolis: University of Minnesota Press.

Lidicker, W. Z., Sage, R. D. & Sage, D. G. (1981). Biochemical variation in northern sea lions from Alaska. In *Mammalian Population Genetics*, eds M. H. Smith & J. Joule, pp. 231–241. Athens, GA: University of Georgia Press.

Lindemann, W. (1956). Transplantation of game in Europe and Asia. *Journal of Wildlife Management* **20**: 68–70.

Lindenmayer, D. B., Lacy, R. C., Burgman, M. A., Akçakaya, H. R. & Possingham, H. P. (1995). A review of the generic programs ALEX,RAMAS/space and VORTEX for modelling the viability of wildlife metapopulations. *Ecological Modelling* **82**: 161–174.

Lindsay, I. M. & Macdonald, D. W. (1985). The effects of disturbance on the emergence of Eurasian badgers in winter. *Biological Conservation* **34**: 289–306.

Lindström, E. (1983). Condition and growth of red foxes (*Vulpes vulpes*) in relation to food supply. *Journal of Zoology (London)* **199**: 117–122.

Lindström, E. (1989). Food limitation and social regulation in a red fox population. *Holarctic Ecology* **12**: 70–79.

Lindström, E. (1992). Diet, reproduction, recruitment and growth of the red fox (*Vulpes vulpes*) in relation to population density – the sarcoptic mange event in Scandinavia. In *Wildlife 2001: Populations*, eds D. R. McCullough & R. H. Barrett. Elsevier Science, London.

Lindström, E., Andren, H., Angelstam, P. & Widen, P. (1986). Influence of predators on hare populations in Sweden: a critical review. *Mammal Review* **16**: 151–156.

Lindström, E., Andren, H., Angelstam, P., Cederlund, G., Hornfeldt, B., Jaderberg, L., Lemnell, P., Martinsson, B., Skold, K. & Swenson, J. E. (1994). Disease reveals the predator: sarcoptic mange, red fox predation, and prey populations. *Ecology* **75**: 1042–1049.

Lindström, E., Brainerd, S. M., Helldin, J. O. & Overskaug, K. (1995). Pine marten-red fox interactions: a case of intraguild predation? *Annales Zoologici Fennici* **32**: 123–130.

Lindström, E. & Mörner, T. (1985). The spreading of sarcoptic mange among Swedish red foxes (*Vulpes vulpes* L.) in relation to fox population dynamics. *Revue d'Écologie (La Terre et la Vie)* **40**: 211–216.

Lindzey, F. G. (1971). *Ecology of Badgers in Curlew Valley, Utah and Idaho, with Emphasis on Movement and Activity Patterns*. MS thesis, Utah State University, Logan.

Lindzey, F. G. (1987). Mountain lion. In *Wild Furbearer Management and Conservation in North America*, eds M. Novak, J. A. Baker, M. E. Obbard & B. Malloch, pp. 657–668. North Bay: Ontario Trappers Association.

Lindzey, F. G. & Meslow, E. C. (1980). Harvest and population characteristics of black bears in Oregon (1971–74). *International Conference on Bear Research and Management* **4**: 213–219.

Lindzey, F. G., Thompson, S. K. & Hodges, J. I. (1977). Scent station index of black bear abundance. *Journal of Wildlife Management* **41**: 151–153.

Lindzey, G. G., Van Sickle, W. D., Ackerman, B. B., Barnhurst, D., Hemker, T. P. & Laing, S. P. (1994). Cougar population dynamics in southern Utah. *Journal of Wildlife Management* **58**: 619–624.

Linhart, S. B. & Knowlton, F. F. (1975). Determining the relative abundance of coyotes by scent station lines. *Wildlife Society Bulletin* **3**: 119–124.

Linn, I. J. & Chanin, P. R. F. (1978a). Are mink really pests in Britain? *New Scientist* **77**: 560–562.

Linn, I. J. & Chanin, P. R. F. (1978b). More on the mink menace. *New Scientist* **79**: 38–40.

Linnell, J. D. C., Aanes, R., Swenson, J. E., Odden, J. & Smith, M. E. (1997). Translocation of carnivores as a method for managing problem animals: a review. *Biodiversity and Conservation* **6**: 1245–1257.

Linscombe, G., Kinler, N. & Aulerich, R. J. (1982). Mink. In *Wild Mammals of North America: Biology, Management, and Economics*, eds J. A. Chapman & G. A. Feldhamer, pp. 629–643. Baltimore, MD: Johns Hopkins University Press.

Liss, K. (1998). Trapping. In *Encyclopedia of Animal Rights and Animal Welfare*, ed. M. Bekoff, pp. 338–340. Westport, CT: Greenwood Press.

Little, P. (1994). The link between local participation and improved conservation: a review of issues and experiences. *In Natural Connections: Perspectives in Community-based Conservation*, eds D. Western & R. Wright, R. pp. 347–372. Washington DC: Island Press.

Litvaitis, J. A., Beltran, J. F., Delibes, M., Moreno, A. & Villafuerte, R. (1996). Sustaining felid populations in human-dominated landscapes. *Journal of Wildlife Research* 1: 292–296.

Litvaitis, J. A. & Harrison, D. J. (1989). Bobcat–coyote niche relationships during a period of coyote population increase. *Canadian Journal of Zoology* 67: 1180–1188.

Litvaitis, J. A. & Kane, D. M. (1994). Relationship of hunting technique and hunter selectivity to composition of black bear harvest. *Wildlife Society Bulletin* 22: 604–606.

Litvaitis, J. A. & Villafuerte R. (1996). Intraguild predation, mesopredator release, and prey stability. *Conservation Biology* 10: 676–677.

Litvinenko, N. M. (1993). Effects of disturbance by people and introduced predators on seabirds in the northwest Pacific. In *The Status, Ecology, and Conservation of Marine Birds of the North Pacific*, eds K. Vermeer, K. T. Briggs, K. H. Morgan & D. Siegel-Causey, pp. 227–231. Ottawa, Ontario: Canadian Wildlife Service Special Publication.

Lloyd, H. G. (1975). The red fox in Britain. In *The Wild Canids*, ed. M. W. Fox, pp. 207–215. New York: Van Nostrand Reinhold Co.

Lloyd, S. (1995). Environmental influences on host immunity. In *Ecology of Infectious Diseases in Natural Populations*, eds B. T. Grenfell & A. P. Dobson, pp. 327–361. Cambridge: Cambridge University Press.

Lodge, D. M. (1993). Biological invasions: lessons for ecology. *Trends in Ecology and Evolution* 8: 133–137.

Logan, K. A., Irwin, L. L. & Skinner, R. (1986). Characteristics of a hunted mountain lion population in Wyoming. *Journal of Wildlife Management* 50: 648–654.

Logan, K. A., Sweanor, L. L., Ruth, T. K. & Hornocker, M. G. (1996). *Cougar of the San Andres Mountain, New Mexico. Final Report, Project W-128R*. Santa Fe: New Mexico Department of Game and Fish.

Lombard, A. T. (1995). The problems with multi-species conservation: do hotspots, ideal reserves and exisiting reserves coincide? *South African Journal of Zoology* 30: 145–163.

Long, C. A. (1995). Stone marten (*Martes foina*) in southeast Wisconsin, U.S.A. *Small Carnivore Conservation* No. 13. Gland: IUCN-World Conservation Union.

Long, J. A., Wildt, D. E., Wolfe, B. A., Critser, J. K., De Rossi, R. V. & Howard, J. G. (1996). Sperm capacitation and the acrosome reaction are compromised in teratospermic domestic cats. *Biology of Reproduction* 54: 638–646.

Loope, L. L., Sanchez, P. G., Tarr, P. W., Loope, W. L. & Anderson, R. L. (1988). Biological invasions of arid land nature reserves. *Biological Conservation* 44: 95–118.

Lopez, B. H. (1978). *Of Wolves and Men*. New York: Charles Scribner's Sons.

Lord, R. D., Vilches, A. M., Maiztegui, J. I. & Soldini, C. A. (1970). The tracking board: a relative census technique for studying rodents. *Journal of Mammalogy* 51: 828–829.

Lovallo, M. J. & Anderson, E. M. (1995). Range shift by a female bobcat (*Lynx rufus*) after removal of neighboring female. *American Midland Naturalist* 134: 409–412.

Love, J. A. (1992). *Sea Otters*. Golden: Fulcrum Publishing.

Ludwig, D., Hilborn, R. & Walters, C. (1993). Uncertainty, resource exploitation, and conservation: lessons from history. *Science* **260**: 17, 36.

Luikart G., Allendorf, F., Cornuet J.-M. & Sherwin, W. (1998a). Distortion of allele frequency distributions provide a test for recent population bottlenecks. *Journal of Heredity* **89**: 238–247.

Luikart, G. & England, P. (1999). Statistical analysis of microsatellite DNA data. *Trends in Ecology and Evolution* **14**: 253–256.

Luikart, G., Sherwin, W., Steele, B. & Allendorf, F. (1998b). Usefulness of molecular markers for detecting population bottlenecks via monitoring genetic change. *Molecular Ecology* **1998**: 963–974.

Lüthcke, J. (1999). Norwegians to hunt sheep killing wolves. The Baltic and Nordic News Service, 13–01–99.

Lutz, W. (1984). Die Verbreitung des Waschbären (*Procyon lotor*) in mitteleuropäischen Raum. *Zeitschrift für Jagdwissenschaft* **30**: 218–228.

Lyles, A. M. & Dobson, A. P. (1993). Infectious-disease and intensive management – population-dynamics, threatened hosts, and their parasites. *Journal of Zoo and Wildlife Medicine* **24**: 315–326.

Lynch, J. M. (1995). Conservation implications of hybridisation between mustelids and their domesticated counterparts: the example of polecats and feral ferrets in Britain. *Small Carnivore Conservation* **13**: 17–18.

Lynch, M. (1996). A quantitative-genetic perspective on conservation issues. In *Conservation Genetics, Case Histories from Nature*, eds J. C. Avise & J. L. Hamrick, pp. 471–501. New York: Chapman and Hall.

Lynch, M. & Gabriel, W. (1990). Mutation load and survival of small populations. *Evolution* **44**: 1725–1737.

Lynch, M. & Lande, R. (1998). The critical effective size for a genetically secure population. *Animal Conservation* **1**: 70–2.

Lynx Summary (1999). http://www.bouldernews.com/extra/lynx/

Maehr, D. S. (1997). The comparative ecology of bobcat, black bear and Florida panther in South Florida. *Bulletin of the Florida Museum of Natural History* **40**: 1–176.

MacArthur, R. H. & Wilson, E. O. (1967). *The Theory of Island Biogeography*. Princeton, NJ: Princeton University Press.

MacArthur, R. M. (1972). *Geographical Ecology: Patterns in the Distribution of Species*. New York: Harper and Row.

Macdonald, D. W. (1977). On food preference in the red fox. *Mammal Review* **7**: 7–23.

Macdonald, D. W. (1980). *Rabies and Wildlife: A Biologist's Perspective*. Oxford: Oxford University Press.

Macdonald, D. W. (1982). Studies of wildlife rabies in the northern hemisphere and their relevance to southern Africa. *South African Journal of Science* **78**: 416–417.

Macdonald, D. W. (1983). The ecology of carnivore social behaviour. *Nature* **301**: 379–384.

Macdonald, D. W. (1992). *The Velvet Claw: A Natural History of the Carnivores*. London: BBC Books.

Macdonald, D. W. (1993). Rabies and wildlife: a conservation problem? *Onderstepoort Journal of Veterinary Research* **60**: 351–355.

Macdonald D. W. (1996). Dangerous liaisons and disease. *Nature* **379**: 400–401.

Macdonald, D. W. (2000a). *Animal Emotions.* ed. M. Bekoff, pp. 46–49 and 100–103. Discovery Books.

Macdonald, D. W. (2000b). Bartering biodiversity: what are the options? In *Environmental Policy: Objectives, Instruments & Implementation,* ed. D. Helm, pp. 142–171. Oxford: Oxford University Press.

Macdonald, D. W., Artois, M., Aubert, M., Bishop, D. L., Ginsberg, J. R., King, A., Kock, N. & Perry, B. D. (1992). Cause of wild dog deaths. *Nature* **360**: 633–634.

Macdonald, D. W., Barreto, G. R., Ferreras, P., Kirk, B., Rushton, S., Yamaguchi, N. & Strachan, R. (1999). The impact of American mink, *Mustela vison:* as predators of native species in British freshwater systems. In *Advances in Vertebrate Pest Management,* eds D. P. Cowan & C. J. Feare, pp. 5–24. Fürth: Filander Verlag.

Macdonald, D. W. & Barrett, P. (1993). *Collins Field Guide to the Mammals of Britain and Europe.* London: Harper Collins.

Macdonald, D. W. & Carr, G. M. (1995). Variation in dog society: between resource dispersion and social flux. In *The Domestic Dog: Its Evolution, Behaviour and Interactions with People,* ed. J. Serpell, pp. 199–216. Cambridge: Cambridge University Press.

Macdonald, D. W. & Courtenay, O. (1993). Wild and domestic canids as reservoirs of American visceral leishmaniasis in Amazonia. *Symposia of the Zoological Society of London* **65**: 465–479.

Macdonald, D. W. & Johnson, P. J. (1996). The impact of sport hunting: a case study. In *The Exploitation of Mammal Populations: 1994 Symposium of the Zoological Society of London,* eds N. Dunstone & V. L. Taylor, pp. 160–207. London: Chapman & Hall.

Macdonald, D. W. & Johnson, P. J. (2000). Farmers and the custody of the countryside: trends in loss and conservation of non-productive habitats 1981–1998. *Biological Conservation* 1–14.

Macdonald, D. W., Mace, G. M. & Barreto, G. R. (1999). The effects of predators on fragmented prey populations: a case study for the conservation of endangered prey. *Journal of Zoology (London)* **247**: 487–506.

Macdonald, D. W., Mace, G. & Rushton, S. (1998a). *Proposals for Future Monitoring of British Mammals.* London: Department of Environment, Transport, and the Regions.

Macdonald, D. W., Mace, G. M. & Rushton, S. (2000). British mammals: is there a radical future. In *Priorities for the Conservation of Mammalian Diversity,* eds A. Entwhistle & N. Dunstone, pp. 175–205. Cambridge: Cambridge University Press.

Macdonald, D. W. & Newman, C. (in press). Badger (*Meles meles*) population dynamics in Oxfordshire, UK: numbers, density and cohort life histories, and a possible role of climate change in population growth. *Journal of Zoology (London).*

Macdonald, D. W., Reynolds, J. C., Carbone, C., Matthews, F. & Johnson, P. J. (in press). The bio-economics of fox control. In *British Mammals in Farmed Landscapes,* eds F. H. Tattersall & W. Manley, Linnean Society Occasional Publications Series. Otley: Westbury Publishing.

Macdonald, D. W. & Strachan, R. (1999). *Mink and Watervole: Analyses for Conservation.* Oxford: Wildlife Conservation Research Unit.

Macdonald, D. W., Yamaguchi, N. & Passanisi, W. C. (1998b). The health, haematology and blood biochemistry of free-ranging farm cats in relation to social status. *Animal Welfare* 7: 243–256.

Macdonald, I. A. W., Graber, D. M., DeBenedetti, S., Groves, R. H. & Fuentes, E. R. (1988). Introduced species in nature reserves in Mediterranean-type climatic regions of the world. *Biological Conservation* 44: 37–66.

Macdonald, I. A. W., Loope, L. L., Usher, M. B. & Hamann, O. (1989). Wildlife conservation and the invasion of nature reserves by introduced species: a global perspective. In *Biological Invasions: A Global Perspective*, eds J. Drake, H. Mooney, F. di Castri, R. Groves, F. Kruger, M. Rejmanek & M. Williamson, pp. 215–255. Chichester: John Wiley and Sons Ltd.

Macdonald, S. M. & Mason, C. F. (1982). The otter *Lutra lutra* in central Portugal. *Biological Conservation* 22: 207–215.

Mace, G., Collar, N., Cooke, J., Gaston, K., Ginsberg, J., Leader-Williams, N., Maunder, M. & E. J. Milner-Gulland (1993). The development of new criteria for listing species on the IUCN Red List. *Species* 19: 16–22.

Mace G. & Sillero-Zubiri, C. (1997). A preliminary population viability analysis for the Ethiopian wolf. In *The Ethiopian Wolf. Status and Conservation Action Plan*, eds C. Sillero-Zubiri & D. Macdonald, pp. 51–60. Gland: UCN/SSC Canid Specialist Group.

Mace, G. M. & Balmford, A. (2000). Patterns and processes in contemporary mammalian extinction. In *Future Priorities for the Conservation of Mammalian Diversity*. eds A. Entwhistle & N. Dunstone, pp. 27–52. Cambridge: Cambridge University Press.

Mace, G. M. & Lande, R. (1991). Assessing extinction threats: toward a reevaluation of IUCN threatened species categories. *Conservation Biology* 5: 148–157.

Mace, R. D., Waller, J. S., Manley, T. L., Lyon, L. J. & Zuuring, H. (1996). Relationships among grizzly bears, roads, and habitat use in the Swan Mountains, Montana. *Journal of Applied Ecology* 33: 754–72.

MacFadden, B. J. (1992). Interpreting extinctions from the fossil record: methods, assumptions, and case examples using horses (family Equidae). In *Extinction and Phylogeny*, eds M. J. Novacek & Q. D. Wheeler, pp. 17–45. New York: Columbia University Press.

Mack, A. (Ed.) (1999). *Humans and Other Animals*. Columbus: Ohio State University Press.

Macpherson, A. H. (1969). The dynamics of Canadian arctic fox populations. *Canadian Wildlife Service, Report Series* No. 8.

Maddock, A., Anderson, A., Carlisle, F., Galli, N., James, A., Verster, S. & Whitfield, W. (1996). Changes in lion numbers in Hluhluwe-Umfolozi Park. *Lamergeyer* 44: 6–18.

Maddock, A. H. & Mills, G. L. (1993). Population characteristics of African wild dogs, *Lycaon pictus*, in the Eastern Transvaal Lowveld, South Africa, as revealed through photographic records. *Biological Conservation* 67: 57–62.

Maehr, D. S. (1990). The Florida panther and private lands. *Conservation Biology* 4: 167–170.

Maehr, D. S. (1997). *The Florida Panther: Life and Death of a Vanishing Carnivore*. Washington, DC: Island Press.

Maehr, D. S. & Caddick, G. B. (1995). Demographics and genetic introgression in the Florida panther. *Conservation Biology* 9: 1295–1298.

Maehr D. S., Greiner, E. C., Lanier, J. E. & Murphy, D. (1995). Notoedric mange in the Florida panther (*Felis concolor coryi*). *Journal of Wildlife Diseases* 31: 251–254.

Maehr, D. S., Land, E. D. & Roelke, M. E. (1991). Mortality patterns of panthers in southwest Florida. *Proceedings Annual Conference Southeastern Fish and Wildlife Agencies* 45: 201–7.

Maehr, D. S., Layne, J. N., Land, E. D., McCown, J. W. & Roof, J. (1988). Long distance movements of a Florida black bear. *Florida Field Naturalist* 16: 1–6.

Maehr, D. S., McBride, R. T. & Mullahey, J. J. (1996). Status of coyotes in south Florida. *Florida Field Naturalist* 24: 101–107.

MAFF (1998). *Quarantine and Rabies, A Reappraisal. Report by the Advisory Group on Quarantine to the Minster of Agriculture, Fisheries and Food.* London: MAFF Publications.

Magoun, A. J. (1985). *Population Characteristics, Ecology, and Management of Wolverines in Northwestern Alaska.* PhD thesis, University of Alaska, Fairbanks.

Mainka, S., Xianmeng, Q., Tingmei, H. & Appel, M. J. (1994). Serological survey of giant pandas (*Ailuropoda melanoleuca*), and domestic dogs and cats in the Woolong Reserve, China. *Journal of Wildlife Diseases* 30: 86–89.

Major, M., Johnson, M., Davis, W. & Kellogg, T. (1980). Identifying scats by recovery of bile acids. *Journal of Wildlife Management* 44: 290–293.

Makombe, K. (Ed.). (1994). *Sharing the Land: Wildlife, People and Development in Africa.* Harare: IUCN.

Malcolm, J. R. (1980). Food caching by African wild dogs (*Lycaon pictus*). *Journal of Mammalogy* 61: 743–744.

Malcolm, J. R. (1988). Small mammal abundances in isolated and non-isolated primary forest reserves near Manaus, Brazil. *Acta Amazônica* 18: 67–83.

Malcolm, J. R. & van Lawick, H. (1975). Notes on wild dogs (*Lycaon pictus*) hunting zebras. *Mammalia* 39: 231–240.

Maldonado, J. E., Orta-Davila, F., Stewart, B. S., Geffen, E. & Wayne, R. K. (1995). Intraspecific genetic differentiation in California sea lions (*Zalophus californianus*) from southern California and the Gulf of California. *Marine Mammal Science* 11: 46–58.

Malik, S., Wilson, P. J., Smith, R. J., Lavigne, D. M. & White, B. N. (1997). Pinniped penises in trade: a molecular genetic investigation. *Conservation Biology* 11: 1365–1374.

Mano, T. (1995). Sex and age characteristics of harvested brown bears in the Oshima Peninsula, Japan. *Journal of Wildlife Management* 59: 199–204.

Mansfield, T. (1997). What we've learned about lions. *Outdoor California* 57: 4–8.

Maran, T. & Henttonen, H. (1995). Why is the European mink (*Mustela lutreola*) disappearing? A review of the process and hypothesis. *Annales Zoologici Fennici* 32: 47–54.

Maran, T., Kruuk, H., Macdonald, D. W. & Polma, M. (1998). Diet of two species of mink in Estonia: displacement of *Mustela lutreola* by *M. vison*? *Journal of Zoology (London)* 245: 218–222.

Maran, T., Macdonald, D. W., Kruuk, H., Sidorovich, V. & Rozhnov, V. V. (1998). The continuing decline of the European mink *Mustela lutreola*: evidence for the intraguild aggression hypothesis. In *Behaviour and Ecology of Riparian*

Mammals, eds N. Dunstone & M. Gorman, pp. 297–323. Cambridge: Cambridge University Press.

Mares, M. A. (1991). How scientists can impede the development of their discipline: egocentrism, small pool size, and the evolution of sapismo. In *Latin American Mammalogy: History, Biodiversity, and Conservation*, eds M. A. Mares & D. J. Schmidley, pp. 57–75. Norman: Oklahoma Museum of Natural History.

Mares, M. A. & Ojeda, R. A. (1984). Fauna commercialization and conservation in South America. *BioScience* **34**: 580–584.

Margules, C. R., Pressey, R. L. & Nicholls, A. O. (1991). Selecting nature reserves. In *Nature Conservation: Cost Effective Biological Surveys and Data Analysis*, eds C. R. Margules & M. P. Austin, pp. 90–97. Canberra: CSIRO.

Marker-Kraus, L. (1996). Cheetah relocation. *Cheetah Conservation Fund Newsletter* **6**: 5.

Marker-Kraus, L. (1997). History of the cheetah in zoos 1829–1994. *International Zoo Yearbook* **35**: 27–43.

Marker-Kraus, L. & Kraus, D. (1997). Conservation of strategies for the long-term survival of the cheetah *Acinonyx jubatus* by the Cheetah Conservation Fund, Windhoek. *International Zoo Yearbook* **35**: 59–66.

Marker-Kraus, L., Kraus, D., Barnett, D. & Hurlbut, S. (1996). *Cheetah Survival on Namibian Farmlands; the Research Findings From the CCF Farm Survey*. Windhoek: Cheetah Conservation Fund.

Marshall, T., Slate, J., Kruuk, L. & Pemberton, J. (1998). Statistical confidence for likelihood-based paternity inference in natural populations. *Molecular Ecology* **7**: 639–655.

Martin, L. D. (1989). Fossil history of the terrestrial Carnivora. In *Carnivore Behavior, Ecology, and Evolution*, ed. J. L. Gittleman, pp. 536–568. Ithaca: Cornell University Press.

Martin, P. S. & Klein, R. G. (Eds) (1998). *Quaternary Extinctions, A Prehistoric Revolution*. Tucson, AZ: University of Arizona Press.

Martin, P. S. & Wright, H. E. (Eds) (1967). *Pleistocene Extinctions: The Search for a Cause*. New Haven, CT: Yale University Press.

Martin, R. B. & de Meulenaer, T. (1988). *Survey of the Status of the Leopard* (Panthera pardus) *in Sub-Saharan Africa*. Lausanne: CITES Secretariat.

Mason, C. F. & Macdonald, S. M. (1986). *Otters: Ecology and Conservation*. Cambridge: Cambridge University Press.

Mason, C. F. & O'Sullivan, W. M. (1992). Organochlorine pesticide residues and PCs in otters (*Lutra lutra*) from Ireland. *Bulletin of Environmental Contamination and Toxicology* **48**: 387–393.

Masood, E. (1988). Outcry as 'scientific' badger cull is launched to target TB. *Nature* **394**: 821.

Masuda, R., Lopes, J. V., Slattery, J. P., Yuhki, N. & O'Brien, S. J. (1996). Molecular phylogeny of mitochondrial cytochrome b and 12S rRNA sequences in the Felidae: ocelot and domestic cat lineages. *Molecular Phylogenetics and Evolution* **6**: 351–365.

Mattlin, R. H. (1978). *Population biology, thermoregulation, and site preference of the New Zealand fur seal*, Arctocephalus forsteri. PhD Thesis, University of Canterbury, New Zealand.

Mattlin, R. H. (1987). New Zealand fur seal, *Arctocephalus forsteri*: within the New

Zealand region. In *Status, Biology, and Ecology of Fur Seals*, eds J. P. Croxall & R. L. Gentry, pp. 49–51. NOAA Tech Rep NMFS, vol. 51.

Mattson, D. J. (1995). The New World Mine and grizzly bears: a window on ecosystem management. *Journal of Energy, Natural Resources & Environmental Law* 15: 267–293.

Mattson, D. J. (1996). Ethics and science in natural resource agencies. *BioScience* 46: 767–771.

Mattson, D. J. (1997). Wilderness-dependent wildlife: the large and the carnivorous. *International Journal of Wilderness* 3: 34–38.

Mattson, D. J. & Craighead, J. J. (1994). The Yellowstone grizzly bear recovery program. In *Endangered Species Recovery: Finding the Lessons, Improving the Process*, eds T. W. Clark, R. P. Reading & A. L. Clarke, pp. 101–129. Washington DC: Island Press.

Mattson, D. J., Herrero, S., Wright, R. G. & Pease, C. M. (1996a). Science and management of Rocky Mountain grizzly bears. *Conservation Biology* 10: 1013–1025.

Mattson, D. J., Herrero, S., Wright, R. G. & Pease, C. M. (1996b). Designing and managing protected areas for grizzly bears: How much is enough? In *National Parks and Protected Areas: Their Role in Environmental Protection*, ed. R. G. Wright, pp. 133–164. Cambridge: Blackwell Science.

Mattson, D. J., Knight, R. R. & Blanchard, B. M. (1987). The effects of developments and primary roads on grizzly bear habitat use in Yellowstone National Park, Wyoming. *International Conference on Bear Research and Management* 7: 259–273.

Mattson, D. J., Wright, R. G., Kendall, K. C. & Martinka, C. J. (1995). Grizzly bears. In *Our Living Resources: A Report to the Nation on the Distribution, Abundance, and Health of U. S. Plants, Animals, and Ecosystems*, eds E. T. LaRoe, G. S. Farris, C. E. Puckett, P. D. Doran & M. J. Mac, pp. 103–105. Washington, DC: U. S. National Biological Service.

May, R. M. (1988). Conservation and disease. *Conservation Biology* 2: 26–30.

May, R. M. & Anderson, R. M. (1978). Regulation and stability of host-parasite population interactions. II. Destabilizing processes. *Journal of Animal Ecology* 47: 249–267.

May, R. M., Lawton, J. H. & Stork, N. E. (1995). Assessing extinction rates. In *Extinction Rates*, eds J. H. Lawton & R. M. May, pp. 1–24. Oxford: Oxford University Press.

McCallum, H. (1996). Immunocontraception for wildlife population control. *Trends in Ecology and Evolution* 11: 491–493.

McCallum, H. I. & Dobson, A. (1995). Detecting disease and parasite threats to endangered species and ecosystems. *Trends in Ecology and Evolution* 10: 190–194.

McCarthy, T. & Allen, P. (1999). Knitting for snow leopards. *Cat News* 30: 24–25.

McCarthy, T. M. & Munkhtsog, B. (1997). Preliminary assessment of snow leopard sign surveys in Mongolia. In *Proceedings of the Eighth International Snow Leopard Symposium*, eds R. Jackson & A. Ahmad, pp. 57–65. Seattle, WA: International Snow Leopard Trust.

McCracken, H. (1957). *The Beast that Walks Like Man – The Story of the Grizzly Bear*. London: Oldbourne Press.

McCrindle, C. M. E. (1998). The community development approach to animal welfare: an African perspective. *Applied Animal Behaviour Science* **59**: 227–233.

McCullough, D. R. (Ed.) (1996). *Metapopulations and Wildlife Conservation.* Washington, DC: Island Press.

McDonald, J. E., Jr. (1998). *The Effects of Food Supply and Nutrition on Black Bear Reproductive Success and Milk Composition.* PhD Dissertation. University of Massachusetts, Amherst.

McDougal, C. (1987). The man-eating tiger in a geographical and historical perspective. In *Tigers of the World: The Biology, Management and Conservation of an Endangered Species*, eds R. L. Tilson & U. S. Seal, pp. 435–448. Park Ride, NJ: Noye Publications.

McDougal, C. (1988). Leopard and tiger interactions at Royal Chitwan National Park, Nepal. *Journal of the Bombay Natural History Society* **85**: 609–611.

McKenna, M. C. & Bell, S. K. (1997). *Classification of Mammals – Above the Species Level.* New York: Columbia University Press.

McKinney, M. L. (1997). Extinction vulnerability and selectivity: combining ecological and paleontological views. *Annual Reviews in Ecology and Systematics* **28**: 495–516.

McKinney, M. L. (1998). Branching models predict loss of many bird and mammal orders. *Animal Conservation* **1**: 159–64.

McLellan, B. N. (1989). Dynamics of a grizzly bear population during a period of industrial resource extraction. III. Natality and rate of increase. *Canadian Journal of Zoology* **67**: 1865–1868.

McLellan, B. N. (1990). Relationships between human industrial activity and grizzly bears. *International Conference on Bear Research and Management* **8**: 57–64.

McLellan, B. N. (1994). Density-dependent population regulation of brown bears. In *Density Dependent Population Regulation in Black, Brown, and Polar Bears*, ed. M. Taylor, pp. 15–24. International Conference on Bear Research and Management Monograph Series No. 3.

McLellan, B. N. & Shackelton, D. M. (1988). Grizzly bears and resource-extraction industries: effects of roads on behaviour, habitat use, and demography. *Journal of Applied Ecology* **25**: 451–460.

McMahan, L. R. (1986). The International cat trade. In *Cats of the World*, eds S. D. Miller & E. D. Everett, pp. 461–487. Washington DC: National Wildlife Federation.

McNamee, T. (1997). *The Return of the Wolf to Yellowstone.* New York: Henry Holt & Co.

McNutt, J. W. (1995). *Sociality and dispersal in African wild dogs*, Lycaon pictus. PhD Dissertation, University of California, Davis.

McNutt, J. W. (1996). Sex-biased dispersal in African wild dogs, *Lycaon pictus. Animal Behaviour* **52**: 1067–1077.

McOrist, S., Boid, R., Jones, T. W., Easterbee, N., Hubbard, A. L. & Jarrett, O. (1991). Some viral and protozool diseases in the European Wildcat (*Felis silvestris*). *Journal of Wildlife Disease* **27**: 693–696.

McOrist, S. & Kitchener, A. C. (1994). Current threats to the European wildcat, *Felis silvestris*, in Scotland. *Ambio* **23**: 243–245.

McVittie, R. (1979). Changes in the social behaviour of South West African cheetah. *Madoqua* **2**: 171–89.

Meagher, T. & Thompson, E. (1986). The relationship between single parent and parent pair genetic likelihoods in genealogy reconstruction. *Theoretical Population Biology* **29**: 87–106

Mech, L. D. (1966). *The Wolves of Isle Royale*. Fauna Series No. 7. Washington, DC: U.S. National Park Service.

Mech, L. D. (1970). *The Wolf: The Ecology and Behavior of an Endangered Species*. Minneapolis: University of Minnesota Press.

Mech, L. D. (1973a). *Wolf Numbers in the Superior National Forest of Minnesota*. U.S. Forest Service Research Paper NC-97. St. Paul, MN: North Cent. For. Exp. Stn.

Mech, L. D. (1973b). Canada lynx invasion of Minnesota. *Biological Conservation* **5**: 151–152.

Mech, L. D. (1977a). Wolf pack buffer zones as prey reservoirs. *Science* **198**: 320–321.

Mech, L. D. (1977b). Productivity, mortality and population trends of wolves in northeastern Minnesota. *Journal of Mammalogy* **58**: 559–574.

Mech, L. D. (1981). *The Wolf: The Ecology and Behavior of an Endangered Species*. New York: Doubleday.

Mech, L. D. (1987). Age, season, distance, direction, and social aspects of wolf dispersal from a Minnesota pack. In *Mammalian Dispersal Patterns: the Effects of Social Structure on Population Genetics*, eds B. D. Chepko-Sade & Z. T. Halpin, pp. 55–74. Chicago: University of Chicago Press.

Mech, L. D. (1995). The challenge and opportunity of recovering wolf populations. *Conservation Biology* **9**: 270–278.

Mech, L. D., Adams, L. G., Meier, T. J., Burch, J. W. & Dale, B. W. (1998). *The Wolves of Denali*. Minneapolis: University of Minnesota Press.

Mech, L. D., Fritts, S. H. & Paul, W. J. (1988a). Relationship between winter severity and wolf depredations on domestic animals in Minnesota. *Wildlife Society Bulletin* **16**: 269–272.

Mech, L. D., Fritts, S. H., Radde, G. L. & Paul, W. J. (1988b). Wolf distribution and road density in Minnesota. *Wildlife Society Bulletin* **16**: 85–7.

Mech, L. D. & Goyal, S. M. (1995). Effects of canine parvovirus on gray wolves in Minnesota. *Journal of Wildlife Management* **59**: 565–570.

Medina, G. (1997). A comparison of the diet and distribution of southern river otter (*Lutra provocax*) and mink (*Mustela vison*) in southern Chilé. *Journal of Zoology (London)* **242**: 291–297.

Meffe, G. K. Boersma, P. D., Murphy, D. D., Noon, B. R., Pulliam, H. R., Soulé, M. E. & Waller, D. M. (1998). Independent scientific review in natural resource management. *Conservation Biology* **12**: 268–70.

Melish, R. & Foster-Turley, P. (1996). First record of hybridisation in otters (*Lutrinae: Mammalia*), between Smooth-coated otter, *Lutrogale perspicillata* (Geoffroy, 1826) and Asian small-clawed otter, *Aonyx cinerea* (Illiger, 1815). *Zoologische Garten* **66**: 284–288.

Melquist, W. E., & Dronkert, A. E. (1987). River otter. In *Wild Furbearer Management and Conservation in North America*, eds M. Novak, J. A. Baker, M. E. Obbard & B. Malloch, pp. 627–641. North Bay: Ontario Trappers Association.

Melquist, W. E. & Hornocker, M. G. (1979). *Methods and Techniques for Studying and Censusing River Otter Populations*. University of Idaho Forest, Wildlife, and Range Experiment Station, Technical Report No. 8.

Melquist, W. E. & Hornocker, M. G. (1983). Ecology of river otters in west central Idaho. *Wildlife Monographs* **83**: 1–60.

Melville, E. G. K. (1994). *A Plague of Sheep*. Cambridge: Cambridge University Press.

Menotti-Raymond, M. & O'Brien, S. J. (1993). Dating the genetic bottleneck of the African cheetah. *Proceedings of the National Academy of Science* **90**: 3172–3176.

Menotti-Raymond, M., David, V. A., Stephens, J. C., Lyons, L. A. & O'Brien, S. J. (1997a). Genetic individualization of domestic cats using feline STR loci for forensic applications. *Journal of Forensic Science* **42**: 1039–1051.

Menotti-Raymond, M., David, V. A. & O'Brien, S. J. (1997b). Pet cat hair implicates murder suspect. *Nature* **386**: 774.

Mercure, A., Ralls, K., Koepfli, K. P. & Wayne, R. K. (1993). Genetic subdivisions among small canids: mitochondrial DNA differentiation of swift, kit, and arctic foxes. *Evolution* **47**: 1313–1328.

Merrill, S. B., Adams, L. G., Nelson, M. E. & Mech, L. D. (1998). Testing releasable GPS radiocollars on wolves and white-tailed deer. *Wildlife Society Bulletin* **26**: 830–835.

Messick, J. P. (1987). North American badger. In *Wild Furbearer Management and Conservation in North America*, eds M. Novak, J. A. Baker, M. E. Obbard & B. Malloch, pp. 587–597. North Bay: Ontario Trappers Association.

Messick, J. P. & Hornocker, M. G. (1981). Ecology of the badger in southwestern Idaho. *Wildlife Monographs* **76**: 1–53.

Messier, F. (1985). Solitary living and extraterritorial movements of wolves in relation to social status and prey abundance. *Canadian Journal of Zoology* **63**: 239–245.

Messier, F. (1987). Physical condition and blood physiology of wolves in relation to moose density. *Canadian Journal of Zoology* **65**: 91–95.

Messier, F. (1994). Ungulate population models with predation: a case study with the North American moose. *Ecology* **75**: 478–488.

Messier, F. (1995). On the functional and numerical responses of wolves to changing prey density. In *Ecology and Conservation of Wolves in a Changing World*, eds L. N. Carbyn, S. H. Fritts & D. R. Seip, pp. 187–197. Edmonton, Alberta: Canadian Circumpolar Institute.

Meyer, W. B. & Turner, B. L. II. (1992). Human population growth and global land-use/cover change. *Annual Review Ecology and Systematics* **23**: 39–61.

Meyers, N. (1988). Threatened biotas: hotspots in tropical forests. *Environmentalist* **8**: 178–208.

Meyers, N. (1990). The biodiversity challenge: expanded hotspot analysis. *Environmentalist* **10**: 243–56.

Mighetto, L. (1991). *Wild Animals and American Environmental Ethics*. Tucson: University of Arizona Press.

Milberg, P. & Tyrberg, T. (1993). Naïve birds and noble savages – a review of man-caused prehistoric extinctions of island birds. *Ecography* **16**: 229–241.

Miller, A. (1999). *Environmental Problem Solving: Psychosocial Barriers to Adaptive Change*. New York: Springer.

Miller, B., Ralls, K., Reading, R. P., Scott, J. M. & Estes, J. (1999). Biological and technical considerations of carnivore translocation: a review. *Animal Conservation* **2**: 59–68.

Miller, B., Reading, R.P. & Forrest, S. (1996). *Prairie Night: Black-footed Ferrets and the Recovery of Endangered Species.* Washington, DC: Smithonian Institution Press.

Miller, F. L. & Russell, R. H. (1977). Unreliability of strip aerial surveys for estimating numbers of wolves on western Queen Elizabeth Islands, Northwest Territories. *Canadian Field-Naturalist* **91**: 77–81.

Miller, S. D. (1980). *The Ecology of the Bobcat in South Alabama.* PhD thesis, Auburn Univ., Auburn, Alabama.

Miller, S. D. (1990). Impact of increased bear hunting on survivorship of young bears. *Wildlife Society Bulletin* **18**: 462–467.

Miller, S. D. (1993). *Impacts of Increased Hunting Pressure on the Density, Structure and Dynamics of Brown Bear Populations in Alaska's Game Management Unit 13.* Alaska: Department of Fish & Game, Fairbanks.

Miller, S. D. & Ballard, W. B. (1982). Density and biomass estimates for an interior Alaskan brown bear, *Ursus arctos*, population. *Canadian Field-Naturalist* **96**: 448–454.

Miller, S. D., White, G. C., Sellers, R. A., Reynolds, H. V., Schoen, J. W., Titus, K., Barnes, Jr., V. G., Smith, R. B., Nelson, R. R., Ballard, W. B. & Schwartz, C. C. (1997). Brown and black bear density estimation in Alaska using radiotelemetry and replicated mark-resight techniques. *Wildlife Monographs* **133**: 1–53.

Mills, J. A. & Servheen, C. (1991). *The Asian Trade in Bears and Bear Parts.* Washington DC: TRAFFIC USA/WWF.

Mills, L. S., Citta, J. J., Lair, K. P. & Schwartz, M. K. (2000). Estimating animal abundance using non-ivasive sampling: promise and pitfalls. *Ecological Applications* **10**: 283–294.

Mills, L. S. & Knowlton, F. F. (1991). Coyote space use in relation to prey abundance. *Canadian Journal of Zoology* **69**: 1516–1521.

Mills, L. S. & Smouse, P. E. (1994). Demographic consequences of inbreeding in remnant populations. *American Naturalist* **144**: 412–31.

Mills, L. S., Hayes, S. G., Baldwin, C., Wisdom, M. J., Citta, J., Mattson, D. J. & Murphy, K. (1996). Factors leading to different viability predictions for a grizzly bear data set. *Conservation Biology* **10**: 863–73.

Mills, M. G. L. (1985). Hyena survey of Kruger National Park: August–October 1984. Unpublished Report: Kruger National Park.

Mills, M. G. L. (1991). Conservation management of large carnivores in Africa. *Koedoe* **34**: 81–90.

Mills, M. G. L. & Biggs, H. C. (1993). Prey apportionment and related ecological relationships between large carnivores in Kruger National Park. *Symposia of the Zoological Society of London* **65**: 253–268.

Mills, M. G. L., Ellis, S., Woodroffe, R., Madock, A., Stander, P., Rasmussen, G., Pole, A., Fletcher, P., Bruford, M., Wildt, D. E., Macdonald, D. & Seal, U. S. (Eds) (1998). *Population and Habitat Viability Assessment for the African Wild Dog (Lycaon pictus) in Southern Africa.* Apple Valley, MN: IUCN/SSC Conservation Breeding Specialist Group.

Mills, M. G. L. & Gorman, M. L. (1997). Factors affecting the density and distribution of wild dogs in the Kruger National Park. *Conservation Biology* **11**: 1397–1406.

Mills, M. G. L. & Hofer, H. (1998). *Hyaenas – Status Survey and Conservation Action Plan.* Gland: IUCN.

Mills, M. G. L. & Mills, M. E. J. (1982). Factors affecting the movement patterns of brown hyaenas, *Hyaena brunnea*, in the southern Kalahari. *South African Journal of Nature Research* **12**: 111–117.

Mills, M. G. L., Wolff, P., le Riche, E. A. N. & Meyer, I. J. (1978). Some population characteristics of the lion (*Panthera leo*) in the Kalahari Gemsbok National Park. *Koedoe* **21**: 163–171.

Minta, S. C., Kareiva, P. M. & Curlee, A. P. (1999). Understanding the history and theory of carnivore ecology and crafting approaches for research and conservation. In *Carnivores in Ecosystems: The Yellowstone Experience*, eds T. W. Clark, A. P. Curlee, S. C. Minta & P. M. Kareiva, pp. 323–404. New Haven, CT: Yale University Press.

Minteer, B. A. (1998). No experience necessary? Foundationalism and the retreat from culture in environmental ethics. *Environmental Ethics* **7**: 338–348.

Miquelle, D. M., Merrill, W. T., Dunishenko, Y. M., Smirnov, E. V., Quigley, H. B., Pikunov, D. G. & Hornocker, M. G. (1999). A habitat protection plan for the Amur tiger: developing political and ecological criteria for a viable land-use plan. In *Riding the Tiger: Tiger Conservation in Human-Dominated Landscapes*, eds J. Seidensticker, S. Christie & P. Jackson, pp. 273–295. Cambridge: Cambridge University Press.

Mishra, C. (1997). Livestock depredation by large carnivores in the Indian trans-Himalaya: conflict perceptions and conservation prospects. *Environmental Conservation* **24**: 338–343.

Miththapala, S., Seidensticker, J. & O'Brien, S. J. (1996). Phylogeographic subspecies recognition in leopards (Panthera pardus): molecular genetic variation. *Conservation Biology* **10**: 1115–1132.

Mittermeier, R. A., Myers, N. & Thomsen, J. B. (1998). Biodiversity hotspots and the major tropical wilderness areas: approaches to setting conservation priorities. *Conservation Biology* **12**: 516–532.

Mizutani, F. (1993). Home range of leopards and their impact on livestock on Kenyan ranches. *Symposia of the Zoological Society of London* **65**: 425–439.

Mladenoff, D. J. & Sickley, T. A. (1998). Assessing potential gray wolf restoration in the northeastern United States: a spatial prediction of favorable habitat and potential population levels. *Journal of Wildlife Management* **62**: 1–10.

Mladenoff, D. J., Sickley, T. A., Haight, R. G. & Wydeven, A. P. (1995). A regional landscape analysis and prediction of favorable gray wolf habitat in the northern Great Lakes region. *Conservation Biology* **9**: 279–294.

Mochizuki, M., Akuzawa, M. & Nagatomo, H. (1990). Serological survey of the Iriomote cat (*Felis iriomotensis*) in Japan. *Journal of Wildlife Disease* **26**: 236–245.

Molinari, P. (1998). The lynx in the Italian south-eastern Alps. *Hystrix* **10**: 55–64.

Møller A. P., Dufva, R. & Allander, K. (1993). Parasites and the evolution of host social behavior. *Advances in the Study of Behavior* **22**: 65–102.

Mollison D. & Levin, S. A. (1995). Spatial dynamics of parasitism. In *Ecology of Infectious Diseases in Natural Populations*, eds B. T. Grenfell & A. P. Dobson, pp. 384–398. Cambridge: Cambridge University Press.

Mollohan, C. M. & LeCount, A. I. (1989). Problems of maintaining a viable black bear population in a fragmented forest. In *Multiresource Management of Ponderosa Pine Forest*, eds A. Tecle, W. W. Covington & R. H. Hamre, pp. 149–159. U.S. Forest Service General Technical Report RM-185.

Mönkkönen, M. & Reunanen, P. (1999). On critical thresholds in landscape connectivity: a management perspective. *Oikos* **84**: 302–305.

Montgomery, W. I. (1987). The application of capture–mark–recapture methods to enumeration of small mammal populations. *Symposia of the Zoological Society of London* **58**: 25–57.

Mooers, A. Ø. & Heard, S. B. (1997). Evolutionary process from phylogenetic tree shape. *Quarterly Review of Biology* **72**: 31–54.

Moore, D. (1992). Re-establishing large predators. *Re-Introduction News* **4**: 6.

Moore, D. E., III & Smith, R. (1991). In *Beyond Captive Breeding: Reintroducing Endangered Animals to the Wild*, ed. J. H. W. Gipps, Oxford: Clarendon Press.

Moore, D. R. J., Breton, R. L. & Lloyd, K. (1997). The effects of hexachlorobenzene on mink in the Canadian environment: an ecological risk assessment. *Environmental Toxicology and Chemistry* **16**: 1042–1050.

Moore, G. C. & Parker, G. R. (1992). Colonization by the eastern coyote (*Canis latrans*). In *The Ecology and Management of the Eastern Coyote*, ed. Boer, pp. 23–38. Frederickton, New Brunswick: University of New Brunswick Press.

Moore, T. D., Spencer, L. E. & Dugnolle, C. E. (1974). *Identification of the Dorsal Guard Hairs on Some Mammals of Wyoming*. Wyoming Game and Fish Dep. Bull. No. 14.

Mooring M. S. & Hart, B. L. (1992). Animal grouping for protection from parasites: selfish herd and encounter-dilution effects. Animal grouping for protection from parasites – selfish herd and encounter-dilution effects. *Behaviour* **123**: 173–193.

Morán, J. F., Saavedra, D. & Ruiz, J. (1998). Veterinary management in the reintroduction of otter (*Lutra lutra*) in northeastern Spain. In *Proceedings of the Second Scientific Meeting of the European Association of Zoo and Wildlife Veterinarians (EAZWV)*, Chester, UK, May 21–24. p. 459.

Morgan, G. S. & Woods, C. A. (1986). Extinction and the zoogeography of West Indian land mammals. *Biological Journal of the Linnean Society* **28**: 167–203.

Morin, P. A., Wallis, J., Moore, J. J., Chakraborty, R., Woodruff, D. (1993). Noninvasive sampling and DNA amplification for paternity exclusion, community structure, and phylogeography in wild chimpanzees. *Primates* **34**: 347–356.

Morin, P. A. & Woodruff, D. S. (1996). Noninvasive genotyping for vertebrate conservation. In *Molecular Genetic Approaches in Conservation*, eds R. K. Wayne & T. B. Smith, pp. 298–313. Oxford: Oxford University Press.

Moritz, C., Heideman, A., Geffen, E. & McRae, P. (1997). Molecular population genetics of the bilby, *Macrotis lagotis*, a marsupial in decline. *Molecular Ecology* **6**: 925–936.

Moritz, C. (1994). Defining 'evolutionary significant units' for conservation. *Trends in Ecology and Evolution* **9**: 373–375.

Moritz, C. (1999). Conservation units and translocations: strategies for conserving evolutionary processes. *Hereditas (Lund)* **130**: 217–228.

Morrison, D. W., Edmunds, R. M., Linscombe, G. & Goertz, J. W. (1981). Evaluation of specific scent station variables in northcentral Louisiana. *Proceedings of the Annual Conference of the Southeast Association of Fish and Wildlife Agencies* **35**: 281–291.

Moulton, M. P. (1993). The all-or-none pattern in introduced Hawaiian passeriforms: the role of competition sustained. *American Naturalist* **141**: 105–119.

Moutou, F. (1995). Re-introducing the brown bear (*Ursus arctos*) in the Pyrenees, France. *Re-Introduction News* **11**: 15.

Mowat, F. (1984). *Sea of Slaughter*. Toronto: McClelland and Stewart.

Mowbray, E. E., Pursley, D. & Chapman, J. A. (1979). *The Status, Population Characteristics, and Harvest of the River Otter in Maryland*. Md. Dep. Nat. Resour., Wildl. Admin., Pub. Wildl. Ecol. 2. 16 pp.

Muir, D. C. G., Segstro, M. D., Hobson, K. A., Ford, C. A., Stewart, R. E. A. & Olpinski, S. (1995). Can seal eating explain elevated levels of PCBs and organochlorine pesticides in walrus blubber from eastern Hudson Bay (Canada)? *Environmental Pollution* **90**: 335–348.

Mulder, J. L. (1990). The stoat *Mustela erminea* in the Dutch dune region, its local extinction, and a possible cause: the arrival of the fox *Vulpes vulpes*. *Lutra* **33**: 1–21.

Müller-Graf, C. D. M., Woolhouse, M. E. J. & Packer, C. (1999). Epidemiology of an intestinal parasite (*Spirometra spp.*) in two populations of African lions (*Panthera leo*). *Parasitology* **118**: 407–415.

Munson, L. (1993). Diseases of captive cheetahs (*Acinonyx jubatus*): results of the cheetah research council pathology survey 1989–1992. *Zoo Biology* **12**: 105–124.

Munson, L. (1996). Emerging helicobacter diseases. *Proceedings of the American Association of Zoo Veterinarians*. Nov: 3–8.

Murcia, C. (1995). Edge effects in fragmented forest: implications for conservation. *Trends in Ecology and Evolution* **10**: 58–62.

Murombedzi, J. C. (1999). Devolution and stewardshiop in Zimbabwe's CAMPFIRE Programme. *Journal of International Development* **11**: 287–293.

Murphy, K. M. (1983). *Relationships between a mountain lion population and hunting pressure in western Montana*. Unpublished MS thesis, University of Montana, Missoula.

Murphy, K. M. (1998). *The Ecology of the Cougar* (Puma concolor) *in the Northern Yellowstone Ecosystem: Interactions with Prey, Bears, and Humans*. PhD dissertation, University of Idaho, Moscow.

Murray, D. L., Kapke, C. A., Everman, J. F. & Fuller, T. K. (1999). Infectious disease and the conservation of free-ranging large carnivores. *Animal Conservation* **2**: 241–254.

Myers, D. L., Zurbriggen, A., Lutz, H. & Pospischil, A. (1997). Distemper: not a new disease in lions and tigers. *Clinical and Diagnostic Laboratory Immunology* **4**: 180–194.

Myers, N. (1988). Threatened biotas: hotspots in tropical forests. *Environmentalist* **8**: 178–208.

Myers, N. (1990). The biodiversity challenge: expanded hotspot analysis. *Environmentalist* **10**: 243–245.

Myers, N. (1998). Lifting the veil on perverse subsidies. *Nature* **392**: 327–328.

Nader, L. (Ed.) (1996). *Naked Science: Anthropological Inquiry into Boundaries, Power, and Knowledge*. New York: Routledge.

Nakata, H., Tanabe, S. R., Tatsukawa, M., Amano, N., Miyazaki & Petrov, E. A. (1995). Persistent organochlorine residues and their accumulation in Baikal seal (*Phoca sibirica*) from Lake Baikal, Russia. *Environmental Science and Technology* **29**: 2877–2885.

Nakamura, K. (1989). Notes on distribution and disappearance of the Japanese sea lion (*Zalophus californianus japonicus*). *Nihon-no-Seibutsu* **3**: 27–34. (In Japanese.)

National Planning Commission (1992). *Republic of Namibia 1991 – Population and Housing Census, Preliminary Report*. Windhoek: Central Statistics Office.

National Research Council. (1997). *Wolves, Bears, and their Prey in Alaska*. Washington: National Academy Press.

Naves, J., Fernandez, A., Gaona, J. F. & Nores, C. (1996). Use of self-activating cameras to collect faunistic information. *Doñana Acta Vertebrata* **23**: 189–199.

Navidi, W., Arnheim, N. & Waterman, M. S. (1992). A multiple-tube approach for accurate genotyping of very small DNA samples by using PCR: statistical considerations. *American Journal of Human Genetics* **50**: 347–359.

Nee, S. & May, R. M. (1992). Dynamics of metapopulations: habitat destruction and competitive coexistence. *Journal of Animal Ecology* **61**: 37–40.

Nee, S. & May, R. M. (1997). Extinction and the loss of evolutionary history. *Science* **278**: 692–694.

Nee, S., Holmes, E. C., May, R. M. & Harvey, P. H. (1994a). Extinction rates can be estimated from molecular phylogenies. *Philosophical Transactions of the Royal Society of London, series B* **344**: 77–82.

Nee, S., May, R. M. & Harvey, P. H. (1994b). The reconstructed evolutionary process. *Philosophical Transactions of the Royal Society of London, series B* **344**: 305–311.

Neigel, J. E. (1997). A comparison of alternative strategies for estimating gene flow from genetic markers. *Annual Review of Ecology and Systematics* **28**: 105–128.

Nel, L. H., Thomson, G. R. & Von Teichman, B. F. (1993). Molecular epidemiology of rabies virus in South Africa. *Onderstepoort Journal of Veterinary Research* **60**: 301–306.

Nellis, C. H. & Keith, L. B. (1976). Population dynamics of coyotes in central Alberta, 1964–68. *Journal of Wildlife Management* **40**: 389–399.

Nellis, D. W. (1982). Mongoose influence on the ecology of islands. *Transactions of the International Congress of Game Biology* **14**: 311–314.

Nellis, D. W. (1989). *Herpestes auropunctatus*. *Mammalian Species* **342**: 1–6.

Nellis, D. W. & Everard, C. O. R. (1983). The biology of the mongoose in the Caribbean. *Studies on Fauna of Curacao and Caribbean Islands* **64**: 1–162.

Nepal, S. K & Weber, K. E. (1995). The quandary of local people-park relations in Nepal's Royal Chitwan National Park. *Environmental Management* **19**: 853–866.

Nettles, V. F., Shaddock, J. H., Sikes, R. K. & Reyes, C. R. (1979). Rabies in translocated raccoons. *American Journal of Public Health* **69**: 601–602.

Neumann, R. P. (1998). *Imposing Wilderness: Struggles over Livelihood and Nature Preservation in Africa*. Berkeley: University of California Press.

Newman, A., Bush, M., Wildt, D. E., Van Dam, D., Frankenhuis, M. T., Simmons, L., Phillips, L. & O'Brien. S. J. (1985). Biochemical genetic variation in eight endangered or threatened felid species. *Journal of Mammalogy* **66**: 256–267.

Newman, C. (2000). The enigma of beauty. *National Geographic*, January: 94–121.

Newmark, W. D. (1995). Extinction of mammal populations in western North American national parks. *Conservation Biology* **9**: 512–526.

Newmark, W. D., Boshe, J. I., Sariko, H. I. & Makumbule, G. K. (1996). Effects of a highway on large mammals in Mikumi National Park, Tanzania. *African Journal of Ecology* **34**: 15–31.

Newsome, A. E. & Corbett, L. (1982). The identity of the dingo. II. Hybridization with domestic dogs in captivity and in the wild. *Australian Journal of Zoology* **30**: 365–374.

Newsome, A. E., Parer, I. & Catling, P. C. (1989). Prolonged prey suppression by carnivores – predator-removal experiments. *Oecologia* **78**: 458–467.

Nicolaus, L. K., Herrera, J., Nicolaus, J. C. & Gustavson, C. R. (1989). Ethynyles-tradiol and generalized aversions to eggs among free-ranging predators. *Applied Animal Behaviour Science* **24**: 313–324.

Nichols, A. O. (1998). Integrating population abundance, dynamics and distribution into broad-scale priority setting. In *Conservation in a Changing World*, eds G. M. Mace, A. Balmford & J. R. Ginsberg, pp. 251–272. Cambridge: Cambridge University Press.

Nichols, J. D. & Pollock, K. H. (1990). Estimation of recruitment from immigration versus in situ reproduction using Pollock's robust design. *Ecology* **71**: 21–26.

Nielsen, O., Nielson, K. & Steward, R. E. A. (1996). Serologic evidence of *Brucella spp.* exposure in Atlantic walruses. (*Odobenujs rosmarus rosmarus*) and ringed seals (*Phoca hispida*) of Arctic Canada. *Arctic* **49**: 383–386.

Norbury, G. L., Norbury, D. C. & Heyward, R. P. (1998). Behavioral responses of two predator species to sudden declines in primary prey. *Journal of Wildlife Management* **62**: 45–58.

Norell, M. A. (1993). Tree-based approaches to understanding history: comments on ranks, rules and the quality of the fossil record. *American Journal of Science* **293A**: 407–417.

Norstrom, R. J., Simon, M., Muir, D. C. G. & Schweinsburg, R. E. (1988). Organochlorine contaminants in arctic marine food chains: Identification, geogrphical distribution, and temporal trends in polar bears. *Environmental Science and Technology* **22**: 1063–1071.

Norton, B. G. (1987). *Why Preserve Natural Variety?* Princeton, NJ: Princeton University Press.

Norton-Griffiths, M. (1975). *Counting Animals*. Nairobi: African Wildlife Leadership Foundation.

Noss, R. F. (1995). *Maintaining Ecological Integrity in Representative Reserve Networks*. Toronto and Washington DC: World Wildlife Fund, Canada; World Wildlife Fund, USA.

Noss, R. F., O'Connell, M. A. & Murphy, D. D. (1997). *The Science of Conservation Planning*. Washington, DC: Island Press.

Noss, R. F., Quigley, H. B., Hornocker, M. G., Merrill, T. & Paquet, P. C. (1996). Conservation biology and carnivore conservation in the Rocky Mountains. *Conservation Biology* **10**: 949–63.

Novak, M. (1987). Wild furbearer management in Ontario. In *Wild Furbearer Management and Conservation in North America*, eds M. Novak, J. A. Baker, M. E. Obbard & B. Malloch, pp. 1049–1061. North Bay: Ontario Trappers Association.

Novaro, A. J. (1995). Sustainability of harvest of culpeo fox in Patagonia. *Oryx.* **29**: 18–22.

Nowak, R. M. (1976). *The Cougar in the United States and Canada*. New York: New York Zoological Society & U.S. Fish and Wildlife Service.

Nowak, R. M. (1979). *North American Quaternary Canis*. Kansas: Museum of Natural History, University of Kansas.

Nowak, R. M. (1991). *Walker's Mammals of the World*, 5th edn. Baltimore: Johns Hopkins University Press.

Nowak, R. M. (1999). *Walker's Mammals of the World*, 6th edn, vol. 2. Baltimore: The Johns Hopkins University Press.

Nowak, R. M. & Paradiso, J. L. (1983). *Walker's Mammals of the World*, 4h edn. Baltimore: John Hopkins University Press.

Nowell, K. & Jackson, P. (1996). *Wild Cats – Status Survey and Conservation Action Plan*. Gland: IUCN.

Nummi, P. (1996). Wildlife introductions to mammal-deficient areas: the Nordic countries. *Wildlife Biology* 2: 221–226.

Obbard, M. E., Jones, J. G., Newman, R., Booth, A., Satterthwaite, A. J. & Linscombe, G. (1987). Furbearer harvests in North America. In *Wild Furbearer Management and Conservation in North America*, eds M. Novak, J. E. Baker, M. E. Obbard & B. Malloch, pp. 1007–1034. North Bay: Ontario Trappers Association.

O'Brien, S. J. (1994a). A role for molecular genetics in biological conservation. *Proceedings of the National Academy of Sciences* USA 91: 5748–5755.

O'Brien S. J. (1994b). Genetic and phylogenetic analyses of endangered species. *Annual Reviews of Genetics* 28: 467–489.

O'Brien, S. J. (1995). Genomic prospecting. *Nature Medicine* 1: 742–744.

O'Brien, S. J. (1998). Intersection of population genetics and species conservation, the cheetah's dilemma. In *Evolutionary Biology*, eds M. K. Hecht, R. J. Macintyre & M. T. Clegg, pp. 79–91. vol. 30. New York: Plenum Press.

O'Brien, S. J. & Evermann, J. F. (1988). Interactive influence of infectious disease and genetic diversity in natural populations. *Trends in Ecology and Evolution* 3: 254–259.

O'Brien, S. J., Joslin, P., Smith, G. L., Wolfe, R., Heath, E., Otte-Joslin, J., Rawal, P. P., Bhattacherjee, K. & Martenson, J. S. (1987a). Evidence for African origin of founders of the Asiatic lion species survival plan. *Zoo Biology* 6: 99–116.

O'Brien, S. J., Martenson, J. S., Eichelberger, M. A., Thorne, E. T. & Wright, F. (1989). Genetic variation and molecular systematics of the black-footed ferret. In *Conservation Biology and the Black-footed Ferret*, eds U. S. Seal, E. T. Thorne, M. A, M. A. Bogan & S. H. Anderson, pp. 21–33. New Haven, CT: Yale University Press.

O'Brien, S. J., Martenson, J. S., Miththapala, S., Janczewski, D., Pecon-Slattery, J., Johnson, W. E., Gilbert, D. A., Roelke, M., Packer, C., Bush, M. & Wildt, D. E. (1996). Conservation genetics of the Felidae. In *Conservation Genetics: Case Histories from Nature*, eds J. C. Avise & J. L. Hamrick, pp. 50–74. New York: Chapman & Hall.

O'Brien, S. J., Martenson, J. S., Packer, C., Herbst, L., de Vos, V., Joslin, P., Ott-Joslin, J., Wildt, D. E. & Bush, M. (1987c). Biochemical genetic variation in geographic isolates of African and Asiatic lions. *National Geographic Research* 3: 114–124.

O'Brien, S. J. & Mayr, E. (1991). Bureaucratic mischief: recognizing endangered species and subspecies. *Science* 251: 1187–1188.

O'Brien, S. J., Roelke, M. E., Marker, L., Newman, A., Winkler, C. A., Meltzer, D., Colly, L., Evermann, J. F., Bush, M. & Wildlt, D. E. (1985). Genetic basis for species vulnerability in the cheetah. *Science* 227: 1428–1434.

O'Brien, S. J., Roelke, M. E., Yuhki, N., Richards, K. W., Johnson, W.E., Franklin, W. L., Anderson, A. E., Bass, O. L. Jr., Belden, R. C. & Martenson, J. S. (1990). Genetic introgression within Florida panther (*Felis concolor coryi*). *National Geographic Research* **6**: 485–494.

O'Brien, S. J., Wildt, D. E., Bush, M., Caro, T. M., FitzGibbon, C., Aggundey, I., & Leakey, R. E. (1987b). East African cheetahs: evidence for two population bottlenecks? *Proceedings of the National Academy of Science* USA **84**: 508–511.

O'Connell, M. O., Wright, J. M. & Farid, A. (1996). Development of PCR primers for nine polymorphic American mink Mustela vison microsatellite loci. *Molecular Ecology* **5**: 311–312.

O'Donoghue, M., Boutin, S., Krebs, C. J., Murray, D. L. & Hofer, E. J. (1998). Behavioural responses of coyotes and lynx to the snowshoe hare cycle. *Oikos* **82**: 169–183.

O'Donoghue, M., Boutin, S., Krebs, C. J. & Hofer, E. J. (1997). Numerical responses of coyotes and lynx to the snowshoe hare cycle. *Oikos* **80**: 150–162.

O'Farrell, T. P. (1987). Kit fox. In *Wild Furbearer Management and Conservation in North America*, eds M. Novak, J. A. Baker, M. E. Obbard & B. Malloch, pp. 423–431. North Bay: Ontario Trappers Association.

Ogutu, J. O. & Dublin, H. T. (1998). The response of lions and spotted hyenas to sound playbacks as a technique for estimating population size. *African Journal of Ecology* **36**: 83–95.

Ohta, T. & Kimura, M. (1973). A model of mutation appropriate to estimate the number of electrophoretically detectable alleles in a finite population. *Genetic Research* **22**: 201–204.

Ojasti, J. (1993). Utilizacion de la fauna silvestre en America Latina, situacion y perspecitvas para un manejo sostenible. Rome: FAO.

Okarma, H., Jedrzejewski, W., Schmidt, K., Sniezko, S., Bunevich, A. N. & Jedrzejewski, B. (1998). Home ranges of wolves in Bialowieza Primeval Forest, Poland, compared with other Eurasian populations. *Journal of Mammalogy* **79**: 842–852.

Okoniewski, J. C. & Chambers, R. E. (1984). Coyote vocal response to an electric siren and human howling. *Journal of Wildlife Management* **48**: 217–222.

Oli, M. K. (1994). Snow leopards and blue sheep in Nepal: densities and predator: prey ratio. *Journal of Mammalogy* **75**: 998–1004.

Oli, M. K., Taylor, I. R. & Rogers, M. E. (1994). Snow leopard *Panthera unica* predation of livestock: An assessment of local perceptions in the Annapurna Conservation Area, Nepal. *Biological Conservation* **68**: 63–68.

Oliver, W. A., Jr. & Pedder, A. E. H. (1994). Crises in the Devonian history of the rugose corals. *Paleobiology* **20**: 178–190.

Olson, D. & Dinerstein, E. (1998). The Global 200: A representation approach to conserving the earth's most biologically valuable regions. *Conservation Biology* **12**: 502–515.

Olson, D. M. & Dinerstein, E. (1997). *The Global 200: a representation approach to conserving the Earth's distinctive ecoregions*. Washington, DC: Conservation Science Program Publication, World Wildlife Fund.

Olson, T. L., Gilbert, B. K. & Squibb, R. C. (1997). The effects of increasing human activity on brown bear use of an Alaskan river. *Biological Conservation* **82**: 95–99.

O'Neill, R. J., O'Neill, M. J. & Graves, J. A. M. (1998). Undermethylation associated with retroelement activation and chromosome remodelling in an interspecific mammalian hybrid. *Nature* **393**: 68–72.

O'Neill, R. V., Milne, B. T., Turner, M. G. & Gardner, R. H. (1988). Resource utilization scales and landscape pattern. *Landscape Ecology* **2**: 63–69.

Orians., G., Cochrane, P. A., Duffield, J. W., Fuller, T. K., Guiterrez, R. J., Hanemann, W. M., James, F. C., Kareiva, P., Kellert, S. R., Klein, D., McClellan, B. N., Olson, P. D. & Yaska, G. (1997). *Wolves, Bears and their Prey in Alaska*. Washington DC: National Academy Press.

Orlans, F. B. (1997). Ethical decision making about animal experiments. *Ethics and Behavior* **7**: 163–171.

O'Ryan, C., Harley, E., Bruford, M. W., Beaumont, M., Wayne, R. K. & Cherry, M. I. (1998). Microsatellite analysis of genetic diversity in fragmented South African buffalo populations. *Animal Conservation* **1**: 85–94.

Osterhaus, A., vandeBildt, M., Vedder, L. *et al.* (1998). Monk seal mortality: virus or toxin? *Vaccine* **16**: 979–981.

Osterhaus, A. D. M. E., Groen, J., Uytdehaag, F. G. C. M., Visser, G., Van de Bilt, M. W. G., Bergmann, A. & Klingeborn, B. (1989). Distemper virus in Baikal seals. *Nature* **338**: 209–210.

OTA (United States Congress Office of Technology Assessment) (1993). *Harmful Non-indigenous Species in the United States*. OTA-F-565. Washington, DC: U.S. Government Printing Office.

Owens, D. & Owens, M. (1993). *Cry of the Kalahari*. London: Collins.

Packard, R. L. & Bowers, J. H. (1970). Distributional notes on some foxes from western Texas and eastern New Mexico. *Southwestern Naturalist* **14**: 450–451.

Packer, C. (1986). The ecology of sociality in felids. In *Ecological Aspects of Social Evolution: Birds and Mammals*, eds D. I. Rubenstein & R. W. Wrangham, pp. 429–452. Princeton: Princeton University Press.

Packer, C. (1990). *Serengeti lion survey: report to TANAPA, SWRI, Mweka and the Game Department*. Serengeti Wildlife Research Centre, Seronera, Tanzania.

Packer, C., Gilbert, D. A., Pusey, A. E. & O'Brien, S. J. (1991a). A molecular genetic analysis of kinship and cooperation in African lions. *Nature* **351**: 562–564.

Packer, C., Herbst, L., Pusey, A. E., Bygott, J. D., Hanby, J. P., Cairns, S. J. & Borgerhoff-Mulder, M. (1988). Reproductive success in lions. In *Reproductive Success: Studies of Individual Variation in Contrasting Breeding Systems*, ed. T. H. Clutton-Brock, pp. 363–383. Chicago, Illinois: University of Chicago Press.

Packer, C. & Pusey, A. E. (1983). Adaptations of female lions to infanticide by incoming males. *American Naturalist* **121**: 716–728.

Packer, C. & Pusey, A. E. (1993). Dispersal, kinship and inbreeding in African lions. In *The Natural History of Inbreeding and Outbreeding*, ed. N. W. Thornhill, pp. 375–391. Chicago: University of Chicago Press.

Packer, C., Pusey, A. E., Rowley, H., Gilbert, D. A., Martenson, J. & O'Brien, S. J. (1991b). Case study of a population bottleneck: lions of the Ngorongoro Crater. *Conservation Biology* **5**: 219–230.

Paetkau, D., Calvert, W., Stirling, I. & Strobeck, C. (1995). Microsatellite analysis of population structure in Canadian polar bears. *Molecular Ecology* **4**: 347–354.

Paetkau, D., Shields, G. F. & Strobeck, C. (1998b). Gene flow between insular, coastal and interior populations of brown bears in Alaska. *Molecular Ecology* 7: 1283–1292.

Paetkau, D., Waits, L. P., Clarkson, P. L. *et al.* (1998a). Variation in genetic diversity across the range of North American brown bears. *Conservation Biology* 12: 418–429.

Pagel, M. D. (1992). A method for the analysis of comparative data. *Journal of Theoretical Biology* 156: 431–442.

Pain, S. (1997). The plague dogs. *New Scientist* 154: 32–37.

Paloheimo, J. E. & Fraser, D. (1981). Estimation of harvest rate and vulnerability from age and sex data. *Journal of Wildlife Management* 45: 948–958.

Palomares, F. & Caro, T. M. (1999). Interspecific killing among mammalian carnivores. *American Naturalist* 153: 492–508

Palomares, F., Ferreras, P., Fedriani, J. M. & Delibes, M. (1996). Spatial relationships between Iberian lynx and other carnivores in an area of south-western Spain. *Journal of Applied Ecology* 33: 5–13.

Palomares, F., Gaona, P., Ferreras, P. & Delibes, M. (1995). Positive effects on game species of top predators by controlling smaller predator populations: an example with lynx, mongooses and rabbits. *Conservation Biology* 9: 295–305.

Pamilo, P. (1990). Comparison of relatedness estimators. *Evolution* 44: 1378–1382.

Panwar, H. S. (1987). Project tiger: the reserves, the tigers and their future. In *Tigers of the World: The Biology, Biopolitics, Management and Conservation of an Endangered Species*, eds R. L. Tilson & U. S. Seal, pp. 110–117. Park Ridge, NJ: Noyes Publications.

Paquet, P. C. (1991). Winter spatial relationships of wolves and coyotes in Riding Mountain National Park, Manitoba. *Journal of Mammalogy* 72: 397–401.

Paquet, P. C. (1992). Prey use strategies of sympatric wolves and coyotes in Riding Mountain National Park, Manitoba. *Journal of Mammalogy* 73: 337–343.

Paquet, P. C. & Hackman, A. (1995). *Large Carnivore Conservation in the Rocky Mountains.* Toronto, Ontario and Washington, D. C: World Wildlife Fund Canada and World Wildlife Fund U.S.

Parker, G. (1995). *Eastern Coyote: The Story of Its Success.* Halifax, NS: Nimbus Publishing.

Parrish, C. R. (1999). Host range relationships and the evolution of canine parvovirus. *Veterinary Microbiology* 69: 29–40.

Parsons, D. R. (1998). 'Green fire' returns to the Southwest: reintroduction of the Mexican wolf. *Wildlife Society Bulletin* 26: 799–807.

Parsons, D. R. (1999). Re-introduction of the Mexican wolf to the southwestern United States. *Re-introduction News* 18: 6.

Parsons, T. J., Olson, S. L. & Braun, M. J. (1993). Unidirectional spread of secondary sexual plumage traits across an avian hybrid zone. *Science* 260: 1643–1646.

Pascal, M. (1980). Structure et dynamiqué de la population de chats harets de l'archipel des Kerguelen. *Mammalia* 44: 161–182. (In French with English summary.)

Pastoret, J. & Brochier, B. (1998). Epidemiology of rabies in Western Europe. *Veterinary Journal* 156: 83–90.

Pastoret, P. P. & Brochier, B. (1999). Epidemiology and control of fox rabies in Europe. *Vaccine* 17: 1750–1754.

Patterson, J. H. (1907). *The Man-Eaters of Tsavo*. London: Macmillan.

Patton, P. W. (1994). The effect of edge on avian nest success: how strong is the evidence? *Conservation Biology* 8: 17–26.

Paxinos E., McIntosh, C., Ralls, K. & Fleischer, R. (1997). A noninvasive method for distinguishing among canid species: amplification and enzyme restriction of DNA from dung. *Molecular Ecology* 6: 483–486.

Pearce, D. (2000). Cost-benefit analysis and environmental policy. In *Environmental Policy: Objectives, Instruments & Implementation*, ed. D. Helm. Oxford: Oxford University Press.

Pearson, O. P. (1964). Carnivore-mouse preadtion: an example of its intensity and bioenergetics. *Journal of Mammalogy* 45: 177–188.

Pearson, S. M. & Gardner, R. H. (1997). Neutral models: useful tools for understanding landscape patterns. In *Wildlife and Landscape Ecology: Effects of Pattern and Scale*, ed. J. A. Bissonette, pp. 215–230. New York: Springer-Verlag.

Pedersen, N. C., Torten, M., Rideout, B., Sprarker, E., Tonachini, T., Luciw, P. A., Ackley, C., Levy, N. & Yamamoto, J. (1990). Feline leukemia virus infection as a potentiating cofactor for the primary and secondary stages of experimentally induced immunodeficiency virus infection. *Journal of Virology* 64: 598–606.

Pech, R. P., Sinclair, A. R. E., Newsome, A. E. & Catling, P. C. (1992). Limits to predator regulation of rabbits in Australia: evidence from predator-removal experiments. *Oecologia* 89: 102–112.

Pecon-Slattery, J. & O'Brien, S. J. (1998). Patterns of X and Y chromosome DNA sequence divergence during the Felidae radiation. *Genetics* 148: 1245–1255.

Peek, J. M., Pelton, M. R., Picton, H. D., Schoen, J. W. & Zager, P. (1987). Grizzly bear conservation and management: a review. *Wildlife Society Bulletin* 15: 160–169.

Pelton, M. R., Cardoza, J., Conley, B., Dubrock, C. & Lindzey, J. (1978). Census techniques and population indices. In *Proceedings of the 4th Eastern Black Bear Workshop*, ed. R. D. Hugie, pp. 242–250. Greenville, Maine.

Pelton, M. R. & Marcum, L. C. (1977). The potential use of radioisotopes for determining densities of black bears and other carnivores. In *Proceedings of the 1975 Predator Symposium*. eds R. L. Phillips & C. Jonkel, pp. 221–236. Montana Forest and Conservation Experiment Station, Missoula: University of Montana.

Pence, D. B. & Windberg, L. A. (1994). Impact of a sarcoptic mange epizootic on a coyote population. *Journal of Wildlife Management* 58: 624–633.

Pennycuick, C. J. & Rudnai, J. (1970). A method of identifying individual lions, *Panthera leo*, with the analysis of the reliability of identification. *Journal of Zoology (London)* 160: 497–508.

Perry, B. D. (1993). Dog ecology in eastern and southern Africa: implications for rabies control. *Onderstepoort Journal of Veterinary Research* 60: 429–436.

Peterman, R. M. (1990). Statistical power analysis can improve fisheries research and management. *Canadian Journal of Fish and Aquatic Sciences* 47: 2–15.

Peterson, M. J. (1991). Wildlife parasitism, science, and management policy. *Journal of Wildlife Management* 55: 782–789.

Peterson, R. O. (1977). *Wolf Ecology and Prey Relationships on Isle Royale*. U.S. National Park Service Scientific Monograph Series No. 11. Washington, DC.

Peterson, R. O. (1995). Wolves as interspecific competitors in canid ecology. In *Ecology and Conservation of Wolves in a Changing World*, eds L. N. Carbyn, S. H.

Fritts & D. R. Seip, pp. 315–323. Canadian Circumpolar Institute Occasional Publications 35.

Peterson, R. O. & Page, R. E. (1988). The rise and fall of Isle Royale wolves, 1975–1986. *Journal of Mammalogy* **69**: 89–99.

Peterson, R. O., Thomas, N. J., Thurber, J. M., Vucetich, J. A. & Waite, T. A. (1998). Population limitation and the wolves of Isle Royale. *Journal of Mammalogy* **79**: 828–841.

Peterson, R. O., Woolington, J. D. & Bailey, T. N. (1984). Wolves of the Kenai Peninsula. *Wildlife Monographs* **88**: 1–52.

Petren, K. & Case, T. J. (1996). An experimental demonstration of exploitation competition in an ongoing invasion. *Ecology* **77**: 118–132.

Petren, K. & Case, T. J. (1998). Habitat structure determines competition intensity and invasion success in gecko lizards. *Proceedings of the National Academy of Sciences of the United States of America* **95**: 11739–11744.

Pfeifer, M. L., Evermann, J. F., Roelke, M. E., Gallina, A. M., Ott, R. L. & McKeirnan, A. J. (1983). Feline infectious peritonitis in captive cheetah. *Journal of the American Veterinary Medical Association* **183**: 1317–1319.

Philobosin, R. & Ruibal, R. (1971). Conservation of the lizard *Ameiva polops* in the Virgin Islands. *Herpetologica* **27**: 450–454.

Phillips, D. M. (1994). *Social and Spatial Characteristics, and Dispersal of Marten in a Forest Preserve and Industrial Forest.* Unpublished M. S. thesis, University of Maine, Orono.

Phillips, M. (1992). Red wolves (*Canis rufus*) in N. Carolina. *Re-Introduction News* **4**: 7–8.

Phillips, M. K. & Scheck, J. (1991). Parasitism in captive and reintroduced red wolves. *Journal of Wildlife Diseases* **27**: 498–501.

Phillips, M. K. & Smith, D. W. (1996). *The Wolves of Yellowstone.* Stillwater: Voyageur Press.

Phillips, M. K. & Smith, D. W. (1997). *Yellowstone Wolf Project Biennial Report 1995–1996.* Yellowstone Center for Resources Document YCR-NR-97-4. Yellowstone, Wyoming.

Phillips, M. K., Smith, R., Henry, V. G. & Lucash, C. (1995). Red wolf reintroduction program. In *Ecology and Conservation of Wolves in a Changing World*, eds L. N. Carbyn, S. H. Fritts & D. R. Seip, pp. 157–168. Occasional Publication No. 35. Edmonton: Canadian Circumpolar Institute.

Phythian-Adams, E. G. (1939). The Nilgiri Game Association – 1879–1939. *Journal of the Bombay Natural History Society* **41**: 384–396.

Pielou, E. C. (1991). *After the Ice Age.* Chicago: University of Chicago Press.

Pils, C. M. (1975). A case against red fox reduction in Wisconsin. *Procedings of the 1975 Predator Symposium, Mountain, Forestry, and Conservation Experimental Station*, eds R. L. Phillips & C. Jonkel, pp. 87–91. Missoula: University of Montana.

Pils, C. M., Martin, M. A. & Lange, E. L. (1981). *Harvest, Age Structure, Survivorships, and Productivity of Red Foxes in Wisconsin, 1975–78.* Wisconsin Department of Natural Resources Technical Bulletin No. 124.

Pimm, S. L. (1991). *The Balance of Nature? Ecological Studies in the Conservation of Species and Communities.* Chicago: University of Chicago Press.

Pimm, S. L. & Gittleman, J. L. (1990). Carnivores and ecologists on the road to Damascus. *Trends in Ecology and Evolution* **5**: 70–73

Pimm, S. L. & Lawton, J. H. (1998). Planning for biodiversity. *Science* **279**: 2068–9.

Pimm, S. L., Russell, G. J., Gittleman, J. L. & Brooks, T. M. (1995). The future of biodiversity. *Science* **269**: 347–350.

Plon, U. (1999). Sweden: No to wolf hunting. The Baltic and Nordic News Service, 25–03–99.

Plowright, W. (1988). Research on wildlife diseases, is a reappraisal necessary? *Revue Scientifique et Technique de l'Office International des Epizooties* **7**: 783–795.

Pocock, R. I. (1930). The lions of Asia. *Journal of the Bombay Natural History Society* **34**: 638–665.

Pollock, K. H. (1981). Capture-recapture models: a review of current methods, assumptions and experimental design. In *Estimating Numbers of Birds. Studies in Avian Biology*, No. 6, ed. R. J. Raitt, pp. 426–435. Lawrence, Kansas: Allen Press.

Pond, D. B. & O'Gara, B. W. (1994). Chemical immobilization of large mammals. In *Research and Management Techniques for Wildlife and Habitats*, ed. T. A. Bookhout, pp. 125–139. 5th edn. Bethesda: The Wildlife Society.

Poole, K. G. (1994). Characteristics of an unharvested lynx population during snowshoe hare decline. *Journal of Wildlife Management* **58**: 608–618.

Post, E., Peterson, R.O. Stenseth, N.C. & McLaren, B.E. (1999). Ecosystem consequences of wolf behavioural response to climate. *Nature* **401**: 905–907.

Potvin, F. (1987). Wolf movements and population dynamics in Papineau-Labelle reserve, Quebec. *Canadian Journal of Zoology* **66**: 1266–1273.

Powell, R. A. (1982). *The Fisher: Life History, Ecology, and Behavior*. Minneapolis: University of Minnesota Press.

Powell, R. A. & Zielinski, W. J. (1994). Fisher. In *The Scientific Basis for Conserving Forest Carnivores: American Marten, Fisher, Lynx, and Wolverine in the Western United States*, eds L. F. Ruggiero, K. B. Aubry, S. W. Buskirk, L. J. Lyon & W. J. Zielinski, pp. 39–73. General Technical Report RM-254. Fort Collins, CO: US Department of Agriculture, Forest Service, and Rocky Mountain Forest and Range Experimental Station.

Powell, R. A., Zimmerman, J. W., Seaman, D. E. & Gilliam, J. F. (1995). Demographic analyses of a hunted black bear population with access to a refuge. *Conservation Biol.ogy* **10**: 224–234.

Prager, M. H. & Fabrizio, M. C. (1990). Comparison of logistic regression and discriminant analysis for stock identification of anadromous fish, with application to striped bass (*Morone saxtilis*) and American shad (*Alosa sapidissima*). *Canadian Journal of Fisheries and Aquatic Science* **47**: 1570–1577.

Prendergast, J. R., Quinn, R. M., Lawton, J. H., Eversham, B. C. & Gibbons, D. W. (1993). Rare species, the coincidence of diversity hotspots and conservation strategies. *Nature* **365**: 335–337.

Prendergast, J. R., Quinn, R. M. & Lawton, J. H. (1999). The gaps between theory and practice in selecting nature reserves. *Conservation Biology* **13**: 484–492.

Pressey, R. L. (1994). Ad hoc reservations: forward or backward steps in developing representative reserve systems? *Conservation Biology* **8**: 662–668.

Pressey, R. L., Humphries, C. J., Margules, C. R., Vane-Wright, R. I. & Williams, P.

H. (1993). Beyond opportunism: key principles for systematic reserve selection. *Trends in Ecology and Evolution* 8: 124–128.

Pressey, R. L., Possingham, H. P. & Day, J. R. (1997a). Effectiveness of alternative heuristic algorithms for identifying indicative minimum requirements for conservation reserves. *Biological Conservation* 80: 207–19.

Pressey, R. L., Possingham, H. P. & Margules, C. R. (1997b). Optimality in reserve selection algorithms: when does it matter and how much. *Biological Conservation* 76: 259–67.

Prevett, J. P. & Kolenosky, G. B. (1982). The status of polar bears in Ontario. *Nature Canada (Quebec)* 109: 933–939.

Previtali, A., Cassini, M. H. & Macdonald, D. W. (1998). Habitat use and diet of the American mink (*Mustela vison*) in Argentinian Patagonia. *Journal of Zoology (London)* 246: 482–486.

Price, E. O. (1984). Behavioural aspects of animal domestication. *Quarterly Review of Biology* 59: 1–32.

Priklonski, S. G. (1970). Winter transect count of game animals. *International Congress of Game Biologists* 9: 273–275.

Primm, S. A. (1993). *Grizzly Conservation in Greater Yellowstone*. MA Thesis, Department of Political Science, University of Colorado, Boulder.

Primm, S. A. (1996). A pragmatic approach to grizzly bear conservation. *Conservation Biology* 10: 1026–1035.

Primm, S. A. & Clark, T. W. (1996). Making sense of the policy process for carnivore conservation. *Conservation Biology* 10: 1036–1045.

Prodöhl, P., Loughry, W., McDonough, C. M., Nelson, W. S., Thompson, E. A. & Avise, J. C. (1998). Genetic maternity and paternity in a local population of armadillos assessed by microsatellite DNA markers and field data. *American Naturalist* 151: 7–19.

Proulx, G., Kolenosk, A. J., Badry, M. J. Drescher, R. K., Seidel, K. & Cole, P. J. (1994). Post-release movements of translocated fishers. In *Martens, Sables, and Fishers. Biology and Conservation*, eds S. W. Buskirk, A. S. Harestad, M. G. Raphael & R. A. Powell, pp. 197–203. Ithaca: Cornell University Press.

Pukazhenthi, B. S., Wildt, D. E., Ottinger, M. A. & Howard, J. G. (1996). Compromised sperm protein phosphorylation after capacitation, swim-up and zona pellucida exposure in teratospermic domestic cats. *Journal of Andrology* 17: 409–419.

Pulliam, H. R. (1988). Sources, sinks, and population regulation. *The American Naturalist* 132: 652–61.

Purvis, A. (1996). Using interspecific phylogenies to test macroevolutionary hypotheses. In *New Uses for New Phylogenies*, eds P. H. Harvey, A. J. Leigh Brown, A. Maynard, J. Smith & S. Nee, pp. 153–168. Oxford: Oxford University Press.

Purvis, A., Agapow, P-M., Gittleman, J. L. & Mace, G. M. (2000a). Non-random extinction and the loss of evolutionary history. *Science* 288: 328–330.

Purvis, A., Gittleman, J. L., Cowlishaw, G., & Mace, G. M. (2000b). Predicting extinction risk in declining species. *Proceedings of the Royal Society of London, Series B*. 267: 1947–1952.

Purvis, A., Nee, S. & Harvey, P. H. (1995). Macroevolutionary inferences from primate phylogeny. *Proceedings of the Royal Society of London, Series B* 260: 329–333.

Purvis, A. & Rambaut, A. (1994). Comparative analysis by independent contrasts (CAIC), version 2. Oxford: Oxford University.

Purvis, A. & Rambaut, A. (1995). Comparative analysis by independent contrasts (CAIC): an Apple Macintosh application for analysing comparative data. *Computer Applications in Bioscience* **11**: 247–251.

Pusey, A. E. & Packer, C. (1987). The evolution of sex-biased dispersal in lions. *Behaviour* **101**: 275–310.

Putman, R. J. (Ed.) (1989). *Mammals as Pests.* London: Chapman & Hall.

Putnam, R. J. (1996). Ethical considerations and animal welfare in ecological field studies. In *Ecologists and Ethical Judgements,* eds N. S. Cooper & R. C. J. Carling, pp. 123–135. London: Chapman & Hall.

Quammen, D. (1998). *Wild Thoughts from Wild Places.* New York: Scribner.

Qiu, X. M. & Mainka, S. A. (1993). Review of mortality of the giant panda (*Ailuropoda melanoleuca*). *Journal of Zoo and Wildlife Medicine* **24**: 425–429.

Quattro, J. M., Avise, J. C. & Vrijenhoek, R. C. (1992). Mode of origin and sources of genotypic diversity in triploid gynogenetic fish clones (*Poeciliopsis: Poeciliidae*). *Genetics* **130**: 621–628

Queller, D. & Goodnight, K. (1989). Estimating relatedness using genetic markers. *Evolution* **43**: 258–275.

Quenette, P. Y., Alonso, M., Chayron, L., Cluzel, P., Dubarry, E., Dubreuil, D. & Palazon, S. (1997). First transplantation of brown bear (*Ursus arctos*) in Pyrenees (France). the first results. *11th International Conference on Bear Management & Research, European Session, September 1997, Graz, Austria.* Abstract 60.

Quick, H. F. (1944). Habits and economics of the New York weasel in Michigan. *Journal of Wildlife Management* **8**: 71–78.

Quick, H. F. (1956). Effects of exploitation on a marten population. *Journal of Wildlife Management,* **20**: 267–274.

Quigley, H. B. & Crawshaw, P. G. (1992). A conservation plan for the jaguar *Panthera onca* in the Pantanal region of Brazil. *Biology of Conservation* **61**: 149–157.

Quinn, N. W. S. & Parker, G. (1987). Lynx. In *Wild Furbearer Management and Conservation in North America,* eds M. Novak, J. A. Baker, M. E. Obbard & B. Malloch, pp. 683–694. Toronto: Ontario Ministry of Natural Resources.

Quinn, N. W. S. & Thompson, J. E. (1987). Dynamics of an exploited Canada lynx population in Ontario. *Journal of Wildlife Management* **51**: 297–305.

Rabinowitz, A. (1986). *Jaguar: One Man's Struggle to Establish the World's First Jaguar Reserve.* New York: Arbor House.

Rabinowitz, A. (1989). The density and behavior of large cats in a dry tropical forest mosaic in Huai Kha Khaeng Wildlife Sanctuary, Thailand. *Natural History Bulletin of the Siam Society* **37**: 235–251.

Rabinowitz, A. (1991). *Chasing the Dragon's Tail: The Struggle to Save Thailand's Cats.* New York: Doubleday.

Rabinowitz, A. (1993). Estimating the Indochinese tiger (*Panthera tigris corbetti*) population in Thailand. *Biological Conservation* **65**: 213–217.

Rabinowitz, A. (1995). Jaguar conflict and conservation: a strategy for the future. In *Integrating People and Wildlife for a Sustainable Future,* eds J. A. Bissonette & P. R. Krausman, pp. 394–397. Bethesda: The Wildlife Society.

Rabinowitz, A. (1999). The status of the Indochinese tiger: separating fact from fiction. In *Riding the Tiger: Tiger Conservation in Human-Dominated Landscape,* eds J. Seidensticker, S. Christie & P. Jackson, pp. 114–122. Cambridge: Cambridge University Press.

Ralls, K., Ballou, J. D. & Templeton, A. (1988). Estimates of lethal equivalents and the costs of inbreeding in mammals. *Conservation Biology* 2: 185–193.

Ralls, K., Demaster, D. P. & Estes, J. A. (1996). Developing a criterion for delisting the southern sea otter under the US endangered species act. *Conservation Biology* 10: 1528–1537.

Ralls, K. & Eberhardt, L. L. (1997). Assessment of abundance of San Joaquin kit foxes by spotlight surveys. *Journal of Mammalogy* 78: 65–73.

Ralls, K., Harvey, P. H. & Lyles, A. M. (1985). Inbreeding in natural populations of birds and mammals. In *Conservation Biology: The Science of Scarcity and Diversity*, ed. M. E. Soulé, pp. 35–56. Sunderland, MA: Sinauer Associates, Inc.

Ralls, K. & White, P. J. (1995). Predation on San Joaquin kit foxes by larger canids. *Journal of Mammalogy* 76: 723–729.

Rambaut, A., Grassly, N. C., Nee, S. & Harvey, P. H. (1996). Bi-De: an application for simulating phylogenetic processes. *Computer Applications in the Biosciences* 12: 469–471.

Randi, E., Francisci, F. & Lucchini, V. (1995). Mithocondrial DNA restriction-fragment-length monomorphism in the Italian wolf (*Canis lupus*) population. *Journal of Zoological Systematics and Evolutionary Research* 33: 97–100.

Randi, E., Gentile, L., Boscagli, G., Huber, D. & Roth, H. U. (1994). Mitochondrial DNA sequence divergence among some west European brown bear (*Ursus arctos*) populations. Lessons of conservation. *Heredity* 73: 480–489.

Rannala, B. & Mountain, J. (1997). Detecting immigration by using multilocus genotypes. *Proceedings of the National Academy of Sciences of the United States of America* 94: 9197–9201.

Rao, P. V., Portier, K. M. & Ondrasik, J. A. (1981). Density estimation using line transect sampling. In *Estimating Numbers of Birds. Studies in Avian Biology*, No. 6, ed. R. J. Raitt, pp. 441–444. Lawrence, KS: Allen Press.

Rappole, J. H., Lopez, D. N., Tewes, M. & Everett, D. (1985). Remote trip cameras as a means for surveying for nocturnal felids. In *Nocturnal Mammals: Techniques for Study*, ed. R. P. Brooks, pp. 45–53. Pennsylvania State University, School of Forest Resources, University Park.

Rasa, O. A. E. (1983). A case of invalid care in wild dwarf mongoose. *Zeitschrift für Tierpsychologie* 62: 235–240.

Rasmussen, G. S. A. (1999). Livestock predation by the painted hunting dog, *Lycaon pictus* in a cattle ranching region of Zimbabwe: a case study. *Biological Conservation* 88: 133–139.

Rauer, G. (1997). Radio-tracking results of three brown bears released by WWF Austria in the Alps of Lower Austria and Styria. *11th International Conference on Bear Management and Research, European Session, September 1997, Graz, Austria.* Abstract 62.

Raup, D. M. (1991). *Extinction: Bad Genes or Bad Luck?* New York: W.W. Norton & Company.

Raven, P. H. (1988). Tropical floristics tomorrow. *Taxon* 37: 549–560.

Ray, J. (in press). Carnivore biogeography and conservation in the African Forest: a community perspective. In *African Rain Forest Ecology and Conservation.* eds W. Weber, L. J. T. White, A. Vedder & L. Naughton-Treves, New Haven, CT: Yale University Press.

Raymond, M. & Bergeron, J.-M. (1982). Response numerique de l'hermine aux fluctuations d'abondance de Microtus pennsylvanicus. *Canadian Journal of Zoology* **60**: 542–549.

Reading, R. P. & Clark T. W. (1996). Carnivore re-introductions: an interdisciplinary examination. In *Carnivore Behavior, Ecology, and Evolution*, vol. 2, ed. J. L. Gittleman, pp. 296–336. Ithaca: Cornell University Press.

Reading, R. P., Clark, T. W. & Griffith, B. (1997). The influence of valuational and organizational considerations on the success of rare species translocations. *Biological Conservation* **79**: 217–225.

Reading, R. P. & Kellert, S. R. (1993). Attitudes toward a proposed reintroduction of black-footed ferret (*Mustela nigripes*). *Conservation Biology*, **7**: 569–580.

Reading, R. P. & Miller, B. J. (1995). The black-footed ferret recovery program: unmasking professional and organizational weaknesses. In *Endangered Species Recovery. Finding the Lessons, Improving the Process*, eds T. W. Clark, R. P. Reading & A. L. Clarke, pp. 73–100. Washington DC: Island Press.

Reading, R. P., Mix, H., Lkhagvasuren, B. & Seveenmyadag, N. (1998). The commercial harvest of wildlife in Dornod Aimag, Mongolia. *Journal of Wildlife Management* **62**: 59–71.

Redford, K. H. (1992). The empty forest. *BioScience* **42**: 412–422.

Redhead, T. D., Singleton, G. R., Myers, K. & Coman, B. J. (1991). Mammals introduced to southern Australia. In *Biogeography of Mediterranean Invasions*, eds R. Groves & F. di Castri, pp. 293–308. Cambridge: Cambridge University Press.

Reed, J., Tollit, D., Thompson, P. & Amos, W. (1997). Molecular scatology: the use of molecular genetic analysis to assign species, sex and individual identity to seal faeces. *Molecular Ecology* **6**: 225–234.

Reed, J. M. & Blaustein, A. P. (1997). Statistical power analysis and amphibian population trends. *Conservation Biology* **11**: 273–275.

Reed, J. M., Murphy, D. D. & Brussard, P. F. (1998). Efficacy of population viability analysis. *Wildlife Society Bulletin* **26**: 244–251.

Reed, R. A., Johnson-Barnard, J. & Baker, W. L. (1996). Contribution of roads to forest fragmentation in the Rocky Mountains. *Conservation Biology* **10**: 1098–1106.

Reeves, R. R. & Ling, J. K. (1981). Hooded seal – *Cystophora cristata*. In *Handbook of Marine Mammals*, vol. 2, eds S. Ridgeway & R. J. Harrison, pp. 171–194. London: Academic Press.

Regan, T. (1983). *The Case for Animal Rights*. Berkeley: University of California Press.

Reich, D. E., Wayne, R. K. & Goldstein, D. B. (1999). Genetic evidence for a recent origin by hybridization of red wolves. *Molecular Ecology* **8**: 139–114.

Reid, D. G., Krebs, C. J. & Kenney, A. J. (1997). Patterns of predation on non-cyclic lemmings. *Ecological Monographs* **67**: 89–108.

Reid, W. V. (1998). Biodiversity hotspots. *Trends in Evolution and Ecology* **13**: 275–280.

Reig, S., Daniels, M. J. & Macdonald, D. W. (2001) Craniometric differentiation within wild-living cats in Scotland using 3D morphometrics. *Journal of Zoology (London)*. (In press.)

Reijnders, P., Brasseur, P., van der Toorn, J., van der Wolf, P., Boyd, I., Harwood, J., Lavigne, D. & Lowry, L. (1993). *Seals, Fur Seals, Sea Lions, and Walrus. Status Survey and Conservation Action Plan*. Gland: IUCN.

Rejmanek, M. & Richardson, D. M. (1996). What attributes make some plant species more invasive? *Ecology* **77**: 1655–1661.

Rensburg, van. P. J. J., Skinner, J. D. & van Arde, R. J. (1987). Effects of feline leucopaenia on the population characteristics of feral cats on Marion Island. *Journal of Applied Ecology* **24**: 65–73.

Reubel, G. H. Dean, G. A., George, J. W., Barlough, J. E. & Pedersen, N. C. (1994). Effects of incidental infections and immune activation on disease progression in experimentally feline immunodeficiency virus-infected cats. *Journal of Acquired Immune Deficiency Syndromes and Human Retrovirology* **7**: 1003–1015.

Reyers, B., van Jaarsveld, A. S., McGeoch, M. A. & James, A. N. (1998). National biodiversity risk assessment: a composite multivariate and index approach. *Biodiversity and Conservation* **7**: 945–65.

Reynolds, J. (1983). *A Plan for the Re-introduction of Swift Fox to the Canadian Prairies*. Master's thesis, University of Alberta, Calgary.

Reynolds, J. C. & Tapper, S. C. (1996). Control of mammalian predators in game management and conservation. *Mammalian Review* **26**: 127–156.

Rhodes, C. J., Atkinson, R. P. D., Anderson, R. M. & Macdonald, D. W. (1998). Rabies in Zimbabwe: reservoir dogs and the implications for disease control. *Philosophical Transactions of the Royal Society of London, Series B* **353**: 999–1010.

Rhymer, J. M. & Simberloff, D. (1996). Extinction by hybridization and introgression. *Annual Review of Ecology and Systematics* **27**: 83–109.

Richards, J. F. (1990). Land transformation. In *The Earth Transformed by Human Action*, eds B. L. Turner, II, W. C. Clark, R. W. Kates, J. F. Richards, J. T. Mathews & W. B. Meyer, pp. 163–178. Cambridge: Cambridge University Press.

Richardson, L., Clark, T. W., Forrest, S. C. & Campbell, III, T. M. (1985). Snow-tracking as a method to search for and study black-footed ferret. In *Proceedings of the Black-Footed Ferret Workshop*, eds S. H. Anderson & D. B. Inkley, pp. 25.1–25.11. Laramie: University of Wyoming.

Ricklefs, R. E. (1990). *Ecology*, 3rd edn. New York: W. H. Freeman & Co.

Riedman, M. (1990). *The Pinnipeds*. Berkeley: University of California Press.

Rieseberg, L. H., Archer, M. A. & Wayne, R. K. (1999). Transgressive segregation, adaptation and speciation. *Heredity* **83**: 363–372.

Rishi, V. (1988). Man, mask and maneater. *Tigerpaper* **15**: 9–14.

Ritland, K. (1996). Estimators for pairwise relatedness and individual inbreeding coefficients. *Genetic Research* **67**: 175–185.

RMAD (1999). Lynx fact sheet. http://rmad.org/lynxfact.html

Robbins, A. H., Borden, M. D., Windmiller, B. S., Niezgoda, M., Marcus, L. C., O'Brien, S. M., Kreindel, S. M., McGuill, M. W., DeMaria, A. Jr., Rupprecht, C. E. & Rowell, S. (1998). Prevention of the spread of rabies to wildlife by oral vaccination of raccoons in Massachusetts. *Journal of the American Veterinary Medical Association* **213**: 1407–1412.

Robbins, J. (1997). In two years, wolves reshaped Yellowstone. *New York Times* 147, December 30: B13, F1(1).

Robinson, J. G. & Redford, K. H. (Eds) (1991). *Neotropical Wildlife Use and Conservation*. Chicago: University of Chicago Press.

Robinson, J. R. & Bennett, E. L. (1999). *Hunting for Sustainability in Tropical Forests*. New York: Columbia University Press.

Robinson, J. R., Redford, K. H. & Bennett, E. L. (1999). Wildlife harvest in logged tropical forests. *Science* **284**: 595–596.

Rodgers, A. (1974). The lion population of the Eastern Selous Game Reserve. *East African Wildlife Journal* **12**: 313–317.

Rodriguez, A. & Delibes, M. (1992). Current range and status of the Iberian lynx, *Felis pardina* Temminck, 1824 in Spain. *Biological Conservation* **61**: 189–196.

Rodriguez, E. & Wrangham, R. (1993). Zoopharmacognosy: the use of medicinal plants by animals. In *Phytochemical Potential of Tropical Plants*, eds K. R. Downum *et al.*, pp. 89–105. New York: Plenum Press.

Roelke, M. E., Martenson, J. S. & O'Brien, S. J. (1993a). The consequences of demographic reduction and genetic depletion in the endangered Florida Panther. *Current Biology* **3**: 340–349.

Roelke, M. E., Forrester, D. J., Jacobson, E. R., Kollias, G. V., Scott, F. W., Barr, M. C., Evermann, J. F. & Pirtle, E. C. (1993b). Seroprevalence of infectious disease agents in free-ranging Florida panthers (*Felis concolor coryi*). *Journal of Wildlife Diseases* **29**: 36–49.

Roelke-Parker, M. E., Munson, L., Packer, C., Kock, R., Cleaveland, S., Carpenter, M., O'Brien, S. J., Pospischil, A., Hofmann-Lehmann, R., Lutz, H., Mwamengele, G. L. M., Mgasa, M. N., Machange, G. A., Summers, B. A. & Appel, M. J. G. (1996). A canine distemper virus epidemic in Serengeti lions (*Panthera leo*). *Nature* **379**: 441–445.

Roemer, G. W., Garcelon, D. K., Coonan, T. J. & Schwemm, C. (1994). The use of capture-recapture methods for estimating, monitoring, and conserving island fox populations. In *4th California Islands Symposium: Update on the Status of Resources*, eds W. L. Halvorson & G. J. Maender, pp. 387–400. Santa Barbara: Santa Barbara Museum of Natural History.

Rogers, L. L. (1987). Effects of food supply and kinship on social behavior, movements, and population growth of black bears in northeastern Minnesota. *Wildlife Monographs* **97**: 1–72.

Rolley, R. E. (1987). Bobcat. In *Wild Furbearer Management and Conservation in North America*, eds M. Novak, J. A. Baker, M. E. Obbard & B. Malloch, pp. 671–681. North Bay: Ontario Trappers Association.

Rood, J. P. (1987). Dispersal and intergroup transfer in the dwarf mongoose. In *Mammalian Dispersal Patterns: the Effects of Social Structure on Population Genetics*, eds B. D. Chepko-Sade & Z. T. Halpin, pp. 85–103. Chicago: University of Chicago Press.

Rosatte, R. C. (1987). Skunks. In *Wild Furbearer Management and Conservation in North America*, eds M. Novak, J. A. Baker, M. E. Obbard & B. Malloch, pp. 599–613. North Bay: Ontario Trappers Association.

Rose, M. D. & Polis, G. A. (1998). The distribution and abundance of coyotes: the effects of allochthonous food subsidies from the sea. *Ecology* **79**: 998–1007.

Ross, P. I. & Jalkotzy, M. G. (1992). Characteristics of a hunted population of cougars in south-west Alberta. *Journal of Wildlife Management* **56**: 679–690.

Ross, P. S., De Swart, R. L., Timmerman, H. H., Reijinders, P. J. H., Vos, J. G., Van Loveren, H. & Osterhaus, A. D. M. E. (1996). Suppression of natural killer cell activity in harbour seals (*Phoca vitulina*) fed Baltic Sea herring. *Aquatic Toxicology* **34**: 71–84.

Roughton, R. J. (1979). Developments in scent station technology. In *Proceedings of the Midwest Furbearer Conference*, pp. 17–44. Manhattan: Kansas State University, Coop. Ext. Serv.

Roughton, R. J. & Sweeny, M. W. (1982). Refinements in scent-station methodology for assessing trends in carnivore populations. *Journal of Wildlife Management* 46: 217–229.

Roy, C. C. (1997). River otter restoration in Kansas. *15th Midwest Furbearer Workshop, Carbondale, Illinoise*. Abstract.

Roy, M. S., Geffen, E., Smith, D., Ostrander, E. A. & Wayne, R. K. (1994a). Patterns of differentiation and hybridization in North American wolf-like canids, revealed by analysis of microsatellite loci. *Molecular Biology and Evolution* 11: 553–570.

Roy, M. S., Geffen, E., Smith, D. & Wayne, R. K. (1996a). Molecular genetics of pre-1940 red wolves. *Conservation Biology* 10: 1413–1424.

Roy, M. S., Girman, D. J., Taylor, A. C. & Wayne, R. K. (1994b). The use of museum specimens to reconstruct the genetic variability and relationships of extinct populations. *Experientia* 15: 551–557.

Roy, K., Jablonski, D. & Valentine, J. W. (1996b). Higher taxa in biodiversity studies: patterns from eastern Pacific molluscs. *Philosophical Transactions of the Royal Society of London, Series B* 351: 1605–1613.

Royama, T. (1992). *Analytical Population Dynamics*. London: Chapman and Hall.

Rozhnov, V. V. (1993). Extinction of the European mink: ecological catastrophe or natural process? *Lutreola* 1: 10–16.

Rudnai, J. A. (1973). *The Social Life of the Lion*. Wallingford, PA: Washington Square East Publishers.

Rudnai, J. (1979). Ecology of lions in Nairobi National Park and the adjoining Kitengela Conservation Unit in Kenya. *African Journal of Ecology* 17: 85–95.

Ruff, F. J. (1939). Region 8 techniques of wildlife inventory. *Transactions of the North American Wildlife Conference* 4: 542–545.

Rushton, S. P., Barreto, G. W., Cormack, R. M., Macdonald, D. W. & Fuller, R. (2000). Modelling the effects of mink and habitat fragmentation on the water vole. *Journal of Applied Ecology* 37: 1–17.

Russell, D. N. (1971). History and status of the felids of Texas. In *Proceedings of a Sympsoium on the Native Cats of North America – Their Status and Management*, eds S. E. Jorgensen & L. D. Mech, Twin Cities, Minnesota: United States Department of the Interior.

Russell, G. J. (1998). Turnover dynamics across ecological and geological scales. In *Biodiversity Dynamics*, eds M. L. McKinney & J. A. Drake, pp. 377–404. New York: Columbia University Press.

Russell, G. J., Brooks, T. M., McKinney, M. L. & Anderson, C. G. (1998). Present and future taxonomic selectivity in bird and mammal extinctions. *Conservation Biology* 12: 1365–1376.

Russell, J. K. (1979). *Reciprocity in the social behavior of coatis* (Nasua narica). PhD Thesis, University of North Carolina, Chapel Hill.

Russell, R. H. & Zendran, J. (1983). A proposal to re-introduce the swift fox (*Vulpes velox*) to Alberta. *Canadian Wildlife Service Report*.

Rybarczyk, W. B., Andrews, R. D., Klass, E. E. & Kienzler, J. M. (1981). Raccoon spotlight survey technique: a potential population trend indicator. In *Proceedings of the Worldwide Furbearer Conference*, eds J. A. Chapman & D. Pursley, pp. 1413–1430. Frostburg, MD.

Ryder, O. A. (1986). Species conservation and systematics: the dilemma of subspecies. *Trends in Ecology* 1: 9–10.

Rylands, A. B. & Keuroghlian, A. (1988). Primate populations in continuous forest and forest fragments in central Amazonia. *Acta Amazônica* 18: 291–307.

Saberwal, V. K., Gibbs, J. P., Chellam, R. & Johnsingh, A. J. T. (1994). Lion–human conflict in the Gir Forest, India. *Conservation Biology* 8: 501–507.

Sadleir, R. M. F. S. (1969). The role of nutrition in the reproduction of wild mammals. *Journal of Reproduction And Fertility Supplement* 6: 29–48.

Safalsky, N. (1994). Ecological limits and opportunities for community-based conservation. In *Natural Connections: Perspectives in Community-Based Conservation*, eds D. Western & R. Wright, pp. 448–471. Washington DC: Island Press.

Sagor, J. T., Swenson, J. E. & Roskaft, E. (1997). Compatibility of brown bear *Ursus arctos* and free-ranging sheep in Norway. *Biological Conservation* 81: 91–95.

Samuel, M. D., Garton, E. O., Schlegel, M. W. & Carson, R. G. (1987). Visibility bias during aerial surveys of elk in northcentral Idaho. *Journal of Wildlife Management* 51: 622–630.

Sandell, M. (1989). The mating tactics and spacing patterns of solitary carnivores. In *Carnivore Behavior, Ecology, and Evolution*, ed. J. L. Gittleman, pp. 164–182. London: Chapman & Hall.

Sanderson, G. C. (1951a). The status of the raccoon in Iowa for the past twenty years as revealed by fur reports. *Proceedings of the Iowa Academy of Sciences* 58: 527–531.

Sanderson, G. C. (1951b). Breeding habits and a history of the Missouri raccoon population from 1941 to 1948. *Transactions of the North American Wildlife Conference* 16: 445–461.

Sanderson, G. C. (1987). Raccoon. In *Wild Furbearer Management and Conservation in North America*, eds M. Novak, J. A. Baker, M. E. Obbard & B. Malloch, pp. 487–499. Toronto: Ontario Ministry of Natural Resources.

Santiapillai, C. & Jackson, P. (1990). *The Asian Elephant: An Action Plan for its Conservation*. Gland: IUCN/SSC Asian Elephant Specialist Group.

Sargeant, A. B. & Allen, S. H. (1989). Observed interactions between coyotes and red foxes. *Journal of Mammalogy* 70: 631–633.

Sargeant, A. B., Allen, S. H. & Hastings, J. O. (1987). Spatial relationships between sympatric coyotes and red foxes in North Dakota. *Journal of Wildlife Management* 51: 285–293.

Sargeant, A. B., Greenwood, R. J., Sovada, M. A. & Shaffer, T. L. (1993). *Distribution and Abundance of Predators that Affect Duck Production – Prairie Pothole Region*. USDI Fish and Wildlife Service, Resource Publication No. 194.

Sargeant, G. A., Johnson, D. H. & Berg, W. E. (1998). Interpreting carnivore scent-station surveys. *Journal of Wildlife Management* 62: 1235–1245.

Saunders, J. K., Jr. (1961). *The biology of the Newfoundland lynx* (Lynx canadensis subsolanus *Bangs*). PhD Thesis, Cornell Univ., Ithaca, New York.

Sausman, K. (1997). Sand cat: a true desert species. *International Zoo Yearbook* 35: 78–81.

Savidge, J. A. & Seibert, T. F. (1988). An infrared trigger and camera to identify predators at artificial nests. *Journal of Wildlife Management* 52: 291–294.

Schaller, G. B. (1967). *The Deer and the Tiger*. Chicago: University of Chicago Press.

Schaller, G. (1972). *The Serengeti Lion: A Study of Predator–Prey Relations*. Chicago: University of Chicago Press.

Schaller, G. B. (1983). Mammals and their biomass on a Brazilian ranch. *Arquivos de Zoologia* **31**: 1–36.

Schaller, G. B. (1993). *The Last Panda*. Chicago: University of Chicago Press.

Schaller, G. B. (1996). Introduction: carnivores and conservation biology. In *Carnivore Behavior, Ecology, and Evolution*, vol. 2, ed. J. L. Gittleman, pp. 1–10. Ithaca: Cornell University Press.

Schaller, G. B. (1998). *Wildlife of the Tibetan Steppes*. Chicago: University of Chicago Press.

Schaller, G. B. & Crawshaw, Jr. P. G. (1980). Movement patterns of jaguar. *Biotropica* **12**: 161–168.

Schaller, G. B., Hu , J., Pan, W. & Zhu, J. (1985). *The Giant Pandas of Wolong*. Chicago: University of Chicago Press.

Scheel, D. & Packer, C. (1991). Group hunting behaviour of lions: a search for cooperation. *Animal Behaviour* **41**: 697–709.

Scheepers, J. L. & Venzke, K. A. E. (1995). Attempts to reintroduce African wild dogs *Lycaon pictus* into Etosha National Park, Namibia. *South African Journal of Wildlife Research* **25**: 138–140.

Schemnitz, S. D. (1994). Capturing and handling wild animals. In *Research and Management Techniques for Wildlife and Habitats*, 5th edn, ed. T. A. Bookhout, pp. 106–124. Bethesda: The Wildlife Society.

Schitoskey, F. (1975). Primary and secondary hazards of three rodenticides to kit fox. *Journal of Wildlife Management* **39**: 416–417.

Schmidt, G. (1993). *Progress Report of Riding Mountain National Park American Pine Marten Reintroduction Program 1992–93*. Parks Canada, Riding Mountain National Park.

Schmidt, K. (1998). Maternal behaviour and juvenile dispersal in the Eurasian lynx. *Acta Theriologica* **43**: 391–408.

Schmidt, R. H. (1985). Controlling arctic fox populations with introduced red foxes. *Wildlife Society Bulletin* **13**: 592–594.

Schneider, R. R. & Yodzis, P. (1994). Extinction dynamics in the American marten (*Martes americana*). *Conservation Biology* **8**: 1058–1068.

Schoener, T. W. (1977). Competition and the niche. In *Biology of the Reptilia*, vol. 7, *Ecology*, eds A. C. Gans & D. W. Tinkle, pp. 35–136. New York: Academic Press.

Schoener, T. W. (1983). Field experiments on interspecific competition. *American Naturalist* **122**: 240–285.

Schonewald-Cox, C., Aari, R. & Blume, S. (1991). Scale, variable density, and conservation planning for mammalian carnivores. *Conservation Biology* **5**: 491–495.

Schowalter, D. B. & Gunson, J. R. (1982). Parameters of population and seasonal activity of striped skunks, *Mephitis mephitis*, in Alberta and Saskatchewan. *Canadian Field-Naturalist* **96**: 409–420.

Schrag, S. J. & Wiener, P. (1995). Emerging infectious disease: what are the relative roles of ecology and evolution? *Trends in Ecology and Evolution* **10**: 319–324.

Schreiber, A., Wirth, R., Riffel, M. & Van Rompaey, H. (Eds) (1989). *Weasels, Civets, Mongooses and their Relatives. An Action Plan for the Conservation of Mustelids and Viverrids*. Gland: IUCN.

Schubert, C. A., Barker, I. K., Rosatte, R. C., MacInnes, C. D. & Nudds, T. D. (1998a). Effect of canine distemper on an urban raccoon population: an experiment. *Ecological Applications* **8**: 379–387.

Schubert, C. A., Rosatte, R. C., MacInnis, C. D. & Nudds, T. D. (1998b). Rabies control: an adaptive management approach. *Journal of Wildlife Management* **62**: 622–629.

Schwartz, M. (1997). *A History of Dogs in the Early Americas.* New Haven, CT: Yale University Press.

Schwartz, M., Tallmon, D. & Luikart, G. (1998). Review of DNA-based census and effective population size estimators. *Animal Conservation* **1**: 293–299.

Schwartz, M. K., Tallmon, D. A. & Luikart, G. (1999). DNA-based methods for estimating population size: many methods, much potential, uncertain utility. *Animal Conservation* **2**: 321–323.

Schwarzkopf, L. & Rylands, A. B. (1989). Primate species richness in relation to habitat structure in Amazonian rainforest fragments. *Biological Conservation* **48**: 1–12.

Scott, J. M., Murray, D. & Griffith, B. (1999). Lynx reintroduction. *Science* **286**: 49–55.

Scott, R. F., Kenyon, K. W., Buckley, J. L. & Olson, S. T. (1959). Status and management of the polar bear and pacific walrus. *Transactions of the North American Wildlife Conference* **24**: 366–374.

Scott-Brown, J. M., Herrero, S. & Reynolds, J. (1987). Swift fox. In *Wild Furbearer Conservation and Management in North America*, eds M. Novak, J. A. Baker, M. E. Obbard & B. Molloch, pp. 432–441. Toronto: Ontario Ministry of Natural Resources.

Seal, U. S. (Ed.) (1994). *A Plan for Genetic Restoration and Management of the Florida Panther* (Felis concolor coryi). Report to the Florida Game and Fresh Water Fish Commission. Apple Valley, MN: Conservation Breeding Specialist Group.

Seal, U. S., Jackson, P. & Tilson, R. L. (1987). A Global tiger Conservation Plan. In *Tigers of the World*, eds R. L. Tilson & U. S. Seal, pp. 487–498. Park Ridge, NJ: Noyes Publications.

Seber, G. A. F. (1982). *The Estimation of Animal Abundance and Related Parameters*, 2nd edn. London: Griffin.

Seddon, P. J. (1999). Reintroductions, introductions, and the importance of post-release monitorings: lessons from Zanzibar. *Oryx* **33**: 89–90.

Seidel, J. (1998). Clawmarks. *Colorado Division of Wildlife* **1**: 1–4.

Seidel, J., Andree, B., Berlinger, S., Byrne, G., Gill, B., Kenvin, D. & Reed, D. (1998). *Draft Strategy for Conservation and Reestablishment of Lynx and Wolverine in the Southern Rocky Mountains*. Denver: Colorado Division of Wildlife.

Seidensticker, J. (1976). On the ecological separation between tigers and leopards. *Biotropica* **8**: 225–234.

Seidensticker, J. (1986). Large carnivores and the consequences of habitat insularization: Ecology and conservation of tigers in Indonesia and Bangladesh. In *Cats of the World: Biology, Conservation, and Management*, eds S. D. Miller & D. D. Everett, pp. 1–41. Washington: National Wildlife Federation.

Seidensticker, J., Christie, S. & Jackson, P. (Eds) (1999). *Riding the Tiger: Tiger Conservation in Human-Dominated Landscapes*. New York: Cambridge University Press.

Seidensticker, J. C., IV, Hornocker, M. G., Wiles, W. V. & Messick, J. P. (1973). Mountain lion social organization in the Idaho Primitive Area. *Wildlife Monographs* **35**: 1–60.

Seidensticker, J. & Lumpkin, S. (1992). *Great Cats*. London: Merehurst.

Seidensticker, J., Sunquist, M. & McDougal, C. (1990). Leopards living at the edge of Royal Chitwan National Park, Nepal. In *Conservation in Developing Countries: Problems and Prospects*, eds J. C. Daniel & J. S. Serrao, pp. 415–423. Proceedings of the Centenary Seminar of the Bombay Natural History Society. Bombay: Oxford University Press.

Sepkoski Jr, J. J. (1991). Population biology models in macroevolution. In *Analytical Paleobiology: Short Courses in Paleontology*, No. 4, eds N. L. Gilinsky & P. W. Signor, pp. 136–156. A Publication of the Paleontological Society.

Sepkoski Jr, J. J. (1998). Rates of speciation in the fossil record. *Proceedings of the Royal Society of London, Series B* **353**: 315–326.

Serfass, T. L. (1998). River otter *Lutra canadensis* re-introduction: a conflict in New York, USA. *Re-introduction News* **16**: 11–2.

Serfass, T. L., Brooks, R. P., Novak, J. M., Johns, P. E. & Rhodes, O. E., Jr. (1998). Genetic variation among populations of river otters in North America: considerations for reintroduction projects. *Journal of Mammalogy* **79**: 736–746.

Serpell, J. (1991). Playing favourites: analysis of results of BBC Wildlife's favourite animal survey. *BBC Wildlife*, December.

Servheen, C., Herrero, S. & Peyton, B. (1999). *Bears – Status Survey and Conservation Action Plan*. Gland: IUCN.

Seton, E. T. (1909). *The Arctic Prairies*. New York: Charles Scribner's Sons.

Seymour, K. L. (1989). Panthera onca. *Mammalian Species* **340**: 1–9.

Shaffer, M. L. (1981). Minimum population sizes for species conservation. *BioScience* **31**: 131–134.

Shaffer, M. L. (1983). Determining minimum viable population sizes for grizzly bear. *International Conference Bear Research & Management* **5**: 133–139.

Shaffer, M. L. (1992). Keeping the grizzly bear in the American West: a strategy for real recovery. Washington: The Wilderness Society.

Shahi, S. P. (1983). Status of the grey wolf (*Canis lupus pallipes*, Sykes) in India. *Acta Zoologica Fennica* **174**: 283–286.

Shankaranayanan, P., Banjeree, M., Kacker, R. K., Aggarwal, R. K. & Singh, L. (1997). Genetic variation in Asiatic lions and Indian tigers. *Electrophoresis* **18**: 1693–1700.

Shankaranayanan, P. & Singh, L. (1998). Mitochondrial DNA sequence divergence among big cats and their hybrids. *Current Science* **75**: 919–923.

Shaughnessy, P. D. & Fletcher, L. (1987). Fur seals, *Arctocephalus* spp., at Macquarie Island. In *Status, Biology, and Ecology of Fur Seals*, eds J. P. Croxall & R. L. Gentry, pp. 177–187. NOAA Technical Report NMFS. **51**.

Shaugnessy, P. D. & Payne, A. I. L. (1979). Incidental mortality of cape fur seals during trawl fishing activities in South African waters. *Fisheries Bulletin (South Africa)* **12**: 20–25.

Shaw, J. H. (1993). The outlook for sustainable harvest of wildlife in Latin America. In *Neotropical Wildlife Use and Conservation*, eds J. G. Robinson & K. H. Redford, pp. 24–34. Chicago: University of Chicago Press.

Shea, K. & NCEAS Working Group on Population Management (1998). Management of populations in conservation, harvesting and control. *Trends in Ecology and Evolution* **13**: 371–375.

Sherburne, S. S. & Bissonette, J. A. (1994). Marten subnivean access point use: response to subnivean prey levels. *Journal of Wildlife Management* **58**: 400–405.

Shieff, A. & Baker, J. A. (1987). Marketing and international fur markets. In *Wild Furbearer Management and Conservation in North America*. eds M. Novak, J. A. Baker, M. E. Obbard & B. Malloch, pp. 862–877. Toronto: Ontario Trappers Association North Bay.

Shields, G. F. & Kocher, T. D. (1991). Phylogenetic relationships of North American ursids based on analysis of mitochondrial DNA. *Evolution* **45**: 218–221.

Shields, W. M. (1987). Dispersal and mating systems: Investigating their causal connections. In *Mammalian Dispersal Patterns*, eds B. D. Chepko-Sade & Z. T. Halpin, pp. 3–24. Chicago: University of Chicago Press.

Shigesada, N. & Kawasaki, K. (1997). *Biological Invasions: Theory and Practice*. Oxford: Oxford University Press.

Short, J., Bradshaw, S. D., Giles, J., Prince, R. I. T. & Wilson, G. R. (1992). Reintroduction of macropods (*Marsupialia: Macropodoidea*) in Australia – a review. *Biological Conservation* **62**: 189–204.

Short, J., Turner, B., Risbey, D. A. & Carnamah, R. (1997). Control of feral cats for nature conservation. II. Population reduction by poisoning. *Wildlife Research*, 703–714.

Shortridge, G. C. (1934). *The Mammals of South-West Africa*. London: William Heinemann Ltd.

Shriver, M. D., Smith, M. W., Jin, L., Marcini, A., Akey, J. M., Deka, R. & Ferrell, R. E. (1997). Ethnic-affiliation estimation by use of population-specific DNA markers. *American Journal of Human Genetics* **60**: 957–964.

Sidorovich, V. E., Kruuk, H. & Macdonald, D. W. (1999). Body size, and interactions between European and American mink (*Mustela lutreola* and *M. vison*) in Eastern Europe. *Journal of Zoology (London)* **248**: 521–527.

Sidorovich, V., Kruuk, H., Macdonald, D. W. & Maran, T. (1998). Diets of semiaquatic carnivores in northern Belarus, with implications for population changes. In *Behaviour and Ecology of Riparian Mammals*, eds N. Dunstone & M. Gorman, pp. 177–189. Cambridge: Cambridge University Press.

Sillero-Zubiri, C. & Gottelli, D. (1991). Aberdare rhinos: predation versus poaching. *Pachyderm* **14**: 37–38.

Sillero-Zubiri, C. & Gottelli, D. (1993). Population ecology of spotted hyaena in an equatorial mountain forest. *African Journal of Ecology* **30**: 292–300.

Sillero-Zubiri, C. (1996). *Ajag versus Banteng: a Conservation Dilemma*. Unpublished report. Oxford: IUCN/SSC Canid Specialist Group.

Sillero-Zubiri, C., King, A. A. & Macdonald, D. W. (1996). Rabies and mortality in Ethiopian wolves (*Canis simensis*). *Journal of Wildlife Diseases* **32**: 80–86.

Sillero-Zubiri, C. & Macdonald, D. W. (Eds) (1997). *The Ethiopian Wolf: Status Survey and Conservation Action Plan*. Gland: IUCN Canid Specialist Group.

Simberloff, D. (1981). Community effects of introduced species. In *Biotic Crises in Ecological and Evolutionary Time*, ed. M. Nitecki, pp. 53–81. New York: Academic Press.

Simberloff, D. (1986). The proximate causes of extinction. In *Patterns and Processes in the History of Life*, eds D. M. Raup & D. Jablonski, pp. 259–276. Heidelberg: Springer-Verlag.

Simberloff, D. (1995). Why do introduced species appear to devastate islands more than mainland areas? *Pacific Science* 49: 87–97.

Simberloff, D. (1996). Hybridization between native and introduced wildlife species: importance for conservation. *Wildlife Biology* 2: 143–150.

Simberloff, D. & Cox, J. (1987). Consequences and costs of conservation corridors. *Conservation Biology* 1: 63–71.

Simms, D. A. (1979). Studies of an ermine population in southern Ontario. *Canadian Journal of Zoology* 57: 824–832.

Simon, H. A. (1983). *Reason in Human Affairs*. Stanford: Stanford University Press.

Simonetti, J. A. (1995). Wildlife conservation outside parks is a disease-mediated risk. *Conservation Biology* 9: 454–456.

Simonsen, V. (1982). Electrophoretic variation in large mammals. II. the red fox, *Vulpes vulpes*, the stoat, *Mustela putorius*, the pine marten, *Martes martes*, the beech marten, *Martes foina*, and the badger, *Meles meles*. *Hereditas* 96: 299–305.

Simpson, G. G. (1953). *The Major Features of Evolution*. New York: Columbia University Press.

Sinclair, A. R. E. (1989). Population regulation in animals. In *Ecological Concepts*, eds J. Cherrett & A. Bradshaw, pp. 197–241. Oxford: Blackwell Scientific Publications.

Sinclair. A. R. E. (1995). Serengeti past and present. In *Serengeti II: Dynamics, Conservation and Management of an Ecosystem*, eds A. R. E. Sinclair & P. Arcese, pp. 3–30. Chicago: University of Chicago Press.

Sinclair, A. R. E., Pech, R. P., Dickman, C. R., Hik, D., Mahon, P. & Newsome, A. E. (1998). Predicting effects of predation on conservation of endangered prey. *Conservation Biology* 12: 563–575.

Singer, P. (1990). *Animal Liberation*. New York: New York Review of Books.

Singer, S. F. (1994). Problems and strategies in the scientific management of fisheries and marine mammals: from the 'tragedy of the commons' to an era of sustainable development. *Environmental Conservation* 21: 184–185.

Singh, A., Ramachandran, B., Fosnight, G., Chenoweth, S. & Crawford, T. (1998). Biodiversity-Rich Ecoregions in Africa Need Protection. http://grid2.cr.usgs.gov/publications/biodiversity/biodiversity.html

Singh, H. S. & Kamboj, R. D. (1996). Predation pattern of the Asiatic lion on domestic livestock. *Indian Forester* 122: 869–876.

Singh, V. B. (1991). How man-eating started in the Corbett. *Indian Forester* 117: 799–803.

Sjöåsen, T. (1995). Re-introduction of European otters in Sweden. *Re-Introduction News* 10: 10–1.

Sjöåsen, T. & Sandegren, F. (1992). Reintroduction of Eurasian otters (*Lutra lutra*) in southern Sweden 1987–1992: a project in progress. In *Habitat 7: Otterschutz in Deutschland*, ed. C. Reuter, pp. 143–146. Hankensbüttel.

Skalski, J. R. A. & Robson, D. S. (1992). *Techniques for Wildlife Investigations*. London: Academic Press.

Skalski, J. R., Simmons, M. A. & Robson, D. S. (1984). The use of removal sampling in comparative censuses. *Ecology* 65: 1006–1015.

Skead, C. J. (1987). *Historical Mammal Incidence in the Cape Province*, vol. 2, *The Eastern Half of the Cape Province, including the Ciskei, Transkei and East Griqualand*. Cape Town: Department of Nature and Environmental Conservation.

Skinner, C. A., Skinner, P. J. & Harris, S. (1991). An analysis of some of the factors affecting the current distribution of badger *Meles meles* setts in Essex. *Mammal Review* 21: 51–65.

Skinner, J. D. & Smithers, R. H. N. (1990). *The Mammals of the Southern African Subregion*, 2nd edn. Pretoria: Pretoria University Press.

Skirrow, M. B. (1994). Diseases due to *Campylobacter, Helicobacter*, and related bacteria. *Journal of Comparative Pathology* 111: 113–149.

Slate, J., Marshall, T. & Pemberton, J. (2000). A retrospective assessment of the accuracy of the paternity inference program CERVUS. *Molecular Ecology* 9: 801–808.

Slatkin, M. (1985). Gene flow in natural populations. *Annual Review of Ecology and Systematics* 16: 393–430.

Slatkin, M. (1987). Gene flow and the geographic structure of natural populations. *Science* 236: 787–792.

Slatkin, M. (1994). Gene flow and population structure. In *Ecological Genetics*, ed. L. A. Real, pp. 3–17. Princeton: Princeton University Press.

Sleeman, D. P. (1992). Long-distance movements in an Irish badger population. In *Wildlife Telemetry*, eds I. G. Priede & S. M. Swift, pp. 670–676. Chichester: Ellis Horwood.

Sleeman, D. P. & Mulcahy, M. F. (1993). Behavior of Irish badgers in relation to bovine tuberculosis. In *The Badger*. ed. T. J. Hayden, Dublin: Royal Irish Academy.

Slooten, E. & Dawson, S. M. (1995). Conservation of marine mammals in New Zealand Pacific. *Conversation Biology* 2: 64–76.

Slough, B. G. (1994). Translocation of American martens: an evaluation of factors in success. In *Martens, Sables, and Fishers. Biology and Conservation*, eds S. W. Buskirk, A. S Harestad, M. G. Raphael & R. A. Powell, pp. 165–178. Ithaca: Cornell University Press.

Slough, B. G. & Smits, C. M. (1985). *Yukon Marten Management, Progress to August, 1985*. Whitehorse: Yukon Department of Renewable Resources Progress Report.

Smal, C. M. (1991). Population studies on feral mink *Mustela vison* in Ireland. *Journal of Zoology (London)* 224: 233–249.

Smallwood, K. S. (1999). Scale domains of abundance amongst species of mammalian Carnivora. *Environmental Conservation* 26: 102–11.

Smallwood, K. S. & Fitzhugh, E. L. (1993). A rigorous technique for identifying individual mountain lions, *Felis concolor*, by their tracks. *Biological Conservation* 65: 51–59.

Smallwood, K. S. & Fitzhugh, E. L. (1995). A track count for estimating mountain lion, *Felis concolor californica*, population trend. *Biological Conservation* 71: 251–259.

Smallwood, K. S. & Schonewald, C. (1996). Scaling population density and spatial pattern for terrestrial, mammalian carnivores. *Oecologia* 105: 329–335.

Smallwood, K. S. & Schonewald, C. (1998). Study design and interpretation of mammalian carnivore density estimates. *Oecologia* 113: 474–491.

Smith, A. B. (1994). *Systematics and the Fossil Record*. Oxford: Blackwell Scientific.

Smith, A. P. & Quin, D. G. (1996). Patterns and causes of extinctions and decline in Australian conilurine rodents. *Biological Conservation* 77: 243–267.

Smith, B. L. (1995). Education to promote male-selective harvest of grizzly bears in the Yukon. In *Integrating People and Wildlife for a Sustainable Future*, eds J. A. Bissonette & P. R. Krausmann, pp. 156–174. Bethesda: The Wildlife Society.

Smith, D., Meier, T., Geffen, E. *et al.* (1997). Is inbreeding common in wolf packs? *Behavioral Ecology* 8: 384–391.

Smith, D. J., Harris, L. D. & Mazzaotti, F. J. (1996). A landscape approach to examining the impacts of roads on the ecological function associated with wildlife movement and movement corridors: Problems and solutions. In *Proceedings of the Transportation Related Wildlife Mortality Seminar*, eds G. L. Evink, P. Garrett, D. Zeigler & J. Berry, pp. 58–96. FL-ER. Tallahassee: Florida Department of Transportation.

Smith, D. W. & Phillips, M. K. (2000). Gray wolf. In *Endangered Species: Conflict and Context*, eds R. P. Reading & B. J. Miller, pp. Westport, CT: Greenwood Press.

Smith, H. S. (1935). The role of biotic factors in the determination of population densities. *Journal of Economic Entomology* 28: 873–898.

Smith, J. L. D. (1993). The role of dispersal in structuring the Chitwan tiger population. *Behaviour* 124: 165–195.

Smith, J. L. D., Tunhikorn, S., Tanhan, S. & Simcharoen, S. (1999). Metapopulation structure of tigers in Thailand. In *Riding the Tiger: Tiger Conservation in Human-Dominated Landscape*, eds J. Seidensticker, S. Christie & P. Jackson, pp. 114–122. Cambridge: Cambridge University Press.

Smith, J. S. (1989). Rabies virus epitopic variation: use in ecologic studies. *Advances in Virus Research* 36: 215–253.

Smith, J. S., Sumner, J. W., Roumillat, L. F., Baer, G. M. & Winkler, W. G. (1984). Antigenic characteristics of isolates associated with a new epizootic of raccoon rabies in the United States. *Journal of Infectious Diseases* 149: 769–774.

Smith, K. G. & Clark, J. D. (1996). Black bears in Arkansas: characteristics of a successful translocation. *Journal of Mammalogy* 75: 309–20.

Smith, K. G., Clark, J. D. & Gipson, P. S. (1990). History of black bears in Arkansas: overexploitation, near elimination, and successful re-introduction. *Eastern Workshop on Black Bear Research and Management.* 10: 5–14.

Smith, N. J. H. (1976). Spotted cats and the Amazon skin trade. *Oryx* 13: 362–371.

Smith, W. P., Borden, D. L. & Endres, K. M. (1994). Scent-station visits as an index to abundance of raccoons: an experimental manipulation. *Journal of Mammalogy* 75: 637–647.

Smithers, R. H. N. (1983). *The Mammals of the Southern African Subregion*. Pretoria: University of Pretoria.

Smithsonian Institute (1993). Mammal species of the world web site http://nmnhwww.si.edu/cgi-bin/wdb/msw/names

Smouse, P. E. & Chevillon, C. (1998). Analytical aspects of population-specific DNA fingerprinting for individuals. *Journal of Heredity* 89: 143–150.

Smuts, G. L. (1976). Population characteristics and recent history of lions in two parts of Kruger National Park. *Koedoe* 19: 153–164.

Smuts, G. L. (1978). Effects of population reduction on the travels and reproduction of lions in Kruger National park. *Carnivore* 1 (Part 2): 61–72.

Smuts, G. L. (1982). *Lion.* Johannesburg: MacMillan South Africa.

Snyder, J. E. & Bissonette, J. A. (1987). Marten use of clear-cuttings and residual forest stands in western Newfoundland. *Canadian Journal of Zoology* **65**: 169–74.

Sober, E. (1986). Philosophical problems for environmentalism. In *The Preservation of Species: The Value of Biological Diversity*, ed. B. G. Norton, pp. 173–194. Princeton, New Jersey: Princeton University Press.

Soorae, P. S. (1997). Wild dog release in Tsavo National Park, Kenya. *Re-Introduction News* **14**: 11–2.

Soorae, P. S. & Stanley Price, M. R. (1997). Successful re-introductions of large carnivores what are the secrets? *11th International Conference on Bear Research and Management in Graz, Austria, September 1997.*

Sørensen, O. J., Swenson, J. E. & Kvam, T. (1999). Status and management of the brown bear in Norway. In *Bears – Status Survey and Conservation Action Plan*, eds C. Servheen, S. Herrero & B. Peyton, pp. 86–89. Gland: IUCN.

Soulé, M. E. (1983). What do we really know about extinction? In *Genetics and Conservation*, eds C. M. Schonewald-Cox, S. M. Chambers, B. MacBryde & L. Tomas, pp. 111–124. Menlo Park, CA: Benjamin/Cummings.

Soulé, M. E. (1985). What is conservation biology? *BioScience* **35**: 727–34.

Soulé, M. E. (Ed.) (1986). *Conservation Biology: The Science of Scarcity and Diversity.* Sunderland, MA: Sinauer Associates.

Soulé, M. E. (1987). *Viable Populations for Conservation.* Cambridge: Cambridge University Press.

Soulé, M. E. (1990). The onslaught of alien species, and other challenges in the coming decades. *Conservation Biology* **4**: 233–239.

Soulé, M. E., Bolger, D. T., Alberts, A. C., Wright, J., Sorice, M. & Hill, S. (1988). Reconstructed dynamics of rapid extinctions of chaparral-requiring birds in urban habitat islands. *Conservation Biology* **2**: 75–92.

Soulé, M. E. & Noss, R. F. (1998). Rewilding and biodiversity's complementary goals for continental conservation. *Wild Earth* **8**: 18–28.

Soulé, M. E. & Terborgh, J. (1999). *Continental Conservation: Scientific Foundations for Regional Reserve Networks.* Washington, DC: Island Press.

Soulé, M. E. & Wilcox, B. A. (Eds) (1980). *Conservation Biology: An Evolutionary-Ecological Perspective.* Sunderland MA: Sinauer Associates.

Sovada, M. A., Roy, C. C., Bright, J. B. & Gillis, J. R. (1998). Causes and rates of mortality of swift foxes in western Kansas. *Journal of Wildlife Management* **62**: 1300–1306.

Spelman, L. H. (1998). North American river otter (*Lutra canadensis*) translocation in North Carolina 1989–96. *Proceedings of the Second Scientific Meeting of the European Association of Zoo and Wildlife Veterinarians (EAZWV), Chester UK, May 21–24*, pp. 461–465.

Spendelow, J. A., Nichols, J. D., Nisbet, I. C. T., Hays, H., Cormons, G. D., Burger, J., Safina, C., Hines, J. E. & Gochfeld, M. (1995). Estimating annual survival and movement rates of adults within a metapopulation of Roseate Terns. *Ecology* **76**: 2415–2428.

Stahl, P. & Artois, M. (1991). *Status and Conservation of the Wild Cat (Felis silvestris) in Europe and Around the Mediterranean Rim.* Strasbourg: Council of Europe Publishing.

Stander, P. E. (1990). A suggested management strategy for stock-raiding lions in Namibia. *South African Journal of Wildlife Research* **20**: 37–43.

Stander, P. E. (1991a). *Aspects of the Ecology and Scientific Management of Large Carnivores in Sub-Saharan Africa*. PhD dissertation, Cambridge University, Cambridge.

Stander, P. E. (1991b). Demography of lions in Etosha National Park, Namibia. *Madoqua* **18**: 1–9.

Stander, P. E. (1993). Conserving large African carnivores in a developing world. In *Wildlife Ranching: A Celebration of Diversity*, eds W. Van Hoven, H. Ebedes & A. Conroy, pp. 368–372. Pretoria: Promedia.

Stander, P. E. (1998). Spoor counts as indices of large carnivore populations: the relationship between spoor frequency, sampling effort and true density. *Journal of Applied Ecology* **35**: 378–385.

Stander, P. E., Ghau, X., Tsisaba, D. & Txoma, X. (1996). A new method of darting: stepping back in time. *African Journal of Ecology* **34**: 48–53.

Stander, P. E., Haden, P. J., Kaqece & Ghau (1997). The ecology of asociality in Namibian leopards. *Journal of Zoology (London)* **242**: 343–364.

Stanley, H. F., Casey, S., Carnahan, J. M., Goodman, S., Harwood, J. & Wayne, R. K. (1996). Worldwide patterns of mitochondrial DNA differentiation in the harbor seal (*Phoca vitulina*). *Molecular Biology and Evolution* **13**: 368–382.

Stanley, H. F. & Harwood, J. (1997). Genetic differentiation among subpopulations of the highly endangered Mediterranean monk seal. In *The Role of Genetics in Conserving Small Populations*, eds T. E. Tew, T. J. Crawford, J. W. Spencer, D. P. Stevens, M. B. Usher & J. Warren, pp. 97–101. Proceedings of a British Ecological Society Symposium.

Stanley, S. M. (1979). *Macroevolution: Pattern and Process*. San Francisco: W. H. Freeman.

Stanley Price, M. R. (1991). A review on mammal re-introductions, and the role of the Re-introduction Specialist Group of IUCN/SSC. *Symposia of the Zoological Society of London* **62**: 9–25.

Starfield, A. M., Shiell, J. D. & Smuts, G. L. (1981). Simulation of lion control strategies in a large game reserve. *Ecological Modelling* **13**: 17–38.

Stauffer, D. & Aharony, A. (1992). *Introduction to Percolation Theory*, 2nd edn. London: Taylor and Francis.

Steadman, D. W., Pregill, G. K. & Olson, S. L. (1984). Fossil vertebrates from Antigua, Lesser Antilles: evidence for late Holocene human-caused extinctions in the West Indies. *Proceedings of the National Academy of Sciences of the United States of America* **81**: 4448–4451.

Stearns, B. P. & Stearns, S. C. (1999). *Watching from the Edge of Extinction*. New Haven, CT: Yale University Press.

Steinmetz, R. & Mather, R. (1996). Impact of Karen villages on the fauna of Thung Yai Naresuan Wildlife Sanctuary: A participatory research project. *Natural History Bulletin of the Siam Society* **44**: 23–40.

Stenseth, N. C., Falck, W., Chan, K. S., Bjornstad, O. N., O'Donoghue, M., Tong, H., Boonstra, R., Boutin, S., Krebs, C. J. & Yoccoz, N. G. (1998). From patterns to processes: phase and density dependencies in the Canadian lynx cycle. *Proceedings of the National Academy of Sciences of the United States of America* **95**: 15430–15435.

Stephenson, R. O., Ballard, W. B., Smith, C. A. & Richardson, K. (1995). Wolf biology and management in Alaska, 1981–1992. In *Ecology and Conservation of Wolves in a Changing World*, eds L. N. Carbyn, S. H. Fritts & D. R. Seip, pp. 43–54. Canadian Circumpolar Institute, Occasional Publication No. 35.

Sterling, B., Conley, W. & Conley, M. R. (1983). Simulations of demographic compensation in coyote populations. *Journal of Wildlife Management* **47**: 1177–1181.

Steun, S., Have, P., Osterhaus, A. D. M. E., Arnemo, J. M. & Moustgard, A. (1994). Serological investigation of virus infections in harp seals (*Phoca groenlandica*) and hooded seals (*Cystophora cristata*). *Veterinary Record* **134**: 502–503.

Stirling, I., Calvert, W. & Andriashek, D. (1980). *Population Ecology Studies of the Polar Bear in the Areas of Southeastern Baffin Island*. Canadian Wildlife Service, Occasional Paper No. 44.

Stirling, I., Pearson, A. M. & Bunnell, F. L. (1976). Population ecology studies of polar and grizzly bears in northern Canada. *Transactions of the North American Wildlife and Natural Resources Conference* **41**: 421–430.

Stone, P. A., Snell, H. L. & Snell, H. M. (1993). Behavioural diversity as biological diversity: introduced cats and lava lizard wariness. *Conservation Biology* **8**: 569–573.

Stoneberg, R. (1995). Tracking re-introduced black-footed ferrets in Montana, USA. *Re-Introduction News* **11**: 17–8.

Storm, G. L., Andrews, R. D., Phillips, R. L., Bishop, R. A., Siniff, D. B. & Tester, J. R. (1976). Morphology, reproduction, dispersal and mortality of midwestern fox populations. *Wildlife Monographs* **49**: 5–82.

Stover, J. & Evans, J. (1984). Interspecies embryo transfer of gaur (*Bos gaurus*) to domestic Holstein cattle (*Bos taurus*) at the New York Zoological Park. *Tenth International Congress on Animal Reproduction and Artificial Insemination* **2**: 243.

Strachan, C., Jefferies, D. J., Barreto, G. R., Macdonald, D. W. & Strachan, R. (1998). The rapid impact of resident American mink on water voles: case studies in lowland England. *Symposia of the Zoological Society of London* **71**: 339–357.

Strachan, R. & Jefferies, D. J. (1993). *The Watervole* Arvicola terrestris *in Britain 1989–1990: Its Distribution and Changing Status*. London: The Vincent Wildlife Trust.

Strachan, R. & Jefferies, D. (1996). *Otter Survey of England 1991–1994. A Report on the Decline and Recovery of the Otter in England and on its Distribution, Status and Conservation in 1991–1994*. London: Vincent Wildlife Trust.

Strachan, R., Jefferies, D. J.& Chanin, P. R. F. (1996). *Pine Marten Survey of England and Wales, 1987–1988*. Peterborough: Joint Nature Conservation Committee.

Strickland, M. A. & Douglas, C. W. (1984). *Results of Questionnaires Sent to Trappers of Fisher and Marten in the Algonquin Region (Ontario) in Five Consecutive Years 1979 to 1983*. Toronto: Ontario Ministry for Natural Resources Report.

Strickland, M. A. & Douglas, C. W. (1987). Marten. In *Wild Furbearer Management and Conservation in North America*, eds M. Novak, J. A. Baker, M. E. Obbard & B. Malloch, pp. 531–546. Toronto: Ontario Ministry of Natural Resources.

Stuart, C. & Stuart, T. (1991). The feral cat problem in southern Africa. *African Wildlife* **45**: 13–15.

Sturtevant, B. R., Bissonette, J. A. & Long, J. N. (1996). Temporal and spatial dynamics of boreal forest structure in western Newfoundland: silvicultural implications for marten management. *Forest Ecology and Management* **87**: 13–25.

Sumner, P. W. & Hill, E. P. (1980). Scent stations as indices of abundance in some furbearers of Alabama. *Proceedings of the Annual Conference of Southeastern Association of Fish and Wildlife Agencies* **34**: 572–583.

Sunquist, M., Karanth, K.U. & Sunquist, F. (1999). Ecology, behaviour and resilience of the tiger and its conservation needs. In *Riding the Tiger – Tiger Conservation in Human-Dominated Landscapes*, eds J. Seidensticker, S. Christie & P. Jackson, pp. 5–18. Cambridge: Cambridge University Press.

Sutherland, W. J. (1998). The importance of behavioural studies in conservation biology. *Animal Behaviour* **56**: 801–809.

Swank, W. G. & Teer, J. G. (1989). Status of the jaguar – 1987. *Oryx* **23**: 14–21.

Swanson, T. M., Barbier, E. B. (Eds) (1992). *Economics for the Wilds: Wildlife, Wildlands, Diversity and Development*. London: Earthscan Publications.

Swart, J. A. A., Reijnders, P. J. H. & Van-Delden, W. (1996). Absence of genetic variation in harbor seals (*Phoca vitulina*) in the Dutch Wadden Sea and the British Wash. *Conservation Biology* **10**: 289–293.

Swenson, J., Gerstl, N., Dahle, B. & Zedrosser, A. (1999). *Final Draft Action Plan for Conservation of the Brown Bear* (Ursus arctos) *in Europe*. Strasbourg: Council of Europe Publishing.

Swenson, J. E., Sandegren, F., Bjarvall, A., Soderberg, A., Wabakkan, P. & Franzen, R. (1994). Size, trend, distribution and conservation of the brown bear *Ursus arctos* population in Sweden. *Biological Conservation* **70**: 9–17.

Swenson, J. E., Sandergren, F., Soderberg, A. *et al.* (1997). Infanticide caused by hunting of male bears. *Nature* **386**: 450–1.

Swinton, J., Harwood, J., Grenfell, B. T. & Gilligan, C. A. (1998). Persistence thresholds for phocine distemper virus infection in harbour seal *Phoca vitulina* metapopulations. *Journal of Animal Ecology* **67**: 54–68.

Swinton, J., Tuyttens, F., Macdonald, D. W., Nokes, D. J., Cheeseman, C. L. & Clifton-Hadley, R. (1997). A comparison of fertility control and lethal control of bovine tuberculosis in badgers: the impact of perturbation induced transmission. *Philosophical Transactions of the Royal Society* **264**: 1–13.

Taberlet, P. & Bouvet, J. (1992). Bear conservation genetics. *Nature* **358**: 197.

Taberlet, P. & Bouvet, J. (1994). Mitochondrial DNA polymorphism, phylogeography, and conservation genetics of the brown bear *Ursus arctos* in Europe. *Proceedings of the Royal Society of London, Series B* **255**: 195–200.

Taberlet, P., Camarra, J-J., Griffin, S. *et al.* (1997). Non-invasive genetic tracking of the endangered Pyrenean brown bear population. *Molecular Ecology* **6**: 869–876.

Taberlet, P., Griffin, S., Goossens, B. *et al.* (1996). Reliable genotyping of samples with very low DNA quantities using PCR. *Nucleic Acids Research* **26**: 3189–3194.

Taberlet, P. & Luikart, G. (1999). Non-invasive genetic sampling and individual identification. *Biological Journal of the Linnean Society* **68**: 41–55.

Taberlet P., Swenson, J. E., Sandegren, F. & Bjarvall, A. (1995). Localization of a contact zone between two highly divergent mitochondrial DNA lineages of the brown bear Ursus arctos in Scandinavia. *Conservation Biology* **9**: 1255–1261.

Taberlet, P. & Waits, L. P. (1998). Non-invasive genetic sampling. *Trends in Ecology and Evolution* **13**: 26–27.

Taberlet, P., Waits, L. P. & Luikart, G. (1999). Non-invasive genetic sampling: look before you leap. *Trends in Ecology and Evolution* **14**: 321–325.

Tabor, J. E. & Wight, H. M. (1977). Population status of river otter in western Oregon. *Journal of Wildlife Management* **41**: 692–699.

Talbot, L. M. (1959). *A Look at Threatened Species – A Report on Some Animals of the Middle East and Southern Asia which are Threatened with Extermination*. Washington DC: Fauna Preservation Society.

Talbot, S. L. & Shields, G. F. (1996). Phylogeography of brown bears (*Ursus arctos*) of Alaska and paraphyly within the Ursidae. *Molecular Phylogenetics and Evolution* **5**: 477–594.

Tapper, S. (1979). The effect of fluctuating vole numbers (*Microtus agrestis*) on a population of weasels (*Mustela nivalis*) on farmland. *Journal of Animal Ecology* **48**: 603–617.

Tapper, S. & Reynolds, J. (1996). The wild fur trade: historical and ecological perspectives. In *The Exploitation of Mammal Populations*, eds V. Taylor & N. Dunstone, pp. 28–44. London: Chapman and Hall.

Taylor, K. (1984). The influence of watercourse management on moorhen breeding biology. *British Birds* **77**: 144–148.

Taylor, R. H. (1979). How the Macquarie Island parakeet became extinct. New Zealand *Journal of Ecology* **2**: 42–45.

Taylor, R. H. (1984). Distribution and interactions of introduced rodents and carnivores in New Zealand. *Acta Zoologica Fennica* **172**: 103–105.

Taylor, V. J. & Dunston, N. (1996). The exploitation, sustainable use, and welfare of wild animals. In *The Exploitation of Mammal Populations*, eds V. J. Taylor & N. Dunston, pp. 3–15. London: Chapman and Hall.

Temple, S. A. (1990). The nasty necessity: eradicating exotics. *Conservation Biology* **4**: 113–115.

Templeton, A. R. (1990). The role of genetics in captive breeding and reintroduction for species conservation. *Endangered Species Update* **8**: 14–17.

Templeton, A. R. & Read, B. (1994). Inbreeding: one word, several meanings, much confusion. In *Conservation Genetics*, eds V. Loeschcke, J. Tomiuk & S. K. Jain, pp. 91–105. Birkhäuser Verlag, Basel.

Terborgh, J. (1974). Preservation of natural diversity: the problem of extinction prone species. *BioScience* **24**: 715–22.

Terborgh, J. (1988). The big things that run the world – a sequel to E. O. Wilson. *Conservation Biology* **2**: 402–403.

Terborgh, J., Estes, J. A., Paquet, P., Ralls, K., Boyd-Heger, D., Miller, B. J. & Noss, R. F. (1999). The role of top carnivores in regulating terrestrial ecosystems. In *Continental Conservation: Scientific Foundations of Regional Reserve Networks*, eds M. E. Soulé & J. Terborgh, pp. 39–64. Washington: Island Press.

Terborgh, J., Lopez, L., Tello, J., Yu, D. & Bruni, A. R. (1997). Transitory states in relaxing ecosystems of land bridge islands. In *Tropical Forest Remnants: Ecology, Management, and Conservation of Fragmented Communities*, eds W. F. Laurance & R. O. Bierregaard, Jr., pp. 256–274. Chicago: University of Chicago Press.

Ternovskij, D. V. (1977). *[Biology of the mustelids (Mustelidae)]*. Novosibirsk: Nauka. (In Russian.)

Tewes, M. E. & Everett, D. D. (1982). Status and distribution of the endangered ocelot and jaguarundi in Texas. In *Cats of the World: Biology, Conservation, and Management*, eds S. D. Miller & D. D, Everett, pp. 147–158. Washington, D.C: National Wildlife Federation.

Tewes, M. E. & Schmidly, D. J. (1987). The neotropical felids: jaguar, ocelot, margay, and jaguarundi. In *Wild Furbearer Management and Conservation in North America*, eds M. Novak, J. A. Baker, M. E. Obbard & B. Malloch, pp. 697–712. Toronto: Ontario Ministry of Natural Resources.

Thapar, V. (1992). *The Tiger's Destiny*. London: Kyle-Cathie.

Thapar, V. (1999). The tragedy of the Indian Tiger: starting from scratch. In *Riding the Tiger: Tiger Conservation in Human-Dominated Landscape*, eds J. Seidensticker, S. Christie & P. Jackson, pp. 296–306. Cambridge: Cambridge University Press.

Thiel, R. P. (1985). The relationship between road densities and wolf habitat suitability in Wisconsin. *American Midland Naturalist* 113: 404–407.

Thiel, R. P. & Ream, R. R. (1995). Status of the gray wolf in the lower 48 United States to 1992. In *Ecology and Conservation of Wolves in a Changing World*, eds L. N. Carbyn, S. H. Fritts & D. R. Seip, pp. 59–62. Edmonton: Canadian Circumpolar Institute.

Thomas, W. L., Jr. (Ed.) (1956). *Man's Role in Changing the Face of the Earth*. Chicago: University of Chicago Press.

Thomson, A. P. D. (1951). A history of the ferret. *Journal of the History of Medicine* 6: 471–480.

Thomson, G. R. & Meredith, C. D. (1993). Rabies in the bat-eared fox in South Africa. *Onderstepoort Journal of Veterinary Research* 60: 3399–3403.

Thomson, P. C., Rose, K. & Kok, N. E. (1992). The behavioral ecology of dingoes in northwestern Australia. IV. Social and spatial organization, and movements. *Wildlife Research* 19: 543–563.

Thompson, E. & Meagher, T. (1987). Parental and sib likelihoods in genealogy reconstruction. *Biometrics* 43: 585–600.

Thompson, I. D. (1994). Marten populations in uncut and logged boreal forests in Ontario. *Journal of Wildlife Management* 58: 272–280.

Thompson, I. D. & Coglan, P. W. (1987). Prey choice by marten during a decline in prey abundance. *Oecologia* 83: 443–451.

Thompson, W. L., White, G. C. & Gowan, C. (1998). *Monitoring Vertebrate Populations*. New York: Academic Press.

Thor, G. & Pegel, M. (1992). *Zur Wiedereinbürgerung des Luchses in Baden-Württemberg*. Aulendorf: Wildforschungsstelle Baden-Württemberg, Band 2.

Thorne, E. T. & Williams, E. S. (1988). Disease and endangered species: the black-footed ferret as a recent example. *Conservation Biology* 2: 66–74.

Thresher, P. (1982). The lion's share. *New Scientist* 15 April.

Thurber, J. A., Peterson, R. O., Woolington, J. D. & Vucetich, J. A. (1992). Coyote coexistence with wolves on the Kenai Peninsula, Alaska. *Canadian Journal of Zoology* 70: 2494–2498.

Thurber, J. M. & Peterson, R. O. (1991). Changes in body size associated with range expansion in the coyote (*Canis latrans*). *Journal of Mammalogy* 72: 750–755.

Tikel, D., Blair, D. & Marsh, H. (1996). Marine mammal faeces as a source of DNA. *Molecular Ecology* 5: 456–457.

Tilman, D. (1986). Resources, competition and the dynamics of plant communities. In *Plant Ecology*, ed. M. J. Crawley, pp. 51–74. Oxford: Blackwell Scientific Publications.

Tilman, D., May, R. M., Lehman, C. L. & Nowak, M. A. (1994). Habitat destruction and the extinction debt. *Nature* 371: 65–66.

Tilson, R. & Nyhus, P. (1998). Keeping problem tigers from becoming a problem species. *Conservation Biology* 12: 261–262.

Tilson, T. L. & Seal, U. S. (1987). *Tigers of the World*. Park Ridge, NJ: Noyes Publications.

Tilson, R., Traylor-Holzer, K. & Miang Jiang, Q. (1997). The decline and impending extinction of the South China tiger. *Oryx* 31: 243–52.

Tischendorf, L., Thulke, L.-H., Staubach, C. Muller, M. S., Jeltsch, F., Gorestzki, J., Selhorst, T., Muller, T., Schluter, H. & C. Wissel (1998). Chance and risk of controlling rabies in large-scale and long-term immunized fox populations. *Proceedings of the Royal Society of London, Series B* 265: 839–846.

Tobias, M. (1998a). *World War III: Population and the Biosphere at the End of the Millennium*. New York: New York Continuum.

Tobias, M. (1998b). *Nature's Keepers: On the Front Lines of the Fight to Save America's Wildlife*. New York: Wiley.

Tocher, M. D., Gascon, C. & Zimmerman, B. L. (1997). Fragmentation effects on a Central American frog community: A ten-year study. In *Tropical Forest Remnants: Ecology, Management, and Conservation of Fragmented Communities*, eds W. F. Laurance & R. O. Bierregaard, pp. 124–137. Chicago: University of Chicago Press.

Todd, A. W. (1985). Demographic and dietary comparisons of forest and farmland coyote, *Canis latrans*, populations in Alberta. *Canadian Field-Naturalist* 99: 163–171.

Todd, A. W. & Keith, L. B. (1983). Coyote demography during a snowshoe hare decline in Alberta. *Journal of Wildlife Management* 47: 394–404.

Todd, A. W., Keith, L. B. & Fischer, C. A. (1981). Population ecology of coyotes during a fluctuation of snowshoe hares. *Journal of Wildlife Management* 45: 629–640.

Todd, N. B. (1978). An ecological, behavioural genetic model for the domestication of the domestic cat. *Carnivore* 1: 52–60.

Toft, C. A., Aeschlimann, A. & Bolis, L. (1991). *Parasite–Host Associations: Coexistence or Conflict?* Oxford: Oxford University Press.

Torres, D. & Aguayo, A. (1971). Algunas observaciones sobre la fauna del archipelago de Juan Fernandez. *Mammals. Bollettino Universidad de Chile* 112: 34–35.

Torres, S. (1997). Mountain lion attacks on humans. *Outdoor California* 57: 10–11.

Tough, S. C. & Butt, J. C. (1993). A review of fatal bear maulings in Alberta, Canada. *Amererican Journal of Forensic Medicine and Pathology*, 14: 22–27.

Towns, D. R. & Daugherty, C. H. (1994). Patterns of range contractions and extinctions in the New Zealand herpetofauna following human colonization. *New Zealand Journal of Zoology* 21: 325–339.

Towns, D. R., Simberloff, D. & Atkinson, I. A. E. (1997). Restoration of New Zealand islands: redressing the effects of introduced species. *Pacific Conservation Biology* 3: 99–124.

Trapp, G. R. 1978. Comparative behavioral ecology of the ringtail (*Bassariscus astutus*) and gray fox (*Urocyon cinereoargenteus*) in southwestern Utah. *Carnivore* 1: 3–32.

Trautman, C. G., Fredrickson, L. F. & Carter, A. V. (1974). Relationship of red foxes and other predators to populations of ring-necked pheasants and other prey, South Dakota. *Transactions of the North American Wildlife Conference* 39: 214–255.

Travaini, A., Laffitte, R. & Delibes, M. (1996). Determining the relative abundance of European red foxes by scent-station methodology. *Wildlife Society Bulletin* **24**: 500–504.

Travaini, A., Palomeres, F. & Delibes, M. (1993). The effects of capture and recapture on space use in large grey mongooses. *South African Journal of Wildlife Research*, **23**: 95–97.

Trent, T. T. & Rongstad, O. J. (1974). Home range and survival of cottontail rabbits in southwestern Wisconsin. *Journal of Wildlife Management* **38**: 459–472.

Trevino, J.C. & Jonkel, C. (1986). Do grizzly bears still live in Mexico? *International Conference on Bear Research and Management* **6**: 11–13.

Trewhella, W. J., Harris, S. & McAllister, F. E. (1988). Dispersal distance, home range size and population density in the red fox (*Vulpes vulpes*): a quantitative analysis. *Journal of Applied Ecology* **25**: 423–434.

Truyen, U. (1999). Emergence and recent evolution of canine parvovirus. *Veterinary Microbiology* **69**: 47–50.

Tully, R. J. (1991). Results, 1991 questionnaire on damage to livestock by mountain lion. In *Mountain Lion–Human Interactions Symposium Colorado Division of Wildlife*, ed. C. L. Braun, pp. 68–74. Denver.

Turner, B. L., II, Clark, W. C., Kates, R. W., Richards, J. F., Mathews, J. T. & Meyer, W. B. (Eds) (1990). *The Earth as Transformed by Human Action*. Cambridge: Cambridge University Press.

Turner, M. (1987). *My Serengeti Years*. London: Elm Tree Books.

Tuyttens, F. A. M. & Macdonald, D. W. (1998). Fertility control: An option for nonlethal control of wild carnivores? *Animal Welfare* **7**: 339–364.

Twichell, A. R. & Dill, H. H. (1949). One hundred raccoons from one hundred and two acres. *Journal of Mammalogy* **30**: 130–133.

Tyndale-Biscoe, C. H. (1994). Virus-vectored immunocontraception of feral mammals. *Reproduction, Fertility and Development* **6**: 281–287.

Underhill, L. G. (1994). Optimal and sub-optimal reserve selection algorithms. *Biological Conservation* **70**: 87.

Usher, M. B. (1988). Biological invasions of nature reserves: a search for generalisations. *Biological Conservation* **44**: 119–135.

US CensusBureau (1998). International Data Base, world wide web: http://www.census.gov/ipc/www/idbprint.html

US CensusBureau (1999). Historical census data, world wide web: http://www.census.gov/population/www/censusdata/hiscensusdata.html

US Fish and Wildlife Service (1993). *Grizzly bear recovery plan*. Missoula: U.S. Fish and Wildlife Service.

US Fish & Wildlife Service (USFWS) (1996). *Reintroduction of the Mexican Wolf Within its Historic Range in the Southwestern U.S. – Final Environmental Impact Statement*. Albuquerque, New Mexico: U.S. Department of the Interior.

US Fish and Wildlife Service (1999). Mexican Wolf Recovery Program Web Page http://ifw2es.fws.gov/mexicanwolf

Valenzuela, D., Ceballos, G. & García A. (2000). Mange epizootic in white-nosed coatis in western Mexico. *Journal of Wildlife Diseases* **36**: 56–63.

Van Aarde, R. J. (1984). Population biology and the control of feral cats on Marion Island. *Acta Zoologica Fennica* **172**: 107–110.

van Apeldoorn, R. C., Oostenbrink, W. T., Vanwinden, A. & van der Zee, F. F.

(1992). Effects of habitat fragmentation on the bank vole, *Clethrionomys glareolus*, in an agricultural landscape. *Oikos* **65**: 265–74.

Van Ballenberghe, V., Erickson, A.W. & Byman, D. (1975). Ecology of the timber wolf in northeastern Minnesota. *Wildlife Monograph* **43**: 1–43.

van de Bildt, M. W. G., Vedder, E. J., Martina, B. E. E., AbouSidib, B., Jiddou, A. B., Barham, M. E. O., Androukaki, E., Komnenou, A., Niesters, H. G. M. & Osterhaus, A. D. M. E. (1999). Morbilliviruses in Mediterranean monk seals. *Veterinary Microbiology* **69**: 19–21.

van der Zee, F. F., Wiertz, J., Ter Braak, C. J. F. & van Apeldoorn, R. C. (1992). Landscape change as a possible cause of the badger *Meles meles* L. decline in the Netherlands. *Biological Conservation* **61**: 17–22.

Van Dyke, F. G., Brocke, R. H. & Shaw, H. G. (1986). Use of road track counts as indices of mountain lion presence. *Journal of Wildlife Management* **50**: 102–109.

Van Gruisen, J. & Sinclair, T. (1992). *Fur Trade in Kathmandu: Implications for India.* New Delhi: TRAFFIC India.

van Heerden, J., Mills, M. G. L., van Vuuren, M. J., Kelly, P. & Dreyer, M. J. (1995). An investigation into the health status and diseases of wild dogs (*Lycaon pictus*) in the Kruger National Park. *Journal of the South African Veterinary Association* **66**: 18–27.

Van Jaarsveld, A.S., Freitag, S., Chown, S.L., Muller, C., Koch, S., Hull, H., Bellamy, C., Krüger, M., Endrödy-Younga, S., Mansell, M.W. & Scholtz, C.H. (1998). Biodiversity assessment and conservation strategies. *Science* **279**, 2106–2108.

Van Orsdol, K. G., Hanby, J. P. & Bygott, J. D. (1985). Ecological correlates of lion social organization. *Journal of Zoology (London)* **206**: 97–112.

Van Sickle, W. D. & Lindzey, F. G. (1991). Evaluation of a cougar population estimator based on probability sampling. *Journal of Wildlife Management* **55**: 738–743.

Van Sickle, W. D. & Lindzey, F. G. (1992). Evaluation of road track surveys for cougars (*Felis concolor*). *Great Basin Naturalist* **52**: 232–236.

Van Strien, A. J., van de Pavert, R., Moss, D., Yates, T. J., Van Swaay, C. A. M. & Vos, O. (1997). Statistical power of two butterfly monitoring schemes to detect trends. *Journal of Applied Ecology* **34**: 817–828.

Van Valkenburgh, B. (1988). Trophic diversity within past and present guilds of large predatory mammals. *Paleobiology* **14**: 156–173.

Van Valkenburgh, B. (1996). Feeding behavior in free-ranging large African carnivores. *Journal of Mammalogy* **77**: 240–254.

Van Valkenburgh, B. (1999). Major patterns in the history of carnivorous mammals. *Annual Review of Earth Planet. Science* **27**: 463–493.

Van Valkenburgh, B. & Janis, C. M. (1993). Historical diversity patterns in North American large herbivores and carnivores. In *Species Diversity in Ecological Communities*, eds R. E. Ricklefs & D. Schluter, pp. 230–240. Chicago: University of Chicago Press.

Van Valkenburgh, B. & Wayne, R. K. (1994). Shape divergence associated with size convergence in sympatric east African jackals. *Ecology* **75**: 1567–1581.

Van Vuren, B. J. & Robinson, T. J. (1997). Genetic population structure in the yellow mongoose, *Cynictis penicillata*. *Molecular Ecology* **6**: 1147–1153.

Van Vuren, D. (1998) Mammalian dispersal and reserve design. In *Behavioral Ecology and Conservation Biology*, ed. T. Caro, pp. 369–393. New York: Oxford University Press.

Vargas, A., Lockhart, M. & Gober, P. (1999). Adapting species management to endangered species management: black-footed ferret recovery, USA. *Re-introduction News* **18**: 17–18.

Varner, G. E. (1998). *In Nature's Interests? Interests, Animal Rights, and Environmental Ethics.* New York: Oxford University Press.

Vedros, N. A., Smith, A. W., Schonewald, J., Migaki, G. & Hubbard, R. C. (1971). Leptospirosis epizootic among California sea lions. *Science* **172**: 1250–1251.

Veeramani, A., Jayson, E. A. & Easa, P. S. (1996). Man-wildlife conflict: cattle lifting and human casualties in Kerala. *Indian Forester* **122**: 897–902.

Veitch, C. R. (1985). Methods of eradicating feral cats from offshore islands in New Zealand. In *Conservation of Island Birds*, ed. P. J. Moors, pp. 125–141. Gland: Technical Publication No. 3 ICBP.

Veitch, C. R. & Bell, B. D. (1990). Eradication of introduced animals from the islands of New Zealand. In *Ecological Restoration of New Zealand Islands*, eds D. R. Towns, C. H. Daugherty & I. A. E. Atkinson, pp. 137–146. Conservation Sciences Publication No. 2, Wellington: NZ Department of Conservation.

Venables, W. N. & Ripley, B. D. (1997). *Modern Applied Statistics with S-plus*, 2nd edn. New York: Springer-Verlag.

Venkataraman, A. B. (1995). Do dholes (*Cuon alpinus*) live in packs in response to competition with or predation by large cats? *Current Science* **69**: 934–936.

Vilà, C. & Wayne, R. K. (1999). Hybridization between wolves and dogs. *Conservation Biology* **13**: 195–198.

Vilà, C., Savolainen, P., Maldonado, J. E., Amorim, I. R., Rice, J. E., Honeycutt, R. L., Crandall, K. A., Lundeberg, J. & Wayne, R. K. (1997). Multiple and ancient origins of the domestic dog. *Science* **276**: 1687–1689.

Vilà, C., Urios, V. & Castroviejo, J. (1995). Observations on the daily activity patterns in the Iberian wolf. In *Ecology and Conservation of Wolves in a Changing World*, eds L. N. Carbyn, S. H. Fritts & D. R. Seip, pp. 335–340. Edmonton: Canadian Circumpolar Institute.

Villafuerte, R., D. F. Luco, C. Gortazer & Blanco, J. C. (1996). Effect on red fox litter size and diet after rabbit haemorrhagic disease in north-eastern Spain. *Journal Zoology (London)* **240**: 764–767.

Vitousek, P. M. (1990). Biological invasions and ecosystem processes: towards an integration of population biology and ecosystem studies. *Oikos* **57**: 7–13.

Voigt, D. R. & Earle, B. D. (1983). Avoidance of coyotes by red fox families. *Journal of Wildlife Management* **47**: 852–857.

Voigt, D. R. & Johnston, D. H. (1992). Control of wildlife rabies in Canada: lessons for Australia. In *Wildlife Rabies Contingency Planning in Australia*, eds P. O'Brien & G. Berry, pp. 311–349. Bureau of Rural Resources Proceedings No. 11. Canberra: Australian Government Publishing Service.

Voigt, D. R. & Macdonald, D. W. (1984). Variation in the spatial and social behaviour of the red fox, *Vulpes vulpes*. *Acta Zoologica Fennica* **171**: 261–265.

Voipio, P. (1990). The Samson fox episode in Finland in the 1930s and 1940s, and the hypothetico-deductive method. *Annales Zoologici Fennici* **27**: 21–27.

von Teichman B. F., Thomson, G. R. Meredith, C. D. & Nel, L. H. (1995). Molecular epidemiology of rabies virus in South Africa: evidence for two distinct virus groups. *Journal of General Virology* **76**: 73–82.

Vucetich, J. A. & Creel, S. (1999). Ecological interactions, social organization, and extinction risk in African wild dogs. *Conservation Biology* **13**: 1172–1182.

Wagner, F. H. & Stoddart, L.C. (1972). Influence of coyote predation on black-tailed jackrabbit populations in Utah. *Journal of Wildlife Management* 36: 329–342.

Waits, L., Luikart, G. & Taberlet, P. (2001). Estimating the probability of identity among genotypes in natural populations: cautions and guidelines. *Molecular Ecology* (In press).

Waits, L. P., Talbot, S. L., Ward, R. H. & Shields, G. F. (1998). Mitochondrial DNA Phylogeography of the north American Brown Bear and implications for conservation. *Conservation Biology* 12: 408–417.

Wallace, R. L. & Clark, T. W. (1999). Solving problems in endangered species conservation: An introduction to problem orientation. *Endangered Species Update* 16: 28–34.

Walsh, P. B. & Inglis, J. M. (1989). Seasonal and diel rate of spontaneous vocalization in coyotes in south Texas. *Journal of Mammalogy* 70: 169–171.

Waples, K. A. & Stagoll, C. S. (1997). Ethical issues in the release of animals from captivity. *BioScience*, 47: 115–121.

Waples, R. (1989). A generalized approach for estimating effective population size from temporal changes in allele frequency. *Genetics* 121: 379–391.

Waples, R. (1991). Genetic methods for estimating the effective size of cetacean populations. *Report of the International Whaling Commission* 13 (Special Issue): 279–300.

Ward, R. M. P. & Krebs, C. J. (1985). Behavioural responses of lynx to declining snowshoe hare abundance. *Canadian Journal of Zoology* 63: 2817–2824.

Waser, P. M. (1996) Patterns and consequences of dispersal in gregarious carnivores. In *Carnivore Behavior, Ecology, and Evolution*, 2nd edn, ed. J. Gittleman, pp. 267–295. Ithaca: Cornell University Press.

Waser, P. M., Creel, S. R. & Lucas, J. R. (1994b) Death and disappearance: estimating mortality risks associated with philopatry and dispersal. *Behavioral Ecology* 5: 135–141.

Waser, P. M., Elliot, L. F., Creel, N. M. & Creel, S. R. (1995). Habitat variation and mongoose demography. In *Serengeti II: Dynamics Management and Conservation of an Ecosystem*, eds A. R. E. Sinclair and P. Arcese, pp. 421–447. Chicago: University of Chicago Press.

Waser, P. M., Keane, B., Creel, S. R., Elliott, L. F. & Minchella, D. J. (1994a). Possible male coalitions in a solitary mongoose. *Animal Behavior* 47: 289–294.

Waser, P. M. and Strobeck, C. (1998) Genetic signatures of interpopulation dispersal. *Trends in Ecology and Evolution* 13: 43–44.

Watt, J. P. (1993). Ontogeny of hunting behavior of otters in a marine environment. *Symposia of the Zoological Society of London* 65: 87–104.

Watts, W. A. (1983). Vegetational history of the eastern United States. In *Late Quaternary Environments of the United States*, vol. 1, *The Late Pleistocene*, ed. S. C. Porter, pp. 294–310. Minneapolis: University of Minnesota Press.

Wayne, R. K. (1992). On the use of molecular genetic characters to investigate species status. *Conservation Biology* 6: 590–592.

Wayne, R. K. (1996). Conservation genetics in the Canidae. In *Conservation Genetics: Case Histories from Nature*, eds J. Avise & J. Hamrick, pp. 75–118. New York: Chapman & Hall.

Wayne, R. K., Geffen, E., Girman, D. J., Koepfli, K. P., Lau, L. M. & Marshall, C. R. (1997). Molecular systematics of the Canidae. *Systematic Biology* 46: 622–653.

Wayne, R. K. & Gittleman, J. L. (1995). The problematic red wolf. *Scientific American* **273**: 26–31.

Wayne, R. K. & Gottelli, D. (1997). Systematics, population genetics and genetic management of the Ethiopian wolf. In *The Ethiopian Wolf, Status Survey and Conservation Action Plan*, eds C. Sillero-Zubiri & D. Macdonald, pp. 43–50. Gland: IUCN.

Wayne, R. K., Gilbert, D. A., Eisenhawer, A., Lehman, N., Hansen, K., Girman, D., Peterson, R. O., Mech, L. D., Gogan, P. J. P., Seal, U. S. & Krumenaker, R. J. (1991). Conservation genetics of the Isle Royale gray wolf. *Conservation Biology* **5**: 41–51.

Wayne, R. K., George, S., Gilbert, D., Collins, P. W., Kovach, S. D., Girman, D. & Lehman, N. (1991). A morphological and genetic study of the island fox, *Urocyon littoralis. Evolution* **45**: 1849–1868.

Wayne, R. K. & Jenks, S.M. (1991). Mitochondrial DNA analysis implying extensive hybridization of the red wolf, *Canis rufus. Nature* **351**: 565–568.

Wayne, R. K., Lehman, N., Allard, M. W. & Honeycutt, R. L. (1992). Mitochondrial DNA variability of the grey wolf: genetic consequences of population decline and habitat fragmentation. *Conservation Biology* **6**: 559–569.

Wayne, R. K., Kat, P. W., Fuller, T. K., Van Valkenburgh, B. & O'Brien, S. J. (1989). Genetic and morphologic divergence among sympatric canids (Mammalia: Carnivora). *Journal of Heredity* **80**: 447–454.

Wayne, R. K. & Koepfli, K. P. (1996). Demographic and historical effects on genetic variation of carnivores. In *Carnivore behavior, Ecology, and Evolution*, vol. 2, ed. J. L. Gittleman, pp. 453–484. Ithaca: Cornell University Press.

WCMC (1998). World Conservation Monitoring Centre, http://www.wcmc.org.uk/protected—areas/data/

Weaver, J. (1993). *Lynx, Wolverine, and Fisher in the Western United States: Research Assessment and Agenda.* Contract report to the US Fish and Wildlife Service, Missoula, MT.

Weaver, J. L., Paquet, P. C. & Ruggiero, L. F. (1996). Resilience and conservation of large carnivores in the Rocky Mountains. *Conservation Biology* **10**: 964–976.

Webb, N. R. (1999). Ecology and ethics. *Trends in Ecology and Evolution* **14**: 259–260.

Webb, S. D. (1984). Ten million years of mammal extinctions in North America. In *Quaternary Extinctions*, eds P. S. Martin & R. G. Klein, pp. 189–210. Tucson: University of Arizona Press.

Weber, D., Weber, J.-M. & Müller, H.-U. (1991). Fischotter (*Lutra lutra*) im Einzugsgebiet Schwarzwassser-Sense: Dokumentation eines erfolglosen Wiederansiedlungsprojekts. *Mitteilungen der Naturforschenden Gesellschaft (Bern)* **48**: 141–152.

Weber, J., Aubry, S., Lachat, N., Meia, J., Mermod, C. & Paratte, A. (1991). Fluctuations and behavior of foxes determined by nightlighting: preliminary results. *Acta Theriologica* **36**: 285–291.

Weber, W. & Rabinowitz, A. (1996). A global perspective on large carnivore conservation. *Conservation Biology* **10**: 1046–1055.

Webster, R. G., Bean, W. J., Gorman, O. T., Chambers, T. M. & Kawaoka, Y. (1992). Evolution and ecology of influenza-a viruses. *Microbiological Reviews* **56**: 152–179.

Weiss, J. A. (1989). The power of problem definition: the case of government paperwork. *Policy Sciences* **22**: 97–121.

Wenger, C. R. & Cringan, A. T. (1977). Biotelemetry in studying responses of coyotes to electronic sirens. *International Conference on Wildlife Biotelemetry* **1**: 126–130.

Wenger, C. R. & Cringan, A. T. (1978). Siren-elicited coyote vocalizations: an evaluation of a census technique. *Wildlife Society Bulletin* **6**: 73–76.

Wentzel, J., Stephens, J. C., Johnson, W. E., Menotti-Raymond, M., Pecon Slattery, J., Yuhki, N., Carrington, M., Quigley, H., Miquelle, D. G., Tilson, R., Manansang, J., Brady, G., Zhi, L., Wenshi, P., Shi-Quiang, H., Johnston, L., Sunquist, M., Karanth, K. U. & O'Brien, S. J. (1999). Subspecies of tigers: molecular assessment using 'voucher specimens' of geographically traceable individuals. In *Riding the Tiger: Tiger Conservation in Human-Dominated Landscapes*, eds J. Seidensticker, S. Christie & P. Jackson, pp. 40–49. Cambridge: Cambridge University Press.

Werdelin, L. (1996). Carnivoran ecomorphology: a phylogenetic perspective. In *Carnivore Behavior, Ecology, and Evolution*, vol. 2, ed. J. L. Gittleman, pp. 582–624. Ithaca: Cornell University Press.

West, E. W. & Rudd, R. L. (1983). Biological control of Aleutian Island arctic fox: a preliminary strategy. *International Journal for the Study of Animal Problems* **4**: 305–311.

Western, D. & Henry, W. (1979). Economics and conservation in third world national parks. *BioScience* **29**: 2764–2769.

Weyer, K., Fourie, P. B., Durrheim, D., Lancaster, J., Haslov, K. & Bryden, H. (1999). Mycobacterium bovis as a zoonosis in the Kruger National Park, South Africa. *International Journal of Tuberculosis and Lung Disease* **3**: 1113–1119.

Whelan, J. B., Raybourne, J., Tibbs, L. & Large, R. M. (1978). Management of black bears (*Ursus americanus*) in state and provincial parks. In *Proceedings of the 4th Eastern Black Bear Workshop*, ed. R. D. Hugie, pp. 365–375. Greenville, Maine.

Whipple, D. (1999). Hot lynx in Vail. *Audubon:* January–February, 14.

Whipple, J. D., Rollins, D. & Schacht, W. H. (1994). A field simulation for assessing accuracy of spotlight deer surveys. *Wildlife Society Bulletin* **22**: 667–673.

White, G. C. (1983). Numerical estimation of survival rates from band recovery and biotelemetry data. *Journal of Wildlife Management* **47**: 716–728.

White, G. C. (1996). NOREMARK: population estimation from mark-resighting surveys. *Wildlife Society Bulletin* **24**: 50–52.

White, G. C., Anderson, D. R., Burnham, K. P. & Otis, D. L. (1982). *Capture–Recapture and Removal Methods for Sampling Closed Populations*. Los Alamos Nat. Laboratory, Los Alamos, New Mexico. Rep. No. LA-8787-NERP.

White, G. C. & Garrott, R. A. (1990). *Analysis of Radio-Tracking Data*. New York: Academic Press.

White, N. (1999). Lynx, free speech tangle at CU. *Boulder Camera:* July 17, 1B, 4B (http://www.buffzone.com/buffzone/news/17cflap.html).

White, P. J. & Garrott, R. A. (1997). Factors regulating kit fox populations. *Canadian Journal of Zoology* **75**: 1982–1988.

White, P. J. & Ralls, K. (1993). Reproduction and spacing patterns of kit foxes relative to change prey availability. *Journal of Wildlife Management* **57**: 861–867.

White, P. J., Ralls, K. & Garrott, R. A. (1994). Coyote–kit fox interactions as revealed by telemetry. *Canadian Journal of Zoology* **72**: 1831–1836.

White, P. J., Vanderbilt White, C. A. & Ralls, K. (1996). Functional and numerical responses of kit foxes to a short-term decline in mammalian prey. *Journal of Mammalogy* **77**: 370–376.

Whitman, K. L. & Packer, C. (1998). The effect of sport hunting on the population dynamics of the African lion. In: *Proceedings of a Symposium on Lions and Leopards as Game Ranch Animals.*, Ondestepoort: South African Veterinary Wildlife Group, Faculty of Veterinary Science.

Whitmore, T. C. (1997). Tropical forest disturbance, disappearance, and species loss. In *Tropical Forest Remnants: Ecology, Management, and Conservation of Fragmented Communities*, eds W. F. Laurance & R. O. Bierregaard, Jr., pp. 3–12. Chicago: University of Chicago Press.

Whitten, A. L, Damanik, S. J., Anwar, J. & Hisyam, N. (1987). *The Ecology of Sumatra.* Yogyakarta: Gadjah Mada University Press.

Wielgus, R. B. & Bunnel, F. L. (1994). Dynamics of a small, hunted brown bear *Ursus arctos* population in southewestern Alberta, Canada. *Biological Conservation* **67**: 161–166.

Wielgus, R. B., Bunnell, F. L., Wakkinen, W. L. & Zager, P. E. (1994). Population dynamics of Selkirk Mountain grizzly bears. *Journal of Wildlife Management* **58**: 266–272.

Wiertz, J. (1993). Fluctuations in the Dutch badger (*Meles meles*) population between 1960 and 1990. *Mammal Review* **23**: 59–64.

Wiklund, C. G., Kjellen, N. & Isakson, E. (1998). Mechanisms determining the spatial distribution of microtine predators on the Arctic tundra. *Journal of Animal Ecology* **67**: 91–98.

Wikramanayake, E. D., Dinerstein, E., Robinson, J. G., Karanth, U., Rabinowitz, A., Olson, D., Mathew, T., Hedao, P., Conner, M., Hemley, G. & Bolze, D. (1998). An ecology-based method for defining priorities for large mammal conservation: The tiger as case study. *Conservation Biology* **12**: 865–878.

Wilcove, D. S., Rothstein, D., Dubow, J., Phillips, A. & Losos, E. (1998). Quantifying threats to imperiled species in the United States. *BioScience* **48**: 607–615.

Wildt, D. E. (1994). Endangered species spermatozoa: diversity, research and conservation. In *Function of Somatic Cells in the Testis:* ed. A. Bartke, pp. 1–23. New York: Springer-Verlag.

Wildt, D. E., Brown, J. L., Barone, M. A., Cooper, K. A., Grisham, J., Bush, M. & Howard, J.G. (1993). Reproductive status of cheetahs (*Acinonyx jubatus*) in North American zoos: the benefits of physiological surveys for strategic planning. *Zoo Biology* **12**: 45–80.

Wildt, D. E., Brown, J. L. & Swanson, W. F. (1998). Reproduction in cats. In *Encyclopedia of Reproduction:* eds E. Knobil & J. Neill, pp. 497–510. New York: Academic Press.

Wildt, D. E., Bush, M., Goodrowe, K. L., Packer, C., Pusey, A. E., Brown, J. L., Joslin, P., & O'Brien, S. J. (1987a). Reproductive and genetic consequences of founding isolated lion populations. *Nature* **329**: 328–31.

Wildt, D. E., Bush, M., Howard, J. G., O'Brien, S. J., Meltzer, D., van Dyk, A., Ebedes, H. & Brand, D. J. (1983). Unique seminal quality in the South African cheetah and a comparative evaluation in the domestic cat. *Biology of Reproduction:* **29**: 1019–25.

Wildt, D. E., Bush, M., Morton, C., Morton, F. & Howard, J. G. (1989). Semen characteristics and testosterone profiles in ferrets kept in long-day photoperiods, and the influence of hCG timing and sperm dilution on pregnancy rate after laparoscopic artificial insemination. *Journal of Reproduction and Fertility:* **86:** 349–358.

Wildt, D. E., Donoghue, A. M., Johnston, L. A., Schmidt, P. M. & Howard, J. G. (1992). Species and genetic effects on the utility of biotechnology for conservation. In *Biotechnology and the Conservation of Genetic Diversity:* eds H. D. M. Moore, W. V. Holt & G. M. Mace, pp. 45–61. Oxford: Clarendon Press.

Wildt, D. E., O'Brien, S. J., Howard, J. G., Caro, T. M., Roelke, M. E., Brown, J. L. & Bush, M. (1987b). Similarity in ejaculate-endocrine characteristics in captive versus free-ranging cheetahs of two subspecies. *Biology of Reproduction* **36:** 351–360.

Wildt, D. E., Phillips, L. G., Simmons, L. G., Chakraborty, P. K., Brown, J. L. & Howard, J. G. (1988). A comparative analysis of ejaculate and hormonal characteristics of the captive male cheetah, tiger, leopard and puma. *Biology of Reproduction* **38:** 245–255.

Wildt, D. E., Rall, W. F., Critser, J. K., Monfort, S. L. & Seal, U. S. (1997). Genome resource banks: 'living collections' for biodiversity conservation. *BioScience:* **47:** 689–98.

Wildt, D. E., Schiewe, M. C., Schmidt, P. M., Goodrowe, K. L., Howard, J. G., Phillips, L. G., O'Brien, S. J. & Bush, M. (1986). Developing animal model systems for embryo technologies in rare and endangered wildlife. *Theriogenology* **25:** 33–51.

Wildt, D. E. & Wemmer, C. (1999). Sex and wildlife: the role of reproductive science in conservation. *Biodiversity and Conservation* **8:** 965–976.

Wilkie, D. S., Curran, B., Tshombe, R. & Morelli, G. A. (1998). Managing bushmeat hunting in Okapi Wildlife Reserve, Democratic Republic of Congo. *Oryx* **32:** 131–144.

Wilkins, L., Arias-Reveron, J. M., Stith, B. M., Roelke, M. E. & Belden, R. C. (1997). The Florida panther *Puma concolor coryi*: a morphological investigation of the subspecies with acomparison to other North and South American cougars. *Bulletin of the Florida Museum of Natural History* **40:** 221–269.

Wilkinson, T. (1998). *Science Under Siege*. Boulder, CO: Johnson Books.

Willard, A. R. & Norchi, C. (1993). The decision seminar as an instrument of power and enlightenment. *Political Psychology* **14:** 575–606.

Willers, B. (1994). Sustainable development: a new world deception. *Conservation Biology* **8:** 146–1148.

Williams, E. S. & Thorne, E. T. (1996). Infectious and parasitic diseases of captive carnivores, with special emphasis on the black-footed ferret (*Mustela nigripes*). *Revue Scientifique et Technique de l'Office National des Epizooties* **15:** 91–114.

Williams, E. S., Thorne, E. T., Appel, M. J. G. & Belitsky, D. W. (1988). Canine distemper in black footed ferrets (*Mustela nigripes*) from Wyoming. *Journal of Wildlife Diseases* **24:** 385–398.

Williams, E. S., Thorne, E. T., Kwiatkowski, D. R. & Oakleaf, B. (1992). Overcoming disease problems in the black-footed ferret recovery program. *Trans. 57th N.A. Wildl. & Nat. Res. Conf.* pp. 474–485.

Williams, O. & McKegg, J. (1987). Nuisance furbearer management programs for urban areas. In *Wild Furbearer Management and Conservation in North America*, eds M. Novak, J. A. Baker, M. E. Obbard & B. Malloch, pp. 156–163. North Bay: Ontario Trappers Association.

Williams, P. (1998). Key sites for conservation: area-selection methods for biodiversity. In *Conservation in a Changing World*, eds G. M. Mace, A. Balmford & J. R. Ginsberg, pp. 211–250. Cambridge: Cambridge University Press.

Williams, P., Gibbons, D., Margules, C., Rebelo, A., Humphries, C. & Pressey, R. (1996). A comparison of richness hotspots, rarity hotspots, and complementary areas for conserving diversity of British birds. *Conservation Biology* **10**: 155–174.

Williamson, M. (1993). Invaders, weeds and the risk from genetically modified organisms. *Experientia* **49**: 219–214.

Williamson, M. (1996). *Biological Invasions*. London: Chapman & Hall.

Williamson, M. (1999). Invasions. *Ecography* **22**: 5–12.

Williamson, M. & Brown, K. C. (1986). The analysis and modelling of British invasions. *Philosophical Transactions of the Royal Society B* **314**: 505–522.

Williamson, M. & Fitter, A. (1996). The varying success of invaders. *Ecology* **77**: 1661–1666.

Wilson, D. E., Cole, F. R., Nichols, J. D., Rudran, R. & Foster, M. S. (Eds) (1996). *Measuring and Monitoring Biological Diversity: Standard Methods for Mammals*. Washington, DC: Smithsonian Institution Press.

Wilson, D. E. & Reeder, D.-A. (Eds) (1993). *Mammal Species of the World*. Washington DC: Smithsonian Instituion Press.

Wilson, E. O. (1987). Causes of ecological success: the case of ants. *Journal of Animal Ecology* **56**: 1–9.

Wilson, P. J., Grewal, S., Lawford, I. D., Heal, J. N. M., Granacki, A. G., Pennock, D., Theberge, J. B., Theberge, M. T., Voigt, D. R., Waddell, W., Chambers, R. E., Paquet, P. C., Goulet, G., Cluff, D., White, B. N. (2000). DNA profiles of the eastern Canadian wolf and the red wolf provide evidence for a common evolutionary history independent of the gray wolf. *Canadian Journal of Zoology* **78**: 2156–2166.

Windberg, L. A. (1995). Demography of a high-density coyote population. *Canadian Journal of Zoology* **73**: 942–954.

Winkler, W. G. (1975). Fox rabies. In *The Natural History of Rabies*, eds Baer, G. M., pp. 3–22. New York: Academic Press.

Wise, M. H., Linn, I. J. & Kennedy, C. R. (1981). A comparison of the feeding biology of mink (*Mustela vison*) and otter (*Lutra lutra*). *Journal of Zoology (London)* **195**: 181–213.

With, K. A. & Crist, T. O. (1995). Critical thresholds in species' responses to landscape structure. *Ecology* **76**: 2446–2459.

Wolch, J. & Emel, J. (Eds) (1998). *Animal Geographies: Place, Politics, and Identity in the Nature-Culture Borderlands*. New York: Verso.

Wolf, C. M., Garland, T. & Griffith, B. (1998a). Predictors of avian and mammalian translocation success: reanalysis with phylogenetically independent contrasts. *Biological Conservation* **86**: 243–255.

Wolf, C.M., Griffith, B., Reed, C. & Temple, S.A. (1996). Avian and mammalian translocations: update and reanalysis of 1987 survey data. *Conservation Biology* **10**: 1142–1154.

Wolf, K., Wildt, D. E., Vargas, A., Marinari, P., Williamson, L., Ottinger, M. A. & Howard, J. G. (1998b). Compromised reproductive efficiency in male black-footed ferrets. *Proceedings of the American Society of Andrology*: Abstract 79.

Wolfe, M. L. & Allen, D. L. (1973). Continued studies of the status, socialization, and relationships of Isle Royale wolves, 1967–1970. *Journal of Mammalogy* **54**: 611–635.

Wong J., Stewart, P. D. & Macdonald, D. W. (1999). Vocal repertoire in the European badger (*Meles meles*): structure, context and function. *Journal of Mammalogy* **80**: 570–588.

Woodford, M. H. & Kock, R. A. (1991). Veterinary considerations in re-introduction and translocation projects. *Symposia of the Zoological Society of London* **62**: 101–110.

Woodford, M. H. & Rossiter, P. B. (1994). Disease risks associated with wildlife translocation projects. In: *Creative Conservation: Interactive Management of Wild and Captive Animals*, eds P. Olney, G. Mace & A. Feistner. London: Chapman & Hall.

Wooding, J. B. & Hardisky, T. S. (1990). Coyote distribution in Florida. *Florida Field Naturalist* **18**: 12–14.

Wooding, S. & Ward, R. (1997). Phylogeography and Pleistocene evolution in the North American black bear. *Molecular Biology and Evolution* **14**: 1096–1105.

Woodley, T. H. & Lavigne, D. M. (1993). Potential effects of incidental mortalitites on the Hooker's sea lion (*Phocarctos hookeri*) population. *Aquatic Conservation and Marine and Freshwater Ecosystems* **3**: 139–148.

Woodroffe, R. (1997). The conservation implications of immobilizing, radio-collaring and vaccinating free-ranging wild dogs. In *The African Wild Dog. Status and Conservation Action Plan*, eds R. Woodroffe, J. R. Ginsberg & D. W. Macdonald, pp. 124–138. Gland: IUCN/SSC Canid Specialist Group.

Woodroffe, R. (1999a). Conserving the African wild dog *Lycaon pictus*. I. Diagnosing and treating causes of decline. *Oryx* **33**: 132–142.

Woodroffe, R. (1999b). Conserving the African wild dog *Lycaon pictus*. II. Is there a role for reintroduction? *Oryx* **33**: 143–151.

Woodroffe, R. (1999c). Managing disease threats to wild mammals. *Animal Conservation* **2**: 185–193.

Woodroffe, R. (2000). Predators and people: using human densities to interpret decline of large carnivores. *Animal Conservation* **3**: 165–173.

Woodroffe, R. & Ginsberg, J. R. (1997a). Country-country action plans for wild dog conservation. In *The African Wild Dog: Status Survey and Conservation Action Plan*. eds R. Woodroffe, J. R. Ginsberg & D. W. Macdonald, pp. 118–123. Gland: IUCN.

Woodroffe, R. & Ginsberg, J. R. (1997b). Past and future causes of wild dogs' population decline. In *The African Wild Dog: Status Survey and Conservation Action Plan*, eds R. Woodroffe, J. R. Ginsberg & D. W. Macdonald, pp. 58–74. Gland: IUCN.

Woodroffe, R. & Ginsberg, J. R. (1997c). The role of captive breeding and reintroduction in wild dog conservation. In *The African Wild Dog. Status and Conservation Action Plan*, eds R. Woodroffe, J. R. Ginsberg & D. Macdonald, pp. 100–111. Gland: IUCN/SSC Canid Specialist Group.

Woodroffe, R. & Ginsberg, J. R. (1998). Edge effects and the extinction of populations inside protected areas. *Science* **280**: 2126–2128.

Woodroffe, R. B. & Ginsberg, J. R. (1999). Conserving the African Wild Dog, *Lycaon pictus*. I. Diagnosing and treating causes of decline. *Oryx* **33**: 132–142.

Woodroffe, R. & J. R. Ginsberg (2000). Ranging behaviour and vulnerability to extinction in carnivores. In *Behaviour and Conservation*, eds L. M. Gosling & W. J. Sutherland, pp. 125–140. Cambridge: Cambridge University Press.

Woodroffe, R., Ginsberg, J. & Macdonald, D. (1997). *The African Wild Dog, Status Survey and Conservation Action Plan*. Gland: IUCN.

Woodroffe, R. & Macdonald, D. W. (1993). Badger sociality and models of spatial grouping. *Symposia of the Zoological Society of London* **65**: 145–169.

Woodroffe, R. & Macdonald, D.W. (1995). Female/female competition in European badgers (*Meles meles*): effects on breeding success. *Journal of Animal Ecology* **64**: 12–20.

Woodroffe, R. & Macdonald, D.W. (2000). Helpers provide no detectable benefits in the European badger, *Meles meles*. *Journal of Zoology (London)* **250**: 113–119.

Woods, J., Paetkau, D., Lewis, D., McLellan, B., Proctor, M. & Strobeck, C. (1999). Genetic tagging free-ranging black and brown bears. *Wildlife Society Bulletin:* **27**: 616–627.

Wozencaft, W. C. (1993). Order Carnivora. In *Mammal Species of the World: A Taxanomic and Geographic Reference*, eds D. E. Wilson & D. M. Reeder, pp. 279–348. Washington, DC: Smithsonian.

Wright, M. (1996). *Ecotourism on Otago Peninsula. Preliminary Studies of Two Species, yellow-eyed penguins* (Megadyptes antipodes) *and Hooker's Sea Lion* (Phocarctos hookeri). Wellington: New Zealand Department of Conservation report, Science for Conservation No. 68. N. Z. Department of Conservation.

Wright, S. (1931). Evolution in Mendelian populations. *Genetics* **16**: 97–159.

Wright, S. (1951) The genetical structure of populations. *Annals of Eugenics* **15**: 323–354.

Wright, S. (1977). *Evolution and the Genetics of Populations: Experimental Results and Evolutionary Deductions*. Chicago: The University of Chicago Press.

Wright, S. J., Gompper, M. E. & DeLeon, B. (1994). Are large predators keystone species in neotropical forests – the evidence from Barro-Colorado Island. *Oikos* **71**: 279–294.

Wydeven, A. P., Schultz, R. N. & Thiel, R. P. (1995). Monitoring of a recovering gray wolf population in Wisconsin, 1979–1991. Pp. 147–156 In *Ecology and Conservation of Wolves in a Changing World*, eds L. N. Carbyn, S. H. Fritts & D. R. Seip, pp. 147–156. Edmonton, Alberta: Canadian Circumpolar Institute.

WRI (1998). *World Resources 1998–1999*. Oxford: Oxford University Press.

Yadvendradew, V. J. (1994). Predation on blackbuck by wolves in Velvadar National Park, Gujarat, India. *Conservation Biology* **7**: 874–881.

Yahner, R. H. (1988). Changes in wildlife communities near edges. *Conservation Biology* **2**: 333–339.

Yahnke, C. J., Johnson, W. E., Geffen, E., Smith, D., Hertel, F., Roy, M. S., Bonacic, C. F., Fuller, T. K., Van Valkenburgh, B. & Wayne, R. K. (1996). Darwin's fox: a distinct endangered species in a vanishing habitat. *Conservation Biology* **10**: 366–375.

Yalden, D. W. (1982). When did the mammal fauna of the British Isles arrive? *Mammal Review* **12**: 1–57.

Yalden, D. W. (1993). The problems of reintroducing carnivores. *Symposia of the Zoological Society of London* **65**: 289–306.

Yamazaki, K. (1996). Social variation of lions in a male-depopulated area in Zimbabwe. *Journal of Wildlife Management* **60**: 490–497.

Yaska (1997). *Wolves, Bears and Their Prey in Alaska*. Washington, DC: National Academy Press.

Yazan, Y. & Knorre, Y. (1964). Domesticating elk in a Russian national park. *Oryx* **14**: 301–304.

Young, S. P. & Goldman, E. A. (1944). *The Wolves of North America*. Washington DC: The American Wildlife Institute.

Young, T. P. (1994). Natural die-offs of large mammals – implications for conservation. *Conservation Biology* **8**: 410–418.

Yuhki, N. & O'Brien, S. J. (1990). DNA variation of the mammalian major histocompatibility complex reflects genomic diversity and population history. *Proceedings of the National Academy of Sciences, USA* **87**: 836.

Zalewski, A., Jedrezejewski, W. & Jedrzejewska, B. (1995). Pine marten home ranges, numbers and predation on vertebrates in a deciduous forest (Bialowieza National Park, Poland). *Annales Zoologici Fennici* **32**: 131–144.

Zhang, Y. P. & Ryder, O. A. (1994). Phylogenetic relationships of bears (the Ursidae) inferred from mitochondrial DNA sequences. *Molecular Phylogenetics and Evolution* **3**: 351–359.

Zielinski, W. J. (1995). Track plates. In *American Marten, Fisher, Lynx, and Wolverine: Survey Methods for their Detection*, USDA For. Serv. Gen. Tech. Rep. PSW-GTR-157. eds W. J. Zielinski & T. E. Kucera, pp. 67–89.

Zielinski, W. J. & Stauffer, H. B. (1996). Monitoring *Martes* populations in California: survey design and power analysis. *Ecological Applications* **6**: 1254–1267.

Zielinski, W. J. & Truex, R. (1995). Distinguishing tracks of closely-related species. *Journal of Wildlife Management* **59**: 571–579.

Zimen, E. (1984). Long range movements of the red fox *Vulpes vulpes* L. *Acta Zoologica Fennica* **171**: 267–270.

Zimmerman, M. E., Callicott, J. B., Sessions, G., Warren, K. J. & Clark, J. (Eds.) (1993). *Environmental Philosophy: From Animal Rights to Radical Ecology*. New York: Prentice-Hall.

Index

Note: Page numbers in **bold** refer to Figures and Tables

Lightning Source UK Ltd.
Milton Keynes UK
UKOW01f1628260416

272990UK00001BA/52/P